Biology and Management of Invasive Quagga and Zebra Mussels in the Western United States

Biology and Management of Invasive Quagga and Zebra Mussels in the Western United States

EDITED BY

Wai Hing Wong • Shawn L. Gerstenberger

CRC Press
Taylor & Francis Group
Boca Raton London New York

CRC Press is an imprint of the
Taylor & Francis Group, an **informa** business

CRC Press
Taylor & Francis Group
6000 Broken Sound Parkway NW, Suite 300
Boca Raton, FL 33487-2742

First issued in paperback 2020

ISBN 13: 978-0-367-57575-5 (pbk)
ISBN 13: 978-1-4665-9561-3 (hbk)

Library of Congress Cataloging-in-Publication Data

Biology and management of invasive quagga and zebra mussels in the western United States / edited by Wai Hing Wong and Shawn L. Gerstenberger.
    pages cm
Includes bibliographical references and index.
ISBN 978-1-4665-9561-3
    1. Quagga mussel. 2. Zebra mussel. 3. Quagga mussel--Control--United States 4. Quagga mussel--Control--West (U.S.) 5. Zebra mussel--Control--United States 6. Zebra mussel--Control--West (U.S.) 7. Introduced organisms--Control--United States. I. Wong, Wai Hing, editor. II. Gerstenberger, Shawn L., 1968- editor.

QL430.7.D8B56 2015
594'.4--dc23
                                  2014041685

Visit the Taylor & Francis Web site at
http://www.taylorandfrancis.com

and the CRC Press Web site at
http://www.crcpress.com

This book is dedicated to the many men and women who work tirelessly to prevent the spread of aquatic invasive species in the field, whether they reside in the heat of the southwestern desert or in the bitter cold of the upper Midwest or upper Northeast.

This book is dedicated to the many men and women who work tirelessly to prevent the spread of aquatic invasive species in the field, whether they reside in the heat of the southwestern desert or in the bitter cold of the upper Midwest or upper Northeast.

Contents

Section III Detection

Section IV Prevention

Section V Policy

Section VI   Monitoring

Section VII   Control

Preface

The speed at which quagga mussels have spread throughout the southwestern United States is unprecedented. Quagga mussels (*Dreissena rostriformis bugensis*) were found in Lake Mead (Arizona–Nevada) in January 2007 and were also identified in reservoirs in San Diego (California) in the same year. By contrast, in the eastern United States, zebra mussels were discovered in 1993 in New York's Hudson River but were not detected in Massachusetts until 2008. Although the distance between these sites is comparable, the time it took the mussels to move this distance is not even on the same scale (i.e., 11 months vs. 15 years). Apart from the Lower Colorado system, quagga mussels have also been found in lakes and reservoirs in Arizona, California, Colorado, Nevada, and Utah. To address this emerging issue in the Southwest, federal, state, and local agencies began to establish policies to attenuate or stop the spread of quagga mussel populations in the western United States. An *ad hoc* interagency quagga mussel committee was established, and they worked collaboratively to establish an interagency monitoring action plan (I-MAP), which outlines agency objectives related to quagga mussel monitoring and provides approaches to realize these objectives. I-MAP team members and their respective agencies provide technical, logistical, and financial support in monitoring quagga mussels and their environmental impacts to Lake Mead. The members include a variety of federal, state, and regional agencies, including the National Park Service–Lake Mead National Recreation Area, the U.S. Bureau of Reclamation, Southern Nevada Water Authority, Nevada Department of Wildlife, Arizona Game and Fish Department, the University of Nevada–Las Vegas, the Metropolitan Water District of Southern California, the U.S. Fish and Wildlife Service, and the U.S. Geological Survey. The U.S. Fish and Wildlife Service, working in concert with the Western Regional Panel of the Aquatic Nuisance Species Task Force, developed the "Quagga–Zebra Mussel Action Plan (QZAP) for Western U.S. Waters" to respond to the westward spread of quagga and zebra mussels. Many western states have passed legislation requiring boat inspection and decontamination to keep invasive mussels out of their waters.

Valuable tools such as books and other publications exist for individuals and groups in the West to find information on coping with these invasive mussels (e.g., the classic zebra mussel book from Nalepa and Scholesser [1993; 2013]). However, the mussels in the western United States are different from those colonizing the Great Lakes or European waters. For example, not only are quagga mussel veligers uniquely present in Lake Mead/Lake Mohave year round, but pediveligers can also survive up to 27 days during transport in the semiarid Southwest. Therefore, the corresponding containment strategies of these invasive mussels in the West can be different in various respects. Significant work has been undertaken in the western United States since the 2007 initial detection of invasive quagga mussels in Lake Mead. In order to contain the mussels more effectively and efficiently, information from different agencies was exchanged and good experience shared. Our hope is that boat owners and the general public become more actively engaged in this issue.

This book provides information on biology, detection, prevention, regulation, monitoring, and control of invasive quagga and zebra mussels in the West. Due to proactive actions and beneficial coordination by the western states, technologies and policies developed and implemented to contain and manage invasive mussels in western waters are leading in the country. For example, boat inspection and decontamination have been mandatory in many lakes and reservoirs in the West since 2008, while the first water body east of the Mississippi River (Lake George in New York) to adopt such a mandatory program did not do so until May 2014. This book can facilitate communication among different agencies and the general public. It also summarizes technologies in mussel containment in the West. Overall, this book is a synthesis of the progress made by scientists and managers working on invasive quagga/zebra mussels in the western United States and will specifically benefit individuals and groups who are fighting invasive mussels in the field. This book documents many efforts, both successful and unsuccessful, of individuals and agencies after dreissenid mussels invaded the West. The information presented in this book will be valuable to other U.S. regions and European countries and offers an opportunity for scientists and lake managers worldwide to compare successful strategies relevant to their unique situation.

Wai Hing Wong
Shawn L. Gerstenberger

Acknowledgments

We appreciate the many authors who worked hard to contribute to this book. It has been an honor to participate in this long process with you. Your great work and willingness to candidly share your successes and failures will help advance our abilities to effectively deal with invasive quagga and zebra mussels. We also express our thanks to about 100 individual chapter reviewers; your time and expertise as volunteers significantly improved the quality of this book. We greatly appreciate the advice from John Sulzycki and Jennifer Ahringer of CRC Press. We are grateful to our parents, mentors, colleagues, and students for their inspiration and dedication to preventing the spread of aquatic invasive species. Finally, we express our appreciation to our wives, Ting and Linda, for their long-time understanding and support of our careers.

Editors

Wai Hing Wong (David) is an assistant professor in the Department of Biology and a research scientist in the Biological Field Station, State University of New York (SUNY) at Oneonta. He has 18 years of experience in studying invasive mussels and other mollusks in North America and Asia. He has a PhD in ecology and master's and bachelor's degrees in aquaculture and fisheries. His research interests include invasive species, pollution ecology, and aquatic ecosystems. He has been a principal investigator of more than 30 projects related to invasive quagga/zebra mussels' early detection, prevention, containment, control, and management. He is also a cofounder of the Technical Center for Aquatic Nuisances. He is an associate editor for three international journals, including *Aquatic Invasions*. He is also a review panelist for the National Science Foundation and a committee member for the federal Aquatic Nuisance Species Task Force Research Committee.

Shawn L. Gerstenberger is the Lincy Professor of Public Health and dean, School of Community Health Sciences, University of Nevada–Las Vegas. He has facilitated a quarterly interagency quagga mussel meeting in the Lower Colorado Region between 2007 and 2012. He has a PhD and a master's degree in aquatic toxicology and a bachelor's degree in reclamation. He is the founder of the Nevada Healthy Homes Partnership and has multiple collaborative grants with key community partners such as the Southern Nevada Health District, City of Henderson, and the Nevada State Health Division. He has done many invasive mussel projects and other environmental health–associated projects funded by the 100th Meridian, National Park Service, South Nevada Water Authority, the U.S. Fish and Wildlife Service, the Pacific States Marine Fisheries Commission, the Centers for Disease Control and Prevention, the U.S. Department of Housing and Urban Development, and the Environmental Protection Agency.

Contributors

Kumud Acharya
Division of Hydrologic Sciences
Desert Research Institute
Las Vegas, Nevada

Eric Anderson
Idaho House of Representatives (District 1)
Priest Lake, Idaho

Mark Anderson
Glen Canyon National Recreation Area
National Park Service
Page, Arizona

Amber Barenberg
Department of Fish and Wildlife Sciences
University of Idaho
Moscow, Idaho

Sarah Barnard
Department of Biology
Texas Christian University
Fort Worth, Texas

Beth A. Bear
Aquatic Invasive Species Program
Wyoming Game and Fish Department
Cheyenne, Wyoming

Kevin Bloom
Environmental Applications and Research
    Group
U.S. Bureau of Reclamation
Denver, Colorado

David K. Britton
Fisheries and Aquatic Conservation
U.S. Fish and Wildlife Service
Arlington, Texas

Andrew J. Brooks
Marine Science Institute
University of California, Santa Barbara
Santa Barbara, California

Scott Bryan
Central Arizona Project
Phoenix, Arizona

Andrea M. Caires
Department of Natural Resources and
    Environmental Science
University of Nevada, Reno
Reno, Nevada

Timothy J. Caldwell
Department of Natural Resources and
    Environmental Science
University of Nevada, Reno
Reno, Nevada

Jamie Carmon
GEI Consultants, Inc.
Denver, Colorado

Sudeep Chandra
Department of Natural Resources and
    Environmental Science
University of Nevada, Reno
Reno, Nevada

Earl W. Chilton II
Texas Parks & Wildlife Department
Austin, Texas

Steve Chilton
Lake Tahoe and Northern Nevada
U.S. Fish and Wildlife Service
Zephyr Cove, Nevada

Sean Comeau
School of Medicine
University of Nevada, Reno
Reno, Nevada

Ernest Couch
Department of Biology
Texas Christian University
Fort Worth, Texas

Chad L. Cross
Department of Environmental and Occupational
    Health
University of Nevada, Las Vegas
Las Vegas, Nevada

Carolynn S. Culver
Marine Science Institute
University of California, Santa Barbara
Santa Barbara, California

Daniel Daft
Public Utilities Department
City of San Diego
La Mesa, California

Larry B. Dalton
Wildlife Reflections Consulting
and
Utah Division of Wildlife Resources
Salt Lake City, Utah

Clinton J. Davis
Department of Natural Resources and
    Environmental Science
University of Nevada, Reno
Reno, Nevada

Patricia Delrose
U.S. Bureau of Reclamation
Department of the Interior
Boulder City, Nevada

Debra L. DeShon
Mussel Dogs®
Denair, California

Sandee Dingman
Lake Mead National Recreation Area
National Park Service
Boulder City, Nevada

Joseph DiVittorio (retired)
U.S. Bureau of Reclamation
Department of the Interior
Denver, Colorado

Susan R. Ellis
Invasive Species Program
California Department of Fish and Wildlife
Sacramento, California

Amy Ferriter
CPS Timberland
Boise, Idaho

Shawn L. Gerstenberger
Department of Environmental and Occupational
    Health
University of Nevada, Las Vegas
Las Vegas, Nevada

Mark D. Hatcher
Sweetwater Authority
Spring Valley, California

G. Chris Holdren
Environmental Applications and Research
    Group
U.S. Bureau of Reclamation
Denver, Colorado

Denise M. Hosler
Environmental Applications and Research
    Group
U.S. Bureau of Reclamation
Denver, Colorado

Richard S. Ianniello
Sequoia National Park
and
Kings Canyon National Park
Three Rivers, California

and

Department of Environmental and Occupational
    Health
University of Nevada, Las Vegas
Las Vegas, Nevada

Ashley J. Jensen
Mussel Dogs®
Denair, California

Matthew Kappel
Department of Environmental and Occupational
    Health
University of Nevada, Las Vegas
Las Vegas, Nevada

Carolyn Link
Marrone Bio Innovations
Davis, California

Scott W. McClelland
Sweetwater Authority
Spring Valley, California

Robert F. McMahon
Department of Biology
The University of Texas at Arlington
Arlington, Texas

Bobbi Jo Merten
U.S. Bureau of Reclamation
Department of the Interior
Denver, Colorado

Michael J. Misamore
Department of Biology
Texas Christian University
Fort Worth, Texas

Christine M. Moffitt
Department of Fish and Wildlife Sciences
University of Idaho
Moscow, Idaho

Bryan Moore
Lake Mead National Recreation Area
National Park Service
Boulder City, Nevada

Harry Nelson
Fluid Imaging Technologies, Inc.
Scarborough, Maine

Dominique T. Norton
Invasive Species Program
California Department of Fish and Wildlife
Sacramento, California

Scott O'Meara
Environmental Applications and Research
  Group
U.S. Bureau of Reclamation
Denver, Colorado

Tanviben Y. Patel
Department of Environmental and Occupational
  Health
University of Nevada, Las Vegas
Las Vegas, Nevada

Sherri Pucherelli
Denver Federal Center
U.S. Bureau of Reclamation
Denver, Colorado

Sarahann Rackl
Marrone Bio Innovations
Davis, California

Sherril R. Rahe
Aquatic Invasive Species Program
Wyoming Game and Fish Department
Cheyenne, Wyoming

Stephen Phillips
Pacific States Marine Fisheries Commission
Portland, Oregon

Michael R. Rosen
U.S. Geological Survey
Carson City, Nevada

Allen D. Skaja
U.S. Bureau of Reclamation
Department of the Interior
Denver, Colorado

Ben Spaulding
Fluid Imaging Technologies, Inc.
Scarborough, Maine

Jim Steele
Lake County Department of Water Resources
Lakeport, California

Kelly A. Stockton
Department of Fish and Wildlife Sciences
University of Idaho
Moscow, Idaho

Catherine L. Sykes
BluBridge Inc.
Bentonville, Arkansas

Cynthia Tait
U.S. Forest Service
Ogden, Utah

Melissa Thaw
Division of Hydrologic Sciences
Desert Research Institute
Las Vegas, Nevada

David Tordonato
U.S. Bureau of Reclamation
Department of the Interior
Denver, Colorado

Kent Turner
Lake Mead National Recreation Area
National Park Service
Boulder City, Nevada

Brian Van Zee
Texas Parks & Wildlife Department
Waco, Texas

Martha C. Volkoff
Invasive Species Program
California Department of Fish and Wildlife
Sacramento, California

Barnaby J. Watten
U.S. Geological Survey
S.O. Conte Anadromous Fish Research Center
Turners Falls, Massachusetts

Ashlie Watters
Department of Environmental and Occupational
   Health
University of Nevada, Las Vegas
Las Vegas, Nevada

and

Lake Mead National Recreation Area
National Park Service
Boulder City, Nevada

Daniel Webster
U.S. Geological Survey
Reston, Virginia

Wade D. Wilson
Southwestern Native Aquatic Resources and
   Recovery Center
U.S. Fish and Wildlife Service
Dexter, New Mexico

Wai Hing Wong
Department of Biology
and
Biological Field Station
State University of New York at Oneonta
Oneonta, New York

Bill Zook
Pacific States Marine Fisheries Commission
Portland, Oregon

Section I

Introduction

1

History of Western Management Actions on Invasive Mussels

David K. Britton

CONTENTS

ABSTRACT Once introduced into North America in the 1980s, zebra mussels (*Dreissena polymorpha*) spread quickly through an artificial connection from the Great Lakes into the Mississippi River system. From there, navigable waterways provided conduits for invasion into much of the eastern United States. Conservation agencies in the West expected zebra mussels to threaten their waters too, but limited resources, personnel, and other factors constrained prevention efforts. Meanwhile, quagga mussels (*Dreissena bugensis*), which were not notorious for spreading beyond the Great Lakes, slipped past our defenses and entered the lower Colorado River sometime before 2007. Once introduced, spread was rapid, again, through artificial water connections. The Colorado River Aqueduct brought quagga mussels rapidly into Southern California, while the Central Arizona Project carried them into Arizona. The extent of the current quagga and zebra mussel distribution is largely a reflection of the navigable waterways in the East and the aqueduct and canal systems in the southwest. These pathways are largely uncontrollable. Western agencies have focused efforts, instead, on preventing overland transport of invasive mussels on boats carried on trailers. Only five states in the West have found evidence for dreissenid mussel introductions: California, Utah, Nevada, New Mexico, and Colorado. The introductions in Colorado, New Mexico, and Nevada (except for the original introductions in the lower Colorado River) have not developed into successful invasions. By 2013, all Western states have recognized the serious nature of existing and potential invasions. Many states have enacted new laws and stepped up prevention and containment programs. This chapter highlights some of the management actions taken in response to quagga and zebra mussel discoveries in the West and some of the obstacles that we have faced. As the invasion continues, attention from the public and from our state and federal authorities has grown. Our abilities to prevent further spread are improving, which is essential because there is no cure for this problem.

How Mussels Spread to North America

The St. Lawrence Seaway, an artificial series of locks and canals, opened the Great Lakes to the Atlantic Ocean in 1959 and allowed transoceanic ships to transfer goods to and from other parts of the world. It also allowed the exchange of harmful invasive species between these very same locations (Transportation Research Board 2008). In the 1980s, ships from the United States supplied grains and other products to Europe. On return, however, these ships carried contaminated ballast water (anywhere from 1,200 to 24,500 m³ per vessel), instead of cargo (Transportation Research Board 2008). When released into the Great Lakes, this unregulated ballast water introduced new invaders, including zebra and quagga mussels, to America's largest freshwater lakes.

Zebra and quagga mussels are members of the molluscan family Dreissenidae, a family of bivalves that have a microscopic larval stage called a veliger. At this life stage, the organism is relatively delicate. Few veligers survive naturally occurring conditions (Thomas and Lamberti 1999), but some do. These settle, grow, and mature, and, after reaching adult size, release sperm or eggs into open waters in order to successfully reproduce (Sprung 1993). Sperm and eggs must come together, of course, to produce the next generation of offspring. A complicated and delicate series of events must unfold without problems to create a new generation of dreissenid mussels (Padilla 2005). Some have argued that aquatic organisms must be in close proximity for successful reproduction to occur (Levitan and Peterson 1995). Otherwise, a next generation is unlikely.

It has been commonly accepted that zebra and quagga mussels were inadvertently transported to North America as larvae in the ballast tanks of transoceanic ships, as described earlier (Carlton 1993). This may be true, but there might be a more plausible vector. The likelihood of individual veliger larvae surviving a transatlantic journey is low, but possible, especially considering that individual mussels can potentially release hundreds of thousands of eggs in one reproductive event (Sprung 1990). If the veligers did survive a transatlantic journey, they would have been dispersed with the ballast water when released into open waters, in this case, into the largest open waters in North America. Veligers would then have to successfully settle and grow to maturity in sufficient proximity to other individuals such that their microscopic sperm and eggs could eventually meet in these enormous open waters. Despite the widespread acceptance that zebra and quagga mussels came to North America in ballast water, to some, it remains an unlikely scenario for a rapid introduction of these species into a new habitat.

Alternatively, some have argued that adult populations of zebra and quagga mussels came to North America attached to anchors and chains (McMahon et al. 1993). This scenario is more plausible for an intercontinental rapid introduction. Adult zebra and quagga mussels glue themselves to hard substrates in prodigious densities and are much more resilient to varying environmental conditions than are veliger larvae. Many people believe that zebra and quagga mussels stay put once they glue themselves to hard surfaces, like rocks, and pilings—or anchors and chains—but this is a common misconception. Zebra and quagga mussels can detach from their byssal threads and crawl, like snails, to new areas. In fact, they do so regularly. I once dropped a 20 lb mushroom anchor into Lake Oologah, Oklahoma, and retrieved it the next day to find over a dozen adult zebra mussels freshly attached, some on the anchor itself and some on the nylon rope tied to the anchor. I have even witnessed zebra mussels attached to submerged monofilament fishing line in less than a day. The huge anchors and chains used for transatlantic ships provide considerable surface area for zebra and quagga mussels and are almost certainly significant vectors for their introduction to new habitats, especially considering that these adult mussels are much more hardy than veligers and can survive out of water for weeks if conditions are cool and humid (McMahon et al. 1993). This is plenty of time for a transatlantic passage. In fact, inspectors from Arizona and Utah have reported to me that they regularly find quagga mussels attached to anchors found inside otherwise clean, recreational boats that were trailered into their jurisdictions.

Dense clusters (called druses) of zebra mussels could be released when ships drop anchor on return to the Great Lakes from Europe. A cluster of sexually mature mussels transported on an anchor or chain could provide a much more likely scenario for rapid invasion of new habitat. This is especially true because sexually mature adults are already glued together in close proximity—literally on top of each other—ready to broadcast sperm and eggs together once in a new habitat.

Regardless of the actual vector, zebra and quagga mussels came to North America with transoceanic vessels and were unintentionally released into the Great Lakes system to be discovered in 1988 and 1989, respectively, although quagga mussels were not officially discovered until 1991 (Herbert et al. 1989; May and Marsden 1992).

Zebra and quagga mussel populations initially behaved differently. Zebra mussel populations exploded, reaching a vast distribution in the Great Lakes by 1990 and spreading through connected waterways, like metastatic cancer, into the Mississippi drainage, reaching St. Louis, Missouri, by 1991 and New Orleans, Louisiana, by 1993 (USGS 2010). Meanwhile, quagga mussel populations were restricted to the Great Lakes at least originally (Mills et al. 1996). Their densities were at first relatively low. Thus, many of us were more concerned with the propensity for the rapid spread exemplified by zebra mussels (Bossenbroek et al. 2007). We did not expect quagga mussels to precede zebra mussels into the West. Nevertheless, they did. They have also outcompeted zebra mussels in many areas of the Great Lakes in recent years, yet zebra mussels remain a serious threat for invasion by way of trailered watercraft (Karatayev et al. 2013).

How Mussels Spread to the West

Dreissenid mussel larvae are planktonic. That is, they live suspended in the water column, carried by currents. These veligers can remain in the larval stage for weeks (Mackie 1991), which is plenty of time to be carried hundreds of miles downstream. In this way, zebra mussels from the Great Lakes spread rapidly through an artificial waterway from the Great Lakes into the Illinois River and the Mississippi River Drainage. The Chicago Sanitary and Ship Canal, originally designed to move unwanted material (Chicago's sewage) away from Lake Michigan, provided the same function for Great Lake's aquatic invaders (Transportation Research Board 2008). Once zebra mussels were released through the artificial canal into the Illinois River, downstream spread into the Mississippi system was inevitable. Our largest lakes became a source to infest our largest river in North America. The infested Mississippi River then served as a conduit for zebra mussels to reach all areas connected via navigable waterways (O'Neil and Dextrase 1994). Adult zebra mussels, attached to hulls, anchors, chains, and other submerged equipment on boats and barges, spread upstream to other navigable rivers. Since the 1980s, zebra mussels have invaded much of the Midwest and Northeast regions of the United States, including the Mississippi, Missouri, Illinois, Arkansas, Tennessee, Cumberland, Kentucky, Ohio, and St. Lawrence Rivers as well as the Hudson River by way of the Erie Canal (USGS 2010). In fact, most of the freshwaters in the United States known to be infested with zebra mussels are directly connected to the Great Lakes or navigable waterways. Compare a map of the navigable waterways in the United States (Figure 1.1a) to the known distribution of established zebra mussel populations (Figure 1.1b).

Quagga mussels did not spread throughout North America as rapidly as did zebra mussels. They remained restricted mostly within the Great Lakes and the St. Lawrence River, except for an isolated population discovered in the Mississippi River near St. Louis, Missouri, in 1995 (Kraft 1995). Nevertheless, in 2007, quagga mussels were found in the lower Colorado River at Lake Mead in Nevada, approximately 1200 miles from the closest known population. It is likely that quagga mussels came to Nevada on a highway, attached to a trailered houseboat brought from the Great Lakes. Within a couple of weeks of the initial discovery, additional quagga mussels were found in Lakes Mohave and Havasu, also on the lower Colorado River. It remains unclear which location on the lower Colorado River received quagga mussels first. Regardless, spread from there through connected waters was expeditious.

Artificial conduits helped accelerate the invasion of the American southwest. Water is scarce in the West. Without interbasin transfers, large metropolitan areas, including Los Angeles, San Diego, Phoenix, and Tucson, could not provide adequate water to supply the needs of the millions of people who choose

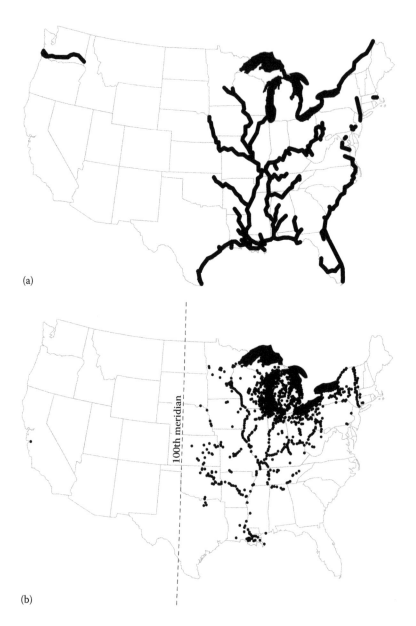

(a)

(b)

FIGURE 1.1   (a) Navigable waterways and (b) the distribution of established zebra mussel populations in the contiguous United States.

to live in these arid regions. Much of Southern California obtains water for municipal supply and irrigation from the Colorado River Aqueduct, which diverts water from the lower Colorado River. Likewise, southern Arizona requires water diverted from the lower Colorado River through the Central Arizona Project. These artificial conduits allowed quagga mussels an immediate pathway to connected waters in California and Arizona, following the flow of water pumped from Lake Havasu on the lower Colorado River. Interbasin transfers of water are like blood transfusions: they may be life-saving, but they can also be disastrous if the source is infected.

Within the same year that quagga mussels were first discovered in the West, they had spread to lakes Murray, Miramar, Sweetwater, Otay, Copper Basin, Dixon, San Vicente, and Skinner in California and Lake Pleasant and the Salt River in Arizona (USGS 2010). By the end of the next year (2008), quagga mussels had spread to lakes Jennings, Rattlesnake, Olivenhain, El Capitan, Romona, Walnut Canyon,

Anaheim, Kraemer, and Poway in California and Mittry Lake and Topock Marsh in Arizona (USGS 2010). Given the copious volume of water transferred across basins and the inability to discontinue these supplies, preventing interbasin spread was impossible.

Quagga mussels now appear to have a permanent foothold in the West and are continuing to spread. Although downstream spread is rapid, largely inevitable, and unpreventable, we are not powerless to control other vectors of spread. Zebra and quagga mussels also reach new habitats by overland transport on trailered boats and equipment. Trailered boats with either zebra or quagga mussels attached have been sighted practically in every Western state. This type of spread occurs more slowly, but steadily, and is entirely preventable.

As of June 2013, five Western states, California, Colorado, Utah, Nevada, and New Mexico, have reported evidence of dreissenid mussel introduction west of the 100th meridian in waters not connected to the lower Colorado River. I discuss each of these in the following.

California

By May 2010, there were 25 water bodies confirmed to have dreissenid mussels in California (USGS 2010). All of these were connected to the lower Colorado River through the Colorado River Aqueduct, except one. A year after quagga mussels were discovered in the lower Colorado River, zebra mussels were found in San Justo Reservoir, near Monterrey Bay, California (all other confirmed sightings of dreissenid mussels in California were for quagga mussels). This was the first occurrence of zebra mussels within the Golden State and the only established population of zebra mussels west of the 100th meridian. Given that San Justo Reservoir has no connection to the lower Colorado River or the Colorado River Aqueduct, it is most likely that zebra mussels entered this reservoir on a trailered boat or other piece of equipment. San Justo Reservoir is part of the U.S. Bureau of Reclamation's San Felipe Division of the Central Valley Project that distributes water through pipes to the Central Valley, the Delta, and up to the Cascade Mountain Range. It is used as for storage, with water usually released only in winter months. According to the San Benito County Water District, a partner that distributes water from San Justo Reservoir, zebra mussels have already infested the Hollister Conduit, which helps fill this reservoir from the Central Valley Project's distribution system. Authorities have suspended recreational boat activities on San Justo Reservoir and are considering implementing a treatment plan to prevent further invasion to other parts of California.

Meanwhile, the State of California and many of its local jurisdictions have begun implementing programs to inspect, interdict, and decontaminate boats before allowing them to enter California's waters. The California Department of Food and Agriculture uses its check stations on key highway entry points to inspect and stop contaminated boats as these enter the state. Some local jurisdictions have prohibited recreational boating in specific water bodies altogether. Others, like authorities for Clear Lake in Lake County, do not allow boats to launch without a prior inspection. Such measures may be inconvenient, but reduce the likelihood of an introduction considerably.

Colorado

So far, there is no evidence for an established population of invasive mussels in Colorado. After quagga mussels were first discovered in the lower Colorado River, a partnership between the Colorado Division of Wildlife and Colorado State Parks (now combined into a single agency, Colorado Parks & Wildlife) began an aggressive sampling campaign to detect and monitor for zebra and quagga mussels within their state. By 2008, Colorado researchers had identified eight reservoirs with evidence of zebra or quagga mussels, including Pueblo, Granby, Shadow Mountain, Willow Creek, Tarryall, Blue Mesa, and Jumbo Reservoirs, as well as Grand Lake (USGS 2010).

As in California, authorities in Colorado recognized the gravity of a dreissenid invasion and acted quickly once invasive mussels were beyond their doorstep. In May 2008, the Colorado General Assembly passed the State Aquatic Nuisance Species Act, making it illegal to possess or transport aquatic nuisance species. This act also authorized peace officers to inspect, quarantine, and decontaminate watercraft, if necessary. The Colorado Division of Wildlife convened a blue-ribbon panel of experts from across the country to help assess their strategy (Anderson et al. 2008). They ranked freshwater bodies based on a prioritization plan, taking into account the relative risk of invasion, and began a sampling schedule with a frequency ranging from every 3 weeks to once a year. Higher-risk waters were sampled more frequently.

Through these efforts, biologists in Colorado found evidence for zebra or quagga mussels at the eight water bodies listed previously. However, continued sampling produced no further evidence of established dreissenids in Colorado, as reported in December 2010 (Colorado Division of Wildlife and Colorado State Parks 2010). Authorities do not know if this means that dreissenids are no longer present in Colorado or if existing populations are lingering below detectable levels. In either case, the State of Colorado has chosen to be proactive in their approach. This should be appreciated beyond the borders of Colorado, whose waters reach 18 other states. The Colorado River flows from Colorado through Utah, Arizona, Nevada, and California. The Rio Grande begins in Colorado and flows through New Mexico and Texas. The Arkansas River flows from Colorado through Kansas, Oklahoma, and Arkansas. The Northern Platte River flows from Colorado through Wyoming and then meets the Southern Platte River (which also begins in Colorado) in western Nebraska; then as the Platte River, these waters empty into the Missouri River near Omaha. The Missouri River then flows through Iowa, Kansas, and Missouri to the Mississippi, which then flows along the state borders of Illinois, Kentucky, Tennessee, Arkansas, Mississippi, and Louisiana. Waters from Colorado also reach the Mexican states of Baja California, Sonora, Chihuahua, Coahuila, Nuevo Leon, and Tamaulipas. Thus, protecting the State of Colorado from aquatic invaders will go a long way toward protecting many other areas from the same.

Utah

During the initial flurry of reaction following the first reports of quagga mussels in the West (at Lake Mead), the State of Utah began an aggressive sampling program. In 2007, an *inconclusive* analysis suggested that quagga mussels may have reached Lake Powell, upriver of Lake Mead on the Colorado River. Subsequent investigations and repeated sampling efforts from 2007 through 2010 have revealed no evidence for zebra or quagga mussels at Lake Powell, a reservoir on the border of Utah and Arizona. Even before quagga and zebra mussels had crossed over the Continental Divide, the National Park Service and especially the staff at Glen Canyon National Recreational Area, which includes Lake Powell, were among the key contributors to the 100th meridian initiative, a partnership of governmental and nongovernmental entities working together to prevent aquatic invasive species from spreading into the West. Glen Canyon National Recreational Area was one of the first Western authorities to introduce proactive measures to prevent dreissenid mussels from entering their waters on trailered boats. Proactive rules insist that boats must be cleaned, drained, and dry before entering the water. These rules follow the recommendations of the national *Stop Aquatic Hitchhikers!* campaign. And although no methods are foolproof, these methods seemed to be working until March 2013, when adult quagga mussels were discovered in Lake Powell. The National Park Service, led by an exemplary program at Glen Canyon National Recreation area, is working with partners to eradicate or control this population at an incipient stage. Although park staff have not dismissed eradication as a possibility, they continue to find mussels at multiple locations near the border of Utah and Arizona.

Utah's state detection and monitoring program has found evidence of zebra or quagga mussels at three additional reservoirs: quagga mussels were found in Red Fleet and Sand Hollow Reservoirs, and zebra mussels were found in Electric Lake (Utah Division of Wildlife Resources 2011). At the end of 2010, the State of Utah listed the status of Red Fleet Reservoir and Electric Lake to be *detected*, which is defined to mean that a plankton sample evidenced a preliminary finding of veligers by way of a microscopic technique, and such a finding has been confirmed by two independent DNA analyses using polymerase chain reaction (PCR) amplification (Larry Dalton, Utah Division of Wildlife Resources, personal communication). Only Sand Hollow was officially recognized as *infested* by Utah authorities. *Infested* in Utah means that the presence of either juvenile or adult mussels has been confirmed by two experts, and two independent PCR methods have verified the dreissenid DNA (Larry Dalton, Utah Division of Wildlife Resources, personal communication). Four other water bodies in Utah remain *inconclusive*, meaning that there is some evidence (by microscopic analysis) that dreissenid larvae may be present, but this has not yet been confirmed by DNA analysis using PCR (Larry Dalton, Utah Division of Wildlife Resources, personal communication).

Nevada

In 2011, evidence for quagga mussels was discovered in Rye Patch Reservoir and Lahontan Reservoir in northern Nevada. In both cases, veliger larvae were collected in plankton samples. Authorities at the Nevada Division of Wildlife have cautioned that these findings are preliminary and need to be confirmed before concluding that a population has established in either of the reservoirs. As of June, 2013, no further evidence has revealed any sustaining population in Rye Patch or Lahontan Reservoirs.

Lake Mead and Lake Mohave, both on the southern Nevada border shared with Arizona, have well-established populations of quagga mussels. These have been a problem since they were discovered in 2007, the first discovery of dreissenid mussels west of the 100th meridian. The National Park Service at Lake Mead National Recreation area, who had been working with partners with prevention efforts before the invasion, had to focus as well on containment, so that invasive mussels would not move from Lake Mead or Mohave to other areas. This has been difficult.

Lake Mead National Recreation Area spans two states, Nevada and Arizona, covers 1.5 million acres, and has approximately 8 million visitors each year. Approximately 3800 boats are moored in water within the park. And, on holiday weekends, 3000–5000 boats may visit the 2 lakes within the park. The Lake Mead National Recreation area now requires anyone with a boat slipped or moored longer than 5 days to provide marina personnel at least 72 h notice prior to pulling the boat from the lake. These boaters are also required to schedule an inspection and a vessel wash. During the inspection, official forms from the State of Nevada or Arizona (depending on which state the boat departs) are completed, signed, and forwarded to the respective state authorities. The states of Nevada and Arizona are responsible for notifying destination states that a potentially contaminated boat is coming.

This system has received considerable criticism from many Western partners because other Western states continued to regularly find boats carrying quagga mussels from the lower Colorado River at inspection stations and interdiction sites, without notice of their departure. In 2011 and 2012, concerned agencies/organizations wrote multiple letters to the director of the U.S. Fish and Wildlife Service, the secretary of the Interior, and members of Congress, calling for improved containment procedures on the lower Colorado River and, especially, at Lake Mead National Recreation Area.

Congress responded in 2012 by directing the U.S. Fish and Wildlife Service to redirect $1 million in existing funds from other priorities to improve containment procedures at federally managed waters in the West at highest risk for spreading invasive mussels. Those who lobbied for change specifically wanted to see the implementation of mandatory inspection and decontamination stations at Lake Mead, as described in the 2010 Quagga/Zebra Mussel Action Plan for Western U.S. Waters.

However, the National Parks Service (and any other federal agency) currently has no legal authority to stop boats carrying quagga mussels. When the U.S. Fish and Wildlife Service listed invasive mussels under the Lacey Act's injurious wildlife provisions, we listed only zebra mussels, specifically *Dreissena polymorpha*. Quagga mussels are a different species, *Dreissena rostriformis bugensis*, and are not covered under the injurious wildlife provisions (Title 18) of the Lacey Act. This was an unfortunate oversight.

Although the superintendent at Lake Mead National Recreation Area had used his Superintendent's Compendium to require slipped and moored boaters to provide notice and stop before leaving for an inspection and a boat wash, the Park Service has no legal authority to stop boaters who choose to ignore these rules. Meanwhile, many boaters do ignore the rules, probably because inspections take considerable time, and boat washes for large boats can cost thousands of dollars.

New Mexico

On May 27, 2011, the State of New Mexico announced a possible invasion into Lake Sumner. Routine analysis of plankton samples revealed DNA from quagga mussels in this New Mexico reservoir, which was temporarily closed to boating. Researchers also found potential evidence for quagga mussel DNA in

Lake Navajo, a reservoir on the New Mexico/Colorado border, later that year. However, as of the writing of this chapter, no established populations of adult mussels have been found in either reservoir. The U.S. Fish and Wildlife Service and the New Mexico Department of Wildlife continue to monitor and strive for early detection of invasive mussels.

What We Are Doing about the Problem

The Western Regional Panel on Aquatic Nuisance Species created a *Quagga-Zebra Mussel Action Plan for Western U.S. Waters*, approved by the federal Aquatic Nuisance Species Task Force in 2010. This plan, affectionately called *QZAP*, identifies priority actions to thwart spread and control existing populations (Western Regional Panel 2010). Specific prevention strategies include implementation of mandatory inspection and decontamination at infested waters, continued development of effective watercraft and equipment inspection and decontamination protocols and standards, adoption of protocols and standards in Western States, establishment and implementation of strong, consistent law enforcement programs, and development of a standardized model and strategy for risk assessment model for water bodies. The QZAP also provides strategies for early detection and monitoring, rapid response, containment and control of existing populations, and outreach and education. Many state and federal agencies have already begun implementing QZAP at all jurisdictional levels, even before it was finalized. However, strong coordination and increased resources are necessary to fully address QZAP priorities. Meanwhile, the public can help by learning about zebra and quagga mussels (and other aquatic invaders) and take the simple steps necessary to prevent unintentional spread. Boaters should always thoroughly clean their equipment, drain all standing water, and dry everything before transport to other waters. If boats are kept moored or slipped in infested waters, a professional cleaning with hot (>140°F), high-pressure water is warranted before transport. A clean boat will also help expedite procedures at inspection stations that are becoming prevalent in the West and may prevent interdiction, impoundment, and/or quarantine.

In 2010, the U.S. Fish and Wildlife Service spent $1.4 million on projects to prevent further invasion in the western United States. A large portion of this funding ($800,000) went to protect Lake Tahoe by funding inspection and decontamination stations on roads leading to the Lake Tahoe region. Approximately $600,000 was also spent to fund some of the highest priorities of the QZAP, including watercraft inspection and decontamination protocols, mussel monitoring technologies, and mussel decontamination technologies, and for early detection activities in the states of Texas, Montana, and Oregon. The U.S. Fish and Wildlife Service also spent an additional $600,000 to help fund state and interstate management plans that had specific actions for preventing the spread of quagga or zebra mussels.

The U.S. Fish and Wildlife Service divided the congressionally directed $1 million in the fiscal year 2012 to several state, federal, and regional agencies for projects devoted to containing quagga mussels in the lower Colorado River Basin. We provided funding to the Lake Mead National Recreation Area to continue and improve their quagga-mussel containment program. The states of Washington, Oregon, California, Nevada, Utah, New Mexico, and Arizona received funding to increase law enforcement activities related to quagga mussels. We also gave the states of Nevada and Arizona funding for outreach and education targeting the lower Colorado River. The U.S. Fish and Wildlife Service provided funding for the Pacific States Marine Fisheries Commission to continue their Level II Watercraft Inspection Training program at Lake Mead. We also provided funding to the Glen Canyon National Recreation Area and New Mexico Game and Fish to conduct boat interdictions and other preventative actions at Lake Powell and Lake Navajo (an interjurisdictional reservoir on the San Juan River, a tributary to the Colorado River and Lake Powell). Additionally, we are working with partners to coordinate the development of a better communication system between states using real-time, on-line databases that could share information on specific, potentially contaminated watercraft.

In the fiscal year 2013, the U.S. Fish and Wildlife Service will also devote approximately $900,000 (reduced by *Sequestration* from $1 million) to containing quagga mussels in the lower Colorado River. In the Spring of 2013, U.S. Representatives Joe Heck (NV-03), Mike Thompson (CA-05), and Mark Amodei (NV-02) introduced a bill that would add quagga mussels to the list of injurious wildlife. This bill, H.R. 1823, called the *Protecting Lakes against Quagga (PLAQ) Act of 2013* would provide federal agencies the ability to take law enforcement actions to prevent the spread of quagga mussels throughout the United States and specifically would allow federal authorities to prosecute boaters who transport live quagga mussels from contaminated waters across state or tribal lines. However, it would not provide federal authority for stopping boaters from simply leaving affected waters if state or tribal boundaries are not crossed. (I am prohibited from influencing potential legislation, so I must be clear that I am neither endorsing nor opposing this or any other bill.)

Early Detection

Many Western states have begun implementing some sort of early detection and monitoring program that specifically targets the planktonic veliger life stage or mussel DNA. This appears to be an improvement over earlier detection programs that looked for settled adults or juveniles on short sections of plastic pipe or bricks suspended into water from a dock or other structure (Kraft 1993; Johnson 1995). Suspending a suitable substrate works well for detecting zebra or quagga mussels once they have already become established in freshwaters, but history has shown that by the time mussels settle on these *samplers*, it is too late to thwart the invasion. In fact, dreissenid mussels may have established sustaining populations in waters for 2 or more years before they are detected on suspended settlement samplers. As a case in point, the National Park Service had monitored settlement samplers deployed in Lake Mead for years before the eventual invasion. Quagga mussels were not initially detected in Lake Mead on these samplers, but on an artificial breakwater during a maintenance operation. To be fair, mussels were eventually found on the Lake Mead samplers. Yet, subsequent searches by the National Park Service recovered from natural substrates mussels large enough to be 2 or more years old (Robert McMahon, The University of Texas at Arlington, personal communication). Thus, if we are interested only in the question of whether a particular water body already has an established population of zebra or quagga mussels, then *settlement samplers* should suffice... eventually. However, if a quick response is warranted, or if we are interested in knowing if zebra or quagga mussels may be present before they have established a sustaining population, then a sampling program that focuses on the microscopic larval stage or DNA is probably a better approach. Especially convincing are the multiple detections of dreissenids in Colorado and Utah by PCR or cross-polarized microscopy, while sightings of adult or juvenile mussels have yet been made. It is possible that monitoring efforts targeting larval stages and DNA are effective enough to detect dreissenids before sustaining populations become established. After an organizational meeting hosted by the U.S. Bureau of Reclamation in Colorado to address early detection and monitoring efforts throughout the West, most Western authorities have chosen to look for larvae or DNA.

Microscopy and PCR

In the West, agencies employ two primary ways for the early detection of zebra or quagga mussels. Both require sampling plankton for larvae. The first technique is simple; a technician can use cross-polarized light microscopy to examine plankton samples for mussel larvae. This technique is tedious and requires an expert eye. With light microscopy, it is not always easy to distinguish between zebra and quagga mussels, or even other microscopic organisms in the sample. There is an automated version of this technique called image flow cytometry that expedites the process by having a robotic machine analyze the samples. The second technique is more complicated but promising. Researchers can analyze plankton samples for molecular markers (DNA) using amplification by PCR. With this method, researchers can easily distinguish between zebra and quagga mussels. However, this technique requires a molecular laboratory and is comparatively expensive.

Many agencies use a combination of these two techniques for early detection. A group of researchers compared traditional cross-polarized light microscopy, image flow cytometry, and PCR assays (Frischer et al. 2011). These researchers found that cross-polarized light microscopy and image flow cytometry had near-perfect accuracy (96.3% and 91.7%, respectively), while PCR assays were accurate only 75.8% of the time. However, researchers in many laboratories across the West are working diligently to improve this technology.

Mandatory Inspections

Across the West, mandatory inspection programs are becoming more ubiquitous. Interdiction or inspection programs exist in California, Idaho, Utah, Oregon, Wyoming, Montana, Washington, Nevada, and Arizona. Lake Mead National Recreation Area and Glen Canyon National Recreation Area are both conducting boat inspections. These programs have all revealed that invasive mussels are common on trailered boats crossing state lines. Transporting live zebra or quagga mussels is illegal in all Western states. As the invasion continues, awareness has elevated. Law enforcement personnel are becoming more responsive to the concerns of those who want to protect their waters. Trailering a boat into the West with invasive species attached is no longer as easy as it used to be.

Consistent Law Enforcement

Enforcing invasive species law is difficult. Some jurisdictions require that those transporting mussels have done so knowingly; otherwise, no violation has occurred. Some jurisdictions require that the mussels are alive while in transport. Such a determination is difficult to make, especially by a law enforcement officer who may have no training in mussel biology. Another problem in inconsistency is related to the infestation status of the jurisdictions (state and federal areas). Those that already have zebra or quagga mussels in their waters appear to be less likely to fine boaters for violating their prohibitions against possessing or moving mussels. Many state agencies with the authority to enforce invasive species laws depend on funds that come, sometimes predominantly, from boat registration fees. In federal areas, recreation fees are often the primary source of funding. Such agencies and areas seem to be averse to biting the hand that feeds them. They do not want to discourage boating or recreational-area visitation. Unfortunately, there is an unspoken conflict of interests between conservation and recreation. Generally, authorities in areas without mussels have asked for our limited funding to be devoted to containment at the source, while authorities in areas with mussels often stress the importance of spending limited funding on prevention at the receiving end of the pathway.

These are problems that the U.S. Fish and Wildlife Service and its partners are addressing. All involved want to prevent the spread of the invasive mussels and are aware of the difficulties that we must overcome. An interagency workshop was convened in Phoenix, Arizona, in 2012 to bring together biologists, law enforcement professionals, and state attorneys general to discuss these issues and build a framework for progress. A second meeting is planned for 2013 in Denver, Colorado. Hopefully, a coordinated approach will minimize overlap and allow us to work cooperatively toward our common goal.

Eradication

Currently, there is no cure for zebra or quagga mussel infestations. Only one eradication attempt has been successful. In Virginia, an isolated small pond was poisoned to remove mussels. A similar attempt in Nebraska was initially thought to be effective, but time revealed a recurrence of zebra mussels in a small lake on Offutt Air Force Base, despite treatment. Poisoning water may not be feasible in larger systems, flowing systems, or in systems where water is used for drinking. The Texas Parks and Wildlife Department attempted to eradicate an incipient population of zebra mussels in Sister Grove Creek, a flowing system in North Texas. State staff used 21,150 lb of potassium chloride, the same chemical that was successful in eradicating the earlier-mentioned population in Virginia. Early indications suggested that the treatment was unsuccessful after biologists found a few living mussels in the creek following treatment (Brian Van Zee, Texas Parks and Wildlife, personal communication). However, a severe drought and an abnormally harsh winter may have finished the job. No mussels have been seen in Sister Grove Creek in over a year.

Unfortunately, there is no magic bullet. Research has focused on controlling these mussels, to keep water flowing, rather than eradicating them in open waters, a more daunting task. With few exceptions, those of us who use waters that currently have zebra or quagga mussels will have to get used to them. Preventing further spread is our best hope at the moment.

The Future

Much has changed in the West since 2007. Before quagga mussels reached the lower Colorado River, very few of us were taking serious precautions against the spread of invasive mussels. Excuses were prevalent from all of our agencies. Limited budgets, limited staff, unmotivated legislators, and other priorities were discussed at our interagency meetings. However, these obstacles quickly disappeared as relentless mussels continued to spread, threatening our waters. In less than a decade, attitudes in the West have changed dramatically as the reality drew closer. Many of the obstacles that hindered us in the past are no longer insurmountable. We have the attention of legislators and governors, and we are doing a fair job at informing the public. Our best strategy is a combination of containment, prevention, and outreach, with law enforcement underlying all three. We may not be able to stop mussels from spreading downstream, but we are making progress in preventing them from reaching new habitat on trailered boats and equipment. Invasion through this pathway is preventable. Until future technologies are developed that will help us control or eradicate these invasive species, our best hope is to slow the spread enough that we can conserve the waters that have not been invaded for as long as possible. A few people have argued with me that invasion is unavoidable. They say it will eventually happen and suggest that this is a reason to do nothing to stop it. I have reminded them that we all will die one day. It is inevitable. But most of us find value in getting up each morning and enjoying our days. We take care of ourselves and try to be optimistic about the future. Giving up is not an option. I believe that many would agree.

Disclaimer

The findings and conclusions in this article are those of the author(s) and do not necessarily represent the views of the U.S. Fish and Wildlife Service.

REFERENCES

Anderson, C., Britton, D., Claudi, R., Culver, M., and Frischer, M. 2008. Zebra/Quagga mussel early detection and rapid response: Blue ribbon panel recommendations for the Colorado division of wildlife. http://wildlife.state.co.us/NR/rdonlyres/010F6AF6-AAF0–41F7-B8CD-BA6722CB7EF9/0/ColoradoBlueRibbonPanelReport.pdf, December 5, 2010.

Bossenbroek, J.M., Johnson, L.E., Peters, B., and Lodge, D.M. 2007. Forecasting the expansion of zebra mussels in the United States. *Conservation Biology* 21:800–810.

California Department of Food and Agriculture. 2009. News release: Agencies ask boaters to 'clean, drain and dry' for a holiday weekend: Help prevent spread of invasive mussels. Release #09-035, www.cdfa.ca.gov/egov/Press_Releases/Press_Release.asp?PRnum = 09-035, January 2011.

Carlton, J.T. 1993. Dispersal mechanisms of the zebra mussel (*Dreissena polymorpha*). In: *Zebra Mussels: Biology, Impacts, and Control.* T.F. Nalepa and D.W. Schlosser (eds.). Lewis Publishers, Ann Arbor, MI. pp. 677–697.

Colorado Division of Wildlife & Colorado State Parks. 2010. State Aquatic Nuisance Species (ANS) Program Summary for Colorado Legislators per SB 08226. www.coloradowater.org/documents/ANSLegislativeReport2010FINAL.pdf, December 12, 2010.

Frischer, M.E., Kelly, K.L., and Nierzwicki-Bauer, S.A. 2011. Accuracy and reliability of *Dreissena* spp. larvae detection by cross-polarized light microscopy, imaging flow cytometry, and polymerase chain reaction assays. *Lake and Reservoir Management* 28:265–276.

Herbert, P.D.N., Muncaster, B.W., and Mackie, G.L. 1989. Ecological and genetic studies on *Dreissena polymorpha* (Pallas): A new mollusc in the great lakes. *Canadian Journal of Fisheries and Aquatic Science* 46:1587–1591.

Johnson, L. 1995. Enhanced early detection and enumeration of zebra mussel (*Dreissena* sp.) veligers using cross-polarized light microscopy. *Hydrobiologia* 312:139–146.

Karatayev, V.A., Karatayev, A.Y., Burlakova, L.E., and Padilla, D.K. 2013. Lakewide dominance does not predict the potential for spread of dreissenids. *Journal of Great Lakes Research* 39:622–629.

Kraft, C. 1993. Early detection of the zebra mussel. In: *Zebra Mussels: Biology, Impacts, and Control.* T.F. Nalepa and D.W. Schlosser (eds.). Lewis Publishers, Ann Arbor, MI. pp. 705–714.

Kraft, C. 1995. Zebra Mussel Update #24. University of Wisconsin-Madison, Wisconsin Sea Grant Institute, Superior, WI.

Levitan, D.R. and Peterson, C. 1995. Sperm limitation in the sea. *Trends in Ecology and Evolution* 10:228–231.

Mackie, G.L. 1991. Biology of the exotic zebra mussel, *Dreissena polymorpha*, in relation to native bivalves and its potential impact in lake St. Clair. *Hydrobiologia* 219:251–268.

May, B. and Marsden, J.E. 1992. Genetic identification and implications of another invasive species of dreissenid mussel in the great lakes. *Canadian Journal of Fisheries and Aquatic Science* 49:1501–1506.

McMahon, R.F., Ussery, T.A., and Clarke, M. 1993. Use of emersion as a zebra mussel control method. U.S. Army Corps of Engineers Contract Report EL-93-1. U.S. Army Engineer Waterways Experiment Station, Vicksburg, MS.

Mills, E.L., Rosenberg, G., Spidle, A.P., Ludyanskiy, M., Pligin, Y., and May, B. 1996. A review of the biology and ecology of the quagga mussel (*Dreissena bugensis*), a second species of freshwater dreissenid introduced to North America. *American Zoologist* 36:271–286.

O'Neil, C.R. and Dextrase, A. 1994. The introduction and spread of the zebra mussel in North America. *Proceedings of the Fourth International Zebra Mussel Conference*, Madison, WI.

Padilla, D.K. 2005. The potential of zebra mussels as a model for invasion ecology. *American Malacological Bulletin* 20:123–131.

Sprung, M. 1990. Costs of reproduction: A study on metabolic requirements of the gonads and fecundity of the bivalve *Dreissena polymorpha*. *Malacologia* 32:267–274.

Sprung, M. 1993. The other life: An account of present knowledge of the larval phase of *Dreissena polymorpha*. In: *Zebra Mussels: Biology, Impacts, and Control.* T.F. Nalepa and D.W. Schlosser (eds.). Lewis Publishers, Ann Arbor, MI. pp. 39–54.

Thomas, G.H. and Lamberti, G.A. 1999. Mortality of zebra mussel, *Dreissena polymorpha*, veligers during downstream transport. *Freshwater Biology* 42:69–76.

Transportation Research Board. 2008. Great Lakes Shipping, Trade, and Aquatic Invasive Species. Transportation Research Board Special Report 291. 148pp. Transportation Research Board. Washington, DC.

U.S. Geological Survey. 2010. *Nonindigenous Aquatic Species Database*, Gainesville, FL. nas.er.usgs.gov, December 10, 2010.

Utah Division of Wildlife Resources. 2011. Utah Division of Wildlife Resources' Attack against the Invasion of Quagga and Zebra Mussels (2010 Boating Season Summary). wildlife.utah.gov/mussels/PDF/ais_summary_annual_2010.pdf, January 3, 2011.

Western Regional Panel on Aquatic Nuisance Species. 2010. Quagga-Zebra Mussel Action Plan for Western Waters. Submitted to the Aquatic Nuisance Species Task Force. www.anstaskforce.gov/QZAP/QZAP_FINAL_Feb2010.pdf, December 2010.

Section II

Biology

2

Invasion by Quagga Mussels (Dreissena rostriformis bugensis Andrusov 1897) into Lake Mead, Nevada–Arizona: The First Occurrence of the Dreissenid Species in the Western United States

Bryan Moore, G. Chris Holdren, Shawn L. Gerstenberger, Kent Turner, and Wai Hing Wong

CONTENTS

ABSTRACT On January 6, 2007, quagga mussels (*Dreissena bugensis*) were found in the Boulder Basin of Lake Mead, the largest reservoir by volume in the United States. This was the first known occurrence of dreissenid species in the western United States, and it was also the first case of large ecosystem being infested by quagga mussels without being previously invaded by zebra mussels, the first dreissenid mussel found in North America. In 2007, the average density of juveniles and adults was 505 mussels/m^2 ranging from 0 to 3368/m^2. There were more mussels in rocky areas than silty areas. Size frequency demonstrated that there were always three to four cohorts for each population with shell lengths ranging from less than 1 mm to 25 mm.

Quagga mussel veligers were present year round. Average veliger numbers from March 2007 to March 2008 were 6.6/L in Boulder Basin, 0.7/L in Overton Arm, 0.3/L in Temple Bar of the Middle of Lake Mead, and 0.04/L in Gregg Basin in upper Lake Mead. In Boulder Basin, veligers peaked in summer (July) and fall (October) of 2007, while in the middle of this reservoir, only one peak was found and there was no peak in the upper Lake Mead. The results of the veliger analyses demonstrated that adult mussels were established in Boulder Basin, but not in the upper lake. This matches the result of abundance and distribution of adults and juveniles.

Shell length structure analysis revealed that the first detectable settlement occurred in mid-2005. It is hypothesized that a reproducing quagga mussel population must have been established in the Boulder Basin by 2003 or 2004 in order to generate enough veliger larvae to settle in 2005. Since 2007, quagga mussels have increased exponentially. These records on the early invasion of quagga mussels into Lake Mead provide baseline information to lake managers and research scientists on monitoring quagga mussels in Lake Mead and the western United States.

The appearance of dreissenid mussels in the Laurentian Great Lakes and their spread into other waters is one of the most significant freshwater invasions in North America. Native to the Black, Caspian, and

Azov Seas, zebra mussel *Dreissena polymorpha* Pallas (1771) was discovered in the United States in 1988 in Lake St. Clair near Detroit, Michigan. It was likely introduced into this area in 1986 through the discharge of ballast water contaminated with planktonic veliger larvae (Hebert et al. 1989; Nalepa and Schloesser 1993). The first detection of zebra mussels is in the western basin of Lake Erie (Ontario, Canada), which was reported on natural gas wellheads and well markers between April and November 1986 (Carlton 2008). The zebra mussels' high fecundity, passively dispersed planktonic veliger larval stage, and ability to attach by proteinaceous byssal threads to boat and barge hulls, nets, buoys, and floating debris allowed it to spread rapidly to the lower Great Lakes, freshwater portions of the St. Lawrence River, and other major basins, such as the Illinois River, Mississippi River, and Hudson River drainages (Ram and McMahon 1996). From 1992 to 2002, however, the average density and biomass of zebra mussels decreased exponentially in Lake Erie (Patterson et al. 2005).

During the same time period, another species, the quagga mussel, *Dreissena rostriformis bugensis* Andrusov (1897), became more dominant in Lake Erie. Although the first sightings of the quaggas in the Great Lakes were in September 1989, when one quagga was found near Port Colborne, Lake Erie (Mills et al. 1993), quagga mussels were first recognized as established in August 1991 in the Erie Canal and Lake Ontario, where they coexisted with more numerous *D. polymorpha* (May and Marsden 1992).

D. rostriformis bugensis was first discovered in the Bug portion of the Dnieper-Bug Estuary near Nikolaev of Ukraine by Andrusov in 1890. Quagga mussels have since spread into the Dnieper River drainage, regions that earlier had only *D. polymorpha* (Mills et al. 1996). Recently, quagga mussels have been found in the Hollands Diep of Western Europe (Molloy et al. 2007). In Lake Erie, the proportion of quagga mussels increased linearly with lake depth (Mills et al. 1993), while in the shallow areas, zebra mussels were more abundant between 1994 and 1998 (Berkman et al. 2000).

As both *Dreissena* species are biofoulers, they have impacted the whole ecosystem and local economy. The economic impacts from *Dreissena* species, such as infestation on power plant water systems, infrastructures, and navigation, are tremendous. The economic loss due to the invasion of zebra and quagga mussels is projected around $1 billion a year in the United States (Pimentel et al. 2005). The zebra/quagga mussel has become arguably the most serious nonindigenous biofouling pest introduced into North American freshwater systems (LaBounty and Roefer 2007) and one of the world's most economically and ecologically important pests (Aldridge et al. 2006).

In North America, quagga mussels were primarily found in the Great Lakes basin with two outside sites on the Mississippi River near St. Louis, Missouri, before 2007 (O'Neill 1995). It was suggested that the quagga mussel would not be able to expand as far south in North America as has the zebra mussel because of its lower upper thermal limit (Spidle et al. 1995). However, quagga mussels appeared in Lake Mead, (Nevada–Arizona) the largest reservoir in the United States by water volume (LaBounty and Roefer 2007). The mussel was first found on January 6, 2007, by an employee at the Las Vegas Boat Marina while he was diving in the Marina area to make repairs following a wind storm that damaged the breakwater and docks. While descending down a cable, he saw a mussel attached to it, pulled it off, and brought it to the surface. Two days later, Wen Baldwin, a National Park Service (NPS) volunteer who has been monitoring for zebra mussels in Lake Mead for several years as part of the 100th Meridian Initiative, was called by the Marina manager. The 100th Meridian Initiative was initiated in 1998 by a group of biologists, wildlife officials, water managers, environmentalists, and others with the goal of preventing invasive species from crossing the 100th meridian, a historical boundary separating eastern and western United States (Britton 2014, Chapter 1 in this book). This mussel was identified as the invasive quagga mussel by the U.S. Wildlife and Fisheries Services (LaBounty and Roefer 2007; Rake 2007). It has been postulated that quagga mussels were introduced into Lake Mead via bilge water (i.e., bait or live wells) carried by a boat from the Great Lakes region (Choi et al. 2013; McMahon 2011; Wong and Gerstenberger 2011).

Once these mussels were found in Boulder Basin, multiple agencies including NPS and Bureau of Reclamation started to assess the extent of the infestation of quagga mussels in Lake Mead. Although there have been several scattered reports on this invasion event to the western United States (LaBounty and Roefer 2007; McMahon 2011; Stokstad 2007; Wong and Gerstenberger 2011), there is no detailed information on the early invasion status of quagga mussels in Lake Mead. In this study, we reported information on the detection of their presence/absence, adult abundance and distribution, population length structure, and veliger abundance during their early invasion stage in Lake Mead.

Methods

Quagga mussels have a planktonic free-swimming larval stage (veligers) and a benthic living stage (attaching to substrates). Therefore, adults, juveniles, and veligers were all monitored beginning in 2007, the first year that quagga mussels were found in Lake Mead. Adults and juveniles were surveyed by SCUBA divers, and veligers were monitored by plankton tows.

Monitoring of adults and juveniles was divided into two phases. Phase I (January 9 to early March 2007) checked for the presence and absence of quagga mussels in different locations of Lake Mead. Lake Mead has four major basins, Boulder Basin, Overton Arm and Virgin Basin, Greg Basin, and Temple Basin. Because these mussels were first found in Las Vegas Boat Harbor Marina in Boulder Basin, the detection of adult and juveniles by divers started from this Marina and other sites in Boulder Basin. The survey was extended to other basins (more information can be found in Table 2.1 and Figure 2.1). While we attempted to count all the mussels found in the study area, only the mussels that were emergent and/or visible after brushing the silt away were counted. Those newly settled mussels with small sizes (<1.5 mm) could not be counted due to lack of visibility. Survey areas deeper than 39.6 m (130 ft) were surveyed with a remotely operated vehicle (ROV, Pro 3 XEGTO made by VideoRay LLC, Phoenixville, Pennsylvania) to detect quagga mussels.

In phase II (late February to the end of 2007), the characteristics of quagga mussel populations, such as density, length frequency, maximum shell length, and growth rate between shell length, were determined at different sites in Lake Mead. To measure the density of quagga mussels in Lake Mead, 15 m transects were installed at nine locations (Table 2.2). Usually, the type of substrates present is a key factor in determining where dreissenid mussels are likely to attach and grow. These mussels mostly settle on hard substrate and tend to prefer dark areas, corners, crevices, and the shells of other mussels (Marsden 1992). Thus, hard substrates such as rocks and stones are expected to have more mussels than those with less compaction such as silt and mud.

The subsurface of Lake Mead is much larger than its surface area, and the sediment composition is heterogeneous. In Lake Mead, the largest proportion of subsurface is rock (Twichell 2005). Therefore, most of the locations selected for setting up transects were rocky areas, although one location was covered with silt and mud (Table 2.2). At each location, a series of three transects at depths of 6.1, 12.2, and 18.3 m were installed. These transects were anchored with large galvanized steel nails (25 cm) pounded into the substrate at both ends, and a piece of yellow polypropylene line affixed to the two nails. The nails were fixed in place with epoxy. In areas composed of cobble, cinder blocks were used to anchor the transect line.

Transects were marked and located with GPS coordinates. A pelican buoy was tossed into the water to mark the GPS location for divers descending to the transect. Once the transect was found, the buoy was tied to the transect for the duration of the sampling to facilitate finding the transect throughout the day. The diver reeled out a meter tape from left to right to mark the area, when oriented facing the shore. Along the transect, the density of quagga mussels was measured and recorded by a SCUBA-equipped observer with a $0.25 \, m^2$ quadrat (PVC), underwater clipboard, and underwater quadrat data sheets. Multiple sites along each transect were sampled at 3 m sampling intervals. The sampling points were recorded on the data sheets.

Due to the amount of silt accumulated on the substrate, divers had to clear it away by fanning the area in the quadrat with their hands in order to get an accurate account. We tried to minimize contact with the bottom as much as possible, as visibility decreased quickly. Newly settled mussels with small sizes could not be observed due to limited visibility. As a result, the density of quagga mussels in this study may be underestimated due to the exclusion of this cohort.

In addition to the transects, rock samples were collected at least 1 m from the transect in order not to alter densities along the transect. Samples were then taken back to the surface where shell length was measured as the distance between anterior and posterior ends of the shell to the nearest 0.1 mm using a vernier caliper.

For each field collection, length–frequency data were grouped into size classes of 1 mm intervals, and the percentage frequency of the whole sample contributed by each size class was plotted as a

TABLE 2.1

Survey on the Presence and Absence of Quagga Mussels in Lake Mead from January to Early March 2007

Date	Location	Basin	Dives	Results
January 9	Las Vegas Harbor Boat Marina	Boulder	2	Mussels were found on concrete anchor blocks, steel cables, and dock structures and breakwater tires from 1 to 50 ft. The majority were concentrated on the concrete anchor blocks.
January 14	Lake Mead Marina	Boulder	2	About 200 individuals in a 0.9 m × 0.9 m area on the underside of a concrete anchor block at 9.8 m. Only a few mussels were found on steel cables and dock structures.
January 18	Overton Beach Marina	Overton	1	No mussels were found in a houseboat, steel dock structures, and concrete anchors.
January 18	Echo Bay Marina	Overton	1	No mussels were found on any of the 10 inspected houseboats and any of the substrates shallow than 13.7 m.
January 19	South Cove Launch Ramp	Gregg	2	No mussels were found on any of the structures and substrates shallower than 27.4 m.
January 26	Flamingo Reef	Boulder	1	Mussels were found on the underside of overhanging rock or in crevices from 12.2 to 30.5 m. No diving past 30.5 m.
January 26	East End Light	Virgin	1	Mussels were found on the underside of overhanging rock or in crevices from 12.2 to 18.3 m. No diving past 30.5 m.
January 28	Middle Point (south and north sides)	Virgin	1 for each side	Dived down to 30.5 m, but no mussels were found on rocks or on top of silt layer.
January 28	Gypsum Bay	Virgin	1	No mussels were found.
January 30	Sandy Point	Gregg	2	No mussels were found in the silt and rocky areas.
February 6	Boulder Wash Cove (east and west sides)	Virgin	1 for each side	No mussels were found in the silt and rocky areas in the eastside (30.5 m), but two mussels were found in the rocky areas in the west side (30.5 m).
February 8	Government Dock	Boulder	1	Lower numbers of mussels were found on all concrete anchor blocks, steel cables, and dock structures.
February 8	Sentinel Island	Boulder	1	Largest number of mussels were found from 0.9 to 30.5 m on the rocks and attached to rocks under silt layer.
February 9	Black Island	Boulder	1	Large number of mussels were found from 3.7 to 30.5 m on the rocks and attached to rocks under silt layer, but smaller than the number of Sentinel Island.
February 9	Las Vegas Bay	Boulder	1	Six mussels were found from 10.7 to 19.8 m in the mud and silt areas, and the maximum diving depth was 26.8 m.
February 10	Boulder Island (Southwest)	Boulder	No dive	A few mussels on a concrete wall were found when an ROV was used to survey the batch plant structures of quagga mussels on concrete, wood, rock, and silt areas up to 43 m in depth.
February 11	Indian Canyon	Boulder	1	Mussels were found from 3 to 16.2 m on rock wall with a maximum diving depth at 16.2 m (computer malfunctioned, so, end diving at 16.2 m).

(Continued)

TABLE 2.1 (*Continued*)

Survey on the Presence and Absence of Quagga Mussels in Lake Mead from January to Early March 2007

Date	Location	Basin	Dives	Results
February 11	Burro Point	Boulder	1	Less number of mussels were found from 6.1 to 14.6 m on the rocks and silt areas with a maximum diving depth at 14.6 m.
February 11	Battleship Rock	Boulder	1	Large number of mussels were found between 7.3 and 16.5 m on rock and silt areas with a maximum diving depth at 16.5 m.
February 11	Water Barge Cove	Boulder	1	Mussels were usually found on rocks on the underside or shaded area, no mussels in silt area, with a maximum diving depth at 12.8 m.
February 19	Wishing Well Cove	The narrows between Boulder and Virgin Basins	1	Mussels were found on rock wall face from 5.5 to 20.4 m with a maximum diving depth at 20.4 m.
February 19	Stewart Cliffs	Virgin	1	Mussels were found on rock wall face from 6.1 to 21.0 m with a maximum diving depth at 21.0 m.
February 24	Overton Arm	Overton	2	No mussels were found on rocks or silt areas.
March 7	Spring Cove	South end of Gregg Basin	1	No mussels were found on rocks or silt areas with a maximum diving depth at 21.9 m.
March 7	The Temple	Temple	1	No mussels were found on rocks or silt areas with a maximum diving depth at 28.3 m.

length–frequency histogram. In order to follow the growth of separate cohorts over the sampling period, individual cohorts (size classes) were separated using the modal progression analysis of Fish Stock Assessment Tool II (FiSAT II, http://www.fao.org/fi/statist/fisoft/fisat). FiSAT II applies the maximum likelihood concept to separate the normally distributed components of size–frequency samples, allowing accurate demarcation of the component cohorts from the composite polymodal population size–frequency distribution. For each cohort identified, mean lengths, with standard deviations, and group sizes (in numbers) were estimated.

Plankton tow were also collected in Lakes Mead and Mohave (downstream of Lake Mead) to look for quagga mussel veligers beginning in March 2007. Samples were taken from the following stations: Sandy Point (N 36.11544, W 114.11563), Echo Bay (N 36.29487, W 114.39855), Temple Bar (N 36.04284, W 114.31030), Hoover Dam (N 36.0193836, W 114.7326641) in Lake Mead and Willow Beach (N 35.86933, W 114.66486), Placer Cove (N 35.70580, W 114.70539), Cottonwood Cove (N 35.49412, W 114.68017), and Katherine Landing (N 35.22015, W 114.57993) in Lake Mohave downstream of Lake Mead.

Monthly samples were collected from each station from March 2007 to 2008. A vertical plankton tow was collected by gently lowering a 15 cm diameter plankton net with a 64 μm mesh size into the water and retrieved at a rate of approximately 1 m/s (a steady and unhurried hand-over-hand motion). Tow lengths were 30 m at the Lake Mead stations and 30 or 1 m above the lake bottom, whichever is less, in Lake Mohave. Usually, the minimum water volume collected was 1000 L. After each tow, the net was washed from top to bottom from the outside with distilled water to rinse veligers into the collection cup. The collection cup side screens were also washed from top to bottom and then emptied into a 500 mL Nalgene bottle. The collection cup was rinsed twice with small amounts of water and emptied into the same bottle. The bottle was labeled with date and location. Sample bottles were kept on ice while in the field. Once sampling was complete, the plankton net was decontaminated by being soaked in a 5 gal bucket filled with white vinegar for about 45 min. The plankton net was thoroughly rinsed with clean water after each soaking (Hosler 2011).

Plankton samples were preserved using ethyl alcohol until it comprised 25% of the final sample volume. Samples were refrigerated until they were ready to be analyzed by cross-polarized light microscopy (CPLM). We originally utilized the Army Corps of Engineers method for veliger enumeration;

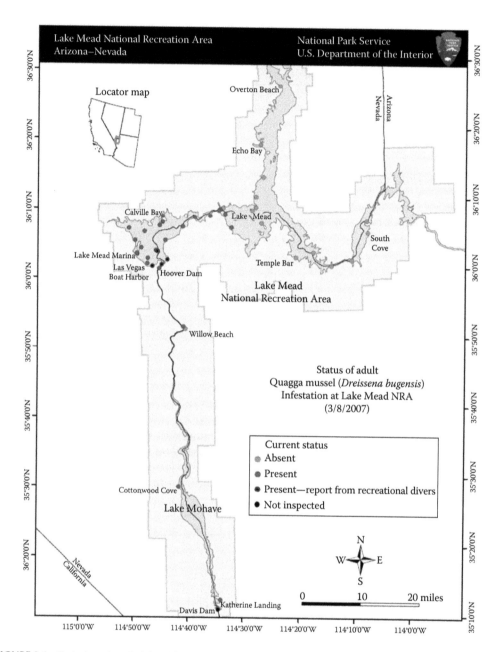

FIGURE 2.1 Early detection of adult and juvenile quagga mussels in Lake Mead, Nevada–Arizona.

however, the method was modified as the concentration of veligers in Lake Mead became too numerous to count. The modified procedure combined Standard Method 10200G, the U.S. EPA Method LG403, Revision 2, and the USACE procedure and yielded consistent quantitative results with the continued benefit of early detection when veliger counts were below five veligers. More detailed information about this method can be found from Hosler (2011) and Carmon and Hosler (2014, Chapter 10 in this book). The sample preparation was modified to allow veligers to settle in an Imhoff cone for a minimum of 24 h, and the results of this modification led to a 98% recovery rate of veligers in the first 15 mL off of the cone. This recovery rate was determined by analyzing successive 15 mL aliquots of the settled sample and examining the contents of each 1 mL aliquot examined under CPLM. Per the USACE method, the mean of five cell counts is used to obtain the mean number of veligers per milliliter in the sample

TABLE 2.2

Locations for Determining Quagga Mussel Density in Lake Mead

Location	Latitude	Longitude	Date	Depth (m)	Substrate
Steward Cliffs	36°08.591 N	114°31.940 W	03/25/07	6.1, 12.2, 18.3	Rock
Sentinel Island (east)	36°06.854 N	114°39.798 W	03/30/07	6.1, 12.2, 18.3	Silt
Sentinel Island (west)	36°03.562 N	114°44.805 W	04/04/07	6.1, 12.2, 18.3	Rock
Black Island	36°06.297 N	114°44.805 W	06/21/07	6.1, 12.2, 18.3	Rock
Indian Canyon Cove	36°06.889 N	114°44.805 W	06/22/07	6.1, 12.2, 18.3	Rock
Steward Cliffs	36°08.591 N	114°44.805 W	03/25/07	6.1, 12.2, 18.3	Rock
		114°16.289 W	08/24/07	6.1,[a] 12.2, 18.3	Rock
Boulder Islands			08/29/07	6.1,[a] 12.2, 18.3	Rock
The Temple	36°02.853 N	114°16.289 W	09/07/07	6.1, 12.2, 18.3	Rock
Cormorant Point	36°13.729 N	114°24.796 W	12/18/07	6.1,[a] 12.2,[a] 18.3	Rock

[a] Samples were not taken from these depths as the original transects were covered by mudslides.

(Hosler 2011). This protocol has been developed by the U.S. Bureau of Reclamation and is currently being used by multiple agencies in the lower Colorado River (Carmon and Hosler 2014, Chapter 10 in this book; Gerstenberger et al. 2011; Hosler 2011).

In addition to microscopy, polymerase chain reaction (PCR), a molecular technique that can be used to provide positive identification of a particular species by confirming the presence of DNA, was also used to check for the possible presence of dreissenids. Plankton samples for PCR analyses were collected using the same methods as the microscopy samples but were preserved to a final ethanol concentration of 75%. The method used by Reclamation for these studies was based on Frischer et al. (2002).

Results

Detection of Adult and Juvenile Quagga Mussels

Three days (i.e., January 9, 2007) after the mussels were found at Las Vegas Boat Harbor Marina, *D. rostriformis bugensis* was confirmed present by diving at that marina (Table 2.1). From January 9 to March 7, 2007, a total of 31 dives and 1 ROV survey were conducted to detect the presence and absence of quagga mussels in Lake Mead (Table 2.1). Generally, mussels were concentrated in the Boulder Basin (lower Lake Mead), while no mussels were found in the upper lake (i.e., Gregg Basin, Temple Bar, Virgin Basin, and Overton Arm) (Figure 2.1). Mussels were found in depths from 0.3 to 43.0 m in concrete anchor blocks, steel cables, dock structures, breakwater tires, crevices, rocks, and some silty areas (Table 2.1). On January 12 and 13, 2007, our survey found no mussels on similar structures (concrete anchor blocks, steel cables, dock structures, and houseboats) at Cottonwood Cove Marina or Willow Beach Gas Dock in Lake Mohave, which is downstream of Lake Mead.

Density

In total, 138 samples were taken from transects (Table 2.2) set up in different locations of the lake (Table 2.3). The average density was 505.7 individuals/m^2, which was close to the density of quagga mussels found in 1992 in Lake Erie (Patterson et al. 2005), where they were first found in August 1991 (May and Marsden 1992). The density of mussels in the rocky area was significantly higher than those in the silty area (T-test, $P < 0.001$). In Sentinel Island (Figure 2.2), there were more mussels in rocky areas than silty areas, and the density in both areas increased with depth (ANCOVA, $P < 0.001$). At Stewart Cliffs, the mean density of quagga mussels at 12.2 and 18.3 m were 20.0 and 17.6 individuals/m^2, respectively, while they increased to 72.8 and 103.2 individuals/m^2, respectively, on August 24, 2007 (Figure 2.3). Analysis of variance showed that the increase is highly significant ($P < 0.001$).

TABLE 2.3

Density of Quagga Mussels Collected from Lake Mead in 2007

Sampling Sites	Density (Individuals/m²)	Number of Sampling Sites	Min.	Max.
Whole lake	505.7	138	0	3368.0
Rocky area	624	108	0	3368.0
Silty area	79.6	30	0	628.0

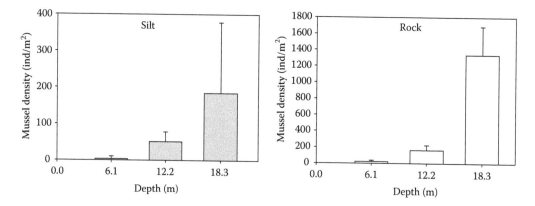

FIGURE 2.2   Quagga mussel density in silty and rocky areas of Sentinel Island, Lake Mead (note: the scales of y-axes are different).

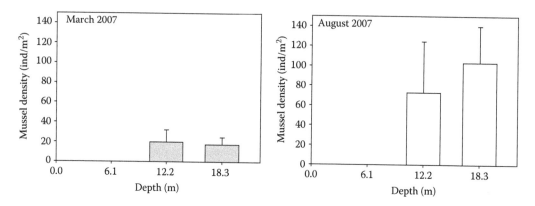

FIGURE 2.3   Quagga mussel density in March and August 2007 at Stewart Cliffs, Lake Mead.

Detection and Abundance of Quagga Mussel Veligers

Although the initial concentrations were low (<0.2 veligers/L), veligers or their DNA were detected at all Lake Mead and Lake Mohave stations beginning in March 2007, the start of veliger monitoring. Microscopy identified veligers at the Temple Bar and Hoover Dam sites in Lake Mead and at all Lake Mohave stations on the first sampling dates in March 2007. For the 10 sampling dates in 2007, veligers were found 7 times at Sandy Point and 9 times at Echo Bay, and these were also found on all sampling dates at both Temple Bar and Hoover Dam and, with the exception of the August 2007 sample at Cottonwood Cove, on all sampling dates at the Lake Mohave stations.

Dreissenid DNA was found by PCR at all stations in both Lakes Mead and Mohave in March 2007. All Lake Mohave samples tested by PCR in 2007 were positive, while 84% of PCR samples in Lake Mead were positive. This compares to positive results in 92% of Lake Mead samples and 98% of Lake Mohave samples by CPLM. Several unexpected difficulties, including inhibition and other interferences, were noted with PCR. As a result of these difficulties, coupled with the now ubiquitous presence of dreissenid DNA in the lower Colorado River and the ability of CPLM to provide quantitative data, PCR analyses were discontinued at the end of September for Lake Mohave samples and at the end of December for Lake Mead samples.

Veligers continue to be present at these stations (Figure 2.5). The highest abundance of veligers was found at the Hoover Dam station (40.2 individuals/L) in October 2007. The average concentration of veligers during the sampling period ranged from 0.04 individuals/L (N = 13 months, standard deviation = 0.06) at South Cove to 6.11 individuals/L at Hoover Dam (N = 13 months, standard deviation = 11.57). The mean concentration of veligers at the upper Lake Mead stations, Sandy Point, Temple Bay, and Echo Bay were all <1.0 individuals/L and were significantly lower than the mean concentration of 6.11 individuals/L at the Hoover Dam station (ANOVA, LSD, $P < 0.05$).

The veliger abundance in stations of Lake Mohave was between 1.0 and 5.0 individuals/L. If Boulder Basin is considered as the lower Lake Mead, other Lake Mead stations are considered as upper Lake Mead, and stations in Lake Mohave are considered as a whole; the order of veliger concentrations is lower Lake Mead (mean = 6.11 individuals/L) ≥ Lake Mohave (2.46 individuals/L) ≥ upper Lake Mead (0.36 Individuals/L) (Figure 2.5). In Boulder Basin, there were two peaks in veliger concentrations, with one in early July and the other one in October. Veliger concentrations at the upper Lake Mead stations also peaked in October, while distributions in Lake Mohave were more variable (Figure 2.5).

Discussion

The discovery of quagga mussels in Las Vegas Boat Harbor Marina on January 9, 2007 (Table 2.1), confirmed that quagga mussels were introduced into Lake Mead. At this same location of Las Vegas Boat Harbor Marina, no juvenile or adult mussel had been detected 10 months prior (Bryan Moore, personal observation). On January 9, 2007, quagga mussels were also discovered in the Nevada Division of Wildlife fish hatchery facilities, which operate with waters from Lake Mead. The initial presence of adult quagga mussels in the Boulder Basin of Lake Mead and their absence in other basins indicated that the Boulder Basin was the origin of invasion for quagga mussels into the largest reservoir in the United States.

Before its discovery in Lake Mead, the quagga mussel was primarily restricted to the Great Lakes region. The presence of quagga mussels in Lake Mead in 2007 was the first confirmed introduction of a dreissenid species in the western United States, and it was also the first time that a large ecosystem was infested by quagga mussels without previous infestation by zebra mussels. It has been postulated that quagga mussels were introduced into Lake Mead via ballast water carried by a boat from the Great Lakes region. Lake Mead National Recreation Area is one of the busiest parks in the United States with more than eight million visitors a year (Holdren and Turner 2010). This discovery of quagga mussels in Lake Mead extended the U.S. range of this nonnative species about 3000 km west of previously known populations in the Great Lakes (Wong and Gerstenberger 2011).

Shell length structure analysis (Figure 2.4) and growth rate recorded in Lake Mead (Wong et al. 2012) revealed that the first detectable settlement occurred in the summer of 2005. It is therefore hypothesized that a reproducing quagga mussel population must have been established in the Boulder Basin by 2003 or 2004 in order to generate enough veliger larvae to settle in 2005. This is in agreement with another independent study by McMahon (2011), who estimated that quagga mussels must have been established during 2003 or 2004.

The lower abundance of veligers in upper Lake Mead is in agreement with the detection of presence/absence results from adult/juveniles in early 2007 where no visible adult/juvenile mussels were found in the upper Lake Mead (Figure 2.5). From April 2008 to March 2009, veligers were present year-round in Lake Mead, with high abundance from September to October (>20 veligers/L) (Gerstenberger et al. 2011). There were also two peaks, one in summer and the second during the fall. However, due to the presence of veligers in upper Lake Mead (this study), it is not surprising that a large cohort of juvenile quagga mussels was found in a rocky area in South Cove in September 2008 with a mean density of 54,242 individuals/m^2 (Figure 2.6), although no adult/juvenile mussels were detected in March 2007 in a nearby area (Sandy Point in the present study). Based on the length frequency data collected in

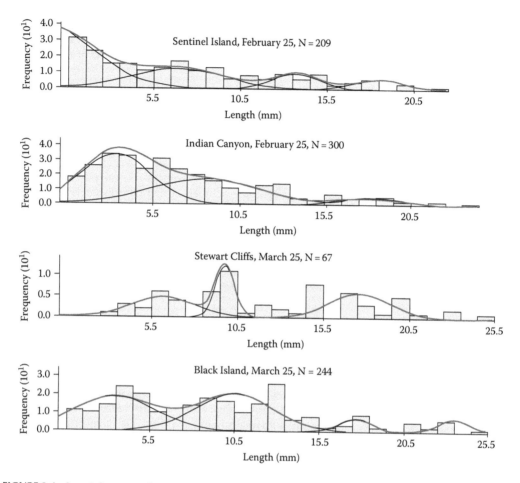

FIGURE 2.4   Length frequency of quagga mussels in Lake Mead in early 2007.

South Cove (Gregg Basin of Lake Mead) and growth rate of quagga mussels in Lake Mead (Wong et al. 2012), the time for the original cohorts to settle should be after January 2008. This is in agreement with the lower number of veligers in Sandy Point (Gregg Basin of Lake Mead) from March 2007 to 2008 (Figure 2.5) because there are not many adults that produce veligers during that period.

Although no mussels were found in our early January 2007 survey in the downstream Cottonwood Cove Marina and Willow Beach Gas Dock, adult mussels were detected in these areas late in 2007. This indicates that Lake Mead was the source providing veligers to these downstream areas where veligers are transported by the flow, and they settle down on available substrates. At the same time, quagga mussels with shell length from 4.0 to 21.2 mm were surprisingly found on houseboat hulls in Katherine Landing, an even further downstream marina, on March 13, 2007 (McMahon 2011). McMahon compared the population structures of quagga mussels collected between January to March 2007 from three sampled sites in Boulder Basin of Lake Mead (Las Vegas Boat Harbor, Callville Bay Marina, and Lake Mead Marina). McMahon (2011) found that it is difficult to speculate Boulder Basin as the original site of quagga mussel introduction due to the essentially equivalent population structure. The largest mussels found in Steward Cliffs and Black Island in March 2007 are both 25 mm in shell length, which is much larger than those found in Lake Mohave (McMahon 2011). It is highly likely that most downstream mussels are produced in Lake Mead in the early invasion of quagga mussels in the Lower Colorado River Basin, based on size frequency analysis and growth rate of quagga mussels in Lake Mead (Wong et al. 2012).

The average quagga mussel density in Lake Mead in 2007 was 506 individuals/m², which was close to the density of quagga mussels in 1992 in Lake Erie (Patterson et al. 2005), where they were first

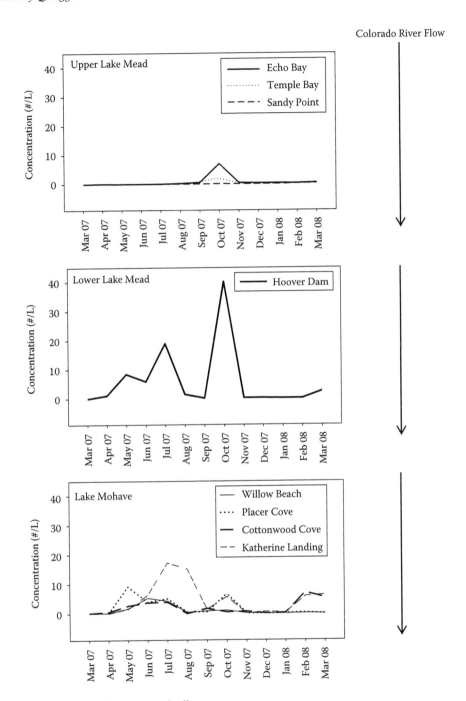

FIGURE 2.5 Abundance of quagga mussel veligers.

found in August 1991 (May and Marsden 1992). The water level in Lake Mead was about 335 m in 2007. The estimated subsurface areas for soft and hard substrates of the Boulder Basin are 55,512,699 and 53,489,199 m^2, respectively (Twichell et al. 1999). Based on the average density of quagga mussels in the present investigation, there were approximately 47,798,551,229 adults and juveniles in the Boulder Basin. At the same water level, the water volume of the Boulder Basin is 6,036,012,620,396 L (Barnett and Pierce 2008). Based on the yearly average concentration of veligers, the estimated number of veligers in this basin is 38,539,940,581,229.

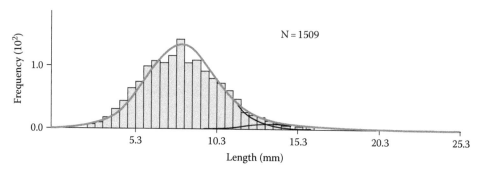

FIGURE 2.6   Length frequency of quagga mussels collected in South Cove, Lake Mead, on September 16, 2008.

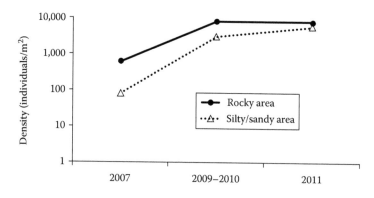

FIGURE 2.7   Quagga mussel density in Lake Mead from 2007 to 2011 (note: the y-axis is in logarithmic scale; the numbers of sampling sites in rocky and silty-sandy areas in 2007, 2009–2010, and 2011 are 108/30, 48/103, 24/34, respectively; only mean density values are presented).

Since 2007, the population size has increased exponentially (Figure 2.7). The overall density of quagga mussel adult and juveniles in Lake Mead in 2010 is 4613 individuals/m², while it was only 506 individuals/m² in 2007. Caldwell et al. (2014, Chapter 3 in this book) also found the same trend of quagga mussels in the soft sediment in Lake Mead. Studies have shown that quagga mussels usually do not reach peak abundance until about 12 years after arrival (Karatayev et al. 2010), but these studies have been conducted in temperate regions such as Europe and Great Lakes.

Why is that quagga mussels grew so fast in such a short time in Lake Mead? Lake Mead is a large subtropical reservoir with thermal and hydrological regimes that are different from water bodies in temperate regions (Holdren and Turner 2010). It provides excellent environmental conditions for quagga mussels to survive, grow, and reproduce (Table 2.4). Therefore, it is not surprising that quagga mussels grew exponentially within 3–4 years. However, the long-term population dynamics in Lake Mead is unclear. It is likely that food could be a potential limiting factor for quagga mussels in Lake Mead as it is a relatively unproductive system, especially in the outer portions of the Boulder and Virgin Basins (Cross et al. 2011). Yet quagga mussels maintain growth at a very low chlorophyll a concentration (0.05 mg/m³) in the Great Lakes (Baldwin et al. 2002). Therefore, long-term monitoring is needed in order to have a better understanding of the population size of quagga mussels in Lake Mead. A long-term standardized quagga mussel monitoring program has been established in Lake Mead (Wong and Gerstenberger 2011). The program tracks changes in mussel size, abundance, and distribution at multiple sites. This program was implemented by lake managers and participating agencies in late 2009 and will help lake managers better understand and accurately assess quagga mussel impacts in the lower Colorado River Basin (Turner et al. 2011).

TABLE 2.4

Environmental Conditions of Lake Mead and Limiting Factors for Dreissenid Mussels

Factors	Threshold	Lake Mead	Limited?
Calcium concentration (mg/L)	<12 (Jones and Ricciardi 2005)	69.1–87.0 (Whittier et al. 2008)	No
Temperature (°C)	>30 (Mills et al. 1996; Spidle et al. 1995)	<28 (LaBounty and Burns 2005)	No
Salinity (ppt)	>4 (Kennedy et al. 1995) or >5 (Strayer 1991)	<1 (LaBounty and Burns 2005)	No
pH	<6.5 (McMahon 1996)	8 (LaBounty and Burns 2005)	No
Oxygen (% of saturation)	<25 (Cohen 2005; Hayward and Estevez 1997; Sorba and Williamson 1997)	>40 (LaBounty and Burns 2005)	No
Secchi depth (m)	<0.1 (Claudi and Mackie 1994; Hayward and Estevez 1997; Sorba and Williamson 1997)	>3.3 (LaBounty and Burns 2005)	No
Chlorophyll a (µg/L)	<2.5 or >25 (Claudi and Mackie 1994; Hincks and Mackie 1997)	>2 (LaBounty and Burns 2005)	No
Flow velocity (m/s)	>1.5 (Claudi and Mackie 1994)	<0.83 (Tracy Vermeyen, personal communication)	No
Total phosphorus (µg/L)	<5 or >35 (Claudi and Mackie 1994; Koutnik and Padilla 1994; Mackie et al. 1989)	10.0–18.0 (LaBounty and Burns 2005)	No

REFERENCES

Aldridge DC, Elliott P, and Moggridge GD (2006) Microencapsulated BioBullets for the control of biofouling zebra mussels. *Environmental Science and Technology* 40: 975–979.

Baldwin BS, Mayer MS, Dayton J, Pau N, Mendilla J, Sullivan M, Moore A, Ma A, and Mills EL (2002) Comparative growth and feeding in zebra and quagga mussels (*Dreissena polymorpha* and *Dreissena bugensis*): Implications for North American lakes. *Canadian Journal of Fisheries and Aquatic Sciences* 59: 680–694.

Barnett TP and Pierce DW (2008) When will lake mead go dry? *Water Resources Research* 44: 1–10.

Berkman PA, Garton DW, Haltuch MA, Kennedy GW, and Febo LR (2000) Habitat shift in invading species: Zebra and quagga mussel population characteristics on shallow soft substrates. *Biological Invasions* 2: 1–6.

Britton DK (2014) The history of western management actions on invasive mussels. In: Wong WH and Gerstenberger S (eds.) *Biology and Management of Invasive Quagga and Zebra Mussels in the Western United States*, CRC Press, Boca Raton, FL, pp. 3–14.

Cadwell TJ, Rosen MR, Chandra S, Acharya K, Caires AM, Davis CJ, Thaw M, and Webster D (2014) Temporal and basin-specific population trends of quagga mussels on soft sediment of a multi-basin reservoir. In: Wong WH and Gerstenberger S (eds.) *Biology and Management of Invasive Quagga and Zebra Mussels in the Western United States*, CRC Press, Boca Raton, FL.

Carlton JT (2008) The zebra mussel *Dreissena polymorpha* found in North America in 1986 and 1987. *Journal of Great Lakes Research* 34: 770–773.

Carmon J and Hosler DM (2014) Understanding dreissenid veliger detection in the western United States. In: Wong WH and Gerstenberger S (eds.) *Biology and Management of Invasive Quagga and Zebra Mussels in the Western United States*, CRC Press, Boca Raton, FL, pp. 123–139.

Choi WJ, Gerstenberger S, McMahon RF, and Wong WH (2013) Estimating survival rates of quagga mussel (*Dreissena rostriformis bugensis*) veliger larvae under summer and autumn temperature regimes in residual water of trailered watercraft at Lake Mead, USA. *Management of Biological Invasions* 4: 61–69.

Claudi R and Mackie GL (1994) Practical manual for zebra mussel monitoring and control. Lewis Publishers, Boca Raton, FL.

Cohen AN (2005) Guide to exotic species of San Francisco Bay. San Francisco Estuary Institute. Oakland, CA. www.exoticsguide.org, (accessed on April 20, 2011).

Cross C, Wong WH, and Che TD (2011) Estimating carrying capacity of quagga mussels (*Dreissena rostriformis bugensis*) in a natural system: A case study of the Boulder Basin of Lake Mead, Arizona–Nevada. *Aquatic Invasions* 6: 141–147.

Food and Agriculture Orgnanization of the United Nations (2011) FISAT II-FAO-ICLARM Stock Assessment Tool. Retrieved from http://www.fao.org/fi/statist/fisoft/fisat on April 20, 2011.

Frischer ME, Hansen AS, Wyllie JA, Wimbush J, Murray J, and Nierzwicki-Bauer, SA (2002) Specific amplification of the 18S rRNA gene as a method to detect zebra mussel (*Dreissena polymorpha*) larvae in plankton samples. *Hydrobiologia* 487: 33–44.

Gerstenberger S, Mueting S, and Wong WH (2011) Abundance and size of quagga mussels (*Dreissena bugensis*) veligers in Lake Mead, Nevada-Arizona. *Journal of Shellfish Research* 30: 933–938.

Hayward D and Estevez E (1997) Suitability of Florida waters to invasion by the zebra mussel, *Dreissena polymorpha*. Florida Sea Grant College Program, Mote Marine Laboratory, Sarasota, FL.

Hebert PDN, Muncaster BW, and Mackie GL (1989) Ecological and genetic studies on *Dreissena polymorpha* (Pallas)—A new mollusk in the Great Lakes. *Canadian Journal of Fisheries and Aquatic Sciences* 46: 1587–1591.

Hincks SS and Mackie GL (1997) Effects of pH, calcium, alkalinity, hardness, and chlorophyll on the survival, growth, and reproductive success of zebra mussel (*Dreissena polymorpha*) in Ontario lakes. *Canadian Journal of Fisheries and Aquatic Sciences* 54: 2049–2057.

Holdren GC and Turner K (2010) Characteristics of Lake Mead, Arizona–Nevada. *Lake and Reservoir Management* 26: 230–239.

Hosler D (2011) Early detection of dreissenid species: Zebra/quagga mussels in water systems. *Aquatic Invasions* 6: 217–222.

Jones LA and Ricciardi A (2005) Influence of physicochemical factors on the distribution and biomass of invasive mussels (*Dreissena polymorpha* and *Dreissena bugensis*) in the St. Lawrence River. *Canadian Journal of Fisheries and Aquatic Sciences* 62: 1953–1962.

Karatayev AY, Burlakova LE, Mastitsky SE, Padilla DK, and Mills EL (2010) Invasion paradox: Why *Dreissena rostriformis bugensis*, being less invasive, outcompete *D. polymorpha*? *International Conference on Aquatic Invasive Species*, San Diego, CA.

Kennedy V, McIninch SP, Wright DA, and Setzler-Hamilton EM (1995) Salinity and zebra and quagga mussels. *The Fifth International Zebra Mussel and Other Aquatic Nuisance Organisms Conference*, Toronto, Canada.

Koutnik MA and Padilla DK (1994) Predicting the spatial distribution of *Dreissena polymorpha* (zebra mussel) among inland lakes of Wisconsin—Modeling with a GIS. *Canadian Journal of Fisheries and Aquatic Sciences* 51: 1189–1196.

LaBounty JF and Burns NM (2005) Characterization of boulder basin, Lake Mead, Nevada-Arizona, USA—Based on analysis of 34 limnological parameters. *Lake and Reservoir Management* 21: 277–307.

LaBounty JF and Roefer P (2007) Quagga mussels invade Lake Mead. *Lakeline* 27: 17–22.

Mackie GL, Gibbons WN, Muncaster BW, and Gray IM (1989) The zebra mussel *Dreissena polymorpha*: A synthesis of european experiences and a preview for North America: A report for the Ontario Ministry of the Environment, Water Resources Branch, Great Lakes Section, Queen's Printer, Toronto, Canada.

Marsden JE (1992) Standard protocols for monitoring and sampling zebra mussels. Illinois National History Survey, Champaign, IL, 40pp.

May B and Marsden JE (1992) Genetic identification and implications of another invasive species of dreissenid mussel in the Great Lakes. *Canadian Journal of Fisheries and Aquatic Sciences* 49: 1501–1506.

McMahon RF (1996) The physiological ecology of the zebra mussel, *Dreissena polymorpha*, in North America and Europe. *American Zoologist* 36: 339–363.

McMahon RF (2011) Quagga mussel (*Dreissena rostriformis bugensis*) population structure during the early invasion of Lakes Mead and Mohave January–March 2007. *Aquatic Invasions* 6: 131–140.

Mills EL, Dermott RM, Roseman EF, Dustin D, Mellina E, Conn DB, and Spidle AP (1993) Colonization, ecology, and population structure of the quagga mussel (Bivalvia, Dreissenidae) in the lower Great Lakes. *Canadian Journal of Fisheries and Aquatic Sciences* 50: 2305–2314.

Mills EL, Rosenberg G, Spidle AP, Ludyanskiy M, Pligin Y, and May B (1996) A review of the biology and ecology of the quagga mussel (*Dreissena bugensis*), a second species of freshwater *Dreissenid* introduced to North America. *American Zoologist* 36: 271–286.

Molloy DP, de Vaate AB, Wilke T, and Giamberini L (2007) Discovery of *Dreissena rostriformis bugensis* (Andrusov 1897) in Western Europe. *Biological Invasions* 9: 871–874.

Nalepa TF and Schloesser DW (1993) *Zebra Mussels: Biology, Impacts, and Control*. Lewis Publishers, Boca Raton, FL, 810pp.

O'Neill C (1995) *Zebra Mussel Clearing House*. SUNY College at Brockport, Brockport, NY, p. 6.

Patterson MWR, Ciborowski JJH, and Barton DR (2005) The distribution and abundance of *Dreissena* species (Dreissenidae) in Lake Erie, 2002. *Journal of Great Lakes Research* 31: 223–237.

Pimentel D, Zuniga R, and Morrison D (2005) Update on the environmental and economic costs associated with alien-invasive species in the United States. *Ecological Economics* 52: 273–288.

Rake L (2007) Lake Mead mussels identified as quagga, not zebra. *Las Vegas Sun*, Las Vegas.

Ram JL and McMahon RF (1996) Introduction: The biology, ecology, and physiology of zebra mussels. *American Zoologist* 36: 239–243.

Sorba EA and Williamson DA (1997) Zebra mussel colonization potential in Manitoba, Canada. Water Quality Management Section, Manitoba Environment, Winnipeg, Manitoba, Canada.

Spidle AP, Mills EL, and May B (1995) Limits to tolerance of temperature and salinity in the quagga mussel (*Dreissena bugensis*) and the zebra mussel (*Dreissena polymorpha*). *Canadian Journal of Fisheries and Aquatic Sciences* 52: 2108–2119.

Stokstad E (2007) Invasive species—feared quagga mussel turns up in western United States. *Science* 315: 453–453.

Strayer DL (1991) Projected distribution of the zebra mussel, *Dreissena polymorpha*, in North-America. *Canadian Journal of Fisheries and Aquatic Sciences* 48: 1389–1395.

Turner K, Wong WH, Gerstenberger SL, and Miller JM (2011) Interagency monitoring action plan (I-MAP) for quagga mussels in Lake Mead, Nevada-Arizona, USA. *Aquatic Invasions* 6: 195–204.

Twichell DC (2005) Seismic architecture and lithofacies of turbidites in Lake Mead (Arizona and Nevada, U.S.A.), an analogue for topographically complex basins. *Journal of Sedimentary Research* 75: 134–148.

Twichell DC, Cross VA, Rudin MJ, and Parolski KF (1999) Surficial geology and distribution of post-impoundment sediment of the western part of Lake Mead based on a sidescan sonar and high-resolution seismic-reflection survey. U.S. Geological Survey Open-File Report.

Whittier TR, Ringold PL, Herlihy AT, and Pierson SM (2008) A calcium-based invasion risk assessment for zebra and quagga mussels (*Dreissena* spp.). *Frontiers in Ecology and the Environment* 6: 180–184.

Wong WH, Gerstenberger S, Baldwin W, and Moore B (2012) Settlement and growth of quagga mussels (*Dreissena rostriformis bugensis*) in Lake Mead, Nevada-Arizona, USA. *Aquatic Invasions* 7: 7–19.

Wong WH and Gerstenberger SL (2011) Quagga mussels in the western United States: Monitoring and management. *Aquatic Invasions* 6: 125–129.

3

Temporal and Basin-Specific Population Trends of Quagga Mussels on Soft Sediment of a Multibasin Reservoir

Timothy J. Caldwell, Michael R. Rosen, Sudeep Chandra, Kumud Acharya, Andrea M. Caires, Clinton J. Davis, Melissa Thaw, and Daniel Webster

CONTENTS

ABSTRACT Invasive quagga (*Dreissena rostriformis bugensis*) and zebra (*Dreissena ploymorpha*) mussels have rapidly spread throughout North America. Understanding the relationships between environmental variables and quagga mussels during the early stages of invasion (2–4 years after detection) will benefit management strategies and allow researchers to predict patterns of future invasions. Quagga mussels were detected in Lake Mead, Nevada, and Arizona in 2007. We monitored the early invasion of mussels in the soft sediment of Lake Mead; this was done at three specific basins (Boulder Basin, Las Vegas Bay, and Overton Arm) and lake wide (average of all basins) biannually from 2008 to 2011. Mean densities of mussels increased during the first year of monitoring and stabilized during the subsequent 2 years on the whole-lake scale (geometric minimum and maximum means = 8–132 individuals $\cdot$ m^{-2}), and in two individual basins, Boulder Basin (73–875 individuals $\cdot$ m^{-2}) and Overton Arm (2–126 individuals $\cdot$ m^{-2}). In Las Vegas Bay, mean mussels densities ranged between 9 and 44 individuals $\cdot$ m^{-2}, significantly lower than the other basins. Low densities in Las Vegas Bay were correlated with high sediment concentrations of metal and warmer (greater than 30°C) water temperatures. Carbon content in sediments increased with depth in Lake Mead, and during some sampling periods, quagga density was also positively correlated with depth, but more research is required to determine the significance of this interaction. Laboratory growth experiments suggested that food quantity may limit mussel growth in

Boulder Basin, which may indicate an opportunity for population expansion if primary productivity were to increase. Overall quagga mussel density in Lake Mead was highly variable and patchy, which suggests temperature, sediment size, sediment metal concentrations, and sediment carbon content contribute to mussel distribution patterns.

Introduction

Invasions of aquatic ecosystems by nonnative species are occurring globally at an increasing frequency (Chandra and Gerhardt 2008; Lodge et al. 2006; Mills et al. 1994; Vanderploeg et al. 2002). Specifically, nonnatives are considered to be one of the largest threats to biodiversity (Sala et al. 2000) and have caused both ecological (Chandra and Gerhardt 2008; Mills et al. 1994; Riccardi and MacIsaac 2000) and economical (Connelly et al. 2007; Leung et al. 2002; Pimentel et al. 2000) damage to regions that are invaded. To reduce the rate of spread and damage to lakes and rivers caused by invasions, it is important for researchers and managers to understand density patterns of invasion during initial colonization.

In the late 1980s and 1990s, zebra mussels (*Dreissena polymorpha*) and later quagga mussels (*Dreissena rostriformis bugensis*, but see Stepien et al. 2013 for more information) colonized waters of the Laurentian Great Lakes (Benson 2013; May and Marsden 1992; Mills et al. 1994; Riccardi and MacIsaac 2000; Strayer et al. 1996) and Europe (Karatayev et al. 1997; Napela and Schloesser 2013). Organisms that colonized North America were likely transported by ballast water from the Ponto-Caspian region of Eurasia (May and Marsden 1992) and have become an invasive aquatic nuisance species in North America (Higgins and Vander Zanden 2010; Karateyev et al. 1997; Ludyanskiy et al. 1993; Strayer et al. 1999). Impacts of dreissenid invasion include shifts in benthic and pelagic community assemblages (Dermott and Kerec 1997; Hoyle et al. 2008; Nalepa et al. 2006, 2007; Ward and Riccardi 2007; Watkins et al. 2007; Wong et al. 2003) and alterations to the flow of nutrients (Barberio et al. 2003; Hecky et al. 2004; Turner 2010) and have been linked to increased toxic cyanobacteria blooms (De Stasio et al. 2008; Knoll et al. 2008). Water treatment centers and hydrologic power plants have been negatively affected by high densities of fouling mussels, which can cause pipes and intake systems of these facilities to clog (MacIsaac 1996; Nalepa and Schloesser 1993). These costs were estimated to be approximately $267 million through 2004 (Connelly et al. 2007), and costs are likely increasing as mussels continue to spread throughout North America.

Dreissenids have the ability to reproduce rapidly (Claxton and Mackie 1997; Mills et al. 1996) and attach to boats, which allows them to be easily transported in water and by trailered boats over land (Wilson et al. 1999). These characteristics make dreissenids difficult to contain and control, and near impossible to eradicate. Thus, despite significant effort to stop the spread of *Dreissena* spp. via state and federal programs (ANSTFSP 2013), quagga mussels recently (2007) invaded the Colorado River system, including Lakes Mead, Havasu, and Mohave, located in Nevada and Arizona (Stokstad 2007).

The Colorado River Basin spans seven states in the United States (Colorado, Wyoming, Utah, Nevada, Arizona, New Mexico, and California), along with portions of Mexico, and provides water to approximately 40 million people located in these regions (Rosen et al. 2012). Within the river system is Lake Mead, which is the largest reservoir by volume in the United States (Bureau of Reclamation 2012). Because human populations in the southwest United States have grown rapidly, the water supply of Lake Mead has become increasingly valuable for drinking and power production (Rosen et al. 2012). In addition, Lake Mead is a recreational area for fishing and watersports and is home to the endangered razorback sucker (*Xyrauchen texanus*), which adds to ecologic, recreational, and economic values of the reservoir (Hickey 2010; Rosen et al. 2012). Recently, climatic changes have lowered reservoir levels (Bernett and Pierce 2008, 2009), and the colonization of the lake by quagga mussels (Stokstad 2007) has threatened the lake's water quality and ecologic integrity.

Quagga mussels are able to colonize both epilimnetic and hypolimnetic waters and have also colonized soft and hard substrates in lakes and reservoirs (Dermott and Munawar 1993; Nalepa et al. 2006, 2007; Roe and MacIsaac 1997; Wittmann et al. 2010), which may spatially separate them from zebra mussels (*D. polymorpha*), which are often limited to hard substrates in epilimnetic waters. Specifically, the distribution of quagga mussels is thought to be regulated by sediment size structure (Berkman et al. 2000), depth, temperature, pH, and certain ion concentrations, especially calcium (Garton et al. 2013; Jones and Riccardi 2005; Spidle et al. 1995). Many studies examining the population dynamics of quagga mussels have been done in the Great Lakes region, where the invader has become well established (e.g., Dermott and Munawar 1993; French et al. 2007; Mills et al. 1993; Nalepa et al. 2006, 2007), and on hard substrates in Lake Mead (McMahon 2011; Wong et al. 2012). Wittmann et al. (2010) described the initial size structure of mussels (Fall 2008, Spring 2009) on soft sediments in Lake Mead in three distinct basins (Boulder Basin, Las Vegas Bay, and Overton Arm). Their data indicated that 2 years after detection, there were five adult size cohorts in both Boulder Basin and Overton Arm. Despite high veliger counts in Las Vegas Bay, there was minimal adult colonization in that basin (Wittmann et al. 2010). Wittmann et al. (2010) suggested that this may have been driven by high anthropogenic inputs of toxic metals (Patiño et al. 2012; Pollard et al. 2007; Rosen and Van Metre 2010), highly turbid waters, and increased summer temperatures, which can cause high mortality of quagga mussels (Spidle et al. 1995).

The objective of this study is to examine the densities of adult (>2 mm) quagga mussels on soft sediments in Lake Mead, including interbasin and temporal dynamics during the first 3 years (2008–2011) of invasion. We examined the populations in three basins (Boulder Basin, Las Vegas Bay, and Overton Arm) and at multiple depth ranges (2.5 to >106 m). Specifically, our objectives were to (1) determine densities of quagga mussels over time at a whole-lake scale and within each basin; (2) understand relationships between mussel density, depth, and environmental variables (substrate size, toxin concentrations, and sediment carbon content); and (3) examine basin-specific growth rates of mussels, in Lake Mead.

Study Site

Lake Mead is a reservoir on the Colorado River located just east of Las Vegas, Nevada, along the border of Arizona (Figure 3.1), which was formed by the construction of the Hoover Dam in 1936. The reservoir has a maximum depth of 180 m and a total surface area of about 66,000 ha (Rosen et al. 2012), making it the largest reservoir by volume in the United States, and an important drinking water and irrigation source for Las Vegas and other lower Colorado River states (i.e., Arizona, California; LaBounty and Burns 2005). Three basins of Lake Mead with unique limnological considerations were monitored: (1) Boulder Basin, (2) Las Vegas Bay, and (3) Overton Arm. These three basins represent the three major inflow areas of Lake Mead (see later) and, while not quantitatively measured, make up approximately 50% of the lake's surface area. Given the size of Lake Mead sampling, the entire lake was not logistically feasible for this project. Soft substrate dominates the benthic habitat of Lake Mead, while thin (1 m), it is evenly distributed throughout the Lake (Rosen et al. 2012).

Las Vegas Bay is the discharge area from the highly urban-influenced Las Vegas Wash and has impacts on water quality in that basin (Bevans et al. 1996; LaBounty and Horn 1997; Patiño et al. 2003; Rosen and Van Metre 2010; Rosen et al. 2010). Additionally, Las Vegas Bay has higher average temperatures and is generally more productive than the deep and cooler Boulder Basin and Overton Arm basins (LaBounty and Burns 2005). Boulder Basin receives a major input from Las Vegas Bay, and its trophic status is likely influenced by this wash; however, the majority of its water by volume is from the Colorado River, which may dilute some effects from Las Vegas Bay (LaBounty and Burns 2005). Overton Arm is the arm furthest north and has major inputs from the Virgin and Muddy Rivers (Figure 3.1).

The Colorado River is the main input of water to the reservoir and impacts temperature, oxygen, and nutrient profiles in the lake (Tietjen et al. 2012). Lake Mead has an average minimum temperature of approximately 11°C in the surface waters during January and can exceed 30°C during summer (Wittmann et al. 2010). Lake levels fluctuate annually due to highly variable snowmelt inputs and consistent losses to evaporation in an arid climate. During the first sampling period of our study (Fall 2008), the surface of the lake was at 338 m above sea level, while the lowest surface level (329 m above sea level) was recorded in November 2010. The following year, the lake rose to a high of 343 m by November 2011 (Rosen et al. 2012).

Lake Mead is designated as a national recreation area making it a popular destination for recreational boaters and anglers. The reservoir is a popular sport fishery, containing various catfish species, large- and

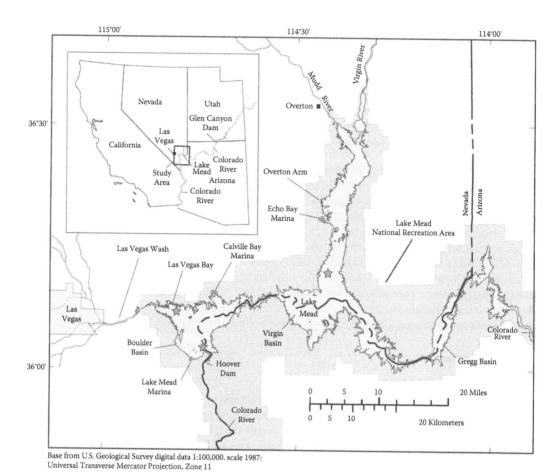

FIGURE 3.1   Map of Lake Mead with basins labeled. The sampling area for each basin is denoted by a star in its respective basin. Marinas sampled for growth experiments (Calville Bay Marina, Lake Mead Marina, and Echo Bay Marina) are labeled by a gray circle and described text within the figure. (This figure is modified from Wittmann, M.E. et al., *Lake Reserv. Manage.*, 26, 316, 2010.)

smallmouth bass (*Micropterus* spp.), and striped bass (*Morone saxatilis*), and holds a smaller population of the endangered razorback sucker (*X. texanus*). Gizzard shad (*Dorosoma cepedianum*), threadfin shad (*Dorosoma petenense*), and centrarchids make up the majority of the forage fish community, and there is a population of common carp (*Cyprinus carpio*; Chandra et al. 2012). Zebra mussels (*D. polymorpha)* have not been detected, while Asian clams (*Corbicula fluminea*) and New Zealand mudsnails (*Potamopyrgus antipodarum*) are two nonnative invertebrates found in Lake Mead (Wittmann et al. 2010). The majority of the native benthic community is made up of oligochaetes and chironomids.

Methods

Sampling

To examine the population of quagga mussels in Lake Mead, surveys were conducted during the fall and spring (Fall 2008 to 2011, excluding Fall 2010). A petite-Ponar grab sampler (Wildlife Supply Company, 231 cm² sample area, 2.4 L volume) was used to sample quagga mussels and sediment. Soft substrate was sampled in three basins of the lake (Boulder Basin, Las Vegas Bay, and Overton Arm) to examine interbasin differences. We calculated the *whole lake* as a mean of the three basins. During each sampling period, samples were collected from 2 to 106 m at approximately 10 m intervals in each basin.

Soft sediments were subjectively classified according to Wentworth's substrate guide (Wentworth 1922). Any samples that had greater than 50% cobble, gravel, or rock were deemed hard substrate dominated and excluded from further analysis. Because our study was focused on the macroinvertebrate community, samples were screened (500 µm mesh) and elutriated (if necessary), and mussels were handpicked for preservation. All mussels were enumerated, identified, and preserved in 70% ethanol.

Mussels were enumerated as the number of individuals $\cdot$ sample^{-1} and divided by the area of the Ponar (0.023 m^2) to determine densities (individuals $\cdot$ m^{-2}). Samples were grouped by five depth intervals (0–10 m, 10–20 m, 20–30 m, 30–40 m, and >50 m) for analysis through the physically stratified layers and data presentation. Shell length was measured as the distance from the umbo to the posterior margin of the shell to the nearest 0.01 mm using a certified digital caliper (Fisher Scientific, Model 14-648-17). Data from Wittmann et al. (2010) suggest that individuals less than 1 mm are not reproductively active. Because our analyses were focused on adult populations only, our study excluded individuals smaller than 2 mm. This size was chosen so that our analysis would be conservative in its classification of adults.

Sediments

We analyzed sediment mineralogy, particle size, metal concentrations, and carbon (inorganic/organic) content at three sites per basin and varied depths during Spring 2011. Sediment samples were collected from the same sample as the quagga mussel collections (see earlier). Additional samples collected before quagga invasion in 2006 were collected using a standard Ponar sampler in Las Vegas Bay and Overton Arm. These samples were used as a pre-quagga comparison to the sediment collected in 2011.

Quantitative particle size analysis was performed on a total of 20 splits of Lake Mead sediment samples. As part of the quality control and quality assurance of the laboratory performing the analysis, four of these samples were blind duplicates. Samples were freeze-dried to obtain workable dry samples for particle size analysis. All samples were prepared by removing organic matter and calcium carbonate using 30% H_2O_2 and 15% HCl, respectively. Particle size was determined using a Malvern Mastersizer (2000) laser analyzer (Malvern Instruments Ltd., Worcestershire, United Kingdom). Quantitative sediment metal analysis was done using inductively coupled plasma atomic emission spectroscopy at the Goldwater Environmental Laboratory, Arizona State University. Total carbon in the sediment was determined using an automated carbon analyzer (Jackson et al. 1987). Inorganic carbon was measured in the sediment as carbonate carbon and was determined as carbon dioxide, CO_2, by coulometric titration. Total organic carbon was determined by subtracting inorganic carbon from the total carbon.

Sediments from March 2011 were analyzed to identify the mineral composition of the sediment using an x-ray diffractometer (Moore and Reynolds 1989). X-ray diffraction analysis was performed on a Panalytical Xpert x-ray generator (Moore and Reynolds 1989). Minerals were characterized with Panalytical Data Collector and High Score Identification software (Panalytical B.V., the Netherlands). The software searches and matches diffractograms of unknown material against reference analyses in a database of standard minerals compiled and licensed by the International Centre for Diffraction Data.

Mussel Growth Rates

To determine the possibility of food limitation of quagga mussels in Lake Mead, we examined the growth rates of quagga mussels from two different basins (Boulder Basin and Overton Arm). Two sets of laboratory growth experiments were conducted in 2011. The first experiment was designed to compare growth rates of mussels collected at two locations (two replicates) within Boulder Basin (Lake Mead Marina and Calville Bay; Figure 3.1) and maximum growth rates (algae supplemented). The second experiment was designed to determine differences of growth rates between basins (Boulder Basin and Overton Arm) and to maximum growth rates against ambient growth rates. All experiments were done under controlled conditions in the laboratory. We did not conduct any experiments using Las Vegas Bay mussels because this is the most productive region of Lake Mead and likely not limited by food.

For experiments, adult mussels were collected from the lake by using a spackle knife to cut the mussels from pier flaps. Mussels were rinsed in lake water and placed in covered containers with air holes, which were filled with Lake Mead water and immediately transported to the laboratory. Mussels were rinsed with deionized water to remove large pieces of algae, plankton, and detritus. Dead mussels were cut away and removed from live mussels. The experimental mussels were then placed in aerated aquariums to acclimate to laboratory conditions, between 24 and 72 h.

Lake water was collected and transported from the same sites that mussels were collected from, including Lake Mead Marina, Calville Bay (Boulder Basin sites), and Echo Bay (Overton Arm site). All lake water was filtered using a 35 μm mesh filter to remove plankton, large pieces of algae, and sediment; water was stored in aerated, lightly covered 5 gal buckets.

Aquariums, kept at room temperature, were aerated, fitted with sponge filters and aquariums lights. Aquarium lights were set by timers for 12 h of light and 12 h of darkness to maintain normal algae growth in the tanks; a layer of bio-balls at the top of each tank added surface area for bacteria growth. Quaggas were not administered additional food other than the algae present in filtered Lake Mead water during the acclimation period. Filtered Lake Mead water was used for all aquariums and experimental containers. During acclimation periods, 25% of the water from each aquarium was exchanged, daily, with fresh, filtered Lake Mead water; during this process, pseudofeces were removed from the aquariums through siphoning to reduce resuspension.

After selecting and measuring live mussels from the acclimation aquariums, groups of similar-sized mussels were placed in 1.5 L lidded, aerated plastic containers, containing 1 L of filtered Lake Mead water. The mussels were kept in an incubator (VWR Signature Diurnal Growth Chamber, Model 2015) at 20°C. The incubator lights remained off during experiments to avoid uncontrolled amounts of algae growth in the experimental containers and to prevent interference with natural Lake Mead food condition. The aerator tubes' air was filtered with a 0.2 μm air filter to keep water aerated and to keep Lake Mead seston (food and other particles in water) in suspension. Filtered Lake Mead water was replaced in each container every other day to every third day depending on the experiment to maintain adequate food supply.

Algae, *Nannochloris coccoides,* was cultured in the laboratory and used to supplement filtered lake water for those mussel groups considered the *maximum growth* group in each experiment. *Nannochloris* spp. is a green, coccoid algae with a cell diameter of approximately 1–3 μm and grows with small to minimal clumping. A Shimadzu UV-1700 spectrophotometer was used to determine the algae density at wavelengths 664.0 and 750.0 nm (Acharya et al. 2006). A regression curve of absorbance, at each wavelength, and cell density in cell·mm^{-3}, was used to determine the amount of the supplemental algae administered to the mussels. Algae was frozen for storage and defrosted to room temperature before administering to the mussels.

In the first experiment, mussels taken from Lake Mead Marina and Calville Bay (two marinas within Boulder Basin) were used. In this experiment, 36 mussels from each location were selected and separated into groups of 6 per container. An additional group of 36 mussels were selected from Boulder Basin and were administered supplemental *Nannochloris* algae to represent the maximum growth. Mussel length was selected based on the average size of those collected during this time period. In this case, mussels with lengths between 8.0 and 8.99 mm were selected to ensure that the range in size remained below 10 mm. Over the course of 6 weeks, one mussel from each container was removed for wet weight measurement to determine the growth rate (i.e., change in body mass over time). To measure wet weight, each mussel was blotted dry with a Kim wipe and allowed to drain for approximately 10 min before weighing. Somatic growth rates were calculated using the following formula (Acharya et al. 2006; Baldwin et al. 2002; MacIsaac 1994a) for all three measurements (length, width, and wet weight):

$$\mu = \ln\left(\frac{\left(\text{final weight/intial weight}\right)}{\text{weeks}}\right)$$

For the second experiment, mussels with lengths between 7.00 and 7.99 (based on the same criteria as in experiment 1) mm were selected and separated into groups of six mussels per container. Six replicate containers were set up to represent each group: Boulder Basin, Overton Arm, and the *maximum growth* group. Over the course of 6 weeks, one mussel from each container was removed for measurement. Mussels were measured using methods similar to that of experiment 1.

Analysis

We used a general linear model (GLM) with density as the response and basin, depth, and time as the predictor variables to test for interactions; the only significant interaction detected was between time and density. Thus, we treated each basin separately and tested for differences over time and depth within each basin. A GLM was selected because of our unbalanced sample design. We tested for significant differences in density over time and depth in each individual basin using a one-way analysis-of-variance (ANOVA) with density as the response with depth and time as the factor. For all ANOVA analysis, Tukey's HSD was used to make statistical groupings. For statistics, all data were log-10 transformed to achieve equal variances, data were then back-transformed, and the geometric means (+SE −SE) were calculated and presented in this Chapter (cf. Caldwell and Wilhelm 2012). Quagga mussel density was regressed against sediment size, mineral content, metal concentration (e.g., As, Pb, and Mg), and carbon concentrations to determine any significant relationships. To determine significant differences between ambient growth rates and maximum growth rates, a one-way ANOVA and Tukey's HSD were used for both experiments. All statistics were performed in Mini-tab® version 16 (Minitab Inc., Pennsylvania, United States).

Results

Population Trends

In general, the lake-wide geometric mean density of quagga mussels in Lake Mead increased between 2008 and 2011 (Table 3.1). Mean densities were significantly lower (ANOVA, $F_{[5,527]} = 16.84$, $p < 0.001$) in Fall 2008 and Spring 2009, then the four other sampling periods (Fall 2009, Spring 2010, Spring 2011, and Fall 2011; Figure 3.2). In Boulder Basin, mean quagga mussel densities increased from 73 ind·m^{-2} (Fall 2008) to a maximum mean density of 875 ind·m^{-2} during Spring 2010 and remained high until Fall 2011. Boulder Basin had the highest overall mean density (with all depths grouped) in the lake (Table 3.1; Figure 3.2). In Overton Arm, mean density was lower than in Boulder Basin, but a similar pattern in temporal variation of quagga density was observed (Figure 3.2). Mean densities in Overton Arm ranged from 2 ind·m^{-2} (Fall 2008) to 126 ind·m^{-2} (Fall 2011), average density of Fall 2009 through 2011 was significantly higher (ANOVA, $F_{[5,284]} = 21.27$, $p < 0.001$) than in Fall 2008 and Spring 2009 (Figure 3.2). The lowest mean density of quagga mussels occurred in Las Vegas Bay and ranged from 9 + 9 − 4 ind·m^{-2} during Fall 2008 to 44 + 32 − 18 ind·m^{-2} Fall 2011. While the population did vary over time in Las Vegas Bay, these fluctuations were not significant (ANOVA, $F_{[5,102]} = 1.58$, $p = 0.173$). Overall, Boulder Basin had the highest densities of mussels, and populations in Boulder Basin, Overton Arm, and the whole lake increased significantly from Spring 2009 to Fall 2009 and stabilized, while the population in Las Vegas Bay remained low. The probability of mussel detection in a sample increased as average densities increased throughout the study.

Quagga density also varied by depth (Table 3.1). In Boulder Basin, the highest mean densities were consistently observed in the 20–30 m depth range (3672 ind·m^{-2} = maximum density in Spring 2010), while the lowest densities were observed in the 0–10 m depth range. In Las Vegas Bay, the highest mean density (182 ind·m^2) occurred at the greater than 50 m depth category with lower densities at depths <40 m. In Overton Arm, the highest mean density observed was 6179 ind·m^{-2} at 40–50 m during Spring 2010; however, there were only two samples in this depth category. The relationship between mean density and depth was often difficult to determine and was not significant in most cases, but analysis may have been limited by low sample size for each depth category. Relationships between specific sample depth and density were not significant ($p > 0.05$, regression). However, in some cases on a whole-lake scale (e.g., Fall 2008 and 2011), mean density increased with each depth category (Table 3.1).

Sediment Analysis

Mineralogy of the sediments in all basins is dominated by quartz and carbonate (calcite and dolomite). The mineralogy of the less than 1 μm separates (clay fraction) is dominated by magnesium-rich clays (smectite, sepiolite, and palygorskite). Illite and kaolinite are also present in subequal amounts in all

TABLE 3.1

Geometric Mean and Standard Error (Individuals m^{-2} + SE −SE) of Densities of Quagga Mussels in Three Basins (Boulder Basin, Las Vegas Bay, Overton Arm, and Whole Lake) at Different Depths and Times in Lake Mead

Depth (m)	Fall 2008	Spring 2009	Fall 2009	Spring 2010	Spring 2011	Fall 2011
Boulder Basin						
0–10	5 + 8 − 3[a]	27 + 60 − 19	N/A	N/A	N/A	227 + 4573 − 216
n	6	4	0	0	0	3
10–20	136 + 200 − 81[a,b]	15 + 204 − 14	31 + 65 − 21	354 + 125 − 92	373 + 440 − 202	957 + 1201 − 533
n	4	2	6	2	4	5
20–30	152 + 370 − 108[a,b]	996 + 1683 − 626	722 + 2186 − 542	3672 + 8720 − 2584	1322 + 920 − 543	2556 + 8936 − 1988
n	2	4	3	2	5	3
30–40	102 + 243 − 72[a,b]	123 + 1462 − 113	449 + 439 − 222	1227 + 1406 − 655	920 + 1688 − 595	590 + 665 − 313
n	5	3	4	3	4	3
40–50	175[a,b]	17 + 40 − 12	1516 + 6007 − 1210	1187 + 162 − 142	182 + 40 − 33	576 + 92 − 79
n	1	5	2	2	3	3
>50	204 + 223 − 107[b]	60 + 47 − 26	97 + 93 − 48	560 + 296 − 193	51 + 158 − 39	511 + 77 − 67
n	10	15	9	6	4	7
Las Vegas Bay						
0–10	0	4 + 6 − 3	65 + 88 − 37	41 + 47 − 22	113 + 63 − 40	22 + 22 − 11
n	3	3	3	3	3	5
10–20	0	4 + 9 − 3	9 + 79 − 8	0	3 + 5 − 2	4 + 6 − 2
n	3	3	2	3	3	6
20–30	0	4 + 9 − 3	9 + 79 − 8	0	3 + 5 − 2	4 + 6 − 2
n	3	1	3	3	3	3
30–40	N/A	N/A	N/A	N/A	N/A	N/A
n	N/A	N/A	N/A	N/A	N/A	N/A
40–50	20 + 372 − 19	N/A	N/A	N/A	292 + 460 − 179	91 + 20 − 16
n	2	0	0	0	3	3
>50	83 + 137 − 52	23 + 500 − 22	78 + 108 − 45	23 + 24 − 12	151 + 46 − 35	182 + 422 − 127
n	8	2	9	8	5	6

(Continued)

TABLE 3.1 (*Continued*)

Geometric Mean and Standard Error (Individuals m^{-2} + SE −SE) of Densities of Quagga Mussels in Three Basins (Boulder Basin, Las Vegas Bay, Overton Arm, and Whole Lake) at Different Depths and Times in Lake Mead

Depth (m)	Fall 2008	Spring 2009	Fall 2009	Spring 2010	Spring 2011	Fall 2011
Overton Arm						
0–10	1 + 0 − 0	5 + 6 − 3	65 + 88 − 37	41 + 47 − 22	113 + 66 − 40	22 + 21 − 11[a]
n	19	11	15	13	22	19
10–20	2 + 1 − 1	3 + 3 − 2	200 + 355 − 128	120 + 239 − 80	313 + 296 − 152	225 + 109 − 73[a,b]
n	9	11	10	6	9	12
20–30	2 + 2 − 1	5 + 9 − 3	144 + 255 − 92	156 + 251 − 96	87 + 92 − 45	193 + 80 − 57[a,b]
n	8	9	10	8	15	12
30–40	5 + 9 − 3	14 + 25 − 9	8 + 10 − 5	60 + 148 − 43	312 + 428 − 180	493 + 676 − 285[b]
n	9	13	8	8	2	6
40–50	15 + 204 − 14	N/A	0	6179 + 2344 − 1699	35 + 54 − 21	5919 + 4821 − 2657[b]
n	2	0	1	2	5	2
>50	N/A	4 + 9 − 3	127 + 55 − 39	3697	N/A	1685 + 2663 − 1032[a,b]
n	0	3	3	1	0	2
Whole Lake						
0–10	2 + 0 − 0[a]	7 + 6 − 3	53 + 55 − 27	33 + 30 − 16	86 + 54 − 33	23 + 20 − 11[a]
n	28	18	19	16	25	27
10–20	5 + 4 − 2[a]	4 + 3 − 2	77 + 83 − 40	40 + 60 − 24	135 + 124 − 65	108 + 79 − 46[a,b]
n	16	16	18	11	16	23
20–30	3 + 3 − 2[a]	20 + 34 − 13	77 + 106 − 45	79 + 128 − 49	151 + 107 − 63	257 + 129 − 86[b]
n	13	14	16	13	23	18
30–40	14 + 20 − 8[a,b]	21 + 33 − 13	31 + 39 − 17	137 + 234 − 86	641 + 693 − 333	524 + 430 − 236[b]
n	14	16	12	11	6	9
40–50	27 + 78 − 20[a,b]	175	132 + 1663 − 122	2708 + 1742 − 1060	98 + 72 − 41	517 + 451 − 241[b]
n	5	1	3	4	11	8
>50	137 + 110 − 61[b]	36 + 27 − 15	92 + 53 − 34	117 + 100 − 54	93 + 80 − 43	397 + 262 − 158[b]
n	18	20	21	15	9	15

Means and standard errors were log$_{10}$ transformed for statistical analyses and are presented in the back-transformed form; thus, positive and negative errors are unequal. Statistical groupings based on depth within basins and seasons during years with significant differences (Tukey's HSD, p < 0.05) are shown by superscript letters.

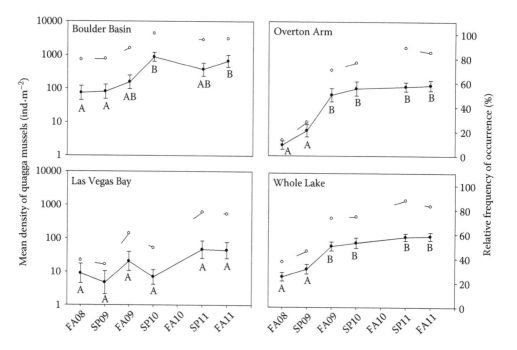

FIGURE 3.2 Geometric means and standard errors (• and vertical lines, respectively) of quagga mussel densities and percentage of samples with quagga mussels present (○) in each basin over the study period (FA = fall, SP = spring). Statistical groupings of the means (Tukey's pairwise comparisons, $p < 0.05$) by time are represented by letters below the data points.

three basins. Palygorskite is present in Las Vegas Bay and Boulder Basin, but not in Overton Arm, and sepiolite is present only in Las Vegas Bay, usually less than 10% of the greater than 1 μm fraction. There were no significant relationships ($p > 0.05$) between mineralogy of each basin and quagga densities in Lake Mead.

Sediment particle size analysis revealed that all sediments were less than 2 mm and varied between depths and basins. In Boulder Basin and Las Vegas Bay, sand-size particles (63–2000 μm) were the dominant substrate size in the shallower depths (less than 20 m), contributing 53.03% and 65.03%, respectively. Silt was the dominant size at the deeper depths in Boulder Basin (55.59% at 34 m, 60.17 at 74 m) and at Las Vegas Bay (72.99% at 15 m and 69.35% at 22 m). In Overton Arm, sand content ranged from 14.10% to 19.95%, and silt was dominant at all depths (71.18% at 5 m, 67.47% at 28 m, and 59.28% at 44 m). In all basins, percent clay increased with depth, with maximums of 29.53% in Boulder Basin (74 m), 20.34% in Las Vegas Bay (22 m), and 20.77% in Overton Arm (44 m). No significant statistical relationships were detected between quagga densities and sediment particle sizes using regression analysis ($p > 0.05$).

Three metals (As, Mg, Pb) out of the twenty-five metals (Al, As, Ba, Be, Bi, Ca, Cd, Co, Cr, Cu, Fe, Hg, K, Mg, Mn, Mo, Na, Ni, Pb, Se, Sr, Th, Ti, V, Zn) analyzed were significantly different between basins (Table 3.2). For all three of these metals, the highest metal contents were observed in Las Vegas Bay. These three metals (As, Mg, and Pb) had significant negative relationship to quagga density.

Total carbon increased as a function of depth on a lake-wide scale ($p = 0.01$, $R^2 = 52.6\%$) and in Las Vegas Bay and Boulder Basin, but not in Overton Arm (Figure 3.3). Increase in total carbon was driven equally by increases in both organic and inorganic carbon (carbonate carbon) with depth. While there were no significant relationships between quagga mussel densities and carbon contents ($p > 0.05$), some observations of high quagga densities did correspond to increased percent carbon in substrates. This was especially noticeable in Overton Arm and Las Vegas Bay, where quagga densities (Table 3.1) and percent carbon generally increased with depth (Figure 3.3).

TABLE 3.2

Arithmetic Mean Concentrations of Elements in the Sediments (mg·g⁻¹) in Each Basin and *p* Values from One-Way ANOVA Comparisons, with Concentration as the Response Variable and Basin as the Predictor Variable (Values n = 3)

	ANOVA				Regression	
Element	BB	LVB	OA	*p*	R² (%)	*p*
Al	5.16	7.33	6.8	0.146	29.8	0.129
As	0.008	0.022	0.012	<u>0.024</u>	52.8	<u>0.027</u>
Ba	0.437	0.213	0.233	0.047	26.2	0.659
Be	6.3×10^{-4}	7.0×10^{-4}	8.0×10^{-4}	0.408	2.2	0.705
Bi	6.7×10^{-4}	6.7×10^{-4}	1.0×10^{-3}	0.63	0.5	0.856
Cd	8.7×10^{-4}	9.7×10^{-4}	7.3×10^{-4}	0.851	3.8	0.616
Co	4.1×10^{-3}	3.4×10^{-3}	4.3×10^{-3}	0.799	0.2	0.919
Cr	0.011	0.017	0.013	0.112	31.1	0.118
Cu	0.018	0.02	0.019	0.949	9.1	0.429
Fe	7.55	10.42	9.69	0.15	21.2	0.213
Hg	2.9×10^{-4}	1.1×10^{-4}	4.3×10^{-4}	0.603	0.0	0.978
K	2.04	2.99	2.72	0.105	40.4	0.066
Mg	6.87	11.73	10.07	<u>0.015</u>	47.3	<u>0.041</u>
Mo	3.3×10^{-4}	7.6×10^{-4}	2.0×10^{-4}	0.076	33.3	0.100
Na	0.5	0.62	0.53	0.677	26.6	0.220
Ni	0.015	0.018	0.017	0.711	13.0	0.341
Pb	0.013	0.047	0.009	<u>0.004</u>	45.9	<u>0.046</u>
Se	3.6×10^{-3}	1.0×10^{-3}	1.6×10^{-3}	0.401	1.5	0.753
Sr	0.46	0.45	0.36	0.827	0.7	0.833
Th	0.014	0.019	0.017	0.627	25.3	0.168
V	0.016	0.019	0.017	0.627	21.7	0.206
Zn	0.04	0.065	0.04	0.063	28.4	0.139

BB, Boulder Basin; LVB, Las Vegas Bay; OA, Overton Arm.

Regression statistics are log quagga density (ind·m⁻²) as a function of element concentration. Underlined numbers indicate statistically significant differences (ANOVA) and interactions (regression). Ca, Mn, and Ti had errors in their analysis associated with nondetection or negative reporting values and are not presented in this table.

Growth Rates

Laboratory experiments showed positive growth rates of mussels; however, the differences varied between the experiments and the methods (Figure 3.4). The most consistent or measurable growth was seen in changes of wet body weight over time compared to changes in shell dimensions (length and width). All experiments lasted approximately 6 weeks, and this duration of time was, perhaps, not long enough to detect measurable differences in shell size.

In the first experiment, the difference in growth rate within Boulder Basin (Lake Mead Marina and Calville Bay) was not significant, while a difference in growth rate was observed between mussels receiving supplemental food and those grown at ambient algal densities ($p < 0.01$; Figure 3.4).

In the second experiment, somatic growth rates of mussels from Boulder Basin and Overton Arm varied (Figure 3.4). The growth rate (change in wet body weight) of mussels from Overton Arm was higher than the growth rate of mussels from Boulder Basin, with somatic growth rates of 0.077 and 0.058 week⁻¹, respectively. The difference in growth rates between mussels from Overton Arm and the maximum growth rate was not statistically significant.

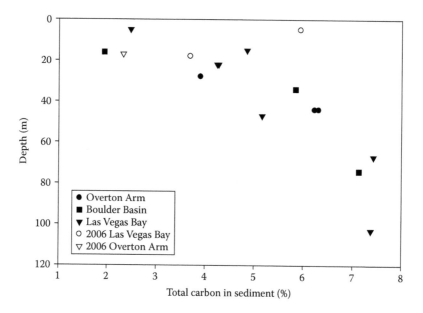

FIGURE 3.3 Total percent carbon in the sediment as a function of depth in Lake Mead; solid shapes represent samples taken during 2011.

Discussion

Mussel Densities

Densities of quagga mussels varied both temporally and between basins. On a lake-wide basis, densities of mussels increased between initial discovery of mussels in 2007 (Stokstad 2007) and until Fall 2009, after which densities appeared to stabilize, specifically in Boulder Basin and Overton Arm. Initial lag times in density increases, followed by exponential population growth, have been documented in invasion dynamics elsewhere (Sakai et al. 2001), and specifically in populations of dreissenids in both the Laurentian Great Lakes (Mills et al. 1996; Strayer et al. 1996) and Europe (Karatayev et al. 1997). We hypothesized to see this expansion pattern throughout Lake Mead; however, densities in some basins (Las Vegas Bay) were significantly lower than the other two basins (Boulder Basin, Overton Arm) and did not change statistically over the course of this study. Interestingly, the highest densities of quagga mussel veligers (free-floating juvenile form) were detected in Las Vegas Bay during 2008 and 2009 (Wittmann et al. 2010), suggesting that there were factors that limited establishment on soft sediments in this basin.

Research on dreissenid mussels indicates that they may be limited by physical parameters, including temperature (Karatayev et al. 1998; Mills et al. 1996; Spidle et al. 1995). Specifically, the upper limit of the quagga mussels is 25°C, while zebra mussels (*D. polymorpha*) have an upper limit of 30°C (Spidle et al. 1995). Because shallow water temperatures during the summer (August) in Las Vegas Bay can approach and pass the lethal limit of 25°C (Veley and Moran 2011) for quagga mussels (Spidle et al. 1995), we suggest that this may be important in the inhibition of quagga colonization in Las Vegas Bay. Colonization by quagga mussels expanded in Boulder Basin and Overton Arm, where water temperatures are often several degrees (2°C–4°C) below the lethal limit (Veley and Moran 2011). Temperature may especially affect mussels in shallower depths of Lake Mead, where in warmer years, mussel die-offs have occurred in the upper 5–6 m (K. Acharya, personal observations). These shallow-water die-offs are largely associated with hard substrates in the littoral nearshore environments; however, these areas are quickly recolonized by mussels. We hypothesize that veliger release by mussels on soft sediments in deeper waters of the lake may help rebuild nearshore populations that

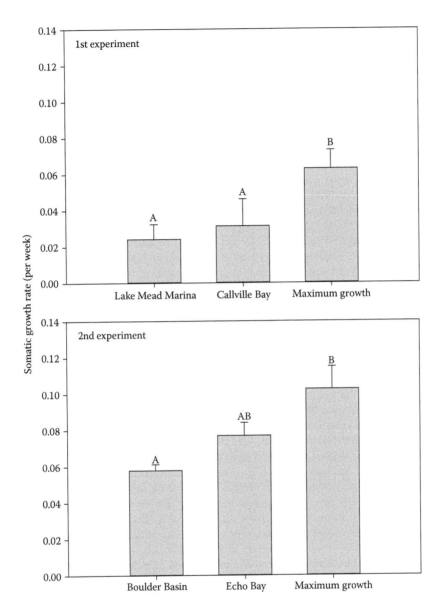

FIGURE 3.4 Somatic growth rate (per week over 6 weeks) for quagga mussels in various basins of Lake Mead. The first experiment was conducted at two sites (Lake Mead Marina and Calville Bay) within Boulder Basin, and the second experiment represents mussels from Boulder Basin and Echo Bay (Overton Arm). Both experiments compared ambient growth rates with maximum growth rates (see methods).

experience such mortality. Wittmann et al. (2010) suggest that quaggas are not limited by temperature alone. We suggest that high temperatures may increase the effect of other stressors (i.e., turbidity, toxins, high sedimentation rates) associated with Las Vegas Bay.

In other water bodies around the world, the literature has suggested that a multitude of variables (other than temperature) have been shown to limit and help explain mussel densities (Jones and Riccardi 2005; Karatayev et al. 1998; Mills et al. 1996). For example, reviews of literature by Mills et al. (1996) and Karatayev et al. (1998) indicate that substrate size is an important regulator of mussel populations. While hard substrates (i.e., large rocks, docks, etc.) are generally most hospitable to quagga mussels in Lake Mead (McMahon 2011; Wittmann et al. 2010; Wong et al. 2012), our study focused on the soft sediment population; thus samples that contained large rocks or hard substrate were not included.

Soft substrates (i.e., sand and silt) are the dominant portion of the lakebed in Lake Mead (Rosen et al. 2012). Our analysis quantified soft sediment particle sizes (<2000 μm); although statistical correlations were not apparent (likely due to a small sample size), quagga densities were higher in deeper waters (>20 m) where there is less sand, more clay, and higher organic and inorganic carbon content of substrates. This was especially apparent in Overton Arm, where the highest densities of quaggas were consistently observed in deeper (>40 m) waters. The higher clay content of deeper parts of the lake, especially in Boulder Basin and Las Vegas Bay, may make substrates more stable for veliger settlement, while higher carbon and inorganic carbon may have provided food and ions for shell growth.

Bivalves have the unique ability to filter large quantities of water (Kryger and Rusgard 1988), allowing them to accumulate high levels of toxins in a short period of time (Kraak et al. 1993; Naimo 1995). Las Vegas Bay is fed by Las Vegas Wash (the urban runoff from nearby city Las Vegas; LaBounty and Horn 1997); thus, anthropogenic influences have been associated with increased concentrations of heavy metals such as As, Cr, Mn, Ni, Pb, and Zn (Patiño et al. 2012; Rosen and Van Metre 2010). These high metal concentrations are hypothesized to prevent quagga mussels from colonizing Las Vegas Bay (Wittmann et al. 2010). Our data support this hypothesis and revealed significant negative correlations between concentrations of arsenic, magnesium, and lead in sediments and quagga densities. Prasada Rao and Khan (2000) present data that suggest that high temperatures may increase effects and concentrations of toxins (copper in this specific research) in mussel tissue, a result of increased respiration rates caused by the stress of high temperatures. Given the higher temperatures and metal concentrations present in Las Vegas Bay, when compared to the other areas studied, it is likely that a similar mechanism may limit colonization of mussels in this particular basin in Lake Mead.

Top-down control is often important in regulating species abundance (Carpenter et al. 1985; McQueen et al. 1989) in lakes and reservoirs. Aside from chemical and physical variables, and anthropogenic inputs to Las Vegas Bay, predation and competition should be considered as a mechanism in controlling quagga populations. Freshwater drum (*Aplodinotus grunniens*), common carp (*C. carpio*), round goby (*Neogobius melanostomus*), northern clearwater crayfish (*Orconectes propinquus*), and a variety of diving duck species all have been found to utilize dreissenid mussels as a food resource in the Great Lakes region (French and Bur 1993; Hamilton et al. 1994; Kuhns and Berg 1999; MacIsaac 1994a; Tucker et al. 1997). However, previous studies on the food web of Lake Mead indicate that the majority of fishes are more reliant on benthic organisms such as oligochaetes and chironomids (Umek et al. 2010) along with smaller forage fish (e.g., centrarchides and threadfin shad), which suggests that fish predation is not an important mechanism in the dynamics of quagga mussels in Lake Mead. Although examination of some common carp (*C. carpio*) stomachs has shown quagga remains (Rosen, unpublished data), additional data are required to determine if carp predation can significantly influence the population of quaggas.

Food quantity and quality may be important in the colonization strategy of quagga mussels. In the Great Lakes, quagga populations have been limited by food quantity (Dermott and Munawar 1993; Karatayev et al. 1998; Mills et al. 1996). In Lake Mead, Wittmann et al. (2010) suggested that future data collection should focus on collecting carbon content of sediment to determine if quagga mussels are food limited. The total carbon content of sediment in Lake Mead was similar between all basins and did not relate significantly to quagga density, suggesting that sediment carbon is either not utilized by quagga mussels or not limiting in Lake Mead. Given that bivalves are filter feeders and they derive the majority of their energy from the sediment water interface, we suggest the former. However, it should be noted that sediment carbon concentrations (both organic and inorganic carbon) increased with depth in all basins of Lake Mead, and in some instances, so did mean quagga density (e.g., Las Vegas Bay during all years and lake wide during Fall 2011). It may be possible for quagga veligers to settle on sediment with higher carbon concentration, but a better understanding of veliger ecology would be needed to confirm this hypothesis.

Samples taken from Las Vegas Bay and Overton Arm in approximately 18 m of water in 2006, before detection of quagga mussels, show similar amounts of total carbon, organic carbon and inorganic carbon. Although the 2006 sample size is limited, carbon concentrations indicate that the quagga mussel invasion does not yet appear to have affected sediment composition. This also suggests that quagga mussels either do not affect/utilize carbon associated with sediments, or they are efficient at recycling nutrients back to sediments.

Mussel Growth Rates

Laboratory experiments determined ambient growth rates for Boulder Basin and Overton Arm, and for the maximum potential growth rates in each basin. Our results imply that quagga mussels are growing at a faster rate in Overton Arm than in Boulder Basin. Because growth rates of mussels in Boulder Basin (at both sites) were significantly lower than the maximum growth rate, we propose that quaggas in Boulder Basin are not growing at their maximum potential rate. Wittmann et al. (2010) suggested that density dependence and competition for food resources may be limiting quagga growth in Boulder Basin, a pattern observed in other quagga populations (Karatayev et al. 1998; Mills et al. 1993). The highest densities in our study were recorded in Boulder Basin, which also correspond to some of the lowest median shell lengths recorded (Caldwell et al. unpublished data); thus, our data also support this hypothesis. We argue that with higher amounts of food, the mussel population in Boulder Basin could continue its expansion until another variable becomes limiting. Conversely, growth experiments from Overton Arm indicate that mussels are growing at their maximum potential in that basin. A stabilized population size indicates that mussels in Overton Arm may be limited by something other than food resources; this supports the hypothesis put forth by Wittmann et al. (2010).

The majority of growth rates reported in the literature use shell length as a measure of growth (Dorgelo 1993; Karateyev et al. 2006; Wong et al. 2012), which can be tracked over long-term studies. We used a laboratory setting in a relatively short-term experiment and quantified somatic growth on a per-week basis, making it difficult to compare our results to other systems. Additionally, Baldwin et al. (2002) report instantaneous growth rates for both quagga and zebra mussels and found that quagga mussels can grow much better in low food conditions than zebra mussels. Our overall growth rate results are somewhat lower than that of Baldwin et al. (2002) and MacIsaac (1994b) studies. One explanation for this is that there could be different nutritional values of the seston (food), and another could be the replenishment of the seston in the flow-through cages in their experiments. While our results do have limitations to their applications, these growth rates provide insight into future population dynamics.

Conclusions

Lake Mead and the Colorado River Aqueduct represent the first invasion in freshwaters of the western United States by quagga mussels and offer a unique opportunity to study their invasion dynamics. Studies done during initial invasion periods in other systems were often focused on populations that inhabit hard substrate (e.g., Nalepa and Fahnenstiel 1995; Strayer et al. 1996; Wong et al. 2012). Thus, the opportunity to study the early invasion dynamics of quagga mussels in a soft sediment environment in the western United States presents new knowledge to the ecology of dreissenids.

Quagga mussel density in Lake Mead varied both spatially and temporally. In Boulder Basin and Overton Arm, mussel populations increased initially but have been relatively stable from 2009 to 2011. However, the density of quaggas in Las Vegas Bay has remained lower than the other basins of the lake and has not changed over time. The lack of colonization by quagga mussels in Las Vegas Bay may be explained by the combination of high temperatures and higher concentrations of heavy metals/toxins when compared to other basins. Carbon content and percent clay content did not vary between basins in Lake Mead; however, both increased with depth in all basins and may help explain greater quagga densities at deeper depth in some cases. Growth rate experiments indicate that mussels in Boulder Basin are not growing at their maximum potential, and increased lake productivity associated with lower water levels and climate change may result in higher growth rates in the future. In contrast, mussels in Overton Arm are growing at or near their maximum potential, as determined by food quantity, but may be limited by other variables.

Our study indicates a highly variable population of quagga mussels; given this variability, determining causes for increased density in some areas while other areas remain sparsely populated is difficult. It is likely that a suite of variables whose effects are compounding upon one another are contributing to the patchiness of this particular population.

Acknowledgments

This project was funded by a grant from the National Park Service through the Great Basin Cooperative Ecosystems Studies Unit (J8R07110002) and the Southern Nevada Public Land Management Act to the U.S. Geological Survey. We give our sincere thanks to the students and staff at the University of Nevada–Reno, Aquatic Ecosystems Analysis Laboratory (Christine Ngai-Ryan, John Umek, Jason Barnes, Trea LaCroix, Robert Barnes, Rob Bolduc, and Marianne Denton). Special thanks to Marion Wittmann for offering her expertise on this subject. We thank Ron Veley for providing access to Lake Mead temperature data and Todd Tietjen for helpful insights on veliger dynamics in Lake Mead. Thanks to Don W. Schloesser and Amy J. Bensen at the USGS and two anonymous reviewers for comments that greatly improved this manuscript.

REFERENCES

Acharya, K., Bukaveckas, P., Jack, J.J., Kyle, M., Elser, J.J., 2006. Consumer growth linked to diet and RNA-P stoichiometry: Response of Bosmina to variation in riverine food resources. *Limnology and Oceanography.* 41: 1859–1869.

ANSTFSP, 2013. Aquatic Nuisance Species Task Force Strategic Plan 2013–2017, 77 FR 46730 (August 6, 2012), pp. 46730–46732.

Baldwin, B.S., Mayer, M.S., Dayton, J., Pau, N., Mendilla, J., Sullivan, M., Moore, A., Ma, A., Mills, E.L., 2002. Comparative growth and feeding in zebra and quagga mussels (*Dreissena polymorpha* and *Dreissena bugensis*): Implications for North American lakes. *Canadian Journal of Fisheries and Aquatic Science.* 59: 1159–1174.

Barbiero, R.P., Rockwell, D.C., Warren, G.J., Tuchman, M.L., 2003. Changes in spring phytoplankton communities and nutrient dynamics in the eastern basin of Lake Erie since the invasion of *Dreissena* spp. *Canadian Journal of Fisheries and Aquatic Sciences.* 63: 1549–1563.

Barnett, T.P., Pierce, D.W., 2008. When will Lake Mead go dry. *Water Resources Research.* 44: W0301.

Barnett T.P., Pierce, D.W., 2009. Sustainable water deliveries from the Colorado River in a changing climate. *Proceedings of the National Academy of Sciences of the United States of America.* 106(18): 7334–7338.

Benson, A.J., 2013. Chronological history of zebra and quagga mussels (Dreissenidae) in North America, 1988–2010. pp. 9–31. *In* Nalepa, T.F. and D.W. Schloesser (eds.), *Quagga and Zebra Mussels: Biology, Impacts, and Control*, 2nd edn. CRC Press, Boca Raton, FL. 775pp.

Berkman, P.A., Garton, D.W., Haltuch, M.A., Kennedy, G.W., Febo, L.R., 2000. Habitat shift in invading species: Zebra and quagga mussel population characteristics on shallow soft substrates. *Biological Invasions.* 2: 1–6.

Bevans, H.E., Goodbred, S.L., Miesner, J.F., Watkins, S.A., Gross, T.S., Denslow, N.D., Schoeb, T., 1996. Synthetic organic compounds and carp endocrinology and histology in Las Vegas Wash and Las Vegas and Calville Bays of Lake Mead, Nevada, 1992 and 1995. U.S. Geological Survey Water Resources Investigative Report 96-4266, Denver, CO. p. 12.

Bureau of Reclamation, 2012. Colorado River basin water supply and demand study. Executive Summary. p. 26. Accessed April 23, 2013 at: http://www.usbr.gov/lc/region/programs/crbstudy/finalreport/Executive%20 Summary/Executive_Summary_FINAL_Dec2012.pdf.

Caldwell, T.J., Wilhelm, F.M., 2012. The life history, growth, and density of *Mysis diluviana* in Lake Pend Oreille, Idaho, USA. *Journal of Great Lakes Research.* 38: 58–67.

Carpenter, S.R., Kitchell, J.F., Hodgson, J.R., 1985. Cascading trophic interactions and lake productivity. *Bioscience* 35: 634–639.

Chandra, S., Abella, S.R., Albrecht, B.A., Barnes, J.G., Engel, E.C., Goodbred, S.L., Holden, P.B. et al., 2012. Chapter 5: Wildlife and biological resources. pp. 69–104. *In* Rosen, M.R., Turner, K., Goodbred, S.L., and Miller, J.M. (eds.), *A Synthesis of Aquatic Science for Management of Lakes Mead and Mohave*, U.S. Geological Survey Circular 1381.

Chandra, S., Gerhardt, A., 2008. Invasive species in aquatic ecosystems: Issue of global concern. *Aquatic Invasions* 3: 1–2.

Claxton, W.T., Mackie, G.L., 1997. Seasonal and depth variations in gametogenesis and spawning of *Dreissena polymorpha* and *Dreissena bugensis* in eastern Lake Erie. *Canadian Journal of Zoology*. 76: 2010–2019.

Connelly, N.A., O'Neill Jr., C.R., Knuth, B.A., Brown, T.L., 2007. Economic impacts of zebra mussels on drinking water and electric power generation facilities. *Environmental Management*. 40: 105–112.

Dermott, R., Kerec, D., 1997. Changes to the deepwater benthos of eastern Lake Erie since the invasion of *Dreissena:* 1979:1993. *Canadian Journal of Fisheries and Aquatic Sciences*. 54: 922–930.

Dermott, R., Munawar, R., 1993. Invasion of Lake Erie offshore sediments by *Dreissena,* and its ecological implications. *Canadian Journal of Fisheries and Aquatic Sciences*. 50: 2298–2304.

De Stasio, B.T., Schrimpf, M.B., Beranek, A.E., Daniels, W.C., 2008. Increased chlorophyll *a*, phytoplankton abundance, and cyanobacteria occurrence following invasion of Green Bay, Lake Michigan by dreissenid mussels. *Aquatic Invasions*. 3(1): 21–27.

Dorgelo, J., 1993. Growth and population structure of the zebra mussel (*Dreissena polymorpha*) in Dutch Lakes differing in trophic state. pp. 79–94. *In* T.F. Nalepa and D.W. Schloesser (eds.), *Zebra Mussels: Biology, Impacts and Control*. Lewis Publishers Inc., Boca Raton, FL. 810pp.

French III J.R.P., Adams, J.V., Craig, J., Stickle, R.G., Nichols, S.J., Fleischer, G.W., 2007. Shell-free biomass and population dynamics of dreissenids in offshore Lake Michigan, 2001–2003. *Journal of Great Lakes Research*. 33(3): 536–545.

French III J.R.P., Bur, M.T. 1993. Predation by the zebra mussel (*Dreissena polymorph*) by freshwater drum in western Lake Erie. pp. 453–464. *In* T.F. Nalepa and D.W. Schloesser (eds.), *Zebra Mussels: Biology, Impacts and Control*. Lewis Publishers Inc., Boca Raton, FL. 810pp.

Garton, D.W., McMahon, R., Stoeckmann, A.M., 2013. Limiting environmental factors and competitive interactions between zebra and quagga mussels in North America. pp. 383–402. *In* Nalepa, T.F. and D.W. Schloesser (eds.), *Quagga and Zebra Mussels: Biology, Impacts, and Control*, 2nd edn. CRC Press, Boca Raton, FL. 775pp.

Hamilton, D.J., Ankney, C.D., Baily, R.C. 1994. Predation of zebra mussels by diving ducks: An exclosure study. *Ecology*. 75(2): 521–531.

Hecky, R.E., Smith, R.E.H., Barton, D.R., Guildford, S.J., Taylor, W.D., Charlton, M.N., Howell, T., 2004. The nearshore phosphorus shunt: A consequence of ecosystem engineering by dreissenids. *Canadian Journal of Fisheries and Aquatic Sciences*. 61: 1285–1293.

Hickey, V., 2010. The Quagga mussel crisis at Lake Mead National Recreation Area, Nevada (U.S.A.). *Conservation Biology*. 24(4): 931–937.

Higgins, S.N., Vander Zanden, M.J., 2010. What a difference a species makes: A meta-analysis of dreissenid mussel impacts on freshwater ecosystems. *Ecological Monographs*. 80(2): 179–196.

Hoyle, J.A., Bowlby, J.N., Morrison, B.J., 2008. Lake whitefish and walleye population responses to dreissenid mussel invasion in eastern Lake Ontario. *Aquatic Ecosystem Health and Management*. 11(4): 403–411.

Jackson, L.L., Brown, F.W., Neil, S.T., 1987. Major and minor elements requiring individual determination, classical whole rock analysis, and rapid rock analysis. pp. G5–G6. *In* Baedecker, P.A. (ed.), *Methods for Geochemical Analysis*. U.S. Geological Survey Bulletin 1770.

Jones, L.A., Riccardi, A., 2005. Influence of physiochemical factors on the distribution and biomass of invasive mussels (*Dreissena polymorpha* and *Dreissena bugensis*) in the St. Lawrence River. *Canadian Journal of Fisheries and Aquatic Sciences*. 62(9): 1953–1962.

Karatayev, A.Y., Burlakova, L.E., Padilla, D.K., 1997. The effects of *Dreissena polymorpha* (Pallas) invasion on aquatic ecosystems in eastern Europe. *Journal of Shellfish Research*. 16(1): 187–203.

Karatayev, A.Y., Burlakova, L.E., Padilla, D.K., 1998. Physical factors that limit the distribution and abundance of *Dreissena polymorpha* (pall). *Journal of Shellfish Research*. 17(4): 1219–1235.

Karatayev, A.Y., Burlakova, L.E., Padilla, D.K., 2006. Growth rate and longevity of *Dreissena polymorpha* (Pallas): A review and recommendations for future study. *Journal of Shellfish Research*. 25(1): 23–32.

Knoll, L.B., Sarnell, O., Hamilton, S.K., Kissman, C.E.H., Wilson, A.E., Rose, J.B., Morgan, M.R., 2008. Invasive zebra mussels (*Dreissena polymorpha*) increase cyanobacterial toxin concentrations in low-nutrient lakes. *Canadian Journal of Fisheries and Aquatic Sciences*. 65: 448–455.

Kraak, M.H.S., Lavy, D., Toussaint, M., Schoon, H., Peeters, W.H.M., Davids, C., 1993. Toxicity of heavy metals to the zebra mussel (*Dreissena polymorpha*). pp. 491–502. *In* T.F. Nalepa and D.W. Schloesser (eds.), *Zebra Mussels: Biology, Impacts and Control*. Lewis Publishers. Inc., Boca Raton, FL. 810pp.

Kryger, J., Rusgard, H.U., 1988. Filtration rate capacities in 6 species of European freshwater bivalves. *Oecologia*. 77: 34–38.

Kuhns, L.A., Berg, M.B., 1999. Benthic invertebrate community responses to round goby (*Neogobius melanostomus*) and zebra mussel (*Dreissena polymorpha*) invasions in southern Lake Michigan. *Journal of Great Lakes Research*. 25(4): 910–917.

LaBounty, J.F., Burns, N.M., 2005. Characterization of boulder basin, Lake Mead, Nevada-Arizona, USA— Based on analysis of 34 limnological patterns. *Lake and Reservoir Management*. 21(3): 277–307.

LaBounty, J.F., Horn, M.J., 1997. The influence of drainage from the Las Vegas Valley on the limnology of Boulder Basin, Lake Mead, AZ-NV. *Lake and Reservoir Management*. 13(2): 95–108.

Leung, B., Lodge, D.M., Finnoff, D., Shogren, J.F., Lewis, M.A., Lamberti, G., 2002. An ounce of prevention or a pound of cure: Bioeconomic risk analysis of invasive species. *Proceedings of the Royal Society of London Series B: Biological Sciences*. 269(1508): 2407–2413.

Lodge, D.M., Williams, S., MacIsaac, H., Hayes, K., Leung, B., Loope, L., Reichard, S. et al., 2006. Biological invasions: Recommendations for policy and management [Position Paper for the Ecological Society of America]. *Ecological Applications*. 16: 2035–2054.

Ludyanskiy, M.L., McDonald, D., MacNeill, D., 1993. Impact of the zebra mussel, a bivalve invader. *Bioscience*. 43: 533–544.

MacIsaac, H.J., 1994a. Size-selective predation on zebra mussels (*Dreissena polymorpha*) by crayfish (*Orconectes propinquus*). *Journal of the North American Benthological Society*. 13(2): 206–216.

MacIsaac, H.J., 1994b. Comparative growth and survival of *Dreissena polymorpha* and *Dreissena bugensis*, exotic mollusks introduced to the Great Lakes. *Journal of Great Lakes Research*. 20: 783–790.

MacIssac, H.J., 1996. Potential abiotic and biotic impacts of zebra mussels on the inland waters of North America. *American Zoology*. 36: 287–299.

May, B., Marsden, J.E., 1992. Genetic identification and implications of another invasive species of dreissenid mussel in the Great Lakes. *Canadian Journal of Fisheries and Aquatic Sciences*. 49: 1501–1506.

McMahon, R.F., 2011. Quagga mussel (*Dreissena rostriformis bugensis*) population structure during the early invasion of Lakes Mead and Mohave January-March 2007. *Aquatic Invasions*. 6(2): 131–140.

McQueen, D.J., Johannes, M.R.S., Post, J.R., Stewart, T.J., Lean, D.R.S., 1989. Bottom up and top-down impacts on freshwater pelagic community structure. *Ecological Monographs*. 59: 289–309.

Mills, E.L., Dermott, R.M., Roseman, E.F., Dustin, D., Mellina, E., Conn, D.B., Spidle, A.P., 1993. Colonization, ecology, and population structure of the "Quagga" mussel (Bivalvia: Dreissenidae) in the lower Great Lakes. *Canadian Journal of Fisheries and Aquatic Sciences*. 50: 2305–2314.

Mills, E.L., Leach, J.H., Carlton, J.T., Secor, C.L., 1994. Exotic species and the integrity of the Great Lakes: Lessons from the past. *BioScience*. 44: 666–676.

Mills, E.L.., Rosenberg, G., Spidle, A.P., Luyanskiy, M., Pligin, Y., May, B., 1996. A review of the biology and ecology of the quagga mussel (*Dreissena bugensis*), a second species of freshwater dreissenid introduced to North America. *American Zoology*. 36: 271–286.

Moore, D.M., Reynolds, R.C. Jr., 1989. *X-Ray Diffraction and the Identification and Analysis of Clay Minerals*. Oxford University Press, New York. 332pp.

Naimo, T.J., 1995. A review of the effects of heavy metals on freshwater mussels. *Ecotoxicology*. 4: 341–362.

Nalepa, T.F., Fahnenstiel, G.L., 1995. *Dreissena polymorpha* in the Saginaw Bay, Lake Huron ecosystem: Overview and perspective. *Journal of Great Lakes Research*. 21(4): 411–416.

Nalepa, T.F., Fanslow, D.L., Foley III A.J., Lang, G.A., Eadie, B.J., Quigley, M.A., 2006. Continued disappearance of the benthic amphipod *Diporeia* spp. in Lake Michigan: Is there evidence for food limitation? *Canadian Journal of Fisheries and Aquatic Sciences*. 63: 872–890.

Nalepa, T.F., Fanslow, D.L., Pothoven, S.A., Foley III A.J., Lang, G.A., 2007. Long term trends in benthic macroinvertebrate populations in Lake Huron over the past four decades. *Journal of Great Lakes Research*. 33: 421–436. 810pp.

Nalepa, T.F., Schloesser, D.W. (eds.), 1993. *Zebra Mussels: Biology, Impacts, and Control*. Lewis Publishers Inc., Boca Raton, FL. 775pp.

Nalepa, T.F., Schloesser, D.W. (eds.), 2013. *Quagga and Zebra Mussels: Biology, Impacts, and Control*. Lewis Publishers Inc., Boca Raton, FL.

Patiño, R., Goodbred, S.L., Draugelis-Dale, R., Barry, C.E., Foott, J.S., Wainscott, M.R., Gross, T.S., Covay, K.J., 2003. Morphometric and histopathological parameters of gonadal development in adult common carp from contaminated and reference sites in Lake Mead, Nevada. *Journal of Aquatic Animal Health*. 15: 55–68.

Patiño, R., Rosen, M.R., Orsak, E., Goodbred, S.L., May, T.W., Alvarez D., Echols, K.R., Wieser, C.M., Ruessler, S., Torres, L., 2012. Patterns of metal composition and morpho-physiological condition and their association in male common carp across an environmental contaminant gradient in Lake Mead National Recreation Area, Nevada and Arizona, USA. *Science of the Total Environment.* 416: 215–224.

Pimentel, D., Lach, L., Zuniga, R., Morrison, D., 2000. Environmental and economic costs of nonindigenous species in the United States. *BioScience.* 50(1): 54–65.

Pollard, J., Cizdziel, J., Stave, K., Reid, M., 2007. Selenium concentrations in water and plant tissues of a newly formed arid wetland in Las Vegas, Nevada. *Environmental Monitoring Assessment.* 135: 447–457.

Prasada Rao, D.G.V., Khan, M.A.Q., 2000. Zebra mussels: Enhancement of copper toxicity by high temperature and its relationship with respiration and metabolism. *Water Environment Research.* 72(2): 175–178.

Riccardi, A., MacIsaac, H.J., 2000. Recent mass invasion of the North American Great Lakes by Ponto-Caspian species. *Trends in Ecology and Evolution.* 15: 62–65.

Roe, S.L., MacIsaac, H.J., 1997. Deepwater population structure and reproductive state of quagga mussels (*Dreissena bugensis*) in Lake Erie. *Canadian Journal of Fisheries and Aquatic Sciences.* 54: 2428–2433.

Rosen, M.R., Alvarez, D.A., Goodbred, S.L., Leiker, T.J., Patiño, R., 2010. Sources and distribution of organic compounds using passive samplers in Lake Mead National Recreation Area, Nevada and Arizona, and their implications for potential effects on aquatic biota. *Journal of Environmental Quality.* 39: 1161–1172.

Rosen, M.R., Turner, K., Goodbred, S.L., Miller, J.L. (eds.), 2012. Lake Mead Circular: A synthesis of aquatic science for management of Lakes Mead and Mohave. USGS Circular 1381. p. 172.

Rosen M.R., Van Metre, P.C., 2010. Assessment of multiple sources of anthropogenic and natural chemical inputs to a morphologically complex basin, Lake Mead, USA. *Palaeogeography, Paleoclimatology, Palaeoecology.* 294: 30–43.

Sakai, A.K., Allendorf, F.W., Holt, J.S., Lodge, D.M., Molofsky, J., With, K.A., Baughman, S. et al., 2001. The population biology of invasive species. *Annual Review of Ecology and Systematics.* 32: 305–332.

Sala, O.E., Chapin III, F.S., Armesto, J.J., Berlow, E., Bloomfield, J., Dirzo, R., Huber-Sanwald, E. et al., 2000. Global biodiversity scenarios for year 2100. *Science.* 287: 1770.

Spidle, A.P., Mills, E.L., May, B., 1995. Limits to tolerance of temperature and salinity in the quagga mussel (*Dreissena bugensis*) and the zebra mussel (*Dreissena polymorpha*). *Canadian Journal of Fisheries and Aquatic Sciences.* 52: 2108–2119.

Stepien, C.A., Grigorovich, I.A., Gray, M.A., Sullivan, T.J., Yerga-Woolwin, S., Kalayci, G., 2013. Evolutionary, biogeographic, and population genetic relationships of dreissenid mussels, with revision of component taxa. pp. 403–444. *In* Nalepa, T.F. and D.W. Schloesser (eds.), *Quagga and Zebra Mussels: Biology, Impacts, and Control,* 2nd edn. CRC Press, Boca Raton, FL. 775pp.

Stokstad, E., 2007. Feared quagga mussel turns up in western United States. *Science.* 315: 453.

Strayer, D.L., Caraco, N.F., Cole, J.T., Findlay, S., Pace, M.L., 1999. Transformation of freshwater ecosystems by bivalves. *Bioscience.* 49: 19–27.

Strayer, D.L., Powell, J., Ambrose, P., Smith, L.C., Pace, M.L., Fischer, D.T., 1996. Arrival, spread, and early dynamics of a zebra mussel (*Dreissena polymorpha*) population in the Hudson River estuary. *Canadian Journal of Fisheries and Aquatic Sciences.* 53: 1143–1149.

Tietjen, T., Holdren, G.C., Rosen, M.R., Veley, R.J., Moran, M.J., Vanderford, B., Wong, W.H., Drury, D.D., 2012. Chapter 4: Lake water quality. pp. 35–68. *In* Rosen, M.R., K. Turner, S.L. Goodbred, and J.M. Miller (eds.), *A Synthesis of Aquatic Science for Management of Lakes Mead and Mohave.* U.S. Geological Survey Circular 1381.

Tucker, J.K., Cronin, F.A., Soergel, D.W., Theiling, C.H., 1997. Predation on zebra mussels (*Dreissena polymorpha*) by common carp (*Cyprinus carpio*). *Journal of Freshwater Ecology.* 11(3): 363–372.

Turner, C.B., 2010. Influence of zebra (*Dreissena polymorpha*) and quagga (*Dreissena rostriformis*) mussel invasions on benthic nutrient and oxygen dynamics. *Canadian Journal of Fisheries and Aquatic Sciences.* 67: 1899:1908.

Umek, J., Chandra, S., Rosen, M., Wittman, M., Sullivan, J., Orsak, E., 2010. Importance of benthic production to fish populations in Lake Mead prior to the establishment of quagga mussels. *Lake and Reservoir Management.* 26: 293–305.

Vanderploeg, H.A., Nalepa, T.F., Jude, D.J., Mills, E.L., Holeck, K.T., Liebig, J.R., Grigorovich, I.A., Ojaveer, H., 2002. Dispersal and emerging ecological impacts of Ponto-Caspian species in the Laurentian Great Lakes. *Canadian Journal of Fisheries and Aquatic Sciences.* 59: 1209–1228.

Veley, R.J., Moran, M.J., 2011. Evaluating lake stratification and temporal trends by using near-continuous water-quality data from automated profiling systems for water years 2005–09, Lake Mead, Arizona and Nevada: U.S. Geological Survey Scientific Investigation Report 2012-5080, 25pp. Accessed November 12, 2013 at: http://pubs.usgs.gov/sir/2012/5080/.

Ward, J.W., Riccardi, A., 2007. Impacts of *Dreissena* invasions on benthic macroinvertebrate communities: A meta-analysis. *Diversity and Distributions*. 13: 155–165.

Watkins, J.M., Dermott, R., Lozano, S.J., Mills, E.L., Rudstam, L.G., Scharold, J.V., 2007. Evidence for remote effects of dreissenid mussels on the amphipod *Diporeia*: Analysis of Lake Ontario benthic surveys, 1972–2003. *Journal of Great Lakes Research*. 33: 642–657.

Wentworth, C.K., 1922. A scale of grade and class terms for clastic sediments. *Journal of Geology*. 30: 377–392.

Wilson, A.B., Naish, K.-A., Boulding, E.G., 1999. Multiple dispersal strategies of the invasive quagga mussel (*Dreissena bugensis*) as revealed by microsatellite analysis. *Canadian Journal of Fisheries and Aquatic Sciences*. 56: 2248–2261.

Wittmann, M.E., Chandra, S., Caires, A., Denton, M., Rosen, M.R., Wong, W.H., Tietjen, T., Turner, K., Roefer, P., Holdren, G.C., 2010. Early invasion population structure of quagga mussel and associated benthic invertebrate community composition on soft sediment in a large reservoir. *Lake and Reservoir Management*. 26: 316–327.

Wong, W.H., Gerstenberger, S., Baldwin, W., Moore, B., 2012. Settlement and growth of quagga mussels (*Dreissena rostriformis bugensis* Andrusov, 1897) in Lake Mead, Nevada-Arizona, USA. *Aquatic Invasions*. 7: 7–19.

Wong, W.H., Levinton, J.S., Twining, B.S., Fisher, N., 2003. Assimilation of micro- and mesozooplankton by zebra mussels: A demonstration of the food web link between zooplankton and benthic suspension feeders. *Limnology and Oceanography*. 48(1): 308–312.

4

Reproductive Biology of Quagga Mussels (Dreissena rostriformis bugensis) *with an Emphasis on Lake Mead*

Michael J. Misamore, Sarah Barnard, Ernest Couch, and Wai Hing Wong

CONTENTS

Introduction

Since their arrival into North America in 1991 (May and Marsden, 1992), quagga mussels (*Dreissena rostriformis bugensis* Andrusov, 1897) have exhibited remarkable ability to spread throughout North American waterways. As seen in Europe, the spread of quagga mussels frequently followed an initial infestation by zebra mussels, with quagga mussels displacing the former in many locations (Mills et al., 1999; Ricciardi and Whoriskey, 2004; Brown and Stepien, 2010; Nalepa et al., 2010). The recent infestation into Lake Mead, Nevada–Arizona, marked the first occurrence of a dreissenid species into the western United States and also marks an unusual situation where solely quagga mussels are present without being proceeded by zebra mussels. It is believed quagga mussels were initially introduced into Lake Mead, quickly spreading downstream of the Hoover Dam, into Lake Mojave, and then continuing onward to their current locations in the southwestern United States. It has been postulated that quagga mussels were introduced into Lake Mead via bilge water (i.e., bait or live wells) carried by a boat from the Great Lakes region (McMahon, 2011; Wong and Gerstenberger, 2011). Thus, the establishment of a reproductive population in Lake Mead has, and continues to play, a critical role in the spread of quagga mussels throughout the southwestern United States. Understanding the reproductive biology of quagga mussels and paying special attention to parameters important to environmental conditions found in the southwest will help provide better understanding of the continued spread into this region.

Like the proverbial first of two children, zebra mussels have garnered considerable attention with regard to nearly every aspect of their biology including reproduction. And like the proverbial second child, quagga mussels have received less intense focus and are often just grouped together with their *older sibling*. This phenomenon holds true with regard to dreissenid reproduction. It is often assumed that reproductive and developmental changes described for zebra mussels may be equally applicable to quagga mussels (Ackerman et al., 1994). While in many regards this assumption is well founded, there are clear differences between these species with regard to reproduction. For example, recent studies have clearly demonstrated differences in spawning activity (Claxton and Mackie, 1998), larval size (Martel et al., 2001), and gamete morphology (Walker et al., 1996) between the two species. The objective of this chapter is to focus specifically on what we know about reproduction in quagga mussels, particularly as it may relate to reproduction in Lake Mead. It is not intended to be a comparison between the two species for two reasons. The first is to avoid the frequent trap of attributing characteristics defined in zebra mussels equally to quagga mussels without definitive evidence supporting such applications. The second is that objective of this text is to focus specifically on quagga mussels and to clearly delineate what we know and do not know about their reproduction. As we progress through the developmental process in quagga mussels, we will occasionally use our knowledge of zebra mussel reproduction to speculate on what may be occurring in quagga mussels. However, we will clearly indicate when this speculation is without direct evidence. Only further research specifically on quagga mussels will allow us to generate a more complete and accurate picture of quagga mussel reproduction.

Broadcast Spawning

Quagga mussels, like zebra mussels and their marine cousins, are broadcast spawners, meaning fertilization and larval development occur in the water column (Ackerman et al., 1994; Nichols and Black, 1994; Mackie and Schloesser, 1996). This differs from most other freshwater North American bivalves, where fertilization and subsequent larval brooding occur within the female (Graf and O' Foighil, 2000; Glaubrecht et al., 2006). To understand the reproductive pattern of quagga mussel spawning, most studies focus on the presence of veligers in the water column or settlement patterns of pediveligers. From this information, time and location of spawning (gamete release) are projected. While this method of projecting spawning patterns relative to larval patterns provides invaluable information, it is important to remember that multiple factors may alter the relationship between spawning patterns and larval distribution patterns (Nichols, 1996). Included is the variability in the length of each larval stage (days to weeks) and the potentially significant transport of larvae from their spawning beds to the site of their collection where environmental conditions may be significantly different.

A more direct method to determine the relationship between environmental factors and reproduction is to view the maturation of the gonads of adult mussels. Histological analysis of gonads can determine the presence, abundance, and maturation of the gametes present in adults from a specific location and time. Furthermore, temporal monitoring of animals can determine when spawning of a particular mussel bed has occurred and the extent to which the animals have released all their gametes. However, few studies have focused on direct monitoring of gonadal development of adult quagga mussels (Claxton and Mackie, 1998). Taking into consideration both strategies (direct monitoring of gonadal development and larval-spawning patterns), we will review our current understanding of quagga mussel spawning particularly as it relates to Lake Mead.

In the Great Lakes, quagga mussel veliger abundance shows seasonal and depth variability. In Lake Michigan, Nalepa et al. (2010) found a distinct spring and fall peak in veliger abundance that varied with depth. At 25 m, the fall spawn was twice as great as the spring spawn, while there was no differ-ence in veliger abundance between spring and fall spawns at 45 m. A later July peak in veliger abun-dance was observed at a depth of 93 m (Nalepa et al., 2010). Roe and MacIsaac (1997) found that quagga mussels collected from deep waters (37, 55 m) exhibited developed gonads based on gonad squashes. They found that over 80% of females exhibited mature oocytes to varying degrees while 20% exhibited spent gonads.

Claxton and Mackie (1998) used histological examination to determine gametogenic development of quagga mussels collected from both the epilimnion and hypolimnion of Lake Erie. They found that males and females from the epilimnion underwent synchronous gametogenesis, and both spawned in late June when water temperature reached 18°C–20°C. Both males and females released the majority of the gametes contained in the gonads. Quagga mussels collected from the hypolimnion (termed profunda) spawned approximately 1 week later. These hypolimnion quagga mussels were able to spawn 9°C–10°C and that spawning was strongly correlated with temperature. In a species comparison, they found that quagga mussels spawned earlier than zebra mussels suggesting quagga mussels may have a lower opti-mal temperature for gametogenesis than their zebra mussel counterpart.

Stoeckmann (2003) directly measured spawning patterns by collecting quagga mussels at vari-ous times and determined reproductive readiness using the ability for serotonin to induce spawning and the total quantity of gametes produced. They found that quagga mussels were able to spawn in April when water temperature was only 9°C with highest levels sampled in mid-June and mid-July. Moreover, they released most of their gametes when they were initially induced to spawn, and few gametes were released following subsequent inductions until no gametes were released by the third induction (Stoeckmann, 2003). Additionally, they did not find a strong relationship between increasing body size and increased fecundity, although this correlation may be attributed to the tim-ing of gamete collection and age/size distribution of the sample population. It is worth noting that in a species comparison, quagga mussels were able to spawn earlier than the zebra mussels, but the female zebra mussels released more eggs than quagga mussel females. There was no difference in sperm release by males. Given the size difference between species, Stoeckmann (2003) concluded that zebra mussels devoted more of their total body mass toward reproduction. Similarly, Ram et al. (2011), using serotonin-induced spawning and gonads squashes to indicate reproductive state, found that quagga mussels had a peak spawn that occurred earlier in the season and at a lower temperature relative to zebra mussels.

In Lake Mead, it has been estimated that quagga mussels may have established a reproductively active population producing larvae as early as 2003 and no later than 2004 (McMahon, 2011). This esti-mate is based on the observed distribution pattern of adult mussels in Lake Mead and downstream loca-tions in 2007. During the initial infestation of Lake Mead, it appears that either a single spawning event occurred or a bimodal spawn took place with minimal midsummer suppression of spawning resulting in the 2006 cohort (McMahon, 2011). In 2009, based on gonadal sampling of adults, McMahon and Lam (2011) observed that adults became gravid during the spring and summer leading to a spawn and subsequent juvenile settlement in early October 2009. Mussels became gravid again in mid-March 2010 leading to a second smaller spawn, which resulted in larval settlement in early May 2010 (McMahon and Lam, 2011).

Based on 60 m vertical plankton samples collected in Lake Mead, Gerstenberger et al. (2011) found veligers in the water column year-round. Two peaks were reported in June and September 2010, with the later peak having three times the number of veligers relative to the June sample (28 veligers/L compared to 9 veligers/L). There were relatively few (1 veliger/L) recorded during January through March 2008. They observed an increase in the mean size of the veligers from August through November, followed by a decrease in January through March. Similarly, they observed an increase in the proportion of pediveli-gers comprising the veliger sample during October through January.

Environmental Cues for Spawning Induction

Numerous factors have been suggested to play a role in the timing of dreissenid gametogenesis and spawning.

Temperature

Temperature is frequently listed as a significant factor in spawning induction (Garton and Haag, 1993; Nichols, 1996; Claxton and Mackie, 1998). In Lake Mead (Nevada–Arizona), temperature showed a significant effect upon quagga mussel reproduction (Ianniello, 2013). Mussels were found to be reproductive year-round, while spawning appears to peak twice in the months of June and September with a smaller third peak in December at all but the deepest collection depths (Ianniello, 2013).

The long-standing limit for zebra mussel spawning required temperatures greater than 12°C (Borcherding, 1991) with fertilization limited to temperatures greater than 10°C (Sprung, 1995). However, several recent studies have shown that quagga mussels are able to undergo gametogenesis and spawn at lower temperatures. Roe and MacIsaac (1997) found that female gonadal development of quagga mussels occurred at depths where temperatures were 4.8°C–6°C. Claxton and Mackie (1998) found that quagga mussels were able to undergo gonadal development and spawning at 9°C. Similarly, Stoeckmann (2003) found that quagga mussels were able to spawn in April when water temperatures were only 9°C. Schwaebe et al. (2013) found that thermal shock (24°C–26°C) was able to induce spawning in quagga mussels, although levels were significantly lower than those achieved by serotonin induction (22% and 77% respectively).

In Lake Mead, the lowest recorded water temperature is 12°C (LaBounty and Burns, 2005). Water temperatures range from 12°C to 12.5°C (hypolimnion), 12°C to 18°C (metalimnion), and 12°C to 27°C (epilimnion) (LaBounty and Burns, 2005), although temperatures of 28°C–30°C have been reported (Rosen et al. 2012). McMahon and Lam (2011) observed that larval settlement occurred after water temperature fell below 23°C and that gonad development and subsequent spawning occurred when temperatures fell below 18°C. Gerstenberger et al. (2011) found that veliger abundance was most strongly correlated with temperature profiles in the metalimnion in Lake Mead. They reported veligers present in the Boulder Basin of Lake Mead year-round.

Algal Content

Numerous studies have shown a relationship between algal concentration and zebra mussel spawning or veligers (Sprung, 1989; Borcherding, 1991; Ram et al., 1992; Hardege et al., 1997). Several studies have shown a direct relationship between zebra mussel gonad development and spawning with algal or chlorophyll concentration (Borcherding, 1995). Both marine and freshwater algal extracts have been shown to directly induce spawning in zebra mussels (Hardege et al., 1997; Kashian and Ram, 2012). Fewer studies have focused on this relationship in quagga mussels. Claxton and Mackie (1998) found that chlorophyll *a* was a significant factor in governing gametogenesis and spawning in quagga mussel from the hypolimnion but not the epilimnion or wave zone. Kashian and Ram (2012) reported that cell-free filtrate of *Chlorella minutissima* stimulated spawning in quagga mussels.

Gamete-Associated Stimuli

Release of gametes has been shown to induce spawning in many broadcast spawning species (Beach et al., 1975; Harrison et al., 1984; Watson et al., 2003; Levitan, 2005). In dreissenids, gonadal slurries have been shown to induce spawning in both zebra mussels (Sprung, 1989; Wright et al., 1996; Hardege et al., 1997) and quagga mussels (Wright et al., 1996; Schwaebe et al., 2013). While evidence suggests that simultaneous spawning is initiated in part by the presence of other gametes, the precise mechanism behind this remains unclear. Kashian and Ram (2012) proposed a generalized model where phytoplankton serves as the primary stimulus to imitate spawning, and early released gametes would induce spawning in neighboring animals to ensure spawning synchrony. These environmental cues induce internal signals that travel through the serotonin pathway resulting in gamete release.

Spawning Induction

Similar to zebra mussels, external application of serotonin has been shown to induce spawning in quagga mussels (Miller et al., 1994; McAnlis et al., 2010; Ram et al., 2011; Schwaebe et al., 2013). Schwaebe et al. (2013) found that greater than 50% of both males and females spawned with females having a slightly lower threshold for serotonin sensitivity than males. The quagga mussels spawned ranged in size from 15 to 28 mm with the highest gamete output from mussels between 15 and 20 mm in shell length. In zebra mussels, serotonin was shown to directly induce germinal vesicle breakdown (GVBD; Fong et al., 1994), and serotonin has been localized to gonadal nerve endings (Ram et al., 1992).

It is presumed that dreissenid sperm become active upon spawning, although some evidence suggests activation might occur just prior to spawning (Mojares et al., 1995; Ciereszko et al., 2001). Zebra mussel sperm show no decrease in mobility after 3 h and are remarkably better at surviving when compared to sperm from other freshwater species (Ciereszko et al., 2001).

Gamete Morphology

Sperm Morphology

The morphology of quagga mussel sperm has been well characterized by several authors (Denson and Wang, 1994; Walker et al., 1996; McAnlis et al., 2010). Like most bivalve sperm studied, quagga mussel sperm is of the primitive type being defined as having a relatively small size, a middle piece containing four to five rounded mitochondria, a single flagellum, and a spherical or conical-shaped nucleus capped by a small acrosome (Franzen, 1983; Walker et al., 1996). As first described in Walker et al. (1996), quagga mussel sperm consist of three regions: (1) a head, (2) a midpiece, and (3) a flagellum (Figure 4.1a and b). The cell body of quagga sperm is approximately 4.6 μm in length × 1.2 μm in width (Walker et al., 1996). The nucleus is conical in shape and slightly curved along its length (Figure 4.1a and b) (Denson and Wang, 1994; Walker et al., 1996; McAnlis et al., 2010). A small conical acrosome lies in a nuclear fossa at the anterior end of the sperm. The acrosome is structurally complex, consisting of an outer thin, electron-dense region widening toward the nucleus and an inner, less electron–dense region. A preformed

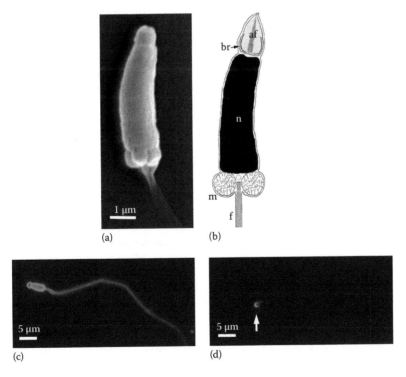

FIGURE 4.1 Sperm morphology. (a) SEM of quagga mussel sperm. (b) Drawing of quagga mussel sperm. (Modified from McAnlis, K.M.K. et al., *J. Shellfish Res.*, 29, 193, 2010. With permission.) (c) Zebra mussel sperm labeled along the entire surface with fluorescently labeled WGA lectin. (d) Quagga mussel sperm labeled only the sperm acrosome (arrow). af, acrosomal filament; br, basal ring; n, nucleus; m, mitochondria; f, flagellum.

acrosomal rod runs the length of the acrosome (Figure 4.1b) (Denson and Wang, 1994; Walker et al., 1996; McAnlis et al., 2010). Like the acrosomes of most bivalves (Brandriff et al., 1978), the acrosome of quagga mussels contains a dense basal ring that is involved in sperm binding to the egg (Figure 4.1b and c) (McAnlis et al., 2010). The sperm middle piece contains four spherical mitochondria and a centriole pair (Walker et al., 1996). Quagga sperm have a single flagellum with typical 9 + 2 microtubule arrangement (Walker et al., 1996). Based on the measurement of 50 sperm from 4 different males, we determined the average length of quagga sperm flagella to be 34 µm.

As stated by Franzen (1983), "There are probably not two bivalve species that have spermatozoa of identical morphology." This holds true for quagga mussel sperm even with regard to the sperm of zebra mussels. As clearly detailed by Walker et al. (1996) and Denson and Wang (1994), quagga sperm are slightly longer, thinner, and slightly curved relative to the shorter (4 µm), thicker (1.5 µm), and straight zebra mussel sperm. There are also biochemical differences between quagga and zebra mussel sperm particularly with respect to carbohydrates associated with the acrosomes (Fallis et al., 2010; McAnlis et al., 2010).

Sperm morphology has been proposed as a method of helping to distinguish individuals between these morphologically similar species (Denson and Wang, 1994; Walker et al., 1996). Furthermore, sperm morphology can be used to separate the fellow dreissenid *Mytilopsis leucophaeata* from both quagga and zebra mussels (Denson and Wang, 1998). Aside from purely morphological differences, there also appear to be biochemical differences between zebra and quagga mussels in regard to carbohydrates associated with the dense basal ring of the acrosome (McAnlis et al., 2010). For example, the lectin wheat germ agglutinin (WGA) labels the entire surface of the sperm of zebra mussels (Figure 4.1c) while labeling only the acrosome of quagga mussels (Figure 4.1d).

Egg Morphology

While several studies on the detailed anatomy of quagga sperm have been conducted, there is little information available regarding quagga mussel eggs. A brief description of gonadal development as indicated by the presence/absence of germinal vesicles has been reported (Claxton and Mackie, 1998; Nalepa et al., 2010). Here, we will provide a comprehensive description of egg morphology. Eggs spawned from Lake Mead quagga mussels ranged in size from 70 to 80 µm with an average size of 75.9 µm (n = 200 eggs from multiple females) (Figure 4.2a). The eggs have a minimal, uniformly distributed yolk and possess a highly transparent cytoplasm similar to zebra mussels (Misamore et al., 1996). The majority of eggs acquired from serotonin-spawned females underwent germinal vesicle breakdown (GVBD) and were arrested in metaphase I (Figure 4.2b). A few released oocytes still possessed intact germinal vesicles. In zebra mussels, Fong et al. (1994) provided a detailed description of GVBD induced by serotonin. They found that GVBD was initiated within 30 min and was complete in most oocytes by 40–50 min.

Surrounding spawned eggs is a prominent jelly layer (Figure 4.2c) (McAnlis et al., 2010). Based on peanut agglutinin lectin labeling, the jelly layer extends approximately 23 µm beyond the egg surface (McAnlis et al., 2010). The jelly layer may extend further but was not detectable. This layer is larger and more durable than the jelly layer reported for zebra mussels (Misamore et al., 1996). Furthermore, the carbohydrate composition of the quagga mussel jelly layer differs from that of zebra mussels (McAnlis et al., 2010). While the precise role of the jelly layer in dreissenids has yet to be determined, several pieces of evidence suggest its role as a sperm chemoattractant, which will be discussed later.

Immediately adjacent to the egg plasma membrane is a vitelline coat (Figure 4.2d and e). Zebra mussels exhibit a similar vitelline coat (Misamore et al., 1996); however, there appear to be biochemical differences between zebra and quagga mussel egg surfaces as they relate to carbohydrate composition and lectin binding (McAnlis et al., 2010). The surface of quagga eggs is covered with microvilli (Figure 4.2d through f) that extend through the vitelline envelope. The microvilli are approximately 0.5 µm in length and are uniform in size and distribution across the entire surface of the egg. The tips of the microvilli contain microvillar tufts (Figure 4.2e) similar to those of other bivalves (Hylander and Summers, 1977). Immediately beneath the egg plasma membrane in the cortex are numerous cortical granules of two general sizes: very small dense granules about 0.1 µm in size and larger less dense granules generally 0.64 µm in diameter (Figure 4.2d and g). Similar cortical granules have been reported in zebra mussels (Misamore et al., 1996).

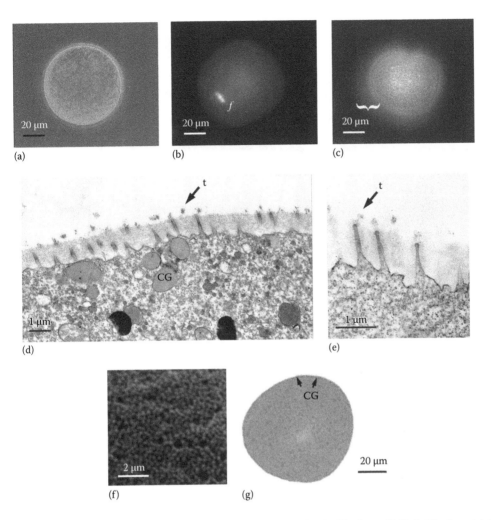

FIGURE 4.2 Quagga egg morphology. (a) Phase-contrast micrograph of unfertilized egg. (b) Unfertilized egg labeled with DNA-specific fluorochrome DAPI showing female DNA (*f*) in metaphase I arrest. (c) Egg jelly layer ({) surrounding an unfertilized quagga mussel egg visualized in an India ink solution. (d and e) TEMs of egg surface illustrating the microvilli (mv), microvillar tufts (t), and cortical granules (cg). (f) SEM of uniform microvilli covering the egg surface. (g) Histological section of unfertilized eggs with numerous cortical granules (cg) along the egg cortex.

Events of Fertilization

The major events of fertilization in zebra mussels have been well documented (Miller et al., 1994; Misamore et al., 1996, 2006; Ram et al., 1996; Luetjens and Dorresteijn, 1998; Misamore and Lynn, 2000; Fallis et al., 2010; McAnlis et al., 2010). However, a comprehensive study on quagga mussel fertilization has not been published. Here, we will present the major events in quagga mussel fertilization while incorporating previous findings from other studies peripherally addressing fertilization.

Motility and Chemotaxis

The importance of chemotaxis, attraction of sperm toward the egg, has been well documented for many broadcast spawning invertebrates (reviewed by Hirohashi et al., 2008). In many broadcast spawning species where chemotaxis has been identified, the chemoattractant that serves to draw the sperm toward the egg is a component of the egg jelly. By creating a diffusion halo around eggs, the chemoattractants

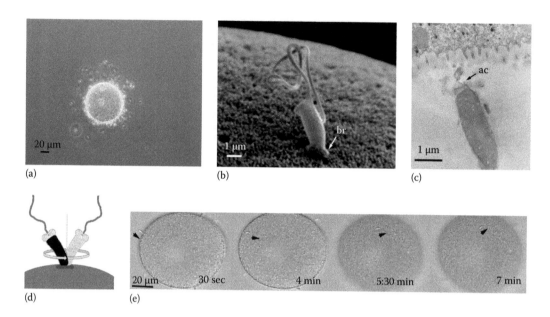

FIGURE 4.3 Early fertilization events. (a) Phase-contrast micrograph of sperm concentrated in jelly layer surrounding the egg. (b) SEM of sperm bound to egg surface (br, basal ring). (c) TEM of sperm bound to egg surface showing opened sperm acrosome (ac). (d) Illustration of fertilizing sperm rotation along an axis (dashed line) perpendicular to egg surface. (e) Time series post insemination of sperm entry and rotation inside the egg cytoplasm (arrowhead indicates orientation of sperm head).

effectively increase the target size of the egg (Jantzen et al., 2001). There are several lines of evidence suggesting that the jelly layer of the quagga egg is involved in sperm chemotaxis. First, Miller et al. (1994) showed that quagga sperm have the ability to undergo chemotaxis. Second, they showed that quagga sperm exposed to whole egg or gonadal extract exhibited chemotaxis (Miller et al., 1994). Third, quagga sperm can be clearly seen collecting in higher densities within the jelly layer relative to the surrounding water (Figure 4.3a). It is worth noting that zebra mussel sperm show remarkable longevity in motility relative to fish sperm. Ciereszko et al. (2001) found that zebra mussel sperm remain motile longer than many other sperm of freshwater species. Zebra mussel sperm showed no decrease in motility up to 3 h after release. Similarly, Quinn and Ackerman (2012b) showed that quagga mussel sperm were able to successfully fertilize eggs 2 h post-spawning.

Acrosome Reaction

In preparation for fusing with the egg, sperm undergo the acrosome reaction. In quagga mussels, this involves the opening of the acrosomal vesicle exposing the inner acrosomal membrane, basal ring, and acrosomal filament to the egg surface (Figure 4.3b and c). As with zebra mussels (Misamore et al., 1996), the preformed acrosomal filament in quagga mussels is now exposed. The filament does not undergo additional elongation or extension as seen in oysters (Niijima and Dan, 1965). The mechanism responsible for inducing the acrosome reaction is unknown in quagga mussels. Possible sources include a component in the jelly layer as seen in sea urchins (Mikamitakei et al., 1991), interaction with the vitelline envelope as seen in other bivalve species (Dan and Wada, 1955), or even binding with receptors on the egg plasma membrane.

Sperm Binding and Entry into the Egg Cytoplasm

Quagga sperm bind uniformly across the surface of the eggs (Figure 4.3a). Sperm bind perpendicular to the egg surface although the curved nature of the cell body changes this angle along the length of the sperm (Figure 4.3b and c). During sperm binding and entry, the flagellum continues to beat, and the sperm rotates around an axis perpendicular to the egg surface (Figure 4.3d). Contact between the acrosomal filament and egg microvilli, as well as contact between the acrosomal basal ring and the egg surface, is made (Figure 4.3c). Sperm enter into the egg cytoplasm by 2 min post insemination (PI). Once in the egg cytoplasm, the sperm rotates, positioning the posterior end toward the egg center (Figure 4.3e), sperm movement inside the egg cytoplasm being about 4 min PI. Through continued ratcheting of the flagellar axoneme as it enters into the egg cytoplasm, the sperm nucleus is translocated across the egg cytoplasm. This sperm flagellar translocation often results in the rapid movement of the sperm across the egg cytoplasm and generates a flow of cytoplasmic particles. This phenomenon was first documented for zebra mussels, but its precise function is unknown (Misamore et al., 1996). Quagga mussels are now the second bivalve species known to exhibit this phenomenon.

Following sperm entry and translocation, quagga mussel sperm nuclei decondense (Figure 4.4a) and form a pronucleus (Figure 4.4b and e). The timing of these events are similar to those reported for zebra mussels (Table 4.1) (Misamore et al., 2006). Quagga mussel pronuclei migrate toward one another, and the pronuclear envelope breaks down prior to pronuclear fusion (Figure 4.4d). Similar to zebra mussels and unlike many fertilization models such as sea urchins, pronuclear fusion does not occur prior to the first cleavage (Luttmer and Longo, 1988).

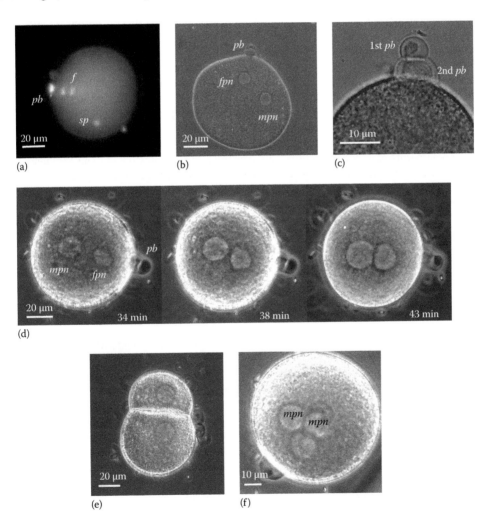

FIGURE 4.4 Pronuclear formation. (a) DAPI-labeled fertilized egg with decondensing sperm nucleus (*sp*) and female DNA (*f*) resuming meiosis creating the first polar body (pb). (b) Female (*fpn*) and male (*mpn*) pronuclei begin to form and enlarge. (c) First and second polar bodies are released. (d) Time series PI showing migration of female (*fpn*) and male (*mpn*) pronuclei. (e) Two-cell stage embryo. (f) Polyspermic egg with multiple male pronuclei (*mpn*).

TABLE 4.1

Summary of Events of Fertilization in Quagga Mussels

PI (min)	Event
1	Sperm binding
2–3	Sperm entry and rotation
3–6	Sperm flagellar translocation
6–8	First polar body formation begins
20	Second polar body formation begins
28–40	Pronuclear formation and migration
70	First cleavage

Timing based on live observations, video microscopy, and fixed sample analysis. Times indicate minutes PI.

Egg Response to Fertilization

Eggs arrested at metaphase I (Figure 4.2b) following spawning resume meiosis during the fertilization process. Polar bodies form as the egg eliminates excess DNA prior to formation of the haploid pronucleus (Figure 4.4a through d). The precise interaction leading to the resumption of meiosis in quagga mussels is unknown. As with many species, the most likely trigger is either sperm–egg binding or sperm–egg membrane fusion (Horner and Wolfner, 2008; Krauchunas and Wolfner, 2013). Approximately 6–8 min PI, the first polar body (Figure 4.4a and b) forms with the initiation of the second polar body (Figure 4.4c) approximately 20 min PI. Although yet to be determined for quagga mussels, cytokinesis required for polar body formation is actin dependent in zebra mussels (Misamore and Lynn, 2000).

Once an egg is fertilized, it must prevent polyspermy (additional sperm from entering the egg). Blocks to polyspermy exist for many broadcast spawning invertebrates (Gould and Stephano, 2003), including several marine bivalve species (Finkel and Wolf, 1980; Alliegro and Wright, 1983; Dufresne-Dube et al., 1983; Togo et al., 1995; Togo and Morisawa, 1999). Precise mechanisms for preventing polyspermy in dreissenids are unknown (Fallis et al., 2010). Under laboratory conditions, quagga mussel eggs can become polyspermic when exposed to high levels of sperm (Figure 4.4f). Similar results have been reported for zebra mussels (Misamore et al., 1996). Blocks to polyspermy typically take two general forms: (1) a *fast block*, which occurs within seconds of fertilization, or (2) a *slow block*, which occurs minutes after fertilization and is longer lasting. Fast blocks frequently involve a change in membrane potential rendering the egg unfertilizable. While a fast block to polyspermy has been suggested for several marine bivalve species (Finkel and Wolf, 1980; Alliegro and Wright, 1983; Dufresne-Dube et al., 1983; Togo et al., 1995; Togo and Morisawa, 1999), no such studies have been conducted on dreissenids.

Slow blocks to polyspermy often involve structural modifications of the egg surface removing nonfertilizing sperm and preventing subsequent binding of late-arriving sperm. Slow blocks often involve the formation of an elevated fertilization envelope and/or enzymatic modification of sperm-binding sites on the egg. Like most mollusks including zebra mussels, quagga mussels do not form a fertilization envelope (Figure 4.3) (Longo, 1973; Hylander and Summers, 1977). The most likely slow block to polyspermy in dreissenids appears to be an enzymatic digestion of sperm-binding sites. In quagga mussels, nonfertilizing sperm that were attached to the egg surface detach and drift away from the egg surface approximately 10–20 min PI (Figure 4.5). In zebra mussels, this sperm detachment occurs at approximately 15 min PI and involves a cleaving of the attachment between egg and sperm by a trypsin-like enzyme between the sperm acrosome and the egg surface (Fallis et al., 2010). In many species, the cleaving of sperm-binding sites typically involves enzymes released by the egg cortical granules (cortical reaction) shortly after fertilization (Gould and Stephano, 2003). However, the absence of a cortical reaction has been observed in many bivalves including oysters, *Mytilus, Spisula,* and zebra mussels (Gould and Stephano, 2003; Fallis et al., 2010). Further work is needed to fully understand the mechanisms preventing polyspermy in dreissenids. This is particularly true in regard to preventing polyspermy prior to the clipping of nonfertilizing sperm.

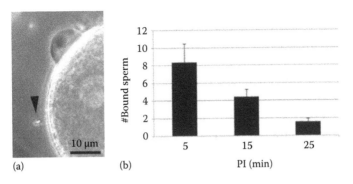

FIGURE 4.5   Sperm detachment. (a) Nonfertilizing bound sperm (arrowhead) detaches from the egg surface. (b) Decrease in the mean number of equatorial bound sperm from 50 eggs per trial (n = 3) from eggs fixed at various time points PI. Bar, standard error.

Sperm–Egg Ratios and Limitations

In many broadcast spawning species, sperm dilution following spawning can negatively affect fertilization success (Levitan, 1993, 2004; Quinn and Ackerman, 2012a). Until recently, few studies have tried to address how dreissenid fertilization occurs *in situ*. Recently, Quinn and Ackerman (2012b) showed that both quagga and zebra mussels were similarly susceptible to sperm limitation. However, both mussels appear to be less susceptible to sperm dilution inhibition of fertilization than other broadcast spawners. Limited fertilization success, as indicated by polar body formation, was achieved with sperm concentrations as low as 10 sperm/mL. Fertilization success increased rapidly from 10^3 to 10^5 sperm/mL, where it leveled off slightly at higher concentrations (Quinn and Ackerman, 2012b). Varying egg concentration had no significant effect on fertilization success. Additionally, they showed that clusters of mussels could alter water flow allowing for greater retention of sperm in downstream eddies, thereby reducing sperm dilution. While we are starting to ascertain a better understanding of dreissenid fertilization under *in situ* conditions, continued work is needed. Based on the aforementioned studies, zebra and quagga mussels appear to be susceptible to sperm limitation but have greater sperm longevity; fertilization success even at low concentrations and mussel cluster fluid dynamics may help limit their vulnerability. Both species must maintain a balance where sufficient sperm concentration is maintained to achieve high rates of fertilization while limiting polyspermy incurred with high sperm concentrations.

Cleavage and Early Development

Little is known about the early developmental stages (pre-trochophore) in quagga mussels. The first cleavage occurs approximately 75 min PI (Figure 4.4e). More detailed observations of early development on zebra mussels have been conducted starting with the initial observations by Meisenheimer (1901). Ackerman et al. (1994) provided a review of the early life history of zebra mussels. Zebra mussels exhibit spiral cleavage (Meisenheimer, 1901) that typically follows a very set pattern of development. In revisiting Meisenheimer's early observations, Luetjens and Dorresteijn (1995) provided the most detailed description of early cleavage events in zebra mussels and found much more variability in early development. They showed that early cell divisions occur approximately every hour for the first four cleavages. However, starting with the fifth cleavage, the cell cycle duration becomes longer and asynchronous. In contrast to Meisenheimer's previous findings, they showed that the pattern of cell division followed multiple alternatives. Blastomeres A–D were arranged in either a clockwise or counterclockwise pattern, and third cleavage chirality was either dextral or sinistral. This variability is uncommon in spiralian embryos (Luetjens and Dorresteijn, 1995). The dorsoventral axis was recognizable at the 16-cell stage with the D-quadrant cells marking the dorsal side. However, the fate of other cells is more variable and depends upon interactions with the D-quadrant cells. They attributed normal larval development within the variable cleavage patterns to a high degree of cell–cell interactions.

Larval Development

In a review of the life history of zebra mussels, Ackerman et al. (1994) provided a summary and detailed description of larval development as follows. Larvae go through a series of developmental changes starting with the trochophore larva, which is a free-swimming larval stage that occurs 6–96 h postfertilization. The trochophore develops into a veliger with the formation of a velum and secretes it first larval (prodissoconch I) shell. This D-shaped veliger persists until 7–9 days postfertilization when the second larval shell (prodissoconch II), characterized by a pronounced umbonal region, forms. This umbonal, or veliconcha, undergoes considerable growth and is the last free-swimming stage. The next larval change, the pediveliger, is characterized by the formation of a foot and behavior changes. Pediveligers swim or crawl along the bottom using their foot until receiving cues to start secreting byssal threads and *settle* between 18 and 90 days postfertilization. The settlement of the pediveliger marks the transition to the postveliger, or plantigrade, mussel, which is characterized by the loss of the velum, formation of the mouth and gills, and secretion of the adult shell (Ackerman et al., 1994).

A comprehensive description of quagga mussel larval development similar to that presented for zebra mussels by Ackerman et al. (1994) has not been reported. While no data were presented, Ackerman et al. (1994) suggested that quagga mussels might exhibit a similar larval development. They did mention that there were subtle differences in size and shape of D-shaped larvae between zebra and quagga mussels.

Martel et al. (2001) provided detailed measurements of quagga mussel veligers from two depths in Lake Erie. They found that quagga veligers had a prodissoconch I shell height of 79.07 μm and a prodissoconch II/settlement size of 256–284 μm for epilimnion/nearshore but were larger for offshore/hypolimnion prodissoconch II/settlement-sized veligers with a mean shell height of 313.64 μm. When compared to zebra mussels, the prodissoconch I size was similar among quagga mussels. However, the prodissoconch II/settlement size was significantly smaller for zebra mussels (236–249 μm) relative to

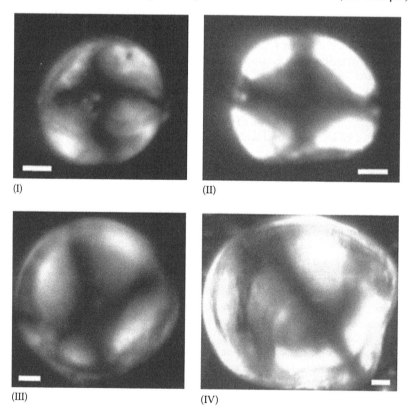

(I)   (II)

(III)   (IV)

FIGURE 4.6   Quagga mussel veligers. Cross-polarized light microscopy image of the four stages of quagga mussel veligers from Lake Mead. (I) trochophore, (II) straight-hinged veliger (D-Stage), (III) umbonal veliger, and (IV) pediveliger. Bar = 20 μm.

TABLE 4.2

Sizes of Different Developmental Stages of Quagga Mussel Veligers

	Length ± Standard Deviation (μm)	Width ± Standard Deviation (μm)	No. of Veligers Measured
Trochophore	102.3 ± 4.9	86.3 ± 4.9	34
D-stage	123.5 ± 9.2	106.3 ± 10.4	39
Umbonal	167.9 ± 21.2	153.8 ± 24.8	41
Pediveliger	234.3 ± 19.8	220.1 ± 23.9	22

quagga mussels (Martel et al., 2001). This size difference was attributed in part to a longer planktonic period observed in the offshore hypolimnion veligers.

In Lake Mead, quagga developmental stages are similar in size and morphology to both zebra mussels (Ackerman et al., 1994) and quagga mussels from other locations (Figure 4.6; Table 4.2). Quagga veligers of various stages are present year-round with peak abundance in September through November with a smaller peak in June (Gerstenberger et al., 2011; Moore et al., 2014). The size of quagga veligers ranged from 78 to 355 μm with the largest veligers present from November through January. The timing of maximum size corresponds with the highest percentage of pediveligers present in September through November (Gerstenberger et al., 2011).

As with other aspects of dreissenid reproduction, assumptions regarding quagga mussel larval development are frequently based on observations of zebra mussels. While many of these assumptions may prove to be correct, there are clearly differences between the species as well. More detailed work focusing specifically on quagga mussels is needed to fully determine the similarities and differences in larval development between zebra mussels and the less documented quagga mussels.

Methodology

The following is a description of the techniques used to spawn, handle gametes, and perform fertilizations, as well as the details of the original research presented in this chapter. Additional observations on techniques are provided to help facilitate improvements and lessons regarding reproductive experiments using quagga mussels.

Collection and Spawning of Animals

Quagga mussels were collected from Lake Mead and Lake Mojave during various times throughout the year by the National Park Service. Animals were maintained in 200 L aquaria containing artificial pond water (PW) modified from Dietz et al. (1994) (0.1 mM KCl, 0.7 mM $MgSO_4$, 0.8 mM $NaHCO_3$, 0.6 mM $CaCl_2$). The mussels were held at 10°C under a 14:10 h light: dark cycle and fed a mixture of marine microalgae (Shellfish Diet 1800, Reed Mariculture) twice weekly. Individual animals ranging from 10 to 35 mm in length were placed in 120 mL specimen cups containing cold PW and allowed to warm to room temperature (~23°C) over a 48 h period prior to spawning induction. This isolation and gradual warming appear to make them more amenable to serotonin-induced spawning and ensures no cross-contamination of gametes between individuals. For exposure to serotonin, isolated animals were transferred to 25 mL culture tubes and submerged in 1 mM serotonin (5-hydroxytryptamine) in PW for 30 min (Ram et al., 1993; McAnlis et al., 2010). At the end of the incubation period, animals were washed twice with deionized water and re-submerged in PW. Males typically started releasing gametes within 30 min, which is evident by a clouding of the culture tube. When sufficient sperm were released, males were removed from the culture tube to limit potential sperm damage due to siphoning. Once females started releasing eggs (typically 1 h after treatment), they were transferred to 70 × 50 mm crystallizing dish containing approximately 50 mL of PW to allow for spawning. The added volume for spawning allowed eggs to disperse upon release thus avoiding clumping and damage.

Fertilizations

For observations of fertilization events, approximately 30,000–40,000 eggs were transferred to 5 mL glass beakers, and a 10^3–10^6 sperm/mL concentration was added. Sperm concentrations were determined by multiple samples counted using a hemocytometer. For morphological observations, samples were fixed at various timing points using one of several fixation protocols described later. To determine numbers of bound sperm for the detachment assay, three independent replicate fertilizations were performed and samples fixed at predetermined time intervals (5, 15, and 25 min PI). To determine numbers of bound sperm, the numbers of sperm bound within an equatorial focal plane using a 40× objective were determined for 50 eggs from each replicate. The equatorial focus allowed for more accurate determination of bound sperm (Misamore and Lynn, 2000; Fallis et al., 2010).

Fixation

Depending on need, one of two fixation protocols was used. Samples required for fluorescent labeling were fixed in 3.2% paraformaldehyde in TAPS–MeOH buffer (5.5 mM TAPS, 0.2 mM KCl, 2 mM NaCl, 1.8 mM Na_2SO_4, 1.35 mM $MgSO_4$, 2.0 mM $NaHCO_3$, 2.25 mM $CaCl_2$, 20% MeOH). Samples for electron microscopy were fixed in 2.5% glutaraldehyde in TAPS–MeOH buffer. Samples were fixed for a minimum of 2 h followed by washes into mussel buffer (5.5 mM TAPS, 0.145 mM KCl, 0.8 mM NaCl, 0.8 mM Na_2SO_4, 0.89 mM $MgSO_4$, 1.32 mM $NaHCO_3$, 1.19 mM $CaCl_2$) (Misamore and Lynn, 2000). Electron microscopy samples were washed twice in 30 mM sodium cacodylate buffer, post-fixed in 0.5% osmium tetroxide for 1 h, dehydrated through an acetone series, and embedded in a modified Spurr's medium (Spurr, 1969; Misamore and Lynn, 2000). Samples for scanning electron microscopy were critical point-dried and sputter-coated with 10 nm of gold/palladium (Misamore et al., 1996).

Microscopy

For observation under light microscopy, inseminated samples were placed on a slide under a cover slip supported by vacuum grease posts to avoid crushing the eggs. Light and fluorescent microscopies were performed using a Zeiss Axiovert 200 or a Nikon Optiphot equipped with phase contrast and epifluorescence. Digital micrographs were captured using a Zeiss AxioCam MRm and Axiovision software. Digital video microscopy was captured using a Hitachi KP-D20 digital video camera and Plextor convertX processor. Both micrograph and video recordings were used to corroborate live observations. Scanning and transmission electron microscopies were performed on a JOEL model JSM-6100 scanning microscope and a Phillips EM300 transmission electron microscope, respectively. Adobe Photoshop was used for final image processing.

REFERENCES

Ackerman, J. D., B. Sim, S. J. Nichols, and R. Claudi. 1994. A review of the early life history of zebra mussels (*Dreissena polymorpha*): Comparisons with marine bivalves. *Canadian Journal of Zoology* 72:1169–1779.

Alliegro, M. C. and D. A. Wright. 1983. Polyspermy inhibition in the oyster, *Crassostrea virginica*. *Journal of Experimental Zoology* 227:127–137.

Beach, D. H., N. J. Hanscomb, and R. F. G. Ormond. 1975. Spawning pheromone in Crown-of-Thorns starfish. *Nature* 254:135–136.

Borcherding, J. 1991. The annual reproductive cycle of the freshwater mussel *Dreissena polymorpha* Pallas in lakes. *Oecologia* 87:208–218.

Borcherding, J. 1995. Laboratory experiment on the influence of food availability, temperature and photoperiod on gonad development in the freshwater *Dreissena polymorpha*. *Malacologia* 36:15–27.

Brandriff, B., G. W. Moy, and V. D. Vacquier. 1978. Isolation of sperm binding from the oyster (*Crassostrea gigas*). *Gamete Research* 1:89–99.

Brown, J. E. and C. A. Stepien. 2010. Population genetic history of the dreissenid mussel invasions: Expansion patterns across North America. *Biological Invasions* 12:3687–3710.

Ciereszko, A., K. Dabrowski, B. Piros, M. Kwasnik, and J. Glogowski. 2001. Characterization of zebra mussel (*Dreissena polymorpha*) sperm motility: Duration of movement, effects of cations, pH and gossypol. *Hydrobiologia* 452:225–232.

Claxton, W. T. and G. L. Mackie. 1998. Seasonal and depth variations in gametogenesis and spawning of *Dreissena polymorpha* and *Dreissena bugensis* in eastern Lake Erie. *Canadian Journal of Fisheries and Aquatic Sciences* 76:2010–2019.

Dan, J. C. and S. K. Wada. 1955. Studies on the acrosome. IV. The acrosome reaction in some bivalve spermatozoa. *Biological Bulletin* 109:40–55.

Denson, D. R. and S. Y. Wang. 1994. Morphological differences between zebra and quagga mussel spermatozoa. *American Malacological Bulletin* 11:79–81.

Denson, D. R. and S. Y. Wang. 1998. Distinguishing the dark false mussel, *Mytilopsis leucophaeata* (Conrad, 1831), from the non-indigenous zebra and quagga mussels, *Dreissena* spp., using spermatozoan external morphology. *Veliger* 41:205–207.

Dietz, T. H., D. Lessard, H. Silverman, and J. W. Lynn. 1994. Osmoregulation in *Dreissena polymorpha*: The importance of Na, Cl, K and particularly Mg. *Biological Bulletin* 187:76–83.

Dufresne-Dube, L., F. Dube, P. Guerrier, and P. Couillard. 1983. Absence of a complete block to polyspermy after fertilization of *Mytilus galloprovincialis* (Mollusca, Pelecypoda) oocytes. *Developmental Biology* 97:27–33.

Fallis, L. C., K. K. Stein, J. W. Lynn, and M. J. Misamore. 2010. Identification and role of carbohydrates on the surface of gametes in the zebra mussel, *Dreissena polymorpha*. *Biological Bulletin* 218:61–74.

Finkel, T. and D. P. Wolf. 1980. Membrane potential, pH and the activation of surf clam oocytes. *Gamete Research* 3:299–304.

Fong, P. P., K. Kyozuka, H. Abdelghani, J. D. Hardege, and J. L. Ram. 1994. In vivo and in vitro induction of germinal vesicle breakdown in a freshwater bivalve, the zebra mussel *Dreissena polymorpha* (Pallas). *Journal of Experimental Zoology* 269:467–474.

Franzen, A. 1983. Ultrastructural studies of spermatozoa in three bivalve species with notes on evolution of elongated sperm nucleus in primitive spermatozoa. *Gamete Research* 7:199–214.

Garton, D. and W. Haag. 1993. Seasonal reproductive cycles and settlement patterns of *Dreissena polymorpha* in western Lake Erie. pp. 111–128 in *Zebra Mussels: Biology, Impacts and Control*, T. F. Nalepa and D. W. Schloesser, eds. CRC Press, Boca Raton, FL.

Gerstenberger, S. L., S. A. Mueting, and W. H. Wong. 2011. Veligers of invasive quagga mussels (*Dreissena rostriformis bugensis*, Andrusov 1897) in Lake Mead, Nevada-Arizona. *Journal of Shellfish Research* 30:933–938.

Glaubrecht, M., Z. Feher, and T. von Rintelen. 2006. Brooding in *Corbicula madagascariensis* (Bivalvia, Corbiculidae) and the repeated evolution of viviparity in corbiculids. *Zoologica Scripta* 35:641–654.

Gould, M. C. and J. L. Stephano. 2003. Polyspermy prevention in marine invertebrates. *Microscopy Research and Technique* 61:379–388.

Graf, D. L. and D. O' Foighil. 2000. The evolution of brooding characters among the freshwater pearly mussels (Bivalvia: Unionoidea) of North America. *Journal of Molluscan Studies* 66:157–170.

Hardege, J., J. Ram, and M. Bentley. 1997. Activation of spawning in zebra mussels by algae-, cryptomonad-, and gamete-associated factors. *Experimental Biology Online* 2:1–9.

Harrison, P. L., R. C. Babcock, G. D. Bull, J. K. Oliver, C. C. Wallace, and B. L. Willis. 1984. Mass spawning in tropical reef corals. *Science* 223:1186–1189.

Hirohashi, N., N. Kamei, H. Kubo, H. Sawada, M. Matsumoto, and M. Hoshi. 2008. Egg and sperm recognition systems during fertilization. *Development Growth and Differentiation* 50:S221–S238.

Horner, V. L. and M. F. Wolfner. 2008. Transitioning from egg to embryo: Triggers and mechanisms of egg activation. *Developmental Dynamics* 237:527–544.

Hylander, B. L. and R. G. Summers. 1977. An ultrastructural analysis of the gametes and early fertilization in two bivalve molluscs, *Chama macerophylla* and *Spisula solidissima* with special reference to gamete binding. *Cell and Tissue Research* 182:469–489.

Ianniello, R. S. 2013. Effects of environmental variables on the reproduction of quagga mussels (*Dreissena rostriformis bugensis*) in Lake Mead, NV/AZ., Doctoral thesis, Department of Environmental and Occupational Health. University of Nevada, Las Vegas, NV. 59pp.

Jantzen, T. M., R. de Nys, and J. N. Havenhand. 2001. Fertilization success and the effects of sperm chemoattractants on effective egg size in marine invertebrates. *Marine Biology* 138:1153–1161.

Kashian, D. R. and J. L. Ram. 2012. Chemical regulation of dreissenid reproduction. pp. 461–469 in *Quagga and Zebra Mussels: Biology, Impacts, and Control*, T. Nalepa and D. Schloesser, eds. CRC Press, Boca Raton, FL.

Krauchunas, A. R. and M. F. Wolfner. 2013. Molecular changes during egg activation. *Current Topic Developmental Biology* 102:267–292.

LaBounty, J. F. and N. M. Burns. 2005. Characterization of boulder basin, Lake Mead, Nevada-Arizona, USA—Based on analysis of 34 limnological parameters. *Lakes and Reservoir Management* 21:277–307.

Levitan, D. R. 1993. The importance of sperm limitation to the evolution of egg size in marine-invertebrates. *American Naturalist* 141:517–536.

Levitan, D. R. 2004. Density-dependent sexual selection in external fertilizers: Variances in male and female fertilization success along the continuum from sperm limitation to sexual conflict in the sea urchin *Strongylocentrotus franciscanus*. *American Naturalist* 164:298–309.

Levitan, D. R. 2005. Sex-specific spawning behavior and its consequences in an external fertilizer. *American Naturalist* 165:682–694.

Longo, F. J. 1973. An ultrastructural analysis of polyspermy in the surf clam, *Spisula solidissima*. *Journal of Experimental Zoology* 183:153–180.

Luetjens, C. M. and A. W. C. Dorresteijn. 1995. Multiple, alternative cleavage patterns precede uniform larval morphology during normal development of *Dreissena polymorpha* (Mollusca, Lamellibranchia). *Rouxs Archives of Developmental Biology* 205:138–149.

Luetjens, C. M. and A. W. C. Dorresteijn. 1998. The site of fertilization determines dorsoventral polarity but not chirality in the zebra mussel embryo. *Zygote* 6:125–135.

Luttmer, S. J. and F. J. Longo. 1988. Sperm nuclear transformations consist of enlargement and condensation coordinate with stages of meiotic maturation in fertilized *Spisula solidissima* oocytes. *Developmental Biology* 128:86–96.

Mackie, G. L. and D. W. Schloesser. 1996. Comparative biology of zebra mussels in Europe and North America: An overview. *American Zoologist* 36:244–258.

Martel, A. L., B. S. Baldwin, R. M. Dermott, and R. A. Lutz. 2001. Species and epilimnion/hypolimnion-related differences in size at larval settlement and metamorphosis in *Dreissena* (Bivalvia). *Limnology and Oceanography* 46:707–713.

May, B. and J. E. Marsden. 1992. Genetic identification and implications of another invasive species of dreissenid mussel in the Great Lakes. *Canadian Journal of Fisheries and Aquatic Sciences* 49:1501–1506.

McAnlis, K. M. K., J. W. Lynn, and M. J. Misamore. 2010. Lectin labeling of surface carbohydrates on gametes of three bivalves: *Crassostrea virginica*, *Mytilus galloprovincialis*, and *Dreissena bugensis*. *Journal of Shellfish Research* 29:193–201.

McMahon, R. F. 2011. Quagga mussel (*Dreissena rostriformis bugensis*) population structure during the early invasion of lakes Mead and Mohave January-March 2007. *Aquatic Invasions* 6:131–140.

McMahon, R. F. and J. P. Lam. 2011. The growth, reproductive and nutritional dynamics of quagga mussels, *Dreissena rostriformis bugensis*, in Lake Mead, NV/AZ. *Journal of Shellfish Research* 30:532.

Meisenheimer, J. 1901. Entwicklungsgeschichte von *Dreissensia polymorpha* Palll. *Zeitschrift fur Wissenschartliche Zoologie* 69:1–137.

Mikamitakei, K., M. Kosakai, M. Isemura, T. Suyemitsu, K. Ishihara, and K. Schmid. 1991. Fractionation of jelly substance of the sea-urchin egg and biological-activities to induce acrosome reaction and agglutination of spermatozoa. *Experimental Cell Research* 192:82–86.

Miller, R. L., J. J. Mojares, and J. L. Ram. 1994. Species-specific sperm attraction in the zebra mussel, *Dreissena polymorpha*, and the quagga mussel, *Dreissena bugensis*. *Canadian Journal of Zoology* 72:1764–1770.

Mills, E. L., J. R. Chrisman, B. Baldwin, R. W. Owens, R. O'Gorman, T. Howell, E. F. Roseman, and M. K. Raths. 1999. Changes in the dreissenid community in the lower Great Lakes with emphasis on southern Lake Ontario. *Journal of Great Lakes Research* 25:187–197.

Misamore, M., H. Silverman, and J. W. Lynn. 1996. Analysis of fertilization and polyspermy in serotonin-spawned eggs of the zebra mussel, *Dreissena polymorpha*. *Molecular Reproduction and Development* 43:205–216.

Misamore, M. J. and J. W. Lynn. 2000. Role of the cytoskeleton in sperm entry during fertilization in the freshwater bivalve *Dreissena polymorpha*. *Biological Bulletin* 199:144–156.

Misamore, M. J., K. K. Stein, and J. W. Lynn. 2006. Sperm incorporation and pronuclear development during fertilization in the freshwater bivalve *Dreissena polymorpha*. *Molecular Reproduction and Development* 73:1140–1148.

Mojares, J. J., J. J. Stachecki, K. Kyozuka, D. R. Armant, and J. L. Ram. 1995. Characterization of zebra mussel (*Dreissena polymorpha*) sperm morphology and their motility prior to and after spawning. *Journal of Experimental Zoology* 273:257–263.

Moore, B., G. C. Holdren, S. Gerstenberger, K. Turner, and W. H. Wong. 2014. Invasion by quagga mussels (*Dreissena rostriformis bugensis* Andrusov 1897) into Lake Mead, Nevada-Arizona: The first occurrence of dreissenid species in the Western United States. in *Chapter 2 in this Book*, Wong W. H. and S. L. Gerstenberer, eds. CRC Press, Baco Raton, FL.

Nalepa, T. F., D. L. Fanslow, and S. A. Pothoven. 2010. Recent changes in density, biomass, recruitment, size structure, and nutritional state of *Dreissena* populations in southern Lake Michigan. *Journal of Great Lakes Research* 36:5–19.

Nichols, S. J. 1996. Variations in the reproductive cycle of *Dreissena polymorpha* in Europe, Russia, and North America. *American Zoologist* 36:311–325.

Nichols, S. J. and M. K. Black. 1994. Identification of larvae: The zebra mussel (*Dreissena polymorpha*), quagga mussel (*Dreissena rosteriformis bugensis*), and Asian clam (*Corbicula fluminea*). *Canadian Journal of Zoology* 72:406–417.

Niijima, L. and J. Dan. 1965. Acrosome reaction in *Mytilus edulis*. 2. Stages in reaction observed in supernumerary and calcium-treated spermatozoa. *Journal of Cell Biology* 25:249–259.

Quinn, N. P. and J. D. Ackerman. 2012a. Biological and ecological mechanisms for overcoming sperm limitation in invasive dreissenid mussels. *Aquatic Sciences* 74:415–425.

Quinn, N. P. and J. D. Ackerman. 2012b. The effect of near-bed turbulence on sperm dilution and fertilization success of broadcast-spawning bivalves. *Limnology and Oceanography Fluids environments* 1:176–193.

Ram, J. L., G. W. Crawford, J. U. Walker, J. J. Mojares, N. Patel, P. P. Fong, and K. Kyozuka. 1993. Spawning in the zebra mussel (*Dreissena polymorpha*): Activation by internal or external application of serotonin. *Journal of Experimental Zoology* 265:587–598.

Ram, J. L., P. Fong, R. P. Croll, S. J. Nichols, and D. Wall. 1992. The zebra mussel (*Dreissena polymorpha*), a new pest in North-America—Reproductive mechanisms as possible targets of control strategies. *Invertebrate Reproduction and Development* 22:77–86.

Ram, J. L., P. P. Fong, and D. Garton. 1996. Physiological aspects of zebra mussel reproduction: Maturation, spawning, and fertilization. *American Zoologist* 36:326–338.

Ram, J. L., A. S. Karim, K. Acharya, P. Jagtrap, S. Purohit, and D. R. Kashian. 2011. Reproduction and genetic detection of veligers in changing *Dreissena* populations in the Great Lakes. *Ecosphere* 2:1–16.

Ricciardi, A. and F. G. Whoriskey. 2004. Exotic species replacement: Shifting dominance of dreissenid mussels in the Soulanges Canal, upper St. Lawrence River, Canada. *Journal of the North American Benthological Society* 23:507–514.

Roe, S. L. and H. J. Macisaac. 1997. Deepwater population structure and reproductive state of quagga mussels (*Dreissena bugensis*) in Lake Erie. *Canadian Journal of Fisheries and Aquatic Sciences* 54:2428–2433.

Rosen, M. R., K. Turner, S. L. Goodbred, and J. M. Miller (eds.). 2012. A synthesis of aquatic science for management of Lakes Mead and Mohave. U.S. Geological Survey Circular 1381, p. 162.

Schwaebe, L., K. Acharya, and M. J. Nicholl. 2013. Comparative efficacy of *Dreissena rostriformis bugensis* (Bivalvia: Dreissenidae) spawning techniques. *Aquatic Invasions* 8:45–52.

Sprung, M. 1989. Field and laboratory observations of *Dreissena polymorpha* larvae—Abundance, growth, mortality and food demands. *Archiv Fur Hydrobiologie* 115:537–561.

Sprung, M. 1995. Physiological energetics of the zebra mussel, *Dreissena polymorpha*, in lakes. 1. Growth and Reproductive Effort. *Hydrobiologia* 304:117–132.

Spurr, A. R. 1969. A low-viscosity epoxy resin embedding medium for electron microscopy. *Journal of Ultrastructure Research* 26:31–43.

Stoeckmann, A. 2003. Physiological energetics of Lake Erie dreissenid mussels: A basis for the displacement of *Dreissena polymorpha* by *Dreissena bugensis*. *Canadian Journal of Fisheries and Aquatic Sciences* 60:126–134.

Togo, T. and M. Morisawa. 1999. Mechanisms for blocking polyspermy in oocytes of the oyster *Crassostrea gigas*. *Journal of Experimental Zoology* 293:307–314.

Togo, T., K. Osanai, and M. Morisawa. 1995. Existence of three mechanisms for blocking polyspermy in oocytes of the mussel *Mytilus edulis*. *Biological Bulletin* 189:330–339.

Walker, G. K., M. K. Black, and C. A. Edwards. 1996. Comparative morphology of zebra (*Dreissena polymorpha*) and quagga (*Dreissena bugensis*) mussel sperm: Light and electron microscopy. *Canadian Journal of Zoology* 74:809–815.

Watson, G. J., M. G. Bentley, S. M. Gaudron, and J. D. Hardege. 2003. The role of chemical signals in the spawning induction of polychaete worms and other marine invertebrates. *Journal of Experimental Marine Biology and Ecology* 294:169–187.

Wong, W. H. and S. L. Gerstenberger. 2011. Quagga mussels in the western United States: Monitoring and management. *Aquatic Invasions* 6:125–129.

Wright, D. A., E. M. Setzler-Hamilton, J. A. Magee, and H. R. Harvey. 1996. Laboratory culture of young zebra (*Dreissena polymorpha*) and quagga (*D. bugensis*) mussels using estuarine algae. *Journal of Great Lakes Research* 22:46–54.

5

Modeling Carrying Capacity of Quagga Mussels in the Boulder Basin of Lake Mead, Nevada

Chad L. Cross and Wai Hing Wong

CONTENTS

ABSTRACT   Dreissenid mussels are known to occupy many freshwater systems throughout the United States. In particular, the quagga mussel (*Dreissena rostriformis bugensis* Anrusov 1897) has been an aggressively invasive species in Lake Mead, Nevada–Arizona, having spread rapidly both in numbers and in distribution. In this chapter, we provide an overview of carrying capacity in general and how it relates to models for mussel culture. Specifically, we examine two facets of carrying capacity for the quagga mussel. The first model is a culture-based model by Incze et al. (1981) and is used to develop an estimate of carrying capacity in terms of total population size. The second model we explore is a box model and is used to simulate the amount of time it would take to reach the carrying capacity estimated in the first model. Using the Incze Model, we predicted a total of approximately 1.02×10^{13} mussels as the carrying capacity in Boulder Basin, with an approximated time to reach this capacity of around 19–21 years as predicted by the box model. Clearly, there are many different models that could potentially be explored, and the models presented herein could certainly be modified, expanded, and reconfigured to answer a myriad of ecological questions (e.g., how circulation or basin water exchange affects the system). There are very few articles or reports that examine a mathematical approach to estimating carrying capacity in nonculture mussel systems. Our hope is that future research will be expanded in understanding invasive mussel species through the use of ecosystem-level and food web-level mathematical models.

Introduction

Dreissenid mussels are known to occupy many freshwater systems throughout the United States. In particular, the quagga mussel (*Dreissena rostriformis bugensis* Anrusov 1897) and the zebra mussel (*Dreissena polymorpha* Pallas 1771) are aggressively invasive species, as evidenced by their rapid spread throughout much of the United States owing to their ability to reproduce rapidly in the absence of natural predators. The economic impact of these invasive mussels is difficult to assess precisely, but their damage to intake pipes, water pump facilities, and electrical generation plants is of particular concern (Britton et al. 2010). Additionally, one aspect of the ecology of invasive species that is of utmost concern is the potential reduction, extirpation, or extinction of native species that lack the ability to outcompete the invaders (Ricciardi et al. 1998).

Because of the potential ecological and economic impacts related to these mussels, a prescient inquiry is to determine the number of mussels that can occupy a given system. This idea is akin to estimating the *carrying capacity* of a system and is simultaneously one of the most important and one of the most challenging ecological determinations for a given species.

The purpose of this chapter is to provide an overview of carrying capacity in general, to discuss aspects of carrying capacity specific to mussel culture and how this relates to carrying capacity models in general, and then to examine two facets of carrying capacity for the quagga mussel. Specifically, we provide two modeling exercises for examining carrying capacity in the Boulder Basin. The first model we explore is used to develop an estimate of carrying capacity in terms of total population size. The second model we explore is used to simulate the amount of time it would take to reach the carrying capacity estimated in the first model.

Concept of Carrying Capacity

Carrying capacity is classically understood to indicate the maximum number of individuals that a system can contain (i.e., *carry*) when adequate food, water, space, and other requirements are available for the species. The concept of carrying capacity enjoys a history exceeding 150 years, having been described by Verhulst in 1838 (*in cit.* Gotelli 1995), and conceptualized most often via the common logistic growth model. In this model, carrying capacity is denoted by the letter "*K*," which is a function of the number of births and deaths in a population over a given, instantaneous growth interval. Functionally, one can express the density-dependent change in the population size over an interval of time with the differential equation:

$$\frac{dN}{dt} = rN\left(1 - \frac{N}{K}\right)$$

It is readily apparent from this equation that the point at which the change in population size as a function of time equals zero is when $N = K$. Hence, carrying capacity is a definable construct in that the population size maximizes at a particular, albeit theoretical, point, above which one would expect there to be ecological damage done to the system that would hinder further growth (Odum 1989; Rees 1992).

Clearly, one can imagine that carrying capacity is neither absolute nor stable. As environmental conditions change, more food is available, or more space is freed, the carrying capacity is expected to change for some period of time before reaching a stable point again. Arguments have been posed by ecologists (Ginzburg 1992; Gabriel et al. 2005) challenging defining carrying capacity using such a simple paradigm, suggesting that the idea of carrying capacity is not the same as the concept of an ecological equilibrium. An interesting argument by Hiu (2006) provides a conceptual clarification by defining carrying capacity as the environment's maximum load—that is, the maximum number of individuals that can be reasonably sustained in an ecological context. The argument brought forth by this concept is that population equilibrium and carrying capacity are separate concepts because an ecosystem may theoretically be able to carry a certain number of individuals; however, the population may reach an equilibrium state lower than the theoretical maximum carrying capacity owing to resource limitations.

Defining Carrying Capacity in Bivalve Culture

A review article by McKindsey et al. (2006) provides a thorough summary of carrying capacity considerations for bivalve culture. Understanding carrying capacity in culture systems is extraordinarily important economically. That is, in a culture system, the idea is to increase bivalve growth to the point at which production is maximized without inadvertently damaging the ecosystem.

Owing to the many constructs of carrying capacity as outlined earlier, McKindsey et al. (2006) explore the four hierarchical definitions of carrying capacity first outlined by Inglis et al. (2000) for understanding marine aquaculture. The first of these is *physical carrying capacity*, which is used to define the geographic area in which the environment has the necessary physical properties to support bivalves. This would include not only physical characteristics such as substrate but also chemical properties such as oxygen and phosphate concentrations.

The second level is the *production carrying capacity*, which defines the level at which mussel production is optimized. Drivers of production carrying capacity are largely related to the availability of resources, such as primary productivity, which is clearly a limiting factor in mussel culture in natural systems. Much of the early work on production carrying capacity has been reported by Bacher et al. (1998, 2003).

The third level is *ecological carrying capacity*, which focuses largely on the broader picture of both the physical and production carrying capacities. Understanding that when aquatic manipulations take place to ensure that adequate physical properties are maintained and that production is maximized by enhancing, say, primary production, it is understood that the larger ecological system as a whole also plays a role in determining how many mussels can be maintained in a system. As discussed earlier, maximizing the size of a population without causing ecological damage is one way to operationally define carrying capacity (see, e.g., Odum 1989; Rees 1992). As one can imagine, considering an entire ecosystem is much more complex, and hence much work to date has been focused on production and physical carrying capacity models (McKindsey et al. 2006).

Finally, the most complex, and hence least studied and understood, is *social carrying capacity*. This final level of carrying capacity includes not only ecological considerations, but also socioeconomic ones and political decision-making. Understanding that ecosystem changes owing to production will occur, one must consider what level of change is acceptable, for example, considering *functional performance indicators* of culture systems (Gibbs 2007), and this could mean a carrying capacity that is quite different from one that is ecologically optimized.

Clearly, the idea of carrying capacity is not as straightforward as estimating a number. Carrying capacity estimation must consider not only the species under investigation, but also the physical, chemical, and ecological contexts in which they live. Many of the models explored here and elsewhere focus on production carrying capacity since the main focus of maximizing population size is to maximize mussel culture for economic reasons. There is an increased interest in developing models addressing ecological carrying capacity, particularly with the recognition that understanding the larger system in which mussels live is important both from an economic and from a sustainability standpoint. Additionally, with the increased availability of computer software and simulation capacity, modeling complexity can lead to a deeper understanding of what drives ecological systems.

Boulder Basin Study Area

The focus of investigation for the present study is the Boulder Basin of Lake Mead, Nevada–Arizona (Figure 5.1). Volumetrically, Lake Mead is one of the largest reservoirs in the Unites States, with a volume of approximately 36.7×10^9 m^3 and an areal extent of 66,000 ha (LaBounty and Burns 2005). The Boulder Basin is where quagga mussels are believed to have first appeared in the Lake, likely through recreational boating, and therefore much fieldwork has been completed in this particular area; hence, it is well suited to address the two main focal points addressed in this chapter, namely, carrying capacity and the length of time it may take to reach this capacity.

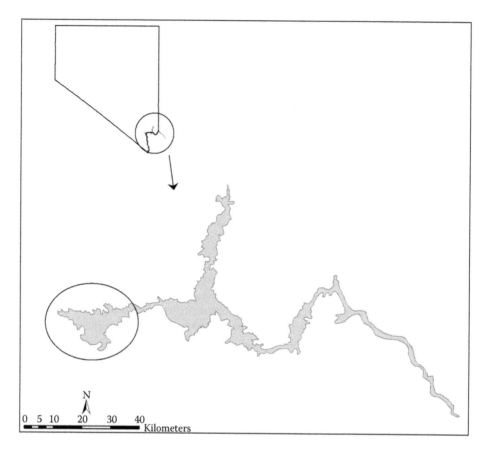

FIGURE 5.1   Graphical representation of Lake Mead Nevada–Arizona. Boulder Basin is circled in the lower left of the figure and is the focus of the data presented in this chapter. (This figure was originally published by Cross, C.L. et al., *Aquat. Invas.*, 6, 141, 2011. The authors retain the copyright to the article, and this figure is reproduced here under a Creative Commons license, http://creativecommons.org/licenses/by/2.0/.)

Key Variables from Boulder Basin

There are several potential variables that may serve as limiting factors for quagga mussels in the Boulder Basin. As a first step in considering an appropriate estimation model, it was necessary to determine if any key variables were beyond the tolerance limits reported for quagga mussels. Tolerance limits for quagga mussels as modeled herein are found in Spidle et al. (1995), McMahon (1996), Mills et al. (1996), and Jones and Ricciardi (2005). Field research reported in work by LaBounty and Burns (2005) and Whittier et al. (2008) provided comparative measures of these variables from Boulder Basin.

The following results were determined from the literature review: (1) calcium concentration ranges between 69.1 and 87.0 mg/L in Boulder Basin, and quagga do poorly when concentrations are <12 mg/L; (2) the average temperature in the Boulder Basin is <28°C, and temperatures >30°C are generally intolerable for quagga mussels; (3) Boulder Basin salinity is <1 psu, and levels >5 psu are not well tolerated; (4) the average pH in Boulder Basin is 8, and the tolerance limit is <6.5; (5) oxygen saturation in Boulder Basin is >40%, and the tolerance limit is <25%; (6) the turbidity in Boulder Basin, as measured by Secchi depth, is >3.3 m, and the turbidity <0.1 m is not well tolerated; and (7) the chlorophyll a concentration in Boulder Basin is 0.9–5.0 µg/L, and the tolerance limit is <2.5 µg/L or >25 µg/L. Hence, a potential limiting factor for determining production carrying capacity in the Boulder Basin is chlorophyll a, as minimum values in the Boulder Basin sometimes fall below the tolerance limit. Therefore, an estimator of carrying capacity that explicitly accounts for suspended food concentration was used to estimate K, the carrying capacity within Boulder Basin.

Estimating Quagga Mussel Carrying Capacity

Culture-Based Model of Carrying Capacity

A culture model first considered by Incze et al. (1981; referred to hereafter at the *Incze Model*) was used to derive an estimate of carrying capacity. The Incze Model was first considered by Che (2010) and Cross et al. (2011) for estimating the carrying capacity of quagga mussels. As with most mussel models described in this document and in the literature, the Incze Model is a cultivation-based model (i.e., it addresses the concept of *production carrying capacity*). In this model, suspended particles flow through the system, the mussels remove these particles, and then the number of mussels either continues to increase or it reaches a point where suspended food particles are reduced below a point that is sustaining in the system, and the culture reaches a relative steady state at a presumed carrying capacity.

The essential feature of the Incze Model is the removal of food particles that are suspended in the water column. If one imagines mussels represented by a *layer* or *tier*, then water with suspended particles that passes through this tier will have an overall reduction in suspended particles, as some concentration of such particles will be filtered out of the water column. If the particles are represented by a limiting factor, such as chlorophyll a, then if n_i chlorophyll a enter the tier, n_j will exit the tier, with $(n_i - n_j)$ representing the difference, or the total number of particles removed by the mussels in a given tier. Here, $i = \{1, 2, 3, ..., k\}$ tiers, with $i < j$. At some point, too few particles will remain in the water column to sustain an additional tier of mussels, and this tier is referred to as tier k, with the total number filtered in the final tier represented as $(n_k - n_{k+1})$, where $k + 1$ is the supremum of the tier distribution, and hence carrying capacity will have been reached in the unit prior to this tier (Figure 5.2).

Nutrients contained in the water column pass through a series of tiers (i.e., layered conglomerations of quagga mussels) and are filtered out by the mussels

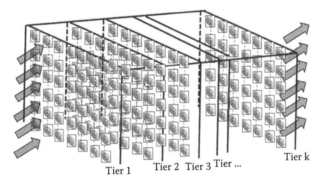

For each tier, n_i nutrients enter and n_j nutrients pass through; the difference $(n_i - n_j)$ represents the total amount of nutrients removed at each tier

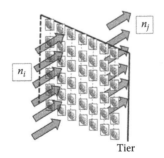

FIGURE 5.2 A graphical representation of the tier system utilized in the Incze Model for estimating carrying capacity. (This figure was originally published by Cross, C.L. et al., *Aquat. Invas.*, 6, 141, 2011. The authors retain the copyright to the article, and this figure is reproduced here under a Creative Commons license, http://creativecommons.org/licenses/by/2.0/.)

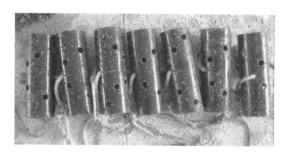

FIGURE 5.3 Photos of the ABS plastic pipes before and after submersion in Boulder Basin. The results of this experiment were used to represent tiers in the Incze Model. (Photos courtesy of W. Baldwin and W. H. Wong.)

An immediate question arises as to how mussels can be represented in a tier system. An experiment carried out by W. Baldwin and W.H. Wong (original data unpublished, but later referenced in Cross et al. 2011) provides a basis for this model. In the Boulder Basin, ABS plastic pipes, approximately 20 cm in length and 6 cm in diameter, were lowered to different depths of the basin. These pipes developed an average of 5079 mussels (adult and juvenile) over a period of 2 years (Figure 5.3).

Additional parameters are needed in the model in order to derive an estimate of carrying capacity. These are the volume of water entering through the first tier (N; L/h), which is a function of water velocity (V; m/h) and area of the system (A; here approximately 40,000 m²). It is assumed that N will be reduced proportionally by the total number of food particles removed from the system such that the particle concentration at a given tier is represented by n_1 and a function of the reduction in N, represented as

$$\frac{n_k}{n_1} = \left\{ \frac{N - \text{number of food particles filtered}}{N} \right\}^{k-1}$$

Next, we must consider the clearance rate (CR; L/h) of the mussels and the total number of mussels/tier (M); the product of CR and M provides an estimate of the number of food particles filtered. Solving the equation for k, which represents the total number of tiers in the system, results in

$$k = \frac{\ln\left(n_k/n_1\right)}{\ln\left((N - CR \times M)/N\right)} + 1$$

Multiplying k by the average number of mussels per tier (5079 as reported earlier) provides the estimated carrying capacity of the system. For a full explication of the model as described here, see Incze et al. (1981) and Cross et al. (2011).

Parameterizing the Incze Model

Further considerations from the literature provide useful results necessary to parameterize the model. We used an estimated filtration rate of 52 L/h/tier, which was reported by Baldwin et al. (2002) for Lake Erie and assumed to be similar in Boulder Basin. We used a chlorophyll estimate of 1.72 µg/L

(W.H. Wong, personal communication) to represent the initial concentration (n_1), and 0.017 μg/L to estimate the concentration in the final tier (a 99% reduction in food particles). Additionally, we partitioned the Boulder Basin into three depths, representing <10 m (mean flow rate at this depth is 329.72 m/h), 10–20 m (mean flow rate at this depth is 111.24 m/h), and >20 m (mean flow rate at this depth is 123.48 m/h). It should be noted that we were concerned primarily with depths at ≤30 m, as colonization below 30 m is rare and quite slow (Meuting 2009; Meuting et al. 2010). Total carrying capacity, then, is the sum across the three depth categories.

Results of the Incze Model

As mentioned earlier, we used an estimated 99% reduction in mean suspended chlorophyll a to represent the point at which carrying capacity may reasonably be assumed to be reached. With this in mind, using the equation and parameters outlined earlier, the calculated carrying capacity in the <10 m depth was 5.96×10^{12} mussels, in the 10–20 m depth was 2.01×10^{12} mussels, and in the >20 m depth was 2.23×10^{12} mussels. Hence, the Incze Model would predict a total of approximately 1.02×10^{13} mussels as the carrying capacity in Boulder Basin.

Estimating the Time to Reach Carrying Capacity

The second major question addressed in this chapter deals with estimating the time it will take for quagga mussels to reach their projected carrying capacity. For this exercise, we turn to the recent literature on box models—so called because they consider the area of estimation to be represented by a box or a series of boxes connected in the natural system; hence, they too conceptualize mussels within the context of a culture system.

Box Models

A large body of research in the last decade has focused on the investigation of ecosystem-level models for aquaculture (Dowd 1997, 2005; Grant et al. 2007; Filgueira and Grant 2009). These models largely address the concept of ecological carrying capacity in that they include the physical resources and the production capacity of the system simultaneously, and hence address the system as a whole.

In ecosystem-level box models, the rate of mussel mass change is considered a function of food availability coupled with feedback circulation of detrital matter and nutrients. Spatial and boundary conditions are responsible for connecting the mussel's local environment to the larger system. The dynamic interplay among zooplankton/phytoplankton, detritus, nutrients, and mussel biomass is represented through a series of differential equations (see, e.g., Grant et al. 2007), with the intent of understanding how energy flows through the system as a function of time. By running simulations with these differential equations, it is possible to examine how energy flow may be related to mussel mass (Filgueira and Grant 2009), for example, through induced nutrient upwelling (Filgueira et al. 2010).

Structure and Mathematical Representation of the Box Model

The standard representation of the box model is provided in Grant et al. (2007). In the basic model, nutrients provide energy to phytoplankton, which in turn are used by zooplankton and mussels. Additional energy flows from detrital matter in the system to both zooplankton and mussels, which in turn flow back into the nutrient pool (principally in the form of nitrate and ammonia). Interestingly, in studies by Filgueira and Grant (2009) and Filgueira et al. (2010), zooplankton played a very minor role in the final outcome of the model simulation. Our simulations showed a very similar outcome, and hence in the present chapter, we did not provide a separate compartment model for zooplankton.

As described earlier, movement among these compartments is estimated through a complex system of differential equations. In their most basic form, designating nutrients as N, phytoplankton as P, detritus as D, and mussels as M, we have the following representative series of differential equations (see Filgueira and Grant 2009):

$$\frac{dN}{dt} = \text{external sources} + \text{mussel excretion} - \text{phytoplankton uptake}$$

$$\frac{dP}{dt} = \text{natural growth} - \text{mortality} - \text{mussel grazing}$$

$$\frac{dD}{dt} = \text{mussel excretion} + \text{phytoplankton mortality} - \text{mussel grazing}$$

$$\frac{dM}{dt} = \text{addtion of new mussels} + \text{net mussel growth} - \text{mortality} - \text{removal}$$

As is evident upon inspection of these equations, several of the variables in each derivative are nonunique, and hence we can envision their relationship to one another such that changes in phytoplankton and detritus, for example, both involve changes owing to mussel grazing. The full representation and subsequent interconnectedness of these equations are much more complex than is represented here, where the interest is for the clarity of presentation and general discussion as opposed to providing a lesson on differential equations. As an example of this complexity, consider one representation of a differential equation for mussels (from Grant et al. 2007):

$$\frac{dM}{dt} = + \left[M_w \left(\frac{\varepsilon_{MP}P + \varepsilon_{MD}\mu_M D}{P + \mu_M D} \right) \times \left(c_{MI} e^{Q_{MI}T} \right) \times \left(\frac{P+D}{k_m + P + D} \right) \times \left(\frac{M_W}{M_{W\,Ref}} \right)^{b_{MI}} \times I_m \right]$$

$$- \left[M_w \times \left(c_{MRS} e^{Q_{MRS}T} \right) \times \left(\frac{M_W}{M_{W\,Ref}} \right)^{b_{MR}} \times \beta_{MRS} \right]$$

$$- \left[M_w \times \sigma_{\gamma M} \varepsilon_M \times \left(c_{MI} e^{Q_{MI}T} \right) \times \left(\frac{P+D}{k_m + P + D} \right) \times \left(\frac{M_W}{M_{W\,Ref}} \right)^{b_{MI}} \times I_m \right]$$

Here there is accounting for mussel growth as a function of feeding efficiency by considering phytoplankton (P) and detritus (D), reduction due to respiration functions, and standard metabolic processes. The rate constants biomass indices in this equation are not important for the general discussion, but full derivations of them can be found in, for example, Dowd (1997) and Grant et al. (2007). The point of showing this equation is not to make the modeling seem overly difficult, but simply to point out that careful attention must be paid to the complex relationships among the modeled compartments in the full model and to emphasize that the models are dependent on a large number of estimates with largely unknown prediction uncertainty.

The box model developed in the current analysis was programmed using Simile Software (v. 5.9; Simulistics Ltd., Midlothian, United Kingdom), which is based on a visual, declarative modeling framework. This software allows for models to be developed quite quickly by using a set of interconnected flow diagrams that connect compartments. Within these compartments, differential equations can be programmed, and the results passed along a flow to other compartment inputs (Figure 5.4). Much more simplified box models can be developed for sensitivity analyses and to investigate various inputs along a potential gradient to gain an understanding of potential outcomes in natural systems (Figure 5.5).

The box model presented here was designed to run a series of simulations to estimate the time it would take to reach the carrying capacity projected by the Incze Model in the previous section. Simulations were run on a Windows 7 Enterprise, 64 bit OS system running at 2.5 GHz with 8 Gb of RAM.

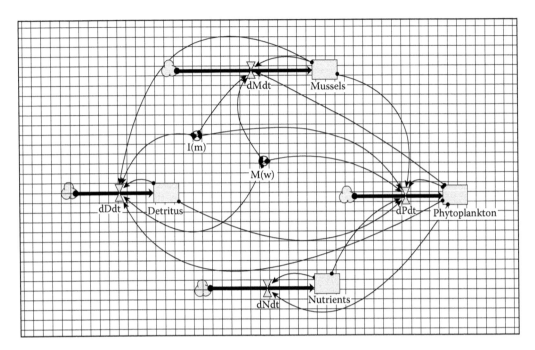

FIGURE 5.4 A representation of one of the simplified box models under development using Simile software. The four major compartments (nutrients, detritus, phytoplankton, and mussels) are defined by differential equations, as described in the text. Complex connections among the compartments show various flows. Other more complex and also more basic box models (as in Figure 5.5) have been examined and are under development.

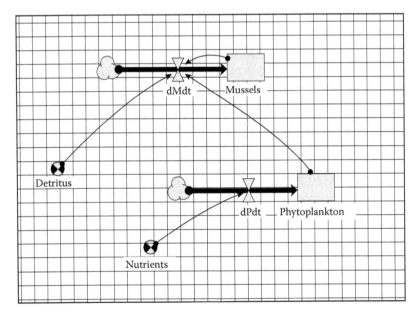

FIGURE 5.5 This figure provides another representation of a box model. In this model, nutrients and detritus are represented as constant values that are connected to phytoplankton and mussel flow diagrams. Models with this level of simplicity can be used to investigate the sensitivity of the model outputs along a gradient of inputs.

TABLE 5.1

Variables and Measurements from Boulder Basin, Lake Mead, and from Published Works Were Used to Parameterize Box Models

Parameter	Value	Source
Starting number of mussels	100,000	—
Mg C per mussel	40.84	O'Brien (2006)
Phytoplankton/detritus assimilation efficiency	0.3	Bayne and Widdows (1978)
Mussel allometric coefficient of ingestion	−0.4	Bayne and Newell (1983)
Phytoplankton concentration (cells/mL)	26,853 (mean 2002–2008)	W.H. Wong
Detritus concentration (cells/mL)	26,854	—
Orthophosphate concentration (µg/L)	2.4 (mean 2005–2008)	W.H. Wong

Source: Grant, J. et al., Ecol. Model., 200, 193-206, 2007. The starting number of mussels and the detritus concentration were unknown, but were used to develop initial model parameters. Other parameters not mentioned here were taken largely from Grant et al. (2007).

Parameterizing the Box Model for Simulation

Many of the parameters for the modeling exercise presented here were taken from those cited in Grant et al. (2007; Table 5.1, p. 199) and from studies done in Boulder Basin by W.H. Wong (personal communication) (Table 5.1). The initial mussel seeding number was 100,000 individuals, and the initial detritus concentration (cells/mL) was assumed to be the same as for phytoplankton since detritus was not directly measured. It should be noted that as the complexity of models increases, sensitivity to minor changes in inputs can lead to demonstrable changes in final results. Hence, the results presented here were simulated several thousand times to ensure their stability for final estimates.

Results of the Box Model

The initial seeding value for mussels and the concentration of detritus were investigated using several different values, but the final simulation results were not demonstrably impacted by changes in these values. Though sensitivity analyses were not a primary outcome for this exercise, the model outcome was relatively robust to input values, suggesting that parameters were within a range in which convergence of the solution was feasible. Ultimately, after several simulation runs, the model predicted that quagga mussels would reach the carrying capacity predicted by the Incze Model in approximately 19–21 years.

Concluding Remarks

In these final few concluding remarks, we examine three additional studies that provide important context for estimating carrying capacity. The use of mathematical modeling and simulation for understanding ecological systems is rapidly growing, and we expect that new techniques and strategies, as outlined in the following studies, will be expanded and further explored.

In an important article by Smaal et al. (1998), a useful definition of carrying capacity is examined in detail. In particular, they define exploitation carrying capacity or that capacity where the available standing stock is maximized so that the marketable cohort is concomitantly maximized. They reviewed several existing models and suggest that minimally, bivalve carrying capacity models should account for organism and population processes and also transport and sediment dynamics. This is certainly in line with many ecosystem box models and fits within the context of what we present for quagga mussels, even though maximizing marketability was not the main goal. However, per their suggestion, we too agree that defining a minimum set of compartments is strategically important for simulation-based models.

Interestingly, modeling of ecosystems has been expanded to include multiple species simultaneously. In a study by Duarte et al. (2003), a multispecies model for scallops and oysters was developed and tested with field data, with some success. The expansion of models to examine polyculture systems is quite complex. It does, however, provide the basis for considering a larger ecosystem-level model, where many organisms are considered as contributing to mussel carrying capacity. As an example, mussels may not be the only species ingesting suspended food particles, and by ignoring other consumers, models of carrying capacity may be too restrictive to provide insights into the larger ecosystem.

Probably the most complex consideration for modeling is to consider food web models, such as that presented by Jiang and Gibbs (2005), where fisheries, as well as mussel culture, are considered. The complexity of such models does make them somewhat difficult to properly parameterize, but the rapidly growing interest in the exploitation of multiple levels of a system simultaneously makes food web models particularly intriguing. Investigating these model structures may be particularly important for invasive species such as quagga mussels since their overall impact on the system as a whole is of utmost interest.

The results of the modeling efforts presented in this chapter are unique in that we strategically combined a basic cultivation-based modeling effort with a fairly complex ecologically based model. The advantage of this approach was that we were able both to estimate carrying capacity computationally and then use these results to consider the ecological context that would allow, and then ultimately limit, population growth of this species. However, we do realize that estimation of carrying capacity using a basic mathematical cultivation model may in fact lead to an overestimation of mussels because it is blinded to ecological context. Clearly, all modeling efforts are ultimately limited in their ability to provide an estimation of reality, and we continue to strive to develop more complexity and logic into our modeling efforts.

The purpose of this chapter was to provide a brief discussion of carrying capacity and to specifically provide a description and illustration of two different methodologies useful for investigating the carrying capacity of quagga mussels. Clearly, there are many different models that could potentially be explored, and the models presented herein could certainly be modified, expanded, and reconfigured to answer a myriad of ecological questions (e.g., how circulation or basin water exchange affects the system). There are very few articles or reports that examine a mathematical approach to estimating carrying capacity in nonculture mussel systems. Our hope is that future research will be expanded in understanding invasive mussel species through the use of ecosystem-level and food web-level mathematical models.

REFERENCES

Bacher, C., Duarte, P., Ferreira, J.G., Heral, M., Raillard, O. 1998. Assessment and comparison of the Marennes-Oleron Bay (France) and Carlingford Lough (Ireland) carrying capacity with ecosystem models. *Aquatic Ecology* 31:379–394.

Bacher, C., Grant, J., Hawkins, A.J.S., Fang, J., Zhu, M., Besnard, M. 2003. Modeling the effect of food depletion on scallop growth in Sungo Bay (China). *Aquatic Living Resources* 16:10–24.

Baldwin, B.S., Mayer, M.S., Dayton, J., Pau, N., Mendilla, J., Sullivan, M., Moore, A., Ma, A., Mills, E. 2002. Comparative growth and feeding in zebra and quagga mussels (*Dreissena polymorpha* and *Dreissena bugensis*): Implications for North American lakes. *Canadian Journal of Fisheries and Aquatic Sciences* 59:680–694.

Bayne, B.L., Newell, J. 1983. Physiological energetics of marine molluscs. In: Saleuddin, S.M., Wilbur, K.M. (Eds.), *The Mollusca*. Part I. *Physiology*, vol. 4. Academic Press, New York, pp. 407–515.

Bayne, B.L., Widdows, R.C. 1978. The physiological ecology of two populations of *Mytilus edulis* L. *Oceologia* 37:137–162.

Britton, D., Brown, E., Heimowitz, P., Morse, J., Norton, D., Pitman, B., Ryce, E., Smith, R., Williams, E., Volkoff, M. 2010. *Quagga–Zebra Mussel Action Plan for Western U.S. Waters. The Western Regional Panel on Aquatic Nuisance Species*. http://www.anstaskforce.gov/QZAP/QZAP_FINAL_Feb2010.pdf (Accessed June 14, 2013).

Che, T. 2010. Estimation of carrying capacity for the quagga mussel (*Dreissena rostriformis*) in Lake Mead, Nevada. MS thesis. University of Nevada, Las Vegas, NV. 16pp.

Cross, C.L., Wong, W.H., Che, T. 2011. Estimating carrying capacity of quagga mussels (*Dreissena rostriformis bugensis*) in a natural system: A case study of the Boulder Basin of Lake Mead, Nevada-Arizona. *Aquatic Invasions* 6:141–147.

Dowd, M. 1997. On predicting the growth of cultured bivalves. *Ecological Modeling* 104:113–131.

Dowd, M. 2005. A bio-physical coastal ecosystem model for assessing environmental effects of marine bivalve aquaculture. *Ecological Modeling* 183:323–346.

Duarte, P., Meneses, R., Hawkins, A.J.S., Zhu, M., Fang, J., Grant, J. 2003. Mathematical modeling to assess the carrying capacity for multi-species culture within coastal waters. *Ecological Modelling* 168:109–143.

Filgueira, R., Grant, J. 2009. A box model for ecosystem-level management of mussel culture carrying capacity in a coastal bay. *Ecosystems* 12:1223–1233.

Filgueira, R. Grant, J., Stand, Ø., Asplin, L., Aure, J. 2010. A simulation model of carrying capacity for mussel culture in a Norwegian Fjord: Role of induced upwelling. *Aquaculture* 308:20–27.

Gabriel, J.P., Saucy, F., Bersier, L.F. 2005. Paradoxes in the logistic equation? *Ecological Modelling* 185:147–151.

Gibbs, M.T. 2007. Sustainability performance indicators for suspended aquaculture activities. *Ecological Indicators* 7:94–107.

Ginzburg, L.R. 1992. Evolutionary consequences of basic growth equations. *Trends in Ecology and Evolution* 7:133.

Gotelli, N.J. 1995. *A Primer of Ecology*. Sunderland, MA: Sinauer.

Grant, J. Curran, K.J., Guyondet, T.L., Tita, G., Bacher, C., Koutitonsky, V., Dowd, M. 2007. A box model of carrying capacity for suspended mussel aquaculture I Lagune de la Grand-Entrée, Iles-de-la-Madeleine, Québec. *Ecological Modeling* 200:193–206.

Hiu, C. 2006. Letter to the editor: Carrying capacity, population equilibrium, and environment's maximum load. *Ecological Modelling* 192:317–320.

Incze, L.S., Lutz, R.A., True, E. 1981. Modeling carrying capacities for bivalve mollusks in open, suspended culture systems. *Journal of the World Aquaculture Society* 12:143–153.

Inglis, G.J., Hayden, B.J., Ross, A.H. 2000. An overview of factors affecting the carrying capacity of coastal embayments for mussel culture. The National Institute of Water and Atmospheric Research. Report CHC00/69. Christchurch, NZ.

Jiang, W., Gibbs, M.T. 2005. Predicting the carrying capacity of bivalve shellfish culture using a steady, linear food web model. *Aquaculture* 244:171–185.

Jones, L.A., Ricciardi, A. 2005. Influence of physicochemical factors on the distribution and biomass of invasive mussels (*Dreissena polymorpha* and *Dreissena bugensis*) in the St. Lawrence River. *Canadian Journal of Fisheries and Aquatic Sciences* 62:1953–1962.

LaBounty, J.F., Burns, N.M. 2005. Characterization of boulder basin, Lake Mead, Nevada-Arizona, USA—Based on analysis of 34 limnological parameters. *Lake and Reservoir Management* 21:277–307.

McKindsey, C.W., Thetmeyer, H., Landry, T., Silvert, W. 2006. Review of recent carrying capacity models for bivalve culture and recommendations for research and management. *Aquaculture* 261:451–462.

McMahon, R.F. 1996. The physiological ecology of the zebra mussel, *Dreissena polymorpha*, in North America and Europe. *American Zoologist* 36:339–363.

Mills, E.L., Rosenberg, G., Spidle, A.P., Ludyanskiy, M., Pligin, Y., May, B. 1996. Review of the biology and ecology of the quagga mussel (*Dreissena bugensis*), a second species of freshwater Dreissenid introduced to North America. *American Zoologist* 36:271–286.

Mueting, S. 2009. Substrate monitoring, contaminant monitoring and public educational outreach on quagga mussels (*Dreissena bugensis*) in Lake Mead, NV. MS thesis. University of Nevada Las Vegas, Las Vegas, NV. 93pp.

Mueting, S.A., Gerestenberger, S.L., Wong, W.H. 2010. An evaluation of artificial substrates for monitoring the quagga mussel (*Dreissena bugensis*) in Lake Mead Nevada-Arizona. *Lake and Reservoir Management* 26:283–292.

O'Brien, M. 2006. Zebra mussels vs. quagga mussels: Survival in oxygen-deficient conditions. *Journal of the U.S. SJWP* 1:59–77.

Odum, E. 1989. *Ecology and Our Endangered Life Support Systems*. Sunderland, MA: Sinauer.

Rees, W.E. 1992. Ecological footprints and appropriated carrying capacity: What urban economics leaves out. *Environment and Urbanization* 4:121–130.

Ricciardi, A., Neves, R.J., Rasmussen, J.B. 1998. Impending extinctions of North American freshwater mussels (Unionoida) following the zebra mussel (*Dreissena polymorpha*) invasion. *Journal of Animal Ecology* 67:613–619.

Smaal, A.C., Prings, T.C., Dankers, N., Ball, B. 1998. Minimum requirements for modeling bivalve carrying capacity. *Aquatic Ecology* 31:423–428.

Spidle, A.P., Mills, E.L., May, B. 1995. Limits to tolerance of temperature and salinity in the quagga mussel (*Dreissena bugensis*) and the zebra mussel (*Dreissena polymorpha*). *Canadian Journal of Aquatic Sciences* 52:2108–2119.

Whittier, T.R., Ringold, P.L., Herlihy, A.T., Pierson, S.M. 2008. A calcium-based invasion risk assessment for zebra and quagga mussels (*Dreissena* spp.). *Frontiers in Ecology and the Environment* 6:180–184.

6

Thermal Tolerance of Invasive Quagga Mussels in Lake Mead National Recreation Area

Matthew Kappel, Shawn L. Gerstenberger, Robert F. McMahon, and Wai Hing Wong

CONTENTS

ABSTRACT Since the introduction and establishment of dreissenid mussels in the Laurentian Great Lakes of North America in 1986, they have invaded water bodies as far west as Southern California. The more commonly found dreissenid mussel in the southwestern United States is the quagga mussel (*Dreissena rostriformis bugensis*). Dreissenid mussels are biofoulers notorious for their ecological and economic impacts. The quagga mussel was first discovered in Lake Mead in January 2007. Because climates and biomes vary from east to west in the United States, thermal tolerance should be determined for dreissenid mussels west of the 100th meridian, where they inhabit warmer water bodies than occur in the northeastern United States. Thermal treatment is an ecologically acceptable approach for control of dreissenid macrofouling (Mackie and Claudi 2010). However, studies of thermal tolerance in dreissenids have focused primarily on populations in the cooler waters of the northeastern United States and particularly on zebra mussels (*Dreissena polymorpha*). The purpose of this study was to determine whether conditions in the southwestern United States affect the acute upper thermal limits of adult quagga mussels. Adult quagga mussels found in Lake Mead (Nevada) could not survive acute exposure to water temperatures greater than or equal to 30°C, above which survival times declined with increasing temperatures up to 40°C. The results of this study revealed that the upper thermal limits of quagga mussels from Lake Mead were similar to those previously recorded for quagga mussels from the northeastern United States.

Introduction

A major aspect of aquatic ecology and environmental health is biological invasions by nonindigenous species. Biological invasions can be expensive in terms of eradication and management cost, manpower and resource allocation, alteration of ecosystem processes, and reduction of biological diversity (Vitousek et al. 1996). Two of the most notable nonindigenous species introduced to North American waters in the 1980s were *Dreissena polymorpha* and *Dreissena rostriformis bugensis*, commonly known as zebra and quagga mussels, respectively (May and Marsden 1992; Claxton et al. 1997). The zebra mussel was discovered in the waters of Lake St. Clair in 1988 (Hebert et al. 1989). It is believed to been released into the Laurentian Great Lakes in 1986 (McMahon et al. 1993). The quagga mussel was discovered a year later in 1989 in the waters of Lake Erie, south of Lake St. Clair (May and Marsden 1992). A more recent paper has provided convincing evidence that it was also present in Lake Erie as early as 1986 (Carlton 2008).

Once a nonindigenous species is introduced into a new environment, it is commonly followed by secondary introductions into other habitats (Kowarik 2003). Both species of dreissenid mussels are believed to have been introduced to North American waters inadvertently by a human-mediated vector, discharged ballast water from transoceanic shipping (Claudi and Mackie 1994; Holeck et al. 2004). Their secondary introductions have been considered to be the result of the natural downstream transport of dreissenid mussels by interconnected bodies of water (Horvath et al. 1996) or larval drift (Griffiths et al. 1991) and overland dispersal on contaminated trailered watercraft (i.e., boats, jet skis, rafts) (Ricciardi et al. 1995).

Dreissenid mussels are macrofoulers capable of colonizing any type of hard substrate and achieving high densities (Boelman et al. 1997). Dreissenid mussels share with marine mytilid mussels the capacity to produce byssal threads for attachment to hard surfaces, external fertilization, and a planktonic larval stage (i.e., veliger stage) (Johnson et al. 2001). Boats that are berthed long enough in mussel-infested waters for dreissenid larvae to settle and develop mussel colonies on their hulls, propulsion units, and other submerged components are vectors for mussel transport when trailered to and launched in other water bodies (Johnson et al. 2001). Since the dreissenid veliger stage is microscopic, they can go undetected in bilge water and be transported overland in trailered watercraft to adjacent unconnected water bodies. Thus, since their introduction and establishment, both dreissenid species have rapidly spread into all of the Great Lakes (Lake Ontario, Lake Michigan, Lake Huron, Lake Erie, and Lake Superior) and adjacent water bodies, as well as into navigable rivers including the Mississippi, Ohio, Tennessee, Arkansas, and Colorado Rivers and their tributaries (McMahon et al. 1995; Ussery and McMahon 1995; Nalepa 2010).

Dreissenid mussels have had serious ecological and economic impacts in North American waters (Bossenbroek et al. 2009). They compete with native freshwater unionacean mussels for phytoplankton leading to a reduction in their abundance and the potential for population extirpations or species extinctions (Ricciardi et al. 1998). Dreissenid mussels foul water intake pipes and raw-water conveyance systems thus costing companies and agencies across the United States billions of dollars per year for repair, maintenance, management, and control (Watters et al. 2013). There have been extensive efforts to monitor, control, and prevent the further spread of these dreissenid mussels. Several methods have been developed to mitigate and prevent the spread of dreissenid mussels. These range from chemical decontamination, heat, hot water/high-pressure washing, freezing, desiccation, and physical removal (Comeau et al. 2011). When considering which method is most appropriate, their effectiveness in mussels control and the impact on health (environmental and anthropogenic) must be taken into consideration. The two most commonly used and widely accepted efficacious methods are the use of molluscicidal chemicals and/or thermal treatments for decontamination (Jenner et al. 1998; Mackie and Claudi 2010). Concerns about chemical by-products arise from the use of molluscicidal agents and their bioaccumulation in dreissenid mussels (i.e., polycyclic aromatic hydrocarbons), which may be toxic to the surrounding environment (Roper et al. 1997).

An accepted nonchemical approach for dreissenid mussel fouling control in raw-water systems is thermal treatment (McMahon and Ussery 1995; McMahon et al. 1995, Mackie and Claudi 2010). Initial studies on North American dreissenid mussel populations have provided baseline data for lethal temperatures of dreissenid mussels when exposed to chronic (long-term) and acute (short-term) temperature stress (Iwanyzki and McCauly 1993; McMahon and Ussery, 1995; McMahon et al. 1995; Spidle et al. 1995). Most studies of thermal tolerance have concentrated on zebra mussels, but similar levels of tolerance have been generally attributed to closely related quagga mussels. Iwanzyki and McCauly (1993) exposed temperature-acclimated adult zebra mussels to acute lethal temperatures without gradually increasing the temperature. McMahon and Ussery (1995) allowed adult zebra mussels to acclimate to specific temperatures and then tested their acute upper thermal limits by exposing specimens to temperature increasing at different rates from the acclimation temperature until an upper lethal temperature was attained.

Spidle et al. (1995) measured the acute and chronic (i.e., long-term) upper thermal limits of both dreissenid species (i.e., zebra and quagga mussels) after acclimation to specific temperatures. Beyer et al. (2011) acutely immersed specimens of both dreissenid species in hot water baths and subjected adult mussels to different exposure durations to determine the effectiveness of combinations of time and temperature for a proposed boat lift at the Rapid Croche dam on the lower Fox River, Wisconsin.

Most watercraft do not stay immersed in water long enough to become fouled by extensive colonies of dreissenid mussels. However, during reproductive periods, juvenile mussels may settle on watercraft even when launched for short periods (i.e., hours or days) in infested water bodies. Small, newly settled juvenile mussels (shell length = 300–500 μm) settle in crevices or corners where they are not readily observed by untrained observers. Thus, individuals who launch or berth boats in mussel-infested water bodies for even short periods must carry out decontamination procedures before relaunching in another water body even though settled mussels are not readily visible on their watercraft. Use of heated water as a means for mussel decontamination is an environmentally safe approach for recreational boats. However, to be effective, the relationship between application time and water temperature should be determined in order to develop truly effective mussel decontamination procedures (Morse 2009a).

Almost all North American studies of dreissenid thermal tolerance have been carried out on northeastern US zebra or quagga mussel populations (Iwanyzki and McCauly 1993; McMahon and Ussery, 1995; McMahon et al. 1995; Spidle et al. 1995). One purpose of this study was to determine the time at which adult quagga mussels acutely immersed at different lethal temperatures attained 100% mortality. There have been no published studies of the acute upper thermal limits of adult quagga mussel populations from warm southwestern water bodies. Previous studies have indicated differences in the upper thermal limits of zebra and quagga mussels from northeastern US waters (Pathy and Mackie 1993; Ricciardi et al. 1995; Mills et al. 1996; Baldwin et al. 2002; Peyer et al.2009). These studies have generally indicated that the upper thermal limit of quagga mussels is lower than that of zebra mussels (Mills et al. 1996). Zebra mussels survive indefinitely at 30°C, but quagga mussels show rapid mortality at 30°C (Spidle et al. 1995; McMahon 1996).

Currently, the quagga mussel is the species invading water bodies in the southwestern United States. There are over 30 lakes and reservoirs throughout Arizona, California, Colorado, Utah, and Nevada (Nalepa 2010), as well as the entire reach of the Colorado River from Lake Powell to the US–Mexico border and the entire Colorado River Aqueduct in Southern California and Central Arizona (Benson 2012), which have been documented for the presence of adult quagga mussels since their introduction into North American waters. In 2011, the Southern Nevada Water Authority estimated that $172,600 has been spent annually for chlorination additions to treat quagga mussels as well as $340,000 to remove quagga mussels from one potable water intake pipe (Nevada Department of Wildlife 2012). Lake Mead supplies approximately 80% of the potable water used in the Las Vegas Valley and adjacent areas (Mueting et al. 2010). The raw-water transfer infrastructure from Lake Mead could be impacted by quagga mussel macrofouling. Lake Mead also provides high-quality recreational activities for boaters, fisherman, and tourists and experiences roughly more than 8 million visitors per year (Wong et al. 2011) whose recreational boats could become vectors for the further introduction of quagga mussels to western water bodies.

Temperatures in the northeastern United States are generally cooler during summer and colder during winter compared to those of southwestern states. Thus, quagga mussel populations in southwestern water bodies could be acclimating and/or adapting to warmer environmental conditions, which could potentially lead to the development of increased upper thermal limits. The development of increased upper thermal limits among quagga mussels in southwestern water bodies could negatively affect the efficacy of thermal decontamination treatments that are based on upper thermal limit data developed from populations in cooler northeastern U.S. water bodies. Indeed, Morse (2009b) reported that zebra mussels isolated in a warm reservoir lake in southern Kansas (maximum summer surface water temperature = 30°C) had a mean incipient upper thermal limit 1.7°C higher than specimens from a cooler lake in New York (maximum surface water temperature = 25°C) when both groups of mussels were acclimated to 25°C. This result suggested that zebra mussels isolated in warm southwestern U.S. water bodies were evolving an increased thermal tolerance, which could impact the temperature/contact time required for decontamination by hot water spray. It is possible that quagga mussels isolated in the warm waters of Lake Mead and other water bodies on the lower Colorado Rivers could also have similarly evolved and increased thermal tolerance, which would impact the temperature/timing procedures required for efficacious decontamination of infested boats and other equipment by hot water sprays. Thus, the purpose of this study was to compare data on the thermal tolerance of quagga mussels from Lake Mead to that reported in previous studies conducted on mussels from populations in the northeastern United States to determine if Lake Mead mussels have developed an increased thermal tolerance that could impact the efficacy of presently accepted hot water spray decontamination procedures.

Methods

Specimen Collection

In June 2011, specimens of quagga mussels (*D. rostriformis bugensis*) were collected along the boat docks at Hemenway Harbor, Lake Mead, Nevada (GPS: 36° 1.7832' N, 114° 46.24788' W). Adult quagga mussels were collected from the tubing of the boat airlifts and anchoring ropes suspended in the lake. Mussels were found no deeper than 3 m below the surface of the lake. Quagga mussels were collected by using a scraper to carefully remove them from the substratum and placed in a 5-gallon bucket filled with lake water. They were then immediately transported to the Nevada Department of Wildlife's Lake Mead Fish Hatchery where they were placed into multiple 10-gallon flow-through aquaria (continuously aerated and circulated with water from Lake Mead). Mussels were habituated to the flow-through tanks for no less than 10 days during which they were fed Instant Algae® *Isochrysis* (Reed Mariculture, Inc.). The average temperature of the water in the holding aquariums was 26.31°C ± 1.24°C during the entire experiment.

Study Design

Subsequent to habituation to holding conditions, individuals with a shell length greater than 14 mm (shell length range = 14.69–28.05 mm) (Table 6.1) were selected for testing of acute upper thermal limits. These mussels were randomly placed into identical 15.2 × 25.4 cm, 800 μm fine mesh media bags (Foster and Smith, Inc.) in groups of ten individuals per mesh bag. There were eight treatment temperatures (26°C, 28°C, 30°C, 32°C, 34°C, 36°C, 38°C, and 40°C) and 20 exposure times (5, 10, 20, 40, 60, 80, 100, 120, 150, 180, 240, 300, 360, 600, 900, 1,200, 1,800, 3,600, 7,200, and 14,400 s) to determine at which time mussels achieved 100% mortality at each temperature (Table 6.2).

Two identical VWR™ Heated Circulators (Model 1127-2P; VWR International, Inc.) were filled with Lake Mead water (used to replicate the natural environment) and heated to one of the desired treatment temperatures. Once the heated water circulators reached their desired temperatures, four 1000 mL Pyrex beakers (VWR International, Inc.) were submersed into the heated circulator and filled with the heated lake water from inside that circulator. The beakers were used to prevent samples from being sucked into the circulation pump. Prior to experimentation, each sample was randomly selected and labeled by its desired exposure time and temperature. Each sample was then immersed inside the beakers for

TABLE 6.1

Shell Length of Adult Quagga Mussels under Different Temperature Conditions

Temperature (°C)	No. of Mussels	Mean (mm)	Min. (mm)	Max. (mm)	Standard Deviation
26	200	18.62	15.01	25.70	2.07
28	200	17.85	15.01	23.75	1.80
30	200	18.56	15.09	24.12	2.00
32	200	19.06	14.69	26.63	2.03
34	200	19.67	14.98	28.05	2.37
36	200	20.37	16.55	26.03	1.93
38	180	20.14	16.06	25.90	1.76
40	170	20.03	16.50	27.59	1.89

TABLE 6.2

Number of Mussels Taken Out from Each Treatment during Thermal Treatment

Time (s)	26°C	28°C	30°C	32°C	34°C	36°C	38°C	40°C
5	Take 10 mussels from the 200 muscles (20 bags)	"	"	"	"	"	"	"
10	Take 10 mussels from the 200 muscles (20 bags)	"	"	"	"	"	"	"
20	Take 10 mussels from the 200 muscles (20 bags)	"	"	"	"	"	"	"
40	Take 10 mussels from the 200 muscles (20 bags)	"	"	"	"	"	"	"
60	Take 10 mussels from the 200 muscles (20 bags)	"	"	"	"	"	"	"
80	Take 10 mussels from the 200 muscles (20 bags)	"	"	"	"	"	"	"
100	Take 10 mussels from the 200 muscles (20 bags)	"	"	"	"	"	"	"
120	Take 10 mussels from the 200 muscles (20 bags)	"	"	"	"	"	"	"
150	Take 10 mussels from the 200 muscles (20 bags)	"	"	"	"	"	"	"
180	Take 10 mussels from the 200 muscles (20 bags)	"	"	"	"	"	"	"
240	Take 10 mussels from the 200 muscles (20 bags)	"	"	"	"	"	"	"
300	Take 10 mussels from the 200 muscles (20 bags)	"	"	"	"	"	"	"
360	Take 10 mussels from the 200 muscles (20 bags)	"	"	"	"	"	"	"
600	Take 10 mussels from the 200 muscles (20 bags)	"	"	"	"	"	"	"
900	Take 10 mussels from the 200 muscles (20 bags)	"	"	"	"	"	"	"
1,200	Take 10 mussels from the 200 muscles (20 bags)	"	"	"	"	"	"	"
1,800	Take 10 mussels from the 200 muscles (20 bags)	"	"	"	"	"	"	"
3,600	Take 10 mussels from the 200 muscles (20 bags)	"	"	"	"	"	"	"
7,200	Take 10 mussels from the 200 muscles (20 bags)	"	"	"	"	"	"	"
14,400	Take 10 mussels from the 200 muscles (20 bags)	"	"	"	"	"	"	"

its designated exposure time determined by stopwatch (SportsTrackLive). At each designated exposure time, the mesh bag containing mussels was removed from the heated water baths and resubmerged into the continuously aerated Lake Mead water flow-through tank at 26.31°C ± 1.24°C and left to recover for no less than 2 days. At the end of the recovery period, mussels were tested for viability using a procedure similar to that of McMahon et al. (1995). If a mussel did not extend its siphons or respond to prodding with blunt end forceps by valve closure, it was considered to be dead.

Statistical Analysis

Analysis of covariance (ANCOVA) was conducted to determine if there was any significant difference in survival rates among the test temperatures while exposure duration (time) was used as a confounding factor (Choi et al. 2013). The significance criterion was set to $\alpha = 0.05$. All statistical analyses were performed using SAS 9.2 (SAS Institute, Inc., Cary, NC).

Results

Adult Quagga Mussels and Thermal Tolerance

A total of 1550 adult quagga mussels were tested (200 adult quagga mussels per temperature, eight temperatures total). Adult quagga mussels showed a clear response to increasing acute exposure temperature. No mortality occurred at either 26°C or 28°C within the 4 h exposure period (Figure 6.1). Adult quagga mussels reached approximately 30% mortality at 30°C, and mortality doubled at 32°C after the 4 h exposure period (Figure 6.1). At 34°C, 100% mortality was obtained after a 4 h exposure (Figure 6.1). The time to achieve 100% mortality declined with increasing exposure temperature. At 36°C, 38°C, and 40°C, it took 30, 10, and 2.5 min, respectively (Figure 6.1). ANCOVA showed that both temperature and exposure time had significant effects on mussel mortality (DF = 15, F = 19.04, P < 0.01). The results indicated that increased temperatures and longer exposure durations resulted in increased mortality.

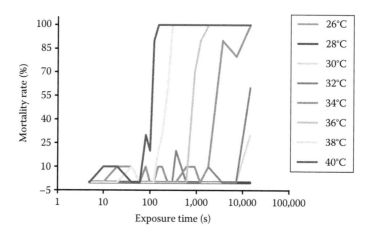

FIGURE 6.1   Mortality rate of quagga mussels exposed to different water temperatures.

Discussion

The results of this study elucidate the combinations of time and acute temperature exposure required to induce 100% mortality in adult quagga mussels collected in the summer of 2011 at Lake Mead. Information on acute upper thermal limits for quagga mussels in North America has been generally surmised from studies conducted on its close relative, the zebra mussel (i.e., Iwanzyki and McCauley 1993; McMahon and Ussery 1995; McMahon et al. 1995; Elderkin and Klerks 2005) with only a few studies specifically on quagga mussels (i.e., Spidle et al. 1995; Beyer et al. 2011). Since the arrival of dreissenid mussels in the Great Lakes, they have found their way as far west as Southern California. In California, the quagga mussel has been the main dreissenid invader of water bodies with zebra mussels established only in San Justo Reservoir and Ridgemark Golf Course (Ram et al. 2012; Pacific States Marine Fisheries Commission 2013). There have been no studies testing the thermal tolerance of quagga mussels west of the 100th meridian.

Climates and biomes differ across the United States, which may in turn provide positive or negative conditions for the survival of dreissenid mussels. Since the quagga mussel was first discovered in Lake Mead in January 2007, there have been enough generations to allow progeny to adapt to the warm thermal regime of Lake Mead, which may have resulted in Lake Mead individuals having a higher thermal tolerance than their counterparts in the northeastern United States as has proven to have been the case for zebra mussels isolated in warm southwestern water bodies (Morse 2009b).

However, unlike the increased thermal tolerance reported for zebra mussels from a warm southern Kansas water body (Morse 2009b), the results of this study did not suggest that Lake Mead (Nevada) adult quagga mussels had a higher thermal tolerance than adult quagga mussels previously tested in the northeastern United States (Spidle et al. 1995; Beyer et al. 2011). Although using slightly different test parameters (i.e., temperature intervals and exposure duration), the results of this study were similar to those of Beyer et al. (2011), who reported that adult quagga mussels from Green Bay (Wisconsin) immersed in 32°C for 20 min resulted in roughly no mortality. Adult quagga mussels found in Lake Mead (Nevada) achieved no mortality after 20 min but reached 60% mortality after 4 h of exposure to 32°C. As the combination of time and temperature were increased, mortality was achieved at a faster rate. After 4 h of exposure to 30°C, quagga mussels achieved 30% mortality, which agreed with the data suggesting the quagga mussel has a lower acute thermal tolerance than the zebra mussel (Spidle et al. 1995; Mills et al. 1996) and cannot survive prolonged periods of time at 30°C (Mills et al. 1996). In this study, adult quagga mussels experienced mortality at or above 30°C. Spidle et al. (1995) found that, at 30°C, one quagga mussel survived longer than 7 days and eventually died on day 14 of the exposure period. Two of the temperatures (26°C and 28°C) tested in this study were below the acute upper lethal temperature of quagga mussels, and as expected, no mortality was observed.

Previous studies have determined the acute upper lethal temperature limits of dreissenid mussels. In these studies, mussels were allowed to acclimate to different temperatures to determine if acclimation temperature induced differences in their acute thermal tolerance. Acclimation to elevated temperatures has been shown to increase the acute upper thermal limits of dreissenid mussels and increase the likelihood of survival at elevated temperatures (McMahon et al. 1993; Spidle et al. 1995). Dreissenid mussels acclimated to lower temperatures (5°C) displayed a lower acute upper thermal limit compared to those acclimated to higher temperatures (15°C–20°C) (McMahon and Ussery 1995; McMahon et al. 1995; Spidle et al. 1995). This study did not measure the acute upper thermal limits of mussels acclimated to different temperatures and did not gradually raise the temperature during immersion. Mussels were acutely immersed in heated water baths at continuous test temperatures near and above the previously determined acute upper lethal temperature for this species (Spidle et al. 1995) to determine the efficiencies of combinations of temperature and exposure times in inducing mussel mortality.

Considering genetic adaptation, higher water temperatures can potentially act as a selective agent (Elderkin and Klerks 2005). Such thermal selection has been demonstrated to have resulted in an increase in the incipient upper thermal limit of specimens of zebra mussels inhabiting a warm southwestern water body (Morse 2009b). Adult individuals able to survive outside of a species' normal environmental temperature limits pass their genes to their progeny, which could result in a change in a gene pool of a dreissenid population inhabiting a water body with elevated temperatures leading to evolution of increased thermal tolerance. The close similarity of the acute upper thermal limits of Lake Mead quagga mussels compared to those from cooler northeastern U.S. water bodies (Spidle et al. 1995; Beyer et al. 2011) suggests that such thermal adaptation has not occurred for quagga mussels in Lake Mead even though it appears to have developed in a population of zebra mussels in a warm water body in the southwestern United States (Morse 2009b).

The main purpose of this study was to develop data relevant to the time–temperature relationship of acute temperature tolerance of North American adult quagga mussels from Lake Mead in the southwestern United States. Thermal tolerance studies have provided baseline data for the development of different thermal treatment methods for control/prevention of dreissenid mussel macrofouling (i.e., decontaminating boats and submerged equipment) prior to overland dispersal. Currently, there is limited data on high-pressurized hot water spray as a means for boat decontamination (Morse 2009a; Comeau et al. 2011). This eco-friendly method of decontamination is cost effective and is based on the acute upper lethal temperatures of dreissenid mussels. This method has been proven to be effective for decontamination of zebra mussels if a contaminated watercraft is exposed to temperatures greater than or equal to 60°C for durations of 10 s or greater (Morse 2009a). The higher the temperature, the less time is required to achieve 100% mortality (Morse 2009a; Comeau et al. 2011; Zook and Phillips 2012). Since our results, like those of others (Spidle et al. 1995; Beyer et al. 2011; Comeau et al. 2011), suggest that quagga mussels have a lower thermal tolerance than zebra mussels, Morse's (2009a) recommendation of a hot water spray of greater than or equal to 60°C for greater than or equal to 10 s would also appear to be sufficient to decontaminate quagga mussels on trailered vessels leaving Lake Mead and other infested water bodies.

It is important to perform studies on local populations of dreissenid mussels to optimize and provide better protocols for treating and managing mussel infestations and preventing their further spread. Pressurized water spray may visibly damage the exterior of an infested watercraft or ruin sensitive equipment (Zook and Phillips 2012). Also pressurized water spray can be applied to only easy-to-reach locations on a boat and on visible mussels. There are locations that are inaccessible to hot water sprays and/or cannot withstand temperatures greater than 54°C (Comeau et al. 2011). Hot water may be poured over hard-to-reach places and used to flush out live wells, intake systems, etc. Life jackets, fishing gear, etc., can be left in hot water to ensure decontamination. One concern with the use of scalding hot water is the possibility of ruining equipment on the watercraft, voiding manufacturer's warranties (Beyer et al. 2011).

The use of hot water immersion for decontamination does not apply solely to adult mussels but can also be used to induce 100% mortality of dreissenid veligers, which are not visible to the human eye. Nevada state law has made it mandatory to decontaminate watercraft before launching into another body of water, and it is a violation to transport aquatic invasive species. Nevada decontamination protocols are as follows: clean, drain, and dry for the day user. It has been estimated that 8 L of residual water may be retained in a watercraft even after it has been efficiently drained (Choi et al. 2013; personal communication, L. Dalton, Utah Division of Wildlife Resources). The residual water left in a watercraft has been shown to provide dreissenid veligers with reasonable conditions for prolonged survival. Quagga mussel veligers from Lake Mead (Nevada) experienced 100% mortality after 5 days in summer conditions at temperatures greater than 30°C (Choi et al. 2013). Similar to adult quagga mussels, quagga mussel veligers cannot survive prolonged exposure to greater than or equal to 30°C, instantaneous death occurring at greater than or equal to 35°C (Craft and Myrick 2011).

If the possibility arises of installing a boat lift and cleaning station, thermal tolerance data for mussels west of the 100th meridian should be taken into account when developing an efficacious decontamination procedure. The results of this study can be considered along with previous data in the design and development of a possible boat decontamination site within the Lake Mead National Recreation area. This data can also be considered by companies in the southwestern United States that are dependent on waterways for their business. The data would be useful in considering their options on maintaining current facilities with respect to the environment and cost-effectiveness. In order to make decontamination times practical, a time–temperature relationship needs to be considered regarding the duration of exposure to different temperatures in order to achieve 100% mussel mortality.

Conclusion

From this study, we can conclude that as the exposure temperature increases, the exposure duration required to achieve 100% mortality of adult quagga mussels decreases. This is supported by previous studies that have generated similar results for both quagga and zebra mussels (McMahon and Ussery 1995; McMahon et al 1995; Spidle et al. 1995). Unfortunately, this study tested only up to 4 h of exposure. Mussels were allowed to thermally habituate in flow-through tanks for up to 10 days prior to experimentation. In future research, use of more prolonged acclimation periods (i.e., 3–4 weeks) to different temperatures reflecting the range experienced in their natural environment and use of test temperatures greater than the maximum of 40°C used in this study (i.e., up to 60°C as used for zebra mussels by Morse 2002a) may provide useful information for the development of efficacious thermally based decontamination procedures for quagga mussels. In this study, single mussels were tested as individuals rather than as druses (i.e., byssally bound clusters of individuals) in which they naturally occur. Druses may provide some form of protection under stressful conditions such as thermal tolerance (Rajagopal et al. 2002).

Acknowledgments

The lead author would like to thank Ashlie Watters for her assistance in the laboratory and field and introduction to fellow colleagues and government employees working with quagga mussels. Appreciation is expressed to a number of individuals who reviewed this manuscript and provided constructive suggestions for its improvement.

REFERENCES

Baldwin, B, Mayer, M S, Dayton, J, Pau, N, Mendilla, J, Sullivan, M, Moore, A, Ma, A, Mills, M L. 2002. Comparative growth and feeding in zebra and quagga mussels (*Dreissena polymorpha* and *Dreissena bugensis*): Implications for North American lakes. *Canadian Journal of Fisheries and Aquatic Science* 59: 680–694.

Benson, A J. 2012. United States Geological Survey, USGS Nonindigenous Aquatic Species "Quagga Mussel Distribution." Last modified 2012. Accessed July 3, 2013. http://nas.er.usgs.gov/taxgroup/mollusks/zebramussel/quaggamusseldistribution.aspx.

Beyer, J, Moy, P, De Stasio, B. 2011. Acute upper thermal limits of three aquatic invasive invertebrates: Hot water treatment to prevent upstream transport of invasive species. *Environmental Management* 47: 67–76.

Boelman, S F, Neilson, F M, Dardeau, E A Jr., Cross T. 1997. *Zebra Mussel (*Dreissena polymorpha*) Control Handbook for Facility Operators*, 1st edn. Technical Report EL-97-1, U.S. Army Engineer Waterways Experiment Station, Vicksburg, MS.

Bossenbroek, J M, Finnoff, D C, Shogren, J F, Warziniack. 2009. Advances in ecological and economic analyses of invasive species: Dreissenid mussels as a case study. In: *Bioeconomics of Invasive Species: Integrating Ecology, Economics, Policy, and Management* (Keller, R P, Lodge, D M, Lewis, M A, Shogren, J F, eds.). Oxford University Press, New York, pp. 244–265.

Carlton, J T. 2008. The zebra mussel *Dreissena polymorpha* found in North America in 1986 and 1987. *Journal of Great Lakes Research* 34: 770–773.

Choi, W J, Gerstenberger, S, McMahon, R F, Wong, W H. 2013. Estimating survival rates of quagga mussel (*Dreissena rostriformis bugensis*) veliger larvae under summer and autumn temperature regimes in residual water of trailered watercraft at Lake Mead, USA. *Management of Biological Invasions* 4: 61–69.

Claudi, R, Mackie, J L. 1994. *Practical Manual for Zebra Mussel Monitoring and Control*. CRC Press, Boca Raton, FL.

Claxton, W T, Martel, A, Dermott, R M, Boulding, E G. 1997. Discrimination of field-collected juveniles of two introduced dreissenids (*Dreissena polymorpha* and *Dreissena bugensis*) using mitochondrial DNA and shell morphology. *Canadian Journal of Fisheries and Aquatic Science* 54: 1280–1288.

Comeau, S, Rainville, S, Baldwin, W, Austin, E, Gerstenberger, S, Cross, C, Wong, W H. 2011. Susceptibility of quagga mussels (*Dreissena rostriformis bugensis*) to hot-water sprays as a means of watercraft decontamination. *Biofouling* 3: 267–274.

Craft, C D, Myrick, C A. 2011. Evaluation of quagga mussel veliger thermal tolerance. *Colorado Division of Wildlife* no. CSU1003: 21pp. Accessed July 11, 2013. http://www.caltrout.org/pdf/Quagga Report - Final.pdf.

Elderkin, C L, Klerks, P L. 2005. Variation in thermal tolerance among three Mississippi River populations of the zebra mussel, *Dreissena polymorpha*. *Journal of Shellfish Research* 24: 221–226.

Griffiths, D W, Schloesser, D W, Leach, J H, Kovalak, W P. 1991. Distribution and dispersal of the zebra mussel (*Dreissena polymorpha*) in the Great Lakes Region. *Canadian Journal of Fisheries and Aquatic Sciences* 48: 1381–1388.

Hebert, P D, Muncaster, N B W, Mackie, G L. 1989. Ecological and genetic studies on *Dreissena polymorpha* (Pallas): A new mollusc in the Great Lakes. *Canadian Journal of Fisheries and Aquatic Sciences* 46: 1381–1388.

Holeck, K, Mills, E L, MacIsaac, H J, Dochoda, M, Colautti, R I, Ricciardi, A. 2004. Bridging troubled waters: Understanding links between biological invasions, transoceanic shipping, and other entry vectors in the Laurentian Great Lakes. *BioScience* 54: 919–929.

Horvath, T G, Lamberti, G A, Lodge, D M, Perry W L. 1996. Zebra mussel dispersal in lake-stream systems: Source-sink dynamics? *Journal of the North American Benthological Society* 15: 564–575.

Iwanyzki, S, McCauley, R Q. 1993. Upper lethal temperatures of adult zebra mussels (*Dreissena polymorpha*). In: *Zebra Mussels: Biology, Impacts, and Control* (Nalepa T F, Schloesser D W, eds.). Lewis Publishers, Boca Raton, FL, pp. 667–673.

Jenner, H A, Whitehouse J W, Taylor, C J L, Khalanski, M. 1998. Cooling water management in European power stations: Biology and control of fouling. *Hydroécologie Appliquée 1–2, Electricité de France, Chatou, Paris* 10: 1–225.

Johnson, L E, Ricciardi, A, Carlton, J T. 2001. Overland dispersal of aquatic invasive species: A risk assessment of transient recreational boating. *Ecological Applications* 11: 1789–1799.

Kowarik, I. 2003. Human agency in biological invasions: Secondary releases foster naturalism and population expansion of alien plant species. *Biological Invasions* 5: 293–312.

Mackie, G L, Claudi, R. 2010. *Monitoring and Control of Macrofouling Molluscs in Freshwater Systems*. CRC Press, Boca Raton, FL. 508pp.

May, B, Marsden, J E. 1992. Genetic identification and implications of another invasive species of dreissenid mussel in the Great Lakes. *Canadian Journal of Fisheries and Aquatic Sciences* 49: 1501–1506.

McMahon, R F. 1996. The physiological ecology of the zebra mussel, *Dreissena polymorpha,* in North America and Europe. *American Zoologist* 36: 339–363.

McMahon, R F, Matthews, M A, Ussery, T A, Chase, R, Clarke M. 1995. Studies of heat tolerance of zebra mussels: Effects of temperature acclimation and chronic exposure to lethal temperatures. Technical Report EL-95-9. U.S. Army Corps of Engineers, Waterways Experiment Station, Vicksburg, MS, 25pp.

McMahon, R F, Ussery, T A. 1995. Thermal tolerance of zebra mussels (*Dreissena polymorpha*) relative to rate of temperature increase and acclimation temperature. Technical Report EL-95-10, U.S. Army Engineer Waterways Experiment Station, Vicksburg, MS, 27pp.

McMahon, R F, Ussery, T A, Clarke, M. 1993. Use of emersion as a zebra mussel control method. Technical Report EL-93-1, U.S. Army Engineer Waterways Experiment Station, Vicksburg, MS, 19pp.

Mills, E L, Rosenberg, G, Spidle, A P, Ludyanskiy, M, Pligin, Y, May, B. 1996. A review of the biology and ecology of the quagga mussel (*Dreissena bugensis*), a second species of freshwater dreissenid introduced to North America. *American Zoologist* 36: 271–286.

Morse, J T. 2009a. Assessing the effects of application time and temperature on the efficacy of hot-water sprays to mitigate the fouling by *Dreissena polymorpha* (zebra mussels Pallas). *Biofouling* 7: 605–610.

Morse, J T. 2009b. Thermal tolerance, physiological condition, and population genetics of dreissenid mussels (*Dreissena polymorpha* and *D. rostriformis bugensis*) relative to their invasion of waters in the western United States. PhD dissertation. The University of Texas at Arlington, Arlington, TX, 279pp.

Mueting, S A, Gerstenberger, S, Wong, W H. 2010. An evaluation of artificial substrates for monitoring the quagga mussel (*Dreissena bugensis*) in Lake Mead, Nevada-Arizona. *Lake and Reservoir Management* 26: 283–292.

Nalepa, T F. 2010. An overview of the spread, distribution, and ecological impacts of the quagga mussel, *Dreissena rostriformis bugensis*, with possible implications to the Colorado River System. In: *Proceedings of the Colorado River Basin Science and Resource Management Symposium* (Melis, T S, Hamill, J F, Bennett, G E, Coggins, L G Jr., Grams, P E, Kennedy, T A, Kubly, D M, Ralston, B E, eds.). U.S. Geological Survey Scientific Investigations Report 2010-5135, Reston, VA, pp. 113–121, 372.

Nevada Department of Wildlife. 2012. Aquatic Invasive Species (AIS) Fact Sheet AB167 2011. Nevada Department of Wildlife. Last modified 2012. Accessed June 20, 2013. http://www.ndow.org/uploadedFiles/ndoworg/Content/public_documents/Commission/AIS_FactSheet_StateLeg2011.pdf.

Pacific States Marine Fisheries Commission. 2013. *Aquatic Nuisance Species News*. Last modified 2013. Accessed June 24, 2013. http://www.aquaticnuisance.org/wordpress/wp-content/uploads/2009/01/CRB-AIS-NEWS-12-19-12.pdf.

Pathy, D A, Mackie, G L. 1993. Comparative shell morphology of *Dreissena polymorpha*, *e*, and the "quagga" mussel (Bivalvia: Dreissenidae) in North America. *Canadian Journal of Zoology* 73: 1012–1023.

Peyer, S M, McCarthy, A J, Lee, C E. 2009. Zebra mussels anchor byssal threads faster and tighter than quagga mussels in flow. *Journal of Experimental Biology* 212: 2027–2036.

Rajagopal, S, Van Der Velde, G, Jenner, H A. 2002. Does status of attachment influence survival time of zebra mussel, *Dreissena polymorpha*, exposed to chlorination? *Environmental Toxicology and Chemistry* 21: 342–346.

Ram, J L, Karim, A S, Banno, F, Kashian, D R. 2012. Invading the invaders: Reproductive and other mechanisms mediating the displacement of zebra mussels by quagga mussels. *Invertebrate Reproduction and Development* 56: 21–32.

Ricciardi, A, Neves, R J, Rasmussen, J B. 1998. Impending extinctions of North American freshwater mussels (Unionoida) following the zebra mussel (*Dreissena polymorpha*) invasion. *Journal of Animal Ecology* 67: 613–619.

Ricciardi, A, Serrouya, R, Whoriskey, F G. 1995. Aerial exposure tolerance of zebra and quagga mussels (Bivalvia: Dreissenidae): Implications for overland dispersal. *Canadian Journal of Fisheries and Aquatic Sciences* 52: 470–477.

Roper, J M, Cherry, D S, Simmers, J W, Tatem, H E. 1997. Bioaccumulation of PAHs in the zebra mussel at Times Beach, Buffalo, New York. *Environmental Monitoring and Assessment* 46: 267–277.

Spidle, A P, May, B, Mills, E L. 1995. Limits to tolerance of temperature and salinity in the quagga mussel (*Dreissena bugensis*) and the zebra mussel (*Dreissena polymorpha*). *Canadian Journal of Fisheries and Aquatic Sciences* 52: 2108–2119.

Ussery, T A, McMahon, R F. 1995. Comparative study of the desiccation resistance of zebra mussels (*Dreissena polymorpha*) and quagga mussels (*Dreissena bugensis*). Technical Report EL-95-6, U.S. Army Engineer Waterways Experiment Station, Vicksburg, MS, 19pp.

Vitousek, P M, D'Antonio, C M, Loope, L L, Westbrooks, R. 1996. Biological invasions as global environmental change. *American Scientist* 84: 468–478.

Watters, A, Gerstenberger, S L, Wong W H. 2013. Effectiveness of EarthTec® for killing invasive quagga mussels (*Dreissena rostriformis bugensis*) and preventing their colonization in the Western United States. *Biofouling* 29: 21–28.

Wong, W H, Gerstenberger, S, Miller, J M, Palmer, C, Moore, B. 2011. A standardized design for quagga mussel monitoring in Lake Mead, Nevada-Arizona. *Aquatic Invasions* 6: 205–215.

Zook, B, Phillips, S. 2012. Uniform Minimum Protocols and Standards for Watercraft Interception Programs for Dreissenid Mussels in the Western United States (UMPS II). Aquatic Nuisance Species Project, Pacific States Marine Fisheries Commission, Portland, OR, 77pp.

7

Potential Impacts of Invasive Quagga Mussels on Diet and Feeding Habitats of Young of the Year Striped Bass in Lake Mohave, Nevada, USA

Richard S. Ianniello, Tanviben Y. Patel, Shawn L. Gerstenberger, and Wai Hing Wong

CONTENTS

ABSTRACT *Morone saxatilis* (striped bass) is a sport fish introduced to Lake Mohave, which developed a self-sustaining population over the late 1900s. With the introduction of *Dreissena rostriformis* (quagga mussels) around 2006 in Lake Mohave, the young of the year (YOY) *Morone saxatilis* began to die off. From July 2011 to April 2012, both quagga mussels and YOY striped bass were collected from Lake Mohave, and stomach content analysis were conducted to compare diet. While results showed some overlap in the diets of both species, they did not necessarily indicate direct competition as the sole reason for the loss of YOY striped bass. It is more likely that the reasons for *Morone saxatilis* YOY mortality are part of more complex trophic interactions.

Introduction

Quagga Mussels

Quagga mussels appeared in Lake Mead, NV–AZ in 2007 near Boulder Basin. Described as one of the most invasive species worldwide, quagga mussels have colonized the majority of Lake Mead since that time. These mussels have caused not only ecological concerns, but economic and recreational damage as well. QM are considered biofoulers and are efficient ecosystem engineers that filter large quantities of water.

Quagga mussels are known to cause a wide variety of problems, including reduced food and oxygen for native fauna, clogging intake pipes, water filtration pumps, and hydroelectric plants. In the Great Lakes where quagga and zebra mussels first appeared, water clarity and composition have been dramatically impacted. Quagga mussels can settle onto both hard and soft sediments causing displacement and competition between native species (Diggins et al. 2004).

These invasive quagga mussels are filter-feeders, removing particles with their cilia from water bodies and consuming particles that pass through a siphon as food. The particles that are not consumed are combined with mucus and excreted as pseudofeces. The main food sources filtered and removed by quagga mussels consist of phytoplankton, zooplankton, and algae. In this way, quagga mussels can greatly affect aquatic ecosystems by removing nutrients and energy from the water column (or pelagic zone) and depositing them into the benthic zone (McMahon 1998). Additional information concerning the ecological impacts of invasive quagga/zebra mussels can be found from Wong et al. (2010).

Lake Mohave

Lake Mohave is a reservoir along the lower Colorado River, which was established with the construction of Davis Dam in 1951. The reservoir is 64 miles long and stretches north through the Black Canyon to the base of the Hoover Dam (Figure 2.1). Water released from Hoover Dam is the only significant source for Lake Mohave, and cold water from Lake Mead makes the upper 20-mile-long Black Canyon section of the reservoir cold year-round, while the lower more open section of the reservoir becomes seasonally warm due to extremely high summer temperatures. Fluctuations in the operations of Davis Dam and more significantly Hoover Dam cause fluctuations in water levels and exchange flows (NDOW 2007). One of the dominant predator species of fish found in the reservoir is the striped bass, but other species such as largemouth bass and channel catfish are also found in the reservoir.

Striped Bass

Striped bass (*Morone saxatilis*) were not intentionally introduced into Lake Mohave. They were first found in the reservoir in the early 1980s and are thought to have entered by means of introduction by an angler or by passing through Hoover Dam from Lake Mead. It is uncommon for striped bass to have self-sustaining populations within a reservoir. This is likely because unless there is sufficient current to keep the eggs suspended in water, they will settle to the bottom and suffocate (Stevens 1966). If there is not sufficient current, then the eggs will need a sandy or rocky surface surrounded by highly oxygenated water in order to survive. Self-sustaining and successfully spawning populations of striped bass are somewhat rare in freshwater ecosystems and are found in only a few places including the lower Colorado River, the Arkansas River, and Lake Marion in South Carolina. Since there are so few sustained populations of the introduced sport fish, it might be expected that striped bass would be particularly sensitive to changes in the chemical, physical, and trophic dynamics of a reservoir.

The striped bass population in Lake Mohave, which began to rise in the late 1980s, dropped severely in 2006 coinciding with the invasion of quagga mussels. From 1990 to 2005 spring, gillnetting surveys produced at least two fish per net night except for in the spring of 2000 and produced an average of around five fish per net night. Since 2006, the per-net-night yield of striped bass with this sampling has been less than 2 and, in 2012, was down to 0.42 fish per net night (Burrell 2012).

Adult striped bass are piscivorous or fish-eating, but the young of the year (YOY) striped bass are not large enough to eat other fish. Instead, they consume zooplankton, insect larvae, invertebrates, and eventually other larval fish. A study in Canton Reservoir in Oklahoma found that striped bass between 10 and 30 mm fed mostly on copepods, a subclass of small crustacean zooplankton, while fish longer than 30 mm fed more commonly on cladocerans, an order of crustacean zooplankton, larger in size than copepods. At around 50–60 mm, the fish became less reliant on crustaceans and fed increasingly on insects and insect larvae. At 100 mm in length, larval gizzard shad began to become an important part of the YOY striped bass diet (Gomez 1970).

Striped bass in Lake Mohave still spawn each year. For this project, YOY striped bass were easily found and caught by sane netting at two different sites in the months of July and September 2011. In November 2011, YOY striped bass were not found by sane netting, but were collected by means of electrofishing. However, after November 2011, no YOY striped bass were able to be found or captured in Lake Mohave. It seems that while new striped bass are spawned each year, very few of the YOY bass survive to eventually become adults. Based on the timing of the beginning of this phenomenon, and the invasion of quagga mussels, we propose to evaluate the possible association between mussels and YOY striped bass.

Methods

For five sampling dates in 2011 and 2012, quagga mussels from Lake Mohave were collected by the removal of substrate (i.e., rocks and sands) by divers at three locations from each of three sites that historically have had striped bass populations: Cottonwood Island Cove, Golden Door Cove, and Orion Cove. The sampling dates were July 19, 2011, September 2, 2011, November 22, 2011, February 3, 2012, and April 19, 2012. While the mussels were originally intended to be sampled with the use of a PONAR sampler, these samplers are effective only in soft sediment. While there are ample soft sediments in Lake Mohave, quagga mussels have a high preference for settling on hard substrate, though mussels have been collected in Lake Mead with the use of a PONAR grab (Wong et al. 2011). However, based on observations, there were very few living mussels in the soft substrates (i.e., muddy or silty areas); accordingly, hard substrates of rocks containing mussels were collected by divers.

Mussels were removed from the rocks by gently scraping them off at each site and then preserved in a 50% ethanol solution immediately to prevent as much loss of stomach contents through digestion and excretion as possible. After being preserved, the mussels were shipped to BSA Environmental Services in Beachwood Ohio for stomach content analysis. The analyses of zooplankton and phytoplankton are standard methods used in this company (http://www.bsaenv.com/aquatics.html).

The mussels that were not removed from the rocks were left on the rocks and frozen in ziplock bags to be later analyzed for mussel density and population structure. This was done by first removing the mussels from the rock using scrapers. The inhabitable surface area of each rock was measured by folding an equivalent area of aluminum foil over the portion of the rock that would not have been in direct contact with the lake bottom. The aluminum foil was then weighed, and a weight ratio was used to calculate the surface area (Strayer and Malcom 2006).

A random sample of 50 of the mussels was taken and individually measured for length, then weighed as a group. Afterward, all mussels removed from the rocks were then weighed. The ratio of the weight of the 50 mussels compared to all of the mussels provided an estimate of the total number of mussels in the sample, and the 50 lengths provided a size distribution for the population.

YOY striped bass were collected by use of a sane net. YOY striped bass were most commonly found at rocky points near the edges of the coves. The sane net was let out to surround a school of striped bass and later pulled into the shore to recover the fish. This process was repeated until at least five fish were captured for each site during the first two sampling events (summer and fall 2011). Stomach content analysis was performed by BSA Environmental Services. During the third sampling (winter 2011), no fish were able to be captured with the use of a sane net. We returned a few days later to attempt electrofishing to capture YOY striped bass. Using electrofishing, we were able to capture striped bass at Orion Cove, but not at Cottonwood Island Cove or Golden Door Cove. However, fish were also captured at Airport Cove, which is near Cottonwood Island Cove in the lake. During the fourth and fifth sampling events, no YOY striped bass were seen or captured.

Benthic samples in the form of soft sediment were also collected at each site with the use of a small PONAR grab. The benthic samples have been preserved and stored for the analysis of benthic organisms.

Results

Stomach content analysis of mussels showed consumed zooplankton such as *Bosmina longirostris*, *Cyclopoid copepodid*, and various Ostracoda species. Mussels also consumed a great deal of their own veligers as well as some amount of *Diptera* heads. The quagga mussel diet was fairly well distributed between these filterable species, but the largest part of their diet consisted of Ostracoda species along with quagga veligers.

The stomach content analysis of the YOY striped bass also commonly showed consumption of zooplankton such as *Bosmina longirostris* and *Cyclopoid copepodid*, along with *Daphnia*. The striped bass also contained a good deal of *Diptera* heads as well as larvae, but by far, using species count as a measurement, *Diptera* pupae constituted the largest part of the diet by an order of magnitude.

TABLE 7.1

Mean of Zooplankton to Mussel Ratio during Season and Location in 2011

Season	Location	Total Zooplankton (Ind/Mussel)
Summer	Cottonwood	0.93
	Golden Door	0.16
	Orion	0.49
Fall	Cottonwood	0.26
	Golden Door	0.32
	Orion Door	0.08

TABLE 7.2

Mean of Phytoplankton to Mussel Ratio during Season and Location in 2011

Season	Location	Total Phytoplankton (Cell/Mussel)
Summer	Cottonwood	79,647
	Golden	26,696
	Orion	49,525
Fall	Cottonwood	46,857
	Golden	32,502
	Orion	49,525

Mussel density at Cottonwood Island Cove averaged around 5000 mussels per square meter, at Orion Cove around 6000 mussels per square meter, and at Golden Door Cove around 7500 mussels per square meter. Mussels averaged 17 mm in length at Cottonwood Island Cove, 18 mm at Golden Door Cove, and 20 mm in length at Orion Cove. Mussels ranged from 3 to 40 mm in lengths.

In Table 7.1, the mean ratio of individual zooplanktons to mussels was highest during the summer season at the Cottonwood Cove location (mean = 0.93) compared to Golden Door with a mean = 0.16 and Orion with a mean = 0.49. During the fall season, Golden Door had the highest mean = 0.32, Cottonwood was close with a mean = 0.26, and Orion Door has a low mean = 0.08.

Table 7.2 displays the phytoplankton cells/mussel mean ratio of 2011 summer and fall season. Cottonwood had the highest (mean = 79,647), while Orion Door had the highest (mean = 49,525) during the fall season. These data could be explained by the drop in water temperature and depths in the various areas.

Conclusion

Based on the data collected from stomach content analysis of YOY striped bass and quagga mussels in Lake Mohave, it appears that direct competition between quagga mussels and striped bass does exist. It is not surprising that dreissenid mussels can consume zooplankton such as rotifers (Wong et al. 2003) and even their own larval veligers (MacIsaac et al. 1991). However, this direct competition is unlikely to be the soul reason for the decline in striped bass population. While both quagga mussels and YOY striped bass do share some of the same food sources, the major food sources for the striped bass, largely *Diptera* pupae, but also *Diptera* larvae and *Daphnia,* were not identified in the stomachs of quagga mussels.

The YOY bass were found easily in both the months of July and September. In July, the size of the striped bass caught averaged around 55–60 mm. According to previous studies (Gomez 1970; Sutton and Ney 2001), this would be the time at which the bass are most inclined to consume zooplankton that would be in direct competition with quagga mussels. In September, the mussels averaged around 80 mm in length. While in other bodies of water YOY striped bass began to consume fish larvae at 60 mm in length, no larval fish were found in the stomachs of the YOY SB. In the month of November when the

YOY striped bass first became uncommon, many of the fish captured by electrofishing were well over 100 mm in length, a point at which larval fish become important in their diet; yet, there was still no evidence of fish in their stomach content. In the study by Gomez (1970), by the time the Oklahoma YOY striped bass averaged 91 mm in length, larval gizzard shad represented 49% of the diet. In a Virginia reservoir, fish including cyprinids and larval sunfish accounted for 85% of the stomach contents of striped bass in between 90 and 140 mm. In Lake Mohave, no YOY striped bass were found after the collection in November.

Lake Mead has not had a decline in striped bass population. The threadfin shad population of Lake Mead has however survived the introduction of striped bass to a greater extent, because threadfin shad can use turbid inflow areas such as where the Virgin and Colorado Rivers enter Lake Mead as refuges against predatory striped bass. Also, in 2007, the same year quagga mussels were discovered in Lakes Mead and Mohave, gizzard shad were found in Lake Mead and have proliferated rapidly (Nielson 2010). It could be that the presence of threadfin and gizzard shad in Lake Mead is allowing for a sustainable population of striped bass in Lake Mead, where the lack of a similar species in Lake Mohave is resulting in a population crash. Loomis et al. (2011) found little significant difference in larval threadfin shad population after the invasion of quagga mussels in Lake Mead, though this could have been complicated by the invasion of gizzard shad, which have larvae indistinguishable from larval threadfin shad.

While all of these factors are possible explanations for the decline in striped bass population, further analysis would need to help develop a definitive answer. A comparison in the stomach content of YOY striped bass in Lake Mohave and Lake Mead, particularly during the late fall months, may be the best next step in trying to explain the loss of YOY striped bass in Lake Mohave at the end of each year.

Acknowledgments

This project is partially funded by Nevada Department of Wildlife. We appreciate Dr. John Beaver's technical support for this project.

REFERENCES

Benson A (2009) Zebra mussel fact sheet. http://fl.biology.usgs.gov (Accessed on November 22, 2013).

Burell MD (2012) Gill netting Lake Mohave 2012 Draft. Nevada Department of Wildlife, Reno, NV.

Burrell MD (2012) The development of the Lake Mohave striped bass (*Morone saxatilis*) fishery and its impact on the Threadfin Shad (*Dorosoma petenense*) population and the stocked rainbow trout (*Oncorhynchus mykiss*) fishery. *Lake Mead Science Symposium*, Las Vegas, NV.

Diggins TP, Weimer M, Stewart KM, Baier RE, Meyer AE, Forsberg RF, Goehle MA (2004) Epiphytic refugium: Are two species of invading freshwater bivalves partitioning spatial resources? *Biological Invasions* 6: 83–88.

Gomez R (1970) Food habits of young-of-the-year striped bass, *Roccus Saxatilis* (Walbaum), in Canton Reservoir. *Proceeding of the Oklahoma Academy of Science* 50: 79–83.

Loomis EM, Sjöberg JC, Wong WH, Gerstenberger SL (2011) Abundance and stomach content analysis of threadfin shad in Lake Mead Nevada: Do invasive quagga mussels affect this prey species? *Aquatic Invasions* 6: 157–168.

MacIsaac HJ, Sprules WG, Leach JH (1991) Ingestion of small-bodied zooplankton by zebra mussels (*Dreissena polymorpha*): Can cannibalism on larvae influence population dynamics? *Canadian Journal of Fisheries and Aquatic Sciences* 48: 2051–2060.

McMahon RF (1996) The physiological ecology of the zebra mussel, in North America and Europe. *Integrative and Comparative Biology* 36(3): 339–363.

NDOW (Nevada Department of Wildlife) (2007) 11. Lake Mohave. http://www.ndow.org/fish/where/waters/south/2007/11_LAKE%20MOHAVE.pdf.

Nielson D (2010) Another new species settles in Lake Mead. *NDOW Press Release*. http://www.ndow.org/about/news/pr/2010/dec_2010/lake_mead_species.shtm.

Stevens RE (1966) Hormone-induced sawning of striped bass for reservoir stocking. *The Progressive Fish-Culturist* 28: 19–28.

Strayer DL, Malcom HM (2006) Long term demography of a zebra mussel (*Dreissena polymorpha*) population. *Freshwater Biology* 51: 117–130.

Sutton TM, Ney JJ (2001) Size-dependent mechanisms influencing first-year growth and winter survival of stocked striped bass in a Virginia mainstream reservoir. *Transactions of the American Fisheries Society* 130: 1–17.

Wong WH, Gerstenberger SL, Miller JM, Palmer CJ, Bryan M (2011) A standardized design for quagga mussel monitoring in Lake Mead, Nevada-Arizona. *Aquatic Invasions* 6: 205–215.

Wong WH, Levinton JS, Twinning BS, Fisher N (2003) Assimilation of micro-mesozooplankton by zebra mussels: A demonstration of the food web link between zooplankton and benthic suspension feeders. *Limnology and Oceanography* 48: 308–312.

Wong WH, Tietjen T, Gerstenberger SL, Holdren GC, Mueting S, Loomis E, Roefer P, Bryan M, Turner K, Hannoun I (2010) Potential ecological consequences of invasion of the quagga mussel (*Dreissena bugensis*) into Lake Mead, Nevada–Arizona. *Lake and Reservoir Management* 26: 306–315.

Section III

Detection

8

High Turbidity Levels as a First Response to Dreissena Mussels Infection in a Natural Lake Environment

Jim Steele and Wai Hing Wong

CONTENTS

ABSTRACT   Invasive mussels are new threats to the environment and economy in the western United States. To better manage these pests, more techniques need to be developed. The present study shows that high turbidity levels can be a used as a first response when a new water body is infected by dreissenid mussels. When quagga mussels (*Dreissena rostriformis bugensis*) collected from Lake Mead (Nevada) were exposed to waters from Clear Lake (California) with different levels of suspended sediment, they were vulnerable. The LC50 of total suspended solids at the fourth day was estimated to be 9.9 g/L. The gill structures of these mussels were damaged by high levels of sediment suspended in the water.

Introduction

Zebra mussels (*Dreissena polymorpha* Pallas, 1771) are indigenous in the Ponto-Caspian area, and quagga mussels (*Dreissena rostriformis bugensis* Andrusov, 1897) originate from the Dnieper River drainage of Ukraine (Mills et al., 1996). In North America, zebra mussel was first sighted in the Great Lakes in 1986 (Carlton, 2008), and the first occurrence of the quagga mussel was documented in 1989 in Lake Erie (Mills et al., 1993). The quagga found its way into lakes/rivers of the east and finally to Lake Mead in Arizona/Nevada of the western United States by 2007 (LaBounty and Roefer, 2007; Mueting et al., 2010; McMahon, 2011).

The introduction of both zebra and quagga mussels into the Great Lakes appears to be from freshwater ballast of ocean-going vessels from European freshwater ports (Mills et al., 1996). The two species are highly polymorphic and prolific with high potential for rapid adaptation that helps with success in different waters. Because the mussels are freshwater mussels with byssal threads that allow attachment to substrates, they are very serious biofouling pests. They have caused considerable damage by fouling pipes, filters, engine intakes, and altering ecological composition (Mills et al., 1996). They are arguably the most serious nonindigenous biofouling pests in North America (LaBounty and Roefer, 2007) and are among the world's most economically and ecologically damaging aquatic invasive species (Aldridge et al., 2006; Connelly et al., 2007).

Sport boating and commercial shipping for all purposes is the most likely threat of introduction to water bodies that are not connected by an aquatic link. Mussels spread by clinging to the outside of boats (or other water-to-water conveyance), and the mussels and larvae also survive in ballast water or wet areas. Mussels are bivalve (two shells) and can close up (retain moisture) when out of a water body for a couple of days depending on temperature and background moisture levels (http://www.100thmeridian.org/emersion.asp). Boats with trailers can be pulled from a water body and transported to a new area within the survival time and ecological parameters for adults clinging to hulls and objects, and adults or larvae in ballast water (Ricciardi et al., 1995).

Interagency consortiums such as the Western Regional Panel to the federal Aquatic Nuisance Species Taskforce, which includes 19 western states, federal agencies, tribes, and others, have developed a mussel action plan that includes inspection, decontamination stations, and compliance enforcement, rapid response to new infestation, and detection and monitoring. This plan is largely unfunded except through local efforts (QZAP, 2010).

Because of the threat, boat inspections in California are sometimes conducted and have created waiting lines at some destination venues that raise the cost of the boating trip by the cost of the inspection (DeLeon S., 2012, pers. comm.). Inspections last approximately 15 min (or more) and cover enough of the boat to establish if they are clean, drained, and dry. Inspectors looking for mussels because of their small size (can be microscopic) and multiple hidden locations on boats have less chance to discover the actual mussel than the clean, drained, and dry conditions so focus on this attribute (DeLeon S., 2012, pers. comm.).

An additional concern is that boats having surface mussels cleaned off at a point of departure may leave those in harder-to-find locations, yet will appear clean. Also, some boats appearing *clean, drained, and dry* have locations that are inaccessible to either inspection or decontamination. Flooded hull (keel) sailboats (e.g., McGregor) and some wake board boats (with bladders) are notoriously wet for long periods and difficult or impossible to inspect (DeLeon S., 2012, pers. comm.). These problems increase the uncertainty of contamination into new water bodies.

Some infected water bodies will be affected at a high level because of background conditions. For example, a natural water body such as Clear Lake, California, is a highly eutrophic water body, cannot be drained, has no structural facilities to manipulate or treat (e.g., concrete channels or inlets), and therefore cannot prevent passing the mussels to downstream users. An infestation there will have unlimited nutrient supply and could affect dozens of important aquatic species as the mussel population progresses. There are no known rapid response procedures to use if an outbreak is discovered. The changes in ecosystem will change fish populations and the tourist economics and raise odor levels from cyanobacteria blooms that affect pleasure boater attitudes, raise drinking water treatment costs, and lower land values. The economic losses to Lake County, which depend on lake visitors and second homes, will be significant compared to the county economic base (Giusti, 2009).

Clear Lake is located at the headwaters of Cache Creek and is tributary to the Sacramento River and thence the Sacramento Delta. These are major fishery resources and both agriculture and domestic water supplies. Once in the delta, mussels spreading to other water bodies throughout the Sacramento River system are almost a certainty (Cohen, 2007). Clear Lake formed by tectonic movement from underlying faults has been in continuous existence for several hundred thousand years and by some accounts may be the oldest lake in Northern America. It is a warm water shallow lake averaging 9 m and has a fine sediment substrate measured at over 500 m deep. This benthos although consolidated at deeper layers is mostly fine particulates and easily stirred up at the top layer down to several centimeters (Lake County, 2012).

Habitat Requirements

There have been a few studies on the habitat requirements and tolerances of quagga and zebra mussels in the western states. Based on a review of the American Great Lakes region and elsewhere, quagga and zebra mussels have similar environmental requirements with quagga handling a broader range.

TABLE 8.1

Water Characteristics of Clear Lake, California

Parameters	Surface Water			Bottom Water		
Total alkalinity (mg/L as CaCO$_3$)	112.3	±	50.0	112.1	±	49.9
Total calcium (mg/L)	19.6	±	8.7	19.4	±	8.6
Hardness (mg/L as CaCO$_3$)	105.1	±	46.8	104.6	±	46.4
Total magnesium (mg/L)	13.7	±	6.1	28.4	±	40.4
Total phosphorus (µg/L)	31.4.0	±	25.4	38.6	±	25.4
Total dissolved solids (mg/L)	132.0	±	58.9	132.7	±	59.1
Total suspended solids (mg/L)	2.0	±	1.6	3.9	±	1.8
Conductance (µS/cm)	240.3	±	106.9	240.1	±	106.8
Dissolved oxygen (mg/L)	7.8	±	3.5	7.0	±	3.3
Water temperature (°C)	13.5	±	6.2	12.7	±	5.9
Turbidity (NTU)	2.5	±	1.5	3.3	±	1.8
pH	6.9	±	3.1	6.8	±	3.0
Secchi depth (meter)	2.6	±	1.2			

These data (mean ± 1 standard deviation) were collected by US EPA on April 19 and May 24, 2011, from three Clear Lake monitoring stations (CLEAR LK 15-UP ARM CL-1, CLEAR LK LO ARM CL3, and CLEAR LK 23 OAKS ARM CL4). Surface water was collected at a depth of 0.5 m and bottom water was collected from 6 to 12 m.

Both require a condition where salinity is low (<5 ppt), calcium is high (>12 mg/L), pH is somewhat alkaline (7.4–9.5 with a low limitation of 6.5), turbidity is higher than 0.1 m, and water flow is slow (<2 m/s) (Cohen, 2007). Quagga can handle lower temperatures and deeper depths. Both require at least 25% oxygen saturation to grow and reproduce (Culver et al., 2009). Based on water properties of Clear Lake such as calcium, dissolved oxygen, turbidity, pH, and water temperature (Table 8.1), quagga mussels should be able to survive and establish a population in this ecosystem.

Dreissena mussels respire and feed by pumping water through tubes across ciliated gill structures that also clear the water column for food particles. This requirement supports a mussel attachment preference for hard surfaces rather than silted benthic areas. Mussels are often found in association with hard substrates like rocky habitat or other shelled organisms. Otherwise, hard structure in the water column is a preferred attachment point (Cross et al., 2011).

Studies on gill damage by suspended particles on the green-lipped mussel *Perna viridis* indicate that manipulating the normal environment using a similar method to upset an early established colony of recently introduced mussels may be effective. Suspended sediments can be considered a nontoxic stressor with both direct and indirect effects to *Dreissena* mussels. Some effects include mechanical abrasion of gills, reduced ability for attachment to structures for veligers, and gill fouling (Shin et al., 2002; Cheung and Shin, 2005). Alexander et al. (1994) observed significant but unexplained decreases in *Dreissena* respiration when animals were exposed to a moderate (31 mg/L) concentration of bentonite clay. Studies have indicated a reduction of feeding rates and scope for growth, increased oxygen consumption, and a smothering effect for small individuals and larvae (Richardson, 1985; Doeg and Milledge, 1991; Newcomb and MacDonald, 1991; Leverone, 1995; Schneider et al., 1998).

Rapid Response Strategy Question

Currently, there are no methods or plans available for Lake County to respond to an early discovered infection (DeLeon S., 2012, pers. comm.). Clear Lake has 22 monitoring stations that are checked monthly by the County of Lake for new colonization by Dreissenid mussels. If one of these stations detect a new colony, can raising the background water column level of fine particulate matter from the local bottom material deter and reduce this infection?

Pumps capable of pumping and discharging significant amounts of bottom sediment are available within the suction gold dredging industry. These pumps are portable, weighing approximately 22 kg and pump bottom gravels and other sediment from river and stream channel bottoms in search of gold. The plan would have boats equipped with these dredges and appropriate length hoses to pump sediment off the bottom and aimed like a fire hose toward structures and hard surfaces in the area of infection.

Study Objectives

Some of the following conditions are assumed prior to the deployment of the rapid response method:

1. Adequate monitoring stations will detect mussels within one breeding season.
2. The area of detection is reasonably contained so that suspended water column turbidity levels can be effective.
3. Equipment is available and ready to deploy.
4. The mussels are not yet acclimated to their new environment and are at their most vulnerability.
5. Mussels are vulnerable to suspended sediment in a concentration that can be repeated in the field.

Test Question for This Study

Are suspended Clear Lake benthic particles lethal to mussels in Clear Lake water and at what concentrations?

The first objective is to find if suspended Clear Lake fine sediment would cause an effect to quagga mussels from Lake Mead. Second, if an effect is detected, quantify the effect so that field trials can determine if this level can be reached. A third objective is to determine if Lake Mead quagga mussels can remain healthy if placed in Clear Lake water during the test.

Materials and Methods

Benthic sediments were collected from the north end of Clear Lake near the City of Lakeport in a plastic bucket. The sediments were filtered with a 500 μm mesh to remove large detritus. One aliquot of wet sediment sample was transferred to a preweighed aluminum disk, dried in an oven (90°C for 24 h), weighed, and then ashed in a muffle furnace (450°C for 6 h) before final weighing. Dry weight of particles and water content (%) of each sediment sample were obtained. Totally five aliquots of sediment (5–9 g) were used to measure the dry weight of particles used to make up the targeted turbidity (0.01–1.25 g/L total suspended particles for experiment one and up to 18.5 g/L particles for experiment two) in the experimental aquaria. The water content of Clear Lake sediment is $87.1 \pm 2.3\%$ (N = 5, mean $\pm$ 1 standard deviation), which means that the dry sediment is about 13%. The fraction of organic materials in the sediment is $14.5 \pm 0.5\%$ (N = 5, mean $\pm$ 1 standard deviation), which means that the inorganic content of the sediment is about 85.5%. Clear Lake water was collected from the same area into a 945 L tank. These materials were transported to the study facilities located in Nevada Department of Wildlife's Lake Mead Hatchery. Before being used for the experiment, the sediments were kept for a week for experiment one and 2 days for experiment two in a refrigerator (4°C) to minimize microbial activity. Water from Clear Lake was kept in a separate storage tank and aerated to maintain freshness. The water was kept for 2 days before being used for the experiment. The total suspended solid concentration of Clear Lake water was estimated as the following: three aliquots (100 mL/aliquot) of water were taken from the storage tank and filtered onto preweighed 47 mm Millipore filters (mesh size of 0.45 μm), rinsed with distilled water, dried in an oven (90°C) for 24 h, weighed, and calculated. The concentration of total suspended solids of Lake Mead water was measured, as was the Clear Lake water. The concentrations of total suspended solids of Clear Lake water and Lake Mead water were 20.1 ± 0.0 mg/L (N = 3, mean $\pm$ 1 standard deviation) and 1.0 ± 1.4 mg/L (N = 2, mean $\pm$ 1 standard deviation), respectively.

Ten 75 L aquaria were set up in a larger fish hatchery raceway that could flow water around the outside of the aquaria to maintain temperatures. Quagga mussels collected from Las Vegas Boat Harbor were acclimated in a flow-through water system provided with raw Lake Mead water for 7 days prior to use. The experimental water (60 L) was filled to each aquarium, and different amounts of sediment were added to achieve widely separated turbidity levels. Two experiments were conducted with the first experiment starting from October 19 to November 23, 2011, and the second experiment starting from December 6, 2011 to January 10, 2012. Each experiment lasted 35 days so that the effects of the lake water and sediments have time to materialize. In experiment one, quagga mussels were exposed to relatively lower particle concentrations to determine response to lower concentrations of suspended particles (likely found in a natural setting). After Clear Lake sediments were added to each of the four experimental aquaria, three aliquots of water (250 mL) were taken and filtered through a preweighed Millipore filter (mesh size 0.45 µm and a diameter of 47 mm). After drying in an oven for 24 h at 90°C, the average sediment concentration in each aquarium was calculated at 0.01, 0.05, 0.25, and 1.25 g/L sediment in experiment one. An additional aquarium with Clear Lake water was used as a control without any sediment addition. In experiment two, quagga mussels were exposed to relatively higher concentrations (simulated saturated levels). After sediment addition to each of the three experimental aquaria, three aliquots of water (20 mL) from each aquarium were taken to a preweighed aluminum disk and left in an oven for 24 h (90°C). The estimated average suspended solid particle concentrations in each experimental aquarium were 3.8, 8.5, and 18.5 g/L. Two additional aquaria were used for Clear Lake water and Lake Mead water only. To maintain suspended particles in a homogeneous suspension in each aquarium with 60 L of water, a water current was generated by installing two submersible Aqueon Circulation Pumps (Model 500 with 1892 lph; A product of central Aquatics™, Franklin, Wisconsin, 53132) on two opposite aquarium walls. The turbidity was measured at the beginning and end of experiment one. In experiment one, water temperature and oxygen were monitored weekly. In experiment two, water temperature, dissolved oxygen, pH, and turbidity were measured daily. The turbidity of water in each aquarium was measured with a Hydrolab equipped with DS5 multiprobe (Loveland, Colorado 80539). The turbidity sensor was calibrated using 2-point standard calibration (0 and 3000 NTU).

For each aquarium, 10 mussels were placed in one loose woven spat bag and three bags placed in each aquarium (A total of 30 mussels in each aquarium). During the experiment, 1 mL of Instant Algae® Isochrysis1800 (Reed Mariculture, Campbell, California) was fed daily to mussels in each aquarium. The mortality rate to date was recorded weekly for experiment one and daily for experiment two. Specimens that did not respond to immediate shell valve closure were then gently stimulated in the area of their inhalant and exhalant siphons using slight finger pressure. Those that did not respond to this latter stimulus by immediate valve closure had their shell valves forcibly closed by finger pressure. If their valves immediately reopened after pressure release, specimens were considered to be dead (Morse, 2009; Comeau et al., 2011). The dead mussels were then completely opened using finger pressure to ensure that they would continue to be counted as dead. The shell length of mussels (i.e., the greatest distance from the anterior tip of the umbos to the posterior shell valve margins measured to the nearest 0.01 mm with digital calipers) used in experiments one and two was recorded. The gill structure of mussels was observed with a stereo microscope (SteREO Discovery V8, Carl Zeiss Inc., Germany).

Data Treatment

Analysis of covariance (ANCOVA) was used to determine if there was any significant difference in different parameters (i.e., temperature, dissolved oxygen, pH, and turbidly) and mortality rates among different groups added different amounts of suspended particles. The experimental time duration was considered as a covariate for ANCOVA. Only the mean value of the mortality rate (N = 3) was used to calculate LC_{50} (the concentration of total suspended solids to reach 50% of mortality of mussels) for experiment two. Analysis of variance (ANOVA) was used to check if there is any significant difference in shell length for mussels in each experiment. The significance criterion was set at $\alpha = 0.05$, and the highly significant criterion was set at $\alpha = 0.01$. All statistical analyses were performed using SAS 9.2 (SAS Institute Inc., Cary, NC, USA).

Results

In experiment one, there is no significant difference in water temperature among the five groups (i.e., Clear Lake water, 0.01, 0.05, 0.25, and 1.25 g/L), while experiment duration has a significant impact (Figure 8.1a; ANCOVA, $F_{9,15} = 3.49$, $P = 0.016$). No difference in dissolved oxygen among the five groups is found (Figure 8.1b; ANCOVA, $F_{9,15} = 0.82$, $P = 0.610$). The turbidity increased exponentially with more suspended particles added to the water (Figure 8.2), but there was no significant change in turbidity between the beginning and the end of the experiment (ANCOVA, $F_{6,3} = 325.5$, $P < 0.01$). The shell length of mussels among treatments was not significantly different (ANOVA, $F_{4,20} = 0.52$, $P = 0.72$) with a mean value of 16.92 mm.

There was no significant difference in mortality rate among treatments at concentrations of 0.01, 0.05, 0.25, 1.25 g/L, and Clear Lake water, but experiment duration time significantly affected mortality rates during the 5-week experiment (ANCOVA, $F_{9,80} = 2.95$, $P = 0.004$). It was noted that mussels from the controls and test levels were found to thrive in Clear Lake water at the concentrations tested with an average of 92% survival rate (Figure 8.3).

For experiment two, the water temperature record shows that there is no significant difference among the five groups, but it was affected by experimental time (ANCOVA, $F_{9,97} = 16.9$, $P < 0.01$; Figure 8.4a).

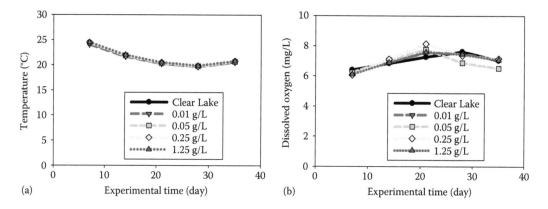

FIGURE 8.1   Water temperature (a) and dissolved oxygen (b) in experiment one.

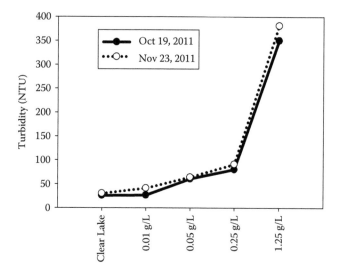

FIGURE 8.2   Water turbidity at the beginning and the end of experiment two.

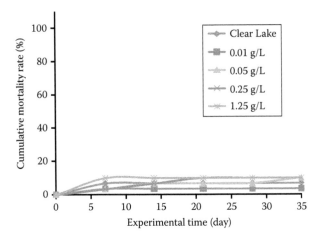

FIGURE 8.3 Mussel mortality of experiment one.

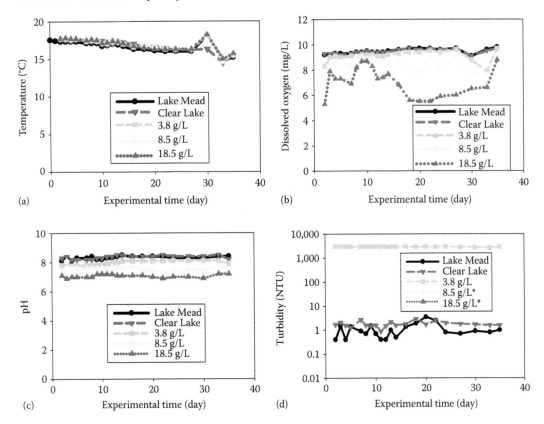

FIGURE 8.4 Water temperature (a), dissolved oxygen (b), pH (c), and turbidity of experiment two (d).

However, for dissolved oxygen, significant difference is found among groups without any impact from experimental time (ANCOVA, $F_{9,95} = 26.9$, P < 0.01). The higher the total suspended solids, the lower the dissolved oxygen (Figure 8.4b). Both suspended solids and experimental time impacted pH values (ANCOVA, $F_{9,95} = 205.3$, P < 0.01) and turbidity (ANCOVA, $F_{9,95} = 288,098$, P < 0.01) significantly. Lower pH values were recorded in the higher suspended particle groups (Figure 8.4c). Higher suspended solids resulted in higher turbidity (Figure 8.4d). The 3.8 g/L group had turbidity around 2954 NTU, while the turbidity of the two groups with 8.5 and 18.5 g/L of suspended solids is beyond the threshold (3000 NTU)

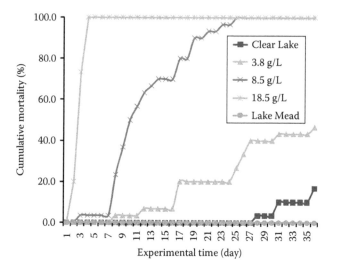

FIGURE 8.5   Mussel mortality of experiment two.

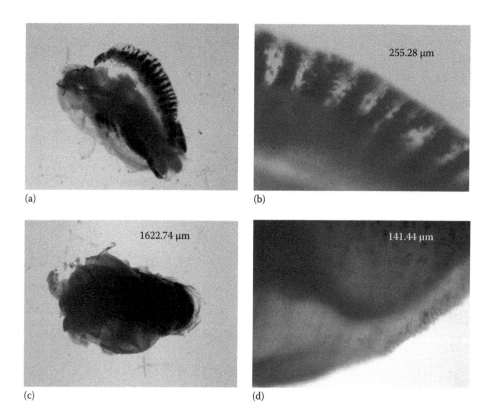

FIGURE 8.6   Gills of mussels exposed to higher suspended solids (a and b) and mussels in natural Clear Lake water (c and d) (photos a and c: 10×; photos b and d: 80×).

of the Hydrolab used in the present experiment. It also shows that the higher the concentration of total suspended solids, the lower the dissolved oxygen (Figure 8.4b) and pH values (Figure 8.4c).

The shell length of mussels among different treatments in experiment two did not differ significantly (ANOVA, $F_{4,72} = 0.78$, $P = 0.54$), and the mean shell length was 15.32 mm. Both concentrations of total suspended solids and experimental time significantly affect mussels' mortality (ANCOVA, $F_{170,9} = 268.2$, $P < 0.01$). Mussel mortality was delayed for each concentration consistent with the level of concentration (Figure 8.5). For the 3.8 g/L sediment concentration, the delay in mortality was 26 days with mortality increasing to 40%. For the 8.5 g/L concentration, it was 7 days with a steeper curve to 100% mortality in 25 days. The 18.5 g/L concentration found mussel mortality the first day with 100% mortality in 3 days. The LC_{50} of total suspended solids at the fourth day was estimated to be 9.9 g/L. There is no mortality recorded in the aquarium with Lake Mead water; for the aquarium with Clear Lake water, no mortality was recorded until the fifth week of the experiment (Figure 8.5).

Microscopic examination of gill structures showed sediment packing in the gill filaments and damage to the gill structure of dead mussels (Figure 8.6), while the healthy mussels had normal gill filament structure.

Discussion

This study is a pilot to determine the potential for looking at this line of thinking. The results of this study demonstrated that at saturation levels (highest possible) of turbidity, 100% mortality can be achieved in a relatively short period of time. The cause for this lethality was not the subject of this study, but damage to the gill structure indicated that mechanical abrasion caused by direct contact with suspended particles resulted in the loss of gill cilia. Cilia are responsible for water flow through the gills or demibranchs (McMahon, 1991). The sediment fouling included sediment-packed gill filaments, which may have been due to an overwhelmed mucous rejection mechanism found in mussels (Seed and Richardson, 1999). In addition, lower dissolved oxygen concentrations and lower pH values were recorded in high-concentration treatments (Figure 8.4b and c), which may also had negative impacts on the healthy status of quagga mussels. It has been addressed that low oxygen and/or low pH can be sublethal or lethal to mussels (McMahon, 1996; Claudi et al., 2012).

The values of oxygen and pH found in the present experiments may alone be sublethal. These lower values of oxygen and pH were probably caused by the higher dissolved suspended solids (Table 8.1) and the higher proportion of organic materials in the added suspended solids, which in turn resulted in higher microbial activities that consumed sufficient amounts of oxygen in the water column. However, dead mussels may also have contributed to the lower oxygen levels, and the study should have included aquaria without mussels to determine the effects of the sediment alone. Mortality should be mainly caused by an inability to gain oxygen or food from the water column, or a combination of both. This is clearly shown in Figure 8.6 that the gill structure of dead mussels has been severely damaged.

It was also determined in both tests that quagga mussels can survive in Clear Lake waters with the turbidity levels thought to be found naturally in Clear Lake. Although the first test estimated background levels based on levels normally found following storm runoff events, they indicate that if mussels were to be brought into the lake environment, they are likely to survive. All parameters (listed in Table 8.1) in Clear Lake also indirectly prove that it is a suitable aquatic ecosystem for quagga mussels.

Water columns supersaturated with sediment are unlikely in any normal situation on a natural shallow water lake. So it's not surprising that quagga mussels have not adapted to this unlikely occurrence. The introduction and mechanical suspension of sediment to the level of saturation in the second tests are thought possible in the natural soft bottom Clear Lake environment using readily available goldfield suction dredges developed for shallow streams and rivers.

A secondary note for these tests is that the high level of organics in the Clear Lake bottom material caused the oxygen levels to drop in the aquaria. In a field test, this effect may be replicated and can drop the background levels of water column oxygen making the method more effective. Further field tests are needed to determine if the laboratory-produced turbidity levels can be reached in the field using suction dredging equipment. This will require cooperation with control agencies such as the Regional Water Quality Control Board, CA Department of Fish and Wildlife, and the public.

REFERENCES

Aldridge DC, Elliott P, Moggridge GD. 2006. Microencapsulated BioBullets for the control of biofouling zebra mussels. *Environmental Science & Technology* 40:975–979.

Alexander JE. Jr., Thorp JH, Fell RD. 1994. Turbidity and temperature effects on oxygen consumption in the zebra mussel (*Dreissena polymorpha*). *Canadian Journal of Fisheries and Aquatic Sciences* 51:179–184.

Carlton JT. 2008. The zebra mussel *Dreissena polymorpha* found in North America in 1986 and 1987. *Journal of Great Lakes Research* 34:770–773.

Cheung SG, Shin PKS. 2005. Size effects of suspended particles on gill damage in green-lipped mussel *Perna viridis. Marine Pollution Bulletin 51*, pp. 801–810.

Claudi R, Graves A, Taraborelli AC, Prescott RJ, Mastitsky SE. 2012. Impact of pH on survival and settlement of dreissenid mussels. *Aquatic Invasions* 7:21–28.

Cohen AN. 2007. Potential distribution of zebra mussels (*Dreissena polymorpha*) and quagga mussels (*Dreissena bugensis*) in California. San Francisco Estuary Institute, Oakland, CA. A report to the CA Department of Fish and Game, Sacramento, CA.

Comeau S, Rainville S, Baldwin B, Austin E, Gerstenberger SL, Cross C, Wong WH. 2011. Susceptibility of quagga mussels (*Dreissena rostriformis bugensis* Andrusov) to hot-water sprays as a means of watercraft decontamination. *Biofouling* 27:267–274.

Connelly NA, O'Neill CR, Knuth BA, Brown TL. 2007. Economic impacts of zebra mussels on drinking water treatment and electric power generation facilities. *Environmental Management* 40:105–112.

Cross C, Wong WH, Che TD. 2011. Estimating carrying capacity of quagga mussels (*Dreissena rostriformis bugensis*) in a natural system: A case study of the Boulder Basin of Lake Mead, Nevada–Arizona. *Aquatic Invasions* 6:141–147.

Culver C et al. 2009. Early detection monitoring manual for quagga and zebra mussels. CA Sea Grant Extension Program, UC Co-Op extension, Davis, CA.

DeLeon S. 2012. Department Chief, County of Lake, Lakeport, CA.

Doeg TJ, Milledge GA. 1991. Effects of experimentally increased concentrations of suspended sediment on macroinvertebrate drift. *Australian Journal of Marine and Fresh Water Research* 42:519–526.

Giusti G. 2009. Identifying risk factors to strengthen current strategies aimed at minimizing the introduction of quagga and zebra mussels to Lake County California. A report prepared by the Lake County Fish and Wildlife Committee, Lakeport, CA.

LaBounty JF, Roefer P. 2007. Quagga mussels invade Lake Mead. *LakeLine* 27:17–22.

Lake County. 2012. Accessed May 22, 2011 at: http://www.co.lake.ca.us.

Leverone JR. 1995. Growth and survival of caged adult bay scallops (*Argopecten irradians concentricus*) in Tampa Bay with respect to levels of turbidity, suspended solids and chlorophyll a. *Florida Science* 58:216–227.

McMahon RF. 1991. Mollusca: Bivalva. In Thorp JH, Covich AP (eds.), *Ecology and Classification of North American Freshwater Invertebrates*. Academic Press, Inc., New York, pp. 315–399.

McMahon RF. 1996. The physiological ecology of the zebra mussel, *Dreissena polymorpha*, in North America and Europe. *American Zoologist* 36:339–363.

McMahon RF. 2011. Quagga mussel (*Dreissena rostriformis bugensis*) population structure during the early invasion of Lakes Mead and Mohave January–March 2007. *Aquatic Invasions* 6:131–140.

Mills EL, Dermott RM, Roseman EF, Dustin D, Mellina E, Conn DB, Spidle AP. 1993. Colonization, ecology, and population structure of the quagga mussel (Bivalvia, Dreissenidae) in the lower Great Lakes. *Canadian Journal of Fisheries and Aquatic Sciences* 50:2305–2314.

Mills EL, Rosenberg G, Spidle AP, Ludyanskiy M, Pligin Y, May B. 1996. A review of the biology and ecology of the quagga mussel (*Dreissena bugensis*), a second species of freshwater Dreissenid introduced to North America. *American Zoologist* 36:271–286.

Morse JT. 2009. Assessing the effects of application time and temperature on the efficacy of hot-water sprays to mitigate fouling by *Dreissena polymorpha* (zebra mussels Pallas). *Biofouling* 23:605–610.

Mueting SA, Gerstenberger SL, Wong WH. 2010. An evaluation of artificial substrates for monitoring the quagga mussel (*Dreissena bugensis*) in Lake Mead, NV–AZ. *Lake and Reservoir Management* 26:283–292.

Newcomb CP, MacDonald DD. 1991. Effects of suspended sediment on aquatic ecosystems. *North American Journal of Fisheries Management* 11:72–82.

QZAP. 2010. Quagga-zebra mussel action plan for western U.S. waters. Submitted to the Aquatic Nuisance Species Task Force by the Western Regional Panel on Aquatic Nuisance Species. Accessed May 22, 2011 at: http://anstaskforce.gov/QZAP/QZAP_FINAL_Feb2010.pdf.

Ricciardi AR, Serrouya R, Whoriskey, FG. 1995. Aerial exposure tolerance of zebra and quagga mussels (*Bivalvia: Dreissenidae*): Implications for overland dispersal. *Canadian Journal of Fisheries and Aquatic Sciences* 52:470–477.

Richardson BA. 1985. The impact of forest road construction on the benthic invertebrate fauna of a coastal stream in southern New South Wales. *Bulletin of Australian Society of Limnology 10*, pp. 65–88.

Schneider DW, Madon SP, Stoeckel JA, Sparks RE. 1998. Seston quality controls zebra mussel (*Dreissena polymorpha & hair* sp.) energetics in turbid rivers. *Oecologia* 117:331–341.

Seed R, Richardson CA. 1999. Evolutionary traits in *Perna viridis* (Linnaeus) and *Septifer virgatus* (Wiegmann) (Bivalvia: Mytilidae). *Journal of Experimental Marine Biology and Ecology* 239:273–287.

Shin PKS, Yau FN, Chow SH, Tai KK, Cheung SG. 2002. Responses of the green-lipped mussel *Perna viridis* (L.) to suspended solids. *Marine Pollution Bulletin* 45:157–162.

9

Utilizing Canines in the Detection of Quagga (Dreissena rostriformis bugensis) and Zebra Mussels (Dreissena polymorpha)

Debra L. DeShon, Ashley J. Jensen, and Wai Hing Wong

CONTENTS

ABSTRACT   As both quagga and zebra mussels have increased their range throughout North America, the development of new and innovative management practices in an attempt to contain these highly invasive species is necessitated. The primary method of preventing introductions of these dreissenid mussels has been the inspection of transient watercraft moving from one water body to another. However, visual inspection for these mussels alone can be very difficult, and the detection of these mussels' microscopic larval stage cannot effectively be done in the field by human inspectors. Trained scent-detection canines can provide inspection personnel with an additional method for detecting these mussels and have been shown previously to effectively detect dreissenid mussels attached to vessels. In addition to this, we found that properly training and utilizing scent-detection canines may be a viable way to detect the larval form of dreissenid mussels in standing water, further reducing the possibility of an introduction to new waters. The canines increase the effectiveness of inspections in general by being able to access scent in areas that cannot be visually inspected and increase effectiveness by their ability to detect dreissenid larva, which human inspectors cannot.

Conservation Canines

Canines and humans have been partners in many roles for thousands of years. With highly developed olfactory senses, dogs can provide invaluable assistance to humans in detecting scents that humans cannot. Canines are currently used to detect various substances or organisms in schools, at border crossings, airports, and workplaces (Department of Homeland Security 2010).

In 1986, the US Border Patrol launched a pilot program, wherein canines were trained to detect heroin, cocaine, methamphetamine, marijuana, and concealed humans. During the course of this program, discoveries by these canine units accounted for numerous apprehensions and seizures of narcotics and concealed persons (Department of Homeland Security 2010).

Conservation canines were first used in the late 1990s. These dogs are used to locate biological material for conservation purposes, such as game animal carcasses or scat samples (Reed et al. 2011, Richards et al. 2014). Conservation canines can even be used in the case of scat detection, to identify the species from which a particular scat was produced (Wasser et al. 2004). In comparison, scent-based searches by these specially trained detection dogs have been shown to be more effective than corresponding human visual searches (Smith et al. 2005). As seen, dogs were finding four times as many kit fox scats as an experienced person visually searching for the same samples (Ralls et al. 2010).

The California Department of Fish and Wildlife has successfully utilized canines to assist conservation officers for over a decade, stating that they have trained dogs to detect numerous biological components ranging from bear gall bladders to abalone, assisting in the efforts of the state's conservation officers (California Department of Fish and Wildlife 2014). In 2007, officers attempted to expand the abilities of their conservation canines and in turn determined that canines could be trained to detect invasive mussels of the family Dreissenidae (Fonseca 2009).

Invasive Mussel Prevention

Dreissenid mussels consist of many species, but most notable are two of particular concern, the quagga mussel (*Dreissena rostriformis bugensis*) and the zebra mussel (*Dreissena polymorpha*). Introduced to North America in the late 1980s, these dreissenid mussels have spread across a great portion of the continent and have caused numerous hardships both ecologically and financially (Lodge 1993, Connelly et al. 2007). The most common vector for dreissenid mussels to expand their range is to be spread from an infested body of water to an uninfested body of water via trailered watercraft (Padilla et al. 1996). These watercraft transfer either settled mussels attached to the vessel with proteinaceous strands known as byssal threads or the microscopic larvae of the mussel known as veligers. One of the most effective methods to prevent the transfer of these mussels is the institution of consistent protocols and standards for watercraft interception programs (Zook and Phillips 2009). Currently, most preventive measures are centered on visual inspections conducted by the vessel owner themselves or by trained inspectors employed by federal, state, or municipal agencies (Schneider et al. 1998).

Human inspections have many limitations. A human looking for the mussels must physically rub their hand over the surfaces of the boat while visually inspecting all surfaces as well; it is easy to miss mussels due to their small size and inconsistency in the inspectors' experience and available search image (WIT 2013). Additionally, all watercraft are not alike, and these varying watercraft come with complex compartments, angles, surfaces, and other components that make human inspections even more challenging. Human inspections also take a considerable amount of time generally lasting 15–20 min per vessel (WIT 2013). However, there is no effective way to detect the microscopic larvae of the species under the constraints of a field inspection setting. Detection canines are proficient at searching areas for scent that are visually impossible to access and may reduce the total time per inspection from 15–20 min with human inspectors alone to 1–3 min with the addition of trained canines (Dahlgren et al. 2012).

One of the most important factors in the effectiveness of a conservation canine is the training program used to develop the dog (Hurt and Smith 2009, Parker and Hurt 2010, Parker 2011). Generally, conservation canine training programs incorporate training methods from narcotic canine, search and rescue canine, and cadaver canine training programs (Reindl-Thompson et al. 2006). Recent training applications and practices have widened the scope and sophistication of conservation canine training programs, mainly through scent detection and discrimination work (Reed et al. 2011). Conservation canines are now being trained to recover carcasses, locate invasive and endangered species, detect animal scent trails, and identify occupied burrows (Reindl-Thompson et al. 2006, Reed et al. 2011). Conservation canines are becoming increasingly popular partly because it is a noninvasive way to monitor a diverse array of threatened and endangered species around the world (Bozarth et al. 2010).

Canine Detection of Mussels

Canine Mussel Detection Training

There are numerous factors to consider when initially training a canine for mussel detection. Hunting breeds are most commonly selected as conservation canines in general because of their innate motivation to search and find. Second, the dog's personal nature, ability, and willingness to learn must be evaluated, as canines that become effective detection canines are those who want to work, not those you make work (Wagner 1997). The final characteristic necessary is a strong *toy drive*, which is the desire of the canine to be rewarded with their personal toy (Dahlgren et al. 2012). We utilize the canine's innate *toy drive* by selecting canines that have an obsession with a particular toy. Ideal canines tend to demonstrate a frantic desire to possess the toy regardless of distractions (Smith et al. 2001). After an acceptable dog has been selected, they then begin the training process required to detect dreissenid mussels (Table 9.1).

The training process again focuses primarily on exploiting the dogs' toy drive. By utilizing this trait, we essentially create a game of *hide and seek* with the dog in training. In the training process employed for our study regarding the canines' ability to detect veligers, the toy was first hidden, and the canine had to successfully find his or her toy 100% of the time utilizing their sense of smell, as opposed to visually, and indicate the toy's presence by sitting at the source of the toy's odor. The next step in the training process was to scent the toy with the target odor. In our case, we used dreissenid mussel odor, placing the toy in with mussels for a short period of time so that the toy absorbs the odor. We then played with the dog with this toy so he or she now associated this smell with the toy. We then hid the toy and initiated the game of *hide and seek* that the dog was now already familiar with.

This game continued with the amount of odor decreasing; this was done by decreasing the amount of time the toy was placed with the mussels. Eventually, the toy is no longer hidden, and groups of mussels or individual mussels were hidden in its place; the canine was still expected to find and indicate odor by sitting at the source of the scent every time and was then rewarded with their toy. It is very important to fluctuate the number of mussels hidden so that the canine is always searching for varying concentrations of odor. We typically use between 1 and 20 mussels for training purposes. During the course of our

TABLE 9.1

Recommendations for Selection, Training, and Evaluation of Canines for Use to Detect Dreissenid Mussels

Canine Training
Initial training
Evaluate the canine's response to a particular toy
Select a canine with a strong desire to possess that toy
Condition the canine to associate the scent of dreissenid mussels with that toy
Train the canine to indicate by sitting when detecting the odor of dreissenid mussels
Train the canine to ignore odors from other sources
Field training
Evaluate the canines' performance in numerous situations and environments
Select a dog with high motivation and consistent performance in the field
Controlled training
Evaluate canine's ability to stay motivated with repetitive tasks and scenarios
Expose canine to setups that contain odor and non-odor designs
Select canine that performs well in both designs
Field and controlled training
Expose canines to varying concentrations of dreissenid mussels
Establish training and maintenance training schedules
Require experience and certification of trainers and handlers

Source: Modified by Smith, D.A. et al., *Anim. Conservat.*, 6, 339, 2003.

training, we had hidden as few as four mussels of varying sizes in the live well of a boat, and the canine successfully indicated on the outside of the boat where the live well drains. After the dogs had been able to successfully locate and identify the scent of mussels, we then sought to transition the dogs to identifying the scent of larval mussels.

Veliger Detection

In June 2013, in conjunction with the State University of New York Research Foundation and with the participation of canine teams from the California Department of Fish and Wildlife, we conducted a study to determine the possibility of training canines to detect microscopic quagga mussel veligers at Lake Mead. This was intended to be a two-part study with the initial aim of determining if canines previously trained to detect adult mussels could detect veligers without further training, and if they could, then we sought to determine the canines' accuracy in detecting lower densities of veligers. If they could not, then the dogs would be initiated to the scent of veligers and would slowly be trained to detect lesser densities of veligers.

We chose four canines that had previously been trained to detect adult mussels, with similar methods as described in Table 9.1. Canine 1 was trained and certified exclusively to detect quagga and zebra mussels, while canines 2, 3, and 4 were trained to detect quagga and zebra mussels, abalone, deer, bear, and lobster. The latter three were also trained in tracking and article search, in addition to fish/wildlife detection.

Samples of veligers were collected from Lake Mead, a lake known to contain a reproducing population of quagga mussels. The samples were filtered to sequentially decreasing concentrations down to a concentration of zero. This sample containing no veligers was used as a control specimen to account for various non-veliger organisms.

Stage 1 consisted of a blind trial exposure, wherein neither the handler nor the canine was previously aware of the contents of the samples. Canines were individually presented with five buckets, one of which contained the veliger (+) sample. They were then instructed by their handlers to inspect the samples.

After this portion of stage 1 was completed, handlers imprinted the veliger scent on their canines in an effort to improve the ability of the canines to detect the veliger sample. During this stage, the handler was aware of the contents of the bucket samples. When the canine reached the bucket containing veligers, the canine was instructed to sit and was immediately rewarded with its toy when they did so. Once the canine learned to associate the veliger scent with its toy, we conditioned the canine to indicate that it had found the veligers by sitting next to the bucket containing veligers (Smith et al. 2003).

Stage 2 began when it was determined that all canines had *successfully* imprinted on the veliger scent. This stage took place over the course of 3 days, in which the canines each completed a combination of training and blind runs for day 1 and day 2 and four blind runs on the third day. For each training run, the canines' handlers were aware of the location of the veliger positive bucket, so that they could reward their canines when they alerted on the correct bucket. Since this was a training run, if the canines did not alert, or alerted on an incorrect bucket, they were shown the correct bucket and signaled to alert by their handlers. The concentration of veligers in the sample was reduced for each training run so that by the sixth run, the sample contained a very low density of veligers. The seventh run used only the control sample with no veligers present.

By the sixth run on the first day of stage 2, all four canines correctly alerted on the veliger bucket and did not alert during the seventh run, when no veliger larvae were present. On the second day of training, it was clear that the canines needed to re-imprint on the veliger scent, as each of them inaccurately alerted or failed to alert in the first run. The canines were re-imprinted with the veliger scent, as explained earlier. After the canine was reacquainted with the scent, they correctly alerted for the run immediately following, but still failed to alert consistently and correctly at all times. At this time, the training was discontinued to reevaluate the process. It was determined that the canines were becoming fatigued with the process, as the canines' alerts were inconsistent, and all trials were halted for day 2. The third day of stage 2 consisted of four blind trials where neither the handler nor the canine was aware of the contents of each bucket, and a minimal concentration of veligers was used. All four dogs successfully alerted with no inaccurate alerts during all of the blind runs on day 3.

Veliger Detection Study Results

Stage 1 of the trials showed that initially, only Canine 1, which was the only canine that was both exclusively trained to detect quagga and zebra mussels and familiar with the bucket inspection protocol, was able to correctly identify the bucket containing the veliger sample. After increasing the concentration of veligers in the sample bucket, Canines 2, 3, and 4 were also able to correctly identify the bucket containing the veliger sample; however, they were not all consistent in their alerts, and the study moved on to remedial training and stage 2 of the sampling protocol.

Stage 2 showed that with additional training and time to imprint further on the smell associated with veligers, the canines could correctly alert to the veligers. Due to the inability of the dogs to alert correctly and consistently on day 2 after initial success on day 1 of this stage, this trial also showed that fatigue and boredom of the canines can play a factor in their ability to detect veligers. As seen, after rest between days 2 and 3, the canines were able to successfully alert to the veligers, without any false alerts.

Conclusions

The results of stage 1 of our study indicate that conservation canines with only previous training in the detection adult dreissenid mussels may or may not consistently and correctly alert to the presence of veligers. After exposure to a dreissenid mussel veliger detection training regimes, these dogs can be expected to detect varying concentrations of veligers within standing water, and that without participating in this targeted training program, canines cannot be expected to correctly alert to dreissenid mussel veligers.

Stage 2 showed that canines can be trained to detect veligers in standing water, but it may be necessary to frequently refresh this training in order to reduce the occurrence of false-positive and false-negative results during detection exercises. Additionally, stage 2 showed that the ability of canines to correctly identify samples containing veligers may be heavily affected by fatigue of the animals, which may further reduce the ability of the trained canines to correctly identify vessels contaminated with veligers.

Another limitation on utilizing canines for the detection of the dreissenid mussels is the effect of the summer heat on the animals. Environmental factors such as temperature, moisture, humidity, and air movements (wind speed) are all factors that affect the abilities of the detection canines. Higher temperatures increase the rate of panting, which causes a decrease in scent efficiency (Smith et al. 2003). It is essential to allow for adequate rest periods to help decrease the rate of panting. It is also important to protect their feet and keep them adequately hydrated during the hot summer months.

In addition to these findings, cursorily we found that the relationship between the handler and the trained canine is paramount to the success of the detection canine. The need for an experienced trainer and an extensive training maintenance program is essential. Another important variable is the rate and effectiveness of communication between the canine and the handler. The canine needs to be able to successfully communicate its alert, and the handler must be able to correctly recognize and reward the alert (Reed et al. 2011). The canine handler must also be able to properly motivate the dog throughout the search.

The final limitation of note in the use of dreissenid mussel detection canines is the cost. The cost of starting up and implementing a canine detection program from scratch is great. Costs include evaluating and training multiple canines for effectiveness, hiring and training handlers, and canine care (food, board, veterinary services). However, some of these costs can be supplemented via the use of canines already in service for conservation purposes.

As well as the ability to detect dreissenid mussels on contaminated craft, canine inspection teams generate conversation with boaters during watercraft inspections. Boaters are curious about the presence of the canines and often inquire about the dog's purpose and abilities. This is an excellent opportunity to educate the boaters on aquatic invasive species and their responsibility to *clean, drain, and dry* their watercraft. Education is an important component of prevention, which is widely recognized as the most cost-effective method of managing invasive species. Even after years of educating boaters on prevention, there are still boaters that have not been reached. Using canine inspection teams is a unique outreach effort and a way to reach new audiences (Culver et al. 2013).

The limitations of dreissenid mussel detection canines' use in state programs have been largely outweighed by the positives of these canines. The California Department of Fish and Wildlife currently has 18 detection teams throughout California. To date, their teams have had a total of 19 positive indications on watercraft where the presence of mussels was confirmed. We currently have one team and perform inspections and education at Lake Sonoma, Lake Mendocino, and Modesto Reservoir and have yet to document an incident where mussels were present on a boat being inspected, and have not had the opportunity to implement the findings of our study of larval detection in the field as of this time. The use of canines to detect adult dreissenid mussels has expanded to other states, with the Minnesota Department of Natural Resources recently adding three zebra mussel sniffing canine teams to their program (Minnesota Department of Natural Resources 2013). Results of our study indicate that it is possible to train these canines to detect veligers as well, and in principle, these canines should be able to be trained to detect other aquatic invasive species as well.

Acknowledgments

We thank Tiffany Sellars for help in organizing and editing this manuscript and Debbie Farmer of Interquest Detection Canines® for her input and support.

REFERENCES

Aquatic Nuisance Species Watercraft Inspection and Decontamination Interception Training (WIT) for Zebra/Quagga Mussels. http://www.aquaticnuisance.org/wit (Accessed June 4, 2013).

Bozarth, C.A., Y.R. Alva-Campbell, K. Ralls, T.R. Henry, D.A. Smith, M.F. Westphal, and J.E. Maldonado. 2010. An efficient noninvasive method for discriminating among faeces of sympatric North American canids. *Conservation Genetics Resources* 2:173–175.

California Department of Fish and Wildlife. K-9 Program, Sacramento, CA. http://www.dfg.ca.gov/enforcement/K9/ (Accessed April 25, 2014).

Connelly, N.A., C.R. O'Neill Jr, B.A. Knuth, and T.L. Brown. 2007. Economic impacts of zebra mussels on drinking water treatment and electric power generation facilities. *Environmental Management* 40:105–112.

Culver, C., H. Lahr, L. Johnson, and J. Cassell. 2013. Quagga and zebra mussel eradication and control tactics. California Sea Grant technical report (T-076) on project A/EA-AR-33.

Dahlgren, D.K., R.D. Elmore, D.A. Smith, A. Hurt, E.B. Arnett, and J.W. Connelly. 2012. Use of dogs in wildlife research and management. In: Silvy, N.J. (ed.), *Wildlife Techniques Manual: Research* (Vol. I, pp. 140–153). John Hopkins University Press, Baltimore, MD.

Department of Homeland Security. 2010. History of CBP Canine Centers, Washington, DC. http://cbp.gov/xp/cgov/border_security/canine/history_3.xml (Accessed June 4, 2013).

Fonseca, F.E.L.I.C.I.A. 2009. Invasive mussels imperil western water system. *New York Daily News.* http://www.nydailynews.com/news/world/invasive-quagga-mussels-cost-west-coast-millions-main-tenancearticle-1.428560 (Accessed June 4, 2013).

Hurt, A. and D.A. Smith. 2009. Conservation dogs. In: Helton, W.S. (ed.), *Canine Ergonomics: The Science of Working Dogs* (pp. 175–194). CRC Press, Taylor & Francis Group, LLC, Boca Raton, FL.

Lodge, D.M. 1993. Biological invasions: Lessons for ecology. *Trends in Ecology & Evolution* 8:133–137.

Minnesota Department of Natural Resources. 2013. Minnesota DNR conservation officers using dogs to detect zebra mussels. http://www.outdoorhub.com/news/minnesota-dnr-conservation-officers-using-dogs-to-detect-zebra-mussels/ (Accessed June 4, 2013).

Padilla, D.K., M.A. Chotkowski, and L.A. Buchan. 1996. Predicting the spread of zebra mussels (*Dreissena polymorpha*) to inland waters using boater movement patterns. *Global Ecology and Biogeography Letters* 5:353–359.

Parker, M. 2011. Wildlife detection dogs. Wildlife Professional, pp. 47–49.

Parker, M. and A. Hurt. 2010. Canine detection teams and conservation. In: K.H. Redford (ed.), *State of the Wild: A Global Portrait* (pp. 183–188). Island Press, Washington, DC.

Ralls, K., S. Sharma, D.A. Smith, S. Bremner-Harrison, B.L. Cypher, and J.E. Maldonado. 2010. Changes in kit fox defecation patterns during the reproductive season: Implications for noninvasive surveys. *Journal of Wildlife Management* 74:1457–1462.

Reed, S.A., A.L. Bidlack, A. Hurt, and W.M. Getz. 2011. Detection distance and environmental factors in conservation detection dog surveys. *Journal of Wildlife Management* 75:243–251.

Reindl-Thompson, S.A., J.A. Shivik, A. Whitelaw, A. Hurt, and K.F. Higgins. 2006. Efficacy of scent dogs in detecting black-footed ferrets at a reintroduction site in South Dakota. *Wildlife Society Bulletin* 34:1435–1439.

Richards, N.L., S.W. Hall, N.M. Harrison, L. Gautam, K.S. Scott, G. Dowling, I. Zorilla, and I. Fajardo. 2014. Merging wildlife and environmental monitoring approaches with forensic principles: Application of unconventional and non-invasive sampling in eco-pharmacovigilance. *Journal of Forensic Research* 5:225.

Schneider, D.W., C.D. Ellis, and K.S. Cummings. 1998. A transportation model assessment of the risk to native mussel communities from zebra mussel spread. *Conservation Biology* 12(4):788–800.

Smith, D., K. Ralls, B. Davenport, B. Adams, and J.E. Maldonado. 2001. Canine assistants for conservationists. *Science* 291:435.

Smith, D.A., K. Ralls, B.L. Cypher, and J.E. Maldonado. 2005. Assessment of scat-detection dog surveys to determine kit fox distribution. *Wildlife Society Bulletin* 33:897–904.

Smith, D.A., K. Ralls, A. Hurt, B. Adams, M. Parker, B. Davenport, M.C. Smith, and J.E. Maldonado. 2003. Detection and accuracy rates of dogs trained to find scats of San Joaquin kit foxes (*Vulpes macrotis mutica*). *Animal Conservation* 6:339–346.

Wagner, E. 1997. Use of canines in accelerant detection. http://www.tcforensic.com.au/docs/uts/essay2.pdf (Accessed May 12, 2013).

Wasser, S.K., B. Davenport, E.R. Ramage, K.E. Hunt, M. Parker, C. Clarke, and G. Stenhouse. 2004. Scat detection dogs in wildlife research and management: Application to grizzly and black bears in the Yellowhead Ecosystem, Alberta, Canada. *Canadian Journal of Zoology* 82:475–492.

Zook, B. and S. Phillips. 2009. Recommended Uniform Minimum Protocols and Standards for Watercraft Interception Programs for Dreissenid Mussels in the Western United States. http://www.100thmeridian.org/Recommended-Protocols-and-Standards-for-Watercraft-Interception-Programs-for-Dreissenid-Mussels-in-the-Western-United-States.pdf (Accessed January 31, 2013).

10

Understanding Dreissenid Veliger Detection in the Western United States

Jamie Carmon and Denise M. Hosler

CONTENTS

ABSTRACT   The Reclamation Detection Laboratory for Exotic Species (RDLES), Denver, CO, is one of the leading laboratories in the western United States for the detection of invasive species. RDLES's primary role is the detection of invasive dreissenid mussel populations in raw water samples. RDLES focuses on identifying the larval (veliger) life stage, as the veliger is free floating in the water column, allowing for increased discovery before adult populations overtake a water body. Multiple testing methods to determine if raw water samples are free of dreissenid mussel larvae are utilized. Studies performed by RDLES have allowed for the optimization of every aspect of the detection process, from sample collection and preservation to increased genetic testing sensitivity. This chapter will focus on advancements in microscopy, scanning electron microscopy (SEM), polymerase chain reaction (PCR), flow cell cytometry (VeligerCam™) for detection of invasive dreissenids, as well as a discussion of laboratory studies conducted to better understand the prevalence of cross-contamination in the field and laboratory and the likelihood of a contaminate being detected by microscopy or PCR.

Background

Aquatic invasive species can negatively influence water bodies in multiple ways. Dreissenid mussels can alter lake chemistry due to their filtering capability (USACE 2007), and disrupt or damage power generation capabilities at dams by clogging water intakes and other infrastructure (USACE 2007; Hosler 2011). In response to these concerns, RDLES, Denver CO, was created to monitor the spread of dreissenid populations throughout the western United States and for early detection of new mussel populations. RDLES also determines biological and environmental factors that may help predict the next large-scale infestation and monitors and supplies dreissenid life history information of infested water bodies.

Detection of adult dreissenids is difficult in raw water bodies due to their size (approximately 6–45 mm) and because they prefer to attach to surfaces that are in low light areas, making access for divers difficult (USACE 2007; Mackie and Claudi 2010). A single adult female can produce over one million eggs per year, and the larvae that are produced remain suspended in the water column for about a month before settling, which increases the likelihood of discovering the larvae over the adult (Mackie and Claudi 2010). RDLES studies have determined that monitoring for the presence of dreissenids in the veliger stage is an improved method over monitoring for adults for early detection (Hosler 2011) because it may provide facility managers with more time to prepare for mussel impacts.

RDLES has implemented multiple detection methods to improve early detection of veligers, which are difficult to detect because they are microscopic (between 97 and 462 μm). An analytical strategy that includes light microscopy (LM), cross-polarized light microscopy (CPLM), SEM, flow cell cytometry, and PCR has been developed. Early detection of dreissenids in raw water has some inherent issues with variability in sampling techniques and analytical methods (Stefanik 2004), resulting in inconsistent reporting between laboratories. When a single veliger body is discovered by microscopy in a water body with no other findings during the sampling season, it is considered a *variable* result. Single findings of this type generate management dilemmas for classification and management of that water body for the future. RDLES has extensive experience with the detection of invasive species (Greene 1997) and has attempted to reduce *variable* results by performing decontamination and optimization studies of field, laboratory, and PCR procedures.

Field Sampling Procedures

The likelihood of dreissenid mussel population explosions is primarily determined by habitat suitability. Therefore, when early detection samples are collected, it is important also to collect water quality data. Calcium concentration, pH, temperature, and turbidity are important water quality parameters that can be used to assess the susceptibility of a reservoir to a dreissenid mussel infestation. Water quality data collected by Reclamation scientists has confirmed that adult quagga mussels prefer low light, moving water, high calcium concentration, and high pH (Claudi and Mackie 1998; Broderick and Hosler 2013). Dreissenid mussels require a higher calcium concentration than most freshwater mollusks as their shell comprises 40% calcium (Secor et al. 1993; Broderick and Hosler 2013); therefore, water systems that have high calcium levels may be more susceptible to invasive mussel infestation. Dreissenid mussels cannot tolerate turbid water as the sediments may disrupt the mussel's ability to filter water and deter the veligers from settling (Steele and Wong 2014, Chapter 8 in this book). In the western United States, dams release water rapidly downstream, and a rapid drawdown of water from reservoirs seems to discourage the settlement of juvenile mussels (Broderick and Hosler 2013). Collection of water quality data alongside mussel samples provides insight into the biology of the organism and documents environmental parameters that trigger a population explosion. In turn, the data may provide increased integrated pest management opportunities and expand knowledge on dreissenid mussels and their water quality tolerances. After water quality data are collected, plankton tow samples are used to determine the presence or absence of quagga mussels.

Mussel samples are collected with a 64 μm plankton tow net with a weighted cup on the end (Figure 10.1). In order to prevent contamination, separate nets are designated for different water systems, especially for suspect and positive water bodies. To collect a sample, the field technician slowly lowers the net and cup

FIGURE 10.1 Plankton tow net with weighted cod end. (From Carmon, J.L. and Hosler, D.M., Field protocol: Field preparation of water samples for dreissenid veliger detection version 4, Technical Memorandum No. 86-68220-13-01, http://www.usbr.gov/mussels/docs/FieldSOPPreparationandAnalysis.pdf, 2013a.)

to the bottom of the water body, and then pulls the net up through the water column effectively filtering the water and collecting anything that is larger than 64 µm. Vertical tows are preferred by RDLES as veligers tend to stay below the photic zone. The net volume and length of the tow are used to calculate the total volume of the tow and the average number of veligers per liter of water (Carmon and Hosler 2013a). Analysis of Reclamation data determined that 75% of all veligers in a positive water body are captured at marinas (Zehfuss 2009); this led to a sampling protocol where five vertical tows are taken at the inflow, outflow, and mid-lake, additionally sampling at marinas, boat launches, and dam sites.

After the tows are collected, buffer is added to every raw water sample, usually sodium bicarbonate (baking soda), as it is inexpensive, nontoxic, and readily available. The buffer stabilizes the pH and maintains the integrity of the sample by preventing rapid degradation of the organisms, which increases the shelf life of the sample (Carmon and Hosler 2013a; Carmon et al. 2014). The sample is preserved by increasing the sample volume by 20% with alcohol, to comply with overnight shipping regulations. Studies by RDLES have documented that the type, proof, or concentration of alcohol in the water sample does not affect the likelihood of detecting a veliger by microscopy or PCR (Carmon et al. 2014). The sample lid is taped with electrical tape, to prevent leakage, and sent to RDLES in a cooler with ice.

Sample Preparation Procedures

Upon arrival, the sample is logged into the laboratory database and given a unique identification and tracking number. The RDLES database has been expanding since 2007 with over 14,000 samples analyzed for dreissenid mussels in the 17 western United States (Hosler 2013). As pH is critical for

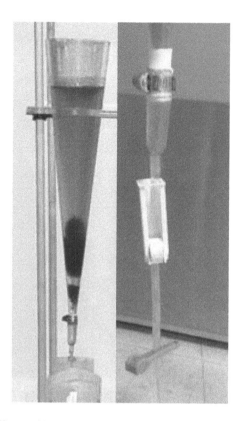

FIGURE 10.2   Modified Imhoff cone with a close-up of passive venoset system. (From Carmon, J.L. and Hosler, D.M., Lab protocol: Preparation and analysis of dreissenid veliger water samples version 4, Technical Memorandum No. 86-68220-13-02, Denver, CO, http://www.usbr.gov/mussels/docs/LabSOPPrepandAnalysis.pdf, 2013b.)

microscopic identification of veligers, sample pH is tested at the time of arrival and adjusted with baking soda if the pH is below 7. After log in, the uniquely numbered raw water sample is poured into a modified Imhoff cone (Figure 10.2) and settled overnight (Carmon and Hosler 2013b). This method was developed from a combination of methods for both zooplankton and veliger detection (Hosler 2011).

The United States Army Corps of Engineers (USACE) (USACE 2007) method was used until 2009, when the veliger counts at Lake Mead were over 500 veligers per milliliter and microscopists had difficulty achieving accurate counts (Hosler 2011). RDLES developed a method combining the USACE method (USACE 2007), Standard Method 10200G (Standard Methods 2001), and the United States Environmental Protection Agency method LG402, Revision 2 (USEPA 2003). In the combined modified procedure, veligers are allowed to settle to the bottom 15 mL of a modified Imhoff cone with a passive venoset system attached. The passive venoset allows for the technician to control the volume of sample delivered from the bottom of the cone (Figure 10.2). Testing of this method demonstrated minimal damage to the veliger, and repeated trials conducted in 2009 demonstrated improved population counts with a 98% recovery rate of veligers in Lake Mead samples (Hosler 2011). To optimize the likelihood of detecting a veliger, RDLES examines the entire 15 mL aliquot of a settled sample (Carmon and Hosler 2013b).

Test Methods for Early Detection in Raw Water Samples

When analyzing raw water samples for microscopic target organisms, many testing methods are used. The preliminary testing method is CPLM and LM. Any suspect organisms are examined further by SEM and/or PCR. Any raw water sample that was positive by microscopy will be analyzed by PCR to determine the presence/absence of any DNA in the sample. Frischer et al. (2011) conducted a double-blind round robin study that found that microscopy is more sensitive than PCR for detecting veligers that have not degraded. Current studies conclude that once the degradation of the veliger occurs, usually due to improper sample preservation techniques, PCR may be the only way to determine the presence of dreissenid mussels (Carmon et al. 2014). Without the microscopic finding, a positive PCR implies only the presence of DNA. In summary, multiple positive results by microscopy with concurrent positive PCR results are needed to consider the possibility of dreissenid mussel establishment (Hosler 2013).

Microscopy

The accepted diagnostic method for the presence of dreissenid veligers is CPLM, where a cross-polarizing filter is used on the objective lens of a dissecting microscope. Organisms with a calcium shell exhibit birefringence character and will shine against a dark background. As light passes through the cross-polarizing filter, a distinctive Maltese-cross pattern appears on the outer shell where the axes of the cross-polarizing filter intersect on the veliger shell (Mackie and Claudi 2010). One milliliter subsamples are analyzed first by CPLM, and if a sample from the water body was previously positive, then the samples are scanned by LM, in search for degraded veligers. A blind round robin study performed in 2010 documented that 98% of veligers will be identified by CPLM through the birefringence of the shell (Frischer et al. 2011). There are interferents that exhibit birefringence under CPLM, such as ostracods, Asian clams, and some particulates. A sample with a high amount of suspended solids and zooplankton may affect microscopic accuracy. RDLES has created an identification flow chart that outlines the morphological differences between ostracods, Asian clams, other bivalve veligers, other birefringent elements in raw water, and dreissenid veligers (Figure 10.3; Carmon and Hosler 2014).

CPLM is the primary tool for detecting veligers; however, the veliger shell is very fragile and can lose some or all of the birefringent character when introduced to acidic conditions (Figure 10.4). RDLES has modified the microscopic techniques so degraded veligers may be discovered using LM in cases where the CPLM result is negative and PCR test result is positive (Carmon and Hosler 2013b).

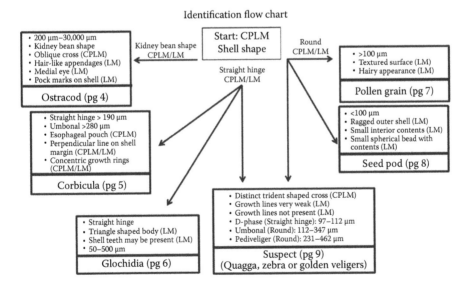

FIGURE 10.3 RDLES identification flowchart for the identification of birefringent particles and organisms (From Carmon, J. and Hosler, D.M., Quality assurance plan for the reclamation detection laboratory for exotic species, Technical Memorandum No. 86-68220-14-06, 2014.)

Veliger morphological changes with
three different microscopy methods

Intact/   Degraded/
not degraded   broken

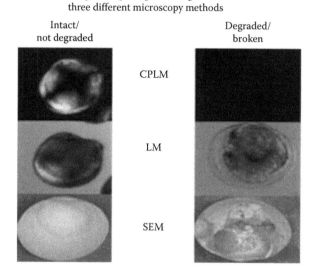

CPLM

LM

SEM

FIGURE 10.4 Intact versus degraded/broken veligers viewed with three microscopy methods: cross-polarized light microscopy (CPLM), light microscopy (LM), and scanning electron microscopy (SEM). (From RDLES, Improving accuracy in the detection of dreissenid mussel larvae, http://www.usbr.gov/mussels/docs/MusselLarvae DetectionReport.pdf, 2013.)

CPLM is the accepted method for the detection of dreissenids; however, if the sample has degraded, then the veliger shell may no longer show birefringence. LM is useful to discover degraded veligers; however, the zooplankton and other interferents can impede detection. SEM gives a close-up view of the veliger shell; however, if the veliger is lost, broken, or degraded, identification can be impossible (Figure 10.4). PCR analysis determines the presence/absence of dreissenid DNA; however, if no veliger is discovered by microscopy, it is impossible to determine if the DNA was a carryover from a predator, a veliger, or a population of live adult dreissenid mussels.

Effect of pH on Veliger Detection

Veliger detection by CPLM is dependent upon the integrity of the veliger shell. If the shell is degraded, the mussel will not show birefringence under cross-polarized light. A veligers shell may be degraded by acidic pH; therefore, a study was conducted to determine if the pH of a sample impacts the integrity of the veliger shell and the detection by CPLM and PCR.

For this study, sets of veligers (100, 50, and 25) were placed in 50 mL tubes containing either deionized (DI) water (pH 5) or buffered DI water (pH 8), buffered per RDLES field standard operating procedures; 0.2 g of baking soda was added per 100 mL of water (Carmon and Hosler 2013a). The samples were analyzed for veliger birefringence loss and PCR detection over the course of 6 weeks. Approximately 80% of the veligers in pH 5 samples were undetectable by CPLM on day 6, and 100% undetected by CPLM by day 14. Detection of veligers by CPLM in buffered samples was greater than the unbuffered samples; however, the veliger body was found to degrade without the addition of alcohol as a preservative. By day 42, over 70% of the veligers in the pH 8 samples were undetectable by CPLM (Figure 10.5; RDLES 2013; Carmon et al. 2014).

Veligers that have lost birefringence could be detected by genetic analysis, because DNA was present in the sample. The PCR results were positive for both buffered and unbuffered samples for the duration of the 6-week study (Figure 10.6; RDLES 2013; Carmon et al. 2014). The results of this study have prompted RDLES to analyze suspect samples or previously positive waters using CPLM, LM, and PCR, since LM and PCR are the only ways to detect degraded veligers (Carmon and Hosler 2013b).

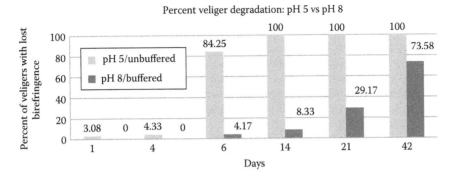

FIGURE 10.5 Percent of veligers with lost birefringence in pH 5 (unbuffered) and pH 8 (buffered) DI water, without alcohol. (From RDLES 2013; Carmon, J., and Hosler, D.M., Quality assurance plan for the reclamation detection laboratory for exotic species . Technical Memorandum No. 86-68220-14-06, 2014.)

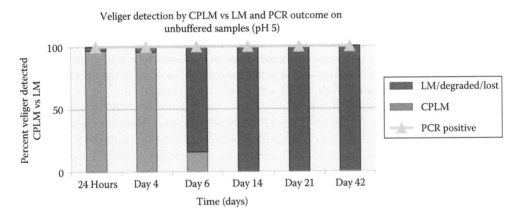

FIGURE 10.6 Veliger detection by CPLM, LM, and PCR in pH 5 (unbuffered) DI water without alcohol. (From RDLES 2013; Carmon, J., and Hosler, D.M., Quality assurance plan for the reclamation detection laboratory for exotic species . Technical Memorandum No. 86-68220-14-06, 2014.)

SEM Procedures

In the event of discovering a suspect by either CPLM or LM, photographic documentation with the SEM is also taken in order to capture the taxonomic features needed to clearly identify the suspect organisms (Hosler 2011). The suspect organism is pipetted onto a glass slide containing the sample location information (Carmon and Hosler 2013b). The sample is analyzed with SEM, and a mollusk taxonomist can determine the species by analyzing taxonomic features and measurements of the prodissoconch (the original larval shell). The prodissoconch generally has a flawless smooth appearance without the presence of growth lines. The dreissenid veliger prodissoconch ranges from 85 to 105 µm (Martel et al. 1995). SEM pictures aid in the clarification between bivalve larvae and Asian clams (Corbicula), which have a prodissoconch that is much larger than a dreissenid veliger. The use of SEM improves diagnostic capability and allows differentiation between varieties of organisms, resolving any questions that arise during initial microscopic identification (Baldwin et al. 1994; Martel et al. 1995). However, even at a high magnification, some taxonomic anomalies such as shell size and shape make identification agreement difficult for taxonomists (Hosler 2011).

Flow Cell Cytometry

Flow cell cytometry, a specialized microscopic technique, is additionally used by RDLES to monitor zooplankton and periphyton communities and to enumerate veligers in samples that have significant numbers of veligers with low debris. More information about Flow cell cytometry can be found in Chapter 11 of this book. As invasive mussels are introduced to a water system, the population dynamics of the micro ecosystem starts to shift as veligers are filter feeders and will decrease zooplankton communities and increase blue-green algae populations (USACE 2007; Mackie and Claudi 2010). These changes can be identified and monitored using a FlowCAM™ to analyze the zooplankton and periphyton communities in samples after they have been analyzed for veligers. RDLES plans to develop a database that will help track ecosystem changes related to mussel infestation over time.

Through a cooperative research and development agreement, the Bureau of Reclamation and Fluid Imaging (Yarmouth, ME) developed a VeligerCam to monitor veligers in water samples. This modified flow cytometry device takes both CPLM and LM pictures and pairs the pictures together on the computer screen. Adjustments of operational parameters, such as flow rate, sample volume, and viscosity with surfactants were made to optimize the VeligerCam performance. RDLES studies concluded that decreasing the flow rate from the factory setting of 1.125 down to 0.875 mL/min increased the likelihood of capturing a photograph of veligers. The sample volume was decreased from 12.5 to 10 mL to decrease sample run time. Adjusting the fluid viscosity with the addition of 40% glycerin increased the likelihood of detecting veligers, as they remain suspended in the liquid for a longer period of time. The glycerin controls the rate at which the veliger passes through the flow cell increasing the likelihood of image capture. The optimization has resulted in capture of 95%–100% of all veligers present in a water sample (Figure 10.7; O'Meara and Bloom 2012).

The VeligerCam is equipped with a 300 μm flow cell requiring that all samples be filtered with a 280 μm sieve. The small flow cell allows for better detection and picture quality of veligers; however, the filtering process decreases the likelihood of detecting veligers that were caught up in debris. Additionally, the use of the 300 μm flow cell will filter out the large pediveligers, altering population numbers for large veligers; therefore, the VeligerCam is routinely used for enumeration of high veliger numbers and not for early detection (O'Meara and Bloom 2012).

In conjunction with enumerating high numbers of veligers, the VeligerCam can be used to determine the effectiveness of control treatments such as turbulence on veligers. The VeligerCam provides detailed images, which can be used to analyze damage caused by a specific treatment. VeligerCam images are especially useful for determining the percent of veligers that have damaged shells. The VeligerCam can process large amounts of veligers in a short amount of time, and analyzing a larger sample size provides a better estimation of the percent damaged by the treatment. By analyzing the experimental control

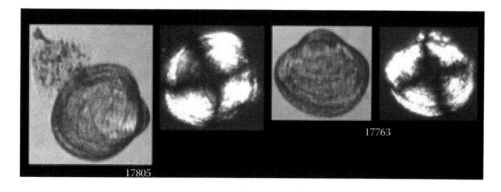

FIGURE 10.7   LM and CPLM paired images taken by the VeligerCam. (From O'Meara, S. and Bloom, K., VeligerCAM protocol optimization: Experiment to increase recovery of dreissenid mussels, Technical Memorandum: 86-68220-12-17, Bureau of Reclamation, 2012.)

sample and a treated sample, minor differences in the VeligerCam photos can be used to analyze the impacts of the control method (O'Meara and Bloom 2012).

PCR Procedures

RDLES performs PCR on previously positive or suspect samples to further detect and establish identity through dreissenid DNA presence. A subsample of the raw water plankton tow sample is analyzed for the presence/absence of dreissenid DNA, by using species-specific primers to amplify a specific gene. The target gene, cytochrome oxidase I (COI), is widely used for the genetic identification of a wide range of organisms, including dreissenid mussels. The COI gene is located in every mitochondrial genome. Due to the critical function of mitochondria, mutations are often lethal; therefore, the COI gene is conserved and differs slightly from species to species making it the ideal gene for species identification. Originally, 710 base pairs of the COI gene were analyzed; currently, RDLES uses a smaller segment of the COI gene, which is 383 base pairs. When performing PCR on raw water samples, a smaller segment of the gene is used due to potential target tissue degradation, which is likely in a raw water sample (Claxton et al. 1997, 1998; Ram et al. 2011; Bellemain 2013; Keele et al. 2013; NPS Lake Powell Water Laboratory 2013).

After a veliger sample has been analyzed by microscopy, the entire raw water sample is agitated to evenly distribute the contents of the sample, and a 40 mL subsample is poured into a Falcon tube and centrifuged for 30 min. The water is decanted off, and 0.5 g of the pellet is collected for DNA extraction. All DNA present in the 0.5 g pellet is extracted from the cells of every organism present in the pellet. The DNA is eluted into 50 mL of an elution buffer solution. PCR is run on the eluted DNA. The subsample of the amplified product is run on an electrophoresis gel that produces a band at a specific size for the target gene. Positive samples are sent to an independent laboratory for gene sequencing to confirm PCR results and for confirmation of the species identity (Keele et al. 2013).

Increasing the sensitivity of the PCR analysis is an ongoing process at RDLES where every step in the DNA extraction and amplification process is analyzed and optimized. Prior to 2012, a blood and tissue DNA extraction kit was used for the detection of dreissenid mussels. However, due to inconsistent results on positive water bodies, it was hypothesized that the lysis step of the blood and tissue kit may not be strong enough to break apart the calcium carbonate matrix of the shell. The raw water pellet is also concentrated with zooplankton and other interferents (such as sand, soil, mud, etc.) and may react more like a soil sample than a blood and tissue sample. One major issue in environmental samples is humic acid and other chemicals that can inhibit the PCR reaction. The soil DNA extraction kits have been designed to remove these inhibitors and thus improve the sensitivity of the PCR assay (Mo Bio Laboratories 2013). The soil kit is optimized for environmental samples and includes an agitation step that helps to break up the calcium carbonate shell releasing the tissue into the sample and increasing the likelihood of a positive PCR result. Optimizing the amplification process usually includes altering magnesium chloride concentrations and/or altering the primer concentration. These are laboratory-specific alterations that are dependent on equipment and reagent variability (Keele et al. 2013).

There are several methods available to determine species identification by PCR. RDLES chose to use agarose gel electrophoresis because it is uncomplicated, sensitive, inexpensive, and produces a visible band. The PCR product can then be sent for gene sequencing, and the species identity can be verified. Other methods, such as qPCR, can be expensive, require specialized equipment, and extensive training. Also qPCR uses a smaller fragment of the gene, which limits the opportunity for gene sequencing (USEPA 2004; Mccord 2011). No matter what method is used, a positive PCR result determines only that dreissenid DNA was present in the sample at the time of collection. Only continued, consistent monitoring can determine if viable populations of mussels are present in the water body. A single positive PCR result, like a single positive microscopy result, does not mean that there is a smoldering population of mussels in the reservoir. Multiple positives at the same location may provide evidence of an undetected population, or it may indicate that multiple introductions are occurring (Hosler 2013).

Cross-Contamination of Samples

For any analytical work, cross-contamination from one source into a sample being analyzed is always a concern. At RDLES, points of cross-contamination have been identified for the field sample collection, laboratory sample preparation, and PCR. The two primary sources of cross-contamination are veliger body contamination and DNA contamination. The first part of this discussion will focus on how a veliger body can be carried from one sample to the next in the field and in the laboratory, followed by a discussion of DNA carryover into other samples and the likelihood of contamination occurring when following RDLES standard operating procedures.

Veliger Contamination

In the Field

Contamination of samples collected in the field is always a concern with biological samples, because of the difficulty of maintaining a clean work space. Field technicians follow the RDLES field standard operating procedures to prevent contamination in the field (Carmon and Hosler 2013a). These procedures include the use of dedicated equipment and decontamination procedures for boats, nets, and equipment.

   Field personnel soak the plankton tow net in vinegar to degrade the veliger shell beyond detection by CPLM in between sites on the same water body (Figure 10.8; Carmon and Hosler 2013a). RDLES laboratory studies concluded that, on average, vinegar will degrade the veliger body beyond detection by CPLM after 3 min, and while the veliger body is still detectable by LM, it would probably remain undetected due to the presence of organic and inorganic materials in the water samples (RDLES 2013). In addition to soaking the net in vinegar in between sites, a bleach rinse is incorporated in between reservoirs to ensure no DNA is transferred from the net to other samples (Carmon and Hosler 2013a). Bleach eliminates DNA cross-contamination, but bleach also brightens the veliger shell and makes it fragile (Figure 10.8). If the net is decontaminated only with bleach, the veliger shell may still be visible under CPLM, resulting in a false-positive finding (RDLES 2013). Bleaching of the net occurs only in between water bodies, so a thorough cleaning in vinegar and water will remove any veligers in the net, while the bleach step will remove any tissue/DNA that may be on the net.

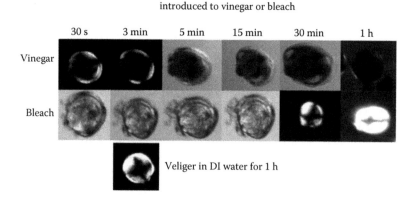

FIGURE 10.8 Morphological changes in veligers over time with different decontamination methods, vinegar, and bleach. (From RDLES, Improving accuracy in the detection of dreissenid mussel larvae, http://www.usbr.gov/mussels/docs/MusselLarvaeDetectionReport.pdf, 2013.)

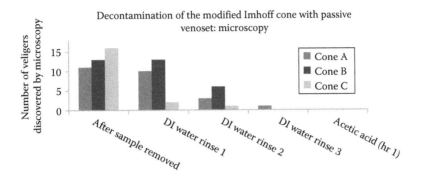

FIGURE 10.9 Modified Imhoff cones with a passive venoset system were tested to determine if veligers stick to the cones after decontamination. Veligers were not detected by microscopy after being in acetic acid for 1 h. (From RDLES, Improving accuracy in the detection of dreissenid mussel larvae, http://www.usbr.gov/mussels/docs/MusselLarvaeDetectionReport.pdf, 2013.)

In the Laboratory

Laboratory contamination is controlled by utilizing good laboratory procedures, new glass pipettes, clean Petri dishes, and surface cleaning of the laboratory table and microscope between water samples (Carmon and Hosler 2013b). To mitigate concerns of cross-contamination occurring in the sample preparation procedure, tests were run to determine if veligers would stick to the modified Imhoff cone (Figure 10.2). RDLES laboratory standard operating procedure requires that after the sample is removed, the cone is triple rinsed, and scrubbed with a designated brush, using DI water. The cone is then soaked in a 5% acetic acid solution for 4 h. Then, the Imhoff cones are scrubbed and rinsed with DI water three times and immersed in a 1% bleach solution for 30 s. DI water is rinsed twice through the cone and then it is set aside to dry (Carmon and Hosler, 2013b; RDLES 2013).

To determine the likelihood of cross-contamination from Imhoff cones that were not properly decontaminated, three 160 mL samples containing greater than 100 veligers per mL were settled in Imhoff cones overnight. After the removal of the sample, 160 mL of DI water was added to the Imhoff cone and swirled around to not break any veligers that may be in the cone, and the scrub bush step was removed to preserve the integrity of the veliger. The entire 160 mL of DI water was analyzed for remaining mussels in 40 mL aliquots. The study found that veligers do not stick to the cone and are, in fact, almost completely washed out by the first hour in acetic acid (Figure 10.9; RDLES 2013).

RDLES has determined that rinsing the Imhoff cones with DI water is sufficient to remove the veliger body from the cone. However, lab technicians follow best management practices and use the standard operating procedures to maintain a high level of QA/QC in sample preparation methods (Carmon and Hosler 2013b). In summary, veliger body contamination in the field is unlikely due to the decontamination procedures in place, which degrades any veligers that may be lingering on the net in between sample locations, so they are not detectable by microscopy. New glass pipettes are used in the laboratory with every sample, and Petri dishes are thoroughly cleaned and decontaminated before use. The settling Imhoff cones could be a prominent source of contamination; however, studies have determined that all veliger bodies wash out of the cone and do not stick to the sides after the wash procedure (RDLES 2013).

DNA Contamination

While the veliger body is a main source of contamination, another source of contamination might be DNA that is transferred from one sample to another. Diligent testing for DNA contamination in the laboratory includes DNA wipe tests, where microscopes and equipment are wiped with a cotton swab to test for the presence of mussel DNA. To date, no free-floating dreissenid mussel DNA has been discovered in the lab (RDLES 2013). RDLES also tested for potential sources of DNA contamination in the field, in the lab, and using PCR methodology. There are three case studies described in the following.

Case Study No. 1: DNA Contamination in the Field

The first case study tested the potential of DNA transfer from one sample bottle to another in the field if proper decontamination procedures are followed. Soaking field equipment in a 5% acetic acid solution (vinegar) and rinsing with bleach to remove lingering DNA is the most effective method of veliger decontamination. Efficacy of field decontamination procedures were tested in the laboratory. Three replicates of 50 veligers were placed in DI water, vinegar, and bleach and analyzed by PCR at 5, 10, 15, and 30 min. One control sample resulted in a negative PCR result. Variability in PCR detection of veligers has always been an issue due to the low concentration of DNA and the quality of the DNA in a single preserved veliger. After 30 min, one vinegar sample was positive, suggesting that a longer vinegar treatment time may be required to degrade all of the tissue in 50 veligers. One bleach sample was positive at 10 min, which could indicate that some veligers were closed tightly enough to retain the DNA in the veliger shell (Figure 10.10; RDLES 2013). This study was performed with 50 veligers per sample, and the likelihood of not detecting a cross-contamination of this size in the microscope first is highly doubtful. Testing indicates that combining vinegar and bleach washes will help to ensure that veliger bodies or DNA are not passed from sample to sample.

Case Study No. 2: DNA Contamination in the Laboratory

The second case study tested the potential of DNA contamination to occur during the sample preparation procedure in the modified Imhoff cones. The previously discussed test for cross-contamination in the modified Imhoff cones (Figure 10.9) was also tested for DNA. At each step, the rinse water was collected and sent for PCR analysis. Results indicate that DNA does not adhere to the Imhoff cones and that all DNA will be rinsed out after the first hour in acetic acid. RDLES follows all decontamination procedures for the settling Imhoff cones, as the acetic acid and bleach steps just further confirm laboratory cleanliness (Figure 10.11; RDLES 2013).

Case Study No. 3: DNA Contamination PCR Testing of Quagga Mussel Models

The third case study was performed to determine whether dead, dried adult tissue still carried enough DNA to obtain a positive result by PCR. During the 2012 sampling season, a raw water sample collected from an isolated spring tested positive for quagga mussel by PCR. It was determined that the plankton tow net used to collect the sample had been placed on top of an adult quagga mussel model (Figure 10.12) in the vehicle during a field sampling event. Dreissenid model displays are created by allowing adult

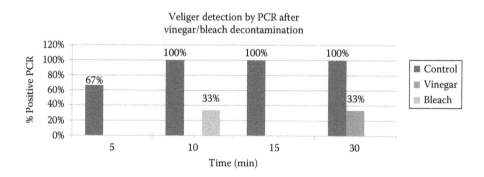

FIGURE 10.10 PCR results for 50 veligers (three replicates) in acetic acid and bleach for a known time period. (From RDLES, Improving accuracy in the detection of dreissenid mussel larvae, http://www.usbr.gov/mussels/docs/MusselLarvaeDetectionReport.pdf, 2013.)

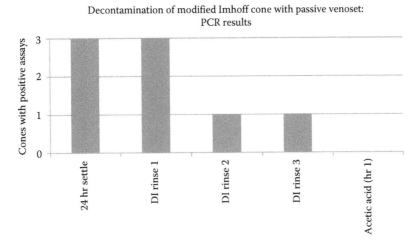

Decontamination of modified Imhoff cone with passive venoset:
PCR results

FIGURE 10.11 PCR assay results (out of three total assays) that were positive for dreissenid DNA after each decontamination step in the modified Imhoff cones with a passive venoset system. (From RDLES, Improving accuracy in the detection of dreissenid mussel larvae, http://www.usbr.gov/mussels/docs/MusselLarvaeDetectionReport.pdf, 2013.)

FIGURE 10.12 Quagga mussel model dried and lacquered. (From Arizona Game and Fish Department, Boaters beware: Citations coming if you don't "clean, drain, dry", Arizona Game and Fish Department: http://azgfd.net/artman/publish/NewsMedia/Boaters-beware-citations-coming-if-you-don-t-Clean-drain-and-dry.shtml, 2013.)

quagga mussels to settle onto an object, and then the object is dried and lacquered. The models are commonly used as an educational tool to demonstrate colonization by dreissenid mussels (Figure 10.12). The positive PCR result was from a water body that had a low probability of mussel inoculation; no boating or fishing occurs on the tested spring. Microscopy was not performed on the sample; therefore, it is unknown if a dreissenid mussel was present in the sample. This event prompted further testing to determine if DNA could be easily transferred from a model display, to the net, and then to subsequent water samples. To test for model contamination, a quagga mussel rope model was shaken, and the debris that fell off was collected and analyzed by PCR. Whole and half-shelled adults were removed from the model and washed with DI water. The following samples were sent for PCR analysis in three replicates: one half shell, three half shells, one crushed whole shell, and three crushed whole shells. Additionally, a wet

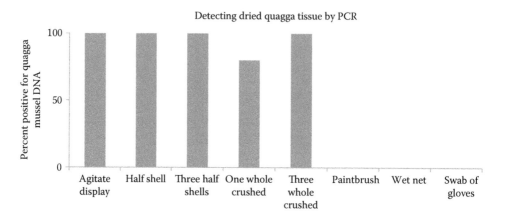

FIGURE 10.13 PCR results from portions of a quagga mussel model. (From RDLES, Improving accuracy in the detection of dreissenid mussel larvae, http://www.usbr.gov/mussels/docs/MusselLarvaeDetectionReport.pdf, 2013.)

paintbrush was brushed over the model and the water was collected for PCR, and then finally a piece of wet plankton tow net was laid over the model for 45 min. The technician's gloves were also swabbed after each step and analyzed by PCR (RDLES 2013).

The results determined that the adult shells (both the half and whole shells) contained tissue and that tissue still contained viable DNA. However, no DNA was detected on the gloves of the laboratory technician, or when the net was wet and laid on the model for 45 min. These results provide evidence that tissue is required for a positive test result to occur by the gel electrophoresis PCR method used by RDLES. PCR is a highly sensitive test; however, the double-blind round robin data indicate that a PCR result is more likely a false negative (Frischer et al. 2011). Based upon the results of this testing and the double-blind round robin, a positive PCR result (with proper QA/QC) indicates that the sample contains dreissenid mussel tissue upon collection, rather than cross-contamination (Figure 10.13; RDLES 2013).

Conclusion

RDLES utilizes various test methods for detection of dreissenid mussels in raw water samples. Early detection of veligers with plankton tow samples offers water managers advanced warning to adjust and modify systems, and plan operation and maintenance activities to reduce negative consequences. Occasionally, early detection methods may yield conflicting results, which creates a dilemma for management decisions that require large dollar expenditures. In the situation where microscopic and DNA testing do not agree, additional testing utilizing additional microscopic imaging, SEM, gene sequencing, and improved laboratory procedures has provided greater confidence in results. Improved sample handling and sample buffering have improved veliger shelf life and PCR results (O'Meara et al. 2013; RDLES 2013; Carmon et al. 2014). Experience with DNA testing to date indicates that tissue from the population must be present in order to achieve a positive PCR result. When consistent DNA tests at the same reservoir reveal a positive result for dreissenids, then the environmental risk assessment becomes the next important management step. Evaluating the environmental conditions gives management some idea of the likelihood of infestation or population explosion (Hosler 2013).

The presence of large quagga and zebra mussel populations has significant impacts on the environment and water-related structures. The economic and functional impacts of an infestation may be overwhelming for water managers (Lovell and Stone 2005). Studies have shown that early detection of dreissenid adults in water systems provides a 3–5 year window before the populations explode to become a real threat to structure and function (Hosler 2011). However, the time frame from detection to infestation in the western United States is not yet well understood. The importance of identifying the presence of dreissenid species early and assessing the risk potential for infestation at vulnerable locations in the facility allows facility managers to make adjustments that will help reduce the economic impacts of an infestation. A risk assessment of the environmental conditions can indicate the likelihood of an infestation; however, speculation cannot be made as to how serious the infestations may become or if the function of the dam components will be affected without the gathering of detailed hydrological, environmental, and biological information (Claudi and Mackie 1998; Lucy 2006; Claudi and Prescott 2007). Early detection of dreissenids, a comprehensive risk assessment of suitability for survival and reproduction at a specific location, combined with an evaluation of vulnerable points in a facility will allow managers to budget and prioritize activities to minimize dreissenid impacts. RDLES has put a great deal of effort into understanding the best methods for early detection of dreissenid mussels. Ongoing monitoring and multiple positive results from a single water body increase confidence of dreissenid presence over time and may predict infestations. The goal of RDLES is that dreissenid species management will be improved with the veliger detection program used in combination with the biologic and environmental data that have been collected for waters of the western United States.

Acknowledgments

The authors would like to thank Curt Brown, Chris Holdren, Fred Nibling, Linnette Ambrosio, Kevin Bloom, Suzanne Brenimer, Sariah Cottrell, Alex Fauble, Tanna George, Rachael Hardinger, Andy Humes, Doug Hurcomb, Brandon Kennedy, Joseph Kubichek, Rachael Lieberman, Judy Lyons, Susan McGrath, Scott O'Meara, Danielle Pucherelli, Sherri Pucherelli, Audry Rager, Miguel Rocha, Jeremiah Root, Benedict Roske, Kyle Rulli, Michael Simonavice, Candice Talbot, Scott Thullen, Francesca Tordanoto, James (Rollie) Williams, Anne Williamson, Dan Williamson, and Dr. Jacque Keele at Reclamation and David Wong at UNLV for the support of the Reclamation dreissenid detection program.

REFERENCES

Baldwin, B. S., Pooley, A. S., Lutz, R. A., Conn, D. B., Kennedy, V. S., and Hu, Y.-P. (1994). Identification of larval and post larval zebra mussels and co-occurring bivalves in freshwater and estuarine habitats using shell morphology. In: *Proceedings: Fourth International Zebra Mussel Conference* (pp. 479–488). Madison, Wisconsin: University Wisconsin Sea Grant Institute.

Bellemain, E. (2013). eDNA barcoding and metabarcoding-general introduction. Belgian Network for DNA Barcoding. http://bebol.myspecies.info/sites/bebol.myspecies.info/files/Bellemain%20Spygen%20eDNA%20intro_1.pdf. Accessed January 06, 2015.

Broderick, S. and Hosler, D. M. (2013). Biological suitability of Lahontan Reservoir for dreissenid mussel infestation: Technical Memorandum 86-68220-13-03.

Carmon, J. and Hosler, D. M. (2014). Quality assurance plan for the reclamation detection laboratory for exotic species. Technical Memorandum No. 86-68220-14-06.

Carmon, J., Keele, J. A., Pucherelli, S. F., and Hosler, D. (2014). Effects of buffer and isopropanol alcohol concentration on detection of quagga mussel (*Dreissena bugensis*) birefringence and DNA. *Management of Biological Invasions*, 5, 151–157. http://www.reabic.net/journals/mbi/2014/2/MBI_2014_Carmon_etal.pdf. Accessed January 06, 2015.

Carmon, J. L. and Hosler, D. M. (2013a). Field protocol: Field preparation of water samples for dreissenid veliger detection version 4. Technical Memorandum No. 86-68220-13-01. http://www.usbr.gov/mussels/docs/FieldSOPPreparationandAnalysis.pdf. Accessed January 06, 2015.

Carmon, J. L. and Hosler, D. M. (2013b). Lab protocol: Preparation and analysis of dreissenid veliger water samples version 4. Technical Memorandum No. 86-68220-13-02, Denver, CO. http://www.usbr.gov/mussels/docs/LabSOPPrepandAnalysis.pdf. Accessed January 06, 2015.

Claudi, R. and Mackie, G. L. (1998). *Zebra Mussel Monitoring and Control*. Boca Raton, FL: Lewis Publishers, Inc.

Claudi, R. and Prescott, T. (2007). Assessment of the potential impact of quagga mussels on Davis Dam and Parker Dam and recommendations for monitoring and control. http://www.usbr.gov/lc/region/programs/quagga/ParkerDavisReport.pdf. Accessed January 06, 2015.

Claxton, T. W., Martel, A., Dermott, R. M., and Boulding, E. G. (1997). Discrimination of field-collected juveniles of two introduced dreissenids (*Dreissena polymorpha* and *Dreissena bugensis*) using mitochondrial DNA and shell morphology. *Canadian Journal of Fish Aquaculture Science*, 54, 1280–1288.

Claxton, T. W., Wilson, A. B., Mackie, G. L., and Boulding, E. G. (1998). A genetic and morphological comparison of shallow- and deep-water populations of the introduced dreissenid bivalve *Dreissena bugensis*. *Canadian Journal of Zoology*, 76, 1269–1276.

Frischer, M., Nierzwicki-Bauer, S., and Kelly, K. (2011). Reliability of early detection of Dreissenid spp. Larvae by cross polarized light microscopy, image flow cytometry and polymerase chain reaction assays. Results of a community double-blind round robin study (Round robin phase II). National Invasive Species Council. http://www.invasivespecies.gov/global/ISAC/ISAC_Minutes/2011/PDF/RRII_Final_Report_(2010).pdf. Accessed January 06, 2015.

Greene, T. (1997). Zebra mussel monitoring research program at the bureau of reclamation: Summary of 1996 monitoring activities. Bureau of Reclamation. Technical Memorandum No. 8220-97-11: http://www.usbr.gov/pmts/eco_research/9711.html. Accessed January 06, 2015.

Hosler, D. M. (2011, June). Early detection of dreissenid species: Zebra/quagga mussels in water systems. *Aquatic Invasions*, 6(2), 217–222. http://www.aquaticinvasions.net/2011/AI_2011_6_2_Hosler.pdf. Accessed January 06, 2015.

Hosler, D. M. (2013). *Reclamation Mussel Detection Program: Lessons Learned from Waters of the West*. Tempe, AZ: SRP Conference: Project Administration Office.

Keele, J. A., Carmon, J. L., and Hosler, D. M. (2013). Polymerase chain reaction: Preparation and analysis of raw water samples for the detection of dreissenid mussels. Technical Memorandum No. 86-68220-13-13. http://www.usbr.gov/mussels/docs/PCRPreparationAnalysisVeligers.pdf. Accessed January 06, 2015.

Lovell, S. J. and Stone, S. F. (2005). The economic impacts of aquatic invasive species: A review of the literature. National Center for Environmental Economics (NCEE), U.S. Environmental Protection Agency. http://yosemite.epa.gov/ee/epa/eed.nsf/54e92d0d1f202a6885256e46007b104c/0ad7644c390503e385256f8900633987/$FILE/2005–02.pdf. Accessed January 06, 2015.

Lucy, F. (2006). Early life stages of *Dreissena polymorpha* (zebra mussel): The importance of long-term datasets in invasion ecology. *Aquatic Invasions*, 1(3), 171–182.

Mackie, G. L. and Claudi, R. (2010). *Monitoring and Control of Macrofouling Mollusks in Fresh Water Systems* (2nd edn.). Boca Raton, FL: CRC Press.

Martel, A., Hynes, T. M., and Buckland-Nicks, J. (1995). Prodissoconch morphology, planktonic shell growth, and size at metamorphosis in *Dreissena polymorpha*. *Canadian Journal of Zoology*, 73, 1835–1844.

McCord, B. (2011). DNA quantitation by real time PCR: Advanced issues. Florida International University, Department of Chemistry. http://dna.fiu.edu/Advanced%20DNA%20Typing%20lectures/An%20introduction%20to%20principles%20of%20QPCR-D.pdf. Accessed January 06, 2015.

Mo Bio Laboratories. (2013). Mo Bio Laboratories. Power SOil DNA Isolation Kit. http://www.mobio.com/soil-dna-isolation/powersoil-dna-isolation-kit.html. Accessed January 06, 2015.

NPS Lake Powell Water Laboratory. (2013). DNA test description. Glen Canyon National Recreation Area, Aquatic Resources Management. http://www.nps.gov/glca/parknews/upload/DNA-Test-20121101a.pdf. Accessed January 06, 2015.

O'Meara, S. and Bloom, K. (2012). VeligerCAM protocol optimization: Experiment to increase recovery of dreissenid mussels. Technical Memorandum: 86-68220-12-17, Bureau of Reclamation.

O'Meara, S., Hosler, D. M., Brenimer, S., and Pucherelli, S. F. (2013). Effect of pH, ethanol concentration, and temperature on detection of quagga mussel (*Dreissena bugensis*) birefringence. *Management of Biological Invasions*, 4(2), 135–138. http://www.reabic.net/journals/mbi/2013/2/MBI_2013_2_OMeara_etal.pdf. Accessed January 06, 2015.

Ram, J. L., Karim, A. S., Acharya, P., Jagtap, P., Purohee, S., and Kashian, D. R. (2011). Reproduction and genetic detection of veligers in changing Dreissena populations in the Great Lakes. *Ecosphere*, 2(1), 1–16.

RDLES. (2013). Improving accuracy in the detection of dreissenid mussel larvae. http://www.usbr.gov/mussels/docs/MusselLarvaeDetectionReport.pdf. Accessed January 06, 2015.

Secor, C., Mills, E., Harshbarger, J., Kuntz, H., Gutenmann, W., and Lisk, D. (1993). Bioaccumulation of toxicants, element and nutrient composition, and soft tissue histology of zebra mussels (*Dreissena polymorpha*) from NY state waters. *Chemosphere*, 26(8), 1559–1575.

Standard Methods. (2001). For the examination counting techniques for phytoplankton(10200F) and zooplankton (10200G). Standard Methods for Examination of Water and Wastewater. http://www.standardmethods. org/store/ProductView.cfm?ProductID=85. Accessed January 06, 2015.

Steele, J. and Wong, W. (2014). *High Turbidity Levels as a First Response to Dreissean Mussels Infection of a Natural Lake Environment*. Boca Raton, FL: CRC Press, In publication.

Stefanik, E. (2004). Summary of zebra mussel monitoring efforts for the upper Mississippi River: 2001 through 2003. http://www.fws.gov/midwest/mussel/documents/zebra_mussel_summary_ 2001–2003. pdf. Accessed January 06, 2015.

USACE. (2007). Aquatic Nuisance Species Information System (ANSIS). http://el.erdc.usace.army.mil/ansrp/ ANSIS/ansishelp.htm. Accessed January 06, 2015.

USEPA. (2003). Standard operating procedure for zooplankton analysis. U.S. Environmental Protection Agency. http://www.epa.gov/greatlakes/monitoring/sop/chapter_4/LG403.pdf. Accessed January 06, 2015.

USEPA. (2004). Quality assurance/quality control guidance for laboratories performing PCR analyses on environmental samples. U.S. Environmental Protection Agency, October. http://www.epa.gov/nerlcwww/ documents/qa_qc_pcr10_04.pdf. Accessed January 06, 2015.

Zehfuss, K. (2009). Probability of a veliger detection. Bureau of Reclamation Data Stewardship Pilot. https:// dosp/techResc/TR/RD/data-stewardship/_layouts/OSSSearchResults.aspx?k=zehfuss&cs=This%20 Site&u=https%3A%2F%2Fdosp%2FtechResc%2FTR%2FRD%2Fdata-stewardship.

11

Automated Method for Dreissenid Veliger Detection, Identification, and Enumeration

Harry Nelson, Ben Spaulding, Scott O'Meara, Kevin Bloom, Sherri Pucherelli, and Denise M. Hosler

CONTENTS

ABSTRACT　In 2008, Fluid Imaging Technologies adapted their imaging particle analyzer, FlowCAM, used to detect and identify zebra and quagga mussel veligers—offering a version of FlowCAM, equipped with cross-polarizing optical filters for the purpose of detecting the natural birefringence seen on veligers. A number of organizations in the Western United States have since been using FlowCAM to monitor veliger activities. Included is the Bureau of Reclamation (BOR), who recently entered into a Cooperative Research and Development Agreement (CRADA) with Fluid Imaging Technologies to further enhance the technology for the express purpose of detecting, identifying, and enumerating dreissenid veligers using a dual camera approach. With input from BOR, engineers at Fluid Imaging Technologies have taken proven FlowCAM technology and improved its ability to detect and identify veligers, while increasing sample volume processing capability.

Introduction: The Problem

Quagga and zebra mussels, at the larval stage, are often very difficult to detect as they range between 80 and 400 µm in size, making them problematic to study. Additionally, it is challenging to determine the difference between a living and a dead mussel. Therefore, Fluid Imaging Technologies, Inc. in Yarmouth, Maine, developed an analytical instrument, the VeligerCAM®.

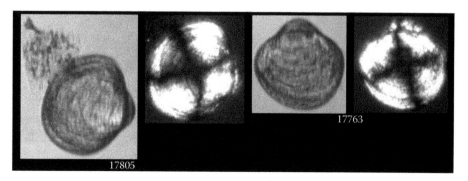

FIGURE 11.1   VeligerCAM images of dreissenid veligers, showing the Maltese cross and the umbonal bump on the shell's hinge.

Materials

Fluid Imaging Technologies offers several options to image particles, including Dynamic Flow Imaging, the FlowCAM® Benchtop/Portable PV Series, FlowCAM XPL, and FlowCAM Dual Camera Analysis (VeligerCAM).

The FlowCAM is an integrated system for rapidly analyzing particles in a moving fluid. The instrument combines selective capabilities of flow cytometry, microscopy, and fluorescence detection. The FlowCAM automatically counts, images, and analyzes particles or cells in a sample or a continuous flow. Additionally, FlowCAM can be customized to accommodate almost any environment or application. Originally developed for oceanographic investigations of organisms and particles in seawater, the FlowCAM provides the user with the capability to rapidly evaluate particulate matter in fluids. The instrument and software provide the user with tools to quickly and efficiently meet challenges, which previously required multiple instruments and many hours of tedious work to complete. The FlowCAM offers the following features and capabilities: high-speed digital imaging, particle size, count and shape, real-time bulk and individual particle analysis, combined benefits of multiple instruments, compact and durable packing, the ability to image particles ranging from 2 μm to 2 mm in diameter, fluorescence detection, and scatter detection for low particle concentrations.

The FlowCAM XPL model (or optional XPL upgrade) utilizes two cross-polarizing filters to properly detect particles that exhibit birefringence when subjected to cross-polarized light. One of the filters is fixed and the other filter is adjustable via an index. With both filters installed, the second may be rotated and adjusted. As this filter is moved, its orientation will allow a varying degree of light to pass through the flow cell and on to the first filter and camera. The screen will fade from light to dark and back again during one complete rotation of the second filter.

The VeligerCAM can provide significant image details so that organisms can be differentiated. Organisms with calcium carbonate shells will display a Maltese cross-pattern when viewed under cross-polarized light (see Figure 11.1). Additionally, these organisms possess other characteristics that vary and are better identified with a standard gray scale/light photo. Therefore, by pairing the images obtained by this dual camera instrument, one can discern which organism is present during evaluation with the side-by-side image display. This instrument provides detailed images of veligers, which can expose minute damages caused by treatment that may not be as apparent as using traditional microscopy. With the assistance of the BOR's series of experiments, protocols, and teamwork, the VeligerCAM now provides for a faster and more reliable technique for identification, enumeration, and method development.

Identification

Samples containing dreissenid veligers may also contain corbicula. Corbicula can be similar in size and shape to the dreissenid mussels, and they also display a Maltese cross-pattern when viewed under cross-polarized light. The main visual difference is that the corbicula has a flat hinge without an umbonal

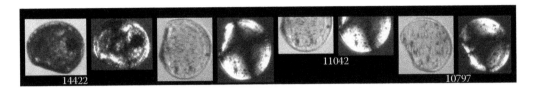

FIGURE 11.2   VeligerCAM images of a corbicula, showing the Maltese cross and the flat hinge.

Eye spot

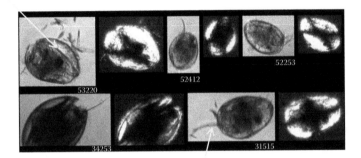

Antenna

FIGURE 11.3   The VeligerCAM images of an ostracod, showing the unique shape, antennae, and eye spot.

bump. Corbicula can be easily distinguished when the VeligerCAM provides a full image in the correct orientation (see Figure 11.2) although partial images may make it difficult to accurately identify the organism.

A third organism that is common in water samples is the ostracod. While the ostracod does display a cross as well, its shell has more of a bean shape than a round shape. Furthermore, when viewing an ostracod under regular light, it is easy to see the eye spot and antennae. See Figure 11.3.

Methods

In order to determine how accurate the VeligerCAM is at detecting veligers, it is important to establish a rate at which veligers are lost during transfer. Many of the protocol optimization tests that have been conducted require veligers to be transferred multiple times between Petri dishes, beakers, disposable pipettes, and test tubes. Dreissenid veligers are small, and it is easy to unknowingly lose them during any one of these transfers. So, the team at the BOR established that the goal was to determine how many veligers were lost during transfer from Petri dish to Petri dish.

Initially, the glassware transfer test was conducted by counting veligers in a glass Petri dish under a cross-polarized light microscope where the Petri dish had a grid. The samples were then transferred into a second clean Petri dish by pouring the liquid and rinsing the dish with a wash bottle containing deionized water. The dish was then rinsed three times, and the sample was recounted under the microscope. The rinse process was repeated for three transfers. A total of three trials were completed using different dreissenid veligers for each trial. Because counting errors do occur, veligers were counted three times after each transfer, and then the counts were averaged. This test was intended to serve as a control and for comparison to VeligerCAM veliger counts.

The results of the earlier study conducted by the BOR can be found in Table 11.1. This table summarizes the veliger counts from the original dish and after each of three transfers for the three trials. In no case did transfer loss seem to be an issue, as counts did not decline. The slight variations recorded were likely due to counting errors.

TABLE 11.1

Average Veliger Recovery after Three Petri Dish Transfers

Trial	Petri Dish	Average Veliger Count	Std. Dev.
1	Original dish	233	2.06
	Transfer 1	236	
	Transfer 2	232	
	Transfer 3	236	
2	Original dish	101	2.38
	Transfer 1	97	
	Transfer 2	98	
	Transfer 3	102	
3	Original dish	91	1.73
	Transfer 1	90	
	Transfer 2	91	
	Transfer 3	94	

Adding a Surfactant

The results of veliger transfer and identification experiments suggest that the veliger recovery rate is not significantly reduced by transfer loss or misidentification. However, significant veliger loss was observed when running spiked samples through the VeligerCAM. In order to reduce this possibility, dish soap was added to the sample to reduce the surface tension of the water. Reducing the surface tension should allow the sample to pass more smoothly through the instrument, leading to better veliger recovery.

Seven samples (20 mL each) with a known number of veligers were analyzed with the VeligerCAM. Approximately three drops of dish soap were added to each sample. All of the veliger recover rates were below 90% (see Table 11.2). The goal was to achieve consistent recovery rates between 95% and 100%. It would appear that adding dish soap did not effectively increase veliger recovery rates.

Altering the Flow Rate

Veligers may also be lost during VeligerCAM processing if they pass through the flow cell too quickly and are not imaged by the camera. The VeligerCAM default rate is 1.125 mL/min, and because veliger recovery was low, slower rates were tested in an attempt to image more veligers. If the flow rate is slowed too much, it is quite possible that multiple images of the same veliger will appear. In the first test, six slower

TABLE 11.2

Percent Recovery of Known Samples from VeligerCAM Tests with the Addition of a Surfactant (Dish Soap)

Flow Rate (mL/min)	Run Time (min: s)	Syringe Size (mL)	Microscope Count	VeligerCAM Count	Percent Recovery (%)
1.125	39:00	12.5	182	137	75.3
1.125	40:25	12.5	202	156	77.2
1.125	33:03	12.5	143	123	86.0
1.125	42:43	12.5	48	38	79.2
1.125	47:20	12.5	77	53	68.8
1.125	46:52	12.5	218	187	85.8
1.125	49:42	12.5	174	152	87.4

TABLE 11.3

Slower Flow Rates Tested, Including the Particle Count and Total Images

Flow Rate (mL/min)	Run Time (min: s)	Particle Count	Fluid Volume Imaged (mL)	Total Images	Frame Rate (Frames per Second)
1.000	10:03	32,503	11.5461	18,113	30
0.875	10:01	35,722	11.5031	18,045	30
0.750	10:03	40,422	11.5407	18,104	30
0.625	10:13	34,181	11.7339	18,407	30
0.500	10:29	41,405	12.0322	18,875	30
0.325	11:09	45,299	12.7952	20,072	30

TABLE 11.4

Percentage of Veligers Recovered at Three Slower Flow Rates, after Two 10 mL Rinses

Flow Rate (mL/min)	Microscope Count	Count after VeligerCAM Run	Count after First 10 mL Rinse	Count after Second 10 mL Rinse	Total Recovery	Percent Recovery (%)
0.625	82	82	24	2	108	132
0.625	28	26	5	2	33	118
0.625	38	45	10	0	55	145
0.750	77	41	20	1	62	81
0.750	46	42	6	4	52	113
0.750	24	20	3	0	23	96
0.875	37	26	5	0	31	84
0.875	56	28	1	0	29	52
0.875	33	15	17	0	32	97

rates were tested to determine which produced double images. The flow rates were slower than the default rate by intervals of 0.125 mL/min, as this was the smallest interval by which the flow rate could be manipulated. Only one run was tested at each flow rate change. In this test, algae were used instead of veligers to determine if double images were occurring. The results of this test can be seen in Table 11.3. Of the six flow rates tested, the only flow rates that had obvious double imaging were 0.500 and 0.325 mL/min. This test helped to narrow down the range of flow rates to be further tested and investigated.

Three of the flow rates that appeared to produce the least amount of double images were further investigated to determine the optimum flow rate for capturing the most veligers. Three replicates were tested for each flow rate. The flow rates tested were 0.625, 0.750, and 0.875 mL/min. For each test, a sample with a known number of veligers was passed through the VeligerCAM, followed by two 10 mL rinses. The veliger images were then counted after the sample run and after each rinse. The results of these tests can be found in Table 11.4. All of the recoveries from the three replicates at 0.625 mL/min were above 100% indicating double imaging. One replicate at 0.750 mL/min was above 100%. The 0.875 mL/min flow rate had good recoveries in two of the three replicates and did not go above 100%; therefore, this flow rate was selected as the optimal flow rate to be used for further testing of the VeligerCAM.

Altering Viscosity

Once the optimal flow rate was determined, there were still some inconsistencies and a small percentage of loss during repetitive runs. Although the flow rate of the water had changed, it is possible that the rate of veliger movement through the VeligerCAM remained the same. Laboratory studies have shown that veligers are negatively buoyant, meaning that veligers will tend to sink to the bottom of a container of water. It was observed that even when the VeligerCAM flow was stopped, the veligers would continue to sink and move through the flow cell. Even with reduced flow, veligers were still moving through the cell too quickly. In order to control how fast the veligers move through the flow cell, the viscosity of the liquid was increased.

Glycerin was selected to increase the viscosity of the samples because it was easy to obtain and is highly soluble in water. The effects of glycerin were first tested at the default flow rate setting of 1.125 mL/min. The percentage of glycerin added to the sample was manipulated until a high recovery was achieved. The goal was to avoid adding too much glycerin as it would increase the volume of the sample, which would increase the duration of the analysis. For each test, a sample with a known number of veligers and glycerin was passed through the VeligerCAM, followed by two 10 mL rinses. The veliger images were counted after the sample run and after each rinse. After testing 20%, 25%, 33%, and 40% glycerin additions, the highest recovery was seen with 40% glycerin (Table 11.5).

Next, the flow rate was altered to determine if recovery would increase to greater than 90% post-glycerin addition. A sample with a known number of veligers and 40% glycerin was analyzed. Counts were conducted after the sample had run and again after each of two 10 mL rinses. Three replicates of each flow rate were tested. The flow rates tested were 1.125, 0.875, and 0.750 mL/min. A flow rate of 0.750 mL/min was too slow and had some double imaging, as all three replicates had recoveries above 100%. The flow rate of 0.875 mL/min still appeared to be the optimal flow rate, even with the addition of 40% glycerin. Recovery rates were 94% and 100% (Table 11.6).

TABLE 11.5

VeligerCAM Recovery Rates with Variable Amounts of Glycerin Added to the Sample

Flow Rate (mL/min)	Microscope Count	Count after VeligerCAM Run	Count after First 10 mL Rinse	Count after Second 10 mL Rinse	Total Recovery	Percent Recovery (%)	Percent Glycerin (%)
1.125	124	83	16	0	99	80	20
1.125	52	38	4	3	45	87	25
1.125	29	24	2	0	26	90	33
1.125	60	53	1	0	54	90	40

TABLE 11.6

Percentage of Veligers Recovered at Three Flow Rates with the Addition of 40% Glycerin

Flow Rate (mL/min)	Microscope Count	Count after VeligerCAM Run	Count after First 10 mL Rinse	Count after Second 10 mL Rinse	Total Recovery	Percent Recovery (%)	Percent Glycerin (%)
1.125	60	53	1	0	54	90	40
1.125	55	44	2	1	47	85	40
1.125	54	44	3	0	47	87	40
0.875	53	27	16	10	53	100	40
0.875	51	26	19	6	51	100	40
0.875	50	21	15	11	47	94	40
0.750	51	57	2	0	59	116	40
0.750	53	33	14	13	60	113	40
0.750	48	45	4	6	55	115	40

TABLE 11.7

Percentage of Veligers Recovered after Reducing the Volume of the Sample to 10 mL from 12.5 mL

Flow Rate mL/min	Microscope Count	Count after VeligerCAM Run	Count after First 10 mL Rinse	Count after Second 10 mL Rinse	Total Recovery	Percent Recovery (%)	Sample Volume mL	Percent Glycerin (%)
0.875	53	27	16	10	53	100	10	40
0.875	51	26	19	6	51	100	10	40
0.875	50	21	15	12	48	96	10	40
0.875	45	40	3	0	42	93	12.5	40
0.875	49	43	2	1	47	96	12.5	40
0.875	50	36	6	1	43	86	12.5	40

Sample Volume

Another way veligers may be lost during analyses is when the syringe is completely full and has to reset and empty its contents to a waste receptacle. When a run resets, images are not being captured, and the rest of the sample in the syringe will continue to empty into the flow cell, resulting in the remaining veligers not being imaged. The volume of the syringe barrel utilized in this application is 12.5 mL. With the aim of reducing the risk of losing veligers during a reset, the veliger recovery rate of a 10 mL sample was compared to a 12.5 mL sample. These tests were run at a 0.875 mL/min flow rate with 40% glycerin. Samples contained a known quantity of veligers and were counted after the run as well as after the two 10 mL rinses. The results of this study can be seen in Table 11.7. Veliger recovery appears to be greater when the volume of the sample is less than the maximum 12.5 mL syringe barrel volume.

Filter Selection

Most of the *real-world* samples that are run through the VeligerCAM are full of algae, plankton, and other debris. These organisms and particles are much larger than the 300 μm flow cell that is used with this application. Samples must be filtered prior to VeligerCAM analysis to prevent clogging of the flow cell. The filter selected for this process must be smaller than the 300 μm flow cell and the detritus, and big enough to allow all of the veligers to pass through.

Initial testing of the VeligerCAM was conducted after running samples through a 300 μm mesh. These samples consistently caused clogging problems in the collar between the tubing and the top of the flow cell. Minor clogs could be cleared by pinching the tubing below the flow cell, allowing the suction pressure to build and then releasing the pinch. Consequently, this created a temporary increase in flow rate and allowed a large number of particles to pass through the flow cell at once. The threat of continual clogging made it necessary for a technician to monitor each sample closely.

Four filters made of four different mesh sizes, and three filter materials were used to determine which was the most appropriate to address the clogging issues. Multiple filter materials were tested because certain filter materials can trap veligers more easily than others. The filters tested included a 240 μm steel filter, a 285 μm Petex filter, a 265 μm Petex filter, and a 280 μm Nitex filter. Five replicates were tested for all four filter types. A sample with a known amount of veligers was passed through the filter and then processed with the VeligerCAM. The number of images recorded by the VeligerCAM was then compared to the initial count; this is summarized in Table 11.8.

All of the filters appeared to work relatively well with recoveries ranging from 90% to 100%. The filter with the greatest average veliger recovery was 280 μm Nitex filter. The average recovery was 98%, and

TABLE 11.8

Percentage of Veligers Recovered after Filtering Samples with Four Filters Made of Variable Materials and Mesh Size

Filter Size (µm) and Material	Run	Start Count	End Count	Percent Recovery (%)	Average (%)
250 Steel	1	21	19	90	96
250 Steel	2	19	18	95	
250 Steel	3	18	18	100	
250 Steel	4	18	17	94	
250 Steel	5	17	17	100	
285 Petex	1	57	56	98	97
285 Petex	2	56	53	95	
285 Petex	3	53	51	96	
285 Petex	4	51	51	100	
285 Petex	5	51	50	98	
265 Petex	1	56	54	96	97
265 Petex	2	54	51	94	
265 Petex	3	51	50	98	
265 Petex	4	50	50	100	
265 Petex	5	50	48	96	
280 Nitex	1	51	47	92	98
280 Nitex	2	47	46	98	
280 Nitex	3	46	46	100	
280 Nitex	4	46	46	100	
280 Nitex	5	46	46	100	

three of the five trials produced 100% recovery. The Nitex filter had the second largest pore size tested, so it is possible that veligers did not become trapped in the Nitex as frequently as in the Petex or steel.

Conclusions

The goal of these tests was to optimize the VeligerCAM process with the intention of recovering the greatest amount of veligers for early detection, monitoring, and research. The results of these tests show that veliger transfer from Petri dish to Petri dish prior to the VeligerCAM processing is not a source of loss. The images provided by the VeligerCAM appear to be of sufficient detail to differentiate dreissenid veligers from other like organisms. Adding a surfactant did not result in greater recovery of veligers; however, increasing the viscosity by adding 40% glycerin did result in a greater recovery. Reducing the flow rate from 1.125 to 0.875 mL/min and reducing the sample volume from 12.5 to 10 mL also resulted in 95%–100% veliger recovery. There are additional challenges faced when processing *real-world* samples due to the possible inclusion of large amounts of debris and organic materials. The results of this study suggest that filtering these samples through a 280 µm Nitex filter should not result in veliger loss. As a result of this research, the VeligerCAM can be used for *real-world* sampling while still providing detailed images of 95%–100% of the veligers in the sample. This instrument and method will significantly aid in the advancement of dreissenid veliger research.

Acknowledgments

The research conducted to write this chapter was performed under Cooperative Research and Development Agreement (CRADA) 10-CR-8-1003 as authorized by 15 USC 3710a. Under the CRADA, the Bureau of Reclamation and Fluid Imaging Technologies worked cooperatively to develop the VeligerCAM

Protocol. Reclamation contributed facilities and research expertise, while Fluid Imaging Technologies contributed equipment, research expertise, and funding to support the effort.

Authors would like to thank Scott O'Meara and Kevin Bloom of the Bureau of Reclamation and Benjamin Spaulding and Harry Nelson from Fluid Imaging Technologies Inc. They would also like to thank Sherri Pucherell of the Bureau of Reclamation for helping with writing and editing this chapter.

Disclaimer

The Bureau of Reclamation co-authorship of this report, and participation in the underlying Cooperative Research and Development Agreement, should not be construed as an endorsement of any product or firm by the Bureau of Reclamation. The Bureau of Reclamation gives no warranties or guarantees, expressed or implied, for the products and results discussed in this report, including merchantability or fitness for a particular purpose.

REFERENCES

Claxton, W.T. and G.L. Mackie. 1998. Seasonal and depth variations in gametogenesis and spawning of *Dreissena polymorpha* and *Dreissena bugensis* in eastern Lake Erie. *Canadian Journal of Zoology* 76: 2010–2019.

Culver, D.A., W.J. Edwards, and L. Babcock-Jackson. 2000. Preventing the introduction of zebra mussels during aquaculture, and fish stocking activities. *Internationale Vereinigung fuer Theoretische und Angewandte Limnologie* 27: 1809–1811.

Edwards, W.J., L. Babcock-Jackson, and D.A. Culver. 2000. Prevention of the spread of zebra mussels during fish hatchery and aquaculture activities. *North American Journal of Aquaculture* 62: 229–236.

Johnson, L.E., A. Ricciardi, and J.T. Carlton. 2001. Overland dispersal of aquatic invasive species: A risk assessment of transient recreational boating. *Ecological Applications* 11:1789–1799.

Klerks, P.L., P.C. Fraleigh, and J.E. Lawniczak. 1997. Effects of the exotic zebra mussel (*Dreissena polymorpha*) on metal cycling in Lake Erie. *Canadian Journal of Fisheries and Aquatic Sciences* 54: 1630–1638.

MacIsaac, H.J. 1996. Potential abiotic and biotic impacts of zebra mussels on the inland waters of North America. *American Zoologist* 36(3): 287–299.

Mackie, G.L., W.N. Gibbons, B.W. Muncaster, and I.M. Gray. 1989. *The Zebra Mussel (Dreissena polymorpha): A Synthesis of European Experiences and a Preview for North America*. Ontario Ministry of Natural Resources, Water Resources Branch, Great Lakes Section, Toronto, Ontario, Canada.

Mills, E.L., G. Rosenberg, A.P. Spidle, M. Ludyanskiy, Y. Pligin, and B. May. 1996. A review of the biology and ecology of the quagga mussel (*Dreissena bugensis*), a second species of freshwater dreissenid introduced to North America. *American Zoologist* 36: 271–286.

Snyder, F.L., M.B. Hilgendorf, and D.W. Garton. 1997. *Zebra Mussels in North America: The Invasion and Its Implications*! Ohio Sea Grant, Ohio State University, Columbus, OH. http://www.sg.ohio-state.edu/f-search.html.

Spaulding, B.W., L. Brown, and H. Nelson. 2009. Field report: Early detection can help eradicate invasive mussels. *Journal of the American Water Works Association (AWWA)* 19–20.

Waller, D.L., S.W. Fisher, and H. Dabrowska. 1996. Prevention of zebra mussel infestation and dispersal during aquaculture operations. *Progressive Fish-Culturist* 58: 77–84.

Section IV

Prevention

12

Challenges for Wildland Firefighters: Preventing the Spread of Aquatic Invasive Species during Fire Operations

Cynthia Tait and Sandee Dingman

CONTENTS

ABSTRACT With the alarming proliferation of aquatic invasive species (AIS) in the West, wildland fire agencies are increasing their capacity to keep fire equipment free of AIS contamination. Firefighting equipment that contacts water not treated to kill AIS may not only spread unwanted organisms across large geographic areas but also, in some cases, be rendered inoperable during emergency situations due to fouling. Interagency wildland fire organizations have accepted responsibility for keeping gear clean, and guidelines for curtailing AIS transmission are beginning to appear in some directives and manuals. Work is underway on national-level, science-based best management practices that will offer consistent, comprehensive, and effective solutions to prevent AIS contamination. Needed also is a national AIS database with regularly updated information on location and extent of infestations. Such a database could be incorporated into the Wildland Fire Decision Support System (WFDSS) and allow fire management organizations to make risk-informed, strategic, and tactical decisions regarding AIS and wildland fires.

Introduction

With the discovery of quagga mussels (*Dreissena rostriformis bugensis*) in the Lower Colorado River, United States, in January 2007, containment of this notorious invasive species became a high priority for many western state and federal agencies. Most of the focus was on recreational boats, widely recognized vectors for spread between hydrologically disconnected water bodies (Rothlisberger et al. 2010). Many effective containment strategies have been developed in the Great Lakes region and eastern waterways to control mussel transport by boaters (e.g., in Minnesota; Minnesota Department of Natural Resources 2015), and consistent educational messaging is available (U.S. Fish and Wildlife Service 2011). However, the distribution of this highly invasive and destructive organism into western waters highlights a vector rarely considered by the AIS community: wildland firefighting operations and equipment.

Between 2000 and 2012, 91 million acres burned in nearly a million wildland fire incidents in the United States (National Interagency Fire Center 2013a), most of which were unplanned ignitions on western public lands. With potential impacts from climate change, such as longer fire seasons and lower fuel moisture conditions, coupled with fuel accumulations from decades of fire suppression in fire-adapted ecosystems, the prognosis is for more and larger fires in the West (Westerling et al. 2006, National Wildfire Coordinating Group 2009). An increase in fire frequency and the number of fire incidents will mobilize ever-greater numbers of personnel and equipment, leading to heightened risk of AIS transmittal by fire operations and equipment. Similarly, larger fires mobilize more type I (national) and type II (national or regional) fire management organizations over large geographic areas. In addition, extreme fire seasons can trigger wide-ranging international responses to wildland fire incidents, for example, Australia and the United States have offset fire seasons and share personnel and equipment (National Interagency Fire Center 2013b), including Erickson Skycranes, which are routinely deployed overseas (Erickson Aviation 2015). Although there have been no documented cases of AIS infestation of a water body caused by fire operations, a rise in fire frequency increases opportunities for fire incident operations and equipment to spread these organisms.

Water is a key fire suppression tool. During a fire incident, most water is delivered aerially and by engines and hoses to either extinguish flames or to wet fuels ahead of the flame front and curtail fire growth. There are no statistics on the amount of water used annually on wildland fires, but if chemical retardant use is an indicator, the amount is substantial. For example, in 2008, nearly 21 million gallons of chemical retardants were delivered in 13,000 drops from helicopters and fixed-wing aircraft (USDA Forest Service 2009).

Possible Mechanisms of Invasion during Fire Operations

A source for aquatic invasive organisms can be the raw water used in firefighting operations, from either a natural (a river or lake) or a human-made water body (a reservoir or stock tank) that has not been treated for municipal use or human consumption. Municipal water distributed via hydrants may also be used in firefighting operations but, because it is treated, is not considered a reservoir for invasive species. Raw water sources may harbor a variety of AIS, including quagga mussels (*D. r. bugensis*), zebra mussels (*D. polymorpha*), New Zealand mud snails (*Potamopyrgus antipodarum*), whirling disease (*Myxobolus cerebralis*), didymo (*Didymosphenia geminata*), hydrilla (*Hydrilla verticillata*), Eurasian watermilfoil (*Myriophyllum spicatum*), and giant salvinia (*Salvinia molesta*), as well as many vertebrate species. In some cases, the occurrence of AIS in a water body is well documented, but for many western waters, such information is incomplete or nonexistent.

Vectors can be the actual pieces of firefighting equipment that contact and/or transport raw water, such as portable pumps (including floatable pumps), portable tanks, helicopter buckets, and internal tanks of fire engines, water tenders, helicopters, and fixed-wing aircraft. Typically, components of the equipment that cannot be drained and dried completely are most likely to harbor invasive organisms and thus serve as vectors. Residual water left in incompletely drained tanks in equipment moved between fire incidents is especially worrisome: quagga mussel larvae are able to survive 5 days in summer and 28 days in autumn in residual water contained within undrained boats (Choi et al. 2013), a time interval that is well within the redeployment period for most firefighting equipment.

Pathways of invasion operate on multiple scales within the context of wildland fire incident response. Within an incident, raw water is routinely moved between watersheds and sometimes between basins (as when a fire straddles the Continental Divide or the Pacific Crest). Typically, large water bodies, such as reservoirs, serve as primary sources to fill various types of firefighting equipment, which then transport and disperse that water to other parts of the fire. Rarely, water loads may be dropped well away from the fire when an aircraft aborts its mission and must jettison its load prior to landing. In many fire incidents, helicopters equipped with snorkels and internal tanks or buckets draft or dip from untreated water sources, then transport to higher elevation areas affected by fire. In this way, infested downstream water could be carried to upstream sites and dispersed across the landscape. While most of the water is applied to fire, vegetation, or soil, some may enter natural waters and unintentionally introduce aquatic invasive organisms.

Sharing firefighting equipment between incidents provides pathways for invasion across large geographic areas, and thousands of pieces of firefighting equipment are managed by dozens of organizations during a fire season. The well-established interagency coordination of firefighting on federal, tribal, and state lands provides an efficient and effective framework to quickly move equipment and personnel where needed. Because of this, a piece of equipment contaminated by AIS in the Lower Colorado River could be deployed on a different fire incident in the Columbia River drainage within days or hours, or fire engines and tankers released from an incident could transport AIS to the home unit in a different state. Dalton and Cottrell (2013) estimated that quagga mussel larvae could survive long enough in contained water to be transported to nearly every region of the United States. While highway and boat launch checkpoints intercept potentially contaminated boats transported overland, firefighting equipment is not typically intercepted through those mechanisms.

Risks posed by AIS to utility infrastructure, ecosystems, and recreation are well documented (see, e.g., Schloesser et al. 2006, Nalepa et al. 2009, Strayer 2010, Moore et al. 2012). What is less recognized is the risk to firefighting equipment itself. Organisms such as dreissenid mussels and New Zealand mudsnails will adhere to interior surfaces of tanks, pumps, and hoses that are not drained and dried completely. If equipment retains raw water containing quagga or zebra mussel larvae, colonizing mussels may clog plumbing and coolant systems, leading to equipment failure (U.S. Army Corps of Engineers 2002). Because fire equipment is primarily used in emergency situations and is in high demand, the potential risk of inoperable equipment due to AIS fouling is a concern to the firefighting community.

Preventing the Spread of Aquatic Invasive Species during Fire Operations

The U.S. Forest Service, Intermountain Region, recognized the risk of AIS transmission during fire incidents and created an internal guidance document entitled "Preventing Spread of Aquatic Invasive Organisms Common to the Intermountain Region, Technical Guidelines for Fire Operations," which was updated in 2013 (USDA Forest Service 2013). In 2009, the interagency Southwest Coordinating Group released similar guidelines for use in the southwestern United States (Southwest Coordinating Group 2009). However, preventing the spread of AIS should not be consigned to individual agencies or regions, and is best addressed from an interagency and national perspective. The National Wildfire Coordinating Group (NWCG) offers leadership to the wildland fire community regarding training, standards, equipment, and other wildland fire functions, and provides a means for agencies to coordinate programs, constructively work together, and avoid duplication of efforts. Land management members of NWCG include the U.S. Forest Service, the Bureau of Indian Affairs, Bureau of Land Management, National Park Service, and U.S. Fish and Wildlife Service, as well as the National Association of State Foresters. NWCG's Equipment Technology Committee formed an Invasive Species Subcommittee, chartered in 2010, to develop recommendations for policies, standards, and procedures to mitigate the risk of transporting and spreading invasive species in all activities related to fire management and incident response. In 2011, the Invasive Species Subcommittee facilitated the insertion of Operational Guidelines for Aquatic Invasive Species into the Interagency Standards for Fire and Fire Aviation Operations (the Redbook) (National Interagency Fire Center 2013c), which serves as the primary policy implementation guidance for federal firefighting agencies in the United States. These guidelines, while general, provide consistent direction to prevent the spread of AIS in firefighting operations and are expected to be updated periodically as new information or methods become available. Furthermore, these requirements are included in the training curriculum for both state and federal fire management personnel.

Procedural and Mechanical Options to Prevent AIS Transmission

The Redbook Operational Guidelines emphasize procedural and mechanical methods to prevent contact with and the spread of AIS. The 2013 Redbook (Chapter 11; National Interagency Fire Center 2013c) states, in part, the following:

> In order to prevent the spread of aquatic invasive species, it is important that fire personnel not only recognize the threat aquatic invasive species pose to ecological integrity, but how our fire operations and resulting actions can influence their spread. Each local land management unit may have specific guidelines related to aquatic invasive species. Therefore, it is recommended that you consult established local jurisdictional guidelines for minimizing the spread of aquatic invasive species and for equipment cleaning guidance specific to those prevalent areas and associated species. To minimize the potential transmission of aquatic invasive species, it is recommended that you:
>
> - Consult with local biologists, Resource Advisors (READ) and fire personnel for known aquatic invasive species locations in the area and avoid them when possible;
> - Avoid entering (driving through) water bodies or saturated areas whenever possible;
> - Avoid transferring water between drainages or between unconnected waters within the same drainage when possible;
> - Use the smallest screen possible that does not negatively impact operations and avoid sucking organic and bottom substrate material into water intakes when drafting from a natural water body;
> - Avoid obtaining water from multiple sources during a single operational period when possible; and
> - Remove all visible plant parts, soil and other materials from external surfaces of gear and equipment after an operational period. If possible, power wash all accessible surfaces with clean, hot water (ideally >140°F) in an area designated by a local READ.

The Redbook urges fire personnel to consult local jurisdictional guidelines and contact local regulatory agencies (usually state departments of environmental quality) for guidance on equipment cleaning to minimize the spread of species specific to the area. However, local guidelines can be contradictory and often do not exist, and fire crews currently must rely on these minimal Redbook best management practices.

Of crucial importance in combating the spread of invasives is locating known AIS infestations in the incident area so as to avoid drafting from contaminated waters. The state of Minnesota, *land of ten thousand lakes*, has been able to inhibit AIS transmission by mapping infestations and supplying fire aviation pilots with GPS coordinates and computer apps pinpointing those uninfested water bodies suitable for drafting (Minnesota DNR 2013). For other parts of the United States, data that track AIS locations are not available in a single, comprehensive (i.e., all aquatic invasive taxa) and nationwide database. Although the U.S. Geological Survey provides AIS spatial data on its Nonindigenous Aquatic Species website (http://nas.er.usgs.gov), it does not track aquatic invasive plants or pathogens. Currently, agencies must pull together species occurrence data on local or regional scales. For example, the Intermountain Region (U.S. Forest Service), which includes national forests in Utah, Nevada, Idaho, and Wyoming, periodically assembles spatial AIS information from state, federal, and academic sources, and compiles it into a downloadable database on a public website (http://www.fs.usda.gov/detail/r4/landmanagement/resourcemanagement/?cid=fsbdev3_016101).

The other procedural and mechanical mitigation measures presented in the Redbook are simple yet likely effective in reducing the risk of spread. Avoidance behaviors (e.g., avoiding infested waters, avoiding transferring water between drainages) are more proactive than cleaning contaminated equipment, assuming situational exigencies allow those behaviors to occur. When equipment does contact untreated water or is contaminated, surfaces must be sanitized. Cleaning external surfaces of gear and equipment is relatively straightforward. Contact with hot water ≥140°F for a few minutes is effective for physically destroying an array of AIS (e.g., Johnson et al. 2003, Kilroy et al. 2007, Comeau et al. 2011) and is relatively safe and inexpensive to provide. In the future, NWCG's Invasive Species Subcommittee plans to explore the feasibility of a hot water option for a variety of firefighting equipment types and situations. High-pressure vehicle cleaning units (e.g., weed washers) with thermal capability or smaller hot water pressure washers could be purchased by fire organizations or vendors and easily deployed during incidents. A primary advantage of hot water sanitation is that containment and disposal of chemical disinfectant are not necessary.

Chemical Disinfectant Options to Prevent AIS Transmission

Although power washing will greatly reduce the likelihood that any target aquatic invasives are present on external surfaces, ensuring that internal tanks of water tenders, engines, and aviation equipment are free from invasive organisms is more problematic. Some internal tanks, such as those in Canadair CL 215 scooper planes, can be opened and are easily accessible. In contrast, aviation and ground-based equipment usually have tanks with baffles that preclude the insertion of hot water nozzles to sterilize the interior space. Thorough drying of internal tank surfaces will effectively kill invasive organisms, but tank equipment frequently arrives to an incident wet and in an unknown, possibly contaminated, state of cleanliness. Until better technology or methods are developed, the best sanitizers for hard-to-access interior spaces are quaternary ammonium disinfectants.

Quaternary ammonium compounds, or *quats*, are common disinfectants with many uses, from killing algae in swimming pools to sanitizing workout equipment at the gym. They are relatively nontoxic, inexpensive, and do not damage fabric or gaskets (McBride 2013), unlike chlorine bleach. There are hundreds of quat compounds, but most research for their effectiveness against AIS has focused on one of the alkyl dimethyl benzylammonium chlorides, abbreviated as ADBAC, and didecyl dimethyl ammonium chloride, or DDAC. ADBAC readily kills whirling disease (Hedrick et al. 2008), and DDAC is effective against chytrid fungus (Johnson et al. 2003). The commercial formulation *Green Solutions High Dilution 256®* (formerly *Sparquat 256®*) contains both ADBAC and DDAC, and so will destroy whirling disease and chytrid fungus. In addition, *Green Solutions®* is effective against New Zealand mudsnails and quagga mussel larvae (Schisler et al. 2008, Britton and Dingman 2011). *Green Solutions* at recommended concentrations will kill many if not all of the waterborne invasives likely to be encountered by firefighting equipment.

Emerging Issues with Chemicals

Disposal

Although household users of quaternary ammonium products can easily discard the used cleaning solution down the drain, disposal of large volumes, as would be generated in wildland fire situations to disinfect equipment, is problematic. While wastewater treatment plants can process small volumes of used quat solutions, large volumes (hundreds of gallons) could overwhelm a wastewater treatment facility, especially ones in small towns (Ron Cook, Spartan Chemical Co., pers. comm.). Quat compounds are toxic to aquatic organisms, so used solution cannot be dumped into streams or lakes, or on areas where it can migrate into any water body. Quat compounds are quickly bound to soil and immobilized, but if soil with bound quat enters water, some of the chemical may be released and become bioactive (Pat Durkin, Syracuse Environmental Research Associates, Inc., pers. comm.). Federal, state, and local regulations for disposal may vary, but in some cases, local regulations allow used solution to be disposed over open land or on roadways where there is no potential for runoff into waterways, storm drains, or sensitive habitats (Idaho Department of Agriculture; Arizona Department of Environmental Quality, Water Quality Division; Utah Department of Environmental Quality).

Corrosion

A recent corrosion study found that, although a 1.8% solution of *Green Solutions High Dilution®* was safe for use on nonmetallic materials (e.g., sealants, gaskets, liners), it did not meet the intergranular corrosion requirements for aluminum, a metal common in aircraft (McBride 2013). Until further review by manufacturers has occurred, *Green Solutions* or other quat compounds should not be used to disinfect fixed-tank helicopters, single-engine fixed-wing air tankers, or multiengine fixed-wing air tankers. Quat compounds remain safe for use on ground-based equipment.

Labeling

Quaternary ammonium compounds appear to be suitable for disinfecting wet internal tanks and inner surfaces of firefighting equipment. Although current EPA labeling does not specify such a use on firefighting equipment *per se*, the label does indicate that *Green Solutions High Dilution 256* is formulated to disinfect hard, nonporous, inanimate, environmental surfaces such as metal, stainless steel, and plastic (rubber), and includes vehicles and transportation equipment in its purview. An environmental risk assessment for *Green Solutions High Dilution 256*, needed to ensure its safe and effective use on federal lands, was initiated in 2013 and is in review by the U.S. Forest Service.

New Directions

Cleaning Internal Tanks

The Invasive Species Subcommittee is currently exploring alternative nonchemical options for safe, efficient, and effective decontamination of fire equipment. Of particular focus are inaccessible internal tanks, especially those with baffles as are found in Erickson Skycrane helicopters and ground-based water tenders, which are difficult to sanitize without chemicals. One strategy will be to assemble a group of invasives species biologists, fire personnel, and equipment contractors/manufacturers to discuss options for cleaning these tanks. Contractors may have ideas that those not as familiar with their equipment have not considered. For example, a Canadair CL-215 scooper plane vendor deploys 25 gal hot water pressure washers along with his planes to decontaminate tanks, which are accessible to spray (Aeroflite Inc, pers. comm). In addition, alternative technologies might be applicable, such as steam injection technology used to sterilize brewer's vats and other large tanks used for food. These steam systems reach 300°F and kill all organisms, and may be an option for quickly sanitizing large aviation and ground-based tanks.

National Wildland Fire Guidebook to Prevent Transmission of Invasive Species

Although some useful regional protocols and guidelines for fire operations exist, there is need for a standardized, interagency, national-level set of operating procedures for the prevention of AIS transmission. In 2013, the NWCG began work on a national AIS prevention guidebook, which will compile up-to-date, scientific decontamination protocols and techniques, with an emphasis on those species that can potentially be transported through fire operations.

AIS Equipment Cleaning Stipulations for Firefighting Equipment Contracts

While at times fire equipment originates on the jurisdictional unit where the fire occurs, frequently, equipment is brought in from off-site by vendors or contractors. Various types of fire aviation contracts exist, including *exclusive use*, which commit aircraft for a season, and *call when needed*, which are for special short-term situations. In addition, small equipment, such as type 3 helicopters, is usually provided through local contracts at the regional or forest/district level. Larger equipment, such as Erickson Skycranes and air tankers, are national contracts, with stipulation language that often differs between agencies.

Currently, few contracts require vendors to decontaminate their equipment or to certify that decontamination has occurred. The complexity and diversity of contract types and verbiage across agencies warrant an interagency approach to the development and insertion of appropriate requirements to address AIS decontamination and documenting that protocols have been followed.

Conclusion and Need for Future Work

With the alarming proliferation of AIS in the West, wildland fire agencies are increasing their capacity to keep fire equipment free of contamination. Interagency wildland fire organizations have accepted responsibility for keeping gear clean, and limited technical guidelines for fire operations on methods to curtail AIS transmission are appearing in Redbook and in direction provided by agencies at local scales. Work is under way on national-level, science-based best management practices that will offer consistent, comprehensive, effective solutions to AIS contamination. In addition, manufacturers and providers of firefighting equipment must be included in the dialogue to brainstorm safe, innovative, and practical technical solutions to tank decontamination. New methodologies, such as emphasis on hot water sterilization, would eliminate the risks of retaining AIS in residual tank water. Needed also is a national AIS database with regularly updated information on location and extent of infestations. Such a database could be incorporated into the WFDSS, a web-based system used to analyze, integrate, and manage data for wildland fire incidents and allow fire management organizations to make risk-informed, strategic, and tactical decisions regarding AIS and wildland fires.

In addition, further research is needed to better understand the biology of various AIS that may be encountered by firefighting equipment, including lethal drying times, thermal tolerances, and contact time needed for hot water sanitation.

In the interim, fire incident resource advisers should be aware of the potential for the spread of aquatic invasive organisms during fire operations and, where possible, incorporate measures to either avoid the potential for spread or sanitize equipment exposed to raw water prior to demobilization of crews after the incident.

REFERENCES

Britton, D.K. and Dingman, S. Use of quaternary ammonium to control the spread of aquatic invasive species by wildland fire equipment. *Aquatic Invasions* 6(2) (2011): 169–173.

Choi, W.J., Gerstenberger, S., McMahon, R., and Wong, W.H. Estimating survival rates of quagga mussel (*Dreissena rostriformis bugensis*) veliger larvae under summer and autumn temperature regimes in residual water of trailered watercraft at Lake Mead, USA. *Management of Biological Invasions* 4(1) (2013): 61–69.

Comeau, S., Rainville, S., Baldwin, W. et al. Susceptibility of quagga mussels (*Dreissena rostriformis bugensis*) to hot-water sprays as a means of watercraft decontamination. *Biofouling* 27(3) (2011): 267–274.

Dalton, L. and Cottrell, S. Quagga and zebra mussel risk via veliger transfer by overland hauled boats. *Management of Biological Invasions* 4(2) (2013): 129–133.

Erickson Aviation, Portland, OR. http://ericksonaviation.com/industries/firefighting (2015).

Hedrick, R., McDowell, T., Mukkatira, K., MacConnell, E., and Petri, B. Effects of freezing, drying, UV, chlorine, and quaternary ammonium treatments on the infectivity of myxospores of *Myxobolus cerebralis* [whirling disease]. *Journal of Aquatic Animal Health* 20 (2008): 116–125.

Johnson, M.L., Berger, L., Philips, L., and Speare, R. Fungicidal effects of chemical disinfectants, UV light, desiccation and heat on the amphibian chytrid *Batrachochytrium dendrobatidis*. *Diseases of Aquatic Organisms* 57 (2003): 255–260.

Kilroy, C., Lagerstedt, A., Davey, A., and Robinson, K. Studies on the survivability of the invasive diatom *Didymosphenia geminata* under a range of environmental and chemical conditions. NIWA Client Report: CHC2006-116, Biosecurity New Zealand, Fall Church, NZ (2007).

McBride, G. Effects of the quaternary ammonium disinfectant *Green Solutions High Dilution*© (1.8%) on metals and non-metallic materials used in fire operations. U.S. Forest Service Technology & Development Project, Missoula Technology and Development Center, Missoula, MT (2013).

Minnesota Department of Natural Resources. Infested waters and waters of special concern—Aircraft operating plan. Division of Forestry, St. Paul, MN (2013).

Minnesota Department of Natural Resources, Protect your waters. http://www.dnr.state.mn.us/invasives/index_aquatic.html (2015).

Moore, J., Herbst, D., Heady, W., and Carlson, S. Stream community and ecosystem responses to the boom and bust of an invading snail. *Biological Invasions* 14 (2012): 2435–2446.

Nalepa, T.F., Fanslow, D.L., and Lang, G.A. Transformation of the offshore benthic community in lake Michigan: Recent shift from the native amphipod *Diporeia* spp. to the invasive mussel *Dreissena rostriformis bugensis*. *Freshwater Biology* 54 (2009): 466–479.

National Interagency Fire Center. Fire information—Wildland fire statistics (2013a). http://www.nifc.gov/fireInfo/fireInfo_statistics.html (Accessed June 2013).

National Interagency Fire Center. International support in wildland fire suppression (2013b). http://www.nifc.gov/fireInfo/fireInfo_international.html (Accessed June 2013).

National Interagency Fire Center. Interagency standards for fire and fire aviation operations. NFES 2724. Standards for Fire and Fire Aviation Operations Task Group, Boise, ID (2013c).

National Wildfire Coordinating Group. Quadrennial fire review. National Interagency Fire Center, Boise, ID (2009).

Rothlisberger, J., Chadderton, W.L., McNulty, J., and Lodge, D. Aquatic invasive species transport via trailered boats: What is being moved, who is moving it, and what can be done. *Fisheries* 35(3) (2010): 121–132.

Schisler, G., Vieira, N., and Walker, P. Application of household disinfectants to control New Zealand mudsnails. *North American Journal of Fisheries Management* 28 (2008): 1172–1176.

Schloesser, D.W., Metcalfe-Smith, J.L., Kovalak, W.P., Longton, G.D., and Smithee, R.D. Extirpation of freshwater mussels (Bivalvia: Unionidae) following the invasion of dreissenid mussels in an interconnecting river of the Laurentian great lakes. *American Midland Naturalist* 155 (2006): 307–320.

Southwest Coordinating Group. Preventing spread of aquatic invasive species in the Southwest—Guidance for fire operations. Operations Committee, Albuquerque, NM (2009).

Strayer, D. Alien species in fresh waters: Ecological effects, interactions with other stressors, and prospects for the future. *Freshwater Biology* 55(Suppl. 1) (2010): 152–174.

U.S. Army Corps of Engineers. Zebra mussel information system. U.S. Army Engineer Research and Development Center, Vicksburg, MS (2002). http://el.erdc.usace.army.mil/ansrp/ANSIS/ansishelp.htm (Accessed June 2013).

U.S. Fish and Wildlife Service. 100th meridian initiative. http://100thmeridian.org (2011).

USDA Forest Service. Retardant use at forest service, Bureau of Land Management, and State Bases—2008 fire season. Fire and Aviation Management, Boise, ID (2009). http://www.fs.fed.us/fire/retardant/index.html (Accessed June 2013).

USDA Forest Service. Preventing spread of aquatic invasive organisms common to the intermountain region: Technical guidelines for fire operations. Intermountain Region, Ogden, UT (2013). http://www.fs.usda.gov/detail/r4/landmanagement/resourcemanagement/?cid=fsbdev3_016101 (Accessed June 2013).

Westerling, A., Hidalgo, H., Cayan, D., and Swetnam, T. Warming and earlier spring increase western U.S. forest wildfire activity. *Science* 313 (2006): 940–943.

13

Boat Decontamination with Hot Water Spray: Field Validation

Sean Comeau, Richard S. Ianniello, Wai Hing Wong, and Shawn L. Gerstenberger

CONTENTS

ABSTRACT The efficiency of hot water spray to decontaminate boats exposed to quagga mussels was evaluated in the field. The water temperature and exposure time needed to result in complete mortality were determined. The time needed for three types of boat areas (easy access, hard access, and specialized areas [i.e., bait and live wells]) to reach the lethal temperature for the required exposure time was determined. Field validation tests were conducted in the summer and winter for the easy access and hard access areas. Summer tests of both areas resulted in 100% mortality, while seven out of eight winter tests resulted in 100% mortality. For bait and live wells, the time needed to reach lethal temperature in winter was much longer than in summer time.

Introduction

The establishment and subsequent invasion of certain nonindigenous species has proven to have profound negative economic, environmental, and even human health impacts (Keller et al. 2007). In the United States alone, the cost and damages associated with aquatic invasive species (AIS) are estimated to be over $7 billion annually (Pimentel et al. 2005). Two AIS of particular importance are the zebra mussel (*Dreissena polymorpha*) and the quagga mussel (*Dreissena rostriformis bugensis*). They are commonly referred to as dreissenid mussels and are two of the most devastating AIS to invade North American freshwater systems (Western Regional Panel on Aquatic Nuisance Species 2010). Many authorities believe that the ongoing spread of these mussels to uncontaminated bodies of freshwater in North America can be ascribed to involuntary transportation of AIS from a contaminated body of water by watercraft (Bossenbroek et al. 2001; Johnson et al. 2001; Leung et al. 2006). While it is possible that some of the spread of the dreissenid mussels could be intentional, most cases of AIS translocation are most likely unintentional with the invasive organisms unknowingly present somewhere in or on the trailered vessel during overland transport (Johnson et al. 2001; Puth and Post 2005). There are many possible transport locations for AIS present on watercraft. These include undrained bait buckets, live wells, and bilge water, all of which provide favorable conditions for possible extended viability. They may also be present to some extent on the hull or entrained on boat exteriors, such as entangled on propellers and trailers (Rothelisberger et al. 2010) or attached to other entangled organisms (Johnson et al. 2001). Any watercraft, such as trailered vessel, kayaks, and canoes, that makes contact with an AIS-contaminated body of water, should be treated as a potential vector for AIS.

In order to prevent further infestations, new research and innovative approaches must be used to generate and revise uniform minimum protocols and standards for watercraft decontamination programs. Protocols regarding safe and inexpensive procedures, such as hot water sprays, which result in the 100% mortality of these dreissenid mussels, need to be established. This would increase the receptivity of the boating public to the established protocols and more likely to follow them.

Current Boat Contamination Protocols

The spread of dreissenid mussels to several previously uncontaminated inland bodies of water in the western United States (Benson 2011) has caused many federal, state, regional, and local agencies to initiate watercraft interception programs to prevent further infestations from occurring by implementing watercraft decontamination protocols (Zook and Phillips 2009). The objective of decontamination is to kill and remove all mussels (any stage of development) present on or in the watercraft. Many of these agencies have protocols that commonly decontaminate watercraft with a pressurized hot water spray exceeding 140°F/60°C (Zook and Phillips 2009; 2014 Chapter 14 in this book). This temperature is based on acute (short-term) upper thermal limit data generated for continuously immersed mussels (Morse 2009). The first data set on the use of hot water spray for mitigation of emersed zebra mussel fouling, which is closer to the field situation where sprays are applied to watercraft, was generated by Morse (2009). Morse found that the survivorship of mussels was affected by two major factors: spray water temperature and exposure duration. Water sprayed at greater than or equal to 140°F/60°C for 10 s or 176°F/80°C at greater than or equal to 5 s was 100% lethal to zebra mussels, which indicates that current decontamination recommendation of spray temperature of greater than or equal to 140°F/60°C may not result in 100% mortality of the mussels if the exposure duration is less than 10 s (Morse 2009).

The data from Morse (2009) can be potentially applied to watercraft areas where the spray directly contacts the fouled areas (Category I areas in Table 13.1). Concurrently, there are also areas on watercraft that hot water sprays cannot directly reach. These decontamination areas can be divided into three categories: (1) areas easy to access, (2) areas difficult to access, and (3) special areas (Table 13.1). These three categories of areas should each be treated differently in order to achieve 100% dreissenid mussel mortality for complete watercraft and equipment decontamination. It is important to evaluate the susceptibility of quagga mussels to hot water sprays to determine if they are more or less susceptible than zebra mussels. This information will be helpful in establishing standards for watercraft interception programs in areas susceptible to quagga mussel colonization (Zook and Phillips 2009).

Protocols for boat decontamination using hot water spray are established mainly for dreissenid mussels in easily accessible areas of watercraft (i.e., surface areas) (Morse 2009; Comeau et al. 2011). At the same time, standards for inaccessible areas and areas with special temperature requirements of watercraft can also be created with proper research and evaluation. Watercraft decontamination areas can be divided into three categories: (I) areas easy to access (e.g., the hull), (II) areas difficult to access (e.g., gimbal areas), and (III) special heat-sensitive areas (e.g., ballast tanks/bladders) (Comeau et al. 2011). These three categories of areas should be treated differently to achieve 100% mussel mortality for legitimate watercraft and equipment decontamination. For category II areas, tests need

TABLE 13.1

Accessibility Categories for Various Decontamination Areas

Category	Characteristics	Areas
I	Easy access surface areas	Hull, transducer, through hull fittings, trim tabs, zincs, centerboard box and keel (sailboats), foot-wells, lower unit, cavitation plate, cooling system intakes (external), prop, prop shaft, bolt heads, engine housing, jet intake, paddles and oars, storage areas, splash wells under floorboards, bilge areas, drain plug, anchor, anchor and mooring lines, PFDs, swim platform, inflatables, downriggers and planing boards, ice chests, fishing gear, bait buckets, stringers, trailer rollers and bunks, light brackets, cross-members, license plate bracket, fenders, spring hangers
II	Hard access areas	Gimbal areas, engine, generator, and AC cooling systems (internal)
III	Special areas that require water temperature ≤ 130°F for decontamination	Ballast tanks/bladders, wash-down systems, bait and live wells, internal water systems

to be conducted to determine how long hot water must be applied to these locations to ensure that they reach the determined lethal temperatures. It will likely take longer time because heat would be lost to conduction across metal and other materials, and will probably vary depending on ambient outside air temperatures. For category III areas, the determined 100% mortality rates for temperatures may be used to prevent heat-associated damage from occurring to these sensitive areas. These field data will also assist policy makers in developing minimal thresholds for associated decontamination and inspection parameters.

Hot Water Spray Decontamination of Quagga Mussels

The use of hot water spray as a method of watercraft decontamination for dreissenid mussels is widely accepted by many federal agencies. These agencies most commonly decontaminate watercraft with pressurized hot water spray temperatures exceeding 60°C. This temperature is based on acute (short-term) upper thermal limit data generated for spray for mitigation of immersed mussels. The first study regarding the use of hot water spray for mitigation of emersed zebra mussel fouling, which is closely related to a field situation where sprays are applied to watercraft, was by Morse (2009), whose results were previously highlighted. Morse's (2009) findings are particularly useful because it was the first study to test thermal spray treatments on emersed mussels and, as such, provides a solid starting point for determining effective field application for watercraft decontamination. There are, however, several important aspects that needed to be addressed regarding species-specific application. This is essential because some inland bodies of water may be infested with only zebra mussels, quagga mussels, or both. In the western United States, quagga mussels are of particular importance, as they are currently the most widespread dreissenid species, whereas only one water body in California is infested by zebra mussels (Benson 2011). Previous studies have shown that there are differences between these two dreissenid species (Pathy and Makie 1993; Mills et al. 1996; Ricciardiand Rasmussen 1998; Peyer et al. 2009). It was important to determine if the quagga mussel is more or less susceptible than the zebra mussel to hot water spray. Studies have also shown that the upper thermal limit of the quagga mussel is lower than that of the zebra mussel (Mills et al. 1996). Zebra mussels survive indefinitely at 30°C, but quagga mussels show rapid mortality at 30°C (Spidle et al. 1995; McMahon 1996). Quagga mussels are also reported to have thinner shells (Zhulidov et al. 2006), less tightly sealing shell valves (Zhulidov et al. 2006), and lower byssal thread synthesis rates in higher flows (Peyer et al. 2009). Therefore, quagga mussels may be more susceptible to death by hot water sprays at a lower temperature than zebra mussels, and the application of hot water spray to these two dreissenid species may be different.

To be effective and efficient in mitigating biofouling by invasive quagga mussels in the western United States, hot water spray thresholds are needed to be evaluated specifically for quagga mussels. In order to accurately determine the temperatures and exposure times necessary to attain 100% mortality of specimens of quagga mussels following exposure to a hot water spray, the present study investigated the lethal effect of hot water sprays on emersed specimens of quagga mussels at water temperatures ranging from 20°C to 80°C and exposure durations of 1, 2, 5, 10, 20, 40, 80, and 160 s. The field data were then compared to existing data regarding zebra mussels to determine if there was any difference in susceptibility regarding the two dreissenid species. The data were also used in an evaluation of the necessary time needed to reach and sustain the lethal temperatures in inaccessible areas (Category II) and heat-sensitive areas (Category III), respectively (Comeau et al. 2011).

Field Tests on Emersed Quagga Mussels

Specimens of adult *D. rostriformis bugensis* (≥12 mm in length) were collected from the hull of an encrusted National Park Service boat that was stationed in Lake Mead, Nevada–Arizona, USA, in 2009. The individuals were then divided among 60 mesh spat bags (~75 in each) and acclimated to the lake water in a boat slip within the Las Vegas Bay Marina (N 36°01.764, W 114°46.400) for 2 weeks prior to experimentation.

After acclimation, the adult mussels were randomly divided into 60 subsamples (n = 50, Table 13.2) and placed into 60 identical pre-labeled 3.0 mm spat bags (Aquatic Eco-Systems Inc., Apopka, FL). Each bag was then suspended over one of two identical open Polyscience Programmable heated circulator wash baths with a 28 liter capacity during the thermal spray treatment (VWR International Inc.). The purpose of using two water baths was to increase the efficiency and speed at which the tests could be conducted by allowing limited water temperature variation. Each mesh spat bag containing a test subsample was held horizontally 20 cm above the heated water bath to prevent any difference in ambient air temperature, which may have resulted from the heated water in the open water baths. Treatment spray was then applied to the samples at a flow rate of approximately 900 mL/min through a fan-shaped nozzle. The distance above each sample at which the spray was applied was modified each time in order to maintain the constant test temperature used for each specific subset. This was done because the environmental field conditions, that is, wind, rain, and ambient air temperature, would affect the contact water temperature if there was a set distance. The specific distances prior to each spray were determined using a ruler and a fast-reacting remote water temperature probe (Pace Scientific Model XR440 Pocket Logger with four temperature probes). The distance between the spray nozzle and the contact point of the water at the necessary test temperature was then calculated. The Pace Scientific Model XR440 Pocket Logger was calibrated prior to use, and an NIST traceable certificate of validation was included from the manufacturer. Temperature readings obtained from the temperature probes were also verified by the use of a

TABLE 13.2

Amount of Adult Quagga Mussels Tested per Treatment Group (n = 50 per Group)

Temperature		1 s	2 s	5 s	10 s	20 s	40 s	80 s	160 s
°F	°C								
68	20	50	50	50	50	50	50	50	50
104	40	50	50	50	50	50	50	50	50
122	50	50	50	50	50	50	50	50	50
130	54	50	50	50	50	50	50	50	50
140	60	50	50	50	50	50	50	50	50
158	70	50	50	50	50	50	50	50	50
176	80	50	50	50	50	50	50	50	50

Raytek MT4 noncontact mini infrared thermometer. The thermal spray was immediately applied to the specific subset at the specifically calculated distance. Each subset of mussels was positioned within the spat bag to form a horizontal line not exceeding 5 cm in width in order to allow the hot water spray to be equally distributed over all of the mussels. The polyethylene mesh of the spat bags allowed the water spray to pass over them without additional pooling or heat transfer beyond that would normally occur from direct exposure to the spray (Morse 2009). Each sample of mussels was separately exposed to thermal spray treatments at 20°C, 40°C, 50°C, 54°C, 60°C, 70°C, and 80°C and exposure durations of 1, 2, 5, 10, 20, 40, 80, and 160 s. Therefore, 56 combinations on temperature by exposure duration were treated (Table 13.2). Four bags that were not treated with hot water spray were used as controls.

Following treatment, each spat bag containing the treatment specimens was then attached to one of the seven 1 cm braided nylon lines (one for each temperature set) spanning the boat slip for over 10 days. These lines were attached to a grid composed of ABS pipe, which was positioned on either side of the slip to allow easy access to the samples. Each line holding the spat bags was approximately 1.5 m out of the water, and the mussels within the bags were kept at a depth of approximately 2 m.

Results from Field Tests on Emersed Mussels

After analysis of the data, it was found that there was a trend that indicated that the higher temperatures induced greater mortality following the same exposure duration (Figure 13.1, Table 13.3). Spray exposures of 1 or 2 s were not found to induce 100% mortality at any of the test temperatures (Table 13.3). However, a 5 s spray exposure did result in 100% mortality (≥ 60°C). The other temperature and time combinations that resulted in 100% mortality were 54°C for 10 s, 50°C for 20 s, and 40°C for 40 s. Estimated LT_{50} values for 1, 2, and 5 s indicate that the temperature to kill 50% of the mussels was between 47.2°C and 47.9°C (Table 13.4), while the estimated LT_{99} with these exposure durations varied significantly from >80°C at 1 s and 2 s to 58.8°C at 5 s (Table 13.4).

The continuously immersed control samples (11.86°C ± 1.60) and the samples exposed to the 20°C spray treatments exhibited high survival rates. The combined four groups of controls exhibited 97% survival (ranging from 94% to 100% (Figure 13.1a)), and the eight 20°C spray treatment subsamples displayed a mean 98% survival rate (ranging from 94% to 100%) with no apparent correlation to duration time (Figure 13.1b). Survival was also high for 40°C at spray exposures of 1 s (98% survival), 2 s (98% survival), 5 s (92% survival), 10 s (88% survival), and for 50°C at 1 s (90% survival).

Evaluation of Category II Watercraft Areas

For areas of the watercraft that are not directly exposed to hot water sprays, the time necessary to preheat these locations to the lethal thermal temperature was evaluated because heat loss can occur during the water flow to these areas. The gimbal unit was tested on an uncontaminated boat, and the contact temperature (internal temperature) was monitored until it reached the lethal temperature. The temperature of the water exiting the gimbal unit was monitored by the use of a fast-reacting remote water temperature probe (Pace Scientific Model XR440 Pocket Logger with four temperature probes). Temperature readings from the contact water at the temperature probes were verified by the use of a Raytek MT4 noncontact mini infrared thermometer. The data from this test can be applied to Category II decontamination areas (Table 13.1). Since weather conditions, especially ambient temperature, could be a confounding factor affecting the surface temperature of these areas, the experiment was conducted twice, once in winter and again in the summer. The summer and winter evaluation experiments were conducted on September 1, 2010, and January 21, 2011, respectively. The longest durations determined from the evaluation of the Category II areas depending on the season were as follows (minimum of three replications): 43 s for summer and 2 min and 7 s for winter. There was a separate validation for both categories for summer and winter.

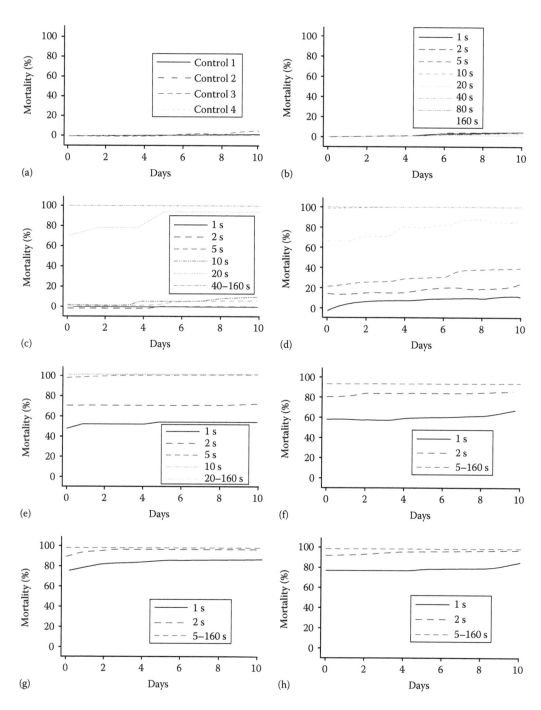

FIGURE 13.1 Mortality (%) of quagga mussels in Lake Mead after hot water spray treatment. (a) Control (11.86°C), (b) 20°C, (c) 40°C, (d) 50°C, (e) 54°C, (f) 60°C, (g) 70°C, (h) 80°C. Note that (c) and (d) share the same symbol and line styles. (Modified from Comeau, S. et al., *Biofouling*, 27(3), 267, 2011.)

TABLE 13.3

Quagga Mussel Mortality (%) under Different Treatments at Day 10

Temperature (°C)	1 s (%)	2 s (%)	5 s (%)	10 s (%)	20 s (%)	40 s (%)	80 s (%)	160 s (%)
20	4	4	6	0	0	2	2	0
40	2	2	8	12	94	100	100	100
50	10	22	36	82	100	100	100	100
54	54	72	98	100	100	100	100	100
60	72	92	100	100	100	100	100	100
70	88	98	100	100	100	100	100	100
80	86	98	100	100	100	100	100	100

Note: The mortality of control (n = 4) was 3%.

TABLE 13.4

Estimated LT_{50} and LT_{99} Values (in Bold) and Their 95% Confidence Limit for Hot Water Spray Treatments on Quagga Mussels at 1, 2, and 5 s Application Durations (n = 350 for Each Duration)

Duration (s)	LT_{50} (°C)	LT_{99} (°C)	SM_{100} (°C)[a]
1	44.1 < **47.9** < 52.5	>80	>80
2	44.0 < **47.8** < 52.3	>80	>80
5	43.5 < **47.2** < 51.7	54.1 < **58.8** < 64.4	60

[a] The SM_{100} is the temperature observed in the experiment that induced 100% mortality.

Evaluation of Category III Watercraft Areas

In some areas of watercrafts, the temperature cannot exceed 130°F, such as ballast tanks and bladders (Zook and Phillips 2009; Comeau et al. 2011). Therefore, the necessary time it takes to reach and maintain a lethal water temperature needed to be evaluated for these areas. The contact temperature (internal temperature) was monitored on the live wells and bait wells of an uncontaminated boat until it reached the lethal temperature. The temperature of the water exiting these areas was monitored by the use of a fast-reacting remote water temperature probe (Pace Scientific Model XR440 Pocket Logger with four temperature probes). Temperature readings from the contact water at the temperature probes were verified by the use of a Raytek MT4 noncontact mini infrared thermometer. Since weather conditions, especially ambient temperature, could be a confounding factor, this experiment was conducted twice: once in winter and again in the summer. The time data from this test would be applied to Category III (Table 13.1) decontamination areas.

For areas on watercrafts that cannot withstand temperatures above 130°F/54°C, flush tests with 54°C water were ran on the live and bait wells of a Cobia 296 boat. Just as in the evaluation of Category II areas, there were tests conducted during the winter and summer, which showed a difference between the necessary application times in the differing ambient conditions (Tables 13.5 through 13.8).

TABLE 13.5

Evaluation of Lethal Temperature Recommendations on the Bait Wells during Summer Weather Conditions

Hot Water Temperature (°F)	Hot Water Temperature (°C)	Air Temperature (°F)	Air Temperature (°C)	Target Temperature (°F)	Target Temperature (°C)	Attempt	Time to Reach Target Temperature (min)
130	54	98	36.7	130	54	1	00:33.1
130	54	98	36.7	130	54	2	00:34.4
130	54	98	36.7	130	54	3	00:33.9

TABLE 13.6

Evaluation of Lethal Temperature Recommendations on the Bait Wells during Winter Weather Conditions

Hot Water Temperature (°F)	Hot Water Temperature (°C)	Air Temperature (°F)	Air Temperature (°C)	Target Temperature (°F)	Target Temperature (°C)	Attempt	Time to Reach Target Temperature (min)
130	54	38	3.3	130	54	1	01:06.9
130	54	38	3.3	130	54	2	01:05.4
130	54	39	3.9	130	54	3	01:03.1

TABLE 13.7

Evaluation of Lethal Temperature Recommendations on the Live Wells during Summer Weather Conditions

Hot Water Temperature (°F)	Hot Water Temperature (°C)	Air Temperature (°F)	Air Temperature (°C)	Target Temperature (°F)	Target Temperature (°C)	Attempt	Time to Reach Target Temperature (min)
130	54	98	36.7	130	54	1	01:10.1
130	54	98	36.7	130	54	2	00:58.2
130	54	98	36.7	130	54	3	01:06.9

TABLE 13.8

Evaluation of Lethal Temperature Recommendations on the Live Wells during Winter Weather Conditions

Hot Water Temperature (°F)	Hot Water Temperature (°C)	Air Temperature (°F)	Air Temperature (°C)	Target Temperature (°F)	Target Temperature (°C)	Attempt	Time to Reach Target Temperature (min)
130	54	38	3.3	130	54	1	01:51.4
130	54	38	3.3	130	54	2	01:41.9
130	54	38	3.3	130	54	3	01:20.5

Boat Decontamination Validation Results

In order to accurately determine if the results regarding the susceptibility of quagga mussels to hot water spray from the previous experiments was applicable as a means of watercraft decontamination, it was necessary to conduct actual field experiments on watercrafts that were encrusted with quagga mussels. Category I areas were sprayed with hot water at a temperature of 60°C for a duration of 5 s because it was the lowest temperature capable of 100% mortality at 5 s, and higher temperatures could be seen a threat to human health (Morse 2009). Category II areas were also sprayed with hot water at a temperature of 60°C, but used the longest duration determined from the evaluation of the areas depending on the season: 43 s for summer and 2 min and 7 s for winter. There was a separate validation for both categories for summer and winter.

Summer Validation

The summer validation experiment took place on September 28, 2010, with an ambient air temperature averaging 95°F/35°C. There were seven replicates for the Category I assessment and three controls. These were located on various freely accessible areas on the boat and were sprayed with 60°C water for a duration of 5 s. For each of the replicates, 100% mortality was achieved immediately after testing (Table 13.9).

TABLE 13.9

Number of Mussels, Percent Mortality, and Average Shell Length of Experimental Groups and Controls for Category I Areas (Summer)

Group	Number of Mussels Present	Number of Mussels Dead	Mortality (%)	Average Shell Length (mm)
1	121	121	100	7.28 ± 0.99
2	163	163	100	8.07 ± 1.47
3	271	271	100	8.04 ± 1.48
4	30	30	100	6.77 ± 1.13
5	39	39	100	4.99 ± 1.50
6	77	77	100	5.62 ± 1.80
7	35	39	100	5.25 ± 1.37
Control 1	126	4	3	8.60 ± 1.84
Control 2	111	19	17	8.15 ± 1.97
Control 3	146	8	6	8.64 ± 1.85

TABLE 13.10

Number of Mussels, Percent Mortality, and Average Shell Length of Experimental Groups and Controls for Category II Areas (Summer)

Group	Number of Mussels Present	Number of Mussels Dead	Mortality (%)	Average Shell Length (mm)
1	109	109	100	7.12 ± 1.18
2	94	94	100	7.92 ± 2.31
Control	57	18	32	10.79 ± 2.83

There were two replicates and one control for the Category II assessment, which evaluated the encrusted gimbal unit of the watercraft. The gimbal unit was flushed with 60°C water for a duration of 48 s (including 5 s of duration to ensure the predetermined lethal duration was met). For each of the replicates, 100% mortality was achieved immediately after testing (Table 13.10). There was a significant difference in the percent mortality of the experimental groups and the controls for each of the tested categories.

Winter Validation

The winter validation experiment took place on January 27, 2011, with an ambient air temperature averaging 50°F/10°C. There were six replicates for the Category I assessment and three controls. These were located on various freely accessible areas on the boat and were sprayed with 60°C water for a duration of 5 s. For each of the replicates, mortality was assessed after 10 days of immersion in Lake Mead after treatment, and 100% mortality was achieved for each replicate (Table 13.11). There were two replicates

TABLE 13.11

Number of Mussels, Percent Mortality, and Average Shell Length of Experimental Groups and Controls for Category I Areas (Winter)

Group	Number of Mussels Present	Number of Mussels Dead	Mortality (%)	Average Shell Length (mm)
1	55	55	100	16.04 ± 4.02
2	43	43	100	18.73 ± 3.85
3	48	48	100	14.78 ± 3.55
4	34	34	100	15.24 ± 4.93
5	64	64	100	13.57 ± 3.25
6	37	37	100	17.03 ± 4.58
Control 1	134	2	2	12.59 ± 4.54
Control 2	107	0	0	12.35 ± 4.78
Control 3	83	0	0	12.35 ± 3.31

TABLE 13.12

Number of Mussels, Percent Mortality, and Average Shell Length of Experimental Groups and Controls for Category II Areas (Winter)

Group	Number of Mussels Present	Number of Mussels Dead	Mortality (%)	Average Shell Length (mm)
1	77	77	100	13.70 ± 3.07
2	55	53	96	13.70 ± 3.73
Control	125	2	2	13.12 ± 4.78

and one control for the Category II assessment, which evaluated the encrusted gimbal unit of the watercraft. The gimbal unit was flushed with 60°C water for a duration of 2 min and 12 s (adding 5 s of duration to ensure that the predetermined lethal duration was met). Only one replicate from the gimbal unit had a resulting 100% mortality, while the other exhibited 96% mortality (Table 13.12). There was a significant difference in the percent mortality of the experimental groups and the controls for each of the tested categories.

Discussion

The data obtained from the evaluation of hot water sprays as a method of decontamination for quagga mussels mirror the reported species-specific characteristic of the upper thermal limit of quagga mussels being lower than that of zebra mussels (Spidle et al. 1995; McMahon 1996; Mills et al. 1996). This vulnerability could be exploited by management agencies in regard to developing a more adaptable and efficient boat decontamination protocol, which may be more apt for recreational boaters to follow due to the less time needed to apply hot water sprays greater than or equal to 60°C to ensure 100% quagga mussel mortality when compared to zebra mussels. There are many areas of boats and other various watercrafts that are capable of being subjected to direct thermal spray (i.e., hull, trim tabs). These areas would require only hot water application of greater than or equal to 5 s at temperatures of greater than or equal to 60°C, instead of the greater than or equal to 10 s contact duration necessary to kill zebra mussels at the same temperature. Though this may not seem like a tremendous difference in application time, a vast majority of the boat area (i.e., hull, deck) would have the treatment time regarding boat decontamination reduced by half. This would appeal to both recreational boaters and government agencies because less money would be spent on the necessary time of labor required to conduct the entire decontamination procedure, and it would allow boaters to leave freshwater recreation areas more quickly. The use of species-specific guidelines for boat decontamination procedures would be more agreeable to monitoring agencies in the western United States, where water bodies are heavily infested specifically by quagga mussels (Benson 2011). In cases where the water body is infested by only zebra mussels or both zebra and quagga mussels could possibly be involved in fouling a boat, a duration of greater than or equal to 10 s at temperature greater than or equal to 60°C should be implemented. Freshwater regions with active surveillance of their specific dreissenid populations will be able to employ species-specific decontamination procedures most effectively as they can determine and use the hot water decontamination standard most applicable toward their particular invasive mussel population.

Field Validation of Category I and Category II Watercraft Areas

Although the new information regarding the increased susceptibility of quagga mussels to thermal spray compared to zebra mussels will be quite useful in revising and developing watercraft decontamination standards and procedures where applicable, the direct application of these data can be used only to readily accessible areas of the watercraft capable of receiving the contact spray directly (Category I areas in Table 13.1). Dreissenid mussels do display a tendency to settle in particularly well-sheltered areas of watercraft such as motors, anchors, intake and outlets, trim tabs, and centerboard slots (Morse 2009), where they may not be able to receive a direct hot water spray and/or may come in contact with sprayed water as runoff from other surfaces where it may have cooled below the lethal temperatures. For these reasons, it was necessary to conduct experiments to evaluate the amount of time necessary for hot water to be applied to these inaccessible areas (Category II in Table 13.1) in order to reach the most efficient and safe temperature resulting in 100% quagga mussel mortality. The inboard/outboard motor gimbal units of two separate boats were evaluated for this experiment, one in the summer and one in the winter. This was done because depending on the ambient temperature and conditions, the surface temperature of the gimbal units may vary, meaning the amount of time necessary to reach the lethal temperature in differing conditions may also vary. As expected, it took significantly longer than the Category I recommended duration of 5 s with 60°C hot water for the top flush of the gimbal unit to reach the target lethal temperature at the bottom of the gimbal unit. The amount of time needed to achieve the target lethal temperature also varied with the specific season: a maximum of 43 s for the summer flush and a maximum of 2 min and 7 s for the winter flush. This was probably due to the different surface area temperatures present between the two seasons.

In addition to areas that are inaccessible to hot water sprays on watercraft, there are also areas that are not capable of withstanding temperatures in excess of 54°C. These areas may be made of materials that could be susceptible to heat-associated damage such as thick plastics or tubing. For these areas (Category III on Table 13.1), the determined 100% mortality rates for temperatures less than or equal to 54°C may be used to prevent such damage from occurring. For the evaluation of Category III areas, the live and bait wells of the same recreational boat were tested in both summer and winter. The temperature of 54°C was used because it would require the least amount of additional contact duration to ensure 100% mussel mortality. As in the evaluation of the gimbal unit, the amount of time necessary to flush the live wells and bait wells was significantly longer than the necessary time regarding Category I areas.

In order for the information obtained from these experiments to be put to practical use in the real world, it was necessary to conduct field tests on actual boats encrusted with quagga mussels. The data regarding the most effective and least hazardous lethal temperature and duration regarding Category I areas could be applied directly to any area of the boat encrusted with mussels that could receive a direct spray. This was determined to be hot water at a temperature of 60°C for a duration of 5 s. The same standard was used for both the winter and summer experiments because this spray would be directly contacting the mussels allowing the lethal temperature to heat the soft tissues of the mussels completely through heat conduction across the shell valves without conduction from outside materials that may be inferring and protecting the mussels. For both winter and summer experiments, all Category I groups tested at this specific temperature and time combination had a 100% quagga mussel mortality result (Tables 13.6 and 13.8). The density of the experimental groups did vary between winter and summer, allowing many more smaller mussels to be killed per experimental group in the summer (mean n = 105) than the winter (mean n = 47). The average shell size between the summer and winter experimental groups also varied at 6.57 ± 2.48 mm and 15.70 ± 4.27 mm, respectively. The 100% mortality within the larger winter group offered confirmation that this specific standard can be used to ensure 100% mortality among some of the hardiest of quagga mussels.

The validation experiments concerning the Category II areas, specifically the quagga encrusted gimbal units of the contaminated boats that were tested, were treated differently than the Category I areas because the hot water spray could not directly contact the mussels colonized deep within the gimbal unit. For these experiments, a combination of the data obtained from the field test on emersed mussels and the evaluation of time needed to reach and sustain lethal temperatures in Category II areas was used.

For the summer validation, a hot water flush (60°C) was applied to the top of the gimbal unit for a total of 48 s: 43 s required to heat the entire unit to the necessary lethal temperature and an additional 5 s to ensure 100% quagga mussel mortality. The results of the experiment showed that the application of hot water to the gimbal unit for this amount of time did ensure 100% quagga mussel mortality in the two experimental groups. For the winter validation, the same technique was used on the gimbal unit regarding the flush, but the amount of time was increased in order to make sure the unit was heated to the necessary lethal temperature in the colder conditions. The hot water flush (60°C) lasted for a duration of 2 min and 12 s: 2 min and 7 s to ensure the unit would be heated to the lethal temperature and 5 s to ensure 100% quagga mussel mortality. Of the two experimental groups, only one displayed 100% mortality, while the other displayed 96% mortality. Since 100% mortality was not achieved in the second experimental group, this current combination of duration and lethal temperature should not be used for Category II areas in winter conditions. One aspect to examine why this specific combination of duration and temperature did not work is in regard to the structure of the gimbal unit. Applying a hot water to only the top of the unit was shown not to be effective. A hot water flush applied only to the top of the gimbal unit may not reach all of the settled quagga mussels within the sides of the hollow cylindrical structure. Therefore, a 2 min and 12 s rinse should have been conducted at both the top and the sides of the gimbal unit in order to make certain that all of the parts are heated to the necessary lethal temperature. For Category III areas (which may vary in size and volume depending on the watercraft), it is recommended that the temperature of the hot water flush be monitored until a temperature of 54°C is reached. After this target temperature is reached, it is necessary to maintain a constant flush of that temperature for at least 10 s in order to ensure 100% quagga mussel mortality.

Conclusion

According to the data obtained from the study testing the susceptibility of quagga mussels to hot water spray as a means of watercraft decontamination, it is recommended that hot water sprays at 60°C for a duration of 5 s can be utilized to ensure 100% quagga mussel mortality under experimental and differing field conditions (winter and summer). If the water temperature is lower than this, 100% mortality cannot be achieved for that specific duration. A temperature of 60°C rather than a higher temperature is recommended because it is reported to have the same efficacy at the same durations, and higher temperatures may be hazardous to human health (Morse 2009). The 60°C/5 s standard is to be used only for readily accessible areas of the watercraft and only used for mitigation of the quagga mussel. For other areas of watercraft (Category II and Category III), it is necessary to verify all surface areas are heated to the correct predetermined lethal temperature for the required amount of time to ensure 100% quagga mussel mortality. The results of the study attempting to validate a time standard for a specific watercraft area (i.e., the gimbal unit) show that developing a specific time standard may not be entirely effective for larger parts and under different weather conditions. Further research needs to be conducted regarding different areas on specific watercraft so that decontamination procedures can be developed depending on the type and model of boat contaminated with quagga mussels.

REFERENCES

Benson AJ. 2011. Quagga mussel sightings distribution. USGS. Retrieved June 20, 2013 from: http://nas.er.usgs.gov/taxgroup/mollusks/zebramussel/quaggamusseldistribution.aspx.

Bossenbroek JM, Kraft, CE, Nekola, JC. 2001. Prediction of long-distance dispersal using gravity models: Zebra mussel invasion of inland lakes. *Ecological Applications* 10:1778–1788.

Comeau S, Rainville S, Baldwin W, Austin E, Gerstenberger S, Cross C, Wong WH. 2011. Susceptibility of quagga mussels (*Dreissena rostriformis bugensis*) to hot-water sprays as a means of watercraft decontamination. *Biofouling* 27(3):267–274.

Johnson LE, Ricciardi A, Carlton JT. 2001. Overland dispersal of aquatic invasive species: A risk assessment of transient recreational boating. *Ecological Applications* 11:1789–1799.

Keller RP, Drake JM, Lodge DM. 2007. Fecundity as a basis for risk assessment of nonindigenous freshwater molluscs. *Conservation Biology* 21:191–200.

Leung B, Bossenbroek JM, Lodge DM. 2006. Boats, pathways, and aquatic biological invasions: Estimating dispersal potential with gravity models. *Biological Invasions* 8:241–254.

McMahon RF. 1996. The physiological ecology of the zebra mussel, *Dreissena polymorpha*, in North America and Europe. *American Zoologist* 36:339–363.

Mills EL, Rosenberg G, Spidle AP, Ludyanskiy M, Pligin Y, May B. 1996. A review of the biology and ecology of the quagga mussel (*Dreissena bugensis*), a second species of freshwater dreissenid introduced to North America. *American Zoologist* 36(3):271–286.

Morse JT. 2009. Assessing the effects of application time and temperature on the efficacy of hot-water sprays to mitigate fouling by *Dreissena polymorpha* (zebra mussels Pallas). *Biofouling* 23:605–610.

Pathy DA, Makie GL. 1993. Comparative shell morphology of *Dreissena polymorpha*, *Mytilopsis leucophaeata*, and the "quagga" mussel (Bivalvia: Dreissenidae) in North America. *Canadian Journal of Zoology* 73:1012–1023.

Peyer SM, McCarthy AJ, Lee CE. 2009. Zebra mussels anchor byssal threads faster and tighter than quagga mussels in flow. *Journal of Experimental Biology* 212:2027–2036.

Pimentel D, Zuniga R, Morrison D. 2005. Update on the environmental and economic costs associated with alien-invasive species in the United States. *Ecological Economics* 52:273–288.

Puth LM, Post DM. 2005. Studying invasion: Have we missed the boat? *Ecology Letters* 8:715–721.

Ricciardi A, Rasmussen JB. 1998. Predicting the identity and impact of future biological invaders: A priority for aquatic resource management. *Canadian Journal of Fisheries and Aquatic Sciences* 55:1759–1765.

Rothelisberger JD, Chadderton LW, McNulty J, Lodge DM. 2010. Aquatic invasive species transport via trailered boats: What is being moved, who is moving it, and what can be done. *Fisheries* 35:121–132.

Spidle AP, Mills EL, May B. 1995. Limits to tolerance of temperature and salinity in the quagga mussel (*Dreissena bugensis*) and the zebra mussel (*Dreissena polymorpha*). *Canadian Journal of Fisheries and Aquatic Sciences* 52:2108–2119.

Western Regional Panel on Aquatic Nuisance Species 2010. http://www.fws.gov/answest/documents.html (accessed on April 20, 2011).

Zhulidov AV, Pavlov DF, Nalepa TF, Scherbina GH, Zhulidov DA, Gurtovaya TY. 2006. Relative distribution of *Dreissena bugenisis* and *Dreissena polymorpha* in the Lower Don River System, Russia. *International Review of Hydrobiology* 89:326–333.

Zook B, Phillips S. 2009. Recommended uniform minimum protocols and standards for watercraft interception programs for dreissenid mussels in the western United States. Western Regional Panel on Aquatic Nuisance Species [Internet]. [Accessed June 20, 2013]. Available from: http://www.aquaticnuisance. org/wordpress/wp-content/uploads/2009/01/Recommended-Protocols-and-Standards-for-Watercraft-Interception-Programs-for-Dreissenid-Mussels-in-the-Western-United-States-September-8.pdf.

14

Uniform Minimum Protocols and Standards for Watercraft Interception Programs for Dreissenid Mussels in the Western United States

Bill Zook and Stephen Phillips

CONTENTS

ABSTRACT   Watercraft and equipment interception programs have been implemented by more than 75 state, federal, tribal, and local government agencies and organizations in the western United States since 2007 in an effort to prevent the overland transport and range expansion of dreissenid mussels (Zook and Phillips 2009b). To maximize the effectiveness of these programs, the Pacific States Marine Fisheries Commission (PSMFC), working in cooperation with the Western Regional Panel (WRP) of the national Aquatic Nuisance Species Task Force (ANSTF) and its many partners, has developed uniform minimum protocols and standards (UMPS) for watercraft, seaplane, and water-based equipment interception based on the best science and information currently available. This chapter details those protocols and standards and provides the scientific justification and summarizes the practical application experience gained from nearly a decade of research and field trials on this issue. It also makes a case for increased coordination, cooperation, and consistency in the application of these protocols and standards across jurisdictional boundaries and encourages reciprocity between jurisdictions to meet the challenge that threatens boundless water supply, recreation, and natural resource values in the West.

Introduction

The primary goal of any watercraft interception program must be to prevent the transfer of quagga and zebra mussels (referred to here as dreissenid mussels) and other aquatic invasive species (AIS) on trailered watercraft, seaplanes, and equipment in order to safeguard natural resources, water supply, recreation, and other important water-dependent values. We also believe that one objective of any long-term mussel interception program should be to keep public and private waters open to boating and seaplane use to the extent possible. It may only take one infested watercraft, seaplane, or piece of water-based equipment (pump, construction barge, navigation buoy, dock, hoist, etc.) to establish a dreissenid mussel population, but the vast majority of watercraft and equipment intercepted by these programs are not moving directly from contaminated waterways and therefore pose little risk for mussel transfer. By following common sense guidelines, a watercraft interception program can be established that readily identifies high-risk watercraft and equipment so that more aggressive strategies can be focused where they are most critically needed to prevent any further range expansion of dreissenid mussels.

We also realize the inherent difficulty in implementing regionally consistent watercraft interception programs. The large number of programs already in place and the wide range of agency/organization capacity (funding, authority, access control, and political will) to implement them make consistency across jurisdictional boundaries difficult to achieve. But the fact remains, interjurisdictional coordination and cooperation will be the key to preventing the range expansion of dreissenid mussels in the western United States. Watercraft interception programs do not in of themselves guarantee that dreissenid mussels will not find their way into protected waterways. A recent case in point is the discovery of quagga mussels in Lake Powell, where, despite a very aggressive and comprehensive watercraft interception program being in place since 2008, quagga mussels were apparently introduced by trailered watercraft or equipment.

Changes to regulations at the local, state, tribal, and federal levels may be necessary to implement a comprehensive multijurisdictional program in the West. We therefore encourage continued discussion, and exchange and support additional cooperation, communication, and coordination among agencies and organizations engaged in these programs in the western United States. *Adopting these protocols and*

FIGURE 14.1   Typical watercraft inspection stations in the western United States. (Photo courtesy of Bill Zook, 2011.)

standards for these programs is one step toward achieving that goal and increasing the overall effectiveness of these programs (Figure 14.1).

Protocols and standards recommended here are products of

1. An extensive literature search and review
2. The results of decontamination protocol research funded by the United States Fish and Wildlife Service (USFWS) and conducted by the University of Nevada Las Vegas (USLV) in 2010–2011 (Wong et al. 2011)
3. Countless personal interviews (see list of all personal communications on pp. 60–62) with program administrators, AIS specialists, private equipment manufacturers, alternative technology proponents, recreational boaters, seaplane pilots, and commercial watercraft haulers
4. Results from a WRP survey of watercraft/equipment interception programs in the 20 western states completed in February 2009 (Zook and Phillips 2009b)
5. A survey of commercial watercraft transport providers completed in 2010 (Zook and Phillips 2010)
6. A cooperative effort with the National Seaplane Pilots Association (SPA) to develop inspection, cleaning, and general operation guidelines for seaplanes completed in 2011 (Zook, personal communication)
7. A November 2010 survey of watercraft interception program managers in the western United States to determine what changes were needed to the 2009 UMPS document (Zook and Phillips, personal communication)
8. A review of individual agency/organization policies, procedures, and standards (Baldwin et al. 2008) (See list of agency/organization manuals on pp. 62–63)
9. The experience and feedback gained from more than 65 watercraft inspection and decontamination training classes delivered to over 3500 individuals representing 180 different agencies, organizations, and water-dependent businesses in 17 western states from 2008 to 2011, and the extensive contact network and ongoing interaction established through that Watercraft Interception Training (WIT) program (http://www.aquaticnuisance.org/wit)

The UMPS report is a living document and will continually evolve as new information and science becomes available. This version (2012) includes many updates from the original document completed in 2009 (Zook and Phillips 2009a), but the basic principles and program elements remain substantially unchanged.

Economics of Prevention

Establishing and implementing a comprehensive dreissenid mussel prevention program can be expensive. Individual state, federal, tribal, and local agencies and organizations in the western United States spend somewhere between an estimated $50,000 and $5.0 million annually on dreissenid mussel prevention programs, including investments for risk assessment, education and outreach, watercraft and equipment interception, early detection monitoring, and response planning. Of these, watercraft interception is normally the highest cost item (Zook and Phillips 2009b).

The Quagga and Zebra mussel Action Plan (QZAP) (Western Regional Panel 2010) developed by the National ANSTF estimated the annual cost to implement mandatory inspection and decontamination of all watercraft at infested waterways in the western United States to be $20 million (plus a one-time initial estimated setup and equipment cost of $25 million) and associated costs for research and development of protocols and standards and enforcement to be about $12 million annually. In addition, QZAP estimated that $31 million would be required annually to fund the implementation of state plans that include state, local, and regional watercraft and equipment inspection programs on uninfested waterways in 19 western states (Western Regional Panel 2010).

Since being first observed in the Great Lakes in 1988, the economic impact of dreissenid mussels has resulted in billions of dollars being spent on control measures for power producers, municipal water suppliers, and other water users. The economic impact of quagga mussels since the infestation in Lake Mead occurred in 2007 has also been significant. The Bureau of Reclamation at its Lower Colorado projects (Hoover, Parker, and Davis Dam) spends approximately $1 million annually on quagga mussel control (Willett, personal communication). The Metropolitan Water District of Southern California has spent over $30 million in the past 5 years for quagga mussel prevention-related operations and maintenance and capital costs in the Colorado River Aqueduct and associated facilities (De Leon, personal communication).

Additionally, an assessment by the Independent Economic Analysis Board (IEAB) (2010) for the Northwest Power and Conservation Council reviewed prevention efforts in the Columbia and Colorado basins, Minnesota, Lake Tahoe, and other locations where economic assessments have been completed. They concluded that existing prevention efforts in the Columbia Basin are underfunded and that an investment on the order of the QZAP estimates to prevent or delay mussel establishment *seems appropriate* given the high cost/benefit ratio (Independent Economic Analysis Board 2010).

The IEAB further concluded that even if dreissenid mussels were to eventually become established, there is great value in delaying establishment because any delay would allow important scientific advances to occur, which may help prevent an introduction, contain an introduction, or eradicate a newly established population, and because the annual cost saving for each year of delay would be substantial and far exceed the cost of implementing a comprehensive prevention program as envisioned by QZAP (Independent Economic Analysis Board 2010).

Existing Watercraft Interception Programs

There are at least 75 jurisdictions in 19 western states that currently employ some form of watercraft interception program on over 400 water bodies (Zook and Phillips 2009b).

Seventy-two of these agencies and organizations received an online survey designed to identify the key elements of each program and gauge support for developing uniform protocols and standards in January 2009 (Zook and Phillips 2009b). Of the 69 entities completing this survey (96% return), nearly 90% favored the development and implementation of more consistent protocols and standards for watercraft interception programs that could be applied across jurisdictional boundaries.

Alternative Decontamination Technologies

Because manually applied hot water spray decontamination is not always 100% effective in *removing* all mussels from hidden areas found on some types of watercraft and/or equipment and because a question remains as to the survivability of attached mussels in some areas of watercraft where visual confirmation of mortality is difficult, alternative methods to hot water spray have been actively pursued and considered for a number of years. Some of the most promising alternative systems include drying time acceleration, dry ice blasting, and semiautomated wash systems. Each of these systems has unique features that may be suitable for wider use in the future.

Seaplanes

For more than a decade, water resource managers throughout North America have been concerned about seaplane activity as a pathway for the spread of aquatic vegetation, dreissenid mussels, and other AIS. In 1998, the Great Lakes Panel of the national ANSTF (Randall 1999) developed *generic* voluntary guidelines for seaplanes that were adopted by the ANSTF as national guidelines in April 1999. Those guidelines still serve as the national standard even though some local jurisdictions have recently expanded on them and, in a couple of cases, made them mandatory (Lake Tahoe and Clear Lake, CA).

While the primary focus of most water resource managers has been and will continue to be on the potential threat of AIS proliferation via the overland transport of watercraft and equipment, the seaplane pathway has been receiving more attention recently as significant progress is being made with other types of more traditional watercraft and equipment interdiction. As dreissenid mussels and invasive aquatic plant species continue to spread throughout North America, individual jurisdictions with relatively high seaplane use are beginning to consider and, in some cases, implement more aggressive regulation of this activity (Figure 14.2).

According to the National SPA, there are an estimated 35,000 seaplane-rated pilots and about 1,500 new seaplane ratings issued each year in the United States (Zook 2010). The Federal Aviation Administration does not distinguish between airplanes with floats, wheels, or skis so the exact number of seaplanes operating in the United States is not known. The SPA estimates that there are between 5,000 and 10,000 seaplanes currently in use in the United States (Windus, personal communication).

Commercial Watercraft and Equipment Haulers

The overland transfer of dreissenid mussels and other AIS on large watercraft and equipment transported by commercial haulers has undoubtedly contributed to their range expansion in North America. Watercraft and equipment that require the services of a commercial hauler tend to be larger, more structurally and functionally complex and more likely to have been in the water for an extended period of time. Those factors elevate the level of risk for having attached mussels, mussel larvae, or other invasive species onboard when these vessels are moved from contaminated to uncontaminated waterways.

A survey and report on the Commercial Watercraft Hauling industry was completed in the fall of 2010 (Zook and Phillips 2010).

The same watercraft interception protocols and standards that apply to smaller watercraft should be used for large vessels and equipment that are commercially hauled. However, large watercraft that are typically commercially hauled generally require more time, effort, and focus because of their large surface areas and complex raw water storage systems (Figure 14.3).

FIGURE 14.2 Screening interview prior to inspecting seaplane. (Photo courtesy of Bill Zook, Long Lake, WA, 2009.)

FIGURE 14.3　Decontamination of a commercially hauled watercraft. (Photo courtesy of Bill Zook, 2009.)

Water-Based Equipment

A variety of water-based equipment are routinely moved between waterways and present the same risks as watercraft and seaplanes. Construction equipment used to build and repair bridges, dredge navigation channels, and install docks and breakwaters can move mussels from contaminated to uncontaminated waterways if not decontaminated before moving between waters. Navigation buoys, boat hoists, and lifts are also a potential source of contamination, as witnessed by the discovery of zebra mussels on a boat hoist moved from Lake of the Ozarks to Smithville Reservoir in Missouri in 2010 (Banek, personal communication), requiring emergency lake treatment (Figure 14.4).

Equipment used to sample fish populations, collect water samples, survey aquatic vegetation, stock fish, and even to sample for dreissenid mussels can also be a pathway for mussels or other AIS to be moved between waters. All agencies and organizations engaged in this type of activity should adopt internal policies and procedures for equipment cleaning and decontamination, especially when working in waterways known or suspected of harboring dreissenid mussels or other aquatic nuisance species (ANS).

The Unites States Bureau of Reclamation has developed an excellent decontamination manual for the handling and cleaning of equipment that can serve as a model for the development or adoption of internal equipment cleaning policies (DiVittorio et al. 2010). The Reclamation Equipment Inspection and Cleaning Manual can be accessed at http://www.usbr.gov/mussels/prevention/docs/Equipment InspectionandCleaningManual2012.pdf.

FIGURE 14.4　Construction equipment typically transported between waterways. (Photo courtesy of Bill Zook 2009.)

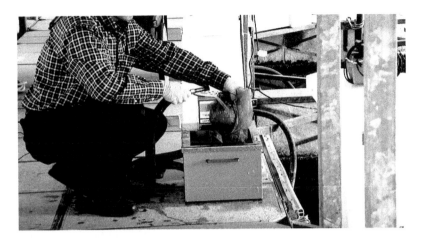

FIGURE 14.5　Testing effect of hot water spray on live quagga mussels at Lake Mead. (Photo courtesy of Bill Zook, Lake Mead Marina, 2009.)

Decontamination Efficacy Research

The first published study to look at the efficacy of hot water sprays to mitigate dreissenid fouling was conducted by Morse (2009). The Morse study showed that zebra mussels required a 10 s spray time with 140°F water to achieve 100% mortality. Morse noted that "it is of interest that quagga mussels are reported to have thinner shells and less tightly sealed shell valves than zebra mussels (Claxton et al. 1997) which may make them more susceptible to hot water sprays." However, this supposition requires experimental confirmation.

The PSMFC with QZAP funding contracted with the UNLV to conduct research to test the efficacy of current UMPS (2009) decontamination protocols and standards. This research was completed by UNLV at Lake Mead, Nevada, in 2011 (Wong et al. 2011). The results of this research determined the following about current decontamination protocols and standards (Figure 14.5):

1. *Hull of watercraft*: Applying hot water spray at *140°F for 5 s* or longer results in 100% quagga mussel mortality. An application time of 10 s at 140°F is still recommended for zebra mussels.

2. *Gimbal area*: In those areas where hot water *flushing* is required and where direct contact with flushed hot water of the entire surface or area is not always possible, the duration of hot water application is increased. *An application of 140°F hot water flush of 130 s or longer* is required to achieve 100% mortality of both quagga and zebra mussels in those areas. (For the gimbal area, it is important to do both a top flush and a side flush {both sides} of 130 s to ensure 100% mortality.)

3. *Live wells/bait wells*: A hot water *flushing* of bait/live wells at *130°F for 70 s* (*note on temperature*: Some pumps are rated only at *120°F*; please refer to your user manual to avoid damaging equipment).

In 2011, another study by UNLV and the Kansas Department of Wildlife, Parks and Tourism was initiated (using QZAP funding) to test the efficacy of hot water sprays on zebra mussels at Lake Wilson in Kansas (Comeau et al. 2011). Initial results from this study indicate that applying hot water spray for 10 s at 130°F and 140°F both resulted in 100% zebra mussel mortality.

Watercraft Interception Training

The PSMFC with funding from the USFWS and Bonneville Power Administration developed the Watercraft Inspection Training Program in 2004. It was originally designed as a 90 min training to enlist the voluntary help of boating law enforcement officers in the western United States to educate boaters and inspect high-risk watercraft in the course of their normal boater safety duties. Three of these trainings were delivered in Oregon, Idaho, and the Lower Colorado River Basin in the fall of 2004 and spring of 2005 training about 150 officers from five western states.

When quagga mussels were discovered at Lake Mead in January 2007 and soon afterward in a number of downstream waterways connected by the Colorado River and Central Arizona aqueduct systems in Southern California and Arizona, the WIT program underwent major reconfiguration and change of direction. The training program was expanded to 5–6 h, and the target audience changed from boating law enforcement officials to state, federal, tribal, and local water, land, and wildlife resource management agencies and organizations as they struggled to come to grips with the looming invasion. The WIT program offered an immediate opportunity for agencies and organizations in the West to train their staff on the dreissenid mussel issue including prevention strategies and watercraft interception protocols and standards (Figure 14.6).

From the early spring of 2007 through the fall of 2008, 25 WIT training programs were delivered in nine western states to about 1200 people. Since late 2008, an additional 40, of what is now referred to as Level One WIT Trainings, have been delivered to an additional 2300 people and nine more western states. There have been 65 Level 1 trainings delivered since 2007 in 17 western states to more than 3500 people representing over 180 different state, federal, and local agencies, tribes, utilities, and other water resource organizations.

In 2008, a Level 2 Watercraft Inspection Training program was developed. Level 2 WIT training is an intensive 2-day training for 10–12 individuals held at Lake Mead, Nevada. This training is designed for those people who will be responsible for developing or managing watercraft interception programs for their agency, tribe, or organization. Level 2 graduates are certified as Level 2 trainers and as first responders. The training focuses on hands-on inspection and decontamination of watercraft and equipment actually infested with live quagga mussels. To date (2012), 19 Level 2 trainings have been delivered producing over 200 certified personnel representing over 50 agencies and organizations from 14 western states (Figure 14.7).

FIGURE 14.6   Level 1 training classes. (Photo courtesy of Wen Baldwin, 2007.)

FIGURE 14.7   Level 2 training class at Lake Mead. (Photo courtesy of Wen Baldwin, 2009.)

Level 2 graduates have delivered an estimated 500 Level 1 trainings in addition to those identified earlier, and several states, most notably Colorado, have established independent training programs based on the WIT training template (Brown, personal communication).

We strongly recommend that all watercraft/equipment inspectors and those performing watercraft/equipment decontamination maintain proficiency through periodic recertification and by taking advantage of continuing education and training opportunities.

Intrajurisdictional Cooperation and Consistency

Achieving a greater level of consistency by adopting UMPS for all watercraft interception programs across the western United States benefits water and resource managers and the boating public in a number of important ways, including the following:

1. Increased effectiveness by ensuring that all programs utilize the best practical science and technology currently available
2. Establishing a high level of confidence in the effectiveness of their own programs and trust in the programs employed by others
3. Reducing the amount of staff time and funding required of all programs by avoiding unnecessary duplication of effort while increasing effectiveness and public acceptance
4. Making it easier for the public to understand, anticipate, and comply with watercraft/seaplane/equipment interception and prevention programs

Not every federal, state, tribal, and local agency or organization currently has the authority or resources (capacity) to implement all of the minimum protocols and standards identified here. In those cases where capacity is lacking, we urge those entities to seek the regulatory authority and resources necessary to stop, inspect, decontaminate, quarantine, or exclude high-risk watercraft in order to ensure protection of the natural resource, economic, public health, and cultural assets that are threatened by this invasion.

In the past several years, many states including Washington, Oregon, Idaho, Montana, Utah, Wyoming, Arizona, Colorado, and Nevada have approved new legislation granting broader authority to intercept watercraft and equipment in transit. In addition, federal agencies like the National Park Service and local government agencies and organizations like the Tahoe Resource Conservation District (USACE 2009) have passed regulations establishing that authority within their respective jurisdictions (Klett, personal communication).

While the protocols and standards recommended in this document are directed at preventing the inadvertent transfer of dreissenid mussels from areas where they are currently present to unaffected waters on trailered watercraft, seaplanes, and water-based equipment, their application will help prevent the spread of other AIS as well. The screening, inspection, decontamination, and quarantine/drying measures described here to reduce the risk of mussel transfer are also effective for reducing the risk of overland transport of invasive aquatic vegetation, fish, disease pathogens, plankton species, and other AIS (Baldwin et al. 2008).

Recommended Program Levels

Many agencies and organizations do not have the capacity to implement state-of-the-art watercraft/equipment interception programs. Funding limitations, lack of access control or authority, and/or the level of political understanding will all play a role in determining whether a water or resource management agency decides to become proactive enough to implement a watercraft interception program and how extensive that program will be. However, in those situations where the risk is high, the potential savings from preventing a dreissenid mussel introduction always far exceed the cost of implementing even the most comprehensive interception program (University of California Agriculture and Natural Resources 2010).

Because of funding/staffing or authority limitations, a number of western agencies and organizations employ only random, periodic, or peak-time interception programs. These programs have obvious limitations so it is vitally important that agencies and organizations implementing this type of watercraft/equipment interception program also complete risk assessments on all major water bodies and use that information to direct those limited resources to waters with the highest risk for mussel introduction.

It is also important that, to the extent practical, all programs should follow the UMPS for all elements of their interception programs and consider adopting more inclusive low-cost programs like volunteer or mandatory self-inspection while seeking more public, political, and financial support for expanded programs as the threat continues to increase with each new mussel discovery in the West.

It is the responsibility of water and resource managers to determine the level of acceptable risk and which type of watercraft interception program most closely reflects the mission, values, and capacity of their agency or organization. However, consideration for the investments made by neighboring water and resource managers should not be overlooked when seeking support for interception programs. A common concern raised by 2009 survey recipients and current WIT training program attendees is that upstream or neighboring managers aren't doing enough to protect those systems, putting their own considerable investments and resources at risk. Regional coordination groups such as the 100th Meridian Initiative, Western Regional Panel (WRP), and basin Aquatic Invasive Species (AIS) teams help promote common approaches to AIS prevention in the western United States to prevent this situation from occurring.

DETERMINING INDIVIDUAL WATER BODY RISK LEVEL

High-Risk Water Body: The determination of a *high-risk water body* is the prerogative of the responsible management entity. Some of the factors that should be used to determine risk potential include the following:

Whether water quality parameters (e.g., calcium and pH level, food supply, and summer water temperatures) will support the survival, growth, and reproduction of dreissenid mussels (these parameters may often vary seasonally and even by location within a large water body)

The amount and type of watercraft activity and where it's coming from

Proximity to dreissenid positive or suspect waters

When the water in question is a headwater, water, or power supply system, or supports species listed under the Endangered Species Act

We recommend employing one of the following three program levels for watercraft/equipment interception programs depending on the risk level and individual agency/organization capacity:

Level 1 (Self-Inspection)

Relatively low-cost program for low-risk waters or on high-risk waters where organizational or physical capacity prevents a more aggressive approach.

As an example, we recommend either a voluntary or mandatory self-inspection program similar to the one developed by the Utah Division of Wildlife Resources and in use at over 100 secondary risk waters in that state (Utah Division of Wildlife Resources 2009). Mandatory programs work best if the authority to enforce provisions of the program (e.g., authority to require that all watercraft operators complete and post self-certification form) is in place. In the absence of that authority, a voluntary program should be implemented.

This type of program involves the dissemination of an inspection form, which can be made available at either an entry station, a kiosk, or a message board with boldly printed instructions for the watercraft/equipment operator to answer all the questions and inspect all designated areas of watercraft, trailers, and equipment. The form is then placed in or on the transport vehicle where it can be easily seen. If the program is mandatory, spot checks by enforcement personnel can be used to enforce compliance.

This type of program has limited effectiveness because it's unmanned, and contaminated watercraft can still be launched unknowingly or otherwise by inexperienced or irresponsible boaters (though it does provide great benefits in terms of public outreach and education). At a relatively low cost, a well-signed self-inspection program essentially equates to having a full-time person (24/7/365) at each location educating boaters and raising their awareness about the consequences of a mussel invasion and the importance of cleaning, draining, and drying watercraft between uses.

Another benefit from this type of program is that it provides a way to overcome political resistance to more *heavy-handed government* approaches by giving the boating public an opportunity to self-regulate and exercise personal responsibility. If the boating public fails to act responsibly, it is much easier then to convince water users and lawmakers that more formal efforts are required to protect water resources and local economies.

Self-inspection programs can be implemented for under $1000/year for individual water bodies. Including staff time for verifying and/or enforcing compliance can add to both effectiveness and cost.

Level 2 (Screening Out High-Risk Watercraft and Equipment)

Moderate to high-risk waterways where budget or other considerations prevent a more comprehensive (Level 3) program.

We recommend a program that includes a screening interview to identify high-risk watercraft and/or equipment followed by a brief inspection to verify interview information. All watercraft that are not clean, drained, and dry, or those that report coming from areas where dreissenid mussels are known to exist within the last 30 days are then excluded from accessing that waterway. This type of program is being employed by several California water and park districts.

This type of program can often be incorporated into an existing entry station operation that is set up to collect access fees, confirm reservations, or provide use information and regulations. Current entry station staff can be easily trained to conduct screening interviews and verifying inspections, and the number of watercraft excluded would normally be expected to be low on waters where this type of program would be implemented. Because a rigorous inspection is not required and no decontamination or quarantine facilities are used, this is a relatively low-cost protection option.

A Level 2 program is designed to exclude all high-risk watercraft where the cost of implementing a more comprehensive program is prohibitive. It maintains boating access for low- risk watercraft (the vast majority) but completely excludes others for lack of comprehensive inspection, decontamination, and/or quarantine capability. Exclusion can have adverse economic, political, and social consequences.

Programs like this typically cost between $2000 and $5000 a year to operate per water body if existing screening facilities and staff are available and are a relatively low-cost option in those situations. However, if those assets are not already in place, the cost can be considerably higher.

NOTE ON LEVEL 1 AND LEVEL 2 PROGRAMS

Level 1 and Level 2 programs are options for local jurisdictions when the capacity to implement more aggressive and effective programs is lacking. These programs, however, *do not* provide the level of security required for any type of cross-jurisdictional reciprocity because they do not offer any assurance that watercraft and/or equipment subjected to either type of program are, to the extent practical, free of mussels or other AIS.

Level 3 (Comprehensive)

High-risk waters, large water bodies, and wherever possible.

We recommend this type of program for all high-risk waters. A Level 3 program should include screening interviews at the point of entry; a comprehensive watercraft/equipment inspection of all high-risk watercraft/equipment performed by trained inspectors; and the decontamination and/or quarantine or exclusion of suspect watercraft and may include vessel certification.

This type of program may require construction or modification of entry facilities, purchase of a hot water power wash and wastewater containment system, hiring and training inspectors and decontamination operators, providing a safe and secure quarantine facility, a good working relationship with law enforcement authorities, and the development of a set of policies and rules that allow all of the earlier actions.

We estimate that about 30 western state, federal, tribal, and local agencies and organizations currently operate Level 3 watercraft intervention programs at the state, regional, tribal, or local level on over 300 high-risk water bodies in the western United States. Programs like this can cost between $50,000 and $1 million per water body per season to operate depending on the size of water involved, type of equipment, facilities used, hours of operation, and the number of access points available to boaters.

Some programs operate border inspection stations. For example, the State of Idaho's program (http://www.idahoag.us/Categories/Environment/InvasiveSpeciesCouncil/Inspection_Stations_2011/Inspection_Stations_2011.php) has 15 mandatory inspection stations located on all major roadways entering the state (7 days a week during the active boating season) at a cost of approximately $850,000 (Amy Ferriter, Idaho Department of Agriculture, personal communication 2011). Idaho also has 11 highway ports of entry whose staff are trained to inspect for contaminated watercraft (particularly commercially hauled watercraft).

Only Level 3 programs offer any opportunity for cross-jurisdictional reciprocity.

UMPS for Watercraft Interception Programs

The term *UMPS* implies that all agencies/organizations should strongly consider adoption of these as integral components of their watercraft interception program. However, because each entity is unique, having different missions, authority, resources, facilities, and governing bodies, it is understood that additional or stricter standards may be implemented and that cross-jurisdictional reciprocity should be left to the discretion of the implementing agency/organization.

These protocols and standards reflect the best currently available science, experience, technology, and understanding. However, we recognize that watercraft interception and decontamination is a rapidly evolving field and that new information may change the way we view these protocols and standards in the future.

There have been few major changes from the 2009 (Zook and Phillips 2009a) to this version of the UMPS regarding specific protocols or standards because the research and experience gained over the past 3 years has confirmed that these protocols and standards still represent the best currently available science and technology. There are, however, many clarifications, elaborations, and updates to the earlier protocols and standards that reflect our growing knowledge base. We will continue to encourage and contribute to the quest to find more effective ways to achieve advancements in efficacy, cost, liability, and delivery of watercraft interception programs in the future.

We recommend the following UMPS for all watercraft interception programs in the western United States.

Self-Inspection Programs (Mandatory or Voluntary)

Self-inspection programs, whether voluntary or mandatory, offer a limited level of protection because compliance and effectiveness are not guaranteed. However, self-inspection programs are very effective boater education tools, provide some level of protection for waters where implemented, and are cost-effective. If a higher level of protection is not available because of insufficient funding, physical site limitations, lack of intervention authority, or the sheer volume of waters needing coverage, the type of program currently implemented by the Utah Division of Wildlife Resources on approximately 100 of their secondary risk waters (Dalton, personal communication) should be considered as a minimal interception tool or *off-hours* adjunct to a more comprehensive program.

Protocols

1. Provide a self-inspection form and clear directions on how to complete the inspection and the form at the point of entry, kiosk, or dedicated check-in area.
2. Require (where a law/rule is in place) or request (when rules are not established) that the form be completed, signed, and posted in clear view on the dash of watercraft/equipment transport vehicle prior to launching.

Standards

Before launching, boaters must confirm that the following conditions have been met by signing and displaying a completed self-inspection form.

1. Watercraft, equipment, trailer have not been in any water known or suspected of having dreissenid mussels in the past 30 days (consider adding a checklist of those water bodies of most concern in your area so boaters can indicate if they have been in any of those specific waters).
2. Watercraft, equipment, trailer have been visually inspected by the operator at the site prior to launching.
3. Watercraft, equipment, and trailer are clean and, to the extent practical, drained, and dried.

Screening Interviews

The screening interview involves collecting information from the vessel operator through a series of questions prior to launching or entry that are designed to determine the level of risk posed by that watercraft based on its recent history of use. This should be an element of every interception program.

In order to be most effective, the screening interview should not rely totally on the responses given, but the person conducting the interview should be attentive enough to make sure that the responses given match the physical evidence available and that they are credible, which may require a brief confirmation inspection.

Protocols

1. Develop and use a standard screening interview form that, at a minimum, includes the following questions:
   a. The home location of the owner/operator
   b. The specific location (water body) where the watercraft or equipment was last used
   c. The date of the last use
   d. If the watercraft/equipment has been cleaned, drained, and dried
2. Verify the responses by checking the license plate or registration (boat ID) number and doing a brief visual inspection.
3. Clarify any inconsistencies between the responses given and the physical evidence before clearing the watercraft or equipment for launch.
4. Screening interviews provide all agencies and organizations implementing interception programs the opportunity to explain the importance of prevention and to educate the boating public on ways they can take personal responsibility for *clean* boating. Use it as an educational opportunity.

Standards

1. Watercraft that have been used in any dreissenid mussel positive or suspect water body in the past 30 days or are not clean, drained, and dry should be subjected to a comprehensive inspection by a trained professional before being allowed to launch, or excluded if inspection or decontamination resources are not available (Figure 14.8).
2. If there is reasonable suspicion of deception on the part of the owner/operator/transporter during the screening interview, the vessel should be subjected to a comprehensive inspection before being permitted to launch.

FIGURE 14.8   Screening interview at Kansas Department of Wildlife, Parks and Tourism Check Station. (Photo courtesy of Bill Zook, 2011.)

Watercraft/Equipment Inspection

Inspecting watercraft and equipment for the presence or likelihood of dreissenid mussels is perhaps the most important and difficult element of a successful interception program. Conducting an effective inspection requires some knowledge of dreissenid mussel identification, life history, and biology; a good understanding of the working parts of a wide range of watercraft types and equipment; and the cooperation of the boat/equipment operator. In addition, watercraft and equipment inspection needs to be systematic and thorough. A checklist should always be used when conducting a watercraft or equipment inspection in order to assure that all areas where mussels and veligers can be found are inspected.

A basic watercraft inspection and decontamination course, like the Level 1 WIT course offered by the PSMFC and certified by the 100th Meridian Initiative (http://www.aquaticnuisance.org/wit), is highly recommended for anyone who will be directly involved in watercraft inspection. An advanced training (WIT Level 2) should be taken by at least one agency or organization representative engaged in or planning to become engaged in watercraft interception. Level 2 training provides the knowledge, tools, and resources necessary to become an in-house Level 1 trainer or interception program manager.

The authority to stop, inspect, decontaminate, and/or quarantine watercraft or equipment varies between jurisdictions. Make sure you understand your authority and exercise it according to the law with regard to search and seizure. It is the inspector's responsibility to know and understand their authority with regard to stopping, inspecting, decontaminating, quarantining, or prohibiting watercraft from launching. It is the jurisdiction having responsibility of access management to inform and train inspectors as to that authority.

Protocols

1. Use an inspection checklist and follow it. The inspection checklist should include (at a minimum) the following information:
    a. The home state or area code where the watercraft or equipment is registered
    b. The vessel ID number
    c. The name and date of the last water visited
    d. A checklist of areas to be inspected includes the following:
        i. *Exterior surfaces*: (at and below the waterline) Hull, transducer, speed indicator, through-hull fittings, trim tabs, water intakes, zincs, centerboard box and keel (sailboats) and foot-wells on personal watercraft (PWCs)
        ii. *Propulsion system*: Lower unit, cavitation plate, cooling system intake, prop and prop shaft, bolt heads, gimbal area, engine housing, jet intake, paddles and oars

    iii. *Interior area*: Bait and live wells, storage areas, splash wells under floorboards, bilge areas, water lines, ballast tanks, and drain plug

    iv. *Equipment*: Anchor, anchor and mooring lines, personal flotation devices (PFD's), swim platform, wetsuits and dive gear, inflatables, down-riggers and planing boards, water skis, wake boards and ropes, ice chests, fishing gear, bait buckets and stringers

    v. *Trailer*: Rollers and bunks, light brackets, cross-members, hollow frame members, license plate bracket, springs and fenders

2. Inspect all high-risk watercraft (see definition on p. 20).

3. Have a systematic and repeatable plan when conducting inspections to ensure complete coverage of every area of the watercraft.

4. If dreissenid mussels are found anywhere on the watercraft or equipment, the inspection can cease and the entire watercraft, trailer, and equipment will need to be decontaminated or quarantined (preferably both) before being allowed to launch.

5. Use the inspection process as an opportunity to educate the boat owner/operator on the importance of prelaunch self-inspection, proper cleaning and drying, and the reasons why all watercraft and equipment operators need to clean, drain, and dry watercraft and equipment when moving between waters. Demonstrate the proper way to conduct a watercraft, seaplane, or equipment inspection.

Standards

1. If attached mussels or standing/trapped water are found on a high-risk vessel, it should not be allowed to launch without first being decontaminated or subjected to the prescribed quarantined/drying time standard or both.

2. If water is found on exposed areas only (rain or wash water), on an otherwise low-risk and clean watercraft, the watercraft should be thoroughly wiped dry first and allowed to launch.

3. If no mussels or water are found following a thorough inspection of the watercraft that is considered high risk because it has been in known mussel waters within the last 30 days, but has been out of the water long enough to be considered safe by applying relevant temperature and humidity drying time standards, it should be allowed to launch, except for watercraft that have ballast tanks or other difficult to access and completely drain raw water storage areas. *Normal drying time standards do not apply when areas that cannot be visually inspected and completely drained are present. These areas need to be treated to kill any mussels or veligers that may be present.*

4. Any watercraft or piece of equipment with attached vegetation (including algae growth) should not be allowed to launch without their complete removal and a reinspection.

5. Any watercraft with enough dirt, calcium, or bio-fouling buildup so as to make inspection for small attached mussels or other AIS difficult should be required to be cleaned and reinspected before being allowed to launch.

NOTE ON LIVE BAIT FISH

If the use of live bait fish is permitted in your jurisdiction and they are found during inspection, remove the bait, place in a bucket of clean water, drain and flush the live bait container with hot water *(130° water*)*, and then return the bait to the clean container. While this process does not assure that mussel veligers or even small settlers are not present on or in the fish themselves, it is the best *minimum* standard for dealing with this situation currently available.

---

* *Note:* If your live or bait well uses a pump, make sure to check your owner's manual for maximum temperature to avoid damaging equipment.

Watercraft/Equipment Decontamination

If, following inspection, a watercraft or piece of equipment transported from one water body to another is confirmed or believed to have mussels on board, three options are available: (1) decontamination, (2) quarantine/drying, or (3) exclusion. Hot water spray decontamination is the only option that kills *and removes* mussels. Since we cannot be sure that all areas of the watercraft and/or equipment have been adequately treated, we recommend that a period of drying be used in conjunction with decontamination for all watercraft confirmed or suspected of having mussels on board.

The best current technology available for watercraft/equipment decontamination is hot water pressure washing. We recommend the exclusive use of hot water (*140°F* or greater *at the point of contact*) and pressure washing equipment with various attachments to *kill and remove* all visible mussels (live and dead) and kill all veligers from every area of the watercraft, engine, trailer, and equipment.

NOTE ON DECONTAMINATION TECHNOLOGY

Recent research (Wong et al. 2011) and current assessment (this report) strongly indicate that hot water pressure washing using the protocols and standards identified here remains the most effective currently available decontamination methodology. We do not believe that relying solely on aerial exposure and desiccation as the primary means of decontamination is sufficient since dead mussels remaining on watercraft/equipment can be moved to other locations where their discovery can cause expensive and unnecessary response. However, we do encourage and support the combination of hot water decontamination and drying time as the most effective means to assure that all mussels are killed and, to the extent practical, all visible mussels are removed.

The objective of decontamination is to *kill and remove*, to the extent practical, all visible mussels. Killing prevents establishment of new populations resulting from watercraft/equipment transfer, but removing dead mussels is also important because a false-positive finding may result from the presence of mussel shells or DNA in samples collected for genetic analysis (polymerase chain reaction). This can result in unnecessary concern and expensive action if unexplained shells drop or are scrapped off the hull and are subsequently discovered at a boat ramp or the lake bottom, or if the watercraft is intercepted in transit. Furthermore, there are no standard protocols in place to easily confirm the viability of attached mussels within the context of a watercraft inspection or decontamination. Therefore, mussels on watercraft or equipment that appear to be dead do not necessarily indicate that those mussels or others not clearly visible settled elsewhere are in fact dead.

DECONTAMINATION SAFETY ADVISORY

Extreme caution should always be used when working in and around watercraft and equipment. This is particularly true when working with the high-pressure equipment and the high water temperatures recommended here.

Protocols

1. Before commencing a decontamination procedure, get the permission of the vessel owner after explaining the options and decontamination process in detail.
2. Consider requesting a liability waiver signature from the vessel owner as a condition of decontamination. Most owners would agree to sign a liability waiver when the option is quarantine or exclusion. Agencies should consult as necessary with their legal staffs on liability issues.
3. Find a location for the decontamination that is away from the water where the runoff and solids from the cleaning process can be contained and will not reenter any water body.

FIGURE 14.9 Portable decontamination unit including wastewater containment pad. (Photo courtesy of Bill Zook, 2011.)

4. If possible, wastewater and solids should be totally contained (low-cost containment systems now exist for this purpose) and directed to an appropriate waste treatment or disposal facility. New guidelines are currently being developed by the EPA for watercraft/equipment decontamination. [*Note*: For further information go to http://water.epa.gov/lawsregs/lawsguidance/cwa/vessel/CBA/about.cfm].

5. If possible, incorporate some cross-training on the care and maintenance of all types of watercraft in preparing staff that will be extensively involved in watercraft decontamination. Consider asking a local marine mechanic to provide some instruction in the basics relating to cooling system and ballast tank flushing and other mechanical elements of commonly encountered watercraft that will help staff better understand the watercraft they will be entrusted to work with, project a more confident approach, and maintain better public relations (Figure 14.9).

NOTE ON KILLING VELIGERS

Ongoing research (Sykes, personal communication) has indicated that killing veligers in water is much more difficult than previously thought and that veligers are resistant to some chemicals currently recommended for that purpose. The research also indicates that veligers that may appear dead immediately following chemical treatment may revive after several hours of recovery time. Research by Craft and Myrick (2011) showed that quagga mussel veliger predicted survival times (in static water baths) ranged from less than a day at 35°C to *at least 24 days at 10°C*.

Standards

1. Use *140°F* or hotter water (at the point of contact) to kill mussels and veligers. Water loses approximately 10°F–15°F/ft of distance when sprayed from a power nozzle, so initial temperature should be increased to account for this heat loss to the point of contact. Monitor water temperature at the point of contact to be sure that equipment is operating as required before initiating decontamination.

2. Use a plastic scraper, brushes, and gloves to remove attached mussels before applying hot water spray to significantly reduce the time required to complete a watercraft decontamination.

3. When using a hot water pressure washer and/or flushing attachment to kill and remove attached mussels from the surface of watercraft/equipment, allow *at least 5 s for quagga mussels* and *10 s for zebra mussels* to elapse from the leading edge of the spray to the

tailing edge when moving the wand across the surface to maintain sufficient *lethal* contact time. If larger mussels are present, it may require more time to remove them from the surface than to kill them. [*Note*: If you are unsure whether you have quagga or zebra mussels, use 10 s to be safe.]

4. Use a power wash unit capable of spraying at least 5 gal/min with a nozzle pressure of 3000 psi or greater (not to exceed 3500 psi) to remove attached visible mussels from all exposed surfaces of the watercraft, piece of equipment, trailer, and engine.

NOTE ON NOZZLE HEAD CONFIGURATION

Be sure to use a nozzle head that directs the water in a fan-like rather than a pinpoint spray. The shape of the spray as determined by the nozzle head used should be 2–3 in. wide 8 in. out from the head to avoid any paint damage and allow a wider spray area of greater lethal contact time. We also recommend a 40° flat fan spray nozzle (anything less than 40° nozzle can cause damage to the surface) and a 12″ standoff to get the maximum coverage and to prevent damage to the vessel.

NOTE ON REMOVING ATTACHED MUSSELS

When attached mussels are allowed to dry for several days and desiccate, their byssal threads begin to decompose. Removing mussels through scraping or by power washing after a period of drying requires considerably less effort and can be accomplished with lower nozzle velocities than those required for live mussels (3000 psi) (Figure 14.10).

5. Use a flushing attachment to flush all hard-to-reach areas and those areas where pressure may damage the watercraft or equipment (such as the rubber boot in the gimbal area). A brush may also be used in conjunction with flushing to remove more mussels from hard-to-access areas.

6. When flushing hard-to-reach and sensitive areas, maintain a contact time of *70 s* to assure that mussels receiving only indirect contact are killed since it may not be possible to remove them from these areas.

7. First drain and then use a flushing attachment and *130° water* to maintain the contact time of *70 s* to flush the live well, bait well, wet storage compartments, and bilge areas, to kill any mussels and veligers that might be present. (*Note*: Alternatively, live/bait well, bilge areas can be filled with 130° water and held for 30 s, and then drained.)

8.

NOTE ON *HIDDEN* MUSSELS

It may not be possible to remove all attached mussels from every area of the watercraft/equipment. The standard is to remove all *visible* mussels. A day or two following a very thorough decontamination, it is not unusual for mussels to appear as byssal threads decompose and mussels slide out of hidden areas to become visible. In addition, there are some areas of almost any watercraft or piece of equipment that cannot be easily accessed to remove dead mussels. If properly treated, these mussels are dead and are in the process of decay. Brushes may be used in conjunction with flushing in some of these areas when doing the initial decontamination to reduce (not eliminate) this from occurring (Figure 14.11).

9. Use appropriate attachment connected to the power wash unit or other hot water source, start the engine, and run for *130 s* to kill mussels in the engine cooling system (Figure 14.12).

FIGURE 14.10 Removing attached mussels with plastic scrapper prior to pressure washing. (Photo courtesy of Bill Zook, 2011.)

FIGURE 14.11 Using a brush to remove and flushing attachment to kill mussels in the gimbal area. (Photo courtesy of Bill Zook, 2011.)

FIGURE 14.12 Flushing live well (note thermometer to maintain lethal temperature). (Photo courtesy of Bill Zook, 2011.)

WARNING ON ENGINE COOLING SYSTEMS

Marine engine cooling system pumps and engines are not designed to operate at less than 7 gallons per minute (gpm) over an extended period, and most current power wash units are not designed to deliver more than 5 gpm. Therefore, when using a power wash unit for this purpose, it is important to limit run-time to *130 s* to avoid any possible engine/pump/impeller damage. No such limitation exists if an outboard is *tank run* in hot water without the use of a power wash unit.

There must be enough volume to properly supply an engine's cooling system in order to keep them from overheating. Five gpm will suffice as long as the engine is idling. In all cases, the operator must watch the temperature gauge during the flushing process.

10. Some ballast system manufacturers have indicated that their pumps and/or other electrical system components are designed for temperatures of no more than 120°. For that reason, we recommend using a 3–4 ft hose extension from the end of the flushing attachment to introduce hot water from the source to the ballast or raw water storage tank. The extension allows the water temperature to cool by an additional 15°–20° in order to reduce effective water temperatures in the bladder or tank to below 120°. To maintain lethal temperatures long enough to achieve 100% mortality, it is important to pump water into the area until the exiting water reaches a temperature of *120°F*. The exiting water temperature can be monitored with a handheld temperature gauge or thermometer. Leaving the water in that area for a minimum of *130 s* will assure 100% mortality (Figure 14.13).

11. Use the flushing attachment to treat PFDs, anchor and lines, paddles and oars, water toys, boat fenders, and other equipment that have been in the water by flushing (or spraying if it will not damage the equipment) with *140°F water* to kill any veligers or mussels present (remember that equipment fouled with settled mussels will require more time to decontaminate, see point (3)).

12. All accessible surfaces in watercraft and equipment trailers should be sprayed with 140° or hotter water. Since trailers are normally out of the water, juvenile and adult mussel are not usually attached to any surfaces; however, mussels can become scraped-off watercraft and equipment during loading and become lodged on the trailer and should be removed with hot water spray. Be sure to drain and flush all hollow frame members (Table 14.1).

FIGURE 14.13   Using hose extension to lower hot water flush temperature to 120°F to treat ballast tank. (Photo courtesy of Bill Zook, 2011.)

TABLE 14.1

Lethal Water Temperatures, Exposure Time, and Mortality Rates at Day 10 for Quagga Mussels as Tested at Lake Mead

Temperatures		Times in seconds							
°C	°F	1 s	2 s	5 s	10 s	20 s	40 s	80 s	160 s
20	68	4	4	6	0	0	2	2	0
40	104	2	2	8	12	**94**	**100**	**100**	**100**
50	*122*	10	22	36	82	**100**	**100**	**100**	**100**
54	*129*	54	72	98	**100**	**100**	**100**	**100**	**100**
60	*140*	72	92	**100**	**100**	**100**	**100**	**100**	**100**
70	158	88	**98**	**100**	**100**	**100**	**100**	**100**	**100**
80	176	86	**98**	**100**	**100**	**100**	**100**	**100**	**100**

Source: Wong, D. et al., Using hot water spray to kill quagga mussels on watercraft and equipment, Final Report to PSMFC and USFWS, Department of Environmental and Occupational Health, University of Nevada, Las Vegas, NV, 2011. Control mortality (n = 4) was 3%. Italic value emphasize the quagga mortality rate of 100%. Bold values emphasize the estimated VHS and spiny flea mortality rate of 100%.

13. When carpeted bunks are present, flush for at least 70 s with 140° water using a slow flush along the bunk that will allow the capillary action to pull enough hot water through the carpet to kill any veligers present. Any dislodged adult or juvenile mussels landing on the bunks will be killed by crushing action so the boat does not need to be removed to access this area.

14. Always use a thermometer or temperature logger to verify and maintain proper water temperatures at the point of contact.

WARNING ON WATERCRAFT/ENGINE DAMAGE CAN OCCUR IF DECONTAMINATION PROTOCOLS ARE NOT FOLLOWED

The most likely place where the decontamination process may cause damage to a watercraft or marine engine are during the cooling system flush where it is critical that engines are run at idle for a maximum of 130 s and that the ear muffs or *fake-a lake* attachment is properly and securely sealed, and in the ballast tank flush (where it is critical that water temperature be reduced to 120°F or less to avoid damage to the electrical components). If these and other protocols are strictly applied, there is little prospect of damage resulting from the application of these protocols and standards (Figure 14.14).

FIGURE 14.14 Flushing life preservers (PFDs). (Photo courtesy of Bill Zook, 2011.)

NOTE ON CHEMICAL TREATMENTS TO KILL DREISSENID MUSSELS

A number of agencies and organizations in the western United States currently recommend and use various chemical compounds for decontamination including potassium chloride (KCL), formalin, vinegar, and other substances. The latest and ongoing research by Catherine Sykes of the U.S. Fish and Wildlife Service being conducted at Willow Beach National Fish Hatchery on the Colorado River indicates that at the recommended concentrations, KCL and formalin are not effective decontamination agents. Vinegar has not yet been tested, but until similar research confirms that it is effective, we have chosen not to recommend any chemical treatments. For the latest information on chemical compounds and their effect on adult and juvenile dreissenid mussels, please contact Catherine Sykes at CatherineSykes@fws.gov.

Quarantine or Drying Time

If watercraft and/or equipment suspected of carrying zebra or quagga mussels cannot be decontaminated for any reason, then they must be held out of water for a period of time necessary to desiccate and kill all mussels and veligers on board through desiccation. The amount of time required to achieve complete desiccation varies depending on temperature, relative humidity, and size of the mussels, and can range from 1 to 30 days (McMahon 2009; Ussery et al. 1995).

Quarantine/drying is likely the most effective way to assure that live mussels are not transported between water bodies on trailered watercraft or equipment (Morse 2009). The biggest concern with quarantine/drying is that it does not remove attached mussels. If mussels remain on the vessel, they will eventually drop off. If that occurs at a boat ramp or beach, the presence of mussel shells can raise concern of a new infestation, triggering alarm and resulting in expensive and unnecessary action. For that reason, we recommend that all visible mussels be removed from quarantined/dried watercraft before they are allowed to launch.

NOTE ON TREATING BALLAST TANKS

Remember, drying time does not apply in the same way to watercraft with ballast tanks or other water storage areas that are not easily accessed for inspection and cannot be completely drained. If these areas maintain water, then the actual time required to achieve 100% mortality either through desiccation or through anoxia will most likely exceed the drying time standards recommended here.

The 100th Meridian Initiative has developed a quarantine time calculator based on the research conducted by Dr. Robert McMahon and others at the University of Texas, Arlington (McMahon and Ussery 1995). That calculator is available on the organization's website http://www.100thmeridian.org/emersion.asp. When practical, we recommend using this standard for determining the length of quarantine or drying time needed to assure that a watercraft or piece of equipment is safe to launch (except when ballast tanks or other inaccessible raw water storage systems are involved). When this level of precision is not practical for field operation, a second, more easily calculated and remembered standard is also recommended in the following.

Protocols

1. Requiring quarantine, drying time, or a waiting period should be applied to all watercraft and equipment that meet the definition of high risk, either in lieu of decontamination or in addition to decontamination as an *insurance policy.*
2. Implementation of this option can take several forms.
   a. Physically quarantining a watercraft or piece of equipment requires providing a safe and secure holding area where it can be *parked* for the amount of time required to desiccate

all mussels onboard. A few agencies/organizations have used this option to take or oversee possession of suspect watercraft (with or without the owner's permission, depending on individual jurisdiction's authority) until they remain out of the water long enough to be considered safe. Establishing and maintaining a dedicated quarantine facility can be expensive and comes with some potential liability issues.

b. When a quarantine facility is not available, then quarantine/drying time can be achieved by banding (secured connection between watercraft and trailer) the watercraft or piece of equipment to the trailer or other means of transport. The operator is advised or required not to launch into any freshwater area until the date indicated on the *band* or an accompanying paper certificate.

c. The final option is simply to require that all high-risk watercraft serve a predetermined drying/waiting period prior to launch (duration determined by risk level and current temperature and humidity conditions). Under this scenario, all high-risk watercraft are prohibited from launching until the required drying time has passed, as determined by the screening interview.

3. All visible mussels should be removed from watercraft or equipment following quarantine or drying period before being allowed to launch.

Standards

1. Where practical, the 100th meridian initiative quarantine time *calculator* should be used to determine the length of quarantine/drying time required (provides the greatest precision but limited availability and predictability for boaters).

2. When the use of the *calculator* is not practical, the following standards should be applied to determine the length of the quarantine/drying time required. (Note: Information provided in the following table was developed in cooperation with Dr. Robert McMahon, University of Texas Arlington.)

Maximum Daily Temperature (°F)	Minimum Days Out of Water	
<30	3	
30–40	28	(4 weeks)
40–60	21	(3 weeks)
60–80	14	(2 weeks)
80–100	7	(1 week)
>100	3	

Note: Add 7 days for temperatures ranging from 32°F to 95°F if relative humidity exceeds 50% (McMahon, personal communication 2009).

3. Watercraft with ballast or other internal water storage tanks that cannot be completely drained should be treated differently with regard to drying time.

Watercraft/Equipment Exclusion (or Sometimes Referred to as Quarantine)

High-risk watercraft that are not decontaminated and/or quarantined whether the result of vessel owner refusal or the lack of available equipment, trained applicators, or facilities should be excluded and not allowed to launch. Exclusion should not be used as a long-term substitute for the development of a more user-friendly and proactive interception program that recognizes the value of recreational boating and seaplane operation to the economy and the legitimate interests and enjoyment of the boating and flying public.

Since dreissenid mussels were first found in the western United States in 2007, some agencies and organizations responsible for water and recreation management have continued to resort to the use of exclusion to protect those resources from the dreissenid mussel threat; however, that number has declined over the past 3 years. The case for using exclusion as a prevention strategy has diminished as agencies and organizations have had time to develop public policy, establish regulations, budget for equipment and manpower, train staff, and purchase equipment needed for more proactive and considerate approaches. However, when exclusion remains the only available option, it should be used to assure that contaminated watercraft are not allowed to enter the waterway.

Protocols

1. High-risk watercraft and equipment that have not been or cannot be decontaminated or meet the quarantined/drying time standard should be excluded from launching.
2. The information obtained from the screening interview used to determine risk level should be shared with the watercraft owner/operator and made available on a real-time basis at all access points to prevent excluded watercraft/equipment from attempting to launch from any other point of access on the same water body.

Standards

1. Watercraft or equipment that were last used in known dreissenid mussel areas within the past 30 days and have not been decontaminated and/or been out of the water for the required time (based on temperature and humidity conditions as determined by either the quarantine time calculator or alternative method recommended here) should be decontaminated if approved facilities are available; placed in self- or on-site quarantine for the required time frame; or excluded.
2. Watercraft that are not clean (having attached vegetation, debris, or surface deposits that can mask the presence of small mussels), drained (having visible water in any live well, bait well, bilge area, engine compartment, floor, or cooler), and dry (been out of the water long enough for attached mussels to desiccate) should be decontaminated and/or quarantined or excluded.

Watercraft Certification/Banding

A growing number of boating and water management agencies and organizations currently offer some form of certification for watercraft or equipment that have passed inspection, been decontaminated, or have remained out of the water long enough to satisfy quarantine/drying time requirements. Certification of this type helps the operator avoid repeated time delays upon reentry and makes it easier for the management agency/organization implementing watercraft/equipment interception programs by reducing workload and processing time and by allowing them to concentrate limited resources on higher-risk watercraft (Figure 14.15).

Some entities currently offer a sticker or paper certificate; however, since there is no way to determine where that watercraft or equipment has been between interceptions, this form of certification offers limited benefit except as an indicator that appropriate fees to support the program have been paid. Many agencies and organizations have addressed this shortcoming by applying *bands* that connect the watercraft/equipment to the trailer so that it cannot be used between interceptions without detection. In some cases, a written certificate is also issued with banding.

If agencies and organizations choose to offer certification, we recommend that the watercraft/equipment be banded in such a manner that it cannot be launched between interceptions without detection. If banding is coordinated between jurisdictions, further action can be expedited (at the discretion of the implementing agency/organization) at the next launch site anywhere in the western United States so long as the band remains intact. Such a system would reduce the amount of staff and equipment time required at interception facilities region-wide, thereby increasing resource protection, saving money, reducing waiting time, and crowding and lowering the frustration level of staff and the boating public alike.

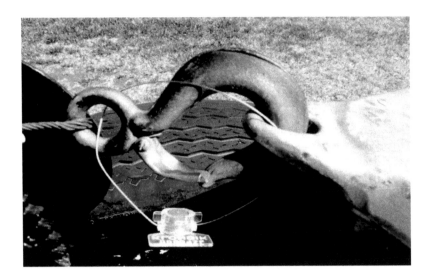

FIGURE 14.15　Banded watercraft. (Photo courtesy of Bill Zook, 2011.)

Protocols

In order to implement a region-wide program that may be acceptable to most agencies and organizations in the western United States, three conditions must be met:

1. The agency/organization placing the band must implement all UMPS to ensure that the best practical science and technology has been employed in certifying the watercraft or equipment.
2. Only those programs that comply with Level 3 inspection and decontamination protocols and standards offer reciprocity opportunity (between jurisdictions).
3. All agencies and organizations participating in the certification program should use a banding system that attaches the watercraft to the trailer that cannot be tampered with or removed without detection. The certification is no longer valid if the band has been tampered with, severed, or removed.
4. While a variety of different *band* styles and materials may continue to be used, all bands should have the following features: This information can either be incorporated into the band or be provided on an accompanying paper receipt or certificate.
   a. The name and contact telephone number of the agency/organization applying the band.
   b. Some way to indicate the basis for certification as one of the following three categories: inspection, decontamination, or quarantine (several options are available including color coding or preprinted number or letter coding).
   c. In the absence of an automated tracking system, the banding date should be indicated on the tag or by providing a dated *paper* certificate.

Standards

1. Only watercraft or equipment that have passed inspection or have been decontaminated or quarantined by trained and certified personnel in accordance with all of the UMPS as adopted should receive certification banding.
2. Certification banding should be applied only by a trained inspector.
3. Watercraft and equipment that have been certified and banded by an agency or organization utilizing these UMPS *may* receive expedited processing at the discretion of the receiving agency/organization in other jurisdictions.

Seaplane Guidelines

1. All seaplane pilots should view the seaplane inspection and cleaning video on the 100th meridian initiative website (http://www.youtube.com/v/luDZptFsQDk?fs=1&hl=en_US) and complete the training course and carry a certificate available online at http://www.aopa.org/asf/video/safety-videos.html?mobileOverride=1

2. Before entering the aircraft
   a. Inspect and remove all aquatic plants or attached mussels, snails, or other animals from all exterior surfaces of floats, wires, cables, transoms, spreader bars, and rudders.
   b. To the extent practical, pump, remove, or otherwise treat (household bleach (one part household bleach to five parts water mixed in a spray bottle) or *140°F* water) all water from floats, wheel wells, and any other compartments or areas of the aircraft that can contain or maintain raw water.

3. Before takeoff
   a. Taxi clear of any aquatic plants.
   b. Reinspect for any visual sign of attached aquatic vegetation.
   c. Raise and lower rudders several times or otherwise remove any aquatic vegetation.
   d. Make sure all floats remain as dry internally as possible during takeoff.

4. After takeoff
   a. Raise and lower rudders several times to free any remaining aquatic vegetation while over the departing water body or dry land.
   b. If aquatic plants persist and are still visible on floats, cables, or rudders, return to the same water body and manually remove them.

5. Storage and mooring
   a. Remove aircraft from the water whenever practical to better facilitate self-inspection, drainage, removal, cleaning, and drying.
   b. Maintain floats and hulls to make sure they remain water tight, including sealing seams, replacing gaskets on inspection covers, and repairing any cracks.

GLOSSARY OF TERMS

Certification: A process whereby watercraft/equipment are determined to present minimal risk based on inspection, decontamination, or quarantine/drying time, and receive some visible form of certification of that fact (e.g., trailer tag, band, paper certificate). It is important to note that it is not possible to certify that watercraft are *free of mussels*, only that the most currently available and effective protocols and standards have been applied to kill and remove all visible mussels.

Clean: Absent visible ANS, attached vegetation, dirt, debris, or surface deposits including mussel shells, byssal threads, or residue on the watercraft, trailer, outdrive, or equipment that could mask the presence of attached mussels.

Drained: To the extent practical, all water drained from any live well, bait well, storage compartment, bilge area, engine compartment, floor, ballast tank, water storage and delivery system, cooler or other watered area of the watercraft, trailer, engine, or equipment.

Dry: No visible sign of water on or in the watercraft, trailer, engine, or equipment. Out of the water long enough to be totally dry.

Decontamination: The process of killing and removing all visible attached mussels and, to the extent practical, killing all veligers and concealed mussels from every area of watercraft, trailer, and equipment.

Exclusion: Not allowing watercraft or equipment to be launched. In extreme cases, exclusion can be applied to all watercraft but, in most cases, is applied to only watercraft and equipment that are considered to be high risk, when other options are not available.

High-Risk Water Body: The determination of *high-risk water body* is the prerogative of the responsible management entity. Some of the factors used to determine risk potential include the following:

Whether water quality parameters will support the survival, growth, and reproduction of dreissenid mussels

The amount and type of boater use

Proximity to dreissenid positive or suspect waters

Whether the water in question is a headwater, water, or power supply system, or supports listed species

High-Risk Watercraft/Equipment: Any vessel or piece of equipment that has operated on or in any water body known or suspected of having zebra or quagga mussels in the past 30 days or any watercraft or equipment that is not clean and, to the extent practical, drained and dry.

Screening Interview: Asking the vessel operator a series of questions prior to launching or entry that are designed to determine the level of risk based on the recent history of use. This should be an element of every intervention program.

Quarantine/Drying Time: The amount of time out of the water required to assure that all mussels and veligers are killed through desiccation. This time requirement varies widely depending on temperature and humidly conditions.

Self-Inspection (Voluntary/Mandatory): A self-inspection program can be implemented alone or as an *off-hour* adjunct to a more direct and comprehensive inspection program. This type of program involves requiring (mandatory) or requesting (voluntary) the cooperation of individual watercraft operators to complete an inspection of their vessel prior to launching by following a set of instructions and completing a checklist provided at an entry station or kiosk.

Reciprocity: The acceptance of watercraft/equipment inspection and/or decontamination by several or all jurisdictions when similar protocols and standards are employed by similarly trained and motivated professionals.

Watercraft/Equipment Inspection: Where all or selected watercraft are subjected to a thorough visual and tactile inspection of all exterior and interior surfaces, areas of standing/trapped water, trailer, and equipment to determine the presence or likelihood of mussel contamination.

Watercraft Interception Program (WIP): Any program that seeks to prevent the spread of dreissenid mussels and other aquatic nuisance species (ANS) on trailered watercraft or equipment by requiring that they be cleaned and, to the extent practical, drained and dried prior to launching.

To access a list of Agencies and Organizations Implementing Watercraft Interception Programs in the western United States (January 2012), *see* http://www.aquaticnuisance.org/wit/reports Attachment #1

REFERENCES

Baldwin, W., E. Anderson, L. Dalton, E. Ryce, S. Ellis, M. Anderson, R. Francis, M. Davis. 2008. Best inspection and cleaning procedures for all water craft owners. 100th Meridian Initiative Report (unpublished). http://wildlife.utah.gov.quagga/pdf/boat_inspection/pdf.

Claxton, W.T., A. Martel, R.M. Dermott, E.G. Boulding. 1997. Discrimination of field-collected juveniles of two introduced dreissenids (*Dreissena polymorpha* and *Dreissena bugensis*) using mitochondrial DNA and shell morphology. *Can J Fish Aquat Sci* 54:1280–1288.

Comeau, S., S. Rainville, W. Baldwin, E. Austin, S. Gerstenberger, C. Cross, W. Wai Hing. 2011. Susceptibility of quagga mussels (*Dreissena rostriformis bugensis*) to hot-water sprays as a means of water craft decontamination. *Biofouling* 27(3):267–274, first published on March 7, 2011 (iFirst).

Craft, C.D., C.A. Myrick. 2011. Evaluation of quagga mussel veliger thermal tolerance. Report prepared for the Colorado Division of Wildlife, Department of Fish, Wildlife, and Conservation Biology, Colorado State University, Fort Collins, CO, 21pp.

DiVittorio, J., M. Grodowitz, J. Snow. 2010. Inspection and cleaning manual for equipment and vehicles to prevent the spread of invasive species. Technical Memorandum No. 86-68220-07-05. U.S. Department of the Interior, Bureau of Reclamation, Denver, CO.

Independent Economic Analysis Board of the Northwest Power and Conservation Council. 2010. Economic risk associated with the potential establishment of zebra and quagga mussels in the Columbia River Basin. IEAB 2010-1. Portland, OR.

McMahon, R.F., T.A. Ussery. 1995. Thermal tolerance of zebra mussels (*Dreissena polymorpha*) relative to rate of temperature increase and acclimation temperature. U.S. Army Corps of Engineers, Waterways Experiment Station, Vicksburg, MS, pp. 1–21.

Morse, J.T. 2009. Assessing the effects of application time and temperature on the efficacy of hot-water sprays to mitigate fouling by *Dreissena polymorpha* (zebra mussels Pallas). *Biofouling* 25(7):605–610.

Randall, W.J. Summer/Fall, 1999. National voluntary ANS guidelines: A strategy to interrupt recreational pathways of spread. *ANS Update* 5(3):1. Minnesota Department of Natural Resources, St. Paul, MN. http://www,gic.org/ans/ansupdate/pdf/ans1099.pdf.

University of California Agriculture and Natural Resources, Cooperative Extension. October 22, 2010. *Minutes of the California Dreissenid Mussel Summit*. Lions Gate Hotel, Sacramento, CA. http://www.terc.ucdavis.edu

USACE. 2009. Lake Tahoe region aquatic invasive species management plan, California—Nevada. 84pp ± Appendices. http://www.trpa.org/wp-content/uploads/01_Updated_Lake-Tahoe-AIS-Management-Plan_Final_July-2014.pdf.

Ussery, T., R.F. McMahon. 1995. Comparative study of the desiccation resistance of zebra mussels (*Dreissena polymorpha*) and quagga mussels (*Dreissena bugensis*). Center for Biological Macrofouling Research, University of Texas at Arlington, Arlington, TX.

Western Regional Panel on Aquatic Nuisance Species. 2009. Quagga-zebra mussel action plan, 45pp. http://www.aquaticnuisance.org.

Wong, D., S. Gerstenberger, W. Baldwin, E. Austin. 2011. Using hot water spray to kill quagga mussels on watercraft and equipment. Final report to PSMFC and USFWS. Department of Environmental and Occupational Health, University of Nevada, Las Vegas, NV.

Zook, W.J. 2010. Recommendations for seaplane inspection and decontamination for aquatic nuisance species. Pacific States Marine Fisheries Commission, Portland, OR, 5pp.

Zook, W.J., S.H. Phillips. 2009a. Recommended uniform minimum protocols and standards for watercraft interception programs for dreissenid mussels in the western United States. Prepared for the Western Regional Panel. Pacific States Marine Fisheries Commission, Portland, OR, 53pp.

Zook, W.J., S.H. Phillips. 2009b. A survey of watercraft intervention programs in the western United States. Report for the Western Regional Panel. Pacific States Marine Fisheries Commission, Portland, OR, 69pp.

Zook, W.J., S.H. Phillips. 2010. Preventing the transfer of dreissenid mussels and other aquatic nuisance species in North America by commercial watercraft and equipment transport providers. Report for the Western Regional Panel. Pacific States Marine Fisheries Commission, Portland, OR, 37pp.

Personal Communications

(Not all specifically cited, but all of whom provided valuable input and/or important document review)

1. K. Smith and J. Foust, Hydro Engineering, Salt Lake City, UT.
2. Dr. D. Britton, USFWS, Arlington, TX.
3. Dr. R. McMahon, University of Texas, Arlington, TX.
4. W. Baldwin, Lake Mead Boat Owners Association, Boulder City, NV.
5. L. Dalton, Utah Division of Wildlife Resources, Salt Lake City, UT.
6. S.E. Anderson and A. Pleus, Washington Department of Fish and Wildlife, Olympia, Washington, DC.
7. R. Billerbeck and G. Seagle, Colorado State Parks, Denver, CO.
8. D. Norton and B. McAlexander, California Department of Fish and Game, Sacramento, CA.
9. T. McMahon and K. Bergersen, Arizona Game and Fish Department, Phoenix, AZ.
10. M. Pike and S. Senti, Quagga Inspection Services, Red Bluff, CA.
11. S. Wickstrum, General Manager, Casitas Municipal Water District, Oak View, CA.
12. S. Smith, United States Geological Service, Seattle, WA.
13. P. Heimowitz, United States Fish and Wildlife Service, Portland, OR.

14. K. Kreif, Lake Kahola Zebra Mussel Committee, Emporia, KS.
15. C. Dearman, Powerwash Plus, Boise, ID.
16. T. McNabb and L. Elgethun, Clean Lakes Inc., Coeur d' Alene, ID.
17. K. Zeile, Prefix Corporation, Rochester Hills, MI.
18. B. Rappoli, EPA, Washington, DC.
19. S. Mangin, ANS Task Force, Arlington, VA.
20. C. Sykes, USFWS Dexter National Fish Hatchery and Technology Center, New Mexico.
21. L. Willett, Bureau of Reclamation, Boulder City, NV.
22. R. De Leon, Metropolitan Water District of Southern California, La Verne, CA.
23. K. Klett, County of Santa Clara, Los Gatos, CA.
24. B. Phillips, Monterey County Water Resources Agency, Salinas, CA.
25. A. Ferriter, Idaho Department of Agriculture, Boise, ID.
26. D. Wong, University of Nevada Las Vegas, Las Vegas, NV.
27. B. Zook and S. Phillips, PSMFC, Portland, OR.
28. W. Windus, National Seaplane Pilots Association, Los Angeles, CA.
29. T. Banek, Missouri Department of Conservation, Jefferson City, MO.

Watercraft interception program details and manuals were used as references in this document from the following: (Not all specifically cited, but many concepts for UMPS taken from the review of these manuals)

1. Arizona Game and Fish Department 2011. Decontamination procedures—Day users and long term use & moored boats. Phoenix, AZ.
2. California Department of Fish and Game. 2008. A guide to cleaning boats and preventing mussel damage. Sacramento, CA, 20pp.
3. Casitas Municipal Water District. 2007. Lake Casitas Recreation Area invasive species contamination threat. Information, training & guidelines for protection of water Quality. Ventura, CA, 35pp.
4. Colorado Division of Wildlife. 2011. *Aquatic Nuisance Species (ANS) Watercraft Decontamination Manual.* Denver, CO, 60pp.
5. Colorado Division of Wildlife. 2009. Aquatic nuisance species (ANS) watercraft inspection handbook, Official State of Colorado watercraft Inspection and decontamination procedures. Denver, CO, 48pp.
6. Colorado State Parks. 2008. *Colorado State Parks Aquatic Nuisance Species (ANS) Inspection and Education Handbook,* Version 2. Denver, CO, 107pp.
7. East Bay Municipal Utility District. 2009. Quagga/zebra mussel prevention program. Oakland, CA.
8. Kahola Homeowners Association. 2009. Zebra mussels information for Kahola. Emporia, KS.
9. Los Angeles Department of Water & Power and Crowley Lake Fish Camp. 2009. Crowley Lake—Boat use survey and vessel inspection certification Form. Los Angeles, CA.
10. Metropolitan Water District of Southern California. 2008. Watercraft and equipment inspection and cleaning procedures for Diamond Valley Lake and Lake Skinner. Los Angeles, CA, 18pp.
11. Nevada Department of Wildlife. 2008. Aquatic nuisance species prevention and disinfection guidelines. Las Vegas, NV, 16pp.
12. Oregon Marine Board. 2010. Angler/Boater Survey Questions and Aquatic Nuisance Species Boat Inspection Form. Salem, OR.
13. Oregon State Marine Board. 2008. Quagga/zebra mussel, Dreissena enforcement strategy & protocol (draft). Oregon State Police, Oregon Department of Fish and Wildlife, and County Sheriff Departments. Salem, OR, 12pp.
14. Palmquist, E., J. Granet, and M. Anderson. 2008. Zebra mussel prevention at Glen Canyon NRA in 2007. National Park Service, Glen Canyon National Recreation Area. Page, AZ, 17pp.
15. Ruth Lake Community Service District. 2009. Watercraft inspection and banding procedures instructions for inspectors. Mad River, CA, 8pp.
16. Tahoe Resource Conservation District. 2010. Screening process for aquatic invasive species and Lake Tahoe aquatic invasive species watercraft inspection form. South Lake Tahoe, CA, 3pp.

17. Utah Division of Wildlife Resources. 2009. How to decontaminate your boat and mussel-free certification. Salt Lake City, UT.

18. Utah Division of Wildlife Resources. 2009. Requirements to prevent the spread of aquatic invasive species (Self certification form for watercraft owners). Salt Lake City, UT.

19. Washington Department of Fish and Wildlife, Fish and Wildlife Enforcement. 2009. Invasive species vessel inspection form. Olympia, WA.

20. Whiskeytown National Recreation Area. Date Unknown. Quagga and zebra mussel-free certification. Whiskeytown, CA, 3pp.

21. State of Wyoming Aquatic Invasive Species (AIS). *Watercraft Inspection Manual.* 2010.

15

Implementing Hazard Analysis and Critical Control Point (HACCP) Planning to Control a Pathway from Spreading Dreissenid Mussels and Other Aquatic Invasive Species

David K. Britton

CONTENTS

ABSTRACT   Hazard analysis and critical control point (HACCP) planning is an easy and effective way to ensure that natural resource managers and staff will do no harm to the environment while performing specific activities or practices. Transporting equipment, conducting research, collecting samples, restoring habitat, and similar activities may provide pathways for nonnative species to invade. Although many feel a responsibility to protect and conserve our environment, attention is often narrowly focused on a specific management objective, an approach that may result in more harm than good when viewed from a broader perspective. This is avoidable with proper planning. HACCP allows natural resource management work to be assessed with respect to the nonnative and potentially invasive species that could be introduced. It allows for the development of a strategy to eliminate or minimize the risk of spread and provides a back-up plan when and if risk-reduction measures fail. HACCP is an effective tool used successfully by the National Aeronautics and Space Administration and the food industry for decades. The National Sea Grant Program adapted HACCP for the baitfish aquaculture industry, and the U.S. Fish and Wildlife Service in partnership with the National Oceanic and Atmospheric Administration expanded this idea by adapting HACCP for all natural resource management activities. This chapter serves as a brief overview of HACCP, using an example where zebra or quagga mussels, as well as other potential invaders, could be inadvertently moved during a common conservation activity. The five steps of HACCP will be discussed to show how the steps work together to create a comprehensive plan to prevent unwanted invasions.

Stewardship

Invasive species alter our native ecosystems, often permanently, causing ecological and economic damage and may negatively impact human health. In the United States, invasive species have impacted nearly half of the species currently listed as threatened or endangered under the U.S. Federal Endangered Species Act (Wilcove et al., 1998). Nonnative species are capable of altering the habitats they invade by reducing the abundance of native species and modifying food webs and other ecosystem processes. These species also threaten economic development by increasing costs of business, decreasing revenue, reducing fishing opportunities, and causing reservoir closings. Throughout history, epidemic diseases have spread using organisms as vectors and reservoirs. Diseases such as West Nile Virus, viral hemorrhagic septicemia (VHS), swine flu (H1N1), and rat lungworm are all examples of health threats carried by invasive species.

Fortunately, these negative impacts can be avoided or minimized by understanding the problem and taking necessary precautions. However, society, in many ways, tends to be reactionary rather than proactive, even though the old adage *an ounce of prevention is worth a pound of cure* is well known. Too often, the lack of a proactive approach has resulted in widespread invasions and costly long-term control efforts. For example, quagga and zebra mussels (*Dreissena polymorpha* and *D. rostriformis bugensis*, respectively) invaded the Great Lakes in the 1980s and have since spread throughout much of the eastern United States and now have a foothold in the Colorado River. They are now a major threat to western regions.

The pathways by which nonnative species arrive to new locations are numerous. Some of the most obvious are the intentional release of captive animals from cages or aquaria, dumping of unwanted plants, and escape from aquaculture farms and gardens that use nonnative plants. The shipping industry is a prevalent pathway as it has unintentionally aided the introduction of numerous species that were transported, on the hulls of ships, in ballast water, and in cargo carried from overseas. For example, zebra and quagga mussels came to the United States in ballast tanks or on anchors and chains of transoceanic ships.

Nonnative species may also be intentionally introduced through stocking, restoration, or farming activities. Examples include the eastern brook trout (*Salvelinus fontinalis*) and rainbow trout (*Oncorhynchus mykiss*) stocked for recreation, Chinese mitten crab (*Eriocheir sinensis*) introduced as a biological control for a different invader, and salt cedar (*Tamarix ramosissima*) planted for erosion control efforts. Even species introduced with good intentions are still capable of spreading and causing widespread damage. Unfortunately, the impacts from introductions are often realized too late and may result in the need for significant long-term control efforts.

It is important to understand that all actions have consequences. Careful planning is necessary to ensure that precautions are being taken to prevent unwanted results. Planning is not always easy, but it is essential. Fortunately, there exists a standard system for planning natural resource management activities, known as hazard analysis and critical control point (HACCP) planning. This system is an international standard (ASTM E2590-09) used by many federal and state agencies for invasive species risk assessment and planning to ensure that all reasonable efforts have been made to minimize the potential for inadvertently moving or introducing unwanted biological organisms.

Traditionally, medical doctors take an oath to do no harm. This is necessary because physicians' actions (or inactions) can permanently damage or destroy an individual's health. This oath requires doctors to act cautiously and remember that their actions may potentially result in irreparable damage. Natural resource managers and other environmental stewards should show a similar level of precaution, as native species, ecosystems, recreational opportunities, and economic health are all at stake. All of these could be quickly impaired or destroyed if prudent caution is not exercised. It is the responsibility of environmental stewards to take care of natural resources by being aware and vigilant and treading lightly on the earth. When it comes to preventing invasions, good stewards must strive to be part of the solution, not part of the problem.

HACCP Overview

Without appropriate planning, natural resource management work could become a potential pathway for spreading invasive species. In 1999, President Clinton signed Executive Order 13112, which stated the following:

> Each federal agency whose actions may affect the status of invasive species shall, to the extent practicable and permitted by law, identify such actions… [and] …not authorize, fund, or carry out actions that it believes are likely to cause or promote the introduction or spread of invasive species in the United States or elsewhere unless, pursuant to guidelines that it has prescribed, the agency has determined and made public its determination that the benefits of such actions clearly outweigh the potential harm caused by invasive species; and that all feasible and prudent measures to minimize risk of harm will be taken in conjunction with the actions. (Clinton, 1999)

Essentially, federal agencies are required to evaluate whether their actions could move invasive species, and if so, either cease performing such actions or take precautions to minimize the risk. Although the order applies only to federal agencies, many other natural resource management entities comply for the sake of good stewardship. The best way to assess and manage the risk from nonnative species is using the HACCP planning process. For this reason, the U.S. Fish and Wildlife Service (USFWS) adopted HACCP as official policy for all of its fisheries offices and hatcheries.

Risk for the purposes here is defined as unwanted exposure to potential negative consequences. HACCP planning is a systematic tool for managing the risk of spreading invasive species. HACCP planning allows natural resource managers and staff to examine routine activities and assess whether anthropogenic pathways may allow for the unintentional transport of nonnative, potentially invasive species. HACCP helps identify where potential risks exist and identify the most effective opportunities to reduce this risk. HACCP includes safeguards such as prescribed ranges, limits, or criteria for specific control measures and prearranged corrective actions to ensure that control is achieved. In short, HACCP evaluates specific tasks for their potential to spread invasive species and allows any such risks to be minimized or eliminated in an efficient and effective way.

HACCP planning has its roots in the food industry. The Pillsbury Company pioneered the concept in the early 1960s in partnership with the National Aeronautic and Space Administration (NASA) as a way to prevent contaminants and pathogens from entering the food supply for the U.S. space program. Pillsbury and NASA considered anything that could negatively impact food products to be *hazards*. Subsequently, HACCP has spread throughout the food industry and is now a compulsory practice required by the Food and Drug Administration and the U.S. Department of Agriculture. Eventually, the National Sea Grant Program recognized the value of HACCP and applied it to the wild baitfish industry to reduce the risk of spreading aquatic invasive species (Gunderson and Kinnunen, 2006). The National Sea Grant had redefined the term *hazard* to include aquatic invasive species. Expanding the idea further, the USFWS adopted HACCP and later partnered with the National Oceanic and Atmospheric Administration (NOAA) to develop a straightforward HACCP planning process applicable to any natural resource management activity (Britton et al., 2012).

Getting Started

A successful HACCP plan depends on the assembly of an effective HACCP team that includes managers, biologists, and field staff. Plans created by a single person are usually less thorough and lack the variety of perspectives necessary to conduct a comprehensive analysis to ensure that possible pathways of species introduction are not overlooked. Diversity in job duties and expertise is important to ensure that the team becomes familiar with the overall management objective, on-the-ground actions, as well as the variety of organisms that could be moved through work conducted. Each of these components is essential to create plan that is both effective and feasible.

In this chapter, the basics of the USFWS/NOAA version of HACCP Planning to Prevent the Spread of Invasive Species will be discussed with an example that relates to quagga and zebra mussels as well as other notorious invaders. This example will encompass all five steps of HACCP planning: (1) activity description, (2) activity flow chart, (3) identify potential non-targets, (4) analyze non-targets, and (5) create a risk action plan. We will address these in order. The term *non-target* will be defined in detail later; for now, think of *non-target* as synonymous with invasive species.

HACCP Step 1: Activity Description

Activities are projects performed in order to meet management objectives. For example, a management objective may be to restore a stream to open it for fish passage. An activity that may be conducted to meet this objective might be *stabilize the stream bank using plants, tree stumps, and logs*. HACCP may be applied to many different kinds of activities. However, HACCP plans work best when applied to very specific rather than general activities. For example, an HACCP plan written for *backpack electrofishing in Northern Arizona Lakes* is more likely to be successful than an HACCP plan written for *fish collection using various methods*. It should be easy to answer the questions *who*, *what*, *when*, *where*, *why*, and *how*. If one finds this difficult, especially if the answer to any of those questions is *it depends*, then it is likely that one is trying to cover more than a single activity with a single HACCP plan. Instead, try to focus on an individual, specific activity. You may need to create more than one HACCP plan. Saving time or paperwork should not be more important than saving our natural resources.

Activities often contain multiple steps. As part of the planning process, HACCP breaks down activities into specific, sequentially completed tasks. For example, the activity of *collecting plankton tow samples from a motorboat* can be broken down into (1) gather gear, (2) load the boat, (3) drive to the field site, (4) launch the boat, (5) conduct sampling, (6) retrieve the boat, and (7) move to the next site of return to the field station. Notice that each task must be completed before the next task begins. That is, these tasks are sequential.

HACCP plans usually are not effective if they include tasks that occur simultaneously or depend on special circumstances. Often, such activities can be broken down into subactivities that are better suited for individual HACCP plans. To illustrate this, it is helpful to look forward to the next step in HACCP planning, which is to diagram activity flow. Step 2 lays out the sequence of tasks in order from start to finish. This order must be linear. If an activity will require a complex (nonlinear) flow chart with *if/then* decisions, then your activity is too general to be served well by a single HACCP plan. As an example, imagine creating an HACCP plan for the activity, *collecting samples*. The broad nature of this activity may include tasks such as *seining*, *electrofishing*, *plankton tows*, *gill-net sampling*, *Surber sampling*, and/or *Ponar grab sampling*. The techniques and order used may vary each time the activity is performed. For example, on a particular day, only plankton samples may be collected, thus tasks are not completed for *seining* or *gill netting*. Some days such tasks may be skipped; other times, there may be another person or crew conducting these tasks simultaneously. In these situations, an activity flow chart for the *collecting samples* activity will likely need some *if/then* decisions. But effective HACCP flow charts should generally not include *if/then* decisions, thus avoiding unnecessary complexity. Squeezing multiple activities into a single HACCP plan may seem like a good idea to reduce paperwork, but it will also add complexity and make analyzing tasks very difficult when trying to minimize the risk of spreading invasive species at the field level. To avoid this, each one of the earlier *tasks* (e.g., *seining* and *electrofishing*) should be treated as an *activity* that has its own set of tasks. In this case, separate HACCP plans are needed for each type of collection method—not one broad plan that covers everything that may or may not occur while collecting samples. It is strongly advised to fight the urge to lump multiple activities under a single HACCP plan in order to save time and paperwork. Doing so will not only make completing the hazard analysis (step 4) very difficult, but also will weaken the effectiveness of the plan in minimizing the risk of spreading unwanted organisms. A good steward would take the time to create specific, effective HACCP plans for all reasonable activities.

A hypothetical HACCP plan for conducting plankton tows on a motorboat is used to demonstrate the HACCP planning process. Figure 15.1 shows the completed first page for this HACCP plan or *Step 1—Activity Description*. In the top sections of the form, a clear, distinct title and management objective for the plan is written. In this example, the HACCP plan is titled "HACCP Plan for Plankton-Net Sampling from a Trailered Motorboat," and "Conduct aquatic plankton-net sampling using a motorboat that is trailered from site to site" has been inserted as the management objective (Figure 15.1). To fill out the pertinent contact information (Figure 15.1), name the person who knows most about the particular HACCP plan as the contact person, usually the HACCP plan team leader or point of contact for any questions regarding the plan. The activity description on the lower portion of the form should concisely explain who, what, when, where, how, and why for the activity. It is important to be clear but concise here. The objective is to allow someone unfamiliar with the activity to be able to understand the context and details of the HACCP plan.

The most difficult part of completing the first step in an HACCP plan is choosing an activity suited for HACCP. To determine what activities are suitable, it may help to browse other HACCP plans to use as examples. A database of HACCP plans is available online at www.HACCP-NRM.org. Once the activity description is complete, it is time to move on to step 2: activity flow chart.

Management Objective & Contact Information	
HACCP Plan Title: HACCP Plan for Plankton-Net Sampling from a Trailered Motorboat	
Management Objective: Conduct aquatic plankton-net sampling using a motorboat that is trailered from site to site.	Contact Person: David Britton
	Phone: 505-366-9565
	Email: david_britton@fws.gov

Activity Description i.e. Who; What; Where; When; How; Why
Field station staff keep trailered boats in a garage near the Hypothetical Field Station in Fort Worth, Texas. These boats are trailered to field sites within 100 miles of Fort Worth and used to conduct dreissenid veliger (larvae) sampling at various rivers, reservoirs, and large ponds. Field sampling is conducted regularly between March and October each year, usually twice a month, sometimes more frequently. Trailered boats may visit multiple sites on each excursion, sometimes in different river basins. Sampling is performed to assess whether zebra or quagga mussels are present in the sampled water bodies.

FIGURE 15.1 HACCP Step 1—Activity description.

HACCP Step 2: Activity Flow Chart

If the activity chosen was specific enough for an individual HACCP plan, then it should contain 10 or less sequential tasks. Step 2 will lay these tasks in order from start to finish. Step 1 described simple tasks necessary to conduct the activity *collecting plankton tow samples from a motorboat*. They were (1) gather gear, (2) load the boat, (3) drive to field site, (4) launch the boat, (5) conduct sampling, (6) retrieving the boat, and (7) move to next site or return to the field station. Since Tasks 1 and 2 may occur simultaneously at the same location, they can be combined. This is permissible if it is known that the tasks are related and that there is no significant risk of moving a nonnative species between two tasks. With this minor change, there are now six total tasks. Continuing on with the hypothetical examples from step 1, Figure 15.2 portrays these tasks in sequence with a brief description regarding what exactly is done during each task.

A keen eye will notice that the last task *move to next site or return to field station* has an *or* in the description (Figure 15.2). It was stated earlier that effective HACCP flow charts should not include conditional decisions. However, this exception has been included to illustrate a point. HACCP plans should not consist of *ifs* that can lead into tasks that will branch the flow chart into subactivities that may, or may not, be performed each time the activity is conducted or may occur in parallel. Doing so would make the activity too complex for the HACCP plan to be effective. This concept will become clearer during the hazard analysis in step 4. For now, notice that the flow chart in the hypothetical plan (Figure 15.2) does not have branches; rather, each task sequentially follows the previous task. If additional field sites are visited, steps 2–5 will be repeated. Sampling occurs during every outing. This creates

| Task 1 | **Title:** Gather Gear and Load Boat |
| | **Description:** Gear is collected from the field station and loaded onto the boat. |

| Task 2 | **Title:** Drive to Field Site |
| | **Description:** Boat is trailered behind a pickup truck to a field site where sampling will be conducted. |

| Task 3 | **Title:** Launch Boat |
| | **Description:** Staff launches the boat into a river, reservoir, or pond. |

| Task 4 | **Title:** Conduct Sampling |
| | **Description:** Staff collects samples at multiple locations within water body by towing a 64 micron plankton net behind a motorboat. Samples are stored in 1 L bottles. |

| Task 5 | **Title:** Retrieve Boat |
| | **Description:** Staff returns the boat to the ramp and is put back on the trailer. |

| Task 6 | **Title:** Move to Next Site or Return to Field Station |
| | **Description:** If multiple sites must be sampled, the staff may continue to the next site. Otherwise, they will return the boat to the field station. |

FIGURE 15.2   HACCP Step 2—Activity description: Outline sequential tasks of activity.

a cyclical flow to the activity chart rather than branching. Branching might occur if we had another task after *Task 6—move to next site or return to field station*. What the field crew would do next would depend on whether they *move to next site* or *return to field station*. However, this example HACCP plan does not branch because it ends without any additional tasks. The idea of applying such actions will be discussed in step 4, analyze non-targets. Prior to that, however, step 3 must be completed to identify and list potential non-target species.

HACCP Step 3: Identify Potential Non-Targets

Targets are anything that is moved purposely, including yourself, gear, and/or any organism that you intend to move. Non-targets are anything that may be unintentionally moved. In the case of biological organisms, a non-target species is one that has reasonable potential to be introduced to new habitats by the activity under consideration. For example, if the activity were stocking rainbow trout, then it is obvious that we would want to intentionally move rainbow trout. In this case, rainbow trout would be our *target* species. Any other species is a *non-target species* and a potential hazard. Non-target species should be identified in advance so that appropriate control measures can be implemented to prevent an unintended invasion.

Sometimes natural resource managers purposely move invasive species. Collecting samples and returning these to the laboratory is an example. In this case, the studied invasive species is intended to be moved and, thus, is not a non-target. It is, in fact, a target. We defined *non-target* earlier to be synonymous with *invasive species*. HACCP plans usually focus on non-target species, but target species can be invasive species as well. Precautions still need to be planned so that the hazard, whether a target or non-target species, is not unintentionally released to new habitat. In the example, dreissenid mussels collected via plankton nets would be targets because the crew intentionally moves them. However, any dreissenid mussels in standing water or attached to the boat and equipment would be non-targets because the crew has no intention to move them. Thus, the concepts of *target* and *non-target* are not mutually exclusive.

It is helpful to categorize non-target species into four general categories: vertebrates, invertebrates, plants, and other organisms (e.g., diseases, pathogens, and parasites). This is recommended, as the control measures that will help minimize or eliminate the risk of non-targets are often specific for each of these categories. Plants, for example, may require different control measures than invertebrates. Although focus may be on one particular group or species (e.g., quagga or zebra mussels), all potential non-target species that have a reasonable probability of being moved during the activity should be considered. If there is uncertainty about whether a species should or should not be listed as a potential non-target, it probably should be listed. To reduce the risk from all non-target species, HACCP follows the precautionary principle. The precautionary principle states that if an action could plausibly cause harm, but there is not yet a scientific consensus that the action is actually harmful, then the burden of proof that such an action is not harmful falls on those taking the action. The precautionary principle implies a social responsibility to protect the environment from exposure to harm when careful consideration has found a plausible risk. This protection can be relaxed only if scientific findings emerge that provide sound, convincing evidence that no harm will result. It is up to the HACCP team to determine how far to carry this principle into the plan. Discussions with local experts and other HACCP planning teams about potential non-target hazards can help focus planning objectives and establish the basic foundation for each HACCP plan. Moreover, discussions during the HACCP planning process may reveal non-target species that were not previously recognized as present in the activity area.

Non-target species can often be lumped together into taxonomic groupings if all species within the group can be controlled effectively with the same methods. For example, all potentially encountered mussels can be combined into one group under the invertebrate category. However, this group should include specific examples of mussels of high concern. For example, *invasive mussels* (e.g., *zebra and quagga mussels*) may be one group in the list of potential non-targets and would be treated as one potential hazard. If for any reason a particular non-target species requires special control measures, this species should be listed separately. Thus, when deciding whether to lump non-target species into categories, it is helpful to consider how they will be eventually controlled.

Figure 15.3 shows the non-target species identified for the hypothetical example. To keep this example simple, we assumed that small fish or other vertebrates were not likely to be moved during sampling. Although this may be a convenient determination for the purposes of example, a real-world HACCP plan should consider this possibility carefully before deciding that there would be no vertebrates moved. Zebra and quagga mussels are potentially moved because the crew would be specifically sampling for them. Both dreissenid species are lumped together in one group, as were aquatic weeds (in a second group), because it is anticipated that all species encountered within the groupings will be effectively controlled using the same control measures. In the box for *Other Organisms*, VHS was listed. Although VHS has not yet been encountered in Texas (where the example activity is hypothetically conducted), it is an imminent threat to the state and, thus, remains a concern. When developing an HACCP plan, it is best to be proactive and have a control measure in place before the hazard is manifested. Justification for including or not including non-target species is added in each box of step 3. Step 4 will describe how and where to add control measures to the HACPP plan.

Nontargets That May Potentially Be Moved/Introduced
Vertebrates: None Justification: Field sampling in a boat using plankton nets is unlikely to move any vertebrates.
Invertebrates: Invasive Mussels (e.g., zebra or quagga mussels) Justification: Some waters in Texas have zebra mussels.
Plants: Aquatic Weeds (e.g., hydrilla, giant salvinia) Justification: Many waters in Texas have aquatic weeds.
Other Organisms (pathogens, parasites, etc.): Viral Hemorrhagic Septicemia (VHS) Justification: Although not yet reported in Texas, VHS is a potential pathogen for many Texas sport fish.

FIGURE 15.3   HACCP Step 3—Identity potential non-targets.

HACCP Step 4: Analyze Non-Targets

The first three steps of HACCP are used to prepare and organize for the most important and challenging parts of HACCP planning. In step 4, the potential hazard (non-target species) will be analyzed with respect to each task listed in step 2. Lumping similar species into groups in step 3 will expedite this analysis by allowing for fewer iterations of the analytical process.

This step is difficult to visualize without referring to the hypothetical example. The non-target analysis worksheet (NTAW) used to complete step 4 (Figure 15.4) is filled out from left to right, then top to bottom. The NTAW has a row for each task. The titles for each task listed in the activity flow chart (Figure 15.2)

1	2	3	4	5	6	7
Tasks (From step 2)	**Potential Nontargets** (From step 3)	**Risk Assessment** Are any non targets significant? Yes or no	**Justification** Justify your answer in column 3	**Control** What control measures can be applied during this task to reduce the risk of nontargets?	**CCP?** Is this task a CCP? Yes or no	**Justification** Justify your answer in column 6
Task # **1** Title: Gather Gear and Load Boat	Vertebrates None	N/A	N/A	N/A	N/A	N/A
	Invertebrates Invasive mussels	Yes	Equipment and boat may not be and clean	Verify proper decontamination has been done or wash boat and equipment with high pressure, hot water and let dry	Yes	Gear could have been stored dirty
	Plants Aquatic weeds	Yes	Equipment and boat may not be dry and clean	Verify proper decontamination has been done or wash boat and equipment with high pressure, hot water and let dry	Yes	Gear could have been stored dirty
	Others VHS	Yes	Equipment and boat may not be dry and clean	Verify proper decontamination has been done or wash boat and equipment with high pressure, hot water and let dry	Yes	Gear could have been stored dirty
Task # **2** Title: Drive to Field Site	Vertebrates None	N/A	N/A	N/A	N/A	N/A
	Invertebrates Invasive mussels	No	Equipment and boat should be dry and clean	N/A	N/A	N/A
	Plants Aquatic weeds	No	Equipment and boat should be dry and clean	N/A	N/A	N/A
	Others VHS	No	Equipment and boat should be dry and clean	N/A	N/A	N/A

FIGURE 15.4 HACCP Step 4—Non-target activity worksheet, page 1.

go into the first column (labeled *Tasks*) (Figure 15.4). The second column (labeled *Potential Non-Targets*) should list all of potential non-target species from step 3 (Figure 15.3).

The third column (labeled *Risk Assessment*) is used to indicate whether each non-target could reasonably be encountered during the specific task in question. *Yes* or *no* should be indicated here. There are many ways to perform a risk assessment to determine the answer to this; in fact, whole volumes have been written on various risk assessment strategies. Such a discussion is beyond the scope of this chapter. Formal risk assessment requires an evaluation of the likelihood of an event occurring and the severity of negative impacts from such an event. With non-target species, the consequences of a new introduction is not always known; however, HACCP employs the precautionary principle that assumes that the introduction of a non-target species will always result in negative, unacceptable impacts. With all impacts assumed to be significantly negative, this leaves only the likelihood of an unintentional introduction to be assessed. Some may want a formal risk analysis, while others may be comfortable with common sense. The HACCP team should determine the level of risk accepted after considering the specific management goals for the activity and the value of the resources at stake. When considering this factor, it is suggested to remember how much an ounce of prevention is worth.

The fourth column on the NTAW (labeled *Justification*) is space to provide justification for our *yes* or *no* answer in the third *Risk Assessment* column (Figure 15.4). That is, this area should be used to explain why a task does or does not pose a significant risk. In the fifth column of the NTAW, an effective control measure should be listed for any non-target that is determined to be a significant risk for the specific task in question. The final two (the sixth and seventh) columns are for identifying critical control points. This is an important topic that will be considered in detail later. For now, only the first five columns from the hypothetical example will be reviewed.

In the hypothetical example, the first task (*Gather Gear and Load Boat*) was listed in the first column (Figure 15.4). In the second column, all the potential non-target species from step 2 were listed. To complete the third column for the first task, the following question should be asked: "Is it reasonably possible for non-target species to be encountered while gathering gear and loading the boat?" The question will be asked and answered for each non-target category (vertebrate, invertebrate, plants, and other). In this example, it was decided that invasive mussels, aquatic weeds, and VHS may be encountered, while gathering gear and loading the boat if the previous crew neglected to clean the gear and let it dry. Accordingly, the answer *yes* was prescribed for each of these non-target categories under the third column (*Risk Assessment*) (Figure 15.4). If it was determined that a risk of movement did not apply to one or all of these non-targets, then the answer *no* would be used in the third column next to the corresponding potential non-target. In either case, a justification is required in the fourth column for the outcome of this risk assessment. It is important for those reading the plan to know why this decision was made. For this example, it was explained in the *Justification* column that invasive mussels, aquatic weeds, and VHS could be found while gathering gear and loading the boat because *equipment and boat may not be dry and clean*. It is best to be proactive rather than assume the risk that whoever used the boat and equipment last did a proper job of decontamination. Zebra and quagga mussels can live out of water for days or weeks if the temperatures are low and relative humidity is high. Some plant fragments can also survive for days or weeks. Therefore, it is essential to check that all equipment is clean and dry prior to use.

The first task (*Gather Gear and Load Boat*) was determined to have significant risks from non-target species (*yes* in the third column); therefore, a control measure must be added under the fifth column, which is labeled *Control* (Figure 15.4). A control measure is any action that can be used during or after a specific task to kill or remove the non-target species so that it can no longer pose a significant risk. For instances where potential hazards are determined to be unlikely, *N/A* for *not applicable* is written under column 5 because a control measure is not needed if a hazard does not exist. Otherwise, the most effective control measure that could be applied during or after this task should be listed. It is possible that no control measure can be applied during this particular task (task 1). If this is the case, *no control measures are possible* should be written in the *Control* column. In this example, however, effective control measures are available. Specifically, one could wash the boat and equipment with high-pressure hot water and then dry everything. However, these control measures may not be necessary if it can be verified that the equipment has already been properly decontaminated. If so, this verification will need to be made each time the gear is gathered and loaded into the boat to avoid application of additional control measures.

Notice that the responses for task 1 are identical in columns for *Risk Assessment, Justification*, and *Control* with respect to invasive mussels, aquatic weeds, and VHS (Figure 15.4). Although identical in the simple hypothetical example, this may not always be the case in other HACCP plans. For example, some stations use eco-friendly disinfectants especially for diseases. Potential risks and effective control measures will vary based on the activities being performed and the list of potential non-target species and should always be considered independently. Often, there are a variety of control measures that could be used. The ones chosen by the HACCP team should be scientifically based and effective for the particular non-target species controlled. Unique circumstances may need to be considered when making these decisions. This is the responsibility of the HACCP team.

Earlier, to maintain simplicity, the sixth and seventh columns used to identify critical control point were skipped. This topic will be discussed in detail later. For now, only the first five columns for the remaining tasks will be reviewed. For task 2, the title (*Drive to Field Site*) was added to the box in the first column and then followed exactly the same procedure from the previous task. Information for *Risk Assessment*, and *Justification*, was completed for each potential non-target category. For this task was determined that there was not a significant risk of encountering non-target species while driving to the field site, such that *N/A* was written under *Control* because control measures are not needed for nonexistent hazards (Figure 15.4).

An effective HACCP plan will include at least one control measure for each potential non-target during at least one of the tasks. In other words, all potential non-targets must have a control measure somewhere. Otherwise, there would be an uncontrolled hazard that could likely introduce an unwanted invader. It is the responsibility of good stewards to prevent this by identifying the best task or tasks where control measures can be applied. These are called *critical control points*, and they form the core of any HACCP plan.

If a control measure could be applied during or immediately after a task, then the task is a control point. Control points may be optional or critical. Critical control points are those tasks where control is possible and necessary to prevent moving a non-target species, and there is no better task for minimizing or eliminating the risk of spread. All other control points are optional. We must determine which task or tasks are critical control points. Every HACCP plan must have at least one critical control point. It is normal for a typical HACCP plan to have one or two critical control points. Tasks where there are no significant risks of encountering a non-target species cannot be critical control points. Likewise, tasks where effective control measures cannot be applied (for whatever reason) cannot be critical control points either. Instead, focus should be given to tasks that have potential hazards and where control measures can be applied. The HACCP planning process includes a decision tree to help decide whether a task is a critical control point or not (Figure 15.7). HACCP plans require that control measures be employed at all critical control points; however, if resources allow, it may be beneficial to implement action at optional control points to further minimize risk well below acceptable levels. This is a decision that should be made by the HACCP planning team.

In the hypothetical example, task 1 (*Gather Gear and Load Boat*) may be a critical control point because there are potential hazards (column 3, Risk Assessment), and control here is possible (column 5, Control). The Critical Control Point Decision Tree (Figure 15.7) is used to determine if this task is a critical control point. Starting at the top of the tree, question 1 is asked by the HACCP team: "Could contamination with the identified non-targets occur or increase beyond acceptable levels?" For task 1, the answer was *yes* for all three non-targets (invasive mussels, aquatic weeds, and VHS), justified by the fact that the equipment and boat may not be dry and clean following its previous use. This answer leads to question 2 in the decision tree: "Do control measures exist during this task or a subsequent task that reduce the occurrence of a non-target to an acceptable level?" The answer, once again, was *yes* for task 1 because an effective control measure was listed for each non-target. The decision tree takes us to the next question (question 4): "Is there a better point within the activity to reduce the non-target to an acceptable level (i.e., a previous task or subsequent task)?" The answer is *no* because if the non-target species is not controlled at this point, then the non-target species may use the equipment or boat as a vector to be carried to the field site. There is no previous task and no other opportunity to remove the hazard before transporting the equipment and boat. Using the decision tree, it was determined that the first task from the hypothetical example is a critical control point. So, the answer *yes* is placed in column 6 (*CCP?*) of the NTAW (Figure 15.4) for all three non-target categories (invasive mussels, aquatic weeds, and VHS).

It is possible in some HACCP plans to see a task that is a critical control point for one non-target species or group, but not for others. Therefore, it is important to assess each non-target species for each task separately. Once it is determined whether the task is a critical control point or not, a justification for this determination is added to column 7 of the NTAW (Figure 15.4). The justification written is *gear could have been stored dirty*. There is no alternative for controlling our hazards in this case. The equipment and boat must be cleaned and allowed to dry before use unless it can be verified that the equipment has already been properly decontaminated.

After finishing task 1, task 2 (*Drive to Field Site*) is analyzed. This task does not contain any potential hazards and therefore is not a critical control point. *N/A* for *not applicable* is placed in the empty boxes under columns 6 and 7 (Figure 15.4). It is important to put a notation of some sort in each box so that others know that areas were not overlooked. Do not leave boxes blank.

Task 3 (*Launch Boat*) is also not a critical control point because, similar to task 2, it was indicated that there is no significant risk of encountering non-target species during this task. Task 4 (*Conduct Sampling*) could be a critical control point because we could reasonably encounter potential non-target species while sampling. However, following the decision tree (Figure 15.7), it can be decided that task 4 is not a critical control point because control measures do not exist for invasive mussels and VHS, and those for aquatic weeds would be better applied during a subsequent task, as it would be difficult to effectively remove aquatic weeds from the boat and equipment while they are still in the water. The answer *no* is written in column 6 for task 3, indicating that this task is not a critical control point and with justification in column 7 (Figure 15.5). Although this task is not a critical control point, it is an optional control point. Removing weeds as they are seen during sampling is good practice and will ease the burden of cleaning during the critical control point.

Since control measures were not applied while sampling (task 4), the boat and equipment used could likely carry unwanted organisms. Therefore, task 5 (*Retrieve Boat*) has a significant risk of encountering non-target species and is a place where control measures can be applied. Since the next task involves transporting the equipment and boat back to the field station or next site, it is important to take actions during or immediately after task 5 to prevent the spread of non-target, potentially invasive, species. Answering the questions from the decision tree helps determine that this is the best place to apply control measures. Task 5 is a critical control point for all hazards—invasive mussels, aquatic weeds, and VHS. This is indicated as such in column 6 of the NTAW, and justification is provided in column 7 (Figure 15.6).

In task 6 (*Move to Next Site or Return to Field Station*), the boat and gear have been decontaminated following task 5. Thus, no hazards are likely encountered, and task 6 is not a critical control point. *N/A* is placed into the appropriate boxes, and the NTAW (Figures 15.4 through 15.6) is complete. The next step is to create a Risk Action Plan.

In the provided example, we did not explore the entire decision tree (Figure 15.7). It is possible to traverse the decision tree and find that significant, unacceptable risks do exist, but there is no opportunity to apply control measures. If such a situation occurs, we have two options. The first option is to modify the activity and tasks so that effective control measures could be applied and then start the HACCP process over. The second option is more dangerous: the risk could be accepted (along with the responsibility), and the activity continued as originally planned without controls for invasive species. However, any federal agency choosing this latter option is compelled by Executive Order 13112 to make public "a determination that the benefits of such actions clearly outweigh the potential harm caused by invasive species; and that all feasible and prudent measures to minimize risk of harm will be taken in conjunction with the actions." A good steward, even if nonfederal, will avoid this option except under extreme (dire) circumstances. If it is at all possible to rework the activity to allow for control measure implementation, that is almost always the better option.

1	2	3	4	5	6	7
Tasks (From step 2)	**Potential Non-targets** (From step 3)	**Risk Assessment** Are any non targets significant? Yes or no	**Justification** Justify your answer in column 3	**Control** What control measures can be applied during this task to reduce the risk of nontargets?	**CCP?** Is this task a CCP? Yes or no	**Justification** Justify your answer in column 6
Task # **3** Title: Launch Boat	Vertebrates None	N/A	N/A	N/A	N/A	N/A
	Invertebrates Invasive mussels	No	Equipment and boat should be dry and clean	N/A	N/A	N/A
	Plants Aquatic weeds	No	Equipment and boat should be dry and clean	N/A	N/A	N/A
	Others VHS	No	Equipment and boat should be dry and clean	N/A	N/A	N/A
Task # **4** Title: Conduct Sampling	Vertebrates None	N/A	N/A	N/A	N/A	N/A
	Invertebrates Invasive mussels	Yes	Adults or larvae could attach to equipment or be in water on board	No effective control measures exist during this task	No	Control measures cannot be applied here
	Plants Aquatic weeds	Yes	Weeds or fragments could attach to equipment or boat	Remove any weeds or fragments that attach to boat or equipment	No	Task 5 is better for applying control
	Others VHS	Yes	Pathogen could be on anything wet	No effective control measures exist during this task	No	Control measures cannot be applied here

FIGURE 15.5 HACCP Step 4—Non-target activity worksheet, page 2.

HACCP Step 5: Create a Non-Target Risk Action Plan (NTRAP)

Once the critical control point or points are identified, they must be addressed. This is the objective of the NTRAP, which clearly identifies what the control measures will be, when they will be applied, who is in charge of making sure these are completed correctly, and what actions should be taken if something goes wrong.

An NTRAP is needed for each unique control measure applied to each critical control point. In the hypothetical example, two critical control points were identified, task 1 (*Gather Gear and Load Boat*) and task 5 (*Retrieve Boat*). For both of these critical control points, the control measure listed in column 5 of the NTAW (step 4) is to use high-pressure hot water to remove or kill invasive mussels, aquatic weeds, and VHS. Since each task has only one control measure, only one NTRAP is needed for each

1	2	3	4	5	6	7
Tasks (From step 2)	**Potential Non-targets** (From step 3)	**Risk Assessment** Are any non targets significant? Yes or no	**Justification** Justify your answer in column 3	**Control** What control measures can be applied during this task to reduce the risk of nontargets?	**CCP?** Is this task a CCP? Yes or no	**Justification** Justify your answer in column 6
Task # **5** Title: Retrieve Boat	Vertebrates None	N/A	N/A	N/A	N/A	N/A
	Invertebrates Invasive mussels	Yes	Adults or larvae could attach to equipment or be in water on board	Wash boat and equipment with high pressure, hot water and let dry	Yes	This is the best opportunity to control mussels
	Plants Aquatic weeds	Yes	Weeds or fragments could attach to equipment or boat	Wash boat and equipment with high pressure, hot water and let dry	Yes	This is the best opportunity to remove weeds
	Others VHS	Yes	Pathogen could be on anything wet	Wash boat and equipment with high pressure, hot water and let dry	Yes	This is the best opportunity to kill VHS
Task # **6** Title: Move to Next Site or Return to Field Station	Vertebrates None	N/A	N/A	N/A	N/A	N/A
	Invertebrates Invasive mussels	Yes	Adults or larvae could be on equipment or in water on board	Wash boat and equipment with high pressure, hot water and let dry	No	Task 5 is a better opportunity to control mussels
	Plants Aquatic weeds	Yes	Weeds or fragments could be on equipment or boat	Wash boat and equipment with high pressure, hot water and let dry	No	Task 5 is a better opportunity to remove weeds
	Others VHS	Yes	Pathogen could be on anything wet	Wash boat and equipment with high pressure, hot water and let dry	No	Task 5 is a better opportunity to kill VHS

FIGURE 15.6   HACCP Step 4—Non-target activity worksheet, page 3.

critical control point. So, for this example, a total of two NTRAP forms are needed. If aquatic weeds were treated with a herbicide during task 1 and task 5, but invasive mussels and VHS were controlled with high-pressure hot water, then four NTRAPs would be needed, including two for task 1 and two for task 5. An NTRAP would be completed for the herbicide application during task 1 and a separate NTRAP for the high-pressure hot water application during task 1. The same would be needed during task 5. Instead, the hypothetical example was written as an example to keep things simple and focus only the components necessary to complete the NTRAP form.

The first four rows of the NTRAP are reminders of management objective, the critical control point (task) that is the focus of this action plan, the significant non-target species that this NTRAP will control,

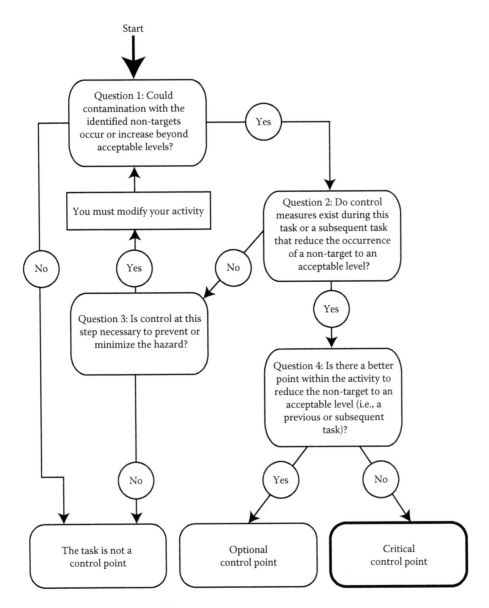

FIGURE 15.7 Critical control point decision tree.

and the method that will be used to control the non-target species. As shown in Figure 15.8, these boxes are filled in with values from previous steps.

The fifth row of the NTRAP calls for *Prescribed Ranges, Limits, or Criteria* for the control measures used in this task. This is a specific measureable attribute that can be used to determine if control measures effectively minimize the risk. This example uses hot water to control non-target species. We do not want ambiguity. So, it is necessary to define how hot the water must be and how long this temperature must be sustained for effective control. This decision should be based on current science. The prescribed ranges, limits, or criteria in this example are specified to be at least 158°F at the contact point for at least 60 s. These prescribed limits for temperature and duration will kill VHS as well as invasive mussels (Zook and Phillips, 2012). It is also specified that the water pressure must be at least 1000 psi and no more than 1500 psi. This is a prescribed range. At least 1000 psi is needed to dislodge weeds and mussels, but more than 1500 psi may damage the equipment or boat. The term *dry* is also defined on the NTRAP such that there is an absence of standing water or wet surfaces.

(Use this form for any "Yes" from Column 6 of HACCP Step 4 - Non-Target Analysis Worksheet) One page for each Critical Control Point	
Mangement Objective From Step 1	Conduct aquatic sampling using a motorboat that is trailered from site to site.
Critical Control Point: Task # 1	**Title:** Gather Gear and Load Boat
Significant Non-Target(s) (Step 4, Column 3)	Invasive Mussels, Aquatic Weeds, and VHS
Control Measure(s) (Step 4, Column 5)	Inspect boat and equipment. Wash boat and equipment with high pressure, hot water if necessary and let dry.
Precribed ranges, limits, or cieria for control measure(s): (PRLC)	Water must be at least 158 °F with 60 sec contact time to kill VHS, this criteria will also kill mussels. Pressure must be at least 1000 psi and no more than 1500 psi. Weeds and fragments will be removed by pressure. Dry means no standing water or wet surfaces anywhere.
Monitoring the Control Measure(s) Who?	The staff member conducting the wash is responsible for monitoring the temperature and pressure.
How?	Temperature should be measured with an IR temperature gun. Pressure is monitored by the settings on the wash unit.
Where?	Decontamination should be done at the field station site before the boat is moved elsewhere.
How often?	Decontamination must be done if anything is wet or there is evidence of invasive mussels or plant fragments.
Corrective Action(s) if Control Measures Fail (or PRLC cannot be met)	If for any reason the boat or equipment cannot be properly cleaned, the staff must postpone field work until proper decontamination can be performed. Under no circumstances should a dirty or wet boat be moved from the station.
Supporting Documents *(For example, Management Plan, Checklist, Decontamination Techniques, SOPs, Scientific Journal Articles, etc.)* Uniform Minimum Protocols and Standards for Watercraft Interception Programs for Dreissenid Mussels in the Western United States, 2012, Pacific States Marine Fisheries Commission. (Control measures for invasive mussels) Technical Operating Procedure 028.0/U.S., Procedures for Decontamination of Equipment to Prevent Spread of Viral Hemorrhagic Septicemia and Other Infectious Diseases and Biota, 2008, U.S. Fish & Wildlife Service. (Control measures for VHS)	
Development Team Members	David Britton
Date Developed: 3/4/2013	**Date(s) Reviewed:** 1/15/2014

FIGURE 15.8   HACCP Step 5—NTRAP for critical control point 1.

The next four rows on the NTRAP form describe who will monitor the prescribed ranges, limits, and criteria, as well as how, where, and when this evaluation will be performed. Monitoring to ensure the control measures is very important. If, for example, the water temperature was only 120°F, then the VHS virus or invasive mussels may not be killed as intended. Decontamination would be ineffective. An effective HACCP plan will clearly identify how monitoring will be performed to minimize the risk of moving non-target species to areas where they may become invasive. Remember, it is the prescribed ranges, limits, and criteria that are monitored here—not just the apparent results of the control measures. Novice HACCP planners often miss this important point.

The next row in the NTRAP calls for *Corrective Actions*. This is a procedure that must be followed if a control measure fails at a critical control point. Even if a control measure appears to be successful, if the prescribed ranges, limits, and criteria are found to be outside their defined parameters, one must conclude that the control measures have failed. A corrective action is the backup plan used as a second line of defense that may be necessary to prevent the spread of a non-target if the first-line defense (a control measure) fails to control it. All effective HACCP plans must have a backup plan, arranged in advance, that can be implemented if the prescribed ranges, limits, and criteria cannot be met or if the control measures fail for some other reason. In the hypothetical example, the corrective actions are to sequester the equipment and boat, and not use them until they can be properly decontaminated.

The lower section of the NTRAP is intended to list any and all supporting documents. HACCP plans should be scientifically based, and the science behind the plan development should be cited. For example, the prescribed ranges, limits, and criteria of 158°F to kill VHS was found with this information in the paper *Technical Operating Procedure 028.0/U.S., Procedures for Decontamination of Equipment to Prevent Spread of Viral Hemorrhagic Septicemia and Other Infectious Diseases and Biota*, 2008, USFWS. By including this reference material in this section, it will show others the information used to develop a valid approach to controlling the pathway.

The hypothetical HACCP plan includes two NTRAPS (Figures 15.8 and 15.9). The first is used for the first critical control point, which was our first task. This was found to be a critical control point because movement of the equipment or boat, if potentially contaminated, would risk spreading non-target species. The NTRAP for the task (Figure 15.8) details the plan to control invasive mussels, aquatic plants, and VHS before the items leave the field station. There is also a corrective action, or backup plan, if the control measure cannot be performed or deviates from the prescribed ranges, limits, and criteria. The second NTRAP (Figure 15.9) explains the control measures to be implemented following task 5. It is very similar to the first NTRAP because the control measures and corrective actions are essentially the same. The HACCP planning process has found and addressed two critical control points. The management objective can now be carried out, knowing confidently, that substantial thought has been put into the activity.

In the end, the purpose for HACCP planning is not for creating paperwork or adding a new layer of compliance tracking—it is for protecting our resources. As stewards of natural resources, it is our responsibility to do our best to protect and preserve. Often that requires us to think from a broader perspective and scrutinize our activities in fine detail. HACCP provides a planning tool to avoid unintentionally spreading invasive species in the course of conservation efforts.

Discussion

The main purpose of HACCP is to embrace the precautions necessary to do no harm. It is the same reasoning doctors use for washing hands before treating a patient. Imagine if doctors carried out their regular actions, examinations, or even surgeries, without taking a moment to ensure that precautions were taken to avoid spreading disease. Because doctors follow a strict set of planned procedures, patients do not have to worry about being in harm's way. In much the same way, natural resource professionals hold the health of natural resources in their hands; this generates a responsibility to understand the consequences of their actions (or inactions). Good stewards of natural resources must protect and preserve.

Vigilance is necessary in a dynamic world. Over time, the environment and work may change, and new challenges must be confronted. Once complete, an HACCP plan begins a journey toward obsolesce. The older the plan, the more likely it is no longer valid. HACCP plans should be revisited regularly to assess if the tasks and control measures used are still relevant or if new non-target species pose a threat. An HACCP plan should be a living document that is reviewed or updated at least once a year.

This overview should provide the information needed to begin the HACCP planning process; however, help is available to those in need of further assistance. Both NOAA and the USFWS staff conduct HACCP training at various locations throughout the year. Several state wildlife agencies also have qualified HACCP trainers who can help provide hands-on learning experiences. Additional information and training inquiries can be found by contacting any regional (USFWS) aquatic invasive species coordinator.

(Use this form for any "Yes" from Column 6 of HACCP Step 4 - Non-Target Analysis Worksheet) One page for each Critical Control Point	
Mangement Objective From Step 1	Conduct aquatic sampling using a motorboat that is trailered from site to site.
Critical Control Point: Task # 5	Title: Retrieve Boat
Significant Non-Target(s) (Step 4, Column 3)	Invasive Mussels, Aquatic Weeds, and VHS
Control Measure(s) (Step 4, Column 5)	Wash boat and equipment with high pressure, hot water and let dry.
Precribed ranges, limits, or citeria for control measure(s): (PRLC)	Water must be at least 158 °F and contact time must be at least 1 minute to kill VHS, this criteria will also kill mussels. Pressure must be at least 1000 psi and no more than 1500 psi. Weeds and fragments will be removed by pressure. Dry means no standing water or wet surfaces anywhere.
Monitoring the Control Measure(s) Who?	The staff member conducting the wash is responsible for monitoring the temperature and pressure.
How?	Temperature should be measured with an IR temperature gun. Pressure is monitored by the settings on the wash unit.
Where?	Decontamination should be done at the launch site before the boat is moved elsewhere.
How often?	Decontamination must be done every time the boat is retrieved from open water, before it is moved to another location or returned to the field station.
Corrective Action(s) if Control Measures Fail (or PRLC cannot be met)	If for any reason the boat or equipment cannot be properly cleaned on site, the staff must return the boat and equipment to the field station until a proper decontamination can be performed. Under no circumstances should a dirty or wet boat be moved from one site to another.
Supporting Documents (For example, Management Plan, Checklist, Decontamination Techniques, SOPs, Scientific Journal Articles, etc.) Uniform Minimum Protocols and Standards for Watercraft Interception Programs for Dreissenid Mussels in the Western United States, 2012, Pacific States Marine Fisheries Commission. (Control measures for invasive mussels) Technical Operating Procedure 028.0/U.S., Procedures for Decontamination of Equipment to Prevent Spread of Viral Hemorrhagic Septicemia and Other Infectious Diseases and Biota, 2008, U.S. Fish & Wildlife Service. (Control measures for VHS)	
Development Team Members	David Britton
Date Developed: 3/4/2013	**Date(s) Reviewed:** 1/15/2014

FIGURE 15.9   HACCP Step 5—NTRAP for critical control point 2.

REFERENCES

Britton, D., Heimowitz, P., Pasko, S., Patterson, M., and Thompson, J. 2012. HACCP: Hazard analysis and critical control point planning to prevent the spread of invasive species. National Conservation Training Center, U.S. Fish & Wildlife Service, Washington, DC (A training manual).

Clinton, W.J. 1999. Executive order 13112 of February 3, 1999: Invasive species. *Federal Register* 64(25):6183–6186.

Gunderson, J.L. and Kinnunen, R.E. 2006. AIS-HACCP: Aquatic invasive species—Hazard analysis and critical control point training curriculum. Minnesota Sea Grant Publication Number: MN SG-F11, Duluth. Michigan Sea Grant Publications Number: MSG-00-400, Ann Arbor.

Wilcove, D.S., Rothstein, D., Dubow, J., Phillips, A., and Losos, E. 1998. Quantifying threats to imperiled species in the United States. *BioScience* 48:607–615.

Zook, B. and Phillips, S. 2012. Uniform minimum protocols and standards for watercraft interception programs for dreissenid mussels in the western United States. A white paper produced for the Western Regional Panel on Aquatic Nuisance Species. Pacific States Marine Fisheries Commission, Portland, Oregon.

16

Equipment Inspection and Cleaning: The First Step in an Integrated Approach to Prevent the Spread of Aquatic Invasive Species and Pests

Joseph DiVittorio
Former Invasive Species/Integrated Pest Management Program Coordinator
U.S. Department of the Interior, Bureau of Reclamation (retired January 2013)

CONTENTS

ABSTRACT   This chapter begins by describing the history of human interaction with transportation, invasive species, and equipment, thereby providing some information for the nonscientist reader on the scope of the invasive species dilemma. The issue of aquatic invasive species and pests is a very public one, and hopefully, this approach is of some value to a reader whose larger interests might be general natural resource issues, environmental public policy, water supply concerns, and budget processes related peripherally to aquatic invasive species and pests. As the chapter progresses, more technical discussion is brought forward from other disciplines, examples, and situations.

Although equipment inspection and cleaning is the first step in an integrated approach to prevent the spread of aquatic invasive species and pests, it must not be an only step. Equipment inspection and cleaning is a proven invasive species and pest prevention tool, but as with many other methods, it has both intrinsic strengths and limitations. These strengths and limitations are discussed to develop a better understanding of the method's prevention potential.

While public outreach remains very important in the campaign against aquatic invasive species and pests, watercraft inspection and cleaning (also referred to as decontamination) as practiced by agencies and the "Clean, Drain, and Dry" equivalent as practiced by the public operate at largely unknown efficacy. For management of the dreissenid mussels in particular, there are indications that recreational watercraft inspection and cleaning does not achieve 100% efficacy 100% of the time. In fact, experience gained through the long-term practice of integrated pest management in other settings demonstrates that no one tactic is capable of producing 100% efficacy.

Equipment inspection and cleaning would be most successful against aquatic invasive species and pests when it is employed in combination with a broader suite of robust tactics than are being currently used. However, even when all of the currently known antidreissenid tactics are integrated and used together, the spread of dreissenid mussels into new sites throughout the western states continues at alarming speed. There are few additional tactics to supplement equipment inspection and cleaning. The failure to research new methods, support innovative ideas, and fund these measures so that they might be enacted in the field is a path to assured and continued aquatic invasion.

Introduction

In developing the *Inspection and Cleaning Manual for Equipment and Vehicles to Prevent the Spread of Invasive Species* (Cleaning Manual) (DiVittorio et al. 2012), three main points of information are provided to equipment users: how to inspect potentially contaminated equipment for various invasive species and pests; how to clean the equipment once contamination is found; and a fast guide of identification and life history information for some of these common species. In addition, depending upon the type of equipment, and the scope and nature of inspection and cleaning requirements, there are several other outstanding works on the topic. Among them, the reader may also consult the U.S. Department of Defense (2004), Cofrancesco et al. (2007), Zook and Phillips (2012, 2014, Chapter 14 in this book), and Britton (2014, Chapter 15 in this book).

There will be limited repetition of the Cleaning Manual in this chapter. For specific equipment inspection and cleaning procedures, the reader is encouraged to refer to the Cleaning Manual. Equipment inspection and cleaning is a proven invasive species and pest prevention tool, but as with many other methods, it has both intrinsic strengths and limitations. This chapter discusses some of these characteristics, providing the natural resource practitioner a fuller understanding of prevention potential. While equipment inspection and cleaning is the first step in an integrated approach to prevent the spread of aquatic invasive species and pests, it must not be an only step. Equipment inspection and cleaning would be most successful against aquatic invasive species and pests when employed in combination with a robust suite of broader tactics than are being currently employed. The failure to research new methods, support innovative, outside-the-box ideas, and fund these measures so that they might be enacted in the field is a path to assured and continued aquatic invasion.

Background

We humans are a traveling lot. We have always been this way, and there is nothing on the horizon to indicate a change in this, one of our most basic behaviors. We are a generalist species untied to strict habitat requirements. Adding to these qualities, humans are also among the few species in the world to use tools (equipment) of some kind for our benefit. We traveled and searched, we moved things; we still do and we always will. Humans and foreign species have intertwined throughout history and will continue to do so into our future. Incidental to human travel is the spread of invasive species and pests via the movement of contaminated equipment.

In our distant past, we drifted with rivers and on the seas and explored and inhabited deserts, tundra, sea ice, mountains, steppes, forests, and any other place that might suit us. When we traveled, we took our belongings—our tools and equipment—with us. Equipment enabled our survival and aided our spread across the world. We generally followed seasonally migrating animal herds that provided not only meat and milk as food but also transportation and products such as bone, sinews, and hides that added to our traveling equipment inventory. What were those seeds hooked by thorns in the animal hides? They accidentally traveled with us also. Along the way, we took advantage of edible plants whenever found and loaded these up for transport to our next stop…wherever that might be. And that is the way it was for the vast extent of human time.

In our recent past, by some accounts about 12,000 years ago, our ancestors discovered that if the animals we pursued and hunted could be tamed a bit, we could manage them and stay awhile at locations that offered water and shelter. And then there are those plants we gathered. Someone must have found that certain parts of those plants when put into the soil could now grow more food plants of the same kind, especially when watered. We made sure those plants had the water they needed because the fall harvest meant the difference between survival and famine. Then as today, of the hundreds of thousands of plant species, subspecies, and varieties the world over, agriculturally capable edible plants number about a hundred species. Animal domestication, the advent of irrigation and seed gathering marked the beginning of agriculture, and it forever changed us. From nomadic groups to agrarian settlement dwellers, in the long trek of human development, agriculture probably ranks as a milestone along with fire use and tool making.

In our modern era, in 1927 aviation great Charles Lindbergh was the first to successfully complete a trans-Atlantic flight. His flight from Long Island, New York to Paris, France, demonstrated the feasibility of intercontinental, transoceanic air travel. Lindbergh's accomplishment ushered in an era of fast, long distance travel, but also unknowingly opened an avenue of global foreign species transport. Today, in a matter of hours, many different species can be moved half a world away. As Lindbergh landed in Paris, an enthusiastic crowd of Parisians numbering in the thousands pushed past barricades to greet him on the runway with hugs, handshakes, and kisses (Berg 1998).

In a more recent aviation achievement, a three-man crew returned from another historic flight, but unlike Lindbergh, there were no personal accolades immediately awarded. This was the return of the successful flight of *Apollo 11*, the first human mission to touchdown on the moon, July 1969. The terms of *Apollo 11*'s return of crew and spacecraft were governed by what was popularly known then as the Extra-Terrestrial Exposure Law. Actually a regulation promulgated into the Code of Federal Regulations (CFR) as 14 CFR Part 1211, the Extra-Terrestrial Exposure Regulation enabled development of protocols by the National Aeronautics and Space Administration to prevent unintended transport of an extraterrestrial species to earth by the Apollo program (Federal Register, multiple citations 1968–1991). Since no extraterrestrial life form had ever been (or has yet to be) identified, there was no information on its biology, and no known tactics to manage such a species if one were inadvertently transported by the mission. Therefore, the protocols were patterned after known earth biology using the best available science involving the use of a chemical disinfectant solution on the returned spacecraft and equipment and isolation of the crew, equipment, and lunar rock and soil samples through quarantine. The Extra-Terrestrial Exposure Regulation was removed from 14 CFR Part 1211 in 1991 and placed into reserved status, pending the next phase of other-world spaceflight.

To be sure, there are examples of species (other than humans) that are the agent of species distribution. The tumbleweed (Russian thistle, *Salsola* spp.), made famous in countless American western movie classics, is pushed by the winds across the rangelands and prairies. Along the way, tumbleweeds both pick up and distribute many plant parts, such as seeds. Not a true thistle, the tumbleweed itself is an introduced species from Eurasia—introduced by human activity to the Americas.

Invasive Species *Perfect Storm*

The relationship of transportation to invasive species and pest is certainly a key point in any discussion of invasive species. But, there are other factors at work that further complicate the invasive species dilemma. Fresh water is a limited and highly managed resource throughout the western states, with numerous and often competing uses. Water is managed for flood control, irrigation, hydroelectric and nuclear power generation, transportation, industrial and manufacturing processes, domestic water supplies, fisheries and riparian habitat, recreation, and other uses. Thinking about the mix of issues regarding water concerns in the western states, I identify the influence of three additional factors in the making of an invasive species perfect storm: (1) declining public sector budgets, (2) climate change, and (3) aging water management infrastructure.

Declining budgets at all levels of government cause invasive species response to be less nimble to rapidly changing conditions. Declining budgets impact invasive species research, reduce funds for invasive

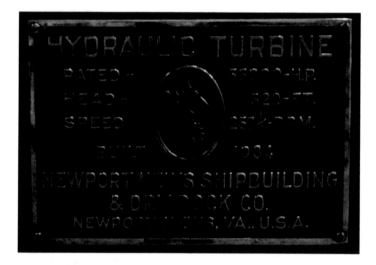

FIGURE 16.1 Turbine nameplate at Hoover Dam. This turbine is an original equipment item in service since its installation in 1934.

species control and management, and early detection and rapid response to new invasive species threats. Uncertainty in budget processes causes difficulty in long-term planning necessary for water infrastructure upgrades needed to combat invasive species attack.

Drought and climate change expressed in other forms can drive some species to become more invasive, while causing other species to be less able to resist invasion. This condition destabilizes habitats, making them yet more vulnerable to invasion (Invasive Species Advisory Committee 2010; Melillo et al. 2014).

The drivers of climate change are outside our immediate control; aging facilities can be mitigated somewhat through continued preventative maintenance and replacement or modification of components, but only as enacted by budget processes.

In the West, two federal agencies, the U.S. Department of the Interior Bureau of Reclamation (Reclamation) and the U.S. Army Corps of Engineers (USACE), plus various state-level agencies, and many municipalities, irrigation districts, and private hydroelectric producers each generally plan, design, construct, and operate water resource infrastructure.

The majority of Reclamation facilities, for example, were planned and constructed between 50 and 100 years ago (Figure 16.1), with some facilities now over 100 years old. Coinciding during this period, the rate of invasive species colonization in the United States sharply increased resulting from large numbers of invasive species arriving from foreign locations. At that time, the impact from invasive species alone could not have been predicted by the planners and designers of water infrastructure as becoming a threat to Reclamation facility operations or mission. However, when combined with aging facilities and periodic drought through climate change, infrastructure that was once designed specifically to deliver water and generate electrical power was not intended to cope with these unforeseen, multiple impacts.

Definitions

> This thing you call language though ... most remarkable. You depend on it for so very much. But is any one of you really its master? (Spock, in *Star Trek* [Bixby 1968])

Several keywords have already been used during this discussion to form the basis of this chapter: equipment use, transportation/travel, human interaction, and the pest/invasive species (Figure 16.2). Each factor has some influence on the other, although the greatest influence is simply that humans move things, mainly equipment.

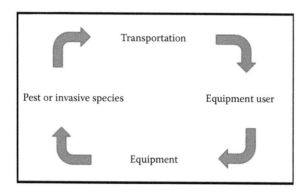

FIGURE 16.2 Interaction between transportation, human interaction of the equipment user, equipment type, and the pest or invasive species.

I have used, and throughout this chapter, I will continue to use the terms pests and invasive species. The Cleaning Manual, though titled as an aid to prevent the spread of invasive species, discusses pests as well throughout its text. It is appropriate to define these terms early on for use in this chapter.

Pest is defined in the Federal Insecticide, Fungicide, and Rodenticide Act, as amended (FIFRA) (FIFRA 1996), as "Any insect, rodent, nematode, fungus, weed, or any other form of terrestrial or aquatic plant or animal life or virus, bacteria, or other micro-organism (except viruses, bacteria, or other micro-organisms on or in living man or other living animals) which the Administrator (of the U.S. Environmental Protection Agency) declares to be a pest...which is injurious to health or the environment."

Invasive species is defined by Executive Order 13112 (Executive Order 13112 1999) as "An alien species whose introduction does or is likely to cause economic or environmental harm or harm to human health."

In some respects though, we seem to be held captive and somehow limited by our own definitions of pests, invasive species, and other terms used. After all, even noted early ecologist Aldo Leopold (Leopold 1924) described the seral succession of one species as invading a site and replacing existing species. The issue here is the invading species was a native, and that succession from one community to the next was occurring. In using the terminology invading/invader, Leopold was not using the equivalent meaning as we use it today (Simberloff et al. 2012). For example, in reading through the literature and in casual conversations with fellow biologists and policymakers, there is a universe of terms in use as we try to describe the complex processes and events associated with pests, invasive species, and invasion biology. The following is a short list of some of these terms, none of which have statutory meaning; simply place the word *species* after each: exotic, nuisance, weed or weedy, introduced, non-native, undesirable, damaging, escaped, unwanted, foreign, invasive native, problem, deleterious, damaging, harmful, and others. Some terms that do have meaning through federal law or the Executive Order: pest, alien species, noxious weed, invasive species, nonindigenous species, and aquatic nuisance species.

The National Invasive Species Council (NISC) Invasive Species Advisory Committee, Definitions Subcommittee submitted an Invasive Species Definition Clarification and Guidance White Paper to help resolve these terminology issues (National Invasive Species Council, Invasive Species Advisory Committee 2006).

Discussion

In spite of our best efforts, there certainly are particular species that seem to defy our traditional definitions. What of the species whose impacts are so far reaching, especially when these species are driven by rapid climate change, that current descriptors are not able to easily convey multiecological effects to legislators, policymakers, and the public? Perhaps a pest or invasive species ought to be classified by its ecosystem impacts rather than where it lives? For example, the mountain pine beetle (*Dendroctonus ponderosae*) might fit this category. This chapter is concerned with the issues of invasive species and pests in aquatic systems, so the reader might wonder why a forest tree pest is being discussed. It is because the mountain pine beetle, although a terrestrial species, and now found throughout the western states, is adversely impacting tens of millions of acres of upper and lower watersheds and their associated aquatic habitats. The mountain pine beetle discussion is included here, if for no other reason, then to pose some flexibility to the definition boundaries of terrestrial and aquatic; of pests and invasive species; of legal and biological definitions; and to observe the expanding human involvement in the spread of biological organisms.

The mountain pine beetle is a species native to the southwestern United States that has become a widespread pest throughout much of the mountainous western states. This insect has historically occupied the pine forests of the warmer, lower latitudes in Arizona and New Mexico. A warming climate however has allowed this species to migrate northward along the Rocky Mountains through Colorado and beyond into southern Canada. Tens of millions of acres of standing dead trees are currently subject to large and nearly impossible to control wildfires along the east and west sides of the Rocky Mountains. As a pest in mountain forests, the implications of the mountain pine beetle infestation are enormous for both upper and lower watersheds. Postfire watershed concerns include erosion loss of mountain soils, increased sediment loads, aquatic habitat loss, water quality issues, watercourse surge flows, and flooding (Figure 16.3). It is startling that an insect, in the adult life stage about the size of a match head (the comparison is intended), is able to inflect such devastating ecosystem-wide impacts.

For example, the High Park Fire was caused by a lightning strike and burned in forest primarily impacted by the mountain pine beetle located west of Fort Collins Colorado on the Poudre River in 2012. It ranked as one of the state's most destructive wildfires at over 87,000 acres of land burned, over 250 buildings destroyed, one fatality, and control costs at over $31 million. Fall rains that year produced flash river flows running soot black and yielding high sediment and ash loads that adversely impacted aquatic habitat, and physical and chemical properties of the river. Poudre River water was not considered suitable for municipal water supplies. Current projections are that these impacts will continue into 2017 (Larimer County Colorado, High Park Fire website 2014).

Individual trees may be successfully treated, but there are no effective control measures for mountain pine beetle infestations when large tracts of forestland are involved. Although a warming climate is the

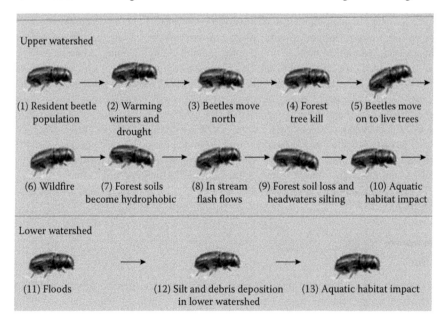

FIGURE 16.3 Connecting the Pest Dots. Progression of the mountain pine beetle leading to adverse impacts in aquatic habitats.

driver, mountain pine beetle spread is not directly linked to human transportation as the beetles will always move on their own to seek susceptible living trees to infest. Human activity has played an indirect role in mountain pine beetle spread due to warming conditions through climate change, enabling the insect to spread ever more northerly.

So, is a new definition needed to describe the type of pest or invasive species whose impacts are so far reaching, such as a terrestrial pest with the power to knock out whole aquatic ecosystems? I will leave that question for the reader to consider. By bringing the mountain pine beetle discussion up in this forum, I hope to highlight the intertwined pest and invasive species complexities impacting the environment and habitats. Policymakers at all levels of government will need the advice of professional resource managers, especially during times of declining budgets, to help resolve these issues.

The introductions of other Eurasian species to the Americas are the familiar aquatic invasive species the zebra and quagga (dreissenid) mussels. These mussels were first introduced into the Great Lakes, then throughout many river systems of the Eastern and Mid-West, and then documented in 2007, into the western watersheds of the United States.

The newest agricultural pest has already arrived in the western states, but it is yet to be fully recognized. This pest does not carry the familiar name of an insect enemy, nor of any fungus, bacterial, or virus disease. This pest is the dreissenid mussel. As the mountain pine beetle is able to damage aquatic systems, the dreissenid mussel has high probability to impact land-based agriculture. Agriculture in the western states is highly dependent on irrigation, and the greatest volume of this water is supplied by surface sources that are vulnerable to dreissenid mussels.

With their capability to rapidly colonize underwater substrates and high reproductive potential, dreissenid mussels can damage irrigation components. Even shells of dead dreissenid mussels can clog irrigation sprinkler nozzles and reduce the overall efficiency of both piped and open-water conveyance systems.

Hydroelectric power producers and municipal and industrial water suppliers have described dreissenid mussel issues in the technical literature. Dreissenid infestation impact on the aquatic ecosystem has also been described in the scientific literature. While the dreissenid mussel has the potential to damage agriculture, agriculture has described few if any dreissenid interactions. Other than their local county agricultural extension agent, growers generally have no system available to report dreissenid impacts to their operations. I feel it will be only a matter of time until irrigated agriculture in the western states is adversely impacted by the dreissenid mussel.

A comparison of Figures 16.4 and 16.5 shows that the most heavily populated traffic routes in the United States correlate with increased zebra and quagga mussel sightings throughout the West (U.S. Department of Transportation 2002; U.S. Geological Survey 2014). For these mussels (Figure 16.6), the general pattern of infestation is overland transport by contaminated watercraft and other in-water equipment via the highway system, then to a water body system, where the invasive species move with downstream water flow. This spread pattern highlights the importance of inspection and cleaning of all types of equipment. Dreissenid mussel prevention in the waters across the western United States is highly dependent upon the 100% successful inspection and cleaning of watercraft and other portable in-water equipment.

In addition to equipment inspection and cleaning, other tactics are also being used to combat aquatic invasive species and pests. Thinking about dreissenid mussel infestation in the western states, these tactics generally include early detection through monitoring, organizational collaboration through public outreach, certain levels of control and management, and research.

Comparison of Invasive Species and Integrated Pest Management Models

National Invasive Species Management Plan and Equipment Inspection and Cleaning

Separately, equipment inspection and cleaning are examples of actions that may be used to deny entry and establishment of an invasive species into an area. Equipment inspection and cleaning as a paired prevention measure coincides with the findings of the NISC and the strategic goals developed in the Invasive Species National Management Plan, 2008–2012 (National Invasive Species Council 2008). The importance of preventing the spread of invasive species is detailed in the Plan

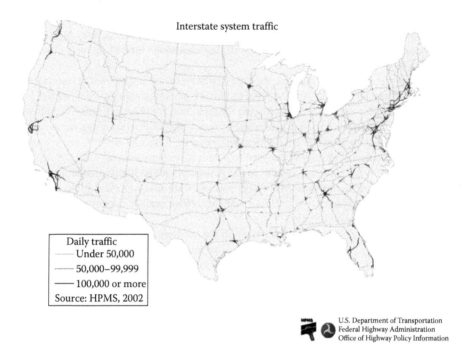

FIGURE 16.4 Daily interstate system traffic in the United States. (From U.S. Department of Transportation, Federal Highway Administration, Daily interstate system traffic in the United States, Washington, DC, http://www.fhwa.dot.gov/, accessed July 2014, 2002.)

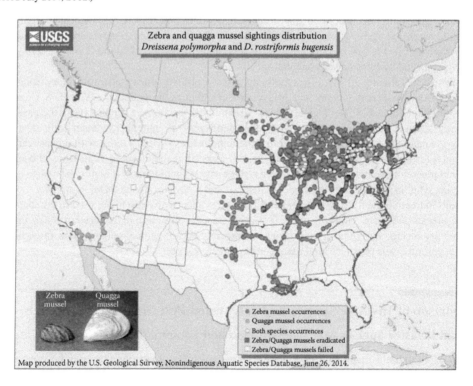

FIGURE 16.5 Zebra and quagga mussel sightings distribution. (From U.S. Geological Survey, Zebra and quagga mussel sightings distribution, Reston, VA, http://nas.er.usgs.gov/taxgroup/mollusks/zebramussel/maps/current_zm_quag_map.jpg, accessed July 2014, 2014.)

FIGURE 16.6 Clusters of adult quagga mussels are covering a rope used in the Lower Colorado River. (From DiVittorio, J. et al., *Inspection and Cleaning Manual for Equipment and Vehicles to Prevent the Spread of Invasive Species*, U.S. Department of the Interior, Bureau of Reclamation, Denver, CO, Technical Memorandum No. 86-68220-07-05, http://www.usbr.gov/mussels/prevention/docs/EquipmentInspectionandCleaningManual2012.pdf, accessed July 2014, 2012.)

where Prevention, Early Detection/Rapid Response, Control and Management, Restoration, and Organizational Collaboration tactics are employed. From the Plan:

Prevention. Prevention is the first-line of defense. The Strategic Goal for Prevention calls for preventing the introduction and establishment of invasive species to reduce their impact on the environment, the economy and health of the United States.

Early Detection Rapid Response. Even the best prevention efforts cannot stop all invasive species. Early Detection, rapid assessment, and Rapid Response (EDRR) may act as a critical second defense. The EDRR Strategic Goal calls for developing and enhancing the capacity in the United States to identify, report, and effectively respond to newly discovered and localized invasive species.

Control and Management. The spread of widely established invasive species can be slowed and their impacts reduced. The Control and Management Strategic Goal calls for containing and reducing the spread of invasive populations to minimize their harmful impacts.

Restoration. Invasive species can severely undermine the ability of plants and animal communities to recover. The Restoration Strategic Goal calls for the restoration of high-value ecosystems to meet natural resource conservation goals by conducting restoration efforts on multiple scales.

Organizational Collaboration. Invasive species cross jurisdictional boundaries, making coordination and collaboration critical to success. The Organizational Collaboration Strategic Goal calls for maximizing organizational effectiveness and collaboration on invasive species issues among international, federal, state, local and tribal governments, private organizations, and individuals.

Integrated Pest Management and Equipment Inspection and Cleaning

Equipment inspection and cleaning can also be considered a sanitation action under the integrated pest management (IPM) defined tools within the cultural control requirement.

IPM is "a sustainable approach to managing pests by combining biological, cultural, physical, and chemical tools in a way that minimizes economic, health, and environmental risks," and "Federal agencies shall use IPM techniques in carrying out pest management activities and shall promote Integrated Pest Management through procurement and regulatory policies, and other activities" (FIFRA 1996).

TABLE 16.1

Comparing National Invasive Species Management Plan Goals to Integrated Pest Management Tactics

National Invasive Species Management Plan		Integrated Pest Management
Prevention	Analog to	Cultural, Physical
Early detection (ED)/rapid response (RR)	Analog to	ED—cultural; RR—chemical, cultural, mechanical, and physical*
Control and management	Analog to	Biological, chemical, cultural, mechanical, and physical
Restoration	Analog to	Cultural
Organizational collaboration	Analog to	Cultural

*Biological controls (usually include predators, parasites, and pathogens) was not included because biological control is a long-term control measure.

Generally, IPM is a process incorporating steps of continuous monitoring, education, and record keeping necessary to maintain pest damage below an unacceptable injury level. However, when considering the dreissenid mussel, nearly any detection will eventually result in an unacceptable level of injury.

Table 16.1 provides a general comparison between the individual National Invasive Species Management Plan goals to a given IPM tactic. Although comparisons from one to the other are somewhat situational, approximations can be made to a corresponding IPM tactic.

Prevention compared to cultural and physical. Equipment inspection and cleaning are examples of sanitation actions, and closely identified as a cultural control tool. Exclusion, a physical tool, has prevention attributes. Prevention communication is usually a cultural control tool.

Early detection/rapid response compared to selected IPM tools. Early detection might include monitoring, surveillance, and similar actions; identifies most closely as a cultural tool. Rapid response involves control in a short time scale to achieve a fast knockdown of the invasive species. IPM tools available include chemical (pesticides), cultural (desiccation), mechanical (harvesting), and physical (barriers, traps, exclusion, etc.). For dreissenid mussel infestations, rapid response is not practically employable. With new detection technologies and increased sampling, incipient populations that have not taken hold yet can be detected. This may provide additional time to harden water management infrastructures and perform other tasks. However, by the time a mussel population is detected, rapid response actions usually fail to control mussel populations in open water situations.

Control and management compared to the full suite of IPM tools. Control and management is usually associated with control actions over the long term. The full range of IPM tools is used.

Restoration compared to cultural. Examples usually include restoration of aquatic and riparian sites using various plant species for soil erosion control and watercourse shading.

Organizational collaboration compared to cultural. Communication with stakeholders, public outreach, billboards, information handouts, etc.

Philosophy of Inspection

Physicist W. Edwards Deming (Deming 1986) studied inspection methods on a statistical basis. His interest area was from the standpoint of business situations and the reliability of manufactured goods. One of Deming's management points faulted reliance of end-of-the-line product inspection for poor quality in finished manufactured items. Inspection, he observed, was labor intensive and costly, and could not find all issues. Yet, Deming did recognize that inspection had its place in certain settings. Equipment inspection and cleaning for invasive species prevention probably fits this category.

Discussing Deming's observations on inspection applied to equipment inspection for invasive species and pests might be a bit *outside the box*. At first glance, it might seem too far a reach to consider a relationship of Deming's work to inspection of equipment contaminated by invasive species. Deming's prescription for product improvement was to increase quality earlier in the manufacturing process and eliminate inspection, but inspecting equipment for invasive species contamination is a distinctly different task than for product inspection. However, the observations made regarding inspection still seem to fit—inspection is labor intensive and costly, inspection can be analyzed statistically, and inspection cannot find all issues.

In a parallel to Deming's observations, the need to constantly improve a given process was recognized. Today, many environmental and natural resource processes use the Deming *plan–do–check–act* system. The inspection cleaning relationship (Figure 16.7) considers a variety of equipment circumstances, situations, and incorporates inspection and cleaning protocols, as well as the plan–do–check–act method of process improvement.

Natural resource practitioners have few additional tactics to supplement inspection for potentially contaminated equipment. We have no control over the last place of use for a privately owned recreational watercraft, its care or cleaning thereafter. There are countless instances that have occurred at our western water bodies where watercraft inspectors must work with the boat in front of them, whatever its condition. We evaluate the costs of inspection, as labor intensive as it is, as a relative bargain when compared to the cost of control and management of an established invasive species, combined with the loss of resource values, for example, damage to recreation, threatened/endangered species, aesthetics, water quality, and infrastructure.

Even after stringent inspection and cleaning controls are put in place at the location of last use, watercraft continue to be found contaminated with dreissenid and other invasive species throughout the western states at new watercraft launch destinations. These watercraft had been professionally inspected, cleaned, and undergone postcleaning inspection and certification by rigorously trained watercraft inspection and cleaning personnel.

Factors Affecting Inspection and Cleaning

Even the most careful inspection and cleaning of any equipment, however, will not guarantee that the equipment is absolutely free of pest or invasive species contamination. Successful inspection and cleaning is dependent upon many factors, such as the amount of care taken during the cleaning operation, the type of cleaning equipment being used, the level of training of the cleaning operator, the type of equipment being cleaned, and the particular invasive species.

Nonhuman Factors

Watercraft are not all alike. There are sailboats, personal-use watercraft, open bow skiffs, so-called bass boats, and many other variations. Nor is in-water equipment all alike, such as construction barges, crane lift-weight testing water bags, and a wide variety of other equipment. With all the complex compartments, angles, surfaces, and components that watercraft and in-water equipment might employ, inspection and cleaning achieving 100% efficacy 100% of the time seems difficult to attain. The standard outboard motor used on so many watercraft is difficult to inspect, but can be treated with 140°F hot water to kill dreissenid mussels hidden in its cooling system (Figure 16.8). The issue here is that not every watercraft equipped with an outboard motor is decontaminated in such a manner, and the outboard motor user could unknowingly transport dreissenid mussels into uncontaminated waters.

Human Factors

Humans are not all alike. Human behavior is certainly a factor in any process. After all, an individual person will be responsible for the eyes-on, hands-on judgment and work of inspection and cleaning equipment. Recalling Figure 16.7, inspection and cleaning can be ultimately broken down into identifiable

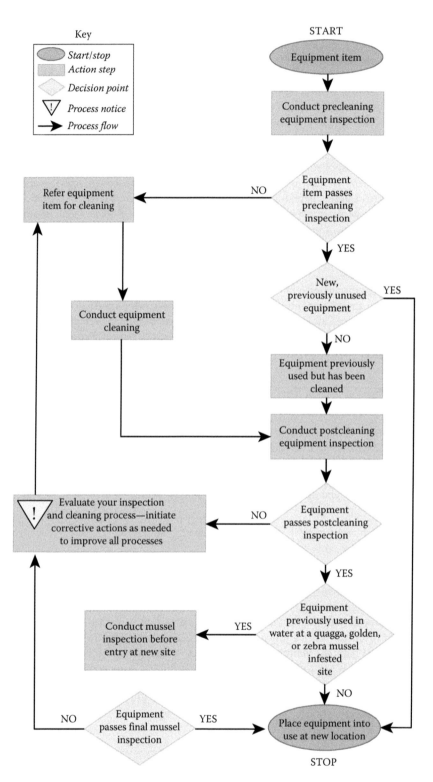

FIGURE 16.7 The inspection–cleaning relationship. (From DiVittorio, J. et al., *Inspection and Cleaning Manual for Equipment and Vehicles to Prevent the Spread of Invasive Species*, U.S. Department of the Interior, Bureau of Reclamation, Denver, CO, Technical Memorandum No. 86-68220-07-05, http://www.usbr.gov/mussels/prevention/docs/ EquipmentInspectionandCleaningManual2012.pdf, accessed July 2014, 2012.)

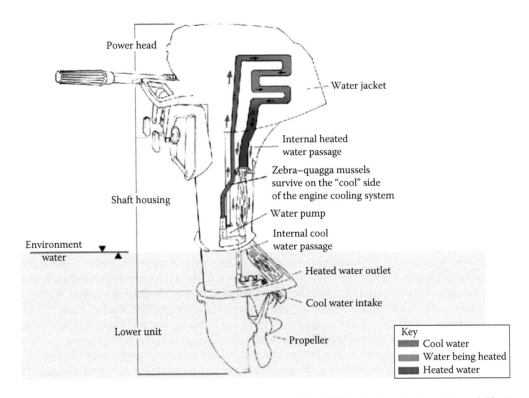

FIGURE 16.8 Simplified cooling system of an outboard motor. (From DiVittorio, J. et al., *Inspection and Cleaning Manual for Equipment and Vehicles to Prevent the Spread of Invasive Species*, U.S. Department of the Interior, Bureau of Reclamation, Denver, CO, Technical Memorandum No. 86-68220-07-05, http://www.usbr.gov/mussels/prevention/docs/EquipmentInspectionandCleaningManual2012.pdf, accessed July 2014, 2012.)

process steps. The expected outcome of successful equipment inspection and cleaning is an equipment item that will not transport an invasive species or a pest. Every one of us, whether we are aware of it or not, however, is subject to development of attitudes and variation from accepted practice in any process, including equipment inspection and cleaning.

Even among experienced, certified, or licensed practitioners of any craft, results many not be always as expected: the automobile that you recently had repaired by a shop still does not run properly; the tooth that you had a root canal performed on 6 months ago is still sensitive; licensed automobile drivers still make driving mistakes; certified pesticide applicators have been known to make errors in pesticide selection and application; buildings and bridges designed by licensed professional engineers have been known to collapse. There are many possible examples to discuss.

Human factors also might influence the outcome during the actual practice of equipment inspection and cleaning work. Fatigue, repetition, boredom, training level, and the pressure to work faster when there are many boaters waiting for inspection prior to launch ... these are just a few instances that can mean the difference between a boat completely free of zebra mussels, and a boat that is still harboring a few living mussels.

Summarizing information from human and nonhuman factors affecting inspection and cleaning, and applying the principles from IPM, demonstrates that no one control tactic is capable of producing 100% efficacy 100% of the time.

Recreational watercraft inspection and cleaning does not always achieve 100% efficacy. There have been countless reports of still contaminated recreational watercraft that have been professionally inspected, cleaned, and certified as dreissenid mussel-free. In these situations, the inspectors and decontamination personnel have been trained under the Watercraft Inspection and Decontamination Interception Training (WIT) For Zebra/Quagga Mussels program. Such boats, after traveling to another launch site, where the watercraft is reinspected, may be found to have live mussels. Such a boat might have had combinations of both human and nonhuman factors involved for passing an infested watercraft. To my knowledge, no scientifically designed study has been yet undertaken to determine why these watercraft are being found infested.

Considering the huge number of watercraft using Lake Mead, for example, on a peak boating day where thousands of boats are on the Lake, it is possible that these instances simply reflect the efficacy limit of watercraft inspection and cleaning when still-contaminated boats get through.

Efficacy of Tactics

This discussion addresses the dreissenid mussels directly. Of the Plan goals (prevention, early detection/rapid response, control and management, restoration, and organizational collaboration), more information will be presented during this discussion on prevention than on the other goals because most attention is focused on prevention in the campaign against mussels. For ease of discussion, we will take each goal in reverse order presented by the Plan.

Organizational collaboration. Organizational collaboration is the communication that takes place with all those concerned about a particular invasive species issue, such the dreissenid mussel issue. Agency meetings including the Aquatic Nuisance Species Task Force, the associated Regional Panels of the Aquatic Nuisance Species Task Force, and the annual National Invasive Species Awareness Week event held in Washington, DC, help foster state and federal agency coordination. The "Clean, Drain, and Dry" message to boaters, boat launch kiosks, handouts, billboard signs, agency press releases, public meetings, and other outreach directly to the public are also a part of organizational collaboration.

As important as public cooperation is in the campaign against zebra and quagga mussels, there is no estimate of its efficacy. Notwithstanding its importance, we just do not know how well it works. This represents a gap in an integrated approach to prevent the spread of aquatic invasive species and pests. For example, surveys conducted of watercraft users show the majority have heard and read of the importance of the "Clean, Drain, and Dry" message. We know that watercraft inspection and cleaning are complex operations where even experienced personnel have had difficulty. Figure 16.7 is understood by professional inspector-decontamination personnel, but likely would not be understood, or even welcomed for that matter, by the public. This is why the short and simple message of "Clean, Drain, and Dry" message is so important to boaters. However, cleaning a watercraft from a boater's point of view might be a task more associated with appearance; while cleaning a watercraft from the point of view of a WIT-certified inspector/decontaminator is a distinctly different task. It is the perception of what *clean* is to different persons that is so elusive. Our earlier discussions regarding definitions and human factors of inspection and cleaning are recalled. Recent modifications of the "Clean, Drain, and Dry" message by some organizations have sought to better define each term to the public.

Restoration. For many terrestrial habitats, restoration actually adds an element of long-term control by bringing stability to the system. Habitats that once were adversely impacted by an invasive species or pest, and successfully restored, usually develop a level of resistance to renewed invasion. However, for all the time the dreissenid mussels have been in North American aquatic ecosystems, there is little success that can be discussed about the restoration goal. Other than two ponds (Milbrook Quarry in Virginia and Base Lake in Nebraska) being treated and with restoration efforts, these mussels continue to invade and overrun aquatic habitats throughout the Country. This represents a gap in an integrated approach to prevent the spread of aquatic invasive species and pests.

Control and management. Control and management is one goal that has shown some limited success toward the dreissenid mussel with a measured efficacy.

Chemical control treatment. Zequanox® is manufactured by Marrone Bio Innovations®. Per company information, Zequanox® is a selective molluscicide (Marrone Bio Innovations® 2014). As of this writing, Zequanox® has shown efficacy only when used in enclosed systems such as hydroelectric plants, cooling water intake systems, and similarly built infrastructure. Open water testing is being conducted currently, but early testing showed little to no efficacy in open water. Other chemical treatments are generally nonselective, such as chlorine. Potash, potassium chloride, and other chemicals have had limited use in several small open water systems. This represents yet another gap in an integrated approach to prevent the spread of aquatic invasive species and pests. Other chemical molluscicides have been used against the dreissenid mussels, but a discussion of these materials in this chapter is beyond the scope of this work. For a more in-depth discussion of chemical molluscicides, see Sprecher and Getsinger (2000).

Mechanical control treatment. Mechanical system retrofit can help protect engineered systems in built infrastructure. These systems include copper ion generators and both low- and high-voltage electrical discharge units, microwater intake filtration, and ultraviolet light exposure. Mechanical control systems function to control dreissenid mussels only in the built water infrastructure, not in open water situations.

Cultural control treatment. Material selection to construct components of water handling infrastructure is an example of a cultural control. Materials research has demonstrated that certain metal alloys, usually those containing copper, are resistant to mussel attachment. It is possible to retrofit various components in built infrastructure. There is no mussel efficacy in open water situations.

Desiccating conditions such as high winds, low humidity, consecutive days of direct sunlight, and high temperatures have been used as a tool in managed water systems such as reservoirs and canals. Canals, for example, might be de-watered to take advantage of desiccating conditions to reduce a local dreissenid population. A similar technique has also been used in reservoir drawdown to attack dreissenids at the upper levels, while using a lowered reservoir pool to force the remaining dreissenid population into reduced oxygen regions in the lower reservoir levels for added efficacy.

Physical control treatment. Coatings may be applied to surfaces to form a barrier between the substrate of an item and the coated outer surface. Coatings work in different modes of action, generally antifouling and foul-release. Antifouling coatings impart a pesticide activity, usually using a metal such as copper, to impede mussel growth. Foul-release coatings offer super-slick outer coating surfaces that inhibit mussel attachment. There is no mussel efficacy in open water situations. For a more in-depth discussion of coatings research, the reader is directed to Skaja (2014, Chapter 29 in this book).

Biological control treatment. Reclamation has conducted limited predation studies using certain fish species for adult quagga mussel control. While this research can be applied to open water situations, the high reproductive potential of dreissenid mussels probably limits the efficacy of this approach. Another example is the co-occurrence of dreissenids and brown gobies in the eastern United States. The brown goby is a voracious dreissenid predator, but has not resulted in substantial dreissenid control. Otherwise, there has been comparatively no research on use of other forms of biological control such as parasites or pathogens against dreissenid mussels. This represents another gap in an integrated approach to prevent the spread of aquatic invasive species and pests.

Early detection/rapid response. Monitoring for dreissenid mussels in western waters has been conducted for some time. Prior to discovery of quagga mussels in Lake Mead in 2007, substrate monitoring was regularly conducted for adult mussels. Adult mussel monitoring continues, with vigorous veliger monitoring at many locations. As discussed earlier, the rapid response portion of *early detection/rapid response* is not a practical tactic against the dreissenid mussels. By the time a mussel population is detected through monitoring, rapid response actions usually fail to control mussel populations in open water situations. But early detection is a valued action because it provides states and federal agencies the advanced warning of possible dreissenid presence. In particular, Reclamation and the USACE use the detection to begin the process to harden water management infrastructure against dreissenid mussel infestation. The federal budget operates on a 3-year planning cycle. So, for example, a positive mussel detection made at a given site because of an early detection effort initiates year one of the cycle, involving submission of a budget request for equipment purchase and contracting action for equipment retrofit to be implemented in budget year three of this process.

Prevention. Prevention has some unique qualities as an IPM tactic. As a physical control, exclusion, for example, can be used on uninfested lakes and reservoirs where outside watercraft are not allowed; only *resident* watercraft, those watercraft that are permanently docked on-site, may be rented for use.

Exclusion has also been used as a quarantine measure of lakes and reservoirs when they are determined to be mussel contaminated. The efficacy of exclusion is obviously high, provided the exclusion of all incoming in-water equipment and management of released infested water is maintained.

By far, most prevention effort dedicated toward prevention through equipment inspection and cleaning. In this, it is helpful to discuss both terrestrial equipment inspection and cleaning, and aquatic equipment inspection and cleaning for a comparison of methods and efficacy.

In a study quantifying the efficacy of equipment inspection and cleaning, Balbach et al. (2008) set up a terrestrial site where wheeled and tracked vehicles ran a designed test course, followed by repeated inspection and cleaning conducted by trained and experienced personnel and using well-maintained and quality wash systems. Soil, plant fragments, and seeds were recovered, separated, weighed, and identified to species. Inspection and cleaning efficacy was determined at approximately the 80%–90% range. Inspection and cleaning efficacy never achieved 100%.

Three interesting points emerge from Balbach's work in achieving the 80%–90% efficacy range at a terrestrial site: (1) trained and experienced personnel always performed the equipment inspection and cleaning tasks; (2) quality wash systems were always used and the wash equipment was more than likely frequently adjusted and repaired quickly; and (3) these personnel probably were not under the same physical and emotional stresses as recreational watercraft inspectors and decontaminators are during a high-volume summer day at the lake. In other words, the wash equipment, and inspection and cleaning personnel involved in the Balbach study performed at 100% efficiency at all times in order to achieve 80%–90% efficacy. In all likelihood, any reduction of performance, either related to human or nonhuman factors, would probably result in lowering efficacy.

It would be difficult to determine how closely these results can be applied to an aquatic equipment inspection and cleaning. There are obvious differences between the terrestrial equipment being tested by Balbach and in-water equipment such as recreational watercraft; as well as differences among terrestrial and aquatic species; as well as the different inspection and cleaning approaches. The results from Balbach's terrestrial work and applying the observed multiple occurrences of infested watercraft that have been passed through point of last use inspection and cleaning, but are later intercepted when crossing into adjacent states, provide an indication that recreational watercraft inspection and cleaning does not always achieve 100% efficacy. In fact, experience gained through the long-term practice of IPM demonstrates that no one control or management tactic is capable of producing 100% efficacy 100% of the time.

For dreissenid mussels in particular, cleaning approaches have looked to water temperatures at 140°F (Morse 2009; Wong et al. 2011) in decontaminating watercraft. When dreissenid mussels become attached to a test surface and are subjected to the specified temperature for the specified period of time, mussel mortality achieved 100%. I do not contest the works by Morse and Wong. On the contrary, these works have done much to strengthen equipment inspection and cleaning methodology. In the Cleaning Manual, I specify that inspection and cleaning methods would assist prevention by limiting the transport of invasive species and pests, but could not guarantee that the treated equipment would be completely free of contamination.

In actual practice, consider the thousands of watercraft being inspected and cleaned at many cleaning check points located throughout the western states by perhaps hundreds of different inspection and decontamination personnel on a weekly basis during the boating season. The current pressurized hot water decontamination approach extrapolates the experimentally observed mussel mortality results of Morse and Wong to mass watercraft and in-water equipment decontaminations, with the expectation the cleaning practice is also achieving 100% mussel mortality. Does the 100% mussel mortality in the experimental testing actually translate into 100% efficacy for all watercraft being cleaned in the field? At this point in time, there have been no peer-reviewed studies to determine the efficacy of watercraft decontaminated on the macro scale. However, we have observed that watercraft formerly used in Lake Mead, and then cleaned, inspected and certified by trained WIT personnel on departure, and upon entering neighboring states, have resulted in repeated interception of contaminated watercraft, causing too high a concern to continue assuming 100% efficacy. Moreover, even when all of the current antidreissenid tactics that have been discussed earlier in this chapter are integrated and used together, the spread of dreissenid mussels into new western sites continues at alarming speed.

We find ourselves in somewhat of a parallel today with dependence of inspection and cleaning as probably the single most important defense against aquatic pests and invasive species as American agriculture was dependent on pesticide control as a single-use tactic more than 40 years ago. The shortcoming of dependence on any single invasive species or pest management tactic is borne out by an unlikely story. It is the case study and history of use of the insecticide dichlorodiphenyltrichloroethane (DDT). DDT was among the first in a family of synthetically produced, chlorinated hydrocarbon organic insecticide materials that showed unequaled efficacy against insect pest populations. DDT was widely used throughout agriculture, residential pest control, and for public health vector control. During the late post–World War II years when DDT was first introduced to the marketplace, and extending into the 1970s when DDT was removed from the market, arthropod pest management was almost entirely pesticide dependent, with little use of nonchemical tools. DDT was most effective early in its use period because insect populations had little to no resistance to the chemical. As large-scale use of DDT progressed during this period, resistance began to develop and increase within target pest populations. As resistance increased, greater DDT dose rates were required, but continued declines in efficacy were observed until DDT was finally removed from the market. The lessons learned from DDT use gave rise to the modern IPM practices that we use today.

Regarding chemical controls, as IPM processes are brought into combined use, a properly functioning IPM program would result in reduced reliance on chemical pesticides. Chemical pesticides will always remain an IPM tool when used in combination with the full suite of other IPM methods. The relative amount of a pesticide used to achieve control of a given pest population can be reduced when used in combination with non-pesticide methods. An important feature of IPM is the relationship of being able to reduce the level of pesticide use while increasing nonchemical IPM methods (Figure 16.9). The product of this relationship is the significant advantage of less pesticide resistance within the pest population. This advantage results in greater pesticide efficacy, and the useful life span of pesticide products can usually be effectively extended (DiVittorio 2008).

Regarding prevention through equipment inspection and cleaning, it is obvious that aquatic invasive species and pests could not develop resistance to inspection and cleaning actions as occurred with a primarily pesticide-dependent approach formerly used in agriculture. The information in Figure 16.9

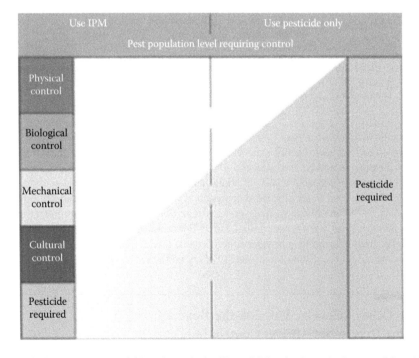

FIGURE 16.9 IPM compared to pesticide-only methods. (From DiVittorio, J. et al., *Integrated Pest Management Manual for Effective Management on Reclamation Facilities*, Bureau of Reclamation, Policy and Program Services, Denver, CO, 2008.)

TABLE 16.2

Summary for Efficacy of Tactics

Plan Goal	IPM Tactic	Action(s)	Gap(s)	Open Water Efficacy	Infrastructure Efficacy
Prevention	CL	Equipment inspection and cleaning	Yes, <100% efficacy; not available at all sites all the time	Operates at <100% efficacy in actual practice	Operates at <100% efficacy in actual practice
Prevention	P	Exclusion	Yes, limited use	100%	100%
Early detection	CL	Monitoring, surveillance	Yes, funding issues	Not a control measure	Not a control measure
Rapid response	C, CL, M, P	None	Yes	No efficacy	No efficacy
Control and management	B	Biocontrol, fish	Yes, limited use; limited biocontrol agents	Low, still experimental	Low, still experimental
Control and management	C	Zequanox®		None, testing in progress	76%–100%*
Control and management	C	Potash, potassium chloride, others	Yes, limited use	Varies	Varies
Control and management	CL	Materials manufacture desiccation	No Yes	No efficacy varies	High, copper alloys varies
Control and management	M	Retrofit-copper ion generators, electrical discharge, microfiltration, UV light	No	No efficacy	Yes, efficacy varies
Control and management	P	Coatings—barrier	No	No efficacy	Probable, still in testing
Restoration	CL	Restore habitat	Yes, not attempted	No efficacy	No efficacy
Organizational collaboration	CL	Communication, outreach, "Clean, Drain and Dry"	Yes, unknown efficacy	Unknown efficacy	Unknown efficacy

*Marrone Bio Innovations®, Marrone Bio Innovations Zequanox®, website http://www.marronebioinnovations.com/products/brand/zequanox/, accessed July 2014.
IPM tactic key: B, biological; C, chemical; CL, cultural; M, mechanical; P, physical.

is presented qualitatively, rather than quantitatively, but demonstrates that an individual control tactic can be rebalanced toward use of a full suite of tactics. We stated earlier in this chapter that equipment inspection and cleaning would be most successfully used against aquatic invasive species and pests when it is employed in combination with a suite of additional, integrated tactics. While equipment inspection and cleaning is the first step in an integrated approach to prevent the spread of aquatic invasive species and pests, it must not be an only step. Widespread recognition is needed that equipment inspection and cleaning does not operate at 100% efficacy all the time. For dreissenid mussels, equipment inspection and cleaning must be supplemented by a broader suite of robust tactics than are being currently employed.

Table 16.2 summarizes the discussion of the "Efficacy of Tactics" section.

Looking to Other Sciences: Outside-the-Box Thinking for Dreissenid Mussel Control

Earlier in the "Efficacy of Tactics" section, we discussed each tactic currently available and noted gaps in efficacy throughout existing tactics. However, where would any additional new control tactics come from that would close these gaps? Research will point the way to the new tactics needed for dreissenid mussel control. Policymakers and legislators will need to fund the pioneering research necessary to develop newer tactics that would be ultimately integrated with equipment inspection and cleaning.

George S. Patton is reported to have said: "If everybody is thinking alike, then somebody isn't thinking." I offer several research approaches to dreissenid mussel control later. The feasibility of these approaches is outcome-dependent on the science. I offer them as possibilities of finding unique approaches that have not yet been submitted for research. My challenge to you, the reader, is a simple one: if you disagree with the approaches presented here, you must publically offer up a new idea, as I have done, to the research community for dreissenid mussel control. Playing it safe by offering a rehash of an attempted or conventional approach, or just remaining silent is not allowed. Innovation is the only way to resolve the mussel problem.

Life cycle manipulation. A dog or cat owner 20 years ago had few options for flea control for their pets. Available to the pet owner at the time were flea collars that emitted a slow-release fumigant pesticide, a shake on pesticide dusting powder, a liquid aerosol pesticide spray labeled for use on pets for flea control, and pet bath soaps and shampoos that contained a pesticide. Each of these product families were designed to bring the pesticide active ingredient in contact with and kill the flea directly.

Novartis® Animal Health began trials on the insect growth regulator lufenuron as an active ingredient for flea control. Lufenuron is a chitin inhibitor that interrupts normal metamorphosis transition from the larval to adult life stage in certain insects. The Novartis® product Program® contains lufenuron and is used widely today on dogs and cats. Fleas not completing metamorphosis cannot survive (Novartis® 2014). Lufenuron has no activity on the adult life stage. In addition to flea control, in some agricultural crop applications, Syngenta® uses lufenuron in a product to control certain insects and mites in a variety of crops (Syngenta® 2014).

Thinking about the dreissenid mussels, we find that zebra and quagga mussels are destructive only in the adult life stage. What if dreissenid mussels were prevented from reaching the adult life stage and remained in any of the pre-adult life stages? The idea that a growth-regulating chemical that could have a similar mode of action in dreissenid mussels as lufenuron has in fleas excites the imagination. Of course, bringing a new pesticide product to market would take years, and there would be a range of complex environmental considerations as well. But, what research has been conducted on the dreissenid hormone systems controlling metamorphosis that parallel the lufenuron work for flea control and agricultural crop protection? None were found during literature search.

Species manipulation. The most destructive feature of dreissenid mussels is found in the adult life stage. The ability to simply attach to a surface using the byssus gives the dreissenids capacity to populate a water body in three dimensions. Without a functioning byssus, zebra and quagga mussels would be relegated along a two dimensional water body substrate. There zebra mussels would probably not be able to survive in high numbers in sediment-covered substrate conditions. The quagga mussel is known to be able to populate a muddier substrate, but its numbers would also be reduced from being deprived of choice locations that offer flowing, food-rich and oxygenated waters. For both species, if the competitive advantage of attachment could be limited, there would be substantial control improvement for both open water and the built infrastructure situations.

But how might such a system work? The genome of *Dreissena polymorpha* and of *D. bugensis* is currently incomplete. Assuming a completed genome is soon to arrive, we can speculate: perhaps the genes controlling the development or function of the byssus can be identified and toggled? Perhaps the gene switches could be manipulated chemically?

In an unpublished work (Misamore et al. 2011), an innovative x-ray experiment was conducted to determine whether irradiated adult quagga mussels might respond to produce sterile males similar to that being used in the screw-worm fly (*Cochliomyia hominivorax*) sterile male release program. Although not within the scope of this experiment, as a member of the Multiagency Irradiation Team chartered by the Utah Division of Wildlife Resources, I wondered whether irradiation might have additional applications. Is it possible that a nonlethal dose of radiation might toggle specific genes in a captive mussel population to disable the byssus? Perhaps such a trait might be passed to progeny, and the treated test population be later released to the wild? Of course, we do not know the answers because no research work has been attempted.

Biological control. The U.S. Department of Agriculture, Animal-Plant Health Inspection Service chartered the Technical Advisory Group for Biological Control Agents of Weeds (TAG) in 1957. The TAG is composed of a multiagency membership that reviews and provides comment to researcher submitted petitions on the proposed importation of biological control agents of weeds. But there is no similar organization for the research of biological controls for animal aquatic invasive species and pests. Generally, biological control research on the dreissenid mussels has lagged and is identified as a gap in IPM tactics.

For these approaches and other new ideas that might be considered, more research will be needed.

Conclusion

Natural resource practitioners have few tactics to supplement equipment inspection and cleaning for potentially contaminated equipment. Equipment inspection and cleaning is the most widely used tactic to prevent the spread of aquatic invasive species and pests, but its efficacy is largely unknown with an operational probably of less than 100%. Equipment inspection and cleaning would be most successful against aquatic invasive species and pests, in particular the dreissenid mussels, when it is employed in combination with a suite of additional, integrated tactics.

For dreissenid mussels, equipment inspection and cleaning must be supplemented by a broader suite of robust tactics than are being currently employed. The results from Balbach's terrestrial work and applying the observed infested watercraft that are interdicted at state line checkpoints provide an indication that recreational watercraft inspection and cleaning does not achieve 100% efficacy. In fact, experience gained through the long-term practice of IPM demonstrates that no one control tactic is capable of producing 100% efficacy.

While equipment inspection and cleaning is the first step in an integrated approach to prevent the spread of aquatic invasive species and pests, it must not be an only step. The failure to research new methods, support innovative, outside-the-box ideas, and fund these measures so that they might be used in combination with equipment inspection and cleaning in the field is a path to assured and continued aquatic invasion.

I completed a car trip heading east from Portland, Oregon, to Denver, Colorado, in November 2011 using the Interstate Highway 84 to 80 to 25 routes through the five states of Oregon, Idaho, Utah, Wyoming, and Colorado. While traveling the nearly 1300 miles, several commercially hauled large pontoon and yacht-styled boats were spotted headed generally west. At that time of the year, most lake inspection check points had already closed down for the season.

Acknowledgments

I thank Dr. Richard Lee, Bureau of Land Management, U.S. Department of the Interior, for his IPM suggestions during the early formative stages of this manuscript. In addition, during the later pre-final version of the manuscript, formal review and comment was provided by Dr. David Britton, Fish and Wildlife Service, U.S. Department of the Interior, and another, but anonymous reviewer. I thank them both for their careful attention to detail and extremely helpful remarks. I also thank Mr. A. Gordon Brown, Invasive Species Program Coordinator, retired, U.S. Department of the Interior for his helpful overall comments. Finally, with his request for this chapter, I am sincerely honored by Dr. David Wong's invitation to participate in the making of *Biology and Management of Invasive Quagga/Zebra Mussels in the Western United States*.

Disclaimer

This chapter, "Equipment Inspection and Cleaning: The First Step in an Integrated Approach to Prevent the Spread of Aquatic Invasive Species and Pests," was prepared by Joseph (Joe) DiVittorio in his personal capacity. The opinions expressed in this article are the author's own and do not necessarily reflect the views or policies of the Bureau of Reclamation, the Department of the Interior, or the United States government.

The information contained herein, regarding commercial products or firms may not be used for advertising or promotional purposes without permission. Mention of firms, brands, or products is not to be construed as an endorsement by the author. Firms, brands, or products that are not listed are not deemed unsatisfactory.

REFERENCES

Balbach, H., L. Rew, and J. Fleming. 2008. *Evaluating the Potential for Vehicle Transport of Propagules of Invasive Species*. U.S. Army ERDC-CERL, Champaign, IL.

Berg, A. S. 1998. *Lindbergh*. The Berkley Publishing Group, New York.

Bixby, J. 1968. "Is There in Truth No Beauty?" *Star Trek*, NBC broadcast, Episode 62, season 3.

Britton, D. K. 2015. Implementing hazard analysis and critical control point (HACCP) planning to control a pathway from spreading dreissenid mussels and other aquatic invasive species. Chapter 15 in this book.

Cofrancesco, Jr., A. F., D. R. Reaves, and D. E. Averett. 2007. *Transfer of Invasive Species Associated with the Movement of Military Equipment and Personnel*. U.S. Army Corps of Engineers—Engineer Research and Development Center, Vicksburg, MS.

Deming, W. E. 1986. *Out of the Crisis*. MIT Press, Cambridge, MA.

DiVittorio, J. 2008. Chapter 9: Regulatory requirements. In DiVittorio, J., D. Eberts, D. Hosler, K. Lair, M. Nelson, F. Nibling, S. O'Meara et al. et al. (eds.), *Integrated Pest Management Manual for Effective Management on Reclamation Facilities*. Bureau of Reclamation, Policy and Program Services, Denver, CO.

DiVittorio, J., M. Grodowitz, and J. Snow. 2012. *Inspection and Cleaning Manual for Equipment and Vehicles to Prevent the Spread of Invasive Species*. U.S. Department of the Interior, Bureau of Reclamation, Denver, CO, Technical Memorandum No. 86-68220-07-05. Website http://www.usbr.gov/mussels/prevention/docs/EquipmentInspectionandCleaningManual2012.pdf, accessed July 2014.

Executive Order 13112, Invasive species. Signed February 3, 1999; Federal Register publication 64 FR 6183, February 8, 1999.

Federal Insecticide, Fungicide, and Rodenticide Act (FIFRA), as amended by the Food Quality Protection Act. 1996. 7 U.S.C. 136.

Federal Register, multiple citations 1968–1991; Doc. 69-8956; 34 FR 12305; Doc. 69-13491; 34 FR 18204; Doc. 70-3182; 35 FR 4674; Doc. 70-17130; 35 FR 19291; Doc. 91-9904; 56 FR 19259.

Larimer County (Colorado) High Park fire. 2014. Website http://larimer.org/highparkfire/, accessed July 2014.

Leopold, A. 1924. Grass, brush, timber, and fire in Southern Arizona. *Journal of Forestry* 22(6):1–10.

Marrone Bio Innovations®. 2014. Marrone Bio Innovations Zequanox®. Website http://www.marronebioinnovations.com/products/brand/zequanox/, accessed July 2014.

Melillo, J. M., T. C. Richmond, and G. W. Yohe (eds.). 2014. *Climate Change Impacts in the United States: The Third National Climate Assessment*. U.S. Global Change Program, Washington, DC, 841pp.

Misamore, M., W. H. Wong, S. L. Gerstenberger, and S. Madsen. 2011. Sterile male release technique (SMRT) using x-ray irradiation as a potential novel method for managing invasive quagga mussels—A laboratory experiment. Report to Utah Division of Wildlife Resources, Salt Lake City, UT, 18pp.

Morse, J. T. 2009. Assessing the effects of application time and temperature on the efficacy of hot-water sprays to mitigate fouling by *Dreissena polymorpha* (zebra mussels Pallas). *Biofouling* 25:7, 605–610.

Invasive Species Advisory Committee, Invasive species and climate change. National Invasive Species Council, December, 2010.

National Invasive Species Council, Invasive Species Advisory Committee. 2006. Invasive species definition clarification and guidance white paper, Washington, DC.

National Invasive Species Council. 2008. Invasive species national management Plan, 2008–2012.

Novartis®. 2014. Website http://www.ah.novartis.com/products/pethealth/cat/index.shtml, Novartis® Animal Health, accessed July 2014.

Simberloff, D., L. Souza, M. A. Nuñez, M. N. Barrios-Garcia, and W. Bunn. 2012. The natives are restless, but not often and mostly when disturbed. *Ecology* 93(3):598–607.

Skaja, A. D. 2015. Coatings for invasive mussel control: Colorado River field study. U.S. Department of the Interior, Bureau of Reclamation, Denver, CO. Chapter 29 in this book.

Sprecher, S. L. and K. D. Getsinger. 2000. *Zebra Mussel Chemical Control Guide*. Publication No. ERDC/EL TR-00-1, U.S. Army Corps of Engineers—Engineer Research and Development Center, Vicksburg, MS.

Syngenta®. 2014. Syngenta Crop Protection®. Website http://www.syngenta.com/country/eg/en/cropprptection/ourproducts/insecticides/Pages/Match50EC.aspx, accessed July 2014.

U.S. Department of Defense. 2004. *Armed Forces Pest Management Board Technical Guide 31, Retrograde Washdowns: Cleaning and Inspection Procedures*. Walter Reed Army Medical Center, Washington, DC.

U.S. Department of Transportation, Federal Highway Administration. 2002. Daily interstate system traffic in the United States. Website http://www.fhwa.dot.gov/, accessed July 2014.

U.S. Geological Survey (USGS). 2014. Zebra and quagga mussel sightings distribution. USGS website http://nas.er.usgs.gov/taxgroup/mollusks/zebramussel/maps/current_zm_quag_map.jpg, accessed July 2014.

Wong, W. H., S. Gerstenberger, W. Baldwin, and E. Austin. 2011. Using hot water spray to kill quagga mussels on watercraft and equipment (final report to PSMFC and USFWS). Department of Environmental and Occupational Health, University of Nevada, Las Vegas, NV.

Zook, B. and S. Phillips. 2012. Recommended uniform minimum protocols and standards for watercraft interception programs for dreissenid mussels in the western United States, revision of 2009. Pacific States Marine Fisheries Commission, Portland, OR. Prepared for the Western Regional Panel on Aquatic Nuisance Species.

Zook, B. and S. Phillips. 2015. Uniform minimum protocols and standards for watercraft interception programs for dreissenid mussels in the western United States. Chapter 14 in this book.

17

Applied Boater Management to Interrupt Population Establishment of Dreissena Species: A Utah Case Study

Larry B. Dalton

CONTENTS

ABSTRACT   Utah supports approximately 250 boatable water bodies, and most of the water bodies evidence limnology suitable for inhabitation by quagga (*Dreissena rostriformis bugensis*) or zebra mussels (*Dreissena polymorpha*). Utah Division of Wildlife Resources demonstrated that programmatic planning coupled with applied management to cause changes in boater behavior likely averted initial inoculations or interrupted early establishing populations of quagga or zebra mussels. Once fully established, quagga or zebra mussel populations can be extremely difficult to eradicate or control. Key to the success in Utah's case study was gaining cooperation and participation by boaters through repetitive outreach strategies. Boaters were educated about the impacts from quagga and zebra mussels along with the process for conducting their own boat inspection and decontamination via the clean, drain, and dry approach in order to kill hitchhiking Dreissena mussels. Routine, random boat inspections and decontaminations when necessary as 140°F water applied via high and low pressure by trained natural resource management professionals served to reinforce Utah's outreach messages to boaters. Success for Dreissena mussel management in Utah was demonstrated: nine water bodies evidenced Dreissena mussels between 2008 and 2013, but only one, Lake Powell, remains infested. Monitoring shows that Dreissena mussels are no longer present in the others, likely due to applied boater management, so the affected water bodies have been reclassified to an undetected status. Utah's other boatable waters remain in the undetected status due in part to the same applied boater management.

Introduction

Utah supports approximately 250 boatable water bodies, defined as being 20 hectares (50 acres) or more in size. And most of those water bodies evidence limnology suitable for inhabitation by quagga (*Dreissena rostriformis bugensis* Andrusov 1897) or zebra mussels (*Dreissena polymorpha* Pallas, 1769). Even the large inflow bays of the Great Salt Lake have acceptable salinity to support the zebra mussel, although the main body of the lake is too saline. Calcium deficiencies in a few high-elevation cirque basin lakes of the Uinta and Boulder mountains may limit Dreissena population establishment. However, low-head concrete dams built at many of those natural lakes by early pioneers to enhance water storage may leach sufficient calcium to sustain a limited Dreissena population (Suflita 2007–2014).

Dreissena mussels, particularly the invasions of the quagga and zebra species, represent a significant economic and ecological impact to the United States. Once established, quagga and zebra mussels are extremely difficult to control (Mackie and Claudi 2010). But, interruption of population establishment through applied boater management techniques, which the author instituted in Utah, can occur, and is the focus of this case study. Applied boater management in Utah urges and even demands via legal consequences that boaters prior to launching self-decontaminate (clean, drain, and dry) their wetted equipment or otherwise secure a professional decontamination that uses 140°F water. Ultimately, this reduces opportunity for Dreissena mussel inoculations. A description of the expected boater sponsored cleaning, draining, and drying promoted in Utah is as follows:

- Clean mud, plants, animals, or other debris from the inside and outside of your boat and equipment.
- Drain ballast tanks, bilge wells, live wells, and motors (lower the foot to let water drain from the lower unit).
- Dry (7 days summer, 18 days spring/fall, and 30 days in winter) or freeze (3 days).

Even after an inoculation occurs, the continued management of boaters reduces additional inoculations, which may reduce the potential for individual mussels getting close enough together to breed, promoting failure for population establishment.

Without question, the management of invasive Dreissena mussels is necessary since they impact aquatic ecosystems; foul water control structures (valves, gates, penstocks, etc.); foul water transport facilities (canals and pipelines); and foul water use apparatus in various industries, including a watercraft's raw water circulation systems (Colorado Division of Wildlife 2011a,b; DiVittorio et al. 2012; Karatayev et al. 2012; O'Neill 1996). Remedial maintenance is significant, costing tens or more millions of dollars per year in individual water conduit systems (AP 2013; Gerstenberger et al. 2011; Pimentel et al. 2005; Wong et al. 2011). Those costs are typically passed on to the public as higher utility or product fees, including need for accelerated boat maintenance of which the cost is borne by the owner.

Conservation entities—state, federal, local governments, and private organizations or water body operators—continue to struggle with management challenges following the arrival of quagga and zebra mussels. Both species arrived in the Great Lakes region in the mid-1980s (O'Neill 1996). Quagga mussels eventually spread to the West's lower Colorado River drainage by January 2007; and zebra mussels were discovered in California's San Justo Reservoir in January 2008 (California Fish and Game Department 2011), which is not associated with the Colorado River drainage. Additional discoveries of Dreissena mussels across the United States continue today (USGS 2014).

Limited control measures for quagga and zebra mussels do exist and more methods continue to be developed, but oft times they are expensive and frequently have significant environmental consequences and associated regulatory limitations (Mackie and Claudi 2010; O'Neill 1996). Thus, water users, aquatic wildlife managers and boaters are eager to find and use inexpensive, but easy and effective control measures, because budgets, either public or private, for management of invasive mussels are lacking or austere at best.

Quagga and zebra mussels are infamous as hitchhikers, traveling from affected water bodies to new locations on or within watercraft, including overland transported recreational boats (Dalton and Cottrell 2013). The mussels accomplish this feat due to their ability to survive out of water for an extended period of time—up to 30 days or longer when ambient air temperature and humidity conditions are favorable (McMahon et al. 1993). Even their larval veliger stage can survive 2 weeks or longer in the stagnant, raw water retained within transported boats (Craft and Myrick 2011). Field tests demonstrate that veligers can survive 5 days in summer and about 27 days in autumn within retained water in boats in the southwest United States (Choi et al. 2013).

Quagga and zebra mussels are also known to be moved long distances in the flow of water between locales (Claudi and Mackie 1993; Dietz et al. 1994; O'Neill 1996). The 541 km long (336 mi) Illinois

Waterway connects Lake Michigan to the Mississippi River; all are infested with quagga and zebra mussels. The Mississippi River then flows another 1910 km (1190 mi) to the Gulf of Mexico. It transports all life stages of live Dreissena mussels downstream to at least New Orleans and likely through the additional 161 km (100 mi) distance to the Gulf of Mexico (Allen 2014). Although zebra mussels breed in the lower Mississippi, annual water temperature impacts to populations cause significant mortality when temperatures of 29°C–30°C occur for a 3-month period (Allen et al. 1999; Cofrancesco 2014). The Central Arizona Project also demonstrates successful movement by quagga mussels in water flow. It delivers quagga mussel infested, raw Colorado River water via a 235 km (146 mi) aqueduct system to Lake Pleasant near Phoenix Arizona. Then the associated Salt River Project aqueduct delivers that water another 306 km (190 mi) to Tucson Arizona (Bryan 2014; McMahon 2014). Additionally, the Colorado River Aqueduct, delivering raw Colorado River water via 390 km (242 mi) of an aqueduct system to Lake Mathews near to Riverside and Lake Skinner near Temecula, both in Southern California, are additional examples for transport of all life stages of live quagga mussels in flow of water (De Leon 2014; Volkoff 2014).

In all cases of Dreissena mussel inoculation, regardless of whether as hitchhikers on boats or carried in the flow of water between water bodies, conservation entities continue to struggle with the mussel's impacts, while assessing options for improved invasive mussel management. Many water body managers have found applied boater management—working with boaters to achieve their cooperation in boat inspection and decontamination—to be the best fit to avert initial inoculations, spread, and resultant Dreissena population establishment. Managers across the nation have found varying degrees of success to achieve funding, personnel, and other resources to fully carry out this effort, and generally, many are underfunded to carry out the needed effort.

Methods

The following approach as a guideline for an impact assessment regarding a Dreissena mussel invasion and ultimate management plan and its implementation may be useful for other states or multistate regions. It certainly proved successful in Utah.

Impact assessment: First and foremost in Utah, we assessed and documented the potential impacts of a full-blown invasion by Dreissena mussels in terms of risk as measured via a water-by-water body suitability for sustaining an invasion based upon limnology and mussel biology and intensity of boater use, along with potential economic costs and ecological damage. A small team of five senior personnel from the Utah Department of Natural Resources (Outreach Chief, Aquatic Research Coordinator, Law Enforcement Captain from Utah Division of Wildlife Resources; Boating Law Administrator from Utah Division of State Parks and Recreation; led by an Engineer from Utah Division of Water Resources) accomplished this task. The impact assessment was the foundation for moving forward with a plan to avoid or mitigate impacts from an invasion; it was prepared as a series of written reports, which were consolidated into an associated PowerPoint presentation. The impact assessment was our initial presentation to agency administrators and political concerns, documenting that a significant, impending problem existed. If we had failed to garner administrative and political support at this point, it is unlikely that significant progress toward impact avoidance or mitigation and an associated budget to manage Dreissena mussels would have resulted. Our goal with the impact assessment was to secure the *go-ahead* to develop a comprehensive Dreissena mussel management plan and reasonable assurance that associated funds for its conduct would be forthcoming.

Management plan: The Utah Division of Wildlife Resources was appointed by the Utah Department of Natural Resources as lead agency for issues dealing with Dreissena mussel management in Utah. We developed a comprehensive, written aquatic invasive species management plan through establishment of a planning team. The author was the chairperson, having sufficient resources and authority to direct the team. The team developed a proposed timeline for plan development with a finite deadline of 1 year or less for plan completion, which included several Utah Division of Wildlife Resources' Regional Advisory Council public hearings and approvals from the Utah Wildlife Board, Utah's Governor, and the national Aquatic Nuisance Species Task Force. Utah's team comprised 10 learned state and federal agency personnel and nonagency stakeholders to provide sufficient intellectual *gray matter* power for planning discussions. Team members provided talent based upon past achievements, particularly as natural resource agency administrators, program coordinators, and managers, along with on-the-ground biologists, law enforcement officers, and outreach professionals. Water users and boaters were helpful stakeholders, too. A primary routine for the team's daily activity was to stimulate in-depth discussions regarding policy, laws, budget, biology, program management, outreach tactics, and state-of-the-art management for Dreissena mussels. It was paramount during the team's formation that too large of a team could become bogged down in conflict, and that even a small team could have its distractions, so we advantaged a professional facilitator whenever possible.

The team decided upon its own decision-making matrices, which were used indiscriminately for varying issues as follows:

1. *Concurrence*—It can be difficult to get, but once achieved the decision was solid.
2. *Majority vote*—It works but leads to an ever changing ad hoc majority frequently overriding another individual's or group's good ideas.
3. *Autocratic decision by the chair*—It is efficient but was seldom used, since the wisdom of the team was not advantaged.

As Utah's team chairperson, the author assigned writing and review responsibilities to appropriate team members. A technical writer assisted Utah's team in the final editing process, giving the plan a singular voice.

The Utah plan's overall mission was defined to be simple—manage aquatic invasive species for a specific geographic area during a selected span of time: *Management Plan for Aquatic Invasive Species in Utah—2009 through 2013*. A 5-year plan seemed to be a good fit. And, state or multistate regional plans meeting guidelines from the national Aquatic Nuisance Species Task Force may achieve limited federal funding assistance (NISA 1996); Utah's plan met those guidelines.

We specified the Utah plan's goal as a future condition we desired to be maintained or achieved: *Keep Aquatic Invasive Species Out of Utah or Contained to Areas Already Inhabited*. Then we developed multiple, time-bound objectives that when accomplished facilitated the goal—for example, *Educate arriving boaters at launch sites during periods when they are in operation about the risks from establishment of a Dreissena mussel population*.

Each objective was supported by one or more proactive strategies: For example, *At launch sites demonstrate to all departing boaters how as an individual they can self-decontaminate their own boat—cleaning, draining, and drying after every use* to preclude population establishment or spread for aquatic invasive species. Additionally, reactive strategies existed as contingencies for inadvertent Dreissena population establishment—for example, *Enlist organized boater groups or individual boaters to aid in the conduct for part of a rapid response, determining the extent of an establishing Dreissena population via beach and near shore searches around a water body*. Topics that were assessed via the Utah plan's objectives and strategies are as follows:

Note: The entire Utah plan with multiple objectives and strategies can be perused at http://wildlife. utah.gov/421-invasive-mussel-plan.html?Itemid=419.

1. Organizational structure and associated costs
   a. Boat interdiction checkpoints: Prioritize water bodies, launch ramps, other checkpoints, and decontamination stations to be staffed based upon boater use—identify the months for staffing per site
   b. Personnel: Identify a coordinator, biologists, outreach specialists, law enforcement officers, technicians, and administrative support—secretary and budget officer—needs and cost
2. Day-to-day operation and associated costs—office and in the field
   a. Office: Space rental, furniture, utilities, communication cost—telephone and Internet, postal mail and shipping, office supplies, etc.

    b. In the field: Fuel and maintenance for trucks, boats and decontamination units, plankton sampling nets, specimen vials and associated preservation chemicals, laboratory supply costs, uniforms, protective clothing, etc.

    c. Training costs: Initial in-house training and annual refreshers

3. Capital costs

    a. Equipment needs: Vehicles, boats, computers, LCD projectors, microscopes, decontamination units, etc.

    b. Contracted support: Laboratory services, advertising or public relations consultant, specialized training, etc.

4. Outreach campaign costs with messages and product

    a. In-reach in order to make sure agency personnel fully understand the plan and its importance.

    b. Media and other promotions: Internet, newspapers, magazines, billboard rental, advertising fees, brochures, posters and signs (include installation cost).

    c. Education for current boater operators—venues for one-on-one outreach (boater check points, boat show, sportsman expos, state and county fairs).

    d. Education for the next generation of boat operators—schools (secondary and university).

5. Early detection monitoring costs

6. Innovative research and implementation of new or evolving technology costs

7. Rapid response strategy and costs to address an invasion, addressing control of the invader and mitigation for impacts that result from the control

8. Monitoring costs for the effectiveness of the plan, making modifications as necessary

9. Develop annual budgets and associated personnel work plans that are linked to the management plan as an implementation schedule

The most notable element of Utah's plan is a requirement for boaters to self-certify that they have inspected and decontaminated their boat prior to every launch. Boaters in Utah under penalty of law must display their self-filed Decontamination Certification in plain view within the launch vehicle each time they launch.

Approval: The Utah plan's purpose is simply to record details of an implementation schedule proposed by the team as influenced by a robust public review. Approval by the Utah Division of Wildlife Resources' Director; then, approval by the Utah Wildlife Board; and finally approval by Utah's Governor was secured. Since securing federal funding assistance was intended, an ultimate review and approval by the Aquatic Nuisance Species Task Force was secured. At times the overall process seemed exhausting, even thankless, but it was well worth the effort to get everyone in agreement. It took approximately 1 year to complete Utah's initial plan. We modified and updated the plan often across its implementation period to meet changing conditions of ecology, economy, politics or improving aquatic invasive species management technology, making it a living plan. Steps are being taken to update it again as a new 5-year plan (Nielson 2014)

Funding: Securing sufficient resources—funds, personnel, and equipment—was a crucial next step after planning. Development of suitable administrative and political support for funding was eased by their earlier involvement in the final review of the plan. All of the good intentions by natural resource management agencies, conservation groups and individuals to *do the best they can, fighting back against the invasion of Dreissena mussels*, are without merit if a management plan coupled with sufficient resources does not exist. Doing the work is the real challenge: It may take years to accomplish, if ever.

Operations: Once the impact assessment and plan development/approval were completed, along with resources being made available, it was time to begin operations. Personnel were hired and trained to meet the 2008 boating season, and the program continues today. Nearly every fiber of work focused upon influencing boater behavior, so they became willing participants in boat inspection and decontamination efforts. In Utah, we attempted to expose boaters to three to seven promotional impressions about boat inspection and decontamination via the clean, drain, and dry protocol on every one of their outings. Boater resistance was minimal, although some occurred. Boaters who cooperated were not compelled to undergo professional boat decontamination; thus, their launch was expedited. Outreach was abundant and repetitive—newspaper, magazine and other written articles; brochures; highway billboards; launch ramp signs; and one-on-one contacts at launch sites; along with personal contacts at boater visited venues (e.g., Boat Show and International Sportsman Expo). On occasion law enforcement tactics and the courts were utilized. Organized boating groups were enlisted to help market the *clean, drain, and dry* message. Politicians, administrators, stakeholders, and personnel were committed to meet success. Monitoring of programmatic success was measured in terms of boater responsiveness, using feedback surveys, supplemented with covert observations of boater behavior. And early detection monitoring for water bodies affected by Dreissena species occurred.

Forty to one hundred Utah water bodies beginning in 2008 were sampled at least once each year for Dreissena veligers using a 64 μm mesh plankton net as per the protocol described by Brown and Hosler (2013). Available funds and workload limited the number of sites and frequency sampled, but once a suspicion or confirmation for existence of Dreissena mussels occurred, samples were taken monthly each year during the mussel's breeding season as predicted by water temperature. Dedicated plankton nets were used at water bodies that indicated any evidence for Dreissena mussels. All nets, dedicated or not, were appropriately decontaminated following a sampling event. Cross-polarized light microscopy, sometimes supplemented with a scanning electron microscope, was utilized to detect veligers in the samples. And endpoint PCR or quantitative PCR was used to substantiate that DNA from sampled veligers was in fact a Dreissena species. Beginning in 2009 water bodies suspicioned or shown as affected by Dreissena species were further assessed via environmental DNA (eDNA). The eDNA samples approximating 500–1000 mL of water was collected along with tow net samples for veligers. Pisces Molecular, LLC in Boulder CO and the US Bureau of Reclamation's Detection Laboratory for Exotic Species in Denver CO conducted the eDNA analysis. Laboratory results as reported by Anderson (2008–2014), Hosler (2008–2014), and Wood (2008–2014) can be perused at http://wildlife.utah.gov/affected-waters.html.

Although use of eDNA is a new and developing segment of molecular biology, it has the potential to dramatically improve our capacity for early detection of aquatic invasive species (Darling and Mahon 2011). Thus, eDNA represents yet another tool, but not the only tool in the biologist's tool bag, and if used properly can aid in decision making. Significant attention was applied to sample collections both in the field and in the laboratory in order to avoid contaminations that can lead to false-positive detections and false-negative detections. Darling and Mahon (2011) fully discuss those issues relative to DNA-based early detection.

Results

Impact assessment: Without an impact assessment, it cannot be fully understood which impacts are trying to be averted or mitigated. Utah's assessment demonstrated likeliness for invasion along with potential economic and ecological threat from aquatic invasive species, particularly quagga and zebra mussels. The impact assessment represented our *sales pitch* for need and authorization to create an aquatic invasive species management plan for consideration of approval and ultimate funding. In Utah, the Nation's second driest state, the Department of Natural Resources' Water Resources Division, assisted by the Utah Department of Natural Resources' State Parks and Recreation Division and Wildlife Resources Division, led the impact assessment. Five key personnel—water resources engineer (chair), boating law administrator, wildlife law enforcement captain, wildlife research coordinator, and conservation outreach chief (the author of this paper)—required 4 months to complete it. We presented the impact assessment to the department's Executive Director, showing that if an invasion occurred, $15 million per year of extraordinary maintenance would be required at Utah's water use industries—water conservancy districts' 6000 mi of pipelines and canals, electric generation plants, culinary water production, manufacturing and agricultural industries, and waste water reclamation. We all agreed that the increased cost would be passed to the end-users. Whereas, fighting the invasion, keeping it at bay, would cost approximately $1.4 million per year. Since all Utahans would benefit from a fight, we opinioned that funding should be from General Funds.

As a result, the department's Executive Director took the presentation to Utah's Governor and leadership within the Utah Legislature's Natural Resources Committee. The politicians agreed that the Utah Department of Natural Resources should address the issue, and via policy, lead responsibility for Dreissena management was assigned to the Utah Division of Wildlife Resources (UDNR 2007).

Management plan, approval, and funding: The author was assigned as Utah's Aquatic Invasive Species Coordinator and led preparation of the Utah Aquatic Invasive Species Management Plan. The Utah Division of Wildlife Resources' Director, Utah Wildlife Board, and Utah Governor each approved the plan in early 2008. Federal approval occurred in January 2009 by the Aquatic Nuisance Species Task Force (UDWR 2009a). Utah's plan displayed an organization structure and strategies facilitating an orchestrated attack against the invasion of aquatic invasive species, particularly quagga and zebra mussels. The 2008 Utah Legislature passed Senate Bill 238, the Aquatic Invasive Species Interdiction Act, providing specific statutory authority for Dreissena management, and appropriated $1.4 million as ongoing General Funds for the Utah Division of Wildlife Resources to carry out the fight. Across the ensuing years as the economy faltered, so the Legislature reduced funding to $1.35 million per year, while multiple stakeholders came forward providing approximately $300,000 per year to the Utah Division of Wildlife Resources, allowing increased management efforts at their individual water bodies.

The 2014 Utah Legislature has again approved $1.35 million ongoing General Funds for Dreissena mussel management, and additional funds for operation of state line checkpoints to interdict boats arriving from affected waters outside of Utah. And stakeholders continue to help fund the program (Nielson 2014).

Operations: The plan's 5-year activities involving the public focused on outreach strategies, monitoring of water bodies, and to a lesser degree on law enforcement (UDWR 2008, 2009b, 2010, 2011, 2012). The author coordinated with Outreach Specialists to design promotional campaigns, messages and outreach product, including product distribution. Additionally, the author coordinated biologists, park rangers, and technicians to monitor the environment and assess data 2008 through 2012 as follows:

- Plankton and water at 40–100 water bodies per year was sampled for early detection of Dreissena mussels. Samples were assessed via cross-polarized light microscopy at the US Bureau of Reclamation's Detection Laboratory for Exotic Species in Denver CO. When a Dreissena or suspect veliger was observed, it was subjected to DNA assessment via PCR, qPCR, and DNA sequencing at either Reclamation's lab or Pisces Molecular's lab in Boulder CO, sometimes at both. Water bodies that had showed positive in the past for Dreissena were also assessed using eDNA. Most samples from Lake Powell were assessed via cross-polarized light microscopy and via PCR at the National Park Service's lab in Page, AZ, while a few samples each year were assessed by Reclamation's and Pisces' labs. Lake Powell samples were also assessed using eDNA by Reclamation's and Pisces' labs (Anderson 2008–2014; Hosler 2008–2014; Wood 2008–2014).

- Personnel interdicted and inspected 1,195,158 boats at water bodies and checkpoints from 2008 through 2012 (an average of 239,032 per year), decontaminating boats when needed and educating boaters about aquatic invasive species using an established protocol (Zook and Phillips 2009). It is apparent that Utah boaters were repeatedly contacted during each year, since an average of only 69,000 registered boats per year existed in Utah from 2008 through 2012; boat numbers increased to 71,499 in 2013 (Hunter 2013; NMMA 2013).

- Personnel working at 41 of Utah's 250 boatable water bodies and other checkpoints decontaminated 28,343 boats and wetted equipment from 2008 through 2012 (an average of 5,669 per year), using an established, professional high pressure, hot water (140°F) protocol (Zook and Phillips 2009).

- It is unknown how many self-inspections and self-decontaminations via the clean, drain, and dry protocol occurred by the boating public in Utah, but it was a significant number since approximately 600,000 launches occur per year. Boater feedback surveys indicate that 75% of the public inspects and decontaminates their boats for every outing. And covert observations by agency personnel of boater behavior showed that at least 50% cleaned and drained their boat on the ramp immediately following recovery from the water. No observational information is available regarding how many let their boat sufficiently dry or how many simply conducted the entire decontamination at home.

In Utah, enactment of needed laws was necessary to allow enforcement of regulations and associated legal prosecution. The author assisted in the development of necessary laws and coordinated with law enforcement's leadership and the Attorney General's Office. Most of the public gladly complied as long as they could understand how the impacts from Dreissena mussels affected them as individuals; how they as boaters could help out; and what the law demanded. In order to deal with the few citizens who were simply *hard-headed*, a cadre of law enforcement officers was essential and available. Successful Conservation Officers employ outreach principles when enforcing regulations. Virtually no enforcement action occurred during 2008 and 2009 other than educational contacts by Conservation Officers with anglers and boaters about aquatic invasive species. Although officers prefer to educate folks about laws and how to comply rather than arrest them, written violations (ranging from informal notices of violation 17,196; warnings 3,764; and notices to appear in court 390) were issued from 2010 through 2012 in Utah. Apprehensions of violators resulted from 151,645 law enforcement contacts in those 3 years (an average of 50,548 per year) that were made specifically for purposes of causing compliance with Utah's aquatic invasive species Rule R657-60 (UDWR 2008, 2009b, 2010, 2011, 2012).

Dreissena mussels have been discovered in some of Utah's water bodies (UDWR 2014a). Immediately upon discovery, the author redirected additional resources (personnel and decontamination units), when needed, to the affected water bodies (UDWR 2014b). Since the author's retirement, the new Aquatic Invasive Species Coordinator has done the same. The purpose for these actions was to more fully engage boaters, compelling decontamination at arrival, if needed, so no new Dreissena mussels could be added to the water body. And compelling decontamination at departure, so existing mussels would not be spread to other water bodies. It is the author's opinion that minimizing the number of mussels in a developing population can interrupt opportunity for them to get close enough together to breed and establish a self-sustaining population.

The following discussion describes Dreissena mussel discoveries in Utah. It also makes limited comparison of Utah's earlier 4-tier affected water body classification (undetected, inconclusive, detected, and infested) to Utah's new 5-tier classification (undetected, inconclusive, suspected, detected, and infested) as defined in Utah's newly modified Aquatic Invasive Species Interdiction Rule R657-60. The new classification system, supported by the Western Region Panel of the Aquatic Nuisance Species Task Force, also includes a de-classification process for affected water bodies (UDWR 2014d). The modified rule became effective March 10, 2014 (http://wildlife.utah.gov/rules-regulations/46-rules/rules-regulations/986-r657-60—aquatic-invasive-species-interdiction.html).

- Suspicion for presence of Dreissena veligers (positive microscopy but negative DNA via PCR from two labs or processes) resulted from single sampling events at five Utah water bodies as follows:

  2008—Suspect veligers at Pelican Lake, Midview Reservoir, Joes Valley Reservoir, and Huntington Reservoir (Hosler 2008–2014; Wood 2008–2014)

  2013—Suspect veligers at Jordanelle Reservoir (Nielson 2014).

- The 2008 observations were termed as *inconclusive*, warranting more samples and limited increased effort to affect boat decontamination. No additional evidence of Dreissena mussels was observed in Pelican Lake, Midview Reservoir, Joes Valley Reservoir, and Huntington Reservoir over the next 2 years, including eDNA assessments, so they were reclassified to an undetected status in late 2010. Regarding Jordanelle Reservoir, it is reported that additional sampling in 2013 failed to show any further evidence for Dreissena mussels, so it remains in the undetected status, too. Today, based upon past experience, the 2008 and 2013 suspicious observations would not likely cause an elevated concern beyond securing more samples. Nor would the standards be met for the new classification of *suspected* (Jordan 2014; UDWR 2014a–c).

- *New suspected status*: Defined as a single sampling event visually discovering any life stage of Dreissena mussel and DNA confirmed by PCR at two independent labs. Declassification can only occur under authority of the Utah Division of Wildlife Resources' Director. Generally, declassification cannot occur until 3 consecutive years of consistent sampling effort shows no evidence for presence of Dreissena mussels or other actions demonstrate successful eradication (UDWR 2014d).

- Detection of Dreissena veligers (positive microscopy and positive DNA via PCR from two labs or processes) was discovered from single sampling events at two Utah water bodies as follows:

  2008—Quagga mussel veligers at Red Fleet Reservoir (Hosler 2008–2014; Wood 2008–2014)

  2008—Zebra mussel veligers at Electric Lake (Hosler 2008–2014; Wood 2008–2014)

- At the time of discovery, those observations were termed as *detected*, warranting more samples and significantly increased effort for boat inspection and decontamination at each water body. Sampling at each water body in succeeding years showed no further evidence of Dreissena mussels, including eDNA assessments, so after 3 years the waters were declassified to an inconclusive status in late 2011. Then after 2 more years of no discoveries for Dreissena mussels the water bodies were declassified to an undetected status in late 2013 (Nielson 2014; UDWR 2014a–c). Today, the observations would again cause an elevated concern for securing more samples and possibly a reassessment of boat decontamination effort, but the standards would not be met for the new classification of detected (UDWR 2014a–c).

- *New detected status*: Defined as two or more consecutive sampling events, detecting Dreissena mussels of any life stage visually identified and DNA confirmed by PCR at two independent labs. Declassification can only occur under the authority of the Utah Division of Wildlife Resources' director. Generally, declassification cannot occur until five consecutive years of consistent sampling effort shows no evidence for the presence of Dreissena mussels or until other actions demonstrate successful eradication (UDWR 2014d).

- Infestation by live juvenile and/or adult quagga mussels has occurred at two Utah water bodies as follows:

  2010–2013—Quagga mussel at Sand Hollow Reservoir. One live, adult quagga mussel was discovered attached to the underside of a boat dock in May 2010; it was removed. No other Dreissena mussels or their life forms have since been discovered at Sand Hollow Reservoir (Hosler 2008–2014; Wood 2008–2014). At the time of discovery, Sand Hollow Reservoir was declared as *infested* by the Utah Wildlife Board since the specimen was visually observed by experts, and DNA tested positive via PCR using two independent labs and methods. An elevated concern existed to secure more samples, and a dramatically elevated effort for boat decontamination occurred. No evidence for veligers was ever observed. Nor was Dreissena eDNA detected, although it was sampled, prior to the single, live adult mussel being discovered. Following the mussel's removal, samples evidenced quagga mussel eDNA up through 2012, but the amount decreased each year, with none found in 2013. Thus, Sand Hollow Reservoir was declassified to the inconclusive status in late 2013 since it did not meet the minimum criteria for detection or infestation. Sand Hollow Reservoir was further declassified to the new undetected status in 2014 (Jordan 2014).

  2012 (April–August)—Quagga mussel eDNA at Lake Powell. Quagga mussel eDNA was discovered via PCR from water samples taken at penstocks in the Glen Canyon Dam (Hosler 2008–2014; Wood 2008–2014).

  2012 (September)—Quagga mussel veligers at Lake Powell. Quagga mussel veligers were discovered via plankton samples at several locales between Antelope Point and the Glen Canyon Dam (Anderson 2008–2014).

2013 (March–June)—Juvenile and adult quagga mussels at Lake Powell. Mussels were discovered living on docks and boats at Antelope Point and Wahweap marinas (Anderson 2008–2014).

2014 (February)—Quagga mussel population expansion at Lake Powell. Quagga mussel population shows expansion to up-lake areas as far north as Bullfrog Bay (Anderson 2008–2014).

- Soon after the discovery of quagga mussel veligers in Lake Powell, it was declared as *infested* by the Utah Wildlife Board since specimens were visually observed by experts, and DNA tested positive via PCR using two independent labs and methods. An elevated concern existed to secure more samples, and a dramatically elevated effort for boat decontamination occurred. Lake Powell remains in the Infested status; it is the only *infested* water body in Utah (UDWR 2014d). And the quagga mussel population in the lake continues to spread. A new Quagga-Zebra Mussel Management Plan for Lake Powell is being implemented. Today, the observations would again cause an elevated concern for securing more samples and increased boat decontamination efforts and the standards would be met for the new classification of infested (Nielson 2014; UDWR 2014a–c).

- *New infested status*: Defined as two or more consecutive sampling events, detecting the presence of multiple age classes of attached Dreissena mussels visually identified and DNA confirmed by PCR at two independent labs. Declassification can only occur under authority of the Utah Wildlife Board. Generally, it cannot occur until seven consecutive years of consistent sampling effort show no evidence for the presence of Dreissena mussels or other actions demonstrate successful eradication (UDWR 2014d).

Collection of statewide plankton samples shows that all other water bodies that have been sampled in Utah are in the undetected status for Dreissena mussels (Hosler 2008–2014; Nielson 2014; Wood 2008–2014). No veligers have been observed via microscopy, nor has Dreissena adults or eDNA been detected. Although not all boatable water bodies have been sampled yet, no evidence exists to show they have been invaded.

Applied management directed at boaters statewide in Utah by marketing required Decontamination Certification (AKA self-certification) and the *clean, drain, and dry* protocol continues to show benefit at interrupting Dreissena mussel inoculations. Many prelaunch boats each year are found with live Dreissena mussels attached. Additionally, aggressive containment management at Dreissena mussel affected water bodies through mandatory inspection of arriving boats with associated education discussion about risk from the mussels, and mandatory professional or self-decontamination for departing boats is believed to have resulted in fewer inoculations. Additionally, routine boaters were purposely exposed in Utah to multiple public outreach messages, with a goal of providing—three to seven promotional messages about cleaning, draining, and drying their boats during any outing, always encouraging them to routinely inspect and decontaminate their boats themselves. Applied management coupled with containment management is the likely cause for initial inoculations to fail. Unregulated use by boaters of water bodies is a disaster waiting to happen. Compliance with Utah's required Decontamination Certification at each launch by boaters has steadily improved, averaging 85% statewide, although compliance is variable on a per water body basis. Only 25% of the boaters reported not conducting a decontamination of their boats after every excursion (so 75% did it), while 91% of the surveyed boaters understood the Dreissena mussel threat (Dolsen and Dalton 2012).

Discussion and Summary

An applied boater management approach is not unique to Utah. Today, most conservation entities strive for a similar approach to manage boaters in order to interrupt establishment of invasive quagga and zebra mussel populations. The management cannot be haphazard, and success requires an impact assessment, leading to a focused, coordinated management plan. Additionally, sufficient resources—funds, personnel, and equipment—are needed to conduct the work, so appropriate administrative and political support is needed. It would be atypical for an agency to secure unlimited resources to do absolutely everything desired; so all elements of the work must be prioritized.

Individual boaters were the targets for applied management in Utah, since Dreissena mussels inadvertently hitchhiked on their boats, successfully staying alive, on or within watercraft while being hauled overland between water bodies. Boaters with boats were interdicted statewide, where they and their friends and families were educated over and over again about the risks from Dreissena mussel establishment. This occurred as one-on-one contacts at launch sites, checkpoints, and other venues. At the same time, boaters were educated about how to inspect and decontaminate their own watercraft or how to secure a professional decontamination. The education focused not only on risk to their boats and their outdoors recreation quests, but also on risk to every use of water possible for mankind. Outreach messages were crafted to be simple with an easily understood call to action (e.g., boaters must clean, drain, and dry their boat and wetted equipment after every excursion). And the Decontamination Certification required for every boat launch reinforced their education about risk and the call to action. The boaters quickly grasped the issue and took action: 91% of boaters understood the risk from Dreissena mussel establishment, while 75% decontaminated their boats after every use. This level of success was achieved in less than 5 years, since decontamination requirements did not exist prior to program establishment in 2008.

Monitoring for success amounted to surveying high priority, boating water bodies each year for the presence/absence of Dreissena mussels. Plankton samples for veligers and collections for juvenile or adult mussels were assessed using microscopy or gross ocular observation, followed by DNA verification using PCR and qPCR. In some cases, eDNA assessment of samples served to show the first signs of an invasion. Evidence for Dreissena mussels were discovered at nine Utah water bodies as follows:

- Pelican Lake, Midview Reservoir, Joes Valley Reservoir, Huntington Reservoir, and Jordanelle Reservoir were each classified as being a *suspected* status. All have been declassified to the undetected status after an appropriate period of monitoring, showing no presence of Dreissena.
- Red Fleet Reservoir and Electric Lake were each classified as being a *detected* status. Both have been declassified to the undetected status after an appropriate period of monitoring, showing no presence of Dreissena.
- Sand Hollow Reservoir and Lake Powell were each classified as being an *infested* status. Sand Hollow Reservoir was declassified to the undetected status, since an appropriate period of monitoring has shown no presence for Dreissena.

Only Lake Powell remains infested with a self-sustaining and expanding population of quagga mussels. It took 5 years for the infestation to occur once quagga mussels were found downstream in the lower Colorado River's impoundments and other water bodies served by aqueducts from those impoundments. In that interim, applied management was working. Boaters moving between Lake Powell and those affected impoundments likely caused the infestation. The number of boaters likely overwhelmed the ability of the National Park Service to inspect and decontaminate, and it is even possible that a boater simply ignored the rules and deviously entered the lake when no one was looking.

Aggressive, applied boater management on a statewide basis, incorporating containment management at affected waters, interrupted the establishment of Dreissena populations. Other factors may have contributed, but how is unknown. The remainder of Utah's 250 boatable water bodies is in an undetected status, too, due to an applied boater management approach. Luck or chance is not a part of the formula. Success resulted from hard work by many dedicated employees, stakeholders, and the unwavering support and cooperation by boaters. The biological mechanism is simple: keep Dreissena mussels from gaining a foothold by reducing their numbers, so their success in breeding is minimized. And boats certainly are a pathway for increasing Dreissena mussel numbers. So, applied management, targeting boaters to eliminate or at least reduce number of inoculations or spreading of the mussels, meets that goal. Fortunately, it is not uncommon for aquatic animal transfers to require several inoculations before a successful, self-sustaining population is established. So, aggressive containment management (mandatory inspections of arriving boats and mandatory decontamination for departing boats) at Dreissena-affected water bodies, coupled with marketing routine Decontamination Certification and self-decontamination via the clean, drain, and dry protocol at Utah's unaffected water bodies, has resulted in fewer inoculations causing all but one initial inoculation to fail.

Recommendations

Effective Dreissena mussel management does not just happen! What has been accomplished in Utah could be replicated or modified for successful implementation in other state or multistate regions. An impact assessment leading to a well-thought-out management plan with a strong promotional outreach strategy, applying management to influence boater behavior, is essential. The impact assessment and plan are critical steps in approaching decision makers for support and assignment of suitable resources—personnel, equipment, and a sustainable budget. Once a plan is approved and funded, operations can begin as follows:

1. Implement an outreach strategy that exposes boaters to multiple promotional expressions on each outing. We strove for—three to seven in Utah.
2. Deliver outreach messages to boaters that are doable, easy to understand, and simple (e.g., clean, drain, and dry your boat after every trip).
3. Based upon available resources, prioritize the water bodies and other venues (launch ramps, checkpoints, boat shows, sportsman expositions, fairs) where you intend to apply management, influencing boaters.
4. Get the boater involved so they are doing their own boat inspections and decontaminations. Their preparation for a next outing begins as the last outing ends, just as the boat is trailered— clean, drain, and dry.
   a. Demand containment management at affected waters (inspect arriving boats, and no boat leaves without cleaning and draining—drying can occur at home). This should become the norm at all waters, whether affected by Dreissena mussels or not.
5. Train and retrain personnel regarding Dreissena mussel management along with boat inspection and decontamination protocols.
6. Conduct monitoring for veligers when water temperature is suitable for breeding to determine the presence/absence of Dreissena species. Use the best available science for detection (e.g., DNA assessment via PCR and eDNA assessment). Shoreline and near-shore monitoring for juvenile and adult mussels can occur anytime there is no ice.
   a. Monitor affected waters at least once per month.
   b. Monitor waters with an undetected status at least once per year or more if budget allows.
7. Stay in-tune with current and upcoming Dreissena management technology.

Do not ever forget that the boater is your customer. Their activities in part pay your salary, and they have the *politician's ear*. The boater's willing participation in your program makes applied management a success. If you give up in frustration, invasive mussels win!

REFERENCES

Allen, Y.C. 2014. Personal communication. Ecologist, US Fish and Wildlife Service, Baton Rouge, LA.

Allen, Y.C., Thompson, B.A., and Ramcharan, C.W. 1999. Growth and mortality rates of the zebra mussel, *Dreissena polymorpha*, in the Lower Mississippi River. *Canadian Journal of Fisheries and Aquatic Sciences*, 56(5): 748–759.

Anderson, M. 2008–2014. Personal communication and laboratory reports regarding cross-polarized light microscopy and DNA identification via end point PCR. Ecologist and Laboratory Supervisor, National Park Service, Page, AZ.

AP. 2013. Associated Press. Invasive quagga mussels may cost the West Coast millions in maintenance. *New York Daily News*. http://www.nydailynews.com/news/world/invasive-quagga-mussels-cost-west-coast-millions-maintenance-article-1.428560. Accessed January 3, 2014.

Brown, C. and Hosler, D. 2013. AARA Invasive Mussel Detection Project. US Bureau of Reclamation, Denver CO. https://www.usbr.gov/mussels/activities/docs/ARRAInvasiveMusselDetectionProjectReporting.pdf. Accessed June 26, 2014.

Bryan, S. 2014. Personal communication. Senior Biologist, Central Arizona Project, Phoenix, AZ.

Californial Fish and Game Department. 2011. Public document on CAFGD web site: http://www.dfg. ca.gov/invasives/quaggamussel/. Dreissenid Mussel Infestations in California. https://nrm.dfg.ca.gov/ FileHandler.ashx?DocumentID=39027. Accessed February 20, 2014.

Choi, W.J., Gerstenberger, S., McMahon, R.F., and Wong, W.H. 2013. Estimating survival rates of quagga mussel (Dreissena rostriformis bugensis) veliger larvae under summer and autumn temperature regimes in residual water of trailered watercraft at Lake Mead, USA. *Management of Biological Invasions*, 4: 669

Claudi, R. and Mackie, G.L. 1993. *Practical Manual for Zebra Mussel Monitoring and Control*, Dispersal Mechanisms (pp. 43). CRC Press, Boca Raton, FL, 240pp.

Cofrancesco, A. 2014. Personal communication. Technical Director, USACE Environmental Laboratory, Vicksburg, MS.

Colorado Division of Wildlife. 2011a. Aquatic nuisance species (ANS) watercraft decontamination manual, 58pp.

Colorado Division of Wildlife. 2011b. Boat compendium for aquatic nuisance species (ANS) inspectors, 42pp.

Craft, C.D. and Myrick, C.A. (2011) Evaluation of quagga mussel veliger thermal tolerance. Colorado Division of Wildlife (CDOW). CDOW Contract Report CSU #53-0555, 21pp.

Dalton, L.B. and Cottrell, S. 2013. Quagga and zebra mussel risk via veliger transfer by overland hauled boats. *Management of Biological Invasions*, 4(2): 129–133.

Darling, J.A. and Mahon, A.R. 2011. From molecules to management: Adopting DNA- based methods for monitoring biological invasions in aquatic environments. *Environmental Research*, 11(7): 978–988.

De Leon, R. 2014. Personal communication. Scientist, Metropolitan Water District of Southern California, La Verne, CA.

Dietz, T.H., Silverman, H., Lessard, D., and Lynn, J.W. 1994. Predicting the spread of zebra mussels: Ionic ratios, magnesium, and selection effect. Paper presented at International Association of Great Lakes Research, June 1994, Windsor, Ontario, Canada.

DiVittorio, J., Grodowitz, M., Snow, J., and Manross, T. 2012. Inspection and cleaning manual for equipment and vehicles to prevent the spread of invasive species. US DOI, Bureau of Reclamation, Technical Memorandum No. 86-68220-07-05, 195pp. http://www.usbr.gov/mussels/prevention/docs/ EquipmentInspectionandCleaningManual2012.pdf. Accessed July 8, 2014.

Dolsen, D.E. and Dalton, L.B. 2012. Utah Division of Wildlife Resources. Utah boater aquatic invasive species survey. Prepared by DISC Information Services Corporation, St Cloud, MN. UDWR Pub. #12–19. 46pp.

Gerstenberger, S.L., Mueting, S.A., and Wong, W.H. 2011. Veligers of invasive quagga mussels (*Dreissena rostriformis bugensis,* Andrusov 1897) in Lake Mead, Nevada–Arizona. *Journal of Shellfish Research*, 30(3): 933–938.

Hosler, D. 2008–2014. Personal communication and laboratory reports regarding cross-polarized light microscopy, scanning electron microscope along with DNA and environmental DNA identification via end point PCR. Ecologist. US Bureau of Reclamation, Reclamation Detection Laboratory for Exotic Species, Denver, CO.

Hunter, T. 2013. Personal communication. Utah Division of Parks and Recreation, Boating Law Coordinator, Salt Lake City, UT.

Karatayev, A.Y, Claudi, R., and Lucy, F.E. 2012. History of Dreissena research and the ICAIS gateway to aquatic invasions science. *Aquatic Invasions*, 7(1): 1–5.

Mackie, G.L and Claudi, R. 2010. *Monitoring and Control of Macrofouling Mollusks in Fresh Water System*. CRC Press, Boca Raton, FL, 508pp.

Maylett, C. 2014. Personal communication. Outreach Coordinator, Utah Division of Wildlife Resources, Salt Lake City, UT.

McMahon, R.F., Ussery, T.A., and Clarke, M. 1993. Use of emersion as a zebra mussel control method. U.S. Army Corps of Engineers Contract Report EL-93-1, Washington, DC, 31pp.

McMahon, T. 2014. Personal communication. Arizona Game and Fish Department, Aquatic Invasive Species Coordinator, Phoenix, AZ.

Nielson, J. 2014. Personal communication. Utah Division of Wildlife Resources, Aquatic Invasive Species Coordinator, Salt Lake City, UT.

NISA. 1996. National Invasive Species Act of 1996, Public Law 104-332, October 26, 1996. 110 Stat. 4073–4092—Reauthorizes and amends the Non-indigenous Aquatic Nuisance Prevention and Control Act of 1990, 16 USC 4701.

NMMA. 2013. National Marine Manufacturers Association. Economic significance of recreational boating—Utah. http://www.nmma.org/assets/cabinets/Cabinet508/Utah_Boating_Economics.pdf. Accessed February 24, 2014.

O'Neill, C.R. Jr. 1996. The zebra mussel—Impacts and control. New York Sea Grant, Cornell Univ. and State University of New York. Cornell Coop. Extension, *Information Bulletin 239*, 62pp.

Pimentel, D., Zuniga, R., and Morison, D. 2005. Update on the environmental and economic costs associated with alien-invasive species in the United States. *Ecological Economics*, 52: 73–288.

Suflita, M. 2007–2014. Personal communication and multiple white papers documenting physical and limnological characteristics of Utah's water bodies, including suitability for Dreissena mussel population establishment. Senior Engineer, UT. Div. Water Resources, Salt Lake City, UT.

UDNR. 2007. Utah Department of Natural Resources, Policy 07-D-11. Salt Lake City, UT.

UDWR. 2008. Utah Division of Wildlife Resources' 2008 annual report presented by the Aquatic Invasive Species Management Program to the 2009 Utah Legislature's Natural Resources Committee. http://wildlife.utah.gov/mussels/PDF/ais_summary_annual_2008.pdf. Accessed February 24, 2014.

UDWR. 2009a. Utah Division of Wildlife Resources' 2009 Aquatic Invasive Species Management Plan. http://wildlife.utah.gov/invasive-mussel-plan.html. Accessed February 24, 2014.

UDWR. 2009b. Utah Division of Wildlife Resources' 2009 annual report presented by the Aquatic Invasive Species Management Program to the 2010 Utah Legislature's Natural Resources Committee. http://wildlife.utah.gov/mussels/PDF/ais_summary_annual_2009.pdf. Accessed February 24, 2014.

UDWR. 2010. Utah Division of Wildlife Resources' 2010 annual report presented by the Aquatic Invasive Species Management Program to the 2011 Utah Legislature's Natural Resources Committee. http://wildlife.utah.gov/mussels/PDF/ais_summary_annual_2010.pdf. Accessed February 24, 2014.

UDWR. 2011. Utah Division of Wildlife Resources' 2011 annual report presented by the Aquatic Invasive Species Management Program to the 2012 Utah Legislature's Natural Resources Committee. http://wildlife.utah.gov/mussels/PDF/ais_summary_annual_2011.pdf. Accessed February 24, 2014.

UDWR. 2012. Utah Division of Wildlife Resources' 2012 annual report presented by the Aquatic Invasive Species Management Program to the 2013 Utah Legislature's Natural Resources Committee. http://wildlife.utah.gov/mussels/PDF/ais_summary_annual_2012.pdf. Accessed February 24, 2014.

UDWR. 2014a. Utah Division of Wildlife Resources. Water bodies affected by quagga or zebra mussels and their status. http://wildlife.utah.gov/affected-waters.html. Accessed February 24, 2014.

UDWR. 2014b. Utah Division of Wildlife Resources. How the DWR discovers and manages water bodies affected by quagga or zebra mussels. http://wildlife.utah.gov/affected-waters/646.html. Accessed February 24, 2014.

UDWR. 2014c. Utah Division of Wildlife Resources. Affected water status definitions. http://wildlife.utah.gov/invasive-mussels/status-definitions.html. Accessed February 24, 2014.

UDWR. 2014d. Utah Division of Wildlife Resources. Rule R657–60 as modified and effective March 10, 2014. http://wildlife.utah.gov/rules-regulations/46-rules/rules-regulations/986-r657–60—aquatic-invasive-species-interdiction.html. Accessed March 11, 2014.

USGS. 2014. US Geological Survey. Zebra and Quagga Mussel Lake Distribution by County. http://fl.biology.usgs.gov/Nonindigenous_Species/Zebra_mussel_distribution/Lakes_by_county/lakes_by_county.html. Accessed February 24, 2014.

Volkoff, M. 2014. Personal communication. Senior Environmental Biologist, Calif. Dept. Fish and Wildlife, Invasive Species Program, Sacramento CA.

Wong, W.H., Gerstenberger, S.L., Miller, J.M., Palmer, C.J., and Moore, B. 2011. A standardized design for quagga mussel monitoring in Lake Mead, Nevada–Arizona. *Aquatic Invasions*, 6(2): 205–215.

Wood, J. 2008–2014. Personal communication and laboratory reports regarding DNA and environmental DNA identification via end point and quantitative PCR. President and Molecular Biologist. Pisces Molecular, LLC, Boulder, CO.

Zook, B. and Phillips, S. 2009. Recommended protocols and standards for watercraft interception programs for dreissenid mussels in the western United States. Prepared by Pacific States Marine Fisheries Commission for the Western Regional Panel on Aquatic Nuisance Species of the Aquatic Nuisance Species Task Force, 49pp.

18

Invasive Mussel Prevention at Lake Powell

Mark Anderson

CONTENTS

ABSTRACT   National Park Service (NPS) management actions at Lake Powell prevented dreissenid infestation for many years despite the lake being at very high risk. Early recognition of the danger, administrative controls, education and outreach, and screening of watercraft to assess risk are credited with keeping mussels out. After invasive mussels were established in the western United States, infestation of Lake Powell was averted through further, even more aggressive actions. An extensive monitoring program was developed for early detection of dreissenid mussels. Indicators of invasive mussels in Lake Powell were detected by the NPS in 2012, 14 years after scientists had predicted Lake Powell would become the first western infested waters.

Introduction

Lake Powell is one of the largest and most scenic man-made reservoirs in North America. Located in southern Utah and northern Arizona, this 200-mile long lake is impounded in deep red-rock canyons and managed by the NPS as a central feature of Glen Canyon National Recreation Area. The park receives over 2 million visitors each year, most of which come for water-based recreation on the lake. This recreation involves the launching of over 100,000 individual privately owns vessels that are transported from across the United States, constituting an enormous risk for spreading aquatic invasive species (AIS).

In 1998, a group of scientists known as the Utah Aquatic Nuisance Species Action Team predicted that Lake Powell would the first western water body to become infested with invasive dreissenid mussels. At that time, only 10 years after the first discovery of dreissenids in North America, the destructive animals had established populations throughout much of the eastern United States. Continued spread was expected despite various efforts, primarily educational and voluntary, to contain the scourge. Lake Powell, if infested, would serve as a significant source for further spread.

Water and environmental conditions throughout most of Lake Powell are suitable for zebra (*Dreissena polymorpha*) and quagga (*Dreissena bugensis*) mussels. High sediments that can exist at times in inflow areas are the only conditions in the lake that do not correspond to a high colonization potential for the mussels (Sorba and Williamson 1997; Vernieu 2009).

Mussel Prevention Efforts 1999–2006

Following the scientists' prediction of Lake Powell's vulnerability, the NPS conducted a simple risk assessment in 1999. Counts were made in marina parking lots of trailer license plates from infested states. NPS judged the risk of infestation as great based on counts as high as 70 empty trailers from infested states on a busy weekend within a single marina. NPS management policies strive to prevent impairment of resources, and so to preserve them for future generations. The threat of mussel infestation was a serious threat to Glen Canyon's best-known resource that draws its largest numbers of visitors.

In 1999, NPS established a prevention program at Lake Powell. The program included education of boating visitors, screening boaters to identify those with risk of spreading mussels, and offering free voluntary decontamination treatments to vessels identified as high risk. High-risk vessels were defined as those that had been used in the prior 30 days in states that have mussel infestations. Early decontaminations were provided by a park concessioner in the interest of cooperation and self-preservation. There were no formal protocols for decontamination in the early years. Decontamination depended on concessioner knowledge of boat systems and NPS guidance to clean all areas that were wet or dirty. The number of high-risk vessels interdicted each year remained consistent at about 50 vessels from eastern states.

Early outreach and education efforts included a park-organized group known as Zebra Mussel Prevention Task Force. The group consisted of concessioners, local businesses, other agencies, and individuals concerned with the mussel threat to Lake Powell. The meetings provided a forum for information exchange and cooperative mussel prevention and education efforts. The first artificial substrate samplers were deployed in Lake Powell by NPS after being constructed by a local business in 2002.

The task force came under internal NPS criticism in 2003 for representing an inappropriate request for aid from the NPS because many of the participants required NPS permits or were otherwise in a position reliant on NPS. The appearance of possible coercion to assist NPS efforts could not be avoided with such obvious NPS leadership of the group. None of the participating entities were able to maintain the group without direct NPS leadership, and the group fell fallow until after mussels were discovered in the western United States in 2007.

The first documented instance of a vessel being transported to Lake Powell with mussels attached was in 2002. The vessel was discovered in a slip at the Bullfrog Marina. Prior to launching in Lake Powell, the vessel had been out of water for 9 months and all of the mussels were dead. This incident prompted the park to require compliance with the program, instead of only promoting voluntary cooperation. The superintendent's compendium of park-specific regulations was altered in 2003 to create a requirement that all vessels identified as high risk must comply with the risk abatement procedures of inspection and decontamination as necessary prior to being launched.

The park developed requirements for agency-controlled functions such as special use and research permits. The permit requirements included mandatory inspections and decontamination, if necessary, of associated equipment and other measures to prevent the spread of AIS. Similar requirements were placed in all contracts and agreements the park entered.

After the initial spread of mussels through the Great Lakes and down the Mississippi River, the spread slowed but continued steadily until Glen Canyon considered all vessels recently used in 23 eastern states and 2 Canadian provinces as high risk in 2006.

Mussel Prevention Efforts 2007–2013

In January of 2007, quagga mussels were discovered in Lake Mead of the Lower Colorado River. At the time of discovery, it was clear from the extent of the population that it had begun several years prior. Lake Powell shares many boaters with Lake Mead. NPS judged the threat to Lake Powell as having increased dramatically.

To facilitate partner education and the park's development of additional prevention elements, the participants of the Lake Powell Zebra Mussel Task Force, which had not formally met since 2003, were contacted to form the Lake Powell Invasive Mussel Prevention Coordination Group, which meets monthly to coordinate mussel prevention efforts for Lake Powell. Members include Lake Powell Resorts and Marinas, Antelope Point Marina, UT Division of Wildlife Resources, AZ Game and Fish Department, Page/Lake Powell Tourism Bureau, Salt River Project, City of Page, Lake Powell Yacht Club, Wilderness Raft Adventures, Navajo Nation Fish and Wildlife, Coast Guard Auxiliary, Bureau of Reclamation, AZ Department of Transportation, and Friends of Lake Powell.

The new level of threat to Lake Powell with dreissenid mussels in the western United States was difficult to characterize. It was clear that there would be many more than 50 high-risk vessels per year. Decontamination treatments prior to 2007 were conducted with standard high-pressure hot water cleaning equipment; temperatures were not consistent in these units, and runoff was not contained. The locations of the equipment were chosen such that runoff from the few treatments prior to 2007 would not reach the lake through the storm water systems. In 2007, NPS secured funding (over $700,000) to develop decontamination facilities at major marinas. These facilities consisted of industrial equipment that consistently produced water at temperatures over 140°F and included water recapture and reuse technology. Water recapture was expected to be necessary with the anticipated volume of high-risk vessels. The locations of the facilities were chosen to be more convenient and appropriate for large numbers of vessels. Formal inspection and decontamination protocols were developed in cooperation with partners including the Pacific States Marine Fisheries Commission (PSMFC). The protocols were consistent with the watercraft inspection training (WIT) later developed by the PSMFC. When WIT was developed, the park adopted those protocols and the training regimen. The park maintains several WIT level II certified trainers on staff for internal training.

Prior to 2007, although high-risk vessels identified were required to be treated, vessels were only screened for risk when entrance booths were open. This approach was effective when the risk was greatest because entrance booths are open during high levels of visitation. Off-season and at night, vessels screening was accomplished through automated fee machines, but compliance was expected to be much lower. With mussels in the West, times of low visitation would present a much greater risk than before. To extend consistent screening to all vessels, the park developed a dashboard certificate requirement. The certificate was required to be displayed through the windshield of the tow vehicle while parked with an empty trailer to indicate that the vessel towed had been screened and determined to be low risk or treated to remove the risk.

The first dashboard certificates involved a self-certification process. The self-certification packet was available on the park's website and at launch ramps. The packet included a flow chart for boaters to assess the risk of their vessels, instructions on how to attain a decontamination treatment if risk was assessed as high, and a certificate to sign, date, and display if risk was assessed as low. While this self-certification approach has been widely criticized as an *honor system* through which many high-risk vessels may be fraudulently certified by their owners, the certificate was and is used at Glen Canyon in cases where, in its absence, there would be no scrutiny of launching vessels. The self-certification process was also seen as an effective education tool. Through the self-certification requirement, every boater to Lake Powell was theoretically exposed to the issue of AIS and invasive mussels and followed a process that showed how risk is assessed. In 2008, Utah developed a similar self-certification process that was required of boaters throughout the state of Utah (Utah Administrative Code R657-60-6 Certification of Decontamination).

With the western discovery of mussels in Lake Mead and the Lower Colorado River, the NPS acted to develop guidance regarding invasive mussel planning for all NPS units with water resources (NPS 2007). Efforts of Glen Canyon National Recreation Area at Lake Powell to prevent mussels, and the response at Lake Mead, were presented as guidance for the many other NPS-managed water bodies across the country threatened by invasive mussels. Three representatives from Glen Canyon participated on the incident command team charged with the rapid development of NPS guidance. The Plan is available online from the National Park Service.

New funding for mussel prevention was developed within Glen Canyon from user fees that allowed a consistent NPS ranger presence to contact visitors about aquatic invasive species as well as inspect and decontaminate visitor watercraft. The funding was sufficient to cover daylight hours at most major marinas. For major marinas where coverage was sufficient, a new certification requirement was created in the superintendent's compendium. Self-certification was no longer allowed, and visitors were required to get an NPS-issued certificate to display through their windshield. To facilitate visitor compliance further, launch ramps were closed to launching at night through the compendium. Some ramps were also closed during low use seasons to control costs. A minimum level of staffing is required to provide consistent visitor services (including inspection and decontamination) and issue certificates at marinas where NPS-issued certificates are required; additional staffing allows for ramp monitoring while the minimum level of staff are inspecting and decontaminating vessels and checking compliance with the certification requirements in parking lots. While monitoring ramps, inspections can be conducted on arriving vessels that have not been identified as high risk. Ramp monitoring significantly increases compliance with the program.

New efforts for outreach about AIS and park requirements were initiated in 2007. These efforts employed the techniques used prior, such as presentations, park newspaper, and website, but now incorporated distribution of brochures to all boaters, radio interviews, and direct visitor contact. The park had employed Stop Aquatic Hitchhikers and Zap the Zebra brochures in the past but now developed a park-specific brochure that included information directly relevant to Lake Powell and the new park requirements for boaters.

NPS staff attending marinas for AIS inspections and decontaminations monitor parking lots daily for vehicles with empty trailers that are not certified by the NPS-issued certificate or the self-certification process as required in each marina. When uncertified tow vehicles are encountered, the AIS ranger collects information on the vehicle and provides an informational card that states "this could have been a ticket" under the windshield wiper. When the compliance checks are finished, the ranger forwards the information on uncertified vehicles to NPS law enforcement staff who exchange the informational card with an actual citation for uncertified launching. Law enforcement rangers use vehicle tire boots when appropriate as determined by officer discretion.

Citations are intended to serve as a deterrent to uncertified launching. Maximum penalties of $5000 and up to 6 months in jail are prominently advertised in educational information, signs, the park paper, and certificates. Actual citations generally carry a mandatory appearance requirement. Park outreach efforts targeting the prosecuting attorneys and presiding judges have led to routine fines of $2500 for individuals disregarding the park AIS regulations. There were typically 300–400 citations written each year from 2009 to 2011; in 2012, when additional funding allowed increased ramp monitoring (estimated 85%), citations were reduced to 150.

In 2009, a vessel with adult mussels was stopped from launching at the Wahweap. The owner refused a decontamination treatment and chose to leave the park. Authority for holding such a vessel was not recognized by the NPS, and the vessel was released. Later discussion with partners generated alarm as to where the vessel had gone after leaving. After this incident, Glen Canyon worked with NPS solicitors to understand authorities that NPS may have to hold contaminated boats in quarantine. It was determined that Lacey Act, supported by the laws of Utah and Arizona, could provide the authority needed for Glen Canyon to quarantine contaminated vessels. The quarantine process that was developed involved providing the boat owners with a form that explains the recognized authority and requests voluntary compliance with a 30-day quarantine to avoid Lacey Act prosecution. After the vessel is decontaminated and in quarantine, a park biologist inspects the vessel, considers the time after decontamination along with the temperature and humidity to recommend a shorter quarantine time if warranted. If the park superintendent accepts the recommendation, the vessel is released prior to the 30 days in accordance with the recommendation. In 2009, 11 watercrafts were found to have adult mussels attached and were quarantined. In 2010 and 2011, there were 14 and 17 quarantines, respectively. In 2012, the number of quarantines increased to 38. The dramatic increase in 2012 was thought to be the result of the mussel infestation in Lake Pleasant, of central Arizona, developing to the point that day-users could readily transport mussels; many of the vessels quarantined had mussels associated with their anchors and anchor wells.

As mussels have spread in the western United States, the risk of infestation to Lake Powell has steadily increased. The challenge of preventing mussel in Lake Powell has become greater each year. Figure 18.1 presents interdiction data from 2006 to 2012, including the number of high-risk vessels identified at Lake Powell, as well as the number of inspections and decontamination conducted.

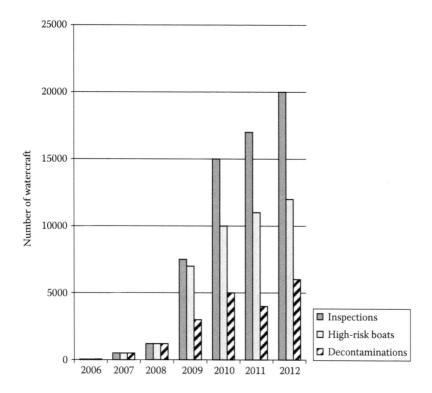

FIGURE 18.1 High-risk vessels, inspections, and decontamination treatments at Lake Powell 2006–2012.

Monitoring

Anecdotal monitoring for adult mussels in Lake Powell began in 1999 with examination of aids to navigation by NPS resource management and facility management staff and passive examination by concessioners on marina structures in the course of other work. Resource management staff developed a random buoy selection process to monitor the aids to navigation and other buoys lake-wide systematically. Buoy and marina structure monitoring only examined areas within a few feet of the water surface. NPS and concessioner divers were educated about mussels and monitored for mussels at greater depths when diving for other purposes.

In 2002, aquatic resources management staff deployed 15 artificial substrate samplers at marinas throughout the lake. These samplers, often referred to as Portland samplers, consisted of PVC or ABS pipe sections of 8–10 in. hung by rope from marina structures to depths ranging from 20 to over 70 ft. The samplers were checked approximately monthly. These samplers were expected to provide a more reliable means of early detection of mussel adults than the anecdotal or random buoy monitoring that preceded them. They have been maintained and monitored to the present.

Routine plankton monitoring has been conducted on Lake Powell for decades, but the first plankton monitoring targeted at early detection of mussels began in 2007. NPS developed laboratory capabilities using cross-polarized light microscopy to detect mussel veligers (larvae). Based on the premise that any control options for a developing mussel infestation would require very early detection, an aggressive sampling regime was developed to collect hundreds of samples each year. Visitor-use zones and data were used to divide the lake into 13 zones for a stratified random sampling program. Samples were collected from each of the zones proportionally with the number of visitors that use each zone. Specific locations within a zone were determined randomly by a computer. Routine plankton sampling for veligers as part of the monitoring program was also conducted monthly at marinas, inflows, and the dam. In 2009, the park acquired a particle image analyzer to supplement the cross-polarized light microscopy, allowing more efficient sample processing and increasing the maximum number of samples that could be processed.

The park developed a polymerase chain reaction (PCR) methodology for potentially even earlier detection of veligers in plankton samples. The several published primers and methodologies were tested for use at Lake Powell. By 2010, the aquatic resources management staff had developed a method on the Ram primer set to detect both zebra and quagga mussels (Ram et al. 2011). In the procedure, when the DNA of the appropriate size is amplified, the DNA is sent to the University of Arizona Genetic Core Laboratory Facility for sequencing to control for occasional false positives. The DNA of mussels was expected to exist to some degree in the absence of viable veligers or an infestation because many high-risk boats were being decontaminated and launched on Lake Powell boat ramps in 2010. To avoid detection of trace amounts of nonviable mussel DNA, the procedure was tuned through experimentation to detect mussel DNA when an amount equivalent to at least one complete veliger or more was present. Upon development of the PCR procedure, the park began collecting two samples at each sampling location. One sample was slated for microscopy and the other to PCR.

Mussel Response at Lake Powell

Plankton monitoring in Lake Powell for veligers by the NPS laboratory had negative results from 2007 until August of 2012. One sample in August 2012 collected near Antelope Point amplified DNA during PCR analysis. While several unknown species in Lake Powell have DNA that will amplify in the NPS PCR procedure, these organisms are typically detected only in winter months. Upon discovery of amplified DNA in the sample, but prior to sequencing and analysis and eventual confirmation of quagga DNA, NPS initiated an intensive sampling effort in the area between Antelope Point and the Glen Canyon Dam. Through the period from late August to September, approximately 150 sample pairs were collected in the area. Of these samples, seven contained a single veliger detected by cross-polarized light microscopy and seven samples contained DNA that amplified in PCR. Assuming a single veliger was present in each PCR-analyzed sample, estimated concentration of veligers in the area between Antelope Point and the dam indicated by the sampling effort was 3×10^{-5} veligers/L or 1 veliger/37,500 L of water.

The veliger detection likely indicated mussel reproduction in Lake Powell. The NPS announced the information to partners and the public in a press release in November 2012. Effort was made to put the low concentrations in perspective as the extremely early detection they likely represented. Included in the release was a commitment to search for adult mussels in the hope of discovering the source of the veligers while feasible control strategies could potentially be successful. The information released included a table of potential control strategies adapted for Lake Powell from the 2007 NPS guidance (NPS 2007).

Through the winter of 2012–2013, the NPS mounted numerous efforts to locate adult mussels on submerged canyon walls in the areas where veligers had been detected. The park owns and operates a remotely operated underwater vehicle (ROV), which is primarily used for retrieving bodies from the lake after fatal boating and other accidents. In ROV surveys of about a dozen locations near the veliger findings, no adult mussels were found.

In the spring of 2013, a commercial boat hauler discovered several mussels on a houseboat that had been retrieved from the Wahweap Marina. The mussels were discovered during an inspection being conducted by the hauler as part of the requirements developed by the park for commercial operations authorized in the park. NPS divers began searching the Wahweap Marina for mussels and discovered adult mussels on a number of houseboats and marinas structures. Surveys of Antelope Point Marina produced similar results. The mussels discovered were in low concentration; all mussels discovered were removed upon discovery. It was expected that additional mussels were present in the marinas. To remove these mussels, the NPS organized a large-scale dive effort. The effort was referred to as the *quagga blitz* and involved approximately 40 divers from various agencies and organizations. The blitz was conducted over 4 days; 411 mussels were found and removed or crushed in place.

Results from the quagga blitz presented an unexpected distribution of mussels. Early detections of adult mussels in Lake Mead were at depths typical of quagga colonization; however, the blitz identified mussels near the surface, but not at depth. Mussels were not evenly distributed, but consisted of localized concentrations of mussels with most boats and marina structures devoid of mussels. The sizes of the mussels were also unexpected, ranging from 3.5 to over 30 mm and clearly not representing a single reproductive event. The larger mussels were expected to have been growing in place for at least 2 years. These findings are difficult to explain, but would be consistent with multiple small introductions of mature veligers. This sort of introduction could plausibly be the result of water containing veligers from watercraft or other sources being introduced in the marina repeatedly over several years in specific locations in the marina. NPS suspected that visitor boats containing contaminated water had entered the marina, moored near the mussel locations, and jettisoned water containing mature veligers; the veligers then attached in the vicinity.

In the summer of 2013, plankton analysis detected additional veligers in the same southern lake region. Veliger concentrations were much higher than in 2012 with a maximum sample concentration of 0.016 veligers/L or 1 veliger in 62 L of water. Approximately 200 other veliger samples collected from areas upstream of Navajo Canyon remained negative throughout the 2013 season.

Routine work on aids to navigation buoys in the summer of 2013 found two adult mussels on a buoy line near the mouth of Wahweap Bay. NPS divers found a concentration of adult mussels about 15 m below the surface. The mussels were not too dense for immediate removal by hand when found; approximately 800 mussels were removed. Additional efforts to determine the full extent of mussels are being planned.

With adult mussels and veligers having been found in Lake Powell, the NPS has begun cooperative efforts with partners and states to prevent the mussels from spreading to other waters. Initial efforts are primarily educational. In 2013, both Utah and Arizona law have been modified to designate Lake Powell as a mussel-affected body of water. These laws require special steps and decontamination of watercraft leaving the reservoir. With the change in mussel status for Lake Powell, in 2013, the NPS embarked on a public planning process to guide the park's actions related to dreissenid mussels through about 2030. The planning effort will focus on monitoring, prevention, containment, and control of mussels at Lake Powell, and funding for these efforts.

REFERENCES

National Park Service. 2007. Quagga/zebra mussel infestation prevention and response planning guide. National Park Service, Ft. Collins, CO: 43pp.

Ram, J.L., A.S. Karim, P. Acharya, P. Jagtap, S. Purohit, and D.R. Kashian. 2011. Reproduction and genetic detection of veligers in changing Dreissena populations in the Great Lakes. *Ecosphere* 2(1): art3.

Sorba, E.A. and D.A. Williamson. 1997. Zebra mussel colonization potential in Manitoba, Canada. Water Quality Management Section, Manitoba Environment, Report No. 97-10. Winnipeg, Manitoba, Canada.

Vernieu, W.S. 2009. Physical and chemical data for water in Lake Powell and from Glen Canyon Dam releases, Utah–Arizona, 1964–2008: U.S. *Geological Survey Data Series* 471: 23pp.

Section V

Policy

19

Protecting California's Environmental Resources and Economic Interests: A Retrospective of the Management Response to Dreissenid Mussels (1993–2013)

Martha C. Volkoff, Susan R. Ellis, and Dominique T. Norton

CONTENTS

ABSTRACT   California is a unique and challenging environment, both in terms of its environmental resources and sociopolitical structure. Responding to, and managing for, the discovery of quagga mussels (*Dreissena rostriformis bugensis*) in 2007 and zebra mussels (*Dreissena polymorpha*) in 2008 has required resource managers to work within the complex intrastate political boundaries and jurisdictional authorities with limited financial resources. Notwithstanding these challenges, there is an imperative to protect vital economic interests intertwined with crucial water conveyance systems, agriculture, and recreation. While there has been universal agreement that dreissenid mussels would, and ultimately, have negatively impacted California, diverse, sometimes competing forces influenced the pace, direction, and outcomes of actions taken to plan for and manage dreissenid mussels in the state. A *post hoc* review of the state's response can offer guidance to others preparing for, or responding to, similar crises. Preplanning, utilization of internal and external expertise, collaboration and partnering, an organized structure for response efforts, development and implementation of a long-term management strategy, and a tailored response to the unique situation and prevailing circumstances have played important roles in the response and ongoing management of dreissenid mussels in the state.

Introduction

Quagga mussels (*Dreissena rostriformis bugensis*) were first discovered west of the Continental Divide in Lake Mead, Nevada, January 6, 2007 (LaBounty, 2007). Subsequent sampling revealed quagga mussels throughout Lake Mead's lower basin and the Colorado River (Figure 19.1; California Department of Fish and Wildlife unpublished report). Water from the Colorado River is diverted into California at Lake Havasu via the Colorado River Aqueduct. Surveys revealed adult quagga mussels throughout the Colorado River Aqueduct and many of the lakes receiving raw Colorado River water. By the end of 2007, quagga mussels had been detected in nine reservoirs in Southern California. As of November 2013, quagga mussels have been detected in an additional 12 waterbodies in Southern California.

Almost exactly a year to the day, on January 10, 2008, a fisherman reported landing a cluster of quagga mussels attached to his fishing line at San Justo Reservoir, San Benito County. San Justo Reservoir is approximately 300 miles northwest of the nearest quagga mussel-infested water, and hydrologically separate. The cluster of mussels was examined by California Department of Water Resources, California Department of Fish and Wildlife, and California Department of Food and Agriculture staff, and determined to be zebra mussels (*Dreissena polymorpha*). The lake was surveyed to determine the extent of the infestation, which revealed zebra mussels throughout. Subsequently zebra mussels were found in the underground conduit system connected to San Justo Reservoir, and a pump for a golf course pond that receives water via the conduit.

It is estimated that within the state, there are over 29,826 miles of rivers and streams and 4,800 lakes and reservoirs (California Fish and Game Commission, 1998). Conditions in aquatic environments throughout a large portion of the state appear suitable for dreissenid mussel survival in terms of dissolved calcium concentration, temperature, and salinity (Cohen and Weinstein, 1998a, 2001; Peterson and Janik, 1997). These suitable conditions overlap a significant portion of the state's ecologically important aquatic environments, including the Sacramento River system, San Joaquin River system, Sacramento–San Joaquin Delta, and the Owens and Lower Colorado rivers. Suitable aquatic conditions for dreissenid mussels exist within many of California's forest, oak woodland, savanna, chaparral, grassland, desert, riparian, wetland, and marsh habitats. Many species of state and federal conservation concern are dependent upon these freshwater habitats for some of part of their life cycle (California Department of Fish and Wildlife, 2013a,b). Many sensitive species that would likely be directly impacted by dreissenid mussels, if they were to co-occur, include invertebrates such as the Shasta crayfish (*Pacifastacus fortis*) and the California freshwater shrimp (*Syncaris pacifica*), as well as numerous fish and amphibian species. In addition, numerous highly desirable nonnative game fish species are, and would be, impacted by dreissenid mussels if they were to become widespread.

The diverse environments found across the state provide abundant recreational opportunities. The activities most directly related to the spread of dreissenid mussels and impacted by their establishment

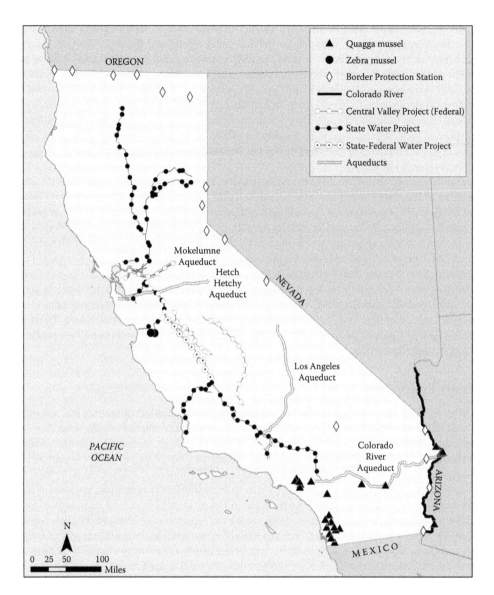

FIGURE 19.1 Map of California showing locations of dreissenid mussel infestations as of July 1, 2013, California Department of Food and Agriculture Border Protection Stations, and the major water distribution systems.

are fishing and boating. Dreissenid mussels can attach to submerged surfaces, such as the hulls of watercraft, and be moved to uninfested waterbodies. In addition, larval dreissenid mussels, or veligers, may be present in water in watercraft that had been in an infested waterbody and transported and released into uninfested water. If environmental conditions are suitable for survival and the released dreissenid mussels are reproductively viable, one or more of these introductions will likely lead to establishment of a new population. Other, more specialized activities, such as freshwater SCUBA diving and gold panning, also have the potential to spread dreissenid mussels or be impacted by their presence.

Outside of their native range, dreissenid mussels have been shown to alter ecosystems by changing the water quality and clarity and outcompeting other species for food and habitat. In addition, they have negatively impacted the enjoyment and aesthetic quality of recreational areas by fouling the hulls and motors of watercraft, encrusting docks, and littering shorelines with vast numbers of sharp shells of dead mussels (Johnson, 2011).

Not only do many of the state's waterways support an abundance of wildlife and diverse recreational opportunities, they also serve as conduits for moving water throughout the state. To meet the growing needs of urban users and agriculture, the state and federal governments developed several water projects in the early to mid-1900s to move water from areas of high abundance to areas of low abundance and high demand, most notably Southern California. The major water projects include the Central Valley Project, State Water Project, Colorado River Aqueduct, Los Angeles Aqueduct, Hetch Hetchy Aqueduct, and the Mokelumne Aqueduct (Figure 19.1). These conveyances, many of which include significant man-made structures (canals, screens, pumping plants, etc.), are vulnerable to dreissenid mussel fouling and would result in widespread distribution of dreissenid mussels throughout the conveyance system if dreissenid mussels were to be present.

Water has historically been central to the economic development and growth of the State of California. Management of the state's water resources, as well as the land, wildlife, and other natural resources, involves many levels of government, including federal, state, and local, with broad ranges of authorities. Management of dreissenid mussels falls within the authority and missions of numerous federal agencies, along with the California Department of Fish and Wildlife (formerly the Department of Fish and Game), California Department of Food and Agriculture, California Department of Parks and Recreation (including the Division of Boating and Waterways, a separate department until July 1, 2013), and California Department of Water Resources. At the local level, cities, counties, municipal water districts, irrigation districts, recreation and park districts, and resources conservation districts manage water deliveries and recreational access, and thus have authorities pertinent to dreissenid mussel management. These diverse authorities result in a mosaic of abutting and overlapping authorities with responsibilities ranging from broad to very limited in scope.

Regulatory Restrictions

As early as 1933, California state law restricted live animal importation of species that were deemed undesirable and a menace to the native wildlife or to the agricultural interests of the state. The original list focused on birds and mammals, but included crayfish, land slugs, and snails. The list of restricted species (Section 671 of Title 14 [Natural Resources], California Code of Regulations) expanded over time through regulatory actions of the California Fish and Game Commission* and provides the legal authority to intercept species determined to pose a threat to native wildlife, the agriculture interests of the state, or to public health or safety. Interceptions may occur at Department of Food and Agriculture inspection stations, as well as in the course of the day-to-day work of Department of Fish and Wildlife wardens.

On July 28, 1993, the Department of Fish and Wildlife submitted an *Initial Statement of Purpose* to the Fish and Game Commission, as part of the regulatory action to amend Section 671 of Title 14. The change added the genus *Dreissena*, thereby adding both zebra and quagga mussels to the list of restricted species. The Department of Fish and Wildlife recognized the possibility that zebra mussels from the Great Lakes could travel overland on trailered watercraft and that they posed a potentially serious threat to fisheries management in the state (California Fish and Game Commission, 1993). With those factors in mind, the Fish and Game Commission approved the addition of *Dreissena* to the restricted species list as a detrimental species, and the regulation became effective on April 6, 1994.

Agricultural Inspection Stations

The Department of Food and Agriculture established its first agricultural inspection station in the early 1920s. Agricultural inspection stations were created to serve as the first line of defense in California's pest and invasive species exclusion efforts (California Department of Food and Agriculture, 2011). Staff at the stations inspect vehicles and commodities to ensure they are pest-free and meet all regulatory requirements. The Department of Food and Agriculture began inspecting watercraft for attached mussels in late

---

* The Fish and Game Commission is composed of five members appointed by the governor, and operates independently of the Department of Fish and Wildlife. The commission provides a forum for public input prior to adopting new regulations or changing existing regulations that will be implemented by the Department of Fish and Wildlife.

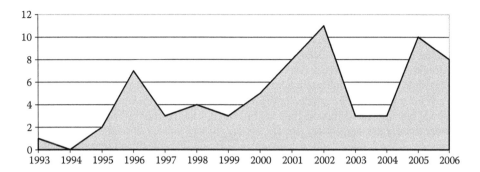

FIGURE 19.2 Dreissenid mussel interceptions at California Department of Food and Agriculture Border Protection Stations 1993–2006.

1992 in response to the 1991 addition of zebra mussels to the federal list of injurious wildlife species. In 1993, the Needles agricultural inspection station intercepted the first vessel with attached zebra mussels. From 1993 to 2006, 68 watercraft with zebra mussels were intercepted at agricultural inspection stations (Figure 19.2), the majority at the Truckee station near Lake Tahoe. All mussels were identified as zebra mussels, with points of origin surrounding the Great Lakes. There was no indication of an infestation in Lake Mead, Nevada, or other points that are more proximate to California.

It is possible that some of these mussels were actually quagga mussels, as the Department of Food and Agriculture did not differentiate between the species at that time. This data does not provide a comprehensive picture of mussel-infested watercraft that were entering California because not all agricultural inspection stations were inspecting watercraft, and of those that did, not all were open full-time. Additionally, inspections primarily consisted of checking watercraft for adult mussels attached to hulls, within the first few inches of engine pipes, and in wet areas within the boat. The limited examinations could have missed larvae or adults attached deeper in the boat (Cohen and Weinstein, 1998a). As of November 2013, the Department of Food and Agriculture operated 16 agricultural inspection stations located on major highways entering the state (Figure 19.1).

Risk Analyses and Early Detection

Between 1998 and 2005, two efforts developed valuable information specific to California that later helped to guide the response to the dreissenid infestation. First, in their 1998 CALFED*-funded report *The Potential Distribution and Abundance of Zebra Mussels in California*, Cohen and Weinstein (1998a) evaluated and mapped the colonization potential of zebra mussels at 160 sites in California, based upon existing available data. They determined that *most of the coastal watersheds in California, the west side of the Sacramento Valley, the San Joaquin River, and southern Delta, provide suitable water chemistry and temperature for colonization* for zebra mussels. They also concluded that many of the state's important water conveyance facilities, including the Los Angeles Aqueduct, the Colorado River Aqueduct, and their associated reservoirs might be suitable for zebra mussel establishment. A companion report, *Methods and Data for Analysis of Potential Distribution and Abundance of Zebra Mussels in California*, provided additional technical information to support their conclusions (Cohen and Weinstein, 1998b).

Second, in 2000, the Department of Water Resources initiated the Zebra Mussel Early-Detection and Outreach Project establishing an early-detection monitoring program for California's Central Valley watershed. This multiyear (2000–2005) effort, also funded by CALFED, included a risk assessment for California, early detection monitoring program, centralized reporting system, rapid response plan, and a public outreach and education program (Messer and Veldhuizen, 2005).

---

* CALFED, created in 1994, was a cooperative venture among 25 state and federal agencies with the mission of improving California's water supply and the ecological health of the San Francisco Bay/Sacramento–San Joaquin River Delta. In 2012, CALFED efforts were absorbed by the Delta Stewardship Council.

Dedicated Staffing and Partnerships

Prior to 2000, the Department of Fish and Wildlife did not have an invasive species coordinator or an invasive species program. This deficiency was addressed when staff was hired in 2000 to participate in the California Non-native Invasive Species Advisory Council (Council) that was established by the CALFED Ecosystem Restoration Program Strategic Plan. The council was comprised of representatives from state and federal CALFED partners for the purpose of providing technical coordination and leadership on invasive species management in the CALFED area, which included the Sacramento–San Joaquin Delta and connected watersheds. While the Department of Fish and Wildlife scientist was hired to focus on the Sacramento–San Joaquin Delta watershed, their role quickly expanded statewide with the heightened concern about zebra mussels and the 2000 discovery of *Caulerpa taxifolia* in Southern California.

In addition, in 2000, the Department of Fish and Wildlife became the state's representative agency to the Western Regional Panel on Aquatic Nuisance Species. At that time, the Western Regional Panel on Aquatic Nuisance Species had seed money available for states interested in developing aquatic nuisance species management plans. The Department of Fish and Wildlife received a portion of these funds and assembled a group of agency scientists from the Department of Water Resources, Department of Food and Agriculture, Department of Boating and Waterways (now within the Department of Parks and Recreation), the State Coastal Conservancy, and the U.S. Fish and Wildlife Service, to identify actions that should be included in the plan. The state plan was drafted in 2002 and finalized in 2008 (California Department of Fish and Game, 2008). Over the course of these 6 years, relationships between staff from the various agencies and departments were formed.

Discovery of Mussels: The Initial Response

Coordinated Effort

On January 6, 2007, quagga mussels were discovered in Lake Mead, Nevada, just east of California (Figure 19.3). The federal representative with the 100th Meridian Initiative* notified the western states invasive species coordinators of the discovery. The same day scientists from the Department of Fish and Wildlife and Department of Water Resources disseminated this information to their managers and colleagues. The Department of Fish and Wildlife, as the public trust and regulatory agency for protecting aquatic habitat and restricting harmful species, convened a briefing on January 12, 2007, with the California Natural Resources Agency and other state agency and department representatives regarding the discovery. That meeting culminated in the formation of the response team, and staff were designated roles and responsibilities for the response. The Department of Fish and Wildlife assumed the lead role and initiated an Incident Command System. The Incident Command System is a standardized, on-scene approach that allows for the integration of facilities, equipment, personnel, procedures, and communications, operating within a common organizational structure; enables a coordinated response among various jurisdictions and functional agencies, both public and private; and establishes common processes for planning and managing resources.

The Incident Command came together quickly, with the Department of Fish and Wildlife's invasive species coordinator serving as the Incident Commander. The position of planning chief was tasked to a Department of Fish and Wildlife fisheries scientist with first-hand knowledge of the Colorado River, and a Department of Food and Agriculture manager acted as operations chief. The Department of Food and Agriculture provided facilities to house the Incident Command Center. All of the participating agencies absorbed the cost of this initial effort. The purposes of the response were to: (1) delineate the infestation, (2) successfully obtain funding for the work of member agencies, (3) establish a closely coordinated interagency scientific and technical consortium, and (4) inform the public and stakeholders of the critical nature of this discovery and to elicit their assistance and cooperation.

---

* The 100th Meridian Initiative is a public and private sector partnership effort led by the U.S. Fish and Wildlife Service to prevent the westward spread of dreissenid mussels and other aquatic invasive species.

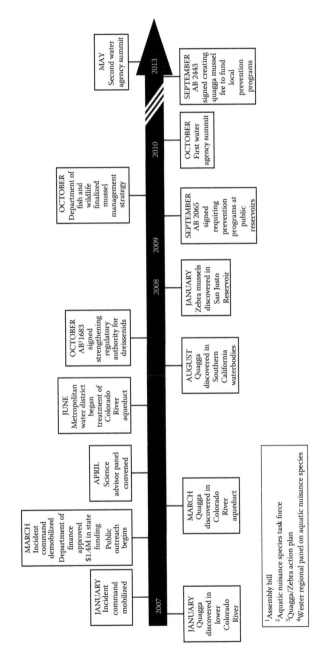

FIGURE 19.3 Timeline of dreissenid mussel efforts post discovery in California.

On Saturday, January 13, 2007, the Quagga Mussel Response Team met for the first time. Staff from the Department of Fish and Wildlife Office of Spill Prevention and Response provided training, shared their Incident Command System experience, and filled many of the key positions needed for a successful emergency response. The Department of Fish and Wildlife, the Department of Food and Agriculture, and the Department of Water Resources provided additional scientific and administrative staff with critical experience and expertise related to emergency response, the state's complex water delivery system, pest exclusion procedures, and aquatic survey techniques. Department of Water Resources staff contacted colleagues at the Metropolitan Water District of Southern California (Metropolitan Water District) and City of San Diego Water Department, who became integral members of the team.

An essential component of the response was the state's continuous coordination with federal agencies and other states in the west. Invasive species staff and Department of Fish and Wildlife leadership held regular conference calls with the National Park Service, Nevada, and Arizona to share information and discuss common goals. This collaboration and the continuous information flow between staff of the 100th Meridian Initiative and members of the Western Regional Panel on Aquatic Nuisance Species served to enhance prevention and outreach efforts. In order to ensure adequate communication within California, the response team compiled daily reports that included progress toward meeting the response objectives and updated information on the infestation, and distributed them to leaders of participating and interested agencies. As the response progressed, the meetings and reports were reduced to weekly, then biweekly, and monthly.

One of the first tasks of the response was to delineate the extent of the infestation by surveying the Colorado River for the presence or absence of mussels. The first mussel detection in California occurred on January 17, 2007, when Metropolitan Water District divers found mussels at the Whitsett Intake Pumping Plant, at the south end of Lake Havasu. Mussels were found on several concrete structures below 30 ft, at a concrete vault at a depth of 20 ft, and at Gene Wash Reservoir, approximately 1.5 miles west of the Whitsett Intake, at a depth of 35 ft. The Whitsett Intake Pumping Plant feeds into Gene Wash Reservoir and into Copper Basin Reservoir, the inception of the 242 mile-long Colorado River Aqueduct, which on average delivers over 6000 acre ft/day of treated and untreated water to Southern California (Metropolitan Water District of Southern California, 2008). On January 19, 2007, the survey discovered quagga mussels at Grass Bay, south of Havasu Landing Resort, and in March, more were discovered in Copper Basin Reservoir, extending the known range of the invasion to about 21 miles into the Colorado River Aqueduct.

Across the state, the Department of Food and Agriculture deployed field crews with experience in surveying for and chemically treating invasive aquatic plants. They were directed to perform visual surface surveys looking for dreissenid mussels. They began surveying on January 30, 2007, and completed surveys on March 19, 2007, surveying 231 reservoirs and finding no dreissenid mussels.

Coincidentally, Metropolitan Water District had scheduled a shutdown of the Colorado River Aqueduct's desert conveyance system for March 2007. This provided the opportunity to inspect the entire system, focusing on dark, sheltered habitats preferred by dreissenid mussels. Metropolitan Water District completed inspections on March 27, shortly after which flows resumed. During the closure, they inspected at least 56 siphons, numerous aqueduct sand traps, Iron Mountain Pumping Plant, and Hines Pumping Plant. They found 778 mussels in siphons between mile 11.98 and 20.96. No mussels were found in or around the two pumping plants or in the open aqueduct. Metropolitan Water District recovered 114 mussels at Whitsett Intake, two at Copper Basin, and 14 in Gene Wash. Monitoring of the Colorado River Aqueduct was extremely important since dozens of waterbodies in Southern California receive raw water from the Colorado River through this delivery system, which serves a population of nearly 19 million in Los Angeles, Orange, San Diego, Riverside, San Bernardino, and Ventura counties. At that time, inspections found no mussels at Diamond Valley Lake, Lake Mathews, Lake Skinner, and many additional raw water recipient reservoirs.

However, less than 6 months later, in August 2007, adult mussels were discovered in Lake Mathews and Lake Skinner reservoirs in Riverside County. Also in August, quagga mussel veligers were found in plankton tow samples collected at the City of San Diego's San Vicente and Lower Otay reservoirs, and dive teams discovered adult mussels attached to a dock and intake structure at the City of Escondido's Dixon Lake. These reservoirs all received water via the Colorado River Aqueduct, and it became clear that the infestation was more widespread than initial surveys suggested.

Initial Control Measures

On the same day as the interagency briefing, the Department of Food and Agriculture took action to enhance interception of trailered watercraft entering the state, with emphasis on those originating from Lake Mead and the Colorado River. The agricultural inspection stations at Yermo, Needles, and Vidal Junction, which saw the most trailered watercraft from Lake Mead and the Colorado River, increased their hours of operation to 24 h a day, 7 days a week. Two days later, they began using an enhanced protocol for inspecting watercraft at the stations. Rather than opportunistically inspecting watercraft for attached mussels, inspectors followed a systematic checklist that entailed a comprehensive inspection, with focus on locations on watercraft most likely to harbor adult mussels. Inspectors also began draining and drying standing water from watercraft, removing attached adult mussels, and decontaminating watercraft when feasible. Once inspected and decontaminated, watercraft with dreissenid mussels were quarantined and allowed to proceed to their destination but prohibited from launching. The Department of Fish and Wildlife staff would follow up to reinspect the watercraft to ensure that it did not pose a threat and release it from quarantine.

In response to the discovery of mussels in their system, Metropolitan Water District implemented weekly shock chlorination treatments in early April at Copper Basin to help reduce veligers in the system. By June 2007, they started to see an increase in mussel density at the Whitsett Intake and found mussels 125 miles into the Colorado River Aqueduct. As a result, they started to chlorinate the system 24 h a day, 7 days a week at the outlet of Copper Basin Reservoir in an effort to control veligers. From October 2 to 5, 2007, Metropolitan Water District had a short shutdown of the Colorado River Aqueduct to evaluate the effectiveness of this continuous chlorination treatment. No small, recently settled mussels were found in a 70-mile stretch of the Colorado River Aqueduct, its pumps, or sand traps. The inspection led Metropolitan Water District to conclude that chlorination was effectively killing veligers in water pumped from Copper Basin Reservoir.

Science Advisory Panel

As part of the Incident Command System, the Department of Fish and Wildlife selected Dr. Andrew Cohen as the department's independent science advisor. He formed a Scientific Advisory Panel that convened on April 4, 2007. Members included preeminent invasive species and *Dreissena* scientists, including Dr. James T. Carlton, Dr. Russ Moll, Dr. Peter Moyle, Dr. Lars Anderson, and Dr. Charles R. O'Neill. Their charge was to develop a report with recommendations for planning the continued response to quagga mussels in California. The panel submitted its report to the Department of Fish and Wildlife in May 2007. The report presented the advisory panel's recommendations in three operational areas: control and eradication in currently infested waters, containment within those waters, and monitoring to detect new infestations (Cohen et al., 2007).

Outreach and Education

As part of the initial response in January 2007, the California Natural Resources Agency, Department of Fish and Wildlife, and Department of Parks and Recreation (Division of Boating and Waterways) began an aggressive statewide public outreach campaign. Emphasis was given to efforts to reach boaters and anglers, and educate them about the issue and how to prevent spreading quagga mussels. At that time, a telephone hotline was created to provide a single point of contact for the response, and a source of information for the public. Informational flyers and posters were developed and distributed in early March 2007 to Department of Food and Agriculture inspection stations, the Department of Fish and Wildlife regional offices, and marinas statewide for dissemination to the public (Table 19.1). During the same time period, the Department of Motor Vehicles, on behalf of the Department of Parks and Recreation—Division of Boating and Waterways, sent a letter to owners of approximately 1.2 million registered watercraft, describing the issue and soliciting their assistance in preventing the further spread of quagga mussels.

TABLE 19.1

Watercraft Inspected at Border Protection Stations for the Period 2007–2012

Year	2007	2008	2009	2010	2011	2012	Total
Number of watercraft inspected	87,840	104,668	162,600	168,773	146,273	122,305	792,459
Number of watercraft cleaned	9,656	9,025	9,445	6,607	4,337	5,019	44,089
Number of watercraft with dreissenid mussels	112	201	230	242	190	162	1137

In order to reach a variety of stakeholder groups, including local water agencies and managers, Department of Water Resources and Department of Fish and Wildlife scientists and managers made presentations throughout the state at both large conferences and smaller venues such as board meetings and local fishing club meetings. To reach an even larger audience, state agency staff participated on radio programs and circulated press releases.

In addition to outreach to the public and stakeholder groups, it was necessary to provide formal training to agency staff to ensure correct species identification as well as effective and consistent watercraft inspection and decontamination. This training was initially provided to the Department of Food and Agriculture inspectors and the Department of Fish and Wildlife staff who were routinely inspecting and decontaminating watercraft. As water agency and reservoir managers established inspection and decontamination programs for watercraft entering their facilities, they began requesting inspection and decontamination training. Between July 2007 and January 2008, Pacific States Marine Fisheries Commission and the Department of Fish and Wildlife provided ten watercraft inspection training classes that focused on identifying dreissenid mussels and techniques and protocols for inspecting and decontaminating watercraft and equipment.

New Authorities

As the scale of the infestation in Southern California was realized, it became clear to the Department of Fish and Wildlife that strategies to prevent the further introduction of quagga mussels into waters or systems not yet affected by the infestation were critical to protecting the state. Although regulations were in place that restricted the movement of adult mussels, the state lacked the authority to interdict or otherwise restrict the movement of veligers in watercraft or other conveyances.

As noted previously, California Code of Regulations Title 14, Section 671, restricted importation, transportation, and possession of dreissenid mussels. However, this regulation did not adequately address the threat posed by veligers contained in bilge water, motors, or flowing water, or adults and veligers in water supply systems. In January 2007, a team of Department of Fish and Wildlife legal, scientific, and enforcement staff began drafting proposed legislation to address these otherwise unregulated aspects of dreissenid mussels in the state. Legislation (AB 1683) intended to close gaps in California's dreissenid mussel regulations was introduced in March, and Governor Schwarzenegger signed it into law on October 10, 2007. This new law added Fish and Game Code Section 2301, enhancing authority for inspections, establishing the authority to quarantine, and placing the responsibility for containment of infested waters on the operators of the water supply systems. Specifically, the Director of the Department of Fish and Wildlife or designee gained the authority to (1) conduct inspections of conveyances that might contain mussels or standing water, (2) order areas of those conveyances that contain water to be drained, dried, or decontaminated, (3) impound or quarantine conveyances for the time needed to ensure that no living mussels persist, (4) inspect waters of the state for mussels, and (5) recommend closure or restrict access to facilities at infested waters. In addition, public or private agencies that operate water supply systems infested with mussels were directed to immediately report any new infestations and prepare plans to either control or eradicate mussels within their facility.

In 2008, additional provisions to prevent the spread of dreissenid mussels in California were approved by the Governor (AB 2065). Section 2302 of Fish and Game Code was enacted and required that managers or owners of reservoirs that are open to the public assess the vulnerability of their reservoir to dreissenid mussel introduction, and develop and implement programs to prevent their introduction.

Funding

In order to pay for the additional workload resulting from the quagga mussel infestation and response, the Department of Fish and Wildlife and the Department of Food and Agriculture submitted emergency deficiency funding requests to the California Department of Finance. Emergency funding is only allocated for expenses incurred in response to conditions of disaster or extreme peril that threaten the immediate health or safety of persons or property of the state. On March 16, 2007, the Department of Finance approved funding totaling $1,459,534 to fund the continued operation of the *Quagga Mussel Emergency Project* through June 30, 2007. The funding was divided between the Department of Food and Agriculture, which was allocated $455,101 to fund additional staffing and for operational expenses to maintain watercraft inspection activities at three agricultural inspection stations, and for visual surveys to assess the extent of the quagga infestations. The Department of Fish and Wildlife was allocated $1,004,433 to staff the Incident Command Center and manage the multiagency response.

After completing its immediate purposes, the Incident Command demobilized on March 16, 2007. The consortium of partners continued to meet on a biweekly basis as the Interagency Team to continue work on actions identified during the response and emerging issues and needs.

After the Incident Command demobilized, the response to the quagga mussel infestation continued. Surveys and research initiated during the response confirmed that eradicating mussels from California was not possible because of the existing level of infestation and lack of tools for eradication. The Interagency Team continued to focus on prevention, containment, and delaying the spread of mussels throughout California. The Department of Fish and Wildlife and the Department of Food and Agriculture submitted budget requests to continue the program and received additional funding for Fiscal Years 2007–2008 and 2008–2009. The Department of Finance approved the addition of $3.2 million for the Department of Fish and Wildlife and $2.5 million for Department of Food and Agriculture.

Due to the size of the state and the number of potential infestation sites, available funding was not adequate for the Department of Fish and Wildlife to shoulder the cost of all the prevention efforts that would be necessary to protect the state. Rather than attempt to implement a statewide prevention program, the Department of Fish and Wildlife determined that the state's role should be advisory and supportive, and focus on collaborative, cost-effective, achievable, adaptive, and efficient elements. The funding enabled the Department of Fish and Wildlife to hire staff to work on prevention, detection, outreach, and education and to develop collaborative partnerships. Six new environmental scientist positions were created so that each of the Department of Fish and Wildlife inland regions would have dedicated staff to lead the program at the local level. Their responsibilities included: following up on quarantined vessels, monitoring high priority waterbodies for mussel infestations, and providing technical and educational resources to local water managers. The Department of Fish and Wildlife headquarters hired staff to develop, implement, and coordinate the statewide program within the Department of Fish and Wildlife, provide oversight for regional activities, continue interagency coordination, and develop consistent scientific and outreach and education components for the program. The Department of Fish and Wildlife also established a laboratory for analyzing water samples to provide early-detection services for the Department of Fish and Wildlife scientists and local water agency monitoring programs.

Living With Mussels: A Long-Term Management Strategy

Once stable funding was established, the Department of Fish and Wildlife began developing a long-term program that would be sustainable. In October 2009, and modeled after the National Parks Service's *Quagga/Zebra Mussel Infestation Prevention and Response Planning Guide* (National Parks Service, 2007), the Department of Fish and Wildlife adopted an internal programmatic guidance document, the *Quagga/Zebra Mussel Management Strategy* (California Department of Fish and Game, 2010). The defined goals of the Department of Fish and Wildlife's dreissenid mussel program were to (1) prevent further introduction of dreissenid mussels into the state, (2) contain mussels within currently infested waters, and (3) to eradicate mussels from infested waters, if feasible.

Strategies and actions for achieving these goals are organized under six broad objectives: (1) coordination, (2) prevention, (3) detection, (4) response, (5) control and eradication, and (6) information dissemination. These objectives were intended to be addressed through a variety of specific actions. To date, significant accomplishments have been made, or are ongoing, and progress has been made in furthering these objectives. The following summarizes the management strategies and some notable accomplishments to date.

Coordination

Before the discovery of quagga mussels in the west, state agencies participated on a variety of invasive species teams. The Lake Mead discovery created a sense of urgency related to aquatic invasive species, and with that, the need to build upon and enhance existing collaborations and partnerships. A core group was formalized from those participating in the response, and over time expanded to include dozens of governmental agencies and stakeholder groups. This ongoing statewide Interagency Team continues to meet regularly to discuss common goals and identify emerging concerns. This network of staff facilitates the sharing of resources, including funding, access to agency expertise, training opportunities, and coordinated outreach.

As previously noted, the Department of Fish and Wildlife dedicated environmental scientist positions to dreissenid mussel management in each of its six inland regions. These scientists continue to work closely with local agencies and reservoir managers to assist with monitoring, provide training, and ensure that local agencies have access to all available resources to implement their prevention and containment programs.

The Department of Fish and Wildlife continues the coordination efforts with other western states that began with the initial response in 2007. These efforts include representation on the Western Regional Panel on Aquatic Nuisance Species, and through coordination with the Invasive Species Coordinators of western states.

Prevention

Preventing the introduction or spread of invasive species is widely recognized as the most cost-effective means for managing invasive species (Buhle, 2005; Leung, 2002), and is the primary focus of California's dreissenid mussel program. For California's purposes, prevention includes stopping new introductions and preventing further spread within the state. A primary means for achieving this objective is through implementation and coordination of watercraft inspections.

In California, watercraft inspections occur both at the borders and throughout the state. Department of Food and Agriculture inspection stations intercept both commercial and private watercraft and inspect them for adult mussels, standing water that may harbor veligers, and attached aquatic vegetation. Watercraft found to carry dreissenid mussels are placed under quarantine and allowed to proceed to a designated destination, but prohibited from launching, or refused entry into the state. After reaching its destination, quarantined watercraft are inspected by Department of Fish and Wildlife staff and released when they are deemed to no longer pose a risk of spreading dreissenid mussels, either as a result of a decontamination and/or dry period. Agricultural inspection stations receive dedicated funding to maximize their hours of operation to inspect watercraft, with the goal of maintaining continuous operation (24 h/day, 7 days/week) at as many of the agricultural inspection stations as possible. In addition to inspections at the border, local agencies across the state implement prelaunch of watercraft inspections.

The new legal authority gained by the Department of Fish and Wildlife in 2007 and 2008 provided additional tools for preventing the spread of dreissenid mussels within the state. In order to prevent dreissenid mussels from moving from infested water supply systems in California, the system operator must develop a plan for either control or eradication of the mussels. These plans include provisions for containing mussels, limiting the release of infested water, managing recreational uses, providing outreach and education, and any other measures necessary to prevent dreissenid mussels from moving outside of their system.

To prevent the introduction of dreissenid mussels into reservoirs, managers or owners of reservoirs that are open to the public and not yet infested with dreissenid mussels are required by Fish and Game Code Section 2302 to assess the reservoir's vulnerability to the introduction of dreissenid mussels and implement a prevention program that includes public education, monitoring, and management of recreational activities.

In 2010, the State Legislature allocated funding to five counties in the San Francisco Bay Area, known as the Bay Area Regional Consortium, to assist in the development and implementation of a pilot cross-jurisdictional dreissenid mussel prevention program. While that funding was temporary, the consortium demonstrated that local agencies desired to enhance prevention programs to protect their waters but lacked adequate funding to do so. They also demonstrated the potential and willingness of local agencies to work together to develop collaborative programs.

With the need for funding at the local level apparent, Legislation (AB 2443) was passed in 2013 to establish a program to help fund local dreissenid mussel prevention programs. The legislation imposed a new fee on vessels registered through the Department of Motor Vehicles and used in freshwater, and established a grant program administered by the Department of Parks and Recreation—Division of Boating and Waterways, to disperse the funds to agencies implementing prevention programs consistent with Section 2302 of the Fish and Game Code.

To the extent possible, the Department of Fish and Wildlife also supports the work of others such as researchers, nongovernmental organizations, and various others, by providing resources and technical assistance for efforts that support or further the department's goals relative to dreissenid mussels.

Detection

Detection of incipient mussel populations is critical to preventing the spread of mussels and for identifying sites for control efforts. In 2012, 255 sites at 95 California waterbodies were monitored for the presence of dreissenid mussels using 1 or more of 3 monitoring methods. Two of those methods, existing surface surveys and dedicated artificial substrates, are used for detection of adult mussels. Water sampling is the third method used to detect dreissenid mussels, and may include collection of veligers and/or genetic material of dreissenid mussels suspended in the water. Sampling for veligers and/or genetic material has the potential to detect an infestation earlier than methods that rely on detection of settled dreissenid mussels. Analysis of water samples requires specialized expertise and equipment, and accurate results are dependent on the use of appropriate methods and adherence to strict procedures during both sample collection and analysis. The Department of Fish and Wildlife developed in-house expertise and facilities for both visual identification and genetic testing for veligers. The Department of Fish and Wildlife lab analyzes water samples collected by Department of Fish and Wildlife staff, as well as those collected by other agencies to support statewide monitoring efforts.

To track monitoring efforts and results, the Department of Fish and Wildlife maintains a statewide database that contains monitoring data from state and federal agencies, as well as water districts and power companies. This database provides the most comprehensive picture of monitoring statewide, allows for identification of gaps in monitoring, and enables the prioritization of future monitoring efforts.

In addition, the Department of Fish and Wildlife has both a reporting website and an information hotline available to the public to report observational information on incipient infestations of dreissenid mussels or other invasive species.

Response

The Department of Fish and Wildlife routinely responds to reports of suspected dreissenid mussel discoveries in California by evaluating photos, reviewing DNA analysis reports, conducting interviews, and participating in follow-up monitoring, in an effort to confirm the presence of suspected infestations. Follow-up monitoring may include dive surveys, targeted surface surveys at watercraft launch ramps and docks, and enhanced water sampling for veligers and/or genetic material.

Since January 2008, dreissenid mussels have been detected in 15 waterbodies and the Department of Fish and Wildlife has assisted waterbody managers in implementing immediate containment via management of exiting watercraft, and subsequently collaborated on the water manager's long-term containment plan.

Control and Eradication

The Department of Fish and Wildlife does not have the authority to direct dreissenid mussel control or eradication efforts in the waters of California, or those under the jurisdiction of other agencies. As such, the department acts in an advisory role and provides assistance relative to outreach, review of environmental documents, and the potential impacts the proposed control or eradication efforts may have on wildlife and habitats. Thus far, dreissenid mussel control or eradication efforts in California have been limited to Metropolitan Water District activities in their water supply system.

Information Dissemination

Outreach and education play an important role in the Department of Fish and Wildlife's prevention efforts. The Department of Fish and Wildlife's approach is twofold: to educate and engage the public and to provide resources to local agencies and other stakeholder groups.

To keep the public informed about dreissenid mussels and their role in preventing their spread, the department maintains a hotline where the public can receive current information or report concerns. The department periodically issues news releases and develops and distributes outreach materials, including posters and pamphlets, to specific stakeholders. Two slogans, namely, *Clean, Drain, Dry* and *Don't Move a Mussel*, have been utilized by the Department of Fish and Wildlife and have aligned with the U.S. Fish and Wildlife Service, *Stop Aquatic Hitchhikers* campaign, to provide consistency with other states in the west.

Over time, the public outreach program evolved from this very basic, traditional approach to include a webpage with current information and links to pertinent resources, on-line reference documents, social media, and integration of emerging electronic media tools.

As part of the Department Fish and Wildlife's outreach and education strategy, the department maintains close working relationships with local agencies and stakeholder groups. Between 2011 and 2013, the department sponsored two stakeholder summits to provide general information on the issue and specifically address two timely items: the practicality or feasibility of a statewide reciprocal watercraft banding program and the process for accessing local assistance funds for prevention programs.

Watercraft inspection training is another important component of the statewide program. Department of Fish and Wildlife staff has provided training to over 2000 individuals and continue to be available to provide training at the request of local water managers.

In order to evaluate the effectiveness of the ongoing outreach program and assess the need for additional outreach, in 2012–2013, the Department of Fish and Wildlife implemented a study to examine the habits and knowledge of anglers and boaters who use popular Southern California quagga mussel-infested waters. This research has not been fully analyzed, but the results of the research will help guide future quagga mussel prevention and management efforts.

Discussion

As of November 2013, efforts to delay the spread of mussels in California have been successful as evidenced by the absence of new dreissenid mussel infestations outside of the water delivery system supplied by the Colorado River Aqueduct. The only populations of quagga mussels discovered in California since 2007 have been found in reservoirs and ponds that received raw water from the Colorado River through the vast and complex Southern California water distribution system. This success can be correlated with many factors, including the drought conditions in California and the national economic downturn during that time, which likely contributed to declines in registered boaters (California Department of Parks and Recreation, unpublished data) and fishing license holders (California Department of

Fish and Wildlife, unpublished data), which in turn may have resulted in fewer boaters traveling from infested waterbodies to uninfested waterbodies. However, a great deal of credit properly goes to state, local, and federal agency managers and scientists, government decision makers, local governments, and the public, for quickly taking action and continuing prevention, containment, and outreach efforts throughout the years.

Agency Staff

Without early recognition of the threat and opportunistic investment in development of in-house expertise and partnerships, California would have been far less prepared and capable of effectively responding to the discovery of dreissenid mussels in 2007. As far back as 1992, multiple California agencies recognized the threat dreissenid mussels posed to the environment, economy, and water conveyance systems and began developing enforceable laws, management strategies, and relationships, as fiscally practical, to prevent the introduction of dreissenid mussels and other invasive species. As large-scale planning efforts like CALFED became the state's new paradigm for resource management, one in which collaboration leads to solutions that satisfy multiple interests, managers began to more fully appreciate the implications of invasive species. In addition to taking a more comprehensive approach to ecosystem management, CALFED brought new financial resources and buy-in from the agencies that would devote staff to the issue. Through CALFED, as well as subsequently during the drafting of the state's Aquatic Invasive Species Management Plan, staff from various agencies came together to identify common interests and goals and established collaborative relationships. These staff, some of whom were dedicated by their agencies to work on invasive species exclusively, and the relationships they established facilitated swift action across multiple agencies when the need to respond to the quagga mussel arose in 2007. These staff immediately understood the gravity of the situation, were effective at communicating with, and mobilizing action by, their executives, and integrating outside experts into the response.

For the initial response, agencies that lacked dedicated staff redirected existing staff as needed. In some cases those were temporary assignments, while in others quagga mussel–related workload became an ongoing component of their duties. These staff have enabled continued coordination and collaboration on quagga mussel management through their participation on the Interagency Team.

While the state could have been more prepared, like many organizations, it was constrained by limited resources. Therefore, the existing financial and staff resources already dedicated to invasive species efforts were critical to the response. Ideally, organizations would have dedicated staff that can watch the horizon for threats; actively engage with the invasive species community locally, regionally, and nationally; and maintain ongoing communication with their management to advise them about what may lie ahead.

Collaboration

The Department of Fish and Wildlife lacked the financial and personnel resources, as well as the authority over local jurisdictions, to implement a sustainable prevention program statewide. Therefore, the Department of Fish and Wildlife opted to invest its limited resources to develop resources to assist local agencies in developing and implementing their own programs, tailored to their available resources and vulnerabilities. In addition to assisting with establishing programs, the Department of Fish and Wildlife worked to develop collaborations among local agencies to improve communication and standardize and interlink prevention programs, thereby strengthening prevention efforts on a regional scale.

Through collaboration, action by many state and local agencies occurred, and a far greater level of effort resulted than could have been achieved by the Department of Fish and Wildlife acting alone. Collaborations enable the sharing of information and pooling of resources, and result in increased benefits for those involved. In some cases, operational or institutional barriers may exist, but the investment in overcoming them, building relationships, and maintaining those relationships is, in the long term, an investment that yields high returns.

Outreach

Both with their compliance with local prevention programs and voluntary implementation of the *Clean, Drain, Dry* message, boaters, anglers, and other recreationists have been instrumental in preventing the spread of dreissenid mussels. A *social marketing* campaign was employed and targeted the largest user groups—boaters and anglers—which were most likely to move dreissenid mussels. The campaign sought to change the behavior of recreational water users and promote environmental stewardship through a sense of collective responsibility. Materials promoted the protection of natural resources, access to recreational opportunities, and avoidance of direct financial impacts (e.g., increased boat maintenance costs, higher water rates). Given limited resources, initial efforts targeted groups believed to pose the greatest threat of spreading dreissenid mussels, and over time expanded to include a wider range of user groups, including fishing guides, bait dealers, etc. Over the years, as new means for reaching the public emerged, the Department of Fish and Wildlife adopted and used these tools. Through such tools as social media, YouTube, and quick response codes, the department continues to extend the reach of its program's messaging.

Even after many years of spreading the message, there are still recreational water users that have not been reached (Department of Fish and Wildlife, unpublished data). Therefore, it is important to continue outreach efforts, seek ways to reach new audiences in meaningful ways, and to continually evaluate efforts and the tools used, and refine the tools and approaches as needed.

Organized Response

In terms of forming and launching a rapid response, the Incident Command System provided an efficient framework for organizing operations, quickly collecting information, evaluating options, and making decisions. This system had the added benefits of temporarily removing organizational and hierarchical barriers, and created an environment without the same level of bureaucratic constraints usually present with government activities. In this case of dreissenid mussels, it also allowed for participation by individuals and organizations outside of state government, which enabled a wider range of expertise and perspective than would have otherwise been represented. Another benefit of the Incident Command System is that it can be, and optimally is, learned before a response is needed. The resources for learning and applying the system are free, and it can be applied to a variety of responses and customized to individual circumstances.

Legal Authorities

To guard against the threat of dreissenid mussels, one of California's first actions back in 1993 was regulatory. While this provided the Department of Fish and Wildlife with the authority to prohibit possession and enforce violations, it was understood that it offered little protection against the more likely means of dreissenid mussel introduction—those that were unintentional. Therefore, regulations alone would not protect the state, and certainly were ineffective at preventing dreissenid mussels from entering the state via the Colorado River. With dreissenid mussels having breached the state's border, additional regulatory authorities were needed.

Laws developed in 2007 in response to the infestation were basic, but gave the Department of Fish and Wildlife new authority specific to stopping, inspecting, and draining conveyances that might be transporting dreissenid mussels, the authority to temporarily close waterbodies, and also required managers of infested waterbodies to contain dreissenid mussels. The following year, owners and managers of reservoirs open to the public were mandated to develop and implement programs to prevent the introduction of dreissenid mussels. While these new laws lacked many specifics compared to those subsequently enacted in other states, they gave the Department of Fish and Wildlife broad authority to address a range of vectors and activities necessary to effectively prevent the further spread of dreissenid mussels, while maintaining its own operational flexibility.

The prevalence of infested waters that lacked managed access, particularly along the Colorado River, played a pivotal role in the state's approach of sharing the responsibility for prevention between managers of infested waters and managers of uninfested waters. To the extent possible, the Department of Fish and Wildlife required managers of infested waters to implement education and inspections (including the draining of all water) of watercraft leaving infested waterbodies. Because watercraft have access at unmanaged, infested waters, managers of uninfested waters still needed to implement prelaunch inspections to be sure every watercraft entering their jurisdiction did not pose a threat for introducing dreissenid mussels.

The reality of unmanaged access to infested waters also meant a statewide watercraft-tracking system, where vessel launches would be documented at managed waters but a vessel could launch without detection or documentation at unmanaged infested waters, was impractical in California. Such a system would create a false sense of security, and be of limited value, because watercraft would still require inspection to ensure they did not carry dreissenid mussels or water. Given those limitations, the state opted not to pursue a statewide watercraft-tracking program.

Because dreissenid mussels were established in the state, watercraft traveling within the state now had the potential to move them to uninfested waters. Therefore, it has been important to continue to engage and involve the public, as well as affected agencies and stakeholder groups, in preventing the introduction and spread of dreissenid mussels and other aquatic invasive species. While Fish and Game Code applies statewide, local agencies lack the authority to enforce it in conjunction with their prevention programs. As a result, some local jurisdictions have developed their own laws that enable them to enforce their prevention programs, cite violators, and quarantine watercraft.

The Department of Fish and Wildlife has opted not to use a heavy hand in enforcing dreissenid mussel laws. Rather, the Department of Fish and Wildlife approaches (all but the most egregious) incidents of noncompliance as opportunities to educate and persuade those who are unaware or indifferent to the law. This is considered to be a far more effective approach to preventing the further spread of dreissenid mussels given the scale of the issue (e.g., number of waters where access is unmanaged, number of watercraft) and the reality that reliance on enforcement would be impossible.

Funding

The ability to formally declare the presence of dreissenid mussels a state of emergency allowed access to the needed financial resources to implement a rapid response. The bar for declaring an emergency is high, and the well-known severity of environmental and economic impacts sustained in the Great Lakes region as a direct result of dreissenid mussels enabled that bar to be met in California. For species without such a notorious history, a strong enough case to substantiate an emergency may be difficult to make. In such cases, it would be prudent to have a rapid response fund set aside that could be used when an invasion warrants prompt action, but does not qualify as an emergency. While identified as a need, California has not yet created a dedicated fund for such efforts.

The emergency funding provided for the initial response, and because the likely further spread of dreissenid mussels was so closely tied to watercraft, the California Department of Finance determined the long-term program would be funded through boat registration fees. This stable funding source has enabled the Department of Parks and Recreation (Division of Boating and Waterways) and the Department of Fish and Wildlife to provide a consistent level of services, despite years in which the state has had financial challenges. While the funding from this source has remained constant for the Department of Food and Agriculture, its other funding sources have sustained reductions, and as a result, agricultural inspection stations have experienced reduced staffing and hours of operation.

While an existing, dedicated funding source is ideal, it may not be available. In some cases, creation of a new revenue-generating program may be necessary to offset the expenses associated with invasive species prevention efforts. In California, this approach was first utilized by local agencies who charged boaters a fee for inspecting their watercraft, and that fee in turn was used to support the inspection program. While not the first state to implement a statewide invasive species fee, in 2014, California began collecting an annual fee along with watercraft registrations to fund state efforts and augment local prevention programs. Whatever its origin, a reliable source of funding improves the potential for long-term stability of any invasive species prevention program.

Management Plan

Long-term financial stability should be paired with long-term goals. California's dreissenid mussel management plan is a general guidance document that identifies achievable actions, which can be realized within the limitations of the available funding. Many of the original actions have been accomplished or are ongoing, but some have not yet been addressed. Full implementation and adaptive management will take a continued commitment by Department of Fish and Wildlife staff and managers.

While not perfect, the strategic management framework addresses key components needed for a successful, sustainable program, including (1) continued efforts to build and enhance collaborations and partnerships, (2) implementation and coordination of effective prevention efforts, (3) monitoring and response to dreissenid mussels throughout the state, and (4) continued development and implementation of outreach to ever-broadening audiences. This management framework allows the department to remain flexible in its specific activities and be able to modify or adapt those activities as needed.

Continuing Challenges

Scientists may take for granted that others share their understanding of the complex impacts invasive species cause to the environment, and that preventing invasive species impacts is a worthwhile investment. While decision makers may not have the same perspective as scientists, scientists can, and must, effectively communicate the issues and consequences in practical, understandable, and persuasive ways if they are to elicit action. They also must be realistic about the importance of environmental concerns when weighted against economic interests and, particularly in California, water interests. In the case of dreissenid mussels, the goals of protecting the environment and economic interests went hand-in-hand. However, this may not always be the case, or at least the perception.

Even when the goal of protecting the state's natural resources is uncontested, conflict can still exist. California has some of the most protective environmental regulations in the country, and they serve important functions in preserving our natural resources, quality of life, and the economic vitality of the state. However, when it comes to managing invasive species, they can also impede quick responses and even mildly aggressive control efforts. The California Environmental Quality Act (CEQA) mandates that all state and local agencies minimize environmental impacts of projects they implement or permit. What is considered an *environmental impact* extends well beyond those that are biological and includes recreational access, air quality, transportation, etc. (Public Resources Code 21000–21177). In addition to CEQA, there are many specific laws that would regulate activities associated with a response to invasive species, including those related to pesticide applications, impacts to water quality, and alteration of streambeds, to name but a few. For these reasons, the response and management of dreissenid mussels in California has largely avoided activities that would trigger an environmental review and approval process. Those that have called for environmental review, such as an eradication effort at San Justo Reservoir or approval of a new pesticide, either labor through the process, fall victim to their own bureaucracy, or both.

Ideally, there would be a means for fast-tracking invasive species control and eradication projects through the environmental regulatory process. In California, this may not be possible, not only because of the regulatory restrictions but also because it would be impossible to predict, much less plan for, all possible scenarios. The most practical alternative, then, is to identify the potential processes, establish relationships with staff that oversee those processes, and develop templates for the most likely needed information. Such preparation could possibly save critical time in a response. Otherwise, it is likely that there will be no rapid responses, and that any response that triggers environmental review will likely take several years as it navigates the approval process.

Conclusion

The short- and long-term management of the dreissenid mussel infestation in California has been shaped, guided, informed, and, in some cases, directed by the structure of governance, the authorities and roles of state agencies, and the environmental conditions in the state. To facilitate an effective and efficient management program, the Department of Fish and Wildlife has assumed an advisory role and, rather than using a heavy-handed regulatory approach, encourages local involvement and public education and participation. California's response to dreissenid mussels has required managers at all levels of government to be open, adaptive, and creative in finding the most effective means for accomplishing their goals within their constraints.

In California, the response to dreissenid mussels created the opportunity for, and in some cases necessitated the formation of, many partnerships and collaborations. The prompting of many local responses formed a truly collaborative venture, facilitated by the existing relationships between each agency, their stakeholders, and the agency scientists. Through the recognition and acknowledgment of various perspectives, California was able to maintain successful, working collaborations. Though collaboration has not always resulted in consensus, focusing discussions and negotiations on underlying common interests has resulted in a rare convergence of interests, based upon mutual gain. Without these local collaborative efforts, the state's response would not have met the same measure of success. Hopefully, the partnerships that have been established through these efforts will persist and not only reduce the likelihood of large-scale dreissenid mussel infestation in California but also keep the threats that invasive species pose to our economy and environment in the forefront of public policy and awareness.

Acknowledgments

Special thanks to Dr. Jeff Janik, Dean Messer, and Tanya Veldhuizen (Department of Water Resources), Dr. Ricardo DeLeon and Dr. Bill Taylor (Metropolitan Water District of Southern California), Dr. Robert Leavitt (Department of Food and Agriculture), and Terry Foreman and Julie Horenstein (Department of Fish and Wildlife) for their contributions to the incident response and continuing efforts. Additional thanks to the many staff from these agencies, as well as the Department of Parks and Recreation, U.S. Fish and Wildlife Service, the Science Advisory Panel, many local agencies, and others who contributed to these efforts. Thanks also go to Mary E. Oliva, Stephen E. Oliva, and Valerie K. Cook Fletcher, and Jeb M. Bjerke for their assistance in preparing this publication, and three anonymous reviewers who provided insightful comments that greatly improved the manuscript.

REFERENCES

Buhle, E. M. (2005). Bang for the buck: Cost-effective control of invasive species with different life histories. *Ecological Economics*: 355–366.

California Department of Fish and Game. (2008). California Aquatic Invasive Species Management Plan. Sacramento, CA: California Department of Fish and Game.

California Department of Fish and Game. (2010). Quagga/Zebra Mussel Management Strategy. Sacramento, CA: California Department of Fish and Game.

California Department of Fish and Wildlife. (2013a). State and Federally Listed Endangered and Threatened Animals of California, January 2013. California Department of Fish and Wildlife, Biogeographic Data Branch. Sacramento, CA: California Department of Fish and Wildlife.

California Department of Fish and Wildlife. (2013b). State and Federally Listed Endangered, Threatened, and Rare Plants of California, April 2013. California Department of Fish and Wildlife, Biogeographic Data Branch. Sacramento, CA: California Department of Fish and Game.

California Department of Food and Agriculture. (2011). California border protection stations (BPS): First line of defense in protecting our environment and resources from invasive species. Retrieved 2013, from California Department of Food and Agriculture, Sacramento, CA, http://www.cdfa.ca.gov/plant/PE/ExteriorExclusion/borders.html.

California Fish and Game Commission. (1993). Final Statement of Purpose for Regulatory Action to Amend Section 671, Title 14, California Code of Regulations. Sacramento, CA: California Fish and Game Commission.

California Fish and Game Commission. (1998). Strategic Plan: An agenda for California's fish and wildlife resources. Sacramento, CA: California Fish and Game Commission.

Cohen, A., Moll, R., Carlton, J., O'Neill, C., Anderson, L., and Moyle, P. (2007). California's response to the zebra/quagga mussel invasion in the West. Quagga/Zebra Mussel Scientific Advisory Panel.

Cohen, A. and Weinstein, A. (1998a). The potential distribution and abundance of zebra mussels in California. Oakland, CA: San Francisco Estuary Institute.

Cohen, A. and Weinstein, A. (1998b). Methods and data for analysis of potential distribution and abundance of zebra mussels in California. Oakland, CA: San Francisco Estuary Institute.

Cohen, A. and Weinstein, A. (2001). Zebra mussel's calcium threshold and implications for its potential distribution in North America. Oakland, CA: San Francisco Estuary Institute.

Johnson, L. E. (2011). Zebra mussel. In: D. Simberloff and M. Rejmanek (Eds.), *Encyclopedia of Biological Invasions* (pp. 709–713). Berkeley, CA: University of California Press.

LaBounty, J. F. (2007). Quagga mussels invade Lake Mead. *LakeLine*, 27: 17–22.

Leung, B. D. (2002). An ounce of prevention or a pound of cure: Bioeconomic risk analysis of invasive species. *Proceedings of the Royal Society.*

Messer, C. and Veldhuizen, T. (2005). Zebra Mussel Early Detection and Public Outreach Program Final Report. Sacramento, CA: California Department of Water Resources.

Metropolitan Water District of Southern California. (2008). Annual report for the fiscal year July 1, 2007 to June 30, 2008. Los Angeles, CA: Metropolitan Water District of Southern California.

National Parks Service. (2007). *Quagga/Zebra Mussel Infestation Prevention and Response Planning Guide.* Fort Collins, CO: National Parks Service.

Peterson, D. and Janik, J. (1997). Keeping zebra mussels out of California. *California Water Plan News*, 2(2): 6–9.

20

Zebra Mussels in Texas: Management Efforts by Texas Parks and the Wildlife Department

Earl W. Chilton II, Robert F. McMahon, and Brian Van Zee

CONTENTS

ABSTRACT   The first live zebra mussel documented in Texas was found in Lake Texoma on April 3, 2009. The North Texas Municipal Water District (NTMWD) has an inter-basin water transfer permit to pump water from Lake Texoma and discharge it into West Prong Sister Grove Creek, a tributary of Lavon Lake that forms part of the headwaters of the Trinity River. Only 4 months after zebra mussels were found in Lake Texoma Texas Parks and Wildlife Department (TPWD) staff found zebra mussels in West Prong Sister Grove Creek. NTMWD discontinued pumping Lake Texoma water into the creek and in 2010 TPWD used KCl treatments in an effort to eradicate zebra mussels from West Prong Sister Grove Creek and help prevent them from spreading through the Trinity River basin. Although KCl treatments were unsuccessful, no young mussels have been observed in the creek since pumping from Lake Texoma stopped. However, zebra mussels have now been found in seven public water bodies including Lakes Belton, Bridgeport, Lavon, Lewisville, Ray Roberts, and Texoma. In 2013, the Texas Legislature gave TPWD additional rule-making authority to require persons leaving or approaching public water to drain from vessels all water resulting from immersion in public water (except salt water), and authority to conduct inspections for compliance.

Background

The first documented case of zebra mussels (*Dreissena polymorpha*) in Texas occurred in 2006 when a boat from Minnesota was transported to Lake Texoma (an 89,000 acre U.S. Army Corps of Engineers project reservoir that straddles the Texas–Oklahoma border, Figure 20.1).

FIGURE 20.1   Map of Lake Texoma showing the Texas–Oklahoma border. The position of the water intake for West Prong Sister Grove Creek is shown as a star in the lower right corner.

An employee at a local marina recognized the mussels on the boat as zebra mussels and notified the local Texas Parks and Wildlife Department (TPWD) Game Warden. The boat was quarantined and cleaned prior to being allowed in the water and TPWD issued a press release on October 30, 2006, entitled *Zebra mussel introduction thwarted on Lake Texoma* regarding the incident. Between 2006 and 2009, four additional boats from out-of-state were intercepted and sanitized prior to being launched in Lake Texoma.

These incidents were serendipitous because there is no formal zebra mussel inspection program in the state. There are two salient reasons why there is no formal zebra mussel inspection program in Texas. First, the cost of setting up an inspection program in a large state such as Texas is prohibitive, and second, it was widely believed that surface waters in Texas would be too warm to support viable zebra mussel populations.

The first live zebra mussel documented in Texas was found in Lake Texoma on April 3, 2009. It was attached to a submerged telephone cable. TPWD notified the Oklahoma Department of Wildlife Conservation and the U.S. Army Corps of Engineers, as well as other state and local governmental entities and industry representatives.

TPWD personnel deployed Portland settlement samplers in an effort to monitor zebra mussel settlement and to determine whether the first documented mussel was an isolated occurrence. In July 2009, live zebra mussels were found in Lake Texoma near the North Texas Municipal Water District (NTMWD) intake structure. NTMWD has an interbasin water transfer permit to pump water from Lake Texoma

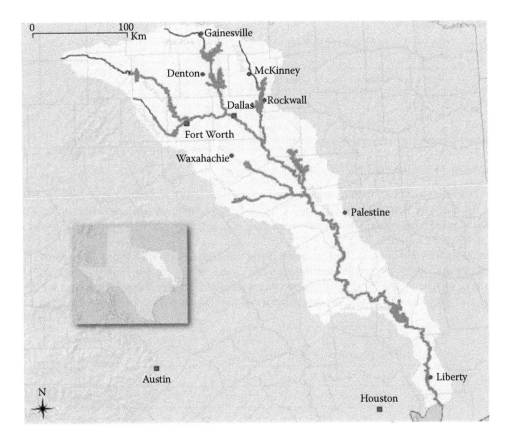

FIGURE 20.2 Trinity River Basin.

and discharge it into West Prong Sister Grove Creek. Sister Grove Creek is a tributary of Lavon Lake and forms part of the headwaters of the Trinity River. On August 3, 2009, TPWD staff inspected the NTMWD outfall area on West Prong Sister Grove Creek and found two live zebra mussels approximately 300 m downstream of where Lake Texoma water entered Sister Grove Creek. Another specimen was found 1.6 km downstream of the outfall area. This was the first detection of zebra mussels in the Trinity River Basin. The Trinity River Basin extends from north of the Dallas-Fort Worth area, southeast to Galveston Bay (Figure 20.2).

Sister Grove Creek Treatments

To reduce the likelihood of the spread of zebra mussels through Sister Grove Creek, TPWD staff determined the best course of action would be to chemically treat West Prong Sister Grove Creek. A decrease in the number of zebra mussels in Sister Grove Creek could reduce the size of a potential source of veligers entering Lake Lavon. Copper and potassium chloride are effective molluscicides for zebra mussels (Boelman et al. 1997; Fisher et al. 1991, 1993; Stubbs 2012; Kennedy et al. 2006; Utah Division of Wildlife Resources 2007; Virginia Department of Game and Inland Fisheries 2013). KCl was chosen for treatments because of significant regulatory hurdles involved in using a copper-based molluscicide on zebra mussels. TPWD staff conducted a series of laboratory experiments to determine the most effective concentration of KCl for zebra mussel mortality. Mussels were collected from Sister Grove Creek and exposed to concentrations of 100, 150, and 200 mg/L KCl and were checked for mortality at 24, 48, and 72 h (Table 20.1).

TABLE 20.1

Percent Mortality of Zebra Mussels Exposed to Different Concentrations of KCl for up to 72 h

Hours	Control (%)	100 mg/L (%)	150 mg/L (%)	200 mg/L (%)
24	0	76.7	100	100
48	0	90.0	NA	NA
72	0	100	NA	NA

Treatment 1: September 20–24, 2010

KCl was used to treat a 51.5 km reach of Sister Grove Creek. The treatment reach extends from where Enloe Road crosses West Prong Sister Grove Creek in Grayson County to the headwaters of Lake Lavon in Collin County (Figure 20.2). During this treatment, a total 9,590 kg of KCl was added to the creek at eight dosing stations (Table 20.2).

Dosing stations 1–4 were located upstream of the confluence with East Prong Sister Grove Creek and dosing stations 5–8 were located between the confluence of the two prongs, and Lake Lavon (Figure 20.3). The reason for this delineation was because of the increase in water flow and volume from East Prong Sister Grove Creek. Estimated travel time was 4–5 days for the treated water to flow

TABLE 20.2

KCl Applied at Each Site during Treatment 1

Site	1	2	3	4	5	6	7	8	Total
KCl (kg)	800	850	900	1100	3800	4450	4350	4550	9590

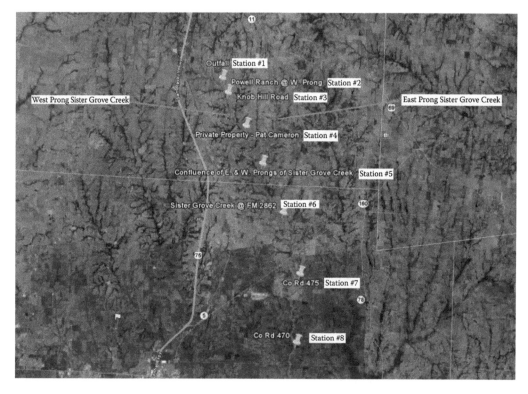

FIGURE 20.3   Dosing stations on Sister Grove Creek during September 2010 KCl treatments.

from one dosing station to the subsequent station downstream. The entire 32 mi of creek was effectively treated in less than 1 week.

KCl was applied over a 48-h period at each dosing station. The target concentration was 175 mg/L of KCl with exposure duration of at least 48 h. QuanTab test strips and conductivity meters were used to determine chloride concentrations and conductance. Chloride and conductance data were used as surrogates to determine the concentration and timing of additional KCl applications at each site.

During a follow-up treatment on September 14, an additional 159 kg of KCl was used to treat the area from Enloe Road to the NTMWD's outfall because of the possibility that section of the creek would stop flowing before treating the rest of the creek the week of September 20.

Prior to treatment, numerous zebra mussels were located in the upper portion of the creek. These mussels were used to determine whether the treatment was effective. No zebra mussels were located downstream of station 4.

During treatment 1, target concentrations were achieved for a 48-h period in the lower portion of the creek (stations 5–8). However, target concentrations were not achieved in the upper portion of the creek (stations 1–4). Subtarget concentrations likely were caused by extremely low flow at stations 1–4 during the 48-h treatment period. As a result, substantially less KCl was used in the upper portion of the creek relative to the lower portion. A large volume of water was likely stored in pools between dosing stations. These pools diluted treated water during downstream transport. Therefore, lower doses of KCl used at the first four stations were diluted as treated water flowed through large pools. Substantially greater amounts of KCl were used at stations 5–8 because of higher flows and water volume below the confluence with the east prong of the creek. Through this reach, pools represented a smaller percentage of total water volume relative to the upper reach (stations 1–4).

Post-treatment evaluations were conducted on September 30 and October 1. More than 40 zebra mussels were found in the upper portion of the creek in pools downstream of dosing stations. Of these, 38 were alive. Therefore, the treatment was not effective in the upper portion of the creek. No mussels were found downstream of the confluence of the east and west prongs of the creek.

Treatment 2: October 6–8, 2010

During September treatments, three Hydrolabs were used to monitor conductivity as a surrogate for KCl concentrations. Conductivity of at least 1000 μmhos/cm indicated KCl concentration in the acceptable range. In the lower section of the creek conductivities, over 1000 μmhos/cm were achieved for at least 48 h, but in the upper section of the creek, conductivities of over 1000 μmhos/cm were not maintained for 48 h. As a result, the second treatment in October of 2010 was focused on the upper section of the creek. The lower section was not retreated since conductivity readings suggested adequate KCl concentrations had been achieved in September, and no zebra mussels were found in that section of the creek. In October, a modified treatment plan was incorporated that utilized two rhodamine dye tests in combination with physical measurements of creek dimensions to estimate the volume of water for the 13.7–km upper reach. Approximately 6280 kg KCl was used to treat the upper reach with a target concentration of 175 mg/L KCl (Table 20.3).

During treatment 2, six dosing stations were used to treat the target reach (Figure 20.4). KCl was applied over a 48-h period at each dosing station (Table 20.3).

Post-treatment evaluations were conducted on October 19 and November 5. Five zebra mussels were found dead and unattached in a pool downstream of station 2. However, both live and dead mussels were found farther upstream and downstream of station 2. Treatment efforts may have been partially effective because dead zebra mussels were found in the treated area after each treatment. However, mortality

TABLE 20.3

KCl Applied at Each Site during Treatment 2

Site	1	2	3	4	5	6	Total
KCl (kg)	1134	1134	1088	975	975	975	6281

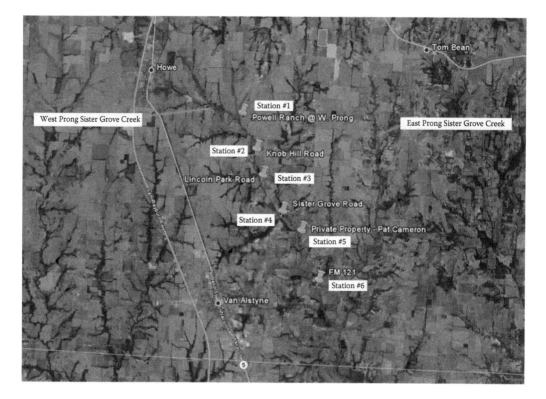

FIGURE 20.4 Dosing stations on Sister Grove Creek during October 2010 KCl treatments.

from other causes could not be ruled out completely. Although KCl treatments did not appear to be as effective as initially hoped, zebra mussels are no longer found in Sister Grove Creek. Shortly after zebra mussels were discovered in Lake Texoma NTMWD discontinued pumping water into West Prong Sister Grove Creek. Since that time no young mussels have been observed and older mussels have died. The last few surveys of the creek have revealed no live zebra mussels.

Monitoring

In 2011, TPWD began working with Dr. Robert McMahon (University of Texas-Arlington [UTA]) to monitor various lakes thought to be at high risk for infestation (either because of proximity to infested lakes or other factors such as high use), to determine both the current status and the likelihood/risk of zebra mussel infestation. Monitoring efforts included analysis of plankton samples and mussel-settlement monitors. Basic physicochemical characteristics were used to determine potential of infestation (Table 20.4). Water bodies were characterized as suitable, unsuitable, or marginal for zebra mussel colonization (Table 20.5). DNA analysis of water samples was added to the monitoring program in 2012.

In 2013, TPWD funded zebra mussel monitoring efforts through UTA. Twenty-three water bodies were monitored during early summer. Monitored lakes include Arlington, Athens, Benbrook, Bridgeport, Cedar Creek, Conroe, Eagle Mountain, Fork, Granbury, Grapevine, Joe Pool, Lavon, Lewisville, Livingston, Marine Creek, Palestine, Possum Kingdom, Purtis Creek, Ray Hubbard, Ray Roberts, Tawakoni, Texoma, and Worth. In 2014, monitoring efforts were expanded to include the following lakes: Austin, Buchanan, Canyon, Georgetown, Graham, Granger, Houston, Hubbard Creek, Inks, LBJ, Richland Chambers, Somerville, Stillhouse Hollow, Travis, Waco, Weatherford, Wheeler Branch, and Whitney. Lakes Bridgeport, Marine Creek, Palestine, Possum Kingdom, Purtis Creek, Ray Roberts, and Texoma will not be monitored during 2014. UTA is also conducting research on Lake Belton where zebra mussels have been found.

TABLE 20.4

Generalized Water Quality Requirements for Zebra Mussels

	Optimum	Minimum	Maximum
Temperature (°F)	62–77	32	88
Spawning Temperature (°F)	≥16	12	25
pH	7.0–8.7	6.6	9.0–9.4
Calcium (Ca^{2+}) (mg/L)	>25	3.6	
Alkalinity ($CaCO^3$) (mg/L)	>50	4.7	
Salinity (ppt OR mg/L)	0–1		12–14

Sources: Johnson, P.D. and McMahon, R.F., *Can. J. Fish. Aquat. Sci.*, 55, 1564, 1998; de Kozlowski, S. et al., Zebra mussels in South Carolina: The potential risk of infestation. South Carolina Zebra Mussel Task Force, 2002; Whittier, T.R. et al., *Front. Ecol. Environ.*, 6(4), 180, 2008.

TABLE 20.5

Risk Analysis Matrix

Lake	O_2	pH	Ca	Temperature	Risk Level
Arrowhead Lake	Suitable	Suitable	Suitable	Suitable	High
Lake Bridgeport	Suitable	Suitable	Suitable	Marginal	Marginal
Eagle Mountain Lake	Suitable	Suitable	Suitable	Suitable	High
Lake Lewisville	Suitable	Suitable	Suitable	Suitable	High
Lake Texoma	Suitable	Suitable	Suitable	Suitable	High
Lake Ray Roberts	Suitable	Suitable	Suitable	Suitable	High
Lake Ray Hubbard	Suitable	Suitable	Suitable	Suitable	High
Lake Lavon	Suitable	Suitable	Suitable	Suitable	High
Lake Tawakoni	Suitable	Suitable	Suitable	Suitable	High
Lake Fork	Suitable	Suitable	Marginal	Suitable	Marginal
Lake Bob Sandlin	Suitable	Suitable	Unsuitable	Unsuitable	Poor
Lake O' the Pines	Suitable	Suitable	Unsuitable	Unsuitable	Poor
Lake Wright Patman	Suitable	Suitable	Suitable	Suitable	High
Caddo Lake	Suitable	Marginal	Unsuitable	Suitable	Poor

Regulations

Zebra mussels were placed on the TPWD list of *Harmful or Potentially Harmful Exotic Fish, Shellfish, and Aquatic Plants* in 1990, long before they were found in Texas. It is an offense for any person to "release into the water of this state, import, sell, purchase, transport, propagate, or possess any species, hybrid of a species, subspecies, eggs, seeds, or any part of any species defined as a harmful or potentially harmful exotic fish, shellfish, or aquatic plant." However, zebra mussels were problematic for law enforcement personnel because one cannot determine if a person is in possession of a zebra mussel veliger without microscopic examination of water in live wells, bilges, other containers, or water systems. Therefore, in March 2012, the Texas Parks and Wildlife (TPW) Commission amended TPWD's regulations to require boats that are operated on Lake Texoma, Lake Lavon, or the Red River be drained (including live wells and bilges) before leaving the water bodies. The regulation change, which took effect in May 2012, was designed to slow the spread of zebra mussels by reducing the potential for overland transport of veligers. In July 2012, the TPWD Executive Director, Carter Smith, signed an emergency order that added Lake Ray Roberts and Lake Lewisville to the existing regulation. In November 2012, the TPW Commission made the emergency order a permanent regulation.

An emergency order was signed in July 2013 that added the West fork of the Trinity River as well as Lakes Bridgeport, Eagle Mountain, and Worth to the list of water bodies covered under the special regulation.

During 2013, the Texas Legislature (HB 1241) granted the TPW Commission authority to adopt rules requiring a person leaving or approaching public water to drain from the vessel or from portable containers on board the vessel any water that was collected from public water (does not apply to salt water). Under the new legislation, authorized TPWD employees may inspect vessels and portable containers for water. In November 2013, the TPW Commission adopted rules requiring anyone leaving or approaching public waters in 17 North Texas counties to drain their boats to prevent the further spread of invasive zebra mussels. In January 2014, the number of counties affected was increased to 30, and in May 2014 the TPWD Commission made the regulations statewide. Statewide regulations went into effect July 1, 2014.

Public Awareness and Outreach

In summer 2011, TPWD and a number of partners including river authorities, municipalities, federal and state agencies initiated a zebra mussel public awareness campaign. What follows is a list of various outlets and promotional materials that were developed:

1. Oversized postcards mailed to 220,000 registered boaters.

2. E-mails sent to 90,000 registered boaters.

3. Seven billboards on key thoroughfares near Lake Texoma.
4. One-hundred-thirty stencils painted on boat ramps and boardwalks on Lake Texoma.
5. Six gas stations participated in *station domination* (cooler door clings, floor stickers, standees, counter mats, and ice bin clings).
6. Twenty-four gas stations near Texoma displayed pump toppers.
7. Twenty-five display banners were distributed to marinas and boat dealers.

Additionally, various media outlets were contacted and promotional materials were distributed including

1. Online ads (Accu-weather, Google, and Facebook)
2. Print ads in TPW magazine, Texas Monthly, and TPWD Outdoor Annual
3. Three 1-min radio features with 72 contracted stations
4. 2000 color posters (11″ × 17″)
5. 600,000 wallet cards (with a grant from APHIS)

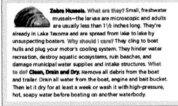

6. 75,000 brochures developed and printed
7. 37 buoys for marinas on Lake Texoma

In May 2012, TPWD and a coalition of partners relaunched the zebra mussel public awareness campaign. Initiation of the campaign coincided with the start of the 2012 boating season and the implementation of new regulations designed to slow the spread of zebra mussels. The theme of the campaign was "Clean, Drain, and Dry" and the goal was to stop the spread of zebra mussels and other invasive species.

Current Status

In Texas, zebra mussels have been confirmed in Lake Texoma, Lake Ray Roberts, Lake Lewisville, Lake Lavon, Lake Bridgeport, and Lake Belton.

REFERENCES

Boelman, S.F., F.M. Neilson, E.A. Dardeau, Jr., and T. Cross. 1997. *Zebra Mussel (*Dreissena polymorpha*) Control Handbook for Facility Operators*, 1st edn. U.S. Army Corps of Engineers, Waterways Experiment Station. Vicksburg, MS.

de Kozlowski, S., C. Page, and J. Whetstone. 2002. Zebra mussels in South Carolina: The potential risk of infestation. South Carolina Zebra Mussel Task Force, South Carolina Department of Natural Resources, Columbia, SC.

Fisher, S.A., S.W. Fisher, and K.R. Polizotto. 1993. Field tests of the molluscistatic activity of potassium to zebra mussel veligers. Report of EPRI, *Third Annual Zebra Mussel Research Conference*, Toronto, Ontario, Canada.

Fisher, S.W., P.C. Stromberg, K.A. Bruner, and L.D. Boulet. 1991. The molluscicidal activity of potassium to the zebra mussel, *Dreissena polymorpha*: Toxicity and mode of action. *Aquatic Toxicology* 20:219–234.

Johnson, P.D. and R.F. McMahon. 1998. Effects of temperature and chronic hypoxia on survivorship of the zebra mussel (*Dreissena polymorpha*) and Asian clam (*Corbicula fluminea*). *Canadian Journal of Fisheries and Aquatic Science* 55:1564–1572.

Kennedy, A.J., R.N. Millward, J.A. Steevens, J.W. Lynn, and K.D. Perry. 2006. Relative sensitivity of zebra mussel (*Dreissena polymorpha*) life-stages to two copper sources. *Journal of Great Lakes Research* 32:596–606.

Stubbs, D. 2012. Section III: Pesticide options for exotic mussel control. In *A Review of the State of Idaho Dreissenid Mussel Prevention and Contingency Plants*, Idaho Department of Agriculture, Aquatic Ecosystem Restoration Foundation, Pacific States Marine Fisheries Commission, Portland, OR.

Utah Division of Wildlife Resources. 2007. http://wildlife.utah.gov/quagga/pdf/boat_inspection.pdf, accessed June 17, 2014.

Virginia Department of Game and Inland Fisheries. 2013. http://www.dgif.virginia.gov/zebramussels, accessed June 17, 2014.

Whittier, T.R., P.L. Ringold, A.T. Herlihy, and S.M. Pierson. 2008. A calcium-based invasion risk assessment for zebra and quagga mussels (*Dreissena* spp.). *Frontiers in Ecology and the Environment* 6(4):180–184.

21

Proactive Approach to the Prevention of Aquatic Invasive Species in Wyoming through Outreach and Watercraft Inspections

Beth A. Bear and Sherril R. Rahe

CONTENTS

ABSTRACT   The Wyoming aquatic invasive species (AIS) program started in 2010 with a focus on legislation, watercraft inspections, and outreach. Legislation has been instrumental in development of a program and in interdiction authority to facilitate the watercraft inspection portion of the program. Legislation continues to help prevent the spread of AIS to Wyoming through mandatory inspections of watercraft entering the state. Since the start of the program, over 200,000 inspections have been conducted and 36 watercrafts with attached dreissenid mussels have been intercepted. Outreach remains the foundation for the program with the goal of educating all water users to drain, clean, and dry their watercraft and equipment to prevent the spread of AIS. This multifaceted approach has so far succeeded in preventing new species from being introduced to Wyoming waters and in slowing the spread of existing species. The program continues to change and adapt as the threat of new AIS appear on the horizon.

Introduction

By taking an aggressive, proactive approach, Wyoming seeks to minimize the risk of new aquatic invasive species (AIS) introductions, such as zebra/quagga mussels, and of spreading AIS that currently exist in the state. Prevention is the best approach to managing AIS and has been shown to be more effective and affordable than dealing with zebra/quagga mussels after they become established (Lee et al. 2007). Over the last

5 years, several western states have launched prevention programs that have succeeded in significantly stopping or slowing the spread of invasive mussels. As such, the Wyoming AIS program was implemented in 2010 to prevent the spread of AIS to Wyoming waters. Three major aspects of the program that will be discussed further include legislation, watercraft inspections, and public outreach and education.

Prior to the formation of the AIS program, limited outreach and watercraft inspections were conducted throughout the state. The Wyoming Game and Fish Department (WGFD) conducted outreach for several years aimed at preventing the illegal introduction or movement of fishes. In 2006, an outreach plan was developed to better educate the public on the issue and threats that AIS pose to Wyoming's waters. Information brochures were distributed to educate anglers and others on the negative impacts of illegal fish stocking. In 2008, the 100th Meridian Initiative funded boater surveys at eight waters in Wyoming that set the stage for the one-on-one education of boaters at ramps that would later develop into a watercraft inspection program.

In 2009, the WGFD determined that because of the increasing presence of zebra/quagga mussels in close proximity to Wyoming (notably Lake Mead, NV, and Pueblo Reservoir, CO), a coordinated, fully funded program aimed at the prevention of these and other AIS to Wyoming was necessary. The WGFD worked with the Wyoming Legislature through 2009 on draft legislation to develop such a program. In 2009, an initial AIS brochure titled "Don't Move a Mussel" was developed and outreach to affected stakeholders began on a larger scale.

After legislation was passed in 2010 and a formal program was formed, the Wyoming Aquatic Invasive Species Management Plan (WGFD 2010) was developed to provide program goals and direction. The plan is the guiding document that serves as a road map for AIS activities in the state. The plan was signed by Wyoming's Governor and approved by the federal Aquatic Nuisance Species Task Force (ANSTF) in November 2010. The plan will be reviewed and updated every 5 years to remain current with changing AIS prevention practices. The plan also enables Wyoming to be eligible for annual ANSTF funding distributed by the US Fish and Wildlife Service.

Specific objectives of the plan and program are

1. To coordinate and implement a comprehensive management program
2. To prevent the introduction of new AIS into Wyoming
3. To detect, monitor, and eradicate AIS in Wyoming
4. To control and eradicate established AIS that have significant impacts on Wyoming waters
5. To educate resource user groups about the risks and impacts of AIS and how to reduce their harmful impacts
6. To support research on AIS in Wyoming and develop efficient systems to disseminate information to research and management communities

---

Legislation

Prior to 2010, legislation pertaining to invasive species consisted of Wyoming Game and Fish Commission (WGFC) regulation Chapter 10 (Regulation for importation, possession, confinement, transportation, sale and disposition of live wildlife) implemented by the WGFD. This regulation prohibits possession of rusty crayfish (*Orconectes rusticus*), New Zealand mudsnail (*Potamopyrgus antipodarum*), and zebra mussel (*Dreissena polymorpha*). The regulation does not address the increasing threat of quagga mussels or numerous other AIS on the horizon. The ability to stop a watercraft to conduct an inspection for AIS (interdiction authority) was also lacking, as was funding for prevention activities.

The Wyoming AIS Act (W.S. 23-4-201-206) was passed by the Wyoming legislature in March 2010 and includes a definition of *AIS*, interdiction authority for watercraft inspections, personnel and funding for the program, and a boater decal fee to partially fund the program. Subsequently, the WGFC developed an AIS regulation (WGFC Chapter 62 Regulation for AIS) to implement the requirements of the new state statute. The regulation includes a detailed list of what species are considered AIS, guidelines for watercraft inspection and decontamination, and the fee structure for the decal.

A 2011 amendment to the AIS act allows for a reciprocity agreement between Wyoming and adjoining states for the purpose of honoring the AIS program fees (decal or sticker) of another state. The only other state that both adjoins Wyoming and currently implements a boater fee for AIS is Idaho, and there is currently no reciprocity agreement in place between the two states.

In 2012, an additional amendment to the AIS statute mandates that all conveyances (motor vehicle, boat, watercraft, raft, vessel, trailer, or any associated equipment or containers, including but not limited to live wells, ballast tanks, bilge areas, and water hauling equipment that may contain or carry an AIS) entering Wyoming by land be inspected for AIS before contacting or entering a Wyoming water. The AIS regulation (Chapter 62) was revised to implement the new watercraft inspection requirement. Guidelines are set in regulation to require this mandatory inspection from March through November each year. At all times of the year, if watercraft has been in a water positive for zebra/quagga mussels within 30 days, it must be inspected prior to launch in Wyoming. Check stations may be placed at Wyoming Department of Transportation port of entries (POE; weigh stations) and rest areas near borders to enable boaters to easily obtain an inspection when entering the state.

Watercraft Inspections

A central idea of the AIS program is the need for consistent and effective protocols for inspecting and decontaminating various types of watercraft that could be transporting AIS. The Pacific States Marine Fisheries Commission was charged with providing trainings and Uniform Minimum Protocols for agencies in the west (Zook and Phillips 2012). Thanks to these protocols and other states that have developed watercraft inspection and decontamination programs, Wyoming moved forward quickly in developing a watercraft inspection and decontamination protocol in 2010. Since the program's inception, there have been many learning opportunities and new information from emerging research, and as a result the program makes annual modifications to protocols (WGFD 2014).

Inspector Certification and Training

As required by state statute, all inspections must be completed by an authorized AIS inspector. The program instituted a training program in 2010 and has since conducted over 40 training courses throughout Wyoming, certifying over 500 inspectors. Inspectors represent state agencies (WY Game and Fish, WY State Parks, WY Department of Agriculture, WY Department of Transportation), federal agencies (National Park Service, US Forest Service), city and county agencies, nonprofit groups, and private entities (Figure 21.1). Training courses are necessary to implement the program and provide an opportunity for interested individuals to be well educated on the subject. Many participants who become an authorized inspector may never do an inspection, yet they are still able to be advocates for the program and provide education to others.

Establishing Inspection Locations

Currently, in Wyoming, there are two types of inspection locations, mandatory check stations and certified inspection locations. A mandatory check station is a location established by WGFD where stopping is mandatory for all watercraft that pass the check station on their route of travel. Certified Inspection Locations are locations where an AIS Inspector is available to conduct inspections and do not require mandatory stops. Inspections can take place at any location where an AIS inspector is available provided the location is high and dry and there is no chance that contaminated water or AIS could come in contact with a water source as a result of the inspection.

From 2010 through 2012, watercraft inspections were primarily conducted at boat ramps and other entrances to major waters in the state. Due to the 2012 statutory change requiring inspection of all conveyances entering the state, watercraft inspections were shifted away from waters and are now conducted at major routes of entry into the state (port of entries, rest areas; Figure 21.2). The program also certifies individuals and their agency or business as a certified inspection location once an individual completes

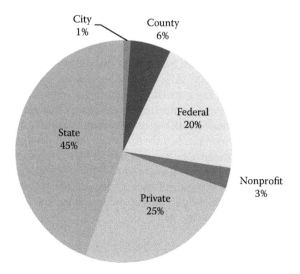

FIGURE 21.1   Association of authorized aquatic invasive species inspectors in Wyoming.

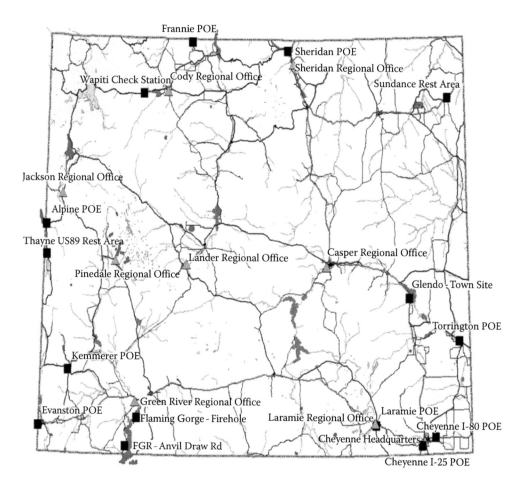

FIGURE 21.2   Watercraft inspection locations including mandatory check stations (square) and WGFD offices (triangle). Over 50 additional private locations are available to conduct inspections but are not indicated on the map as locations may vary frequently.

the training course. This greatly assists the program in providing additional inspection locations for individuals not intercepted at mandatory check stations but still required to have their watercraft inspected prior to launch. In addition to WGFD offices, over 50 certified inspection locations or private authorized inspectors are available statewide to conduct an inspection. Private entities are allowed to charge an inspection fee if they choose to; that aspect is not regulated by the WGFD or other entity.

Inspection and Decontamination Protocols

A training manual was developed in 2010 to explain protocols for the inspection and decontamination of watercraft. This manual is updated annually using current science about invasive mussel survival and watercraft risk (WGFD 2014). The training manual ensures all inspectors are operating under the same set of consistent protocols.

Watercraft inspections include a boat owner or operator interview and a visual and tactile inspection of all parts of the conveyance that may hold or have come in contact with raw water. The interview is focused on determining the watercrafts' previous use, mainly whether it has been used in a water known or suspected to be positive for dreissenid mussels (termed *infested* or *positive* water). Due to the high number of mussel-positive waters in some states and the rapid expansion of zebra/quagga mussels, all waters east of the 100th Meridian are considered positive waters for the purpose of watercraft inspections. Other waters in western states are listed specifically in the inspection manual. The WGFD maintains a list of these waters in the training manual and makes this information available to the public on the Department webpage. The list of mussel positive waters is generated from information in the USGS Nonindigenous Aquatic Species Database and from AIS program coordinators in western states. Based on the watercrafts last use, a standard inspection or high-risk inspection will be conducted.

A standard inspection is a quick inspection of the exterior and interior of the watercraft to ensure the watercraft is drain, clean, and dry. Some of the procedures include lowering the motor of an outboard or inboard/outboard watercraft, removing all drain plugs, and verifying all compartments (storage, live wells, ballast tank) do not contain any standing water. If standing water is found, it is removed by draining completely or removing manually with a sponge, towel, or hand pump. If standing water cannot be drained and presents a high risk, the area is decontaminated by flushing with hot water. Any standing water found in a watercraft used in a water positive for mussels within the last 30 days is considered high risk and requires a mandatory decontamination of that water and associated compartments. Any standing water from watercraft used on waters of a state with known positive waters within the last 30 days requires a high-risk inspection.

A high-risk inspection is a more thorough inspection using both visual and tactile inspection methods. In addition to checking interior compartments, all equipment that may have been used in or exposed to water would require examination. Previous use in a positive water in the last 30 days automatically requires a high risk inspection. Other factors that may trigger a high-risk inspection are standing water from a state with known positive waters, the presence of suspected AIS, and large and complex watercraft.

There are several types of decontaminations performed on watercraft; standing water, motor flush, plant, and full decontamination. A standing water decontamination is advised if water cannot be removed using manual methods (sponge, bilge pump), the watercraft was used within 30 days (increasing the likelihood that veligers are still viable), the water had never been heated to over 90° (effective temperature to kill veligers), or the water is not dirty or oily (making the water inhospitable for veliger survival). Standing water decontamination includes flushing of any internal compartments containing high-risk standing water at 120°F for 2 min.

A motor flush is required when a watercraft was used on a positive water within 30 days, regardless of motor type. Other reasons for performing a motor flush include if the water was used in a state with positive waters and cannot be fully drained from the motor, as is the case with inboard/outboard and inboard motors. A motor flush is conducted by flushing 140°F water through the engine until the water exiting the motor reaches 140°F, indicating the hot water has been circulated throughout the motor system.

A plant decontamination is required if any vegetation on a watercraft cannot be physically removed. For example, vegetation trapped between the watercraft and trailer or wrapped around the propeller or equipment. In these instances, the area containing the plant is flushed with 140°F water for 2 min to kill the vegetation.

A full decontamination is required if zebra/quagga mussels or suspected mussels are found during an inspection. Notification via the state AIS hotline is made at the time of detection and samples are collected and sent to a lab for verification. Full decontamination protocol includes using 140°F water at high pressure (2500 psi) on all exterior surfaces, motor and equipment, and 120°F on all interior compartments including the bilge, storage compartments, live wells, and ballasts.

If zebra/quagga mussels are found on a watercraft and they appear live or their viability is unknown, the watercraft is quarantined to allow additional desiccation to kill any mussels not killed during full decontamination. The 100th Meridian Quarantine Estimator is used to determine the length of quarantine time based on temperature and humidity needed to desiccate mussels.

Lessons Learned

Other Vectors of Spread

AIS can be spread to new waters on watercraft and also on a variety of other vectors. Wyoming statute requires inspection of all conveyances entering the state including all types of watercraft and other equipment that may contain water. Many of these conveyances such as fire equipment, water hauling trucks, and other types of water transportation systems may have a similar level of risk as watercraft of transporting AIS. These systems are often complex and fully draining or flushing with hot water may not be feasible. Some protocols do exist for the treatment of these conveyances to prevent the spread of AIS such as those developed by the US Forest Service for firefighting equipment (USFS 2013). Many fire agency personnel are authorized AIS inspectors in Wyoming to enable the inspection of any firefighting equipment entering the state. Future training programs tailored toward the inspection and decontamination of these other conveyances will help prevent the movement of AIS through these vectors.

Quality Control

In order for inspections to be effective, inspectors must follow protocol and the boating public must comply with and answer honestly regarding their watercraft's last waters of use as this information is critical to conducting an appropriate inspection. To ensure all watercraft receive necessary scrutiny, a standard inspection is conducted on all watercraft to confirm that they are drained, clean, and dry. With respect to quality control of inspections, the program primarily receives feedback on inspections from word of mouth (boater feedback) and staff visits to check stations. As these methods are subjective, in 2013 the program began contracting with a third party to conduct anonymous quality assurance checks at watercraft check stations. These quality assurance checks follow standard evaluation criteria so each check station is graded on the same set of standards. All WGFD operated check stations and select certified inspection locations are evaluated annually to ensure correct and consistent inspections are conducted throughout the state.

Summary

From 2010 through 2014, over 200,000 watercraft inspections were conducted in Wyoming. Of these, 3,763 have been high-risk inspections and 1,578 have resulted in some type of decontamination. Thirty-six watercrafts with dreissenid mussels attached have been intercepted since 2010. On all but one watercraft, the mussels were determined dead at the time of inspection and the watercrafts were decontaminated and allowed to proceed. The watercraft with live mussels was decontaminated and quarantined before being allowed to launch in Wyoming. Boaters have traveled to over 1400 different waters before boating in Wyoming and have been registered in every U.S. state and Canada. Inspectors have contacted over 45,000 individual boaters during inspections and have showed boaters how to inspect their own watercraft and have educated them on the AIS issue and prevention tips of drain, clean, and dry.

Watercraft and equipment inspection and decontamination is an effective method of preventing the spread of AIS. As of December 2014, the state of Wyoming has not detected mussels in any of the 63 waters sampled for dreissenid veligers (for a water to be classified as positive for dreissenid mussels, juvenile or adult mussels must be collected during two or more sampling events and identified with DNA confirmation of the sample, or evidence of veligers is detected via microscopy and by two independent PCR methods during two or more sampling events). In addition, watercraft with high-risk standing water and watercraft containing dreissenid mussels are intercepted and decontaminated as a result of the program. The slow spread of invasive mussels in the west, particularly in states that have instituted watercraft and equipment inspection and decontamination programs, is evidence that these programs are effective at slowing the spread of zebra/quagga mussels to new waters.

Outreach

Developing a Message

Arguably the most important component of an effective AIS program is education and outreach. At the onset, the primary outreach message of the program was "Don't Move a Mussel." This slogan was already in use by several other entities and had some recognition for boaters. The intent of the message was to educate boaters on the impacts of the most damaging AIS in the west, zebra and quagga mussels. The campaign proved successful and boaters began to recognize the slogan and associate it with AIS.

However, many water users such as wading anglers, kayakers, and other nonmotorized users did not relate to this message. Admittedly these vectors are less likely pathways of spread for dreissenid mussels and are more likely to spread New Zealand mudsnail, Asian clam, and invasive plants. To reach a broader audience, the outreach message was changed in 2012, to "Drain, Clean, Dry" to incorporate other vectors of AIS spread. Six outreach brochures were developed aimed at these diverse vectors and included a brochure on general AIS information for all water users, motorized watercraft, nonmotorized watercraft, wading anglers, waterfowl hunters, and water transportation systems (Figure 21.3).

The program uses several strategies to get the AIS message out to water users including newspaper, radio, internet, direct mailing, billboards, highway messaging systems, fishing and watercraft regulation booklets, brochures, posters, etc. Information from a recent boater survey suggests that the most effective outreach method is printed material including brochures and fishing regulations (Bear 2013). Boaters also frequently receive AIS information through one-on-one contact with a WGFD employee that is achieved primarily through watercraft inspections. Outreach to anglers is achieved through presentations to groups such as Trout Unlimited and Walleyes Unlimited and at fishing tournament meetings. Presentations are given to civic groups and booths are setup at fairs and local events around the state to ensure all water users are aware of AIS.

FIGURE 21.3 Outreach brochures developed to target a variety of water users with the "Drain, Clean, Dry" message.

Boater Surveys

Success of educational programs can be analyzed through data on boater behavior and through the use of public opinion surveys. Data collected from inspections suggests boaters are increasingly following drain, clean, and dry guidelines to prevent the spread of AIS. During an inspection, inspectors document whether a boat contains any standing water. In 2010, 9% of boats contained standing water. In 2011, 5% of boats contained standing water, which decreased to less than 1% of boats in 2012. This indicates that boaters are more often following guidelines regarding AIS and are launching with boats drained and dry.

In 2012, a boater survey was conducted on a sample of boaters that purchased the AIS decal the preceding year. The survey was conducted to determine boater's awareness of AIS and experiences at watercraft inspection stations (Bear 2013). Boaters were asked how familiar they were with AIS and 91.4% of boaters responded they were very or somewhat familiar with AIS. In regard to following prevention guidelines, 89.3% of boaters responded that they drained water from their watercraft after each use, with 89.5% of boaters indicating they removed plant, mud, and debris from their watercraft regularly.

Summary

Outreach has been shown to be an effective method for preventing the spread of AIS. In Wyoming, outreach has taken many forms and has so far been successful in increasing awareness among boaters and all water users. The goal of the program in the future will be to continue to change awareness of the issue into prevention action on the part of all users.

Conclusion

By taking a proactive approach, Wyoming continues to be effective at preventing the spread of zebra/quagga mussel to our waters. Legislation enabled the development of a program to implement prevention activities and set forth requirements to limit the movement of water and AIS into the state. By inspecting watercraft potentially transporting AIS, the risk of these species being introduced into the state is greatly reduced. Outreach and education results in a more aware public and more boaters doing their part to protect their waters from harmful AIS by ensuring their boats are drained, clean, and dry.

While the program has been successful in these areas over the last 4 years, there are still many challenges on the horizon and lessons to be learned. Whereas the march of zebra/quagga mussels westward has slowed, invasive mussels have recently been detected in new waters in close proximity to Wyoming. In addition, increased surveying has found other invasive species in Wyoming (Asian clam, curly pondweed) and the challenge will be to prevent their spread to other waters of the state. As the threat of emerging invasive species continues, the program works to remain proactive and adapt to changing species and challenges to protect Wyoming waters.

REFERENCES

Bear, B. A. 2013. Summary of 2012 Wyoming aquatic invasive species boater survey responses. Wyoming Game and Fish Department Administrative Report. Cheyenne, WY.

Lee, D. J., D. C. Adams, and F. J. Rossi. 2007. The economic impact of zebra mussel in Florida. EDIA document FE693. Food and Resources Department, University of Florida. Gainesville, FL.

USFS (United States Forest Service). 2013. Preventing spread of aquatic invasive organisms common to the intermountain region: Operational guidelines for 2013 fire activities. USFS Intermountain Region. Odgen, UT.

WGFD (Wyoming Game and Fish Department). 2010. Wyoming aquatic invasive species management plan. Wyoming Game and Fish Department. Cheyenne, WY.

Wyoming Game and Fish Department. 2014. State of Wyoming aquatic invasive species watercraft inspection and decontamination manual. Wyoming Game and Fish Department. Cheyenne, WY.

Zook and Phillips. 2012. Uniform minimum protocols and standards for watercraft interception programs for dreissenid mussels in the western United States. Pacific States Marine Fisheries Commission Report. Portland, OR.

22

Lake Tahoe Aquatic Invasive Species Program: Successes and Lessons Learned

Steve Chilton

CONTENTS

ABSTRACT   Lake Tahoe is designated a water of extraordinary ecological or aesthetic value. The ongoing decline in Lake Tahoe's water quality is facilitating colonization and expansion of aquatic invasive species (AIS). Action plans set in place provide guidance that delineates appropriate, science-based regulation and monitoring that expressly deals with prevention and management of AIS. Strategic control plans include processes such as subjecting all watercraft inspection and/or decontamination prior to launching in Lake Tahoe. Prevention plans include implementation of legislation, increased certification demand, and increasing awareness and training. Many facets of control and prevention have been implemented and analyzed. Lessons learned show a need for expanded outreach, public relations, and better development of decontamination procedures and data analysis. The effort has many substantive accomplishments to its credit and plans for many more.

Introduction

The Lake Tahoe Aquatic Invasive Species Program has evolved over many years from a relatively minor control approach to voluntary watercraft inspections and to currently a robust program containing outreach, early detection monitoring, rapid response, prevention, and control facets. This has been possible in large part due to collaborative partnerships and passion that Lake Tahoe fosters through the many environmental programs that strive to protect it.

Lake Tahoe is designated an Outstanding National Resource Water (ONRW) under the Clean Water Act (CWA Section 106) due to its extraordinary clarity. However, substantial changes to the Lake Tahoe Region's economy, pristine water quality, aesthetic value, and recreational pursuits are occurring. This is partly due to the harmful impacts of nonnative aquatic plants, fish, invertebrates, and other invaders. These nonnative aquatic organisms are considered aquatic invasive species (AIS) when they threaten the diversity or abundance of native species or the ecological stability of infested waters or commercial, agricultural, aquacultural, or recreational activities dependent upon such waters (NISA 1996; NANPCA90 2000). AIS are commonly spread by activities such as boating, fishing, hatchery releases, and aquarium dumping. The Lake Tahoe Region is not only threatened by new introductions of AIS to Lake Tahoe from other waterbodies, but also the expansion of existing populations within the lake and even as a source of AIS to nearby uninfected waterbodies.

At least 20 nonnative species are established in the Lake Tahoe Region, including aquatic plants, fishes, invertebrates, and an amphibian. As examples, Eurasian watermilfoil (*Myriophyllum spicatum*; an aquatic plant) has been spreading around Lake Tahoe over the last 15–20 years. Curlyleaf pondweed (*Potamogeton crispus*; another aquatic plant) has begun to expand dramatically over the past few years. Beds of Asian clams (*Corbicula fluminea*) are larger and more common than previously known. Populations of warm water fishes such as largemouth bass (*Micropterus salmoides*) and bluegill (*Lepomis macrochirus*) are expanding. Moreover, global climate change has resulted in warmer water temperatures, likely facilitating the establishment of nonnative plants in the nearshore environment and providing increased spawning areas for warm water fishes that compete with desirable species.

The potential economic impact to the Lake Tahoe Region caused by new AIS introductions such as quagga or zebra mussels (*Dreissena rostriformis bugensis* and *Dreissena polymorpha*, respectively) or expanding invasive aquatic plant populations would be substantial. A study found that the combined economic impacts to recreation value, tourism spending, property values, and increased boat/pier maintenance, when evaluated over a 50-year period, was estimated at $417.5 million, with an average annual equivalent value of $22.4 million (US Army Corps of Engineer 2008, but see Tahoefund.org). The largest estimated impacts would be to property values and lost tourism spending, each accounting for 38% of the total estimated AIS damages. Spending for AIS prevention and early eradication produces a higher benefit to cost ratio than postinfestation control programs such that maximum benefits are realized through early and preemptive action.

The 2007 discovery of quagga mussels in the Lower Colorado River—Lakes Mead, Mohave, Havasu, including the Central Arizona Project Aqueduct and Colorado River Aqueduct (LaBounty and Roefer 2007; California Fish and Game Department 2011; Wong and Gerstenberger 2011; Bryan 2014)—prompted rapid cooperation and action by regional, bistate, federal agencies, and nongovernmental organizations in the Lake Tahoe Region. These new threats, coupled with recent studies showing high incidence of Lower Colorado River boat traffic to Lake Tahoe (Environmental Incentives 2013), prompted a tremendous ramping up of education and outreach campaigns. New regulations to prevent accidental introduction of AIS, increased control efforts and research on the biology and distribution of AIS populations occurred, too. Examples of these activities include

- Aquatic Invasive Species Workshop held in April, 2007
- Formation of the Lake Tahoe AIS Working Group (LTAISWG)
- Formation of the Lake Tahoe AIS Coordination Committee (LTAISCC)
- Yearly workshops organized by the LTAISWG to prioritize AIS prevention, monitoring, control, education, and research efforts
- Development and implementation of a Vessel Inspection Program at Lake Tahoe
- Deployment of portable boat washing stations
- Full-time AIS Coordinator hired by US Fish and Wildlife Service (USFWS)
- Increased monitoring for invasive aquatic plants, invertebrates, and warm water fishes
- Use of diver-operated suction and benthic barriers to control invasive aquatic plants

- Evaluation of diver-operated suction and bottom barriers to control Asian clams
- Measurement of warm water fish behavior and diets in and around the Tahoe Keys
- Increased education and outreach activities
- Quagga mussel survivability studies

Geographic Scope: Lake Tahoe Region

The geographic scope of the Lake Tahoe AIS Program encompasses the Lake Tahoe Region (Region) as defined by the Tahoe Regional Planning Agency Compact (TRPA Compact P.L 96-551): The region straddles the California–Nevada border and includes Lake Tahoe and approximately 6 km of the Lower Truckee River below the lake; the adjacent parts of Douglas and Washoe Counties in NV; Carson City, NV; and the adjacent parts of Placer and El Dorado Counties in California. The region drains 63 streams to Lake Tahoe with the Upper Truckee River being the largest. The lake's only outflow, after passing the Lake Tahoe Dam, is the Lower Truckee River at Tahoe City, CA. Beyond the region boundaries, the Truckee River continues to flow approximately 140 mi to its terminus at Pyramid Lake (Murphy et al. 2000). In addition to Lake Tahoe, many smaller lakes and six larger recreation lakes (Fallen Leaf, Echo, and Cascade Lakes in CA; Marlette, Spooner, and Incline Lakes in NV) are located in the region (Figure 22.1).

The majority of the land in the region is owned and managed by public agencies. Approximately 80% of the public lands are managed as the Lake Tahoe Basin Management Unit (USDA-USFS-LTBMU) by the US Department of Agriculture's US Forest Service. There are nine state parks on the California side managed by California Department of Parks and Recreation (CADPR) and the Lake Tahoe Nevada State Park managed by Nevada Division of State Parks (NDSP) on the Nevada side. Also in the Region, the California Tahoe Conservancy (CTC) owns large and small land parcels and the Nevada Division of State Lands (NDSL) owns and manages approximately 500 urban parcels. Most of the private lands are commercially held with most development being in the low-lying areas near Lake Tahoe. The TRPA directs land use and development issues in the Region.

South Lake Tahoe, the only incorporated city in the Tahoe Basin, occupies the south shore of the lake. With respect to AIS, of note is the Tahoe Keys, also on the south shore. The Keys, as it is commonly referred to, is a residential development that includes two marinas. The residential marina is in a western channel and the commercial marina is in an eastern channel, referred to as Tahoe Keys West and Tahoe Keys East, respectively. The Tahoe Keys were constructed within the Upper Truckee Marsh in the mid-1960s when water from the Upper Truckee River was channelized and diverted to allow for significant filling of the marsh for residential roads and building sites. The result is that surface water exchange between the Tahoe Keys and the main body of Lake Tahoe is now limited to the two channels. Water in the keys is shallower, turbid, and warmer, providing habitat for numerous AIS.

Lake Tahoe's water clarity (the depth of light penetration) is one of its most striking features. Lake Tahoe is designated an Outstanding National Resource Water (ONRW) under the federal Clean Water Act (CWA 1972). Likewise, Lake Tahoe is designated a "water of extraordinary ecological or aesthetic value" by the Nevada Division of Environmental Protection. Lake Tahoe has a mean depth of 305 m (maximum 501 m), second only in the United States to the depth of Crater Lake (also designated an ONRW) in Oregon.

Regularly recorded Secchi depths (a measurement of water clarity) have occurred in Lake Tahoe since the late 1960s. Since that time, transparency of up to 41 m has been recorded; however, it has declined up to 0.27 m year^{-1} (Jassby et al. 2003) with recent measurements of 21.4 m (TERC 2008), suggesting a shift in the lake's oligotrophic status (Goldman 1974, 1988). The ongoing decline in Lake Tahoe's water clarity is a result of light scatter from fine sediment particles (primarily particles less than 16 µm in diameter) and light absorption by phytoplankton, resulting in an increased shift in the lake's depth of maximum chlorophyll (LRWQCB 2007). The addition of nitrogen and phosphorus to Lake Tahoe contributes to phytoplankton growth. Fine sediment particles are the most dominant pollutant contributing to the impairment of lake's water, accounting for an estimated two-thirds of the impairment. The decline of

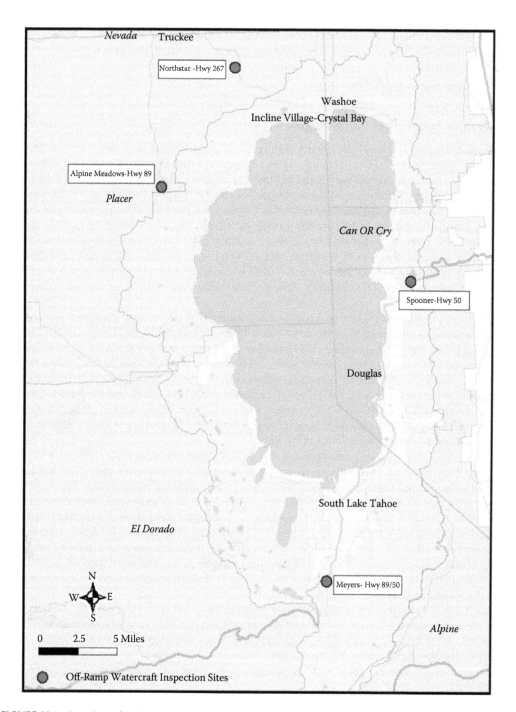

FIGURE 22.1   Locations of on-highway inspection and decontamination stations at Lake Tahoe.

Lake Tahoe's clarity resulted in the listing of Lake Tahoe as impaired for the transparency standard under Section 303(d) of the Clean Water Act. Lake Tahoe's 303(d) listing compelled California and Nevada to develop a total maximum daily load (TMDL). Lake Tahoe's clarity improved in 2012 for the second year in a row, and its waters were the clearest in 10 years, according to University of California, Davis, scientists who study the lake. The 2012 average annual clarity level was 75.3 ft, or a 6.4-ft improvement from 2011(TERC 2012).

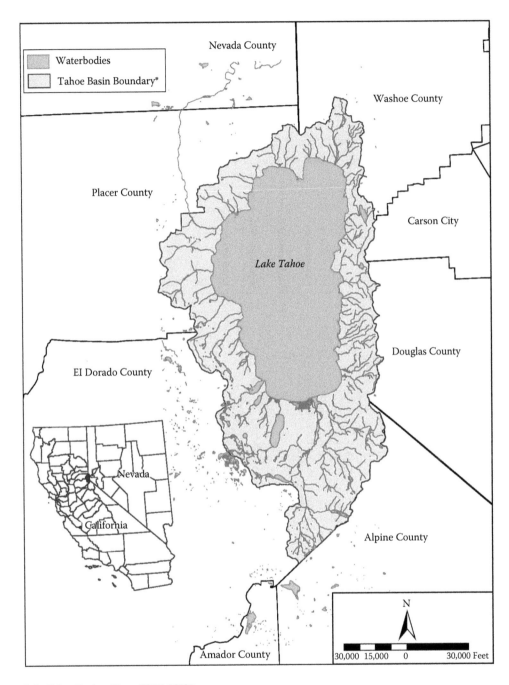

Lake Tahoe Region. (From TRPA 2012.)

Despite its relatively small watershed (812 km²), Lake Tahoe has a surface area of approximately 500 km². This low watershed-to-lake ratio (1.6:1) results in a substantial amount of precipitation falling directly on Lake Tahoe, contributing to its oligotrophic status. It is a subalpine lake (elevation 1897 m) surrounded by mountains over 1200 m above lake level (LRWQCB 2007). Typical surface water temperatures range from 18°C to 21°C during late summer and between 4.5°C and 10°C during the winter. Evidence by Coats et al. (2006), however, strongly suggest increases in the thermal structure of Lake Tahoe, possibly facilitating further colonization and expansion of AIS (UC Davis 2008).

Existing Authorities and Programs

Numerous federal, state, and regional regulations and programs are in place in the Region, to limit the introduction and spread of AIS with no single agency or group responsible for all AIS issues. Table 22.1 lists the various agencies, regulations, and programs associated with AIS in the region. In 2009, the national Aquatic Nuisance Species Task Force approved the Lake Tahoe AIS Management Plan (Plan), a bistate plan (LTAISCC 2009). Governors Schwarzenegger and Gibbons of California and

TABLE 22.1

Federal, State, and Regional Agencies, Regulations, and Programs in the Lake Tahoe Region and Associated AIS Activities

	Control	Coordination	Documentation	Education/Outreach	Eradication	Exportation	Financial Assistance	Importation	Possession	Prevention	Quarantine	Research	Technical Assistance
Federal[a]													
Endangered Species Act of 1973	×				×								
Executive Order 13057		×											
Executive Order 13112		×		×			×			×			×
Lacey Act of 1990 (amended 1998)								×	×				
NANPCA (1990) and NISA (1996)	×	×		×			×			×		×	×
NEPA of 1970			×										
USACE		×					×					×	×
USDA	×	×		×	×		×			×		×	×
USDOI	×	×		×	×		×			×		×	×
State and regional[a]													
CADPR	×	×		×	×					×	×	×	×
CDFA	×	×	×	×	×	×		×	×	×			×
CDFW	×	×		×	×	×		×	×	×	×	×	×
CEQA	×		×		×					×			
CSLC	×	×					×						
CTC				×			×			×			
EIP		×					×					×	
LRWCQB (CRWQCB 2005)	×	×		×			×			×			
LTAISCC	×	×	×	×	×					×		×	×
LTAISWG	×	×		×	×					×		×	×
LTSLT				×				×		×			
NDOW	×			×	×	×		×	×	×	×		×
NDSL		×					×						
NDSP		×		×						×			
Tahoe Area Sierra Club Group				×						×			
Tahoe Science Advisory Group		×											
TKPOA	×												
TRCD	×	×		×	×		×			×			×
TRPA	×	×	×	×	×		×	×	×	×		×	×
TSC		×		×								×	×
UCD—TERC	×			×	×					×		×	×
UNR	×			×						×		×	×

[a] Acronyms listed on Page iv.

Nevada, respectively, also signed the Plan. The Plan guides all aspects of the AIS effort at Lake Tahoe. Management actions must consider the overlapping jurisdictions of the States of California and Nevada as well as the area-wide role of the TRPA.

Federal authority to limit the interstate transport and importation to the United States of prohibited plant species is provided by the Plant Protection Act of 2000, which provides authority to the USDA's Animal and Plant Health Inspection Service for Plant Protection and Quarantine (USDA-APHIS-PPQ). Authority to control prohibited wildlife species is provided to the US Fish and Wildlife Service (USFWS) by the Lacey Act of 1990 as amended in 1998.

In California, the California Department of Fish and Wildlife (CDFW) is responsible for prohibited fish and wildlife resources (CCR, Title 14), and is the lead agency for the California AIS Management Plan (CAISMP). The CAISMP defines invasive species as those

> ...that establish and reproduce rapidly outside of their native range and may threaten the diversity or abundance of native species through competition for resources, predation, parasitism, hybridization with native populations, introduction of pathogens, or physical or chemical alteration of the invaded habitat. Through their impacts on natural ecosystems, agricultural and other developed lands, water delivery and flood protection systems, invasive species may also negatively affect human health and/or the economy.

The purpose of the CAISMP is "to coordinate state programs, create a statewide decision-making structure and provide a shared baseline of data and agreed-upon actions so that state agencies may work together more efficiently." The CAISMP addresses numerous AIS presently established in or threatening introduction to aquatic ecosystems throughout the state. Waterbody types addressed include creeks, wetlands, rivers, bays, and coastal water habitats (CDFW 2008). The CAISMP describes vectors of concern on a statewide-scale including commercial shipping and fishing, recreational equipment and activities, trade in live organisms (e.g., aquarium trade), construction in aquatic environments, and water delivery and diversion systems (CDFW 2008).

The California Department of Fish and Wildlife (CDFW) Code §2301 allows CDFW designated staff (and other authorized state authorities, i.e., CADPR peace officers and California Department of Food and Agriculture [CDFA]) to inspect, impound, or quarantine any conveyance (e.g., watercraft) that may carry dreissenid mussels (i.e., quagga and zebra mussels). CDFA is the lead agency for regulatory activities associated with noxious weeds (CAC Title 3, Sec. 3400). Also in California, the Lahontan Region Water Quality Control Board (LRWQCB) is responsible for region-wide water quality objectives as outlined in the *Water Quality Control Plan for the Lahontan Region North and South Basins* (commonly referred to as the Basin Plan; CRWQCB 2005). With respect to managing AIS, the Basin Plan states that region-wide water quality objectives for pesticides, and related objectives for nondegradation and toxicity, essentially preclude direct discharges of pesticides such as aquatic herbicides.

In Nevada, the Nevada Department of Agriculture (NDA) is the lead agency for regulatory activities associated with noxious weeds, and the Nevada Department of Wildlife (NDOW) is the lead agency for regulatory activities associated with prohibited wildlife. Under NRS Title 14 Chapter 171.123, any peace officer (e.g., NDOW Game Warden, county sheriff deputy, city police agencies) may detain a person that has committed, is committing, or is about to commit a crime (e.g., possession of state listed prohibited wildlife [NAC 503.110] or plant [NAC 555.010] species). Additionally, NDOW Game Wardens (or other Nevada peace officers), as deputies of the USFWS have the authority to uphold provisions of the Lacey Act. Nevada is currently without a comprehensive AIS management plan and instead must rely on the disparate efforts of regional, state, and federal agencies. The state has, however, completed draft guidance to prevent and monitor for AIS, particularly quagga mussel. In 2011, the state passed laws providing for additional boater registration fees that would go to AIS prevention efforts in the state. So, 2013 was the first year of fee collection from these new laws.

The Nevada Board of Wildlife Commissioners has set policy that clearly supports programs that would limit the introduction and impacts of undesirable aquatic species (P-33 Fisheries Management Program). The US Department of Interior's Bureau of Land Management (USDOI-BLM) Nevada State

Office maintains a website for their Invasive Species Initiative for reporting invasive species, but it is not specific to aquatic organisms. Likewise, efforts of the Nevada Invasive Species Council are not focused on AIS. Quagga mussels have been found in Nevada lakes (e.g., Lake Mead, Lahonton, and Rye Patch Reservoirs) that are also popular destinations for Lake Tahoe visitors (LaBounty and Roefer 2007; Wittmann et al. 2010; Wong and Gerstenberger 2011). Presently there is limited mandatory boat inspection or decontamination, when needed, for boats leaving infested waterbodies in Nevada; however, funding has been secured for inspection and decontamination at Lahonton, Rye Patch, and Wildhorse Reservoirs and these begin operations in 2013. Wildhorse Reservoir, near Elko, Nevada, is a headwater of the Columbia River system and holds a significant trout sport fishery.

Region-wide efforts include the designation of TRPA as an area-wide planning agency under Section 208 of the federal CWA. This action is to maintain water quality measures specified in the Water Quality Management Plan for the Lake Tahoe Basin (208 Plan) by limiting the impacts of tourism, ranching, logging, and development on the Lake Tahoe environment and to enforce environmental thresholds. TRPA and its Governing Board have taken an aggressive and proactive role in preventing the introduction of new AIS to Lake Tahoe. The TRPA has the authority to inspect all boats entering Lake Tahoe for AIS or issue penalties starting at $5000 (TRPA Code of Ordinances Chapter 79.3. B). CADPR peace officers (or other state agencies with CDFW Director approval) have the authority to enforce California Code §2301 (related to dreissenid mussel inspections). As of November 1, 2008, all boat launches (public and private) without a trained inspector present are closed (TRPA Code of Ordinances, Chapter 79.3.B (1) and (2)). The Lake Tahoe AIS Program provided funding for gates, where needed, at boat launches in the Region.

TRPA defines an invasive species as

> …both aquatic and terrestrial, that establish and reproduce rapidly outside of their native range and may threaten the diversity or abundance of native species through competition for resources, predation, parasitism, hybridization with native populations, introduction of pathogens, or physical or chemical alteration of the invaded habitat. Through their impacts on natural ecosystems, agricultural and other developed lands, water delivery and flood protection systems, invasive species may also negatively affect human health and/or the economy (TRPA Code of Ordinances, Chapter 79.3).

Evolution of the Program

Prior to 2007, AIS efforts at Lake Tahoe were limited to monitoring and control of Eurasian Water Milfoil, utilizing weed harvest barges to mow. This occurred within the Tahoe Keys residential development and continues to be funded by the homeowners association. The Tahoe Keys is a large area, initially developed by the Dillingham Corporation in the 1960s. The vision of Dillingham was to produce a Venice-like area with lagoons, waterfront properties with lake access, and a large commercial marina from a marsh area that naturally treated the Upper Truckee River as it entered Lake Tahoe. Today and nearly since its inception, the Tahoe Keys is believed to be the primary inoculation point for Water Milfoil into Lake Tahoe, which was carried by power boats. The resultant scatter for fragments of milfoil dispersed and produced populations where suitable habitat exists.

During the 1990s, aquatic pesticides were considered as a means of controlling milfoil within the Tahoe Keys but ultimately were eliminated as a control measure due to regulatory prohibitions that specified a zero detection level for the chemicals (Water Quality Control Plan for the Lahonton Region).

In early 2007, just as maintenance divers in a marina at Lake Mead, Nevada, discovered Quagga Mussels, a workshop on invasive species in the Lake Tahoe Region was in the planning stages. The discovery was a first detection of dreissena mussels in the West, and a call to action for those concerned about the health of Lake Tahoe. Previously, it had been widely thought that these mussels could not survive and produce offspring in the high-temperature waters of the arid southwest. Dreissena mussels, both Quagga and Zebra, soon proved their adaptability and are flourishing in many waters of the southwest (CDFW 2011).

The workshop quickly evolved and placed an emphasis on AIS and in particular quagga and zebra mussels. The prominence of dreissenid mussels on the agenda facilitated a large turnout and lively discussion, primarily surrounding whom and how should the Lake Tahoe Region respond. The numerous Lake Tahoe partners identified AIS as a substantial threat to the economic and environmental health of Lake Tahoe and began taking steps to address the threat. At risk is a substantial investment by private, local, state, and federal interests in this diverse and unique ecosystem. Lake Tahoe is designated an Outstanding National Resource Water (ONRW) under the Clean Water Act (CWA Section 106) due to its extraordinary clarity. Substantial negative changes to the Lake Tahoe Region's economy, pristine water quality, aesthetic value, and recreational pursuits will occur if quagga and zebra mussels become established. The Lake Tahoe Region is not only threatened by new introductions of AIS, but also the expansion of existing populations within the Lake. Lake Tahoe also harbors invasive plants and invertebrates such as the Asian clam that are not found in other lakes within the Tahoe Basin. The effects of climate change amplify these threats to Lake Tahoe and other lakes within the Lake Tahoe Basin.

The AIS effort at Lake Tahoe has expanded substantially in the few, short years since the threat was recognized. In an unprecedented example of cooperation and action, the effort has many substantive accomplishments to its credit and plans for many more.

Funding for the development of a nationally recognized state plan and development of a prevention and control program on the scale needed for Lake Tahoe was difficult at best. Coincidentally there already existed a funding source that had been used primarily to finance the Lake Tahoe Environmental Improvement Program, a large-scale retrofit of numerous water quality, forest health, transportation, and other environmental projects designed to protect and restore the health of the Lake Tahoe ecosystem.

Public Law 105-263, Southern Nevada Public Land Management Act of 1998 as amended (SNPLMA), using authorities granted by the US Congress, allowed the Bureau of Land Management to sell some public lands to private investors. The generated funds were used to fund environmental restoration projects at Lake Tahoe and in Southern Nevada. Through a process similar to grant applications, the Lake Tahoe AIS Program successfully competed for SNPLMA funding. Starting in 2008, operational funds were provided that allowed the program to launch into a watercraft inspection, AIS control, and public outreach program that quickly became endorsed by local, state, and federal agencies. In 2008, SNPLMA funding totaled $450,000. As the program expanded and showed success, funding increased to $500,000 in 2009, $985,000 in 2010, $3,221,000 in 2011, and $3,302,397 in 2012.

Early in the development of the Lake Tahoe AIS Program, it was realized that a strategic plan or management plan was needed to give long-term direction to managing the AIS threats at Lake Tahoe. With a great deal of financial and management assistance from the US Army Corps of Engineers (USACOE), the writing of a Lake Tahoe Region AIS Management Plan (Plan) began in 2008. The Plan was drafted by staff from Tetra Tech, Inc., and was continually reviewed by the LTAIS Coordination Committee. California, Nevada, and the national Aquatic Nuisance Species Task Force (ANSTF) approved the Plan in November 2009. Under the Nonindigenous Aquatic Nuisance Prevention and Control Act of 1990 (P.L. 101-646) (NANPCA), states and regions with approved plans are eligible for federal funds to assist in plan implementation.

The Plan can be perused at http://www.anstaskforce.gov/State%20Plans/Lake_Tahoe_Region_AIS_Management_Plan.pdf. It provides the Lake Tahoe Region with a cohesive, guiding document that prioritizes objectives and identifies lead organizations, specific actions for each organization, and funding sources to combat existing and potential AIS. It also provides guidance that delineates appropriate, science-based regulation and monitoring that expressly deals with prevention and management of AIS. The purpose of the Plan is to facilitate coordination of regional, bistate, state, and federal programs and to guide implementation of AIS prevention, monitoring, control, education and research in the Lake Tahoe Region.

The goals of the Plan are to

- Prevent new introductions of AIS to the Lake Tahoe Region
- Limit the spread of existing AIS populations in the Lake Tahoe Region, by employing strategies that minimize threats to native species, and extirpate existing AIS populations when possible
- Abate harmful ecological, economic, social, and public health impacts resulting from AIS

The implementation of the Plan is structured around seven objectives associated with

- Management plan implementation and updates
- Coordination and collaboration
- Prevention
- Early detection, rapid response and monitoring
- Long-term control and management
- Research and information transfer
- Laws and regulations

To meet these objectives, 23 strategies were identified with respective action items detailing how that objective will be met. The priority of each of the 95 actions included is ranked as low, medium, or high and the lead and cooperating entities are identified. Where applicable, short-term (present through 2010) priorities for action and funding sources are indicated, as are the long-term actions over the 5-year period from 2010 to 2015. In many cases, the LTAISWG or LTAISCC are named as the lead or cooperating entities.

The intent of the Plan is to provide more localized guidance for preventing and managing AIS in the Lake Tahoe Region. It is not in conflict with California's nor Nevada's AIS management plans.

At a minimum, the Plan is reviewed once a year and revised every 5 years by a LTAISCC subcommittee to ensure Plan objectives, strategies, and actions continue to identify and address relevant AIS issues in a timely manner. Individual components of the Plan (e.g., rapid response plans, monitoring plans, and vessel inspection protocols) will be updated more frequently to fully address changing needs in the Lake Tahoe Region.

Summarized in the Plan is the background of nonnative species introductions to the Lake Tahoe Region, the pathways for existing and potential AIS introductions, the types of existing and potential AIS in the Lake Tahoe Region, and short- and long-term priorities for action. Also included (as appendices) are an overview of regulations and programs; the Vessel Inspection Plan; the Small Watercraft Screening Process; an estimate of potential economic impacts from a dreissenid mussel infestation at Lake Tahoe; and an overview of existing and potential AIS life histories, environmental requirements, distributions, and control methods.

Program Oversight

The Lake Tahoe Aquatic Invasive Species Program (LTAISP) is a multiagency bistate, local and federal partnership tasked with the implementation of the Plan. The Plan is one of only a few bistate or regional plans in the nation and serves as the template for aquatic invasive species (AIS) activities in the Tahoe Region. Oversight for state AIS management plans is typically led by a respective state resource agency; however, in the case of bistate or regional plans, oversight is best suited to an organization capable of regulation across state jurisdictions. The Tahoe Regional Planning Agency (TRPA), as created by California, Nevada, and the US Congress, has such regulatory authority. The Plan overlaps with the California state Aquatic Nuisance Species (ANS) plan and the Nevada ANS plan Implementation of the Plan is closely coordinated with the ANS leads of both states to ensure collaborative and complementary efforts.

The Plan is implemented by working groups comprised of various agency representatives. Each working group focuses on a unique challenge and issue presented by individual species or programs (i.e., invasive weeds and motorized watercraft inspections). The working groups are in turn organized into a regional program by the LTAISCC. The LTAISCC is co-chaired by the USFWS and the TRPA and comprised senior managers and senior technical staff from agencies involved in policy, implementation, and regulation of AIS projects and programs. The representatives on the LTAISCC are charged with the development of AIS actions, annual work programs, reporting and policy recommendations.

Funding for the implementation of the Plan and LTAISP is provided by multiple federal, state, local, and private sources. The majority of this funding is provided by the US Fish and Wildlife Service and passed through the TRPA, which acts as the fiscal agent for the Plan. The significant contributions from these multiple partners continue to be a tribute to the success of the Program. The largest single source of funding has been from the Southern Nevada Public Land Management Act (SNPLMA), with funds passed through the USFWS.

Prevention

The Lake Tahoe watercraft inspection program began following the 2007 discovery of quagga mussels in Lake Mead. The 2007 program was a pilot based on voluntary inspections on peak weekends at several public launch facilities conducted by the Tahoe Resource Conservation District (Tahoe RCD).

With the success of the pilot program along with new regulations put in place by the Tahoe Regional Planning Agency (TRPA), making all watercraft subject to inspection prior to launching in Lake Tahoe, the Watercraft Inspection program went into full, year round operation starting May of 2008. In late summer of 2008, the TRPA passed an additional regulation requiring all launch facilities around Lake Tahoe to install gates at the launch ramps to be locked when a certified watercraft inspector is not present. In 2009, the program included greater participation by private marina owners; and the Tahoe RCD and TRPA implemented improvements in inspections and decontamination of watercraft at launch ramps. The watercraft inspection program continues to function under the Lake Tahoe Region Aquatic Invasive Species Management Plan.

Watercraft inspections at Lake Tahoe are conducted to prevent the introduction of new AIS. Because Lake Tahoe has a limited number of AIS, many of the invasive species plaguing other waters in the United States are of concern to Lake Tahoe resource managers. While all new AIS threats, plant, animal, and viral are of high concern, quagga and zebra mussels carry a particular risk of environmental and economic degradation. Quagga and zebra mussels damage the ecosystems they invade. They feed by filtering water and removing large amounts of food, effectively starving native species in infested rivers and lakes. The waste they produce, including pseudo feces, accumulates and degrades the environment, using up oxygen, making the water acidic and producing toxic byproducts. These pollutants can be passed up the food chain if the mussels are consumed.

The primary way invasive mussels and other AIS spread is on overland hauled boats and trailers. If a boat or personal watercraft has been in infested waters, it could be carrying quagga or zebra mussels or other AIS. The mussels microscopic larvae (called veligers) can also be unintentionally transported in water held in live wells, bilges or bait buckets. Since their introduction to the Great Lakes in 1986, dreissena mussels have spread to rivers and lakes throughout the United States. Recently, two northern Nevada reservoirs, Rye Patch and Lahonton, tested positive for quagga mussels. Lahonton is considered positive and Rye Patch is considered suspect at this time. Monthly monitoring of these reservoirs and other lakes in Nevada are conducted by the Nevada Department of Wildlife.

Each year the Tahoe Resource Conservation District (TRCD) conducts watercraft inspector certification classes, certifying over 100 level 1 watercraft inspectors including TRCD, private marina, state park, and public utility district staff. Also, TRCD Level II inspectors have been trained at Lake Mead through the Pacific States Marine Fisheries Commission's (PSMFC) certification process. They are qualified to conduct decontaminations and also conduct the level 1 training sessions.

Through cooperative agreements with local agencies and agreements with private marinas, watercraft inspections have been in place at all operational launch facilities 7 days a week. When launch facilities are not staffed with inspectors, the facility is closed and gated. Each boat is issued a tamper proof inspection band that is attached between the boat and the trailer by staff after the boat has successfully passed the inspection. The band is then removed as the boat is launched. When the boat is later loaded onto its trailer, staff to signify that the boat does not need to be re-inspected attaches a new band. As long as the band is intact, the boat can be launched at a later date without further inspection.

Beginning in May of 2010 the watercraft inspection program was expanded to include inspection and decontamination facilities at four on-highway locations around the Lake Tahoe region. These facilities are intended to be the primary inspection and decontamination locations for watercraft entering the Lake Tahoe Basin. Supplementing these locations, inspections and limited decontaminations are performed at boat ramps and other boat launch facilities on Lake Tahoe, Fallen Leaf Lake, and Echo Lake.

Two of the on-highway watercraft inspection facilities are located on public lands. One is the Alpine Meadows facility, located along highway 89 north of Tahoe, managed by the US Forest Service, Tahoe National Forest, and Truckee Ranger District. The other is the Spooner Summit facility, located in the Nevada Department of Transportation right of way along highway 50 east of Tahoe. The two remaining facilities are on privately owned parcels. One is the Northstar Boulevard facility along highway 267 north of Tahoe, which is owned by Northstar at Tahoe ski resort. And the location in Meyers along US Highway 50 west of Tahoe is owned by a private individual. The property owners at no charge, with the exception of the Meyers facility that is leased, provide all locations.

Each on-highway inspection facility is equipped with a portable decontamination unit, a portable office, and portable restroom facilities. Portable, electronic changeable message signs and other highway signage direct boaters into the stations. In addition to signage at individual locations, the Tahoe AIS Program purchases advertising space on billboards along major routes into the Tahoe region. The Tahoe RCD staffs each facility with three inspector/decontamination personnel (22 seasonal full time equivalent). When a vessel is inspected and/or decontaminated at an on-highway location the vessel is sealed to the trailer, using a tamper-proof wire cable seal, so that when brought to a boat ramp, inspection staff at these locations could verify that it had been inspected and was able to launch. Many vessels are also referred from launch facilities to on-highway stations for decontamination.

The on-highway inspection program was in operation beginning in May 2010 with the start of the normal boating season in the Tahoe Region. Prior to Memorial Day, boating is limited on the lakes of the region due to snow, ice, and low temperatures. Inspections and decontaminations continue at on-highway locations into September of each year. During the winter months, decontaminations are conducted only at the Meyers location. During the period May through September, inspection stations are operated daily between 6 am and 8 pm.

In 2010, the motorized watercraft inspection program conducted a total of 8200 full AIS inspections at Lake Tahoe. Approximately 70% of these inspections were conducted at launch ramps and marinas, and 30% were conducted at roadside inspection stations. An additional 6500 boats were processed as "Tahoe Only." These vessels had been previously inspected by the program and had retained their inspection seal, thereby not requiring a full AIS inspection for the 2010 season. Approximately 90% of these vessels were processed at the launch facilities and marinas, while 10% were fulfilled at the on-highway inspection stations. Approximately 21,400 check-ins were recorded at the launch facilities and marinas, representing previously inspected and sealed motorized watercraft.

The watercraft inspection program in total performed approximately 8000 inspections at all locations, including on-highway, public, and private launch facilities. The majority of these inspections were of vessels registered in California (n = 5704) and Nevada (n = 1607), with the next most frequently represented states being Arizona, Oregon, and Utah. However, vessels registered as far away from Lake Tahoe as Virginia and Florida were also inspected.

The types of watercraft inspected and decontaminated at on-highway facilities ranged from simple pleasure craft under 25 ft, to large, complex cabin cruisers of over 40 ft in length with multiple plumbing systems, all in need of decontamination.

The startup of on-highway inspection and decontamination stations posed many management and logistical challenges. A short planning and contracting window of 90 days required that staff be diligent, creative, and constantly communicating with each other on hiring staff, permitting and acquiring station locations and identifying and purchasing equipment and services.

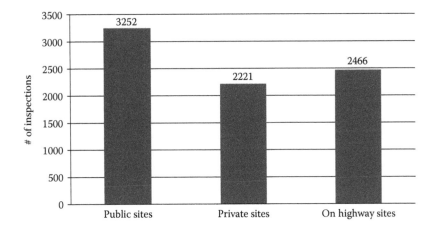

FIGURE 22.2 Number of watercraft inspections performed and the three types of inspection facilities in the Tahoe Region during 2010.

Beginning in 2011, the Program instituted a requirement that all watercraft inspections and decontaminations be conducted at the on-highway stations. The inspection and decontamination data collected for the 2012 inspection year reflect the requirement that no inspections or decontaminations occur at launch ramps. All inspections and decontamination procedures now are limited to the on-highway stations. In 2012, 7443 watercraft inspections were carried out at on-highway stations, which include 3752 watercraft decontaminations. Check-ins at launch facilities and marinas, representing launches of previously inspected and sealed watercraft, totaled 30,569 boats.

The first lesson that was learned from the implementation of on-highway inspection stations was the need for expanded outreach and signage. Because the locations for the on-highway stations were not permanent in 2010, and the limited availability of suitable locations that could be used without significant improvement, most of the on-highway stations were located just off of the main highway rather than in the right of way. The location of the stations combined with the limited temporary signage allowed by state regulations, and the ability of boaters to be inspected at launch facilities led to an underutilization of on-highway stations for inspection (Figure 22.2).

The second lesson that we learned was that the logistics of decontamination was challenging at on-highway locations. Some logistics of decontamination involved providing fresh water and fuel for the portable decontamination machines; and the retrieval and disposal of wastewater. Wastewater from decontaminations could only be disposed of in an unused aeration basin at a sewage treatment plant in the Tahoe Region due its unknown nature of constituents. Wastewater typically contained a high volume of grease and oils form engine bilges and other unknown contaminants from watercraft. Compounding the challenge was the large volume of water used in decontaminating watercraft, particularly large complex vessels of 30 ft and more. Decontamination would often take twice the capacity of a single decontamination unit's 225 gal, which necessitated having two units for those large vessels. This resulted in long waits for decontamination during the busiest times, such as the week leading up to July 4. These logistical issues were resolved by acquiring funding for higher capacity decontamination units that recycle wash water and therefore require less frequent fresh water resupply and shorter wait times. Two large, semi-permanent water recycling decontamination units were purchased during the following season and are in use at the Meyers and Alpine Meadows on-highway stations.

The next lesson learned also involved logistics. In this case, it was the logistics related to data collection and general Internet connectivity. Many of these locations had limited cell phone service and no landline service, which restricted access to the online database that is used to track inspections in the Tahoe Region. The limited access to the database meant that the primary data collection method at many of these locations was on paper datasheets, which slowed down data entry, dramatically reducing our ability to adaptively manage the program. The issue with connectivity has been resolved by either tying into hardline connections or contracting for satellite Internet services.

Finally, there was a lesson learned by the watercraft inspection program that did not just apply to the on-highway inspection and decontamination stations. That lesson is the value of an ongoing quality control program. During the 2010 season, the watercraft inspection program instituted a *secret shopper* program to determine whether inspections were being carried out to protocol. While the majority of these surveys indicated that inspectors were indeed following protocol, some came back indicating that changes to the tracking of inspections and vessels and additional training was needed. The issues that were discovered by the *secret shopper* program have been resolved with a change in the supervisory structure of the watercraft inspection program that allows for more time to be spent by senior inspection staff on quality control and refresher training.

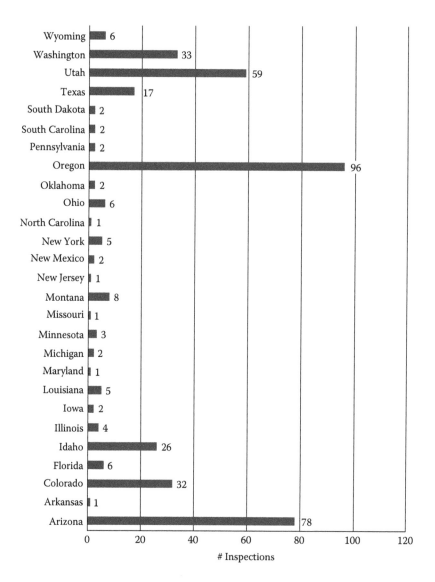

FIGURE 22.3   Number of watercraft inspections performed by state, excluding California (n = 5704) and Nevada (n = 1607) in the Tahoe region during 2010.

Watercraft Decontamination for Dreissenid Mussels

Decontamination procedures for watercraft to reduce the spread of AIS, including quagga and zebra mussels, have been in use for some time and are based on data derived from research from various sources. The Recommended Uniform Minimum Protocols and Standards for Watercraft Interception Programs for Dreissenid Mussels in the Western United States (Zook and Phillips 2009) established that water heated to 140°F should be applied for at least 10 s to kill adult dreissenid mussels, juveniles, and veligers. In addition, Zook and Phillips recommend a combination of high-pressure washing and low-pressure flushing to clean the hull, drive units, and all through-hull fittings, bilges, and other areas that may be exposed to contaminated water. The use of water heated to 60°C (140°F) has been recently tested experimentally and found to be 100% effective at killing adult zebra mussels at an exposure time of at least 10 s (Morse 2009a,b) while it takes at least 5 s at killing adult quagga mussels (Comeau et al. 2011). While this study examined only adult mussels, the shells of adult mussels make them more resistant to heat than juveniles and veligers. Thus, it is reasonable to assume that exposure to water at 140°F for at least 10 s will also kill juveniles and veligers. The use of 140°F water has been shown to be effective and has been employed around the country (Adams and Lee 2010). However, the use of water heated to 140°F in systems such as ballast tanks is not recommended by some boat manufactures as it may damage equipment. So, slightly cooler water left to stand in those areas for longer time will kill the invasive mussels (Comeau et al. 2014, Chapter X in this book).

Lake Tahoe Decontamination Protocol

The Lake Tahoe watercraft inspection program uses a set protocol to determine a vessel's level of risk and then applies a decision tree to prescribe the necessary level of action. The protocol requires decontaminations of high-risk vessels prior to allowing these vessels to launch. The first step in the protocol is to determine the level of risk for an individual vessel using a flow chart. The first step is to determine whether the vessel has a Lake Tahoe Region–issued band sealing the boat to its trailer and indicating that the vessel last launched in a Lake Tahoe Region waterbody. Should a vessel have such a band intact, it is known to be of no risk, allowing its launch. All other vessels are subject to further inspection. The next step is to determine whether the vessel has been in waters infested with invasive species. This is determined from information provided by the operator as well as other clues inspectors are trained to notice. All vessels must undergo a complete inspection regardless.

Any vessel found to be at all wet, including remnant-standing water, even though there is no visible evidence of invasive species, is determined to be a high risk. Wet vessels were flushed with water at 140°F. In cases where only a very limited amount of water was found, bleach has been used. Although bleach is known to be effective at killing veligers, it is difficult to dispose. So, use of bleach was discontinued within the Tahoe program, and full hot water flush decontamination has been used instead since the 2011season.

If a vessel is from a known infected waterbody, regardless of whether water is found on board or not, it is judged to be high risk and at a minimum, a flush is conducted. In situations where an invasive species is found or if a boat from a known infected waterbody is wet or dirty, a full decontamination is performed, consisting of washing any and all surfaces and equipment that may contact the water.

The flushing process used water at 140°F. To assure adequate thermal exposure to kill mussels, water was flushed through flow-thru systems for at least 1 min. This well exceeds the recommended 10 s needed to kill adult invasive mussels, ensuring 100% mortality. On exterior surfaces, 3–5 min of contact time by the 140°F water was used, which is more than sufficient to ensure 100% mortality. Only after the appropriate inspection and decontamination is completed is the vessel allowed to launch.

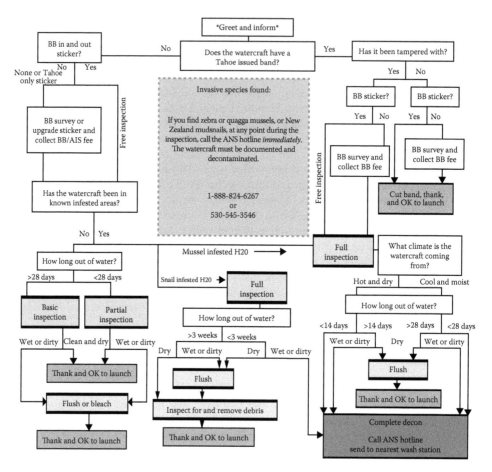

FIGURE 22.4 Flow chart for inspection and decontamination used by the Lake Tahoe watercraft inspection program in 2010.

TRPA implemented watercraft fees in April 2011 to contribute to a long-term funding source for the AIS watercraft inspection program (WIP). The WIP involves two types of stickers—one for boats that are exclusively used on Lake Tahoe (Tahoe Only) and the other for boats that go to other bodies of water as well as Tahoe (Tahoe In and Out). In addition, it also allows for a 7-day pass for a reduced rate. The fee structure provides partial funding for the implementation of watercraft inspections. In addition to the basic inspection fee, additional fees are applied for decontamination. Fees assessed in 2013 are as follows.

Tahoe-Only Stickers	Fee Amount
All sealed vessels	$30.00
Tahoe In and Out Stickers	**Fee Amount**
Personal watercraft (PWC)	$35.00
Vessels 0.1–17.0 ft	$35.00
Vessels 17.1–21.0 ft	$75.00
Vessels 21.1–26.0 ft	$86.00
Vessels 26.1–39.0 ft	$98.00
Vessels 39.1 ft and greater	$121.00

(Continued)

Seven-Day Passes	Fee Amount
Personal watercraft (PWC)	$33.00
Vessels 0.1–17.0 ft	$33.00
Vessels 17.1–21.0 ft	$55.00
Vessels 21.1–26.0 ft	$66.00
Vessels 26.1–39.0 ft	$78.00
Vessels 39.1 ft and greater	$101.00
Upgrades from Tahoe-Only to In and Out	**Fee Amount**
Personal watercraft (PWC)	$5.00
Vessels 0.1–17.0 ft	$5.00
Vessels 17.1–21.0 ft	$45.00
Vessels 21.1–26.0 ft	$56.00
Vessels 26.1–39.0 ft	$68.00
Vessels 39.1 ft and greater	$91.00
7-day pass to in/out	$25.00
Decontamination Fees	**Fee Amount**
Engine and bilge flush	$25.00
Ballast tanks	$10.00
Additional systems	$10.00
Off-site decontamination	$200.00

A nonmotorized watercraft inspection program was instituted in 2009. It has evolved into a project that is now called Tahoe Keepers. The US Forest Service (USFS) began screening nonmotorized watercraft at all beaches and campgrounds staffed by concessionaires in 2009, and Tahoe RCD inspectors also began nonmotorized inspections at Fallen Leaf Lake that same year.

From the inception of the program, the inspection and decontamination of motorized watercraft has taken priority over the inspection of nonmotorized watercraft. There is a lower risk of introduction for many AIS posed by nonmotorized watercraft, and the dispersed nature of launching sites for these craft makes inspection more difficult to manage.

"Tahoe Keepers" is the brand name of the outreach, education, and stewardship campaign launched by the nonmotorized watercraft-working group in 2011. The effort is based on the ability of non-motorized boaters to substantially reduce the risk of AIS introduction or transport through diligent implementation of the Clean, Drain, Dry, and Dispose method of self-inspections and decontamination. The purpose of the Tahoe Keepers campaign is to educate the target user group on the risks, techniques, and laws or ordinances associated with AIS in the Tahoe Region. In 2011, the Non-Motorized Watercraft Working Group developed the Tahoe Keepers campaign concepts; built an online resource, training, and Keeper membership registration portal; launched a prevention outreach campaign targeted to nonmotorized boaters, including public media, public engagement, and social networking (Facebook). In early 2013, there were 727 registered Tahoe Keepers, representing 1282 individual watercrafts.

The season of nonmotorized watercraft inspections begins on May 15 and ends on September 30. Yearly training consists of 2 days of reviewing protocols and procedures, emphasizing customer service skills, and discussing launch locations and kayaking routes. The program objectives are to inform paddlers of the Tahoe Keepers initiative and identify high-risk vessels prior to launching. The inspectors are trained to approach paddlers, informing them of the risk associated with AIS and how they can help stop the spread through self-inspection and decontamination. There is one full time inspector during this time, generally dedicated to Fallen Leaf Lake on the West side of Lake Tahoe. Presently, Fallen Leaf Lake has no known AIS. During 2012, the inspector made 1712 contacts with paddlers.

Control of Existing AIS

The control effort of the LTAISP is focused on the removal of AIS currently found in the Lake Tahoe Region and abating the impacts from these species. The suite of AIS currently in the Lake Tahoe Region includes Eurasian watermilfoil, curlyleaf pondweed, Asian clam, and multiple species of invasive warm water fish such as largemouth bass and small mouth bass. These AIS have been demonstrated to have ecological and economic impacts such as promotion of algal blooms, and impacts to boat access and the native cold water fishery.

Control of AIS implies that populations are present and small enough in population size for control to prevent further increases in the infestation area. Eradication, while less achievable, but appropriate in species-specific and site-specific locations, indicates the elimination of a species at all life stages. Often the methods to control AIS are the same as those to eradicate them; however, the intensity of management may vary greatly from control to eradication. Methods to control or eradicate may overlap between groups of AIS while some methods are specific to a particular AIS.

Asian Clams

Efforts are currently underway in Lake Tahoe to conduct research that will determine the most effective means of controlling Asian clams. The control measures in use or being investigated for Asian clams are not presently aimed at eradication; however, these objectives may change based on research outcomes. Additional research is being conducted on Asian clam populations to better understand responses to control strategies and the costs associated with these efforts.

Asian clam were first documented in Lake Tahoe in 2002. Since the initial clam detection, the populations in South Lake Tahoe increased rapidly from 1 to nearly 200 ac in less than 10 years with densities up to 6000 individuals/m^2 measured. A relatively sparse population was discovered near the mouth of Emerald Bay in 2009. Initial surveys estimated this infestation at approximately 3.5 ac in size just inside and on the south side of the mouth of Emerald Bay, in water depths of 6–30 ft. Surveys in 2011 showed the infestation had spread to approximately 5.5 ac (UC Davis 2011), an increase in infestation size of over 40% in only 2 years. An expansion of the Emerald Bay infestation similar to that seen in South Lake Tahoe would result in considerable ecological and recreation impacts.

The invasion and establishment of Asian clams can lead to a variety of negative impacts. The clams can dominate native benthic invertebrates and plants, which are important components of the food web in Lake Tahoe. The presence of clams may increase the habitat suitability for other nonnative invasive species such as zebra and quagga mussels and promote algal growth. Water quality can be degraded through concentrated nutrient excretion and associated algal blooms. The extended algal blooms supported by clam populations can have far reaching impacts on near shore conditions. When algae die and wash ashore, they decompose and rot on beaches and rocks where they impact aesthetics of the lake and influence water quality and clarity. Localized populations of Asian clams in Lake Tahoe have densities among the highest recorded worldwide, and have been observed growing at greater depths than reported in the literature anywhere else they occur (Brockett 2013).

Options currently available to control Asian clams in Lake Tahoe include physical and mechanical methods. However, these methods are under development and are not yet considered operational lake-wide. Molluscicides are not acceptable. Each control project includes an evaluation of effectiveness and ability to measure success in controlling and/or eradicating the clams. The control program is funded by US Fish and Wildlife Service and Lahontan Regional Water Quality Control Board. Recently additional funding has been obtained through the California Department of Parks and Recreation.

The Asian Clam Working Group (ACWG) comprised members from the Tahoe Resource Conservation District (Tahoe RCD), US Fish and Wildlife Service (USFWS), Lahontan Regional Water Quality Control Board (LRWQCB), Tahoe Regional Planning Agency (TRPA), California Department of Parks and Recreation (CDPR), Nevada Division of State Lands (NDSL), Nevada Division of Wildlife (NDOW), UC Davis Tahoe Environmental Research Center (UCD-TERC), and University of Nevada,

Reno (UNR). The ACWG has been meeting regularly since 2008 to discuss, plan, and implement a pilot project that experiments with nonchemical treatment strategies for Asian clam treatment in Lake Tahoe.

The Asian clam infestation in Emerald Bay is currently a relatively sparse population presumed to still be in the early stages of invasion. This level of infestation offers the opportunity to prevent further spread and contain or control the existing infestation. If the population goes untreated for even a short duration of time, this small and sparse infestation could potentially expand as seen in the populations in South Lake Tahoe and in the 40% increase in infestation size over the 2 years since the infestation was discovered. An expanded infestation will become extremely difficult and expensive to control.

A small-scale experiment by UC Davis researchers in 2009 using ethylene propylene diene monomer (EPDM) pond liners to cover the clams reduced the dissolved oxygen available to them. It resulted in 100% mortality after 28 days at peak summer temperatures. Additional work utilizing this technique has shown that it is a good method with which to cause AC mortality. There is some subsurface water flow through the sill in the mouth of Emerald Bay, so benthic barrier deployment at this site will require barriers to be in place for an extended duration (approximately 12 months) and be augmented with organic material to increase biological oxygen demand beneath the barriers.

In 2012, approximately 5 ac of EPDM pond liner was deployed on the Emerald Bay AC infestation. Those barriers will remain in the water for 1 year. Monitoring will determine the efficacy of these barriers in reducing the AC population. Early indications are that dissolved oxygen (DO) level beneath the barriers is declining, and clam mortality is evident. During the 2012 efforts, the control team made two significant observations:

1. The AC population extended beyond the perimeter of barriers deployed in 2012.
2. Some infested areas at the perimeter of the infestation are strewn with intermittent large to very-large boulders, making deployment of barriers difficult, if not impossible in these locations.

The barriers will remain on the lake bottom until the fall of 2013 when the deployment will be evaluated and future project work will be determined.

Invasive Weeds

Invasive weeds have been detected in numerous locations in Lake Tahoe. The continued spread of Eurasian watermilfoil and curlyleaf pondweed create severe economic and ecological impacts to Lake Tahoe, both from source and satellite populations. The presence of invasive plants provides habitat for undesirable warm water fish and impedes restoration of native fish populations. Funding from local, state, and federal sources have continued the strategy of identification, control, and monitoring of priority sites. The emphasis has been on eliminating source populations and/or satellite populations. The primary source population is in the Tahoe Keys Lagoons. A study plan ("Evaluation of Methods for Aquatic Invasive Species Management in the Tahoe Keys" by Dr. Lars Anderson, USDA-ARS) developed in 2010 that included a cost and task analysis of needed control measures for aquatic invasive weeds for a herbicide-based control strategy. Herbicide use in the Tahoe Keys could be utilized to both continue implementation of the Plan and to implement a management strategy of herbicide application if an application is approved by the California Water Quality Control Board. The California Water Quality Control Board, Lahontan Region (Lahontan Water Board) has been working to adopt procedures for allowing project proposals to come before the Board for review and discretionary authorization of pesticide use. On December 7, 2011, the Lahontan Regional Water Quality Control Board (Lahontan Water Board) adopted Resolution No. R6T-2011-0102 amending the Basin Plan to replace the existing region-wide pesticide water quality objective. This effectively prohibits pesticides in water, with a region-wide waste discharge prohibition with exemption criteria for aquatic pesticide application. The amendment establishes a formal prohibition on pesticides in all waters, but provides a pathway for project proponents to seek an exemption to the prohibition from the Lahontan Water Board for use of aquatic pesticides. The LTAISCC has been working with the Lahontan Water Board so the proposal for use of herbicides to control invasive weeds in the Tahoe Keys lagoon areas is eligible for review and can be considered for discretionary Board authorization if the EPA approves the Basin Plan Amendment.

Early Detection/Rapid Response

Following prevention, early detection and containment and control/eradication of new AIS introductions are equally the second most cost-effective measures to reduce the impacts from AIS. This is accomplished through rigorous monitoring followed by the ability to respond efficiently and aggressively. Effective responses to new or newly spreading AIS require adequate resources, personnel, and procedures in place. Response is facilitated by a collaborative effort between numerous agencies, NGOs, researchers, and other stakeholders.

An Aquatic Invasive Species Early Detection Plan for Zebra and Quagga Mussel Veliger Monitoring was developed in 2010 and implemented in May of 2010 for the Lake Tahoe Region. The Lake Tahoe Basin Interagency Dreissenid Mussel Rapid Response Plan (LTRRP) was written in 2009 and the first workshop was conducted in September of 2012. The purpose of the exercise was to engage local, state, and federal authorities in testing and evaluating the LTRRP and to apply that knowledge to building an effective response to a quagga or zebra mussel detection (Chilton 2009).

Lake Tahoe is fortunate to have one of the most stringent and effective AIS prevention programs in the country. While we trust that a response to a mussel detection will not be needed, it is prudent and responsible to have a tested process in place should the need arise. The results of the workshop provided the LTAISCC with a compilation of the strengths and weaknesses of the Plan and recommendations toward improving our ability to successfully answer such a threat.

Day 1 of the exercise introduced attendees to the Dreissenid mussel risk to Lake Tahoe and the Incident Command System (ICS). ICS is the emergency management system for most AIS rapid response plans. Day 2 reviewed the Lake Tahoe Basin Interagency Rapid Response Plan (Plan) and a response exercise was conducted. The culmination of the day and the workshop was a critique of the Plan and identification of elements to improve the Plan. Workshop objectives included identifying agency responsibilities during a response, applying the organization and principles outlined in the Plan, and identifying training, tools, and planning elements to improve rapid response capabilities and effectiveness.

The purpose of the LTRRP is to provide a framework for an effective rapid response to the discovery of any dreissenid mussel (mussel) AIS in Lake Tahoe.

In the document, *rapid response* means that soon after a detection of a dreissenid mussel (veliger or adult) in Lake Tahoe is discovered, (1) the responsible agency will make a determination of whether it is potentially significant and/or detrimental and (2) if that is the case, the responsible agency will develop and implement a course of action. This also would apply to mussels that are discovered in an adjacent waterway or lake that ultimately enters Lake Tahoe.

The possible courses of action for newly discovered mussels may include an effort to eradicate the species, control its spread, prevent future introductions, minimize or mitigate the damage it causes, or study it further before any other action is taken. Rapid response is the second line of defense after prevention to minimize the negative impacts of AIS on the environment and economy of Lake Tahoe. Once nonnative invasive species become widespread, efforts to control them are typically more expensive and less successful than rapid response measures. The damage caused by AIS that becomes widespread, and the actions that are taken to control it, may be more harmful to the environment than a successful rapid response.

The LTRRP reflects strategies, models, and activities gleaned from a variety of other contingency plans. In particular, it draws from the Columbia River Basin Interagency Invasive Species Response Plan: Zebra Mussels and Other *Dreissenid* Species created in 2008 by the Columbia River Basin Team, 100th Meridian Initiative and the Rapid Response Plan for Aquatic Invasive Species in California created in 2007 by the California Department of Fish & Wildlife, Habitat Conservation Branch, Invasive Species Program.

Science

The Lake Tahoe AIS program includes input from the science community in all aspects of project planning, including project level effectiveness along with status and trend monitoring. While the AIS program to date has not adopted a formal adaptive management system, we have and continue to use the Plan's *do, check, act* approach for refining future actions. The best example of this is the interaction between the agencies and the science community to address Asian clam removal. Staff from agencies and the research community worked hand in hand to plan and carryout many pilot projects to determine the most effective methods for removal of Asian clams. In addition, the science community has been actively involved in the yearly process of project prioritization for determining budgets and scope of work.

Gaps and Challenges

The unique ecological and political landscape of the region presents some policy challenges that could limit the ability of resource managers to achieve management goals. For example, the Nevada Department of Environmental Protection allows for the application of US Environmental Protection Agency (EPA)-approved aquatic herbicides for the control of nuisance aquatic plants. On the California side of Lake Tahoe, however, the LRWQCB's region-wide water quality objectives for pesticides, and related objectives for nondegradation and toxicity, preclude direct discharges of pesticides such as aquatic herbicides. The LRWQCB has recently adopted a Basin Plan amendment to consider proposals for the application of aquatic pesticides in the Region. Pending EPA approvals, limited application of herbicide for the management of invasive aquatic plants could be an additional tool for resource managers.

With the sunset of current federal funding, the AIS Program sought stable funding to avoid the severe economic and ecological consequences of AIS invasion into Lake Tahoe. The Program instigated the development of an AIS Program Funding Strategy Analysis (Environmental Incentives 2013). This funding strategy analysis outlines specific options to secure funding from diverse, reliable, and long-term sources to sustain core elements of the AIS program. The strategy provides an approach and recommendations for funding to implement the Lake Tahoe AIS Management Plan.

The AIS effort at Lake Tahoe has expanded substantially in the few, short years since the threat was recognized. The effort has many substantive accomplishments to its credit and plans for many more. Examples of these accomplishments include

- Development of the Lake Tahoe Region Aquatic Invasive Species Management Plan and adoption of the Plan by the national Aquatic Nuisance Species Task Force
- Formation of the Lake Tahoe AIS Working Group (LTAISWG)
- Formation of the Lake Tahoe AIS Coordination Committee (LTAISCC)
- Development and implementation of a Watercraft Inspection Program at Lake Tahoe
- Requirement that all launch ramps are locked unless there is an AIS inspector present
- Deployment of seven portable watercraft decontamination stations
- Installation of two semipermanent decontamination stations
- Training and certification of more than 100 Watercraft Inspectors
- Hiring an AIS Coordinator through the US Fish and Wildlife Service (USFWS)
- Increased monitoring for invasive aquatic plants, invertebrates, and warm water fishes
- Use of diver-operated suction and benthic barriers to control invasive aquatic plants
- Large-scale invasive plant controls instituted in Emerald Bay
- Commercial-scale techniques developed for the control of Asian clams
- Measurement of warm water fish behavior and diets in and around the Tahoe Keys

- Removal of over 1 ton of invasive warm water fish
- Increased education and outreach activities
- Quagga mussel survivability studies conducted by UNR/UCD

REFERENCES

Adams, D. C. and D. J. Lee. 2010. Selected paper prepared for presentation at the *Agricultural and Applied Economics Association's 2010 AAEA, CAES & WAEA Joint Annual Meeting*, Denver, CO, July 25–27.

Brockett, J. 2013. Emerald Bay Asian Clam Control Project Description, Tahoe Resource Conservation District, South Lake Tahoe, CA.

Bryan, S. 2014. Chapter 27 in this book. Quagga mussel monitoring in the Central Arizona Project: 2009–2012.

Californial Fish and Game Department. 2011. Public document on CAFGD web site: http://www.dfg.ca.gov/invasives/quaggamussel/. Dreissenid Mussel Infestations in California. https://nrm.dfg.ca.gov/FileHandler.ashx?DocumentID=39027 (Accessed February 20, 2014).

Chilton, S. 2009. Lake Tahoe Basin Interagency Dreissenid Mussel Rapid Response Plan. US Fish and Wildlife Service, Washington, DC.

Coats, R., J. Perez-Losada, G. Schladow, R. Richards, and C. Goldman. 2006. The warming of Lake Tahoe. *Climate Change* 76(1–2):1573–1480.

Comeau, S., S. Rainville, B. Baldwin, E. Austin, S. L. Gerstenberger, C. Cross, and W. H. Wong. 2011. Susceptibility of quagga mussels (*Dreissena rostriformis bugensis* Andrusov) to hot-water sprays as a means of watercraft decontamination. *Biofouling* 27:267–274.

Environmental Incentives. 2013. Lake Tahoe AIS Program Funding Strategy Analysis.

Goldman, C. 1988. Primary production, nutrients, and transparency during the early onset of eutrophication in ultra-oligotrophic Lake Tahoe, California–Nevada. *Limnology and Oceanography* 33(6 part 1): 1321–1333.

Goldman, C. R. 1974. Eutrophication of Lake Tahoe emphasizing water quality. Report EPA–660/3–74–034.

Jassby, A. D., J. E. Reuter, and C. R. Goldman. 2003. Determining long-term water quality changes in the presence of climate variability: Lake Tahoe (USA). *Canadian Journal of Fisheries and Aquatic Sciences* 60:1452–1461.

LaBounty, J. F. and Roefer, P. 2007. Quagga mussels invade Lake Mead. *LakeLine* 27:17–22.

LRWQCB and NDEP (Lahontan Region Water Quality Control Board and Nevada Department of Environmental Protection). 2007. Draft Lake Tahoe TMDL Technical Report.

LTAISCC. 2009. Lake Tahoe Region Aquatic Invasive Species Management Plan.

Morse, J. T. 2009a. Assessing the effects of application time and temperature on the efficacy of hot-water sprays to mitigate fouling by *Dreissena polymorpha* (zebra mussels Pallas). *Biofouling* 25(7):605–610.

Morse, J. T. 2009b. Assessing the effects of application time and temperature on the efficacy of hot-water sprays to mitigate fouling by *Dreissena polymorpha* (zebra mussels Pallas). *Biofouling* 23:605–610.

Murphy, D. D. and C. M. Knopp (Eds.). 2000. *Lake Tahoe Watershed Assessment: Volume I.* General Technical Report PSW-GTR-175. Albany, CA: Pacific Southwest Research Station, Forest Service, U.S. Department of Agriculture. 753 pp. http://www.treesearch.fs.fed.us/pubs/26709 (Accessed on July 10, 2014).

NANPAC. 2000. Nonindigenous aquatic nuisance prevention and control act of 1990: As amended through P.L. 106–580, December 29, 2000. http://www.anstaskforce.gov/Documents/nanpca90.pdf (Accessed on July 10, 2014).

NISA. 1996. Public Law 104–332—Oct. 26, 1996. http://www.anstaskforce.gov/Documents/NISA1996.pdf (Accessed on July 10, 2014).

TERC (Tahoe Environmental Research Center). 2012. State of the lake report 2012. http://terc.ucdavis.edu/stateofthelake/StateOfTheLake2012.pdf (Accessed on July 10, 2014).

UC Davis. 2011. Personal communication and unpublished survey reports regarding Asian clam densities in Lake Tahoe. University of California, Davis, CA.

Wittmann, M. E., S. Chandra, A. Caires, M. Denton, M. R. Rosen, W. H. Wong, T. Teitjen, K. Turner, P. Roefer, and G. C. Holdren. 2010. Early invasion population structure of quagga mussel and associated benthic invertebrate community composition on soft sediment in a large reservoir. *Lake and Reservoir Management* 26:316–327.

Wong, W. H. and S. L. Gerstenberger. 2011. Quagga mussels in the western United States: Monitoring and management. *Aquatic Invasions* 6:125–129.

Zook, B. and S. Phillips. 2009. Recommended protocols and standards for watercraft interception programs for dreissenid mussels in the western United States. Prepared by Pacific States Marine Fisheries Commission for the Western Regional Panel on Aquatic Nuisance Species of the Aquatic Nuisance Species Task Force, Portland, OR. 49pp.

23

Idaho's Dreissenid Mussel Prevention Program: Implementing Policy Directives to Protect the State's Resources

Amy Ferriter and Eric Anderson

CONTENTS

ABSTRACT Quagga mussels (*Dreissena rostriformis bugensis*) and zebra mussels (*Dreissena polymorpha*) are dreissenid species native to eastern Europe and western Russia. First discovered in the Great Lakes in the late 1980s, they are now considered invasive species throughout North America. Dreissenid mussels use byssal threads to attach and *hitchhike* between unconnected waterbodies. Resource managers began to recognize the significance of the eastern North American invasion in the early 1990s. It is widely accepted that trailered boats introduced the mussels to western United States waters, where they were first reported in the Lake Mead National Recreation Area in January 2007. The invasive mussels are now found in all western states except Idaho, Oregon, Washington, Montana, Wyoming, Alaska, and Hawaii. The mussels would have a significant economic impact if they are introduced to Idaho's waters and infrastructure systems. In response to this threat, the Idaho Legislature enacted progressive Invasive Species Laws in 2008 to establish state agency authorities and prevent the mussels from being introduced to the state. Through these policies, the state operates highway-based stations to inspect boats that are entering the state. More than 100 mussel-fouled boats have been intercepted by Idaho's program to date. The chances of eradicating a new population of zebra or quagga mussels in an Idaho waterbody will depend directly on the ability of the state to respond quickly. Idaho has developed an exclusion strategy and contingency plan in the event prevention fails.

Background

Quagga mussels (*Dreissena rostriformis bugensis*) and zebra mussels (*Dreissena polymorpha*), closely related dreissenid species native to eastern Europe and western Russia, are thought to have been introduced to North America via the ballast of commercial ships traversing the St. Lawrence Seaway. University researchers first found zebra mussels attached to the surfaces of rocks, piers, and other underwater structures in Lake St. Clair, which is connected to Lakes Huron and Erie, in 1988. It is unclear how long the species was present in US waters before detection.

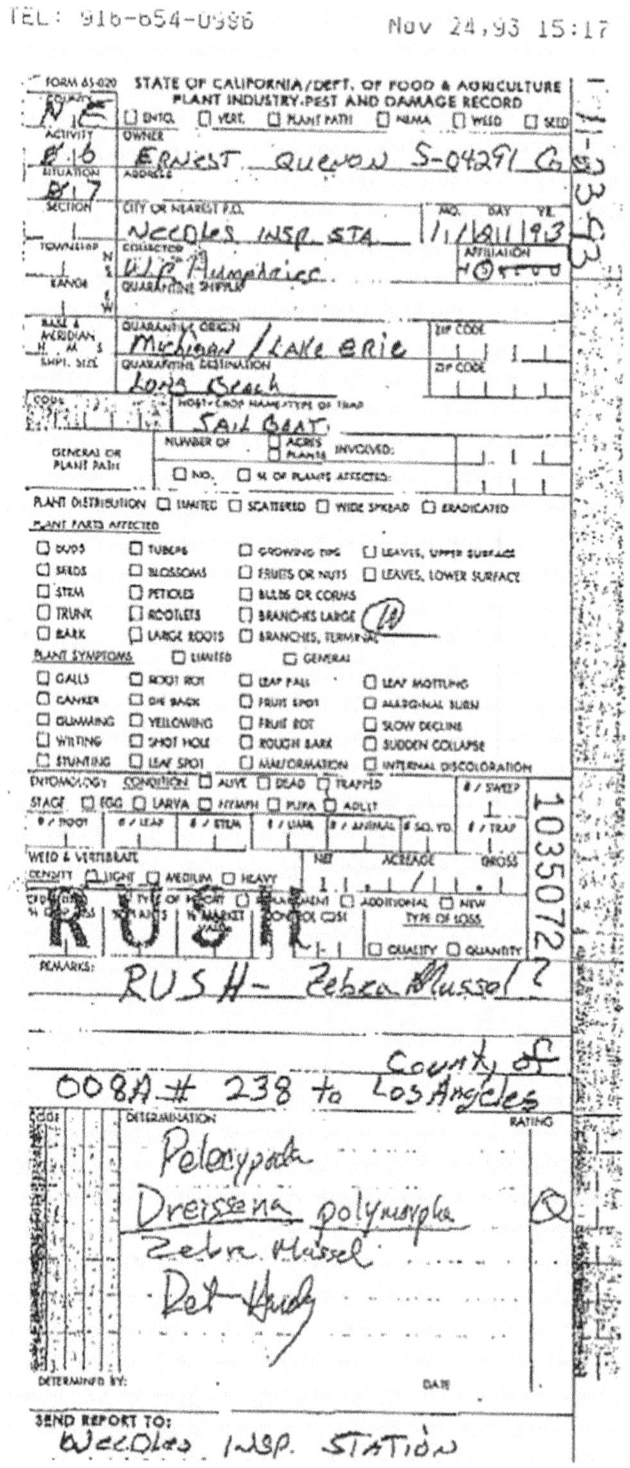

TEL: 916-654-0986 Nov 24,93 15:17

FORM 65-020 STATE OF CALIFORNIA/DEPT. OF FOOD & AGRICULTURE
PLANT INDUSTRY-PEST AND DAMAGE RECORD

COUNTY: NE 1C
ENTO. VERT. PLANT PATH NEMA WEED SEED
ACTIVITY: 816
OWNER: ERNEST QUEVON S-04291 Ca
SITUATION: 817
ADDRESS
SECTION
CITY OR NEAREST P.O.: NEEDLES INSP. STA
MO. DAY YR. 11 18 11 93
TOWNSHIP
COLLECTOR: W.R. Humphries
AFFILIATION
RANGE
QUARANTINE SHIPPER
BASE & MERIDIAN
QUARANTINE ORIGIN: Michigan / LAKE ERIE
ZIP CODE
QUARANTINE DESTINATION: Long Beach
ZIP CODE
HOST-CROP NAME/TYPE OF TRAP: SAIL BOAT
GENERAL OR PLANT PATH
NUMBER OF ACRES PLANTS INVOLVED
NO. % OF PLANTS AFFECTED
PLANT DISTRIBUTION: LIMITED SCATTERED WIDE SPREAD ERADICATED
PLANT PARTS AFFECTED
BUDS TUBERS GROWING TIPS LEAVES, UPPER SURFACE
SEEDS BLOSSOMS FRUITS OR NUTS LEAVES, LOWER SURFACE
STEM PETIOLES BULBS OR CORMS
TRUNK ROOTLETS BRANCHES LARGE
BARK LARGE ROOTS BRANCHES, TERMINAL
PLANT SYMPTOMS LIMITED GENERAL
GALLS ROOT ROT LEAF FALL LEAF MOTTLING
CANKER DIE BACK FRUIT SPOT MARGINAL BURN
GUMMING YELLOWING FRUIT ROT SLOW DECLINE
WILTING SHOT HOLE ROUGH BARK SUDDEN COLLAPSE
STUNTING LEAF SPOT MALFORMATION INTERNAL DISCOLORATION
ENTOMOLOGY CONDITION ALIVE DEAD TRAPPED # / SWEEP
STAGE EGG LARVA NYMPH PUPA ADULT
/ ROOT # / LEAF # / STEM # / LIMB # / ANIMAL # SQ. YD. # / TRAP
WEED & VERTEBRATE NET ACREAGE GROSS
DENSITY LIGHT MEDIUM HEAVY
TYPE OF REPORT ADDITIONAL NEW
TYPE OF LOSS
QUALITY QUANTITY
RUSH
10350722
REMARKS: RUSH- Zebra Mussel ?

008A # 238 to County of Los Angeles

DETERMINATION RATING
Pelecypoda
Dreissena polymorpha Q
Zebra Mussel
Det-Hardy
DETERMINED BY: DATE
SEND REPORT TO: NEEDLES INSP. STATION

FIGURE 23.1 First record of zebra mussels in the western United States.

Unfortunately, containment efforts after the initial discovery in North America were inadequate, largely due to a general failure of officials to recognize the significance of the introduction. Consequently, necessary resources were not allocated. By 1989, zebra mussels had spread to Lake Ontario and the St. Lawrence River. By the early 1990s, the mussels were found throughout the Great Lakes and in major eastern US river systems connected to the Great Lakes watershed via the Chicago Sanitary and Ship Canal. This included the Ohio, the Mississippi, and the Missouri Rivers.

In addition to the passive spread of mature and immature stages of the mussels in flowing water, dreissenid mussels use byssal threads to actively attach to trailered boats, docks, anchors, and related gear, allowing for frequent *hitchhiking* on such equipment between unconnected waterbodies. This particular pathway has intensified the rate of spread in the United States and Canada.

Early efforts to systematically inspect trailered boats began at California Department of Food and Agriculture (CDFA) Border Protection Stations in the early 1990s. Records from the CDFA show that zebra mussels were first intercepted on a sailboat from Lake Erie at Needles, California, in 1993 (Figure 23.1). This was the first record of zebra mussels in the western United States (C. McNabb, personal files).

Alarmed by this interception, which occurred approximately 5 years after the first mussels were detected in the Great Lakes, resource managers began to recognize the significance and potential consequences of the North American Dreissenid invasion. During the early 1990s, the United States Fish and Wildlife Service (USFWS), the National Park Service (NPS), and the United States Bureau of Reclamation (USBR) conducted many interagency meetings and mussel workshops in Denver, Boise, and Chicago where participants struggled to develop strategies to address the issue (C. McNabb, personal files). Interestingly, many of the strategies that were developed in those meetings are still relevant today.

In 1994, the USFWS published a document entitled *Feasibility of Preventing Further Invasion of the Zebra Mussel into the Western United States* (Tyus et al. 1994). This forward-thinking report aimed to determine the feasibility of preventing westward spread and establish the USFWS as the lead agency in this endeavor. The report determined pathways and vectors for westward spread, including detailed information on major highways and other land routes across the Continental Divide, encouraging roadside inspections for all boats traveling west. The authors also recognized the potential for passive spread hydrologically, identifying major east-west rivers. If these strategies had been implemented early in the invasion process, the ecological and economic problems associated with these species in western states would most likely have been avoided.

The popularity of water-based recreation makes this pathway difficult to manage. For example, the Lake Mead National Recreation Area has more than 7 million visitors per year, thousands of which engage in boating activities, attracted by the availability of day use launches and moorage facilities. It is widely accepted that trailered boats were the pathway that introduced dreissenid mussels to Lake Mead sometime prior to January 2007.

Following the detection of dreissenid mussels in Lake Mead in early 2007, the mussels quickly spread to connected waters and reservoirs in Arizona and Southern California via the California Aqueduct and Central Arizona Project. Quagga and/or zebra mussels have also invaded many other hydrologically disconnected waterbodies in the western states of Nevada, Arizona, California, New Mexico, Colorado, Texas, and Utah. To date, invasive mussels have not been found in Idaho, Oregon, Washington, Montana, Wyoming, Alaska, or Hawaii.

The State of Idaho: What Is at Risk?

Congressional researchers estimate that the zebra mussel infestation in the Great Lakes in the period from 1993 through 1999 cost the power industry $3.1 billion, with a total economic impact on industries, businesses, and communities in the area of more than $5 billion. In response to extensive documentation of negative impacts these species have had in the Great Lakes, and the subsequent discovery of the mussels in the Western United States, the state of Idaho (IISC, 2009) conducted an analysis of the potential effects these species would have on Idaho's environment and industries.

The Idaho Invasive Species Council report reviewed existing databases and published manuscripts to generate estimates on possible occurrences in Idaho. The results reflect an estimated cost of direct and indirect impacts on infrastructure and facilities that use surface water. Most of the published data that

TABLE 23.1

Estimated Potential Costs of a Dreissenid Invasion to the State of Idaho

Facility	Number	Estimated Cost Per Unit	Estimated Cost Statewide	Citation
Hydro power	26	$1,817,000	$47,242,000	Phillips et al. (2005)
Other dams	86	$1,730	$148,700	O'Neill (1997)
Drinking water	100	$42,870	$4,287,000	O'Neill (1997)
Golf courses	114	$150	$17,100	O'Neill (1997)
Boat facilities	380	$750	$285,000	O'Neill (1997)
Hatcheries/aquaculture	194	$5,860	$1,136,800	O'Neill (1997)
Boat maintenance	90,000	$265	$23,850,000	Vilaplana and Hushak (1994)
Angler days (4% reduction)	2,917,927	$150	$17,507,500	Vilaplana and Hushak (1994)
Irrigation POD	56,175			Little current published data
Total estimate			$94,474,000	

were reviewed did not report annual costs; however, annual maintenance costs are expected to increase for all categories examined. In some cases, economic impacts could not be estimated. For example, no comparable economic data exist for mussel impacts to irrigation systems; therefore, they are excluded from the potential cost estimates. The estimates are considered conservative and for the most part are reported in 1997 dollars, not having been adjusted for inflation (Table 23.1).

The following categories were examined:

Hydro power: Estimates were based on a study commissioned by the Bonneville Power Administration that examined the estimated hydropower maintenance costs associated with zebra mussel. Costs associated with control of Asian clams at the Bonneville Dam First Powerhouse and a survey of costs of zebra mussel mitigation at other hydropower generation facilities in North America were used. The study estimated the costs for installing sodium hypochlorite systems and applying antifouling paint to 13 federal hydroelectric projects in the Columbia River Basin. The Idaho estimate was based on the Bonneville Power Administration (BPA) average cost per project ($1.8 million) for the 26 hydropower dams in Idaho (Phillips et al. 2005).

Other dams: Data from water impoundment structures not associated with power generation were included in estimates as these structures would most likely incur maintenance costs associated with mussel fouling of pipes and structures. Estimates based on figures from O'Neil (1997) for navigational lock structures ($1700 per structure) applied to 86 structures in the state.

Drinking water intakes: Estimates included information from drinking water facilities that draw surface water for municipal or public drinking water use. Mussels foul intake piping and water processing infrastructure, increasing maintenance costs and degrading water flavor due to mussel waste and decomposition in water lines. Private single family homes with water intakes for drinking and irrigation were not included in this estimate. Estimates based on O'Neill (1997) figures from water treatment facilities ($42,000 per facility) applied to 100 facilities in Idaho.

Golf courses: Golf courses are also at risk for additional maintenance costs for irrigation systems. Fouling of pipes and pumps and clogged sprinklers are projected to increase operating expenses. Estimates based on O'Neill (1997) costs from golf courses ($150 per facility) were applied to 114 Idaho courses.

Boating facilities: Boating facilities included marinas, docks, and boat launches located in Idaho. Increased cost estimates are based on maintenance associated with dock and boat launch fouling. Estimates based on O'Neill (1997) figures from marinas ($750 per facility) were applied to 380 Idaho facilities.

Fish hatcheries and aquaculture: Hatcheries and aquaculture facilities are vulnerable to fouling from dreissenid mussel. Pipes, pumps, and raceway structures would be subject to increased operations and maintenance costs. Estimates based on O'Neill (1997) figures for hatcheries and aquaculture impacts ($5800 per facility) were applied to 163 facilities in Idaho.

Boater costs: More than 90,000 motorized boats were registered in the state of Idaho in 2007. Potential increases in costs to boaters are based on estimates for antifouling paints and increased per-boat maintenance costs. Estimates were based on Vilaplana et al. (1994) for increases in boater maintenance costs ($265 per boat).

Fishing use: Recreational fishing is a $430 million industry in Idaho. Research related to impacts of mussels on fisheries is limited, but reductions of fish numbers are likely. Vilaplana et al. (1994) found a 4% decrease in boater recreation because of introduction of mussels. The Idaho estimate was based on a 4% reduction of use applied to 2,917,972 Idaho fishing trips a year averaging $150 per trip (IDFG 2003).

Irrigation: Approximately 56,175 points of diversion (POD) were identified in Idaho by the Idaho Department of Water Resources. Multiple points of use (POU) may be associated with each POD. Each POD and POU could potentially be affected by dreissenid mussels. The mussels can grow up to 0.5 mm/day under ideal conditions, and could impact water conveyances that are seasonally dry. Fouling and shell production from mussel establishment is cumulative; increased fouling and flow reduction could occur in ditches, pipes, pumps, fish screens, and diversion structures over time. Published research on mussel-related flow reduction in irrigation systems is minimal, but mussel establishment in pipes and pumps is well documented. The true impacts of dreissenid mussel introduction on irrigated agriculture in Idaho are uncertain, but there is a high likelihood that these mussels will increase maintenance costs for operations that rely on surface water for irrigation.

Idaho's Policy

Within months of the discovery of quagga mussels in Lake Mead on January 7, 2007, the Idaho legislature began drafting legislation to address the threat of dreissenid mussels to Idaho's waters. At the time, Idaho did not have an Invasive Species Law. Existing laws and rules such as the Idaho Noxious Weed Law, the Idaho Plant Protection Act, and the Deleterious Animal Rule provided authorities for certain taxa, and groups of species, but the authorities were piecemeal, and the state lacked a comprehensive Invasive Species Law.

The Idaho Invasive Species Law (Title 22, Chapter 19, Idaho Code) was enacted by the Legislature in 2008. The law provides policy direction, planning, and authorities to combat invasive species and to prevent the introduction of new invasive species to the state. This law establishes the duties of the Idaho State Department of Agriculture (ISDA) and its director, authorizes the ISDA director to promulgate rules, and gives the state authorities to conduct inspections as necessary. It also establishes an invasive species fund.

The resultant Idaho Invasive Species Rules (IDAPA 02.06.10) were promulgated by ISDA and underwent extensive negotiation with impacted stakeholders. Engaged stakeholders included the Idaho Water Users, Northside Canal Company, Aberdeen-Springfield Canal Company, Clear Springs Foods, the Aquaculture Association, the Idaho Farm Bureau, the Idaho Department of Fish and Game, the Idaho Conservation League, The Nature Conservancy, Trout Unlimited, Boise Canal Company, the Pend Oreille Basin Commission, Bonner County, and several western states. The rules outline the duties of the department and govern the designation of invasive species, inspection, permitting, decontamination, recordkeeping, and enforcement of regulated species.

The Idaho Invasive Species Act of 2008 created the Idaho Invasive Species Fund, but the fund lacked a dedicated funding source. The Invasive Species Prevention Sticker Rules (IDAPA 26.01.34) were enacted by the legislature in 2009. The rules, which are under the Safe Boating Act (Title 67, Chapter 70, Idaho Code), require motorized and nonmotorized boats to display an invasive species sticker to launch and operate on Idaho's waters (Figure 23.2). The sticker program established annual user fees for resident, nonresident, and nonmotorized vessels. Revenue generated from this program (~$1.7 million/annually) is deposited in the invasive species fund. The fund is administered by the ISDA. With revenue generated by the Invasive Species Prevention Sticker program, ISDA developed a comprehensive statewide prevention

FIGURE 23.2   Idaho's invasive species sticker program is a dedicated funding source for the state's operational inspection stations.

program designed to educate the public about invasive species, monitor Idaho waterbodies for possible introduction of those species, and inspect and decontaminate watercraft that travel to and through Idaho (ISDA 2014).

Idaho's invasive species policy provides basic authorities for mandatory inspections and decontaminations of infested conveyances. It also provides several unique elements that contribute to Idaho's progressive prevention program.

Key elements include the following:

• Idaho's white list
  Section 102 of the rules governing invasive species provides that Idaho has a *white list* for invasive species. New species must be approved by the ISDA prior to import. It states:

  INTRODUCTION OF NEW SPECIES TO THE STATE. Species that are not previously known to occur in Idaho cannot be introduced to the state without a determination from the Department that the subject species is not invasive.

• Invasive species fund
  Section 22-1911 of the Idaho Invasive Species Law provides that the invasive species fund can not only receive revenue but can also accrue and carry forward funds. This is important to the invasive species program, as species outbreaks are inconsistent by nature. The ability to accrue funding and carry balances forward allows the state to spend funding effectively. It states:

  22-1911. INVASIVE SPECIES FUND. There is hereby established in the state treasury an invasive species fund. (1) The fund shall receive such appropriations as deemed necessary by the governor and the legislature to accomplish the goals of this chapter. The fund shall also receive moneys from the collection of reasonable fees for permits or as otherwise required by this chapter or rules promulgated hereunder. The fund may also receive, at the discretion of the director, moneys from any other lawful source including, without limitation, fees, penalties, fines, gifts, grants, legacies of money, property, securities or other assets, or any other source, public or private. (2) Moneys in the invasive species fund are subject to appropriation for the purposes of this chapter. The fund shall be used to support activities related to the prevention, detection, control and management of invasive species in Idaho. (3) All interest or other income accruing from moneys deposited to the fund shall be redeposited and accrue to the fund. Any unexpended balance left in the fund at the end of any fiscal year shall carry forward without reduction to the following fiscal year.

- Idaho's emergency fund
  Section 22-1912 of the Idaho Invasive Species Law authorizes the use of a deficiency warrant in the event of an infestation. ISDA is authorized to spend up to $5 million in unbudgeted general funds annually for control and eradication costs.

  22-1912. CONTROL AND ERADICATION COSTS—DEFICIENCY WARRANTS—COOPERATION WITH OTHER ENTITIES AND CITIZENS. Whenever the director determines that there exists the threat of an infestation of an invasive species on state-owned land or water, private, forested, range or agricultural land or water, and that the infestation is of such a character as to be a menace to state, private, range, forest or agricultural land or water, the director shall cause the infestation to be controlled and eradicated, using such moneys as have been appropriated or may hereafter be made available for such purposes. Provided however, that whenever the cost of control and eradication exceeds the moneys appropriated or otherwise available for that purpose, the state board of examiners may authorize the issuance of deficiency warrants against the general fund for up to five million dollars ($5,000,000) in any one (1) year for such control and eradication. Control and eradication costs may include, but are not limited to, costs for survey, detection, inspection, enforcement, diagnosis, treatment and disposal of infected or infested materials, cleaning and disinfecting of infected premises or vessels and indemnity paid to owners for infected or infested materials destroyed by order of the director. The director, in executing the provisions of this chapter insofar as it relates to control and eradication, shall have the authority to cooperate with federal, state, county and municipal agencies and private citizens in control and eradication efforts; provided, that in the case of joint federal/state programs, state moneys shall only be used to pay the state's share of the cost of the control and eradication efforts. Such moneys for which the state shall thus become liable shall be paid as a part of the expenses of the Idaho state department of agriculture out of appropriations that shall be made by the legislature for that purpose from the general fund of the state. In all appropriations hereafter made for expenses of the department, account shall be taken of and provision made for this item of expense.

State Response

The Idaho invasive species program was the first of its kind in the United States. Several western states have since patterned funding, inspection, and prevention programs after the Idaho model. Nationwide, the majority of aquatic nuisance species laws and authorities are housed within fish and game departments. Idaho is the only state in the country where the lead agency for aquatic nuisance species is a department of agriculture. Departments of agriculture in the United States have a long history of detection, prevention, and quarantine of invasive species—from insect pests to pathogens to noxious weeds. This expertise and the nexus of invasive species threats to irrigated agriculture make ISDA the most logical choice for the inspection program within Idaho state government.

The invasion of dreissenid mussels to western waterbodies has resulted in increased prevention efforts across the region. At the state level, numerous western states have increased efforts in mussel prevention through enhanced monitoring, public outreach, and watercraft inspection programs.

FIGURE 23.3   Trailered boats are a major pathway for dreissenid mussels to the Pacific Northwest.

It is notable that the western watercraft inspection programs are funded with few federal dollars, since nearly all states, including Idaho, fund the programs with state boater license fees, user fees, sticker fees, or general funds. Of particular concern to many western resource managers is the continued state interception of mussel-fouled watercraft originating from federally managed waterbodies in the Lower Colorado River (Figure 23.3).

Idaho's resource managers developed a progressive and proactive prevention program to minimize the risk of introduction to Idaho's waters via mussel-fouled watercraft. Idaho's watercraft inspection program began in 2009. Idaho's inspection stations are placed on major highways at or near the Idaho state line (Figure 23.4). This strategy is designed to maximize contact with boats that are traveling into the state from mussel-infested states. The inspection stations on the southern and eastern borders of the state intercept the majority of the mussel-fouled boats. The Idaho inspection program has inspected boats from every state in the United States (Figure 23.5). A comprehensive summary of Idaho state inspection data can be found on the ISDA website (ISDA 2014).

Boats that have been in mussel-infested states recently (within the last 30 days); watercraft coming from another state (especially commercially hauled boats); boats that show a lot of dirt, grime, or slime below the waterline; or boats that have standing water on board are considered *high risk* to the state of Idaho. High-risk inspections are intense and include a thorough inspection of the exterior and interior parts of the boat.

Idaho's inspections include a thorough and complete visual and tactile examination of all components of the boat, including compartments, bilge, trailer and any equipment, gear, ropes, or anchors (Figure 23.6). If any biological material is found on the vessel or equipment, the inspectors conduct a roadside *hotwash* of the watercraft. This is done to prevent the spread of other invasive species such as New Zealand mudsnail, Eurasian watermilfoil, and hydrilla. Boats that have mussels attached are impounded and decontaminated per ISDA policy.

In 2011, Idaho began issuing voluntary Invasive Species Passports to local boaters (Figure 23.7). This system gives Idaho and Pacific Northwest boaters an expedited *fast pass* when they repeatedly come through Idaho's stations. Boaters are issued a uniquely numbered passport booklet at the beginning of the season. They show the assigned number to inspectors during subsequent inspections throughout the boating season. Inspectors ask boaters if they have left the Pacific Northwest in the last 30 days. If the answer is no, the boat receives an expedited inspection, the passport is stamped

FIGURE 23.4 Idaho's inspection stations are strategically placed on major roadways to intercept boats from high-risk waters.

with the inspection station location, and the boater's information is logged with a handheld data unit. This program dramatically reduces field data collection time and allows for tracking of repeat boaters.

Several stations (such as the one located at I-90 eastbound from the state of Washington) inspect a large volume of boats that travel between the Spokane (WA) area and the lakes of northern Idaho. The passport system allows inspectors to quickly screen boaters based on risk. This is especially critical during busy times when inspectors are able to give low-risk boats an expedited inspection and spend additional time scrutinizing high-risk boats that have come into the region from elsewhere. The system was well received by the boating community and could serve as a model for a regional *Pacific Northwest* (PNW) *Invasive Species Passport*. The concept of a regional passport will become increasingly valuable as other western states establish roadside inspection programs.

More than 154,000 watercraft inspections were conducted in Idaho from 2009 to 2013. Ninety three (93) mussel-fouled boats were intercepted in Idaho during this 5-year time period. The majority of these boats came from federally managed waterbodies in the lower Colorado River system (ISDA 2014).

FIGURE 23.5   Idaho's program inspections boats from every state in the country.

FIGURE 23.6   Inspectors are trained to feel the outsides of the vessels and look closely at the outside and inside of boats entering the state.

What If Prevention Fails in Idaho?

Although the chances of eradicating a new population of zebra or quagga mussels in an Idaho waterbody are small, success depends directly on the ability of the state to respond quickly (and effectively) once a nascent population is detected. There is an urgent need to develop control technologies for dreissenid mussels in Idaho's systems. Water managers in impacted western states (i.e., CA, NV, AZ, and TX)

FIGURE 23.7   Idaho's Invasive Species Passport system began in 2011.

have been forced to scramble to develop control technologies within water delivery infrastructure systems. This work began in 2007, shortly after the discovery of the mussels in the Lake Mead National Recreation Area. Unfortunately, control options for lakes, rivers, and naturally flowing river systems remain poorly developed.

To date, there are no known control technologies available for use outside of closed (infrastructure-type) systems. Applied research is needed to find new tools to eradicate or contain these species in a field response situation in Idaho. Waterbodies such as the Snake River have numerous private and public stakeholders who have access or management authorities. Diversion facilities for irrigation, hydroelectric power generation, municipal water systems, aquaculture, and recreation are just a few of the uses and management influences on the river.

In 2009, the Idaho Invasive Species Council convened a roundtable of stakeholders, including conservation groups, water users, canal companies, irrigation districts, utilities, municipal water companies, and germane state and federal agencies, to determine what steps should be taken to prepare for an invasion of zebra or quagga mussels. Participants were asked to weigh options in the event that these species are discovered in the state. Given the complexities of preventing and treating waterbodies if quagga or zebra mussels are discovered in Idaho, the group recommended that the state develop an *Exclusion Strategy and Contingency Plan*.

The state of Idaho, in cooperation with the Aquatic Ecosystem Restoration Foundation (AERF), assembled a panel of experts to develop the contingency plan. The group's goals were to compile a summary of Idaho's waterbody data, review available control technology options, and assess Idaho's technical and regulatory gaps, including endangered and threatened species concerns. The *Exclusion Strategy and Contingency Plan* was completed in early 2012 (AERF 2012).

The AERF report clearly stated that the discovery of exotic mussels in large river run reservoirs in Idaho would most likely be impossible to eradicate. This conclusion was based in part on the length of time (often weeks or months) between the collection of monitoring samples and subsequent analyses with confirmation. This temporal lag would likely allow mussel populations to reproduce and spread beyond pioneer infestations in marinas or boat moorage locations into free flowing reservoirs and rivers. In addition, an eradication program in large reservoirs would undoubtedly be cost prohibitive.

The AERF panel found that the biology and ecology of dreissenid mussels are well known, as are their pathways of spread. However, data and control technologies that could be used for proactive monitoring and management of mussels are lacking. The panel concluded that there are currently no economic or technical means to control exotic mussels in large river run reservoirs such as the Snake River, Lakes Pend Oreille, Lake Coeur d'Alene, and similar bodies of water in Idaho. Unfortunately, most of the molluscicides that could be used for dreissenid mussel control are toxic to native mussels. In addition, many of these products are toxic to fish, some of which are classified as threatened or endangered.

Additional AERF Panel Recommendations

- The ISDA should share the invasive mussel contingency plan with federal agencies and adjoining states in an effort to foster greater cooperation and to integrate and increase regional prevention efforts.
- Idaho and neighboring states should encourage, insist on, and use all possible means to ensure that states and authorities enforce decontamination of boats leaving mussel-infested waters. Every one of these vessels has the potential to infest other waterways, so it is common sense—from both environmental and economic standpoints—to ensure that departing vessels are decontaminated.
- Increase, improve, and intensify prevention and monitoring programs.
- Educational efforts need to be expanded to include marina owners/operators and commercial boat haulers.
- Engage with a rapid response team (similar to the Columbia River Basin Team that includes WA, OR, MT, and ID) that represents appropriate state, federal, and tribal interests from all neighboring states to evaluate and coordinate the response following any suspected finds of mussels, develop specific treatment plans, and define agency responsibilities and commitments.
- Work with other Idaho state agencies to include mussel exclusion clauses and inspections in state construction contracts when equipment (barges, silt barriers, water tanks, etc.) might be brought into Idaho and placed in waters of the state.
- Collect additional water temperature data on highly vulnerable lakes to optimize the timing of monitoring efforts.

Discussion and Future Needs

Hindsight is always 20/20. Had resource managers initiated effective containment in the Great Lakes in the early 1990s, there is a good chance dreissenid mussels would not have crossed the 100th meridian and invaded western waters. Prevention is an improbable award winner, though. It is extremely difficult to measure (or appreciate) the value of what was prevented, especially when it comes to invasive species.

While states in the Pacific Northwest are doing their best to intercept mussel-contaminated boats coming into the region, inspection and decontamination at the *source*-infested waterbodies is lacking. As evidenced by the watercraft inspection data for the Pacific Northwest, the majority of infested boats entering the region come from the lower Colorado River, and from Lake Mead in particular, which puts the long-term viability of state prevention programs in question. Many western states and organizations such as the Northwest Power and Conservation Council, The Nature Conservancy, the Pacific Northwest Economic Region, and Lake Tahoe have joined Idaho to advocate for inspections and decontaminations of departing watercraft from federally managed mussel-infested waters. In particular, these groups have asked the U.S. Department of the Interior to implement a mandatory inspection and decontamination program for moored watercraft departing the Lake Mead National Recreation Area.

The federal government demonstrated good intentions early in the North American dreissenid invasion curve. Several meetings and workshops were held where innovative strategies were developed in the early 1990s, but decisive action was not taken; dreissenid mussels spread throughout the country's waterways. Learning from this example, Idaho acted quickly to launch a unique invasive species program when the mussels were discovered in the western United States in 2007. The proactive program includes legislation, authorities, a dedicated funding source, and political support.

The ISDA reports (ISDA 2014) that in the program's first 5 years, approximately 100 mussel-fouled boats were intercepted by the state inspection program. Although the state plans to continue this important work, it recognizes that federal partners must also effectively manage mussel-infested waterbodies such as Lakes Mead, Mojave, Havasu, and Powell. For Idaho's program to be successful, federal agencies must do their part to institute parallel and complementary mandatory federal inspection and decontamination programs at *point sources* on the Lower Colorado system. Additionally, states in the Pacific Northwest must work cooperatively as a region instead of focusing on protecting individual jurisdictions. Individual states are connected by highways and by flowing waters; it is critical that all states in the Pacific Northwest recognize their dependency on each other and work cohesively to prevent these species from being introduced to the region. This is the only way to a sustainable and effective containment and prevention strategy for the western United States.

REFERENCES

Aquatic Ecosystem Restoration Foundation 2012, A review of the state of Idaho dreissenid mussel prevention and contingency plans. AERF http://www.aquatics.org/musselreport.pdf, accessed December 15, 2013.

Idaho State Department of Agriculture 2014, Idaho invasive species watercraft inspection program 5 year review 2009–2013. http://www.agri.idaho.gov/Categories/Environment/InvasiveSpeciesCouncil/documents/DataReviewFINAL011514.pdf, accessed March 15, 2014.

IDFG Economic Report 2003, http://fishandgame.idaho.gov/cms/fish/misc/03econstudy/bonner.pdf, accessed March 3, 2009.

O'Neill, C. 1997. Economic impact of zebra mussels: Results of the 1995 zebra mussel information clearinghouse study. *Great Lakes Research Review*, 3(1): 35–42.

Phillips, S., T. Darland, and M. Systma. 2005. Potential economic impacts of zebra mussels on the hydropower facilities in the Columbia river basin. Pacific States Marine Fisheries, Portland, OR.

Tyus, H., P. Dwyer, and S. Whitmore. 1994. Feasibility of preventing the further invasion of the zebra mussel into the western United States. U.S. Fish and Wildlife Service, US Government Printing Office, Report Number 1994-576-764/05146, Washington, DC, 43pp.

Vilaplana, J.V. and L.J. Hushak. 1994. Recreation and the zebra mussel in Lake Erie, Ohio. Technical Summary. OHSU-TS-023. Ohio Sea Grant College Program. Columbus, OH.

Section VI

Monitoring

24

Implementation of a Cost-Effective Monitoring and Early Detection Program for Zebra Mussel Invasion of Texas Water Bodies

Robert F. McMahon

CONTENTS

ABSTRACT   A cost-effective zebra mussel (*Dreissena polymorpha*) risk assessment and monitoring/ early detection system specifically designed for Texas and other southwestern United States water bodies was conducted at 14 Texas lakes during spring and fall 2011 and spring 2012. The water bodies included zebra mussel–infested Lake Texoma (Grayson and Cook counties) and 13 zebra mussel–uninfested water bodies: Lake Arrowhead in the Red River Basin; Lakes Bridgeport, Eagle Mountain, Lewisville, Lavon, Ray Hubbard, and Ray Roberts in the Trinity River Basin; Lake Wright Patman in the Sulfur River Basin; Lakes Tawakoni and Fork in the Sabine River Basin; and Lakes Bob Sandlin, O' the Pines, and Caddo in the Cypress River Basin. Invasion risk assessment was based on published tolerated limits of oxygen saturation, pH, calcium concentration, and upper temperature limits in southwestern zebra mussel populations. Monitoring consisted of cross-polarized light microscopic examination and quantitative polymerase chain reaction (qPCR) testing for mussel DNA in concurrent vertical-tow, near-shore plankton samples along with mussel settlement on scouring pad monitors conducted from marina or privately owned docks, eliminating the need to sample from boats. Five of the water bodies, including infested Lake Texoma, were assigned a high risk for successful invasion as all four tested risk parameters fell within the range for successful invasion. Five were considered of moderate risk because one or more risk factors (primarily mean daily August surface water temperature) fell within a range considered marginal (i.e., approached or periodically occurred outside tolerated limits) for successful invasion and four were considered of low risk because one or more risk factors (primarily calcium concentration and/or August mean daily surface water temperature) fell outside known mussel tolerance limits. Dissolved oxygen (DO) levels were within tolerated limits in all 14 water bodies. Juvenile mussels only occurred on settlement monitors in mussel-infested Lake Texoma. During spring 2011, qPCR detected mussel DNA at high levels in Lake Texoma and at low levels in Lakes Lavon and Ray Roberts, while in fall 2011, it was detected at low levels in Lakes Arrowhead, Bridgeport, Eagle Mountain, Lewisville, Ray Roberts, and Caddo. The fall 2011 increase in the number of water bodies yielding positive qPCR results was attributed to transport of veligers and mussel DNA in water contained in and/or mussels attached to boats during the summer 2011 boating season. A second consecutive qPCR signal occurred only for Lake Ray Roberts in spring 2012 accompanied by detection of a single veliger larva in a plankton sample. A subsequent July 2012 inspection by Texas Parks and Wildlife Department personnel revealed low numbers of settled juvenile mussels in the lake confirming infestation. Thus, the monitoring system (particularly microscopic and qPCR examination of near-shore vertical plankton tow samples) was considered to be effective for early detection of mussel invasion in Texas and other southwestern reservoirs. Of four previously uninfested water bodies later confirmed to harbor mussel larvae and/or settled mussels by January 2014, two (Ray Roberts and Lavon) were previously assigned a *high* risk and two (Lakes Bridgeport and Lewisville) a *moderate* risk of mussel invasion. In contrast, zebra mussel larvae or adults have not been found in the four lakes assigned a *low* invasion risk as of February 2014, suggesting that particularly calcium concentration and mean maximum August surface water temperature are efficacious for evaluating mussel invasion risk in most Texan and southwestern water bodies. The project appeared to attain its objectives of developing an accurate and cost-effective zebra-mussel risk assessment/monitoring system for Texas and other southwestern water bodies and informing the public at large of the economic and ecological threats posed by zebra mussels to Texas and other southwestern states.

Introduction

Invasion of United States Water Bodies by Zebra Mussels

Zebra mussels (*Dreissena polymorpha*), endemic to Europe, were first reported in Lake St. Clair of the Laurentian Great Lakes in 1988 where they were presumed to be introduced via larval transport in the ballast water of commercial ships from the Black Sea or as adults attached to anchors and anchor chains (Mackie and Schloesser 1996). The mussels initially spread rapidly throughout the lower Great Lakes and St. Lawrence River and entered the Mississippi River through the Chicago Sanitary and Ship Channel (i.e., Illinois and Des Plaines Rivers) in 1991. Since then, this species has spread rapidly throughout the Mississippi River and its major tributaries (i.e., the Ohio, Tennessee, Cumberland, Missouri, Arkansas, and Red Rivers) (United States Geological Survey 2010a) mainly due to downstream hydrological transport of its planktonic veliger stage and bidirectional transport of juveniles and adults attached to barge and boat hulls (Mackie and Schloesser 1996). More recently, zebra mussel populations have also been discovered in isolated water bodies and drainages west of the Mississippi River Basin including Lake Offutt, Nebraska, in 2006 (URS Group, Inc. 2009); San Justo Reservoir, California, in 2007 (California Department of Fish and Game 2008); Lake Texoma, Texas/Oklahoma, on the Red River in 2009 (Texas Parks and Wildlife Department 2009a); and Lake Belton, Texas, on the Brazos River (Texas Parks and Wildlife Department 2013a). These new introductions have been presumed to be a result of overland transport of mussel larvae, juveniles, and/or adults on trailered boats or other submerged equipment previously in contact with mussel-infested waters. After initial introduction, zebra mussels can rapidly develop dense populations, which filter phytoplankton and bacterioplankton at high rates, increasing water clarity, stimulating macrophyte productivity, and diverting energy flow from nektonic to benthic communities. Thus, they can have major negative environmental impacts on invaded water bodies (McMahon and Bogan 2001, Qualls et al. 2007, Strayer 2009, Nalepa 2010).

Zebra mussel macrofouling also has major negative economic impacts. Their ability to attach to hard surfaces with proteinaceous byssal threads, along with their sister species, *Dreissena rostriformis bugensis* (quagga mussel), allows them to occlude flow in raw-water systems, negatively impacting operations in power stations, impoundments, potable water treatment plants, irrigation systems, and fire protection systems, among others (Miller et al. 1992, Mussalli et al. 1992, Connelly et al. 2007). They also exacerbate corrosion of metal surfaces reducing raw-water system life spans while increasing maintenance and repair costs (McMahon and Lutey 1996). Zebra mussel infestations also negatively impact sport and commercial fisheries, recreational and commercial boater operations and their sharp-edged shells make beaches and other shallow-water areas unsafe or unsuitable for bathers, all of which have negative impacts on local economies dependent on nearby water bodies or waterways (O'Neill 1996).

In 2004, the combined annual environmental and economic costs of zebra/quagga mussels in the United States were estimated to be approximately one billion dollars (Pimentel et al. 2005). The recent invasion of Lake Mead (Nevada/Arizona) by quagga mussels and its rapid expansion to other water bodies in Southern California, Colorado, and Arizona (United States Geological Survey 2010b) has almost certainly increased the ecological damages and socioeconomic costs associated with zebra and quagga mussels in the United States (Nalepa 2010, Wong and Gerstenberger 2011). Future uncontrolled dispersal of zebra and quagga mussels into other, as yet uninfested western US water bodies will only serve to increase their environmental and economic costs to the United States.

During the initial stages of their U.S. invasion, zebra/quagga mussels were generally restricted to drainages east of the 100th meridian (United States Geological Survey 2010a). Zebra mussels first entered the western tributaries of the Mississippi River in 1992 when they were discovered in the lower Arkansas River (Arkansas). By 1993, they had progressed up the Arkansas River into eastern Oklahoma. Zebra mussels were first discovered in the Missouri River below the Gavin Point and Fort Randall Dams in eastern Nebraska during 2003. After their initial 1992 establishment in the Arkansas River Drainage, zebra mussels remained restricted to the lower portions of the drainage downstream of Tulsa, Oklahoma. Then, beginning in 2003, mussel populations were confirmed at 23 further sites in the Arkansas River Basin extending their distribution to the state's northern border and the lower reaches of the Canadian River near its junction with the Arkansas River (United States Geological Survey 2010c). During that same period, zebra mussels were also first recorded in the upper Arkansas River drainage of Kansas in El Dorado (2003) and Cheney Lakes (2004). Since those initial discoveries, zebra mussels have continued to disperse through the Arkansas River Basin in Kansas being confirmed at 10 additional upstream sites since 2006. An additional two populations were discovered in the Kansas River Basin of northeastern Kansas in 2007 and 2009, respectively (United States Geological Survey 2010c).

Zebra Mussel Invasion of Lake Texoma (TX/OK)

Zebra mussels were discovered at the lower end of Lake Texoma, an impoundment of the Red River, on the border between Texas and Oklahoma in 2009. After discovery of a single specimen near the City of Denison, TX, on April 3, 2009 (Texas Parks and Wildlife Department 2009a), zebra mussels were recorded at three other sites in the lower end of the lake (Texas Parks and Wildlife Department 2009b). Thereafter, the Lake Texoma zebra mussel population expanded rapidly during the summer and fall of 2009 (Texas Parks and Wildlife Department 2009b) such that adult mussels were readily collectable in relatively large numbers in the lower portion of the lake by 2011 (personal observation of the author) where they have been recorded at depths ranging to >10 m (North Texas Municipal Water District [NTMWD], personal communication).

The expansion of zebra mussels into Lake Texoma was unexpected because lake surface ambient water temperatures during late summer (Gido and Matthews 2000, Lienesch and Matthews 2000) were generally equal to or exceeded the mussel's accepted incipient upper thermal limit of 28°C–30°C (McMahon and Tsou 1990, McMahon 1992), which should have made it resistant to infestation. Indeed, a genetic algorithm for rule-set analysis (GARP) model including average annual air temperatures and 11 other environmental factors used to predict the potential distribution of zebra mussels in the United States suggested that predicted mussel invasion probabilities for the eastern portions of Texas, Oklahoma, and Kansas were highly variable and considered unreliable (Drake and Bossenbroek 2004). The thermal resistance of Lake Texoma and other Texas water bodies appeared to be confirmed by the fact that even though boats were being trailered into Texas water bodies from surrounding states with zebra mussel infestations based on Kevin Buch's (2000) boater movement survey in northeastern Texas and later confirmed by the Texas Parks and Wildlife Department (2009a) and Bossenbroek et al. (2007), no dreissenid infestation had been reported for any Texas water body. This was the situation until April 3, 2009, when the first adult zebra mussel was found in the lake (Texas Parks and Wildlife Department 2009b), after which the mussel population rapidly expanded and increased in density.

Invasion Risk to Other Northeastern Texas Water Bodies

It is an open question why it took so long for zebra mussels to invade Lake Texoma in spite of the facts that they rapidly invaded eastern Kansas and Oklahoma water bodies and that mussel-infested recreational boats had been launched in Lake Texoma since mussels first infested the Oklahoma Arkansas River Basin (Buch 2000, Texas Parks and Wildlife Department 2009a). One possibility is that zebra mussel populations genetically isolated in the warm water bodies of upstream portions of infested Kansas and Oklahoma river basins could have been subject to selection for increased thermal tolerance, allowing evolution of increased incipient upper thermal limits that eventually allowed toleration of the elevated summer ambient surface water temperatures of Lake Texoma (Morse 2009, Garton et al. 2014). There is now evidence that this may have been the case. In Lake Oologah, OK, a zebra mussel population thrived and reached very high densities even though surface ambient water temperatures annually approached and periodically exceeded this species' previously accepted 30°C incipient upper thermal limit (Morse 2009). The apparent evolution of increased thermal tolerance by zebra mussel populations isolated in the upper portions of drainages in Kansas and Oklahoma was confirmed for a thriving zebra mussel population in Winfield City Lake (KS), isolated on the Arkansas River drainage near the southeastern Kansas/Oklahoma border. This population was exposed to summer water temperatures reaching 30°C and recorded as having an incipient (i.e., long-term, 28-day) upper thermal limit of 30.7°C, the highest recorded for a natural population of *D. polymorpha* in Europe or North America (Morse 2009). Its elevated incipient upper thermal limit suggested that this population represented an evolved physiological race that could survive in Lake Texoma where surface water temperatures approach but do not exceed 31°C (Gido and Matthews 2000, Lienesch and Matthews 2000, this study). Accordingly, such a thermally tolerant physiological race could be the source population for zebra mussel invasion of other northeastern Texas water bodies whose summer surface water temperatures reach peak values of 30°C–31°C (Ensminger 1999, Espey Consultants, Inc. 2007, this study). Similarly, rapid selection for increased thermal tolerance has been reported in eastern European dreissenid populations isolated in water bodies with elevated ambient water temperatures due to receipt of thermal effluents (Garton et al. 2014).

There is a high level of recreational boat traffic between zebra mussel–infested water bodies in Oklahoma including Lake Texoma and water bodies in northeastern Texas (Buch 2000). If thermally tolerant physiological races of *D. polymorpha* have developed in the isolated warm-water bodies of Kansas and Oklahoma (Morse 2009), allowing establishment of a sustainably reproducing zebra mussel population in Lake Texoma, other Texas water bodies and particularly those in northeastern Texas with annual surface water temperatures similar to that of Lake Texoma appear potentially at grave risk of infestation. In spite of this possibility, as of spring 2011, there had been no organized efforts to either assess the zebra mussel invasion risk potential or implement effective zebra mussel–monitoring programs for major water bodies in northeastern Texas, with the exception of monitoring by the US Geological Service for the presence of veliger larvae (plankton net samples) and scuba diving to search for adult mussels at a limited number of northeastern Texas water bodies serving as raw-water sources for the NTMWD including lakes Texoma, Lavon (initiated in 2010), Ray Hubbard, Lewisville, Grapevine, Ray Roberts, Fork, and Tawakoni (initiated in 2011) and Palestine (initiated in 2012) (Churchill and Baldys 2012). To date, no reports of the results of this monitoring program have been published. With a well-established zebra mussel population now in Lake Texoma and several other Texas water bodies (Texas Parks and Wildlife Department 2014a), it is imperative that an effective dreissenid risk assessment and early-detection monitoring system be established for major water bodies in Texas and throughout the southwestern United States. This report describes the development of and testing outcomes for a cost-effective dreissenid mussel invasion risk assessment and monitoring/early detection program for 14 water bodies in northeastern Texas.

Methods

Water Bodies Chosen for Zebra Mussel Invasion Risk Assessment and Monitoring

Thirteen major northeastern Texas water bodies lying south of and known to receive extensive recreational boat traffic from Lake Texoma (Buch 2000) were chosen for zebra mussel invasion risk assessment and monitoring along with zebra mussel–infested Lake Texoma as a positive control (see Figure 24.1 and Table 24.1 for information on the lakes and their locations). Risk assessments and monitoring for evidence of zebra mussel presence were conducted at each lake during three different periods, spring 2011, fall 2011, and spring 2012. Sampling dates and periods during which juvenile settlement monitors were deployed are listed in Table 24.2.

Invasion Risk Assessment for the Monitored Lakes

During each water body site visit, four physical water parameters considered critical to successful zebra mussel invasion were recorded including surface water temperature, pH, calcium ion concentration, and DO concentration. In addition, surface water temperature was recorded hourly at each site with Hobo® temperature data loggers tethered to weighted nylon lines at a depth of approximately 1–1.5 m. Water temperature was recorded hourly by data loggers from their initial deployment at the beginning of the spring 2011 monitoring period through the end of the spring 2012 monitoring period encompassing a full year of temperature data for each lake (Table 24.2). Hourly water temperature data were used to compute daily mean temperatures and temperature ranges for each site throughout the study period. Mean daily surface water temperature for the month of August 2011 (August was the month when ambient surface

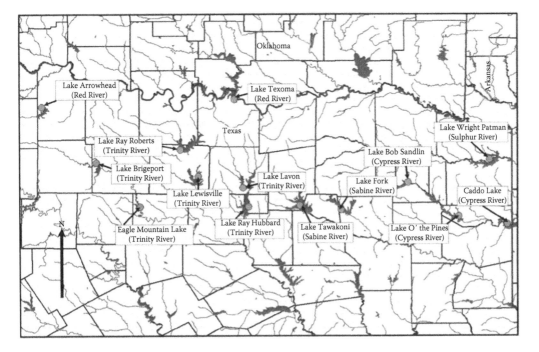

FIGURE 24.1 Map of northeastern Texas showing the location of the 14 water bodies subjected to risk assessment for zebra mussel invasion and monitoring for the presence of zebra mussel veligers and settled juveniles. River basins on which the water bodies are located are shown in parentheses.

TABLE 24.1

List of Water Bodies with Abbreviations, River Basin, Counties Where Was Water Bodies Are Located, Monitoring/Sampling Site, Site Addresses and Geographic Coordinates Where Zebra Mussel Invasion Risk Assessment Conducted Along With Monitoring for Mussel Larva DNA by Quantitative Polymerase Chain Reaction (qPCR), Larval Presence in Plankton Samples, and Juvenile Settlement Observations

Lake	Site Abbreviation	River Basin	Texas Counties Where These Water Bodies Are Located	Monitoring/Sampling Site	Site Address	Geographic Coordinates
Arrowhead Lake	AL	Red River	Archer	Arrowhead Lake State Park	229 Park Road 63, Henrietta, TX 76310	33°45.298′N 98°23.054′W
Lake Bridgeport	LB	Trinity River	Wise	Northside Marina and Resort	180 Private Road 1735, Chico, TX 76431	33°15.858′N 97°52.837′W
Eagle Mountain	EML	Trinity River	Tarrant, Wise	Tarrant Regional Water District Boat Dock	8300 Eagle Mountain Circle Fort Worth, TX 76135	32°52.545′N 97°28.493′W
Lake Ray Roberts	LRR	Trinity River	Denton, Grayson	Lake Ray Roberts Marina	1399 Marina Circle, Sanger, TX 76266	33°22.586′N 97°6.470′W
Lake Lewisville	LLE	Trinity River	Denton	Cottonwood Creek Marina	900 Lobo Lane, Little Elm, TX 75068	33°8.597′N 96°56.565′W
Lake Texoma	LTX	Red River	Grayson, Cook	Eisenhower Yacht Club	2141 Park Road #20, Denison, TX 75020	33°49.240′N 96°36.423′W
Lake Lavon	LLV	Trinity River	Collin	Collin Park Marina	2200 Saint Paul Rd, Wylie, TX 75098	33°2.465′N 96°31.873′W
Lake Ray Hubbard	LRH	Trinity River	Collin, Dallas, Rockwall, Kaufman	Chandler's Landing Marina	1 Harborview Drive, Rockwall, TX 75032	32°52.569′N 96°28.948′W
Lake Tawakoni	LTA	Sabine River	Hunt, Rains	Tawakoni Marina	2201 East Rabbit Cove Road, West Tawakoni, TX 75474	32°53.302′N 96°0.374′W
Lake Fork	LF	Sabine River	Rains, Wood	Popes Landing Marina	195 Private Road 5551, Alba, TX 75410	32°51.004′N 95°37.460′W
Lake Bob Sandlin	LBS	Cypress River	Franklin, Titus	Lake Bob Sandlin Park	Fort Sherman Dam Road, Mount Pleasant, TX 75455	33°5.389′N 95°0.798′W
Lake O' the Pines	LOP	Cypress River	Upshur, Marion	Johnson Creek Marina	440 Johnson Creek Road, Jefferson, TX 75675	32°47.530′N 94°32.375′W
Lake Wright Patman	LWP	Sulphur River	Bowie, Cass	Kelly Creek Marina	4221 CR 1208, Maud, TX 75567	33°17.336′N 94°15.047′W
Caddo Lake	CL	Cypress River	Marion	Private Dock	2119 Ma-CR 3636, Jefferson, TX 75657	32°43.357′N 94°5.215′W

Notes: Water bodies are listed in geographically successive order from the most westerly to the most easterly monitoring sites.

TABLE 24.2

List of Water Bodies with Specific Dates for Zebra Mussel Settlement Monitor Deployment and Recovery with Number of Days That Settlement Monitors Were Deployed at a Depth of Approximately 1–1.5 m

Water Body	Water Body Abbreviation	Spring 2011		Fall 2011		Spring 2012	
		Dates Monitored Day/Month/Year	Days	Dates Monitored Day/Month/Year	Days	Dates Monitored Day/Month/Year	Days
Arrowhead Lake	AL	01/06/11–30/06/11	30	21/10/11–21/11/11	32	26/04/12–04/06/12	40
Lake Bridgeport	LB	01/06/11–30/06/11	30	21/10/11–21/11/11	32	26/04/12–04/06/12	40
Eagle Mountain Lake	EML	01/06/11–30/06/11	30	21/10/11–21/11/11	32	26/04/12–04/06/12	40
Lake Lewisville	LLE	08/06/11–01/07/11	24	22/10/11–19/11/11	29	20/04/12–02/06/12	40
Lake Ray Hubbard	LRH	08/06/11–01/07/11	24	22/10/11–19/11/11	29	27/04/12–02/06/12	37
Lake Lavon	LLV	08/06/11–01/07/11	24	22/10/11–19/11/11	29	27/04/12–02/06/12	37
Lake Ray Roberts	LLR	02/06/11–30/06/11	29	22/10/11–19/11/11	29	20/04/12–02/06/12	37
Lake Texoma	LTX	02/06/11–30/06/11	29	21/10/11–19/11/11	30	20/04/12–01/06/12	36
Lake Tawakoni	LTA	08/06/11–01/07/11	24	22/10/11–19/11/11	29	27/04/12–06/06/12	41
Lake Fork	LF	08/06/11–01/07/11	24	24/10/11–22/11/11	28	27/04/12–06/06/12	41
Lake Bob Sandlin	LBS	07/06/11–05/07/11	29	23/10/11–20/11/11	29	27/04/12–06/06/12	41
Lake O' the Pines	LOP	07/06/11–05/07/11	29	23/10/11–20/11/11	29	28/04/12–06/06/12	40
Caddo Lake	CL	07/06/11–06/07/11	29	23/10/11–20/11/11	29	27/04/12–07/06/12	42
Lake Wright Patman	LWP	07/06/11–05/07/11	29	23/10/11–20/11/11	29	27/04/12–07/06/12	42

Notes: Plankton samples subjected to qPCR (quantitative polymerase chain reaction) analysis for the presence of zebra mussel larval DNA were taken on the date when monitors were deployed. Cross-polarized light microscopy for the presence of mussel veliger larvae was conducted on plankton samples taken on both deployment and recovery dates. Physical data for mussel risk assessment (i.e., surface water calcium concentration, oxygen concentration, and pH) were recorded during deployment and recovery of settlement monitors.

TABLE 24.3

Risk Assessment Parameters for Zebra Mussel Invasion of Texas and Other Southwestern United States
Water Bodies

Physical Parameter	Unsuitable	Marginal	Suitable	References
Average August water temperature	>32°C	31°C–32°C	<31°C	de Kozlowski et al. (2002)
pH	<6.8 or >9.5	6.8–7.4	7.4–9.5	de Kozlowski et al. (2002)
Calcium ion concentration	<12 mg L^{-1}	12–28 mg L^{-1}	>28 mg L^{-1}	Whittier et al. (2008)
Dissolved oxygen as% of air O$_2$ saturation	<30% O$_2$	30%–50% O$_2$	>50% O$_2$	Johnson and McMahon (1998)

water temperatures were maximal at all 14 test water bodies) was determined for all 14 water bodies
from daily mean temperatures and used as a risk factor for mussel invasion (Table 24.3).

Mid-summer measurements of ambient water temperature and DO at Lake Texoma during 2011 revealed
that water temperatures were highly stable throughout the epilimnion (Figure 24.2, data provided by the
Texas Parks and Wildlife Department [TPWD]) such that recording of ambient water temperature at depth
of 1–1.5 m as conducted in this study was reflective of that throughout the epilimnion. Zebra mussels generally are restricted to epilimnetic waters because they cannot tolerate the hypoxic conditions encountered
in the hypolimnion below the thermocline (Figure 24.2, Johnson and McMahon 1998, Garton et al. 2014).

The physical data utilized to develop risk assessments for potential zebra mussel invasion of each
examined water body are detailed in Table 24.3. The four factors listed in Table 24.3 (i.e., oxygen concentration, pH, calcium concentration, and incipient upper thermal limit) were assumed to independently
impact the risk of water body invasion. Thus, if for a specific water body, any of the four parameters listed
in Table 24.3 fell outside the known tolerated limits of zebra mussels, it was taken as indicating that a water
body would be unsuitable or resistant to mussel colonization, giving it a *low* risk of successful invasion.

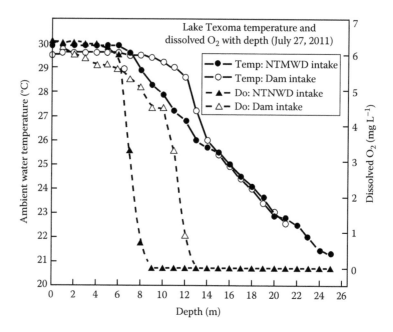

FIGURE 24.2   Ambient water temperature and dissolved oxygen (DO) profiles with depth (horizontal axes) taken in Lake
Texoma, TX, on July 27, 2011 from two locations, the Dam Intake (temperature = open circles, DO = open triangles) and
the North Texas Municipal Water District (NTMWD) Intake (temperature = solid circles, DO = solid triangles). Note that
at both sites surface ambient water temperatures are highly similar with depth throughout the epilimnion and decline rapidly along with DO below the thermocline in the hypolimnion such that ambient surface water temperature at 1 m depth is
representative of water temperature throughout the epilimnion. Zebra mussels do not penetrate cooler hypolimnetic waters
because they cannot tolerate the high levels of hypoxia marked by rapid decline in DO below the thermocline. (Courtesy of
Texas Parks and Wildlife Department, Austin, TX.)

If, for any tested water body, one or more of the four physical parameters listed in Table 24.3 fell within the range known to be marginal for zebra mussel survival and sustainable reproduction, that water body was assigned a *moderate* risk for mussel invasion. Assignment of a moderate invasion risk indicated that a water body could potentially harbor a zebra mussel population; however, it would likely be subject to periodic density reductions due to natural fluctuation of risk factors outside the mussel's tolerated range, preventing development of dense mussel populations. If all four parameters listed in Table 24.3 fell within the range suitable for zebra mussels, a water body was assigned a *high* risk of invasion and assumed to be able to support an extensive, sustainably reproducing population of zebra mussels unlikely to be subject to periodic fluctuations in density due to risk factors falling outside tolerated levels.

Zebra Mussel Settlement Monitoring

Zebra mussel settlement monitors were deployed at depths of approximately 1.0–1.5 m over durations of 24–42 days during the spring 2011, fall 2011, and spring 2012 monitoring periods (Table 24.2). Monitoring sites were selected to be near to high boater usage areas in marinas, state parks, or other relatively secure sites (Table 24.1) where settlement monitors and temperature data loggers would experience minimal disturbance by the public. Settlement monitors and temperature data loggers were suspended in areas with depths of at least 2 m. They were suspended on nylon ropes with a terminal house brick weight. The ropes were tied to the superstructure of floating docks so that the monitors and temperature data loggers remained at the same depth throughout the monitoring period. The one exception was Caddo Lake where, due to the limited number of accessible deep water deployment sites, the settlement monitor and data logger were deployed from a private, fixed-position dock where the water depth was >3 m. Because of its shallow nature, Caddo Lake is not normally subject to extensive water level fluctuation, which prevented immersion of the monitor and data logger throughout the monitoring periods.

Settlement monitors (Figure 24.3) consisted of 15.2 × 20.3 cm, 0.6–0.8 cm thick nylon kitchen scouring pads, previously successfully utilized to monitor zebra mussel juvenile settlement (Martel 1992, Martel et al. 1994). The scouring pads were reinforced by attaching 0.3 cm thick, 1.9 cm wide, 15.2 cm long plexiglass strips to both sides of the shorter, 15.2 cm, ends of the scouring pads with hot glue. A hole was drilled through the plexiglass strips and scouring pad 2.5 cm from each end of the plexiglass strips through which stainless steel machine screws were bolted to further secure the strips to the pad. A further two holes, approximately 0.8 cm apart, were drilled on either side of the center of the plexiglass strips on both ends of the monitor. The upper end of the settlement monitor was secured to the deployment rope with a self-locking nylon wire tie passed through the two central holes in the supporting plexiglass strips and a loop tied in the nylon deployment rope, allowing secure attachment when the self-locking tie was tightened. A second nylon wire tie was used to similarly secure the bottom end of the monitor to the rope. However, this tie was simply passed around the rope so that when the tie was tightened, it cinched the plexiglass supporting bar to the rope. The attachment of the bottom end of the settlement monitor in this fashion allowed it to be stretched out flat on the rope by pulling it downward. Thus, the monitor was held in a vertical position in the water column when deployed on the weighted rope (Figure 24.3). A temperature data logger was similarly attached to a loop tied in the rope just below the bottom end of the settlement monitor (Figure 24.3). Use of nylon wire ties to attach monitors and temperature data loggers to deployment lines allowed their rapid removal and reattachment as attachment ties could be readily cut with wire clippers.

Monitor deployment ranged over periods of 24 to 42 days (Table 24.2) when mussels were reproducing and juvenile settlement was expected (i.e., when spring and fall water temperatures were between 18°C and 25°C; McMahon and Bogan 2001). Settlement monitors were deployed at the beginning of and removed for examination at the end of each monitoring period. On removal, both sides of the settlement monitors were immediately examined in the field with a dissecting microscope at 10× power. If settled mussels were detected, they were either counted in the field, or if juvenile settlement was too dense for accurate counting, the monitor was fixed in 50% ethanol, returned to the laboratory, rinsed clear of debris, and the number of attached juvenile mussels counted in 10 randomly selected 3.142 cm² fields on each side of the monitor. The mean number of mussels in the 20 fields was multiplied by 93.09 to estimate the average number of mussels settled on the 292.5 cm² surface of one side of the monitor.

FIGURE 24.3 Scouring pad settlement monitor attached to deployment rope with temperature data logger and terminal brick weight.

Multiplication of the estimated number of mussels settled on one side of the monitor by 34.2 yielded the estimated settlement of mussels per m².

Plankton Sampling for Zebra Mussel Veliger Larvae

At each site visit (Table 24.2), the water column was sampled for zebra mussel veliger larvae from docks where mussel settlement monitors were deployed using the methodology of Marsden (1992). Veliger sampling was conducted with a 0.91 m long, 63 μm mesh plankton net, with a 0.30 m diameter opening and 500 mL sample bucket (Aquatic Sampling Company®, Buffalo, NY). Sampling consisted of two sets of 4–6 vertical plankton net tows extending from just above the substratum to the water's surface. Depending on available depth, vertical plankton tows ranged from 1.5 to 7.5 m. Samples from each set of plankton tows were combined in a Nalgene® sample bottle. Immediately after collection, one of the two combined samples was preserved for later qPCR testing for the presence of zebra mussel DNA (Claxton and Boulding 1998) by addition of 95% nondenatured ethanol to achieve a concentration of ≈70% ethanol. Sample qPCR testing was conducted by Pisces Molecular, Boulder, Colorado.

Five milliliters of the second plankton sample was pipetted from the bottom of the sample bottle where veliger larvae tended to accumulate and placed in a 6 cm diameter by 1.4 cm high Pyrex® petri dish and immediately microscopically examined in the field for the presence of zebra mussel larvae with a dissecting microscope at 20×. The remainder of the sample was preserved to a final concentration of 70% ethanol, as described earlier for qPCR samples, for later, more extensive, laboratory examination with cross-polarized illumination at 30× power under a dissecting microscope using the methodology of Marsden (1992) and Johnson (1995).

For cross-polarized light examination, a 10 mL sample aliquot was drawn from the bottom of the sample bottle, placed in a 9.0 cm diameter by 0.8 cm high Pyrex petri dish. The sample was initially scanned under cross-polarized illumination for the characteristic birefringent pattern generated by the calcareous larval shell consisting of a darkened central cross separated by four bright peripheral shell sections (Johnson 1995). Any objects exhibiting this pattern of birefringence were further examined under 45× power and normal illumination to confirm that they were a zebra mussel veliger or pediveliger based on the identification manual of Nichols and Black (1994). Examination under normal illumination allowed the identification of other organisms with similarly birefringent calcareous shells, particularly, ostracods (identified by carapace shape, presence of compound eyes and jointed limbs; Johnson 1995), unionacean glochidia larvae (identified by shell shape; Surber 1912, Kennedy and Hagg 2005), and juvenile Asian clams (identified by retention of a straight hinged shell at shell at lengths ≥ 200 µm; Nichols and Black 1994). Four 10 mL aliquots were thus examined for each plankton sample from each water body.

Plankton sampling equipment decontamination followed protocols recommended by the United States National Park Service (2007) and California Department of Fish and Wildlife (2013) to prevent cross-contamination of veligers or other organisms between sampled water bodies. Different plankton nets, tow ropes, and sample buckets were used in each water body during any one sampling trip. Decontaminated nets, tow ropes, and sample buckets were kept in separate plastic bags, preventing contact with other nets or equipment, prior to sampling. Just before water body sampling, plankton nets were rinsed by pulling them vertically through the water column to be sampled several times without the sample bucket attached and the sample bucket thoroughly rinsed in the water body before being reattached to the plankton net. Immediately after sampling, plankton nets, tow ropes, and sample buckets were returned to their storage bags and the bag placed in a container separate from that holding unused, decontaminated nets. The plankton net used in mussel-infested Lake Texoma was always isolated from other nets and never used in another water body.

During the first, spring 2011, round of sampling (Table 24.2), plankton nets, tow ropes, and sample buckets were decontaminated after each use by thoroughly rinsing with City of Arlington, Texas, tap water (CATW) followed by soaking in a 5% acetic acid bath (i.e., white vinegar) for 12 h as recommended by the United States National Park Service (2007) and California Department of Fish and Wildlife (2013). After acetic acid treatment, they were again thoroughly rinsed in CATW and air-dried prior to storage in new plastic bags. The hands and gloves of field personal were thoroughly washed and rinsed between collections to prevent transfer of veligers or other species between sampling sites. Sampling from docks prevented the necessity of entering the water body and having to decontaminate waders, hip boots, or boats. During fall 2011 and spring 2012, plankton nets, tow ropes, and sample buckets were decontaminated by a 30 min submersion in a 10% solution of CATW and commercial bleach (sodium hypochlorite), an exposure concentration and duration greatly exceeding that required to kill mussel larvae and destroy all traces of exogenous mussel DNA (Prince and Andrus 1992, Kemp and Smith 2005) as recommended by the California Department of Fish and Wildlife (2013). Otherwise, all other decontamination procedures were the same as used during spring 2011 sampling.

Results and Discussion

Zebra Mussel Invasion Risk Assessment

Among the 14 examined water bodies, mean surface water calcium concentrations ($n = 6$) fell below the 12 mg Ca L^{-1} generally considered to be the lower limit for successful zebra mussel colonization (Whittier et al. 2008) at Lakes Fork, Lake Bob Sandlin, Lake O' the Pines, and Caddo Lake in the eastern portion of the state (Figure 24.4). One water body, Lake Tawakoni, had a mean Ca concentration of 18 mg Ca L^{-1} (Figure 24.4), which was considered marginal for supporting an extensive zebra mussel infestation (Table 24.3). All nine remaining lakes had Ca concentrations within the range considered highly suitable for sustaining zebra mussel populations (Figure 24.4, Table 24.3).

Means ($n = 6$) of surface water oxygen concentrations (mg O_2 L^{-1}) measured during the spring and fall 2011 and spring 2012 monitoring periods at the 14 monitored water bodies (Figure 24.5a) when converted to mean percent of full air oxygen saturation values were generally well above the 30% of full air O_2 saturation (Figure 24.5b) considered to be the tolerated lower limit for zebra mussels (Table 24.3,

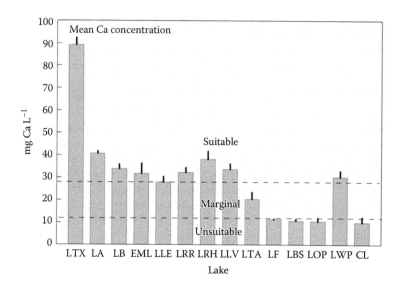

FIGURE 24.4   Mean surface water calcium concentrations (n = 6) taken during the spring and fall of 2011 and spring 2012 at 14 northeastern Texas water bodies as part of a risk assessment for potential invasion by zebra mussels. Horizontal dashed lines delineate thresholds making a water body unsuitable, marginal, or suitable for successful zebra mussel colonization based on mussel calcium tolerance levels listed in Table 24.3. Error bars above histograms represent one standard deviation of the mean. See Table 24.1 for water body names corresponding to abbreviations on the horizontal axis.

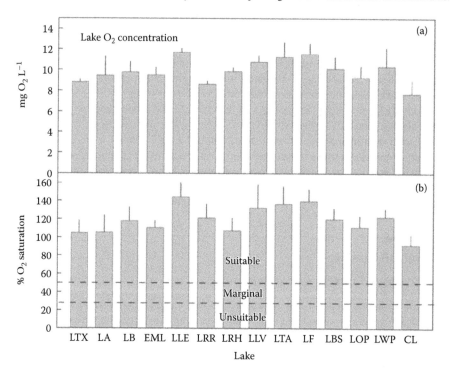

FIGURE 24.5   Mean surface water oxygen concentrations (n = 6) taken during the spring and fall of 2011 and spring 2012 at 14 northeastern Texas water bodies as part of a risk assessment for potential invasion by zebra mussels. (a) Mean raw oxygen concentration values in mg $O_2 L^{-1}$. (b) Mean oxygen concentrations expressed as a percentage of full air O_2 saturation. Horizontal dashed lines in panel b delineate thresholds, making a water body unsuitable, marginal, or suitable for successful zebra mussel colonization based on mussel percent O_2 saturation tolerance levels listed in Table 24.3. Vertical bars above histograms represent standard deviation of the mean. See Table 24.1 for water body names corresponding to abbreviations on the horizontal axis.

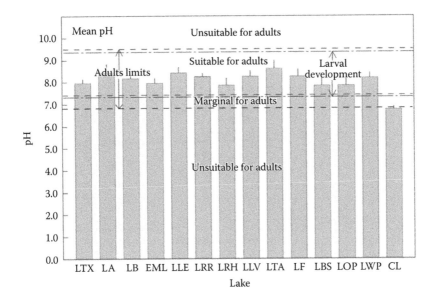

FIGURE 24.6 Mean surface water pH (n = 6) recorded at 14 northeastern Texas water bodies in spring and fall 2011 and spring 2012 as part of a risk assessment for potential invasion by zebra mussels. The horizontal short-dashed lines delineate pH thresholds, making a water body unsuitable, marginal, or suitable for successful zebra mussel colonization based on adult mussel pH tolerance levels listed in Table 24.3. Upper and lower long-short dashed lines represent the upper and lower pH limits for larval development to juvenile mussel settlement. Vertical bars above histograms represent standard deviation of the mean. See Table 24.1 for water body names corresponding to abbreviations on the horizontal axis.

Johnson and McMahon 1998). Air oxygen saturation values are utilized to determine O_2 tolerance levels because, unlike O_2 concentration values, they are independent of temperature and more directly related to an organism's capacity for gas exchange (Johnson and McMahon 1998). Only Caddo Lake had a mean daytime surface water oxygen concentration below 100% of full air O_2 saturation (Figure 24.5b).

Mean surface water pH (n = 6) measured during the spring and fall 2011 and spring 2012 monitoring periods at 13 examined water bodies fell within the range of 7.4–9.5 (Figure 24.6, Table 24.3), suitable for colonization by adult zebra mussels. Only Caddo Lake on the eastern edge of the examined water bodies had a mean pH slightly below the 6.8 threshold considered marginal for adults and well below the 7.3 required for larval development to a settled juvenile (Figure 24.6). These results indicated that only Lake Caddo had a mean surface water pH unsuitable for sustaining a zebra mussel infestation.

Average August 2011 surface water temperatures equaled or exceeded the incipient (long-term) upper 32°C thermal limit estimated for zebra mussels in southwestern US waters (Morse 2009, Table 24.3) at Lakes Bridgeport (LB), Bob Sandlin (LBS), and Lake O' the Pines (LOP) (Figure 24.7), indicating that they were potentially too warm during summer months to be suitable to support a sustainably reproducing zebra mussel population. The remaining 11 water bodies had mean August surface water temperatures below the mussel's 32°C incipient upper thermal limit (Figure 24.7). Of these, Lakes Lewisville (LLE), Ray Hubbard (RH), Tawakoni (LTA), Fork (LF), and Wright Patman (LWP) had mean August surface water temperatures above 31.5°C (Figure 24.7), suggesting that they were marginal zebra mussel habitats (Table 24.3). Lake Texoma, which has harbored and extensive zebra mussel population since 2009, had the lowest August mean surface water temperature of 30.5°C (Figure 24.7), indicating that southwestern water bodies with similar thermal regimes could support extensive zebra mussel populations. Lakes Arrowhead (LA), Eagle Mountain (EML), Lavon (LLV), Ray Roberts (LRR), and Caddo (CL) had mean daily August 2011 surface water temperatures <31.5°C, indicative of a thermal regime suitable for support of an extensive zebra mussel population (Figure 24.7, Table 24.3).

The physical parameters of percent surface water air oxygen saturation, pH, calcium concentration, and mean daily August surface water temperature were used to assess potential zebra mussel invasion risk for the 14 monitored northeastern Texas water bodies (Figures 24.4 through 24.7, Tables 24.2 and 24.3)

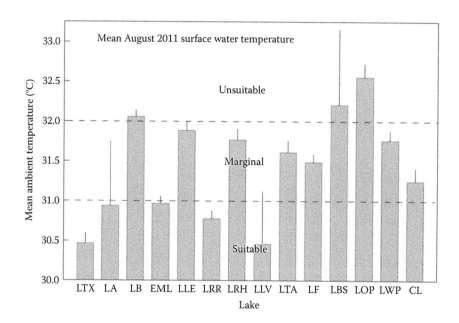

FIGURE 24.7   Mean surface water temperatures (n = 6) during August 2011 recorded from daily mean temperatures at 14 northeastern Texas water bodies as part of a risk assessment for potential invasion by zebra mussels. Horizontal dashed lines delineate thresholds, making a water body unsuitable, marginal, or suitable for successful zebra mussel colonization based on mussel temperature tolerance levels listed in Table 24.3. Vertical bars above histograms represent standard deviation of the mean. See Table 24.1 for water body names corresponding to abbreviations on the horizontal axis.

as *high, moderate,* or *low,* as described in the "Methods" section. Based on these physical parameters, 5 of the 14 monitored water bodies were assigned a *high* risk for successful establishment of an extensive zebra mussel infestation (Table 24.4), including Lake Arrowhead and Lake Texoma both in the Red River Basin and Eagle Mountain Lake, Lake Ray Roberts, and Lake Lavon in the Trinity River Basin (Figure 24.1, Table 24.1). Five water bodies were assigned a *moderate* risk for the establishment of a low density/ephemeral zebra mussel infestation (Table 24.4) including Lake Bridgeport, Lake Lewisville, and Lake Ray Hubbard in the Trinity River Basin; Lake Tawakoni in the Sabine River Basin; and Lake Wright Patman in the Sulfur River Basin (Figure 24.1, Table 24.1) because their mean, ambient daily August surface water temperatures were above 31.5°C, close to the absolute 32°C incipient upper thermal limit of southwestern zebra mussel populations (Morse 2009) (Figure 24.7, Table 24.3). Lake Fork in the Sabine River Basin and Lake Bob Sandlin, Lake O' the Pines, and Caddo Lake in the Cypress River Basin were assigned a *low* risk of zebra mussel invasion (Figure 24.1, Tables 24.1 and 24.4) because they had mean surface water calcium concentrations below the zebra mussel's lower limit of 12 mg Ca L^{-1} (Whittier et al. 2008) (Tables 24.3 and 24.4, Figure 24.4). Further, Lake Bob Sandlin and Lake O' the Pines were also considered unlikely to support zebra mussels because they had daily mean August surface water temperatures greater than the 32°C incipient upper thermal limit for southwestern mussel populations (Morse 2009) (Figure 24.7, Tables 24.3 and 24.4).

Monitoring for Zebra Mussel Invasion of Northeastern Texas Water Bodies

Plankton and Settlement Monitoring

Zebra mussel larvae were present in plankton samples from Lake Texoma throughout the monitoring period corresponding to high levels of mussel DNA detected by qPCR analysis of plankton net samples during spring and fall 2011 and spring 2012 (Figure 24.8).

Juvenile mussels were only found on settlement monitors in Lake Texoma and during spring 2011 and spring 2012 (Figure 24.9). No settlement occurred during fall 2011 even though living zebra mussel larvae were present in plankton samples (Table 24.5). Based on counts of settled juveniles on monitors,

TABLE 24.4

Risk Assessment for Successful Zebra Mussel Invasion and Establishment for 14 Water Bodies in Northeastern Texas Based on Measurements of Dissolved Oxygen Concentration, pH, Calcium Concentration, and Mean August Surface Water Temperatures

Lake	O_2	pH	Ca	Temperature	Risk Level
Arrowhead Lake	Suitable	Suitable	Suitable	Suitable	High
Lake Bridgeport	Suitable	Suitable	Suitable	Marginal	Moderate
Eagle Mountain Lake	Suitable	Suitable	Suitable	Suitable	High
Lake Lewisville	Suitable	Suitable	Suitable	Marginal	Moderate
Lake Texoma	Suitable	Suitable	Suitable	Suitable	High
Lake Ray Roberts	Suitable	Suitable	Suitable	Suitable	High
Lake Ray Hubbard	Suitable	Suitable	Suitable	Marginal	Moderate
Lake Lavon	Suitable	Suitable	Suitable	Suitable	High
Lake Tawakoni	Suitable	Suitable	Suitable	Marginal	Moderate
Lake Fork	Suitable	Suitable	Unsuitable	Marginal	Low
Lake Bob Sandlin	Suitable	Suitable	Unsuitable	Unsuitable	Low
Lake O' the Pines	Suitable	Suitable	Unsuitable	Unsuitable	Low
Lake Wright Patman	Suitable	Suitable	Suitable	Marginal	Moderate
Caddo Lake	Suitable	Marginal	Unsuitable	Suitable	Low

Suitable, condition likely to support the development of a dense zebra mussel population if invaded; Marginal, condition that may allow zebra mussel establishment but is unlikely to support the development of a dense population. Mussel populations could be ephemeral through time depending on variation in water body physical parameters; Unsuitable, condition which falls outside the level required to support a zebra mussel infestation.

settlement was estimated to be 64,618 mussels m^{-2} during spring 2011 declining to 2,750 mussels m^{-2} during spring 2012 (Figure 24.9). Even though larval mussels occurred in Lake Texoma plankton net samples during fall 2011, the lack of juveniles on the settlement monitor at this time (Figure 24.9) suggested that larvae were unable to complete the development to the settlement-competent pediveliger stage. In contrast, fairly extensive settlement of juvenile mussels was recorded on settlement monitors during fall 2012 (peak settlement = 5256 mussels m^{-2}), with peak settlement during spring 2013 further declining to 417.6 mussels m^{-2} (Figure 24.9).

Thus, peak juvenile settlement declined 96% between the springs of 2011 and 2012 and a further 85% between springs of 2012 and 2013 (Figure 24.9). Similar progressive annual declines in zebra mussel settlement rates on concrete panels have been reported in Oologah and Sooner Lakes, OK, from 2003 to 2006 and 2007 to 2010, respectively (Boeckman and Bidwell 2014). At Sooner Lake, OK, mean settled mussel densities declined by 67% between 2007 and 2008, increased by 20% between 2008 and 2009, and then declined by 50% between 2009 and 2010 (Boeckman and Bidwell 2014). The decline in juvenile mussel settlement between spring 2011 and 2012 in Lake Texoma corresponded to an ongoing decline in mussel larval densities since they were first recorded in 2010 (Churchill 2013). Mean peak mussel larval density recorded in Lake Texoma was 42 larvae L^{-1} in 2010, declining to 22 larvae L^{-1} in 2011, further declining to 3.3 larvae L^{-1} in 2012, corresponding to a 48% and 85% reduction in larval density between 2010 and 2011 and 2011 and 2012, respectively (Churchill 2013). The correlation between veliger abundance and juvenile settlement rates on scouring pad monitors recorded in Lake Texoma (Figure 24.9) has also been reported for zebra mussel settlement on similar scouring pad monitors in Lake Erie (Martel et al. 1994).

In addition to that reported for Lake Texoma (Texas/Oklahoma) and Sooner and Oologah Lakes (Oklahoma), major temporal reductions in the density of zebra mussel populations following attainment of maximal densities have been reported in other water bodies to occur over periods of 2–5 years. Mussel densities, size distributions, and condition indices were reduced in the Szczecin Lagoon, Poland, over a period extending from 2001 to 2005 (Wolnomiejski and Woźniczka 2008). A 95% reduction in zebra mussel population density occurred in Lake Mikolajski, Poland, between 1976 and 1977, with the population not recovering through 1989 (Stańczykowska and Lewandowski 1993). After zebra mussels

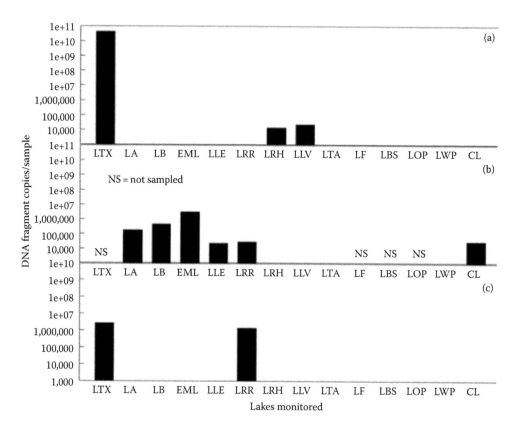

FIGURE 24.8 Results of molecular qPCR testing for the number of replicated zebra mussel DNA fragments derived from plankton samples from 14 monitored water bodies in northeastern Texas. The vertical axis is the exponential number of DNA fragments detected and the horizontal axis gives abbreviations for the water bodies tested as indicated in Table 24.1. *NS* indicates that a water body was not sampled during fall 2011. (a) Results of qPCR testing for spring 2011, (b) fall 2011, and (c) spring 2012.

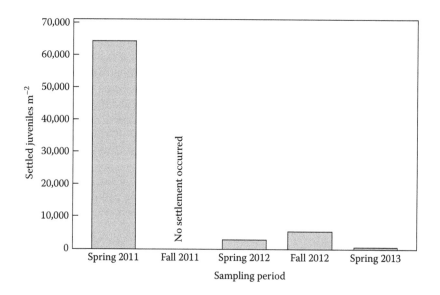

FIGURE 24.9 Estimated zebra mussel settlement rates in juveniles m^{-2} (vertical axis) in Lake Texoma, TX, based on juvenile mussel counts on settlement monitors during the springs of 2011, 2012, and 2013 and falls of 2011 and 2012 (horizontal axis).

TABLE 24.5

Monitoring Outcomes for Zebra Mussels in 14 Texas Water Bodies

Lake	Zebra Mussel Invasion Risk	Spring 2011				Fall 2011				Spring 2012			
		Positive qPCR	Veligers Detected	Juvenile Settlement	Mussels in Lake	Positive qPCR	Veligers Detected	Juvenile Settlement	Mussels Found	Positive qPCR	Veligers Detected	Juvenile Settlement	Mussels Found
Arrowhead (LA)	High					✓							
Bridgeport (LB)	Moderate					✓							
Eagle Mountain (EMB)	High					✓							
Lewisville (LLE)	Moderate					✓							
Ray Hubbard (LRH)	Moderate	✓											
Lavon (LLV)	High	✓											
Ray Roberts (LRR) (infested)	High					✓				✓	✓		✓
Texoma (LTX) (infested)	High	✓	✓	✓	✓	NS	✓			✓	✓	✓	✓
Tawakoni (LTW)	Moderate												
Fork (LF)	Low					NS							
Bob Sandlin (LBS)	Low					NS							
Lake O' the Pines (LOP)	Low					NS							
Caddo (CL)	Low					✓							
Wright Patman (LWP)	Moderate												

Notes: Monitoring occurred during spring 2011, fall 2011, and spring 2012. Zebra mussel invasion risk (see Table 24.4) is listed for each water body determined by measurement of critical parameters (i.e., surface water dissolved O_2, pH, calcium concentration, and mean August ambient water temperature). Check marks indicate positive results for each monitoring category (i.e., positive qPCR recorded for mussel DNA, veligers detected microscopically in a plankton sample, juvenile mussels on a settlement monitor or juvenile/adult mussels discovered in a water body).
NS, not sampled.

were discovered in El Dorado Reservoir (KS) in August 2003, veliger densities peaked at approximately 190 veligers L^{-1} in 2006 followed by a major decline to <12 veligers/L^{-1} in 2009 associated with a concurrent decline in mussel density (Severson 2010). Similarly, Petrie and Knapton (1999) reported that zebra mussels attained peak densities in 1992 at Long Point Bay, Lake Erie, of 2050 mussels m^{-2}, after which the population steadily declined to 606 mussels m^{-2} in 1995. Strayer and Malcom (2006, 2014) reported major annual fluctuations in the density of zebra mussel populations in the Hudson River Estuary between its initial establishment in 1992 until 2010, a period over which they recorded five major peaks in mussel density followed by declines to minimal densities with a periodicity ranging from 2 to 4 years.

A number of causes have been proposed for postinvasion declines in zebra mussel densities. These include predation (Stańczykowska and Lewandowski 1993, van der Velde et al. 1994; Petrie and Knapton 1999, Strayer and Malcom 2006, 2014, Casagrandi et al. 2007), larval and adult food limitation (Stańczykowska and Lewandowski 1993, Strayer and Malcom 2006, 2014, Wolnomiejski and Woźniczka 2008), elevated temperature stress (Morse 2009, Churchill 2013, Boeckman and Bidwell 2014), parasites (Stańczykowska and Lewandowski 1993, Strayer and Malcom 2006), diseases (Stańczykowska and Lewandowski 1993, Strayer and Malcom 2006, 2014), competition for hard substrata with other invasive sedentary species such as the amphipod *Corophium curvispinum* (van der Velde et al. 1994), decline in water levels desiccating emersed mussels (Churchill 2013), and pollution (Stańczykowska and Lewandowski 1993). It is likely that any or all of these factors could negatively impact zebra mussel population densities after initial invasion and that their relative impacts are likely to be specifically dependent on the unique physical, chemical, and biotic characteristics of the invaded water body (Stańczykowska and Lewandowski 1993, Strayer and Malcom 2006, 2014). However, the impacts of physical and biotic factors on zebra mussel population dynamics have not been rigorously documented (Stańczykowska and Lewandowski 1993; Strayer and Malcom 2006, 2014, Churchill 2013).

When the condition of zebra mussels in Winfield City Lake, KS, was measured as dry tissue weight relative to shell length, it was discovered that mussels lost tissue weight throughout the summer when ambient surface water temperatures were >25°C, suggestive of starvation during warm summer months (Morse 2009). The relative degree of tissue mass loss was most pronounced in larger mussels, with a standard individual of 30 mm shell length (SL) losing 53% of its tissue mass while ambient surface water temperatures remained above 25°C. In a controlled laboratory study, Walz (1978) recorded a similar lack of capacity in zebra mussels to maintain a positive scope for growth at temperatures >25°C. Thus, it appears that, in warm southwestern water bodies, zebra mussels must accrue enough tissue biomass during periods when ambient water temperatures are below 25°C to survive an extended period of summer starvation when water temperatures are >25°C.

Maximum zebra mussel tissue biomass in Winfield City Lake occurred in mid-June just prior to ambient water temperature exceeding 25°C (Morse 2009). Just before a zebra mussel population's near-complete extirpation from Lake Oologah (OK) in 2007 (Boeckman and Bidwell 2014), a sample of zebra mussels taken on June 29, 2007, had tissue biomasses that were 32.3%–80.5% lower than that of mussels taken from Winfield City Lake on June 29, 2008, over a size range of 5–30 mm shell length (Morse 2009). The mean daily surface (1.5 m depth) water temperature in August (warmest time of the year) at the Lake Texoma monitoring site was 30.5°C ± 0.71 in 2011 (Figure 24.7), 29.1°C ± 0.95 in 2012, and 29.3°C ± 0.17 in 2013 (RF McMahon, unpublished data), all below the estimated 30.7°C 28-day incipient upper thermal limit reported for zebra mussels in southwestern water bodies (Morse 2009). Very similar August 2011 and 2012 daily mean ambient temperatures have been reported from 1.5 m depth at a site near that used in this study on the southern shore of Lake Texoma (Churchill 2013). The fact that August surface water temperatures in Lake Texoma are not greatly different between years and approached but did not exceed the incipient 30.7°C upper thermal limit of southwestern zebra mussels (Morse 2009) suggests that population crashes rapidly following mussel invasion and establishment in warm southwestern water bodies are not due entirely to elevated water temperatures. Rather, they may result from expanding mussel populations eventually reducing planktonic food availability below levels required for the accrual of enough biomass when ambient water temperatures are <25°C to allow survival through an extended summer period of starvation when they are above the 25°C limit required for *D. polymorpha* to maintain a positive scope for growth.

qPCR Testing

Molecular qPCR testing during spring of 2011 revealed a strong zebra mussel DNA signal in Lake Texoma (TX) as expected due to its extensive mussel population and detection of mussel larvae in a concurrent plankton sample (Figure 24.8). However, two weak qPCR signals for mussel DNA were also recorded from Lake Ray Hubbard (LRH) and Lake Lavon (LLV) (Figure 24.8) where mussel larvae were not found in concurrent plankton samples. These weak signals were associated with (1) launching of a boat infested with zebra mussels at the marina monitoring site on Lake Ray Hubbard several weeks prior to plankton sampling (WFAA 2011) and (2) prior pumping of water into Lake Lavon from mussel-infested Lake Texoma by the NTMWD (Texas Parks and Wildlife Department 2009b).

Somewhat surprisingly, of the 10 water bodies sampled during fall 2011, six including Lake Arrowhead (LA), Lake Bridgeport (LB), Eagle Mountain Lake (EML), Lake Lewisville (LLE), Lake Ray Roberts (LRR), and Caddo Lake (CL) had weak positive qPCR signals for zebra mussel DNA (Figure 24.8) even though mussel larvae were not microscopically detected in concurrent plankton samples. Indeed, Caddo Lake was considered incapable of supporting a zebra mussel infestation due to its calcium concentration being less than the mussel's 12 mg Ca L⁻¹ lower limit (Figure 24.4, Table 24.5). Since all plankton nets were carefully decontaminated of mussel DNA before reuse (see "Plankton Sampling for Zebra Mussel Veliger Larvae" section), these weak positive qPCR results for fall 2011 were highly unlikely to be due to net contamination with mussel DNA. In addition, there were no other indications of zebra mussel infestation in these six water bodies (Table 24.5), suggesting that their weak positive qPCR signals were a result of boaters moving mussel larvae or mussel DNA into them from mussel-infested water bodies (especially Lake Texoma) during the summer boating season.

While mussel larvae or adults may have been carried into these six lakes during the summer and fall of 2011, they were not considered infested because no mussel larvae were found by the microscopic examination of concurrent plankton samples and no juveniles occurred on settlement monitors. Interestingly, the qPCR results for Lakes Ray Hubbard (LRH) and Lavon (LLV), which were weakly positive during spring 2011, were negative during fall 2011, indicating that weak positive qPCR signals for zebra mussel DNA could be ephemeral through time in Texas and other southwestern US water bodies. There have been a number of similar cases reported where a positive qPCR test indicating the presence of zebra or quagga mussel (*Dreissena rostriformis bugensis*) DNA in a water body could not be confirmed by subsequent testing. For instance, in the State of Colorado, positive qPCR DNA tests for quagga mussels reported for six lakes and zebra mussels for one lake could not be confirmed during five subsequent years of testing (United States Geological Survey 2014a,b). In only one lake did subsequent testing reveal the presence of veligers and adult quagga mussels although a sustainable population was never established (United States Geological Survey 2014b).

Detection of Zebra Mussel Invasion of Lake Ray Roberts

Plankton sample qPCR testing was repeated for all 14 lakes during spring 2012 with somewhat surprising results (Figure 24.8, Table 24.5). Only two lakes had positive qPCR signals for zebra mussel DNA, the infested Lake Texoma (LTX) and Lake Ray Roberts (LRR), which also had a much lower positive qPCR result in fall 2011 (Figure 24.8, Table 24.5). The lack of positive tests in spring 2012 for the five other lakes testing positive in fall 2011 supported the hypothesis discussed earlier that these lakes did not harbor sustainably reproducing zebra mussel populations, with their fall 2011 positive qPCR results being an indication of interlake mussel larvae/DNA transport during the summer 2011 boating season.

In contrast, the consecutive and increasingly positive qPCR results (i.e., increasing by over an order of magnitude between fall 2011 and spring 2012 for Lake Ray Roberts (Figure 24.8) suggested that it could have become infested with zebra mussels as it was the only initially uninfested water body of the 13 studied to have two consecutive positive qPCR tests (Figure 24.8). In addition, microscopic examination of a concurrent plankton sample revealed the presence of a single D-shaped early zebra mussel veliger larva (Figure 24.10), confirming the positive qPCR result. TPWD personnel were informed. On July 15, 2012, TPWD personnel found several recently settled juvenile mussels settled on rocks in shallow, near-shore areas of the lake (Figure 24.11), confirming the lake to be in the early stages of a mussel infestation (Texas Parks and Wildlife Department 2012).

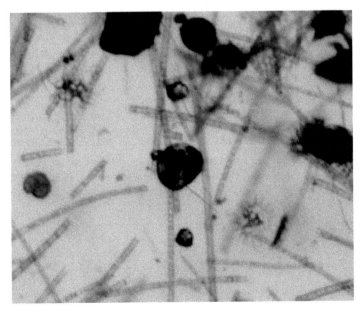

FIGURE 24.10 Photomicrograph of the first zebra mussel veliger larva found in Lake Ray Roberts, Texas, during spring 2012.

FIGURE 24.11   Photograph of a newly settled juvenile zebra mussel found by Texas Parks and Wildlife Personnel in Lake Ray Roberts, on the Elm Fork of the Trinity River, Texas, on July 15, 2012, after spring 2012 qPCR testing indicated the presence of mussel DNA in a plankton net sample. (Photo courtesy of Texas Parks and Wildlife Department, Austin, TX.)

During spring 2013, large adult mussels resulting from mussel spawning and juvenile settlement during summer and fall 2012 were found throughout Lake Ray Roberts. During spring 2013, relatively high densities of zebra mussel veliger larvae occurred in plankton samples and of juveniles on settlement monitors followed by a second major reproductive and settlement period occurring from October 2013 through February 2014 (RF McMahon, unpublished observations). The abundance of adult and juvenile mussels and high densities of planktonic mussel larvae detected at Lake Ray Roberts during 2013 were indicative of a rapidly expanding zebra mussel infestation. Lake Ray Roberts is on the upper end of the East Fork of the Trinity River. Lake Lewisville, rated at *moderate risk* for successful invasion by zebra mussels (Table 24.4), is only 7 km downstream from the outlet of Lake Ray Roberts, suggesting that it could be infested by downstream hydrological transport of mussel larvae in the near future.

The infestation of Lake Ray Roberts in the upper reaches of the Trinity River Basin is of major concern because the basin's drainage feeds a number of impounded reservoirs (Figure 24.1) that are the major source of potable water for the extended Dallas-Fort Worth, TX, area with a population of 6,500,000 people. It is also a concern for the City of Houston, TX, and surrounding municipalities with a population of 6,000,000 people that draw their potable water from Lakes Livingstone and Houston downstream from the Dallas-Fort Worth area on the main stem of the Trinity River (United States Army Corps of Engineers 2013). Zebra mussels can be further spread among the water bodies of the upper Trinity River Drainage Basin not only by recreational boaters and downstream hydrological transport but also by hydrological transport of mussel larvae on water transfers between reservoirs on the river's four different forks (i.e., the Clear, West, Elm, and East Forks) (United States Army Corps of Engineers 2013).

Conclusions and Project Impacts

Effectiveness of the Zebra Mussel Monitoring System

The results of the study indicated that the monitoring system was low cost and effective. Spring 2011 qPCR testing revealed the presence of mussel DNA in two of the 13 investigated uninfested lakes. Subsequent fall 2011 testing indicated that mussel DNA had been transferred to 6 of 10 monitored uninfested water bodies during the summer 2011 boating season. However, examination of concurrently taken plankton samples and juvenile settlement monitors revealed a lack of presence of larval and/or settled juvenile mussels in the eight spring and fall 2011 qPCR positive water bodies. Thus, these water bodies were considered as suspect but not confirmed for zebra mussel infestation. Only in Lake Ray Roberts were two consecutive positive qPCR results obtained, one in fall 2011 and a second in spring 2012. Because signal strength in spring 2012 was more than an order of magnitude greater than recorded in fall 2011, it was indicative of a zebra mussel invasion (Figure 24.8). The positive spring 2012 qPCR was supported by both cross-polarized light microscopic detection of a single veliger larva in a concurrently collected plankton sample (Figure 24.10) and later that summer by discovery of a few recently settled juveniles in the lake proper (Texas Parks and Wildlife Department 2012) (Figure 24.11).

No juvenile mussels occurred on the scouring pad settlement monitors deployed in Lake Ray Roberts during spring 2012 despite juveniles settling on rocks in the lake proper. Settled mussels were found on the settlement monitor only during spring 2013, suggesting that settlement monitors are not as reliable for early detection of a dreissenid invasion as qPCR testing with microscopic examination of plankton samples. Instead, as demonstrated by this project at Lake Texoma, settlement monitors may be more effective for longer-term monitoring of juvenile settlement rates as part of a study of the postinvasion mussel population dynamics (Figure 24.9), allowing determination of the health and growth/decline of a mussel population after its initial establishment (this study, Churchill 2013).

Effectiveness of the Risk Assessment System

In contrast to surface water pH, calcium concentration, and August mean ambient water temperature, surface water DO concentration did not appear to be as effective a risk factor for assessing potential zebra mussel invasion of Texas water bodies because average daytime surface DO in all 14 investigated water bodies generally closely approached or exceeded 100% of full air O_2 saturation, well above the mussel's lower tolerated limit of 30% of full air O_2 saturation (Figure 24.5b). Thus, data on surface water DO appear to be generally less critical for risk assessment of the potential for a southwestern water body to harbor a sustainably reproducing zebra mussel population, and, therefore, appeared to be the least informative of the risk factors addressed in this study.

In contrast, surface water temperature appears to be a critical factor for assessing mussel invasion risk in warm southwestern water bodies relative to more northern US water bodies where it rarely reaches the mussel's incipient upper thermal limit for extended periods of time (the zebra mussel's upper thermal limit has been estimated to be 28°C in northeastern US populations (Spidle et al. 1995, Morse 2009). Thus, mean August surface water temperatures were considered *unsuitable* for zebra mussel infestation at 2 of the 14 tested water bodies and *marginal* for 6 of them (Figure 24.5, Table 24.4). Of the four Texas lakes now considered to be infested with zebra mussels (Texas Parks and Wildlife Department 2014a), three (Lakes Texoma, Lewisville, and Ray Roberts) were considered to have a *suitable* and one (Lake Bridgeport) a *marginal* thermal regime for the development of a sustainably reproducing zebra mussel infestation (Figure 24.7, Table 24.4). In contrast, the two lakes considered to have unsuitable thermal regimes (i.e., Lake Bob Sandlin and Lake O' the Pines) (Figure 24.7, Table 24.4) have yet to be infested with zebra mussels. For this reason, hourly monitoring of surface water temperatures with data loggers during summer months is highly recommended as part of a zebra mussel risk assessment program in warm southwestern and southern water bodies.

Thus, the tested risk assessment system appeared cost-effective and accurate. For the 14 examined northeastern Texas water bodies, the most critical risk assessment factors for mussel invasion proved to be surface water pH, calcium concentration, and August mean ambient surface water temperature. All three factors approached or extended beyond the tolerated range for the successful establishment of a zebra mussel population in some of the 14 studied water bodies, allowing them to be rated as having a *low* (n = 4), *moderate* (n = 5), or high (n = 5) risk of mussel invasion (Table 24.4). Of the four previously uninfested water bodies included in this study, two assessed as having a *high* invasion risk (i.e., Lakes Ray Roberts and Lavon) and two (i.e., Lakes Bridgeport and Lewisville) as having a *moderate* invasion risk (Table 24.4) were subsequently confirmed to harbor zebra mussel larvae and/or settled mussels in 2012 or 2013 (Texas Parks and Wildlife Department 2014a). In contrast, none of the four water bodies assessed as having a *low* invasion risk (i.e., Lake Bob Sandlin, Lake O' the Pines, Lake Fork, and Caddo Lake) (Table 24.4) have been invaded by zebra mussels even though one of them (Caddo Lake) had a weak positive qPCR signal for mussel DNA during fall 2011 (Figure 24.8, Table 24.5).

The development of an accurate, cost-effective, and readily applied zebra mussel invasion risk assessment for Texas water bodies will allow public outreach/education, boat inspections, and monitoring early detection efforts by the TPWD, US Fish and Wildlife Service (USFWS), US Geological Service (USGS), US Army Corps of Engineers (USACE), water utilities, and other concerned entities to be primarily focused on high- and moderate-risk water bodies in order to reduce expenditure of effort and funds for these activities on low-risk water bodies unlikely to become infested with zebra mussels. Thus, water body dreissenid invasion risk assessment programs like that successfully applied under this program could be used to improve the cost-effectiveness of dreissenid invasion prevention efforts and public outreach in water bodies throughout Texas and the southwestern and southern United States.

The monitoring program developed for this project proved effective at early detection of zebra mussel invasion of Texas water bodies. It was the first to detect the presence of both a qPCR signal and veliger larva in Lake Ray Roberts, leading to subsequent detection of settled juveniles during summer 2012. The Tarrant Regional Water District (TRWD), using the same monitoring techniques developed in this project, found zebra mussel larvae in plankton samples taken in Lake Bridgeport on the West Fork of the Trinity River on June 6, 2013 (Texas Parks and Wildlife Department 2013b). This finding was preceded by qPCR signals for zebra mussels being recorded from plankton samples taken from the lake during the falls of 2011 (this study) and 2012 (RF McMahon unpublished observations). Thus, these positive qPCR results for Lake Bridgeport provided an early warning of a potentially developing zebra mussel infestation later confirmed by the finding of mussel larvae by the TPWD. These early detections of zebra mussel infestations have supported the TPWD efforts to develop regulations aimed at preventing further zebra mussel invasions including draining of all water from boats before launching in or leaving any public water body in the State of Texas (Texas Parks and Wildlife Department 2014b).

This study was closely coordinated with the TPWD, allowing project monitoring results to be disseminated to the public through TPWD news releases, which were then further widely disseminated through print, electronic, and televised media. Thus, the project was able to accomplish one of its main goals of increasing public awareness of the zebra mussel invasion threat to Texas water bodies. Dissemination of the project's findings through various media outlets also increased Texas boater awareness of the zebra mussel threat, potentially changing behaviors (i.e., clean, drain and dry boats prior to launching) (Texas Parks and Wildlife Department 2013c) to decrease the likelihood of mussel introduction to uninfested Texas water bodies as well as supporting the development of legally enforceable state regulations and fines directed at containment of mussel infestations by changing boater behavior and awareness of the need to prevent further dispersal of zebra mussels (Texas Parks and Wildlife 2014b). In addition, the qPCR data resulting from this project indicated that zebra mussel larvae are moved between Texas water bodies by boaters placing its many water bodies at risk of mussel invasion. The high levels of zebra mussel larvae transfer between infested and uninfested Texas water bodies elucidated by this study have stimulated independent development of mussel invasion risk assessments for water bodies in other portions of the state including Houston and east-central Texas in order to identify those that are particularly susceptible to mussel invasion. In some cases, identification of *high-risk* water bodies has stimulated the development of independent zebra mussel–monitoring programs, particularly of source water bodies by water utilities. It has also helped stimulate efforts to develop programs designed to prevent zebra mussel introductions to uninfested water bodies by both public and private entities.

Thus, this project appears to have done much to inform the public of the zebra mussel threat in Texas. It also helped to stimulate the development of zebra mussel risk assessments and monitoring programs for Texas water bodies outside those monitored in this study by concerned governmental and nongovernmental entities. In doing so, it is very likely to have slowed further zebra mussel invasion of Texas water bodies. Indeed, the TPWD contracted the author to conduct further zebra mussel invasion risk assessments and monitoring at 23 water bodies throughout eastern Texas during 2013.

Therefore, the project appears to have attained its three major objectives: (1) development and testing of an accurate and cost-effective zebra-mussel risk assessment for Texas water bodies, (2) development and testing of an accurate and economically feasible zebra mussel monitoring system for Texas water bodies, and (3) informing governmental and nongovernmental entities and the public at large of the economic and ecological threats posed by zebra mussels to Texas water bodies, in order to stimulate efforts to protect them from further invasion. These goals appear to have synergistically interacted to increase both public awareness and the state's and concerned entities' efforts to coordinate their programs to prevent further zebra mussel invasion of Texas water bodies (Texas Parks and Wildlife Department 2014a,b). The described program appears to be highly applicable to other water bodies in the southwestern and southern United States.

Acknowledgments

Colette O'Byrne McMahon, Dr. David Britton (USFWS), Mark Ernest and Jenifer Owens (TRWD), and Dr. Michael O'Neill assisted with field collections. Molecular qPCR testing of plankton samples was conducted by Pisces Molecular, LLC, Boulder, Colorado. Dr. David Britton (USFWS) and Dr. Earl Chilton and Brian VanZee (TPWD) provided support and advice. The following people kindly provided access to monitoring/sampling sites: John Ferguson, TPWD (Lake Arrowhead); James and Alan Kennedy (Lake Bridgeport); Mark Ernest and Jennifer Owens, TRWD (Eagle Mountain Lake); Jennifer Morris (Lake Lewisville); Joel Weiner (Lake Ray Hubbard); Joe Castro (Lake Lavon), Bill Williams and Roger Wingo (Lake Ray Roberts); Paul C. Kisel, TPWD, and Maria Boren (Lake Texoma); Larry Wright (Lake Tawakoni); John Goergen (Lake Fork); Judy Barton, Titus County Water Supply District, Lake Bob Sandlin; Sam and Lyda Edwards, Lake O" the Pines; Col. Richard Cary, Caddo Lake, and Leon and Shelly Jennings, Lake Wright Patman. An anonymous reviewer provided criticisms and suggestions that greatly improved the manuscript's final draft. This project was supported by a grant through the Quagga-Zebra Mussel Action Plan from the United States Fish and Wildlife Service to Robert F. McMahon.

REFERENCES

Boeckman CJ and Bidwell JR. 2014. Density, growth, and reproduction of zebra mussels (*Dreissena polymorpha*) in two Oklahoma reservoirs. In TF Nalepa and DW Schlosser (eds.), *Quagga and Zebra Mussels: Biology Impacts, and Control*, 2nd edn. CRC Press, Taylor & Francis Group, Boca Raton, FL, pp. 369–382.

Bossenbroek JM, Johnson LE, Peters B, and Lodge DM. 2007. Forecasting the expansion of zebra mussels in the United States. *Conservation Biology* 21: 800–810.

Buch KL. 2000. Assessing potential for dispersal of aquatic nuisance species by recreational boaters into the fresh waters of the western United States. MS Thesis. The University of Texas at Arlington, Arlington, TX, 90pp.

California Department of Fish and Game. 2008. DGF news release: Zebra mussels found in California reservoir. California Department of Fish and Game, Sacramento, CA, January 16, 2008. Accessed April 8, 2010 at http://www.dfg.ca.gov/news/news08/08005.html.

California Department of Fish and Wildlife. 2013. Zebra and quagga mussel veliger sampling protocol vertical tow. California Department of Fish and Wildlife, Sacramento, CA, 6pp. Accessed December 9, 2013 at https://nrmsecure.dfg.ca.gov/FileHandler.ashx?DoucumentID=4954.

Casagrandi R, Mari L, and Gatto M. 2007. Modelling the local dynamics of the zebra mussel (*Dreissena polymorpha*). *Freshwater Biology* 52: 1223–1238.

Churchill CJ. 2013. Spatio-temporal spawning and larval dynamics of a zebra mussel (*Dreissena polymorpha*) population in a north Texas reservoir: Implications for invasions in the southern United States. *Aquatic Invasions* 4: 389–406.

Churchill CJ and Baldys S, III. 2012. USGS zebra mussel monitoring program for north Texas. U.S. Geological Survey, Fact Sheet 2012-3077, 6pp. Accessed November 15, 2013 at http://pubs.usgs.gov/fs/2012/3077/pdf/fs2012-3077.pdf.

Claxton WT and Boulding EG. 1998. A new molecular technique for identifying field collections of zebra mussel (*Dreissena polymorpha*) and quagga mussel (*Dreissena bugensis*) veliger larvae applied to eastern Lake Erie, Lake Ontario and Lake Simcoe. *Canadian Journal of Zoology* 76: 194–198.

Connelly NA, O'Neill CR Jr, Knuth BA, and Brown TL. 2007. Economic impacts of zebra mussels on drinking water treatment and electric power generation facilities. *Environmental Management* 40: 105–112.

de Kozlowski S, Page C, and Whetstone J. 2002. Zebra mussels in South Carolina: The potential risk of infestation. South Carolina Department of Natural Resources, Columbia, SC, 14pp. Accessed February 16, 2014 at http://www.dnr.sc.gov/invasiveweeds/img/zebramusselassessment.pdf.

Drake JM and Bossenbroek JM. 2004. The potential distribution of zebra mussels in the United States. *BioScience* 10: 931–941.

Ensminger PA. 1999. Bathymetric survey and physical and chemical-related properties of Caddo Lake, Louisiana and Texas, August and September 1998. Report 99-4217, Water-Resources Investigations, US Geological Survey, Reston, VA, 1pp.

Espey Consultants, Inc. 2007. Lake Granbury Water Quality Monitoring Project Phase I. Draft report—Data trend analysis, modeling overview and recommendations. EC Project No. 6025, Brazos River Authority, Waco, TX, 107pp. Accessed April 12, 2010 at http://www.brazos.org/gbWPP/5-22-2007_Granbury_Phase1_ModelingReport.pdf.

Garton DW, McMahon RF, and Stoeckmann AM. 2014. Limiting environmental factors and competitive interactions between zebra and quagga mussels in North America. In TF Nalepa and DW Schlosser (eds.), *Quagga and Zebra Mussels: Biology Impacts, and Control*, 2nd edn. CRC Press, Taylor & Francis Group, Boca Raton, FL, pp. 383–402.

Gido KB and Matthews WJ. 2000. Dynamics of the offshore fish assemblage in a southwestern reservoir (Lake Texoma, Oklahoma–Texas). *Copeia* 2000: 917–930.

Johnson LE. 1995. Enhanced early detection and enumeration of zebra mussel (*Dreissena* spp.) veligers using cross-polarized light microscopy. *Hydrobiologia* 312: 139–146.

Johnson PD and McMahon RF. 1998. Effects of temperature and chronic hypoxia on survivorship of the zebra mussel *Dreissena polymorpha* and Asian clam *Corbicula fluminea*. *Canadian Journal of Fisheries and Aquatic Science* 55: 1564–1572.

Kemp BM and Smith DG. 2005. Use of bleach to eliminate contaminating DNA from the surface of bones and teeth. *Forensic Science International* 154: 53–61.

Kennedy TB and Hagg WR. 2005. Using morphometrics to identify glochidia from a diverse freshwater mussel community. *Journal of the North American Benthological Society* 24: 880–889.

Lienesch PW and Matthews WJ. 2000. Daily fish and zooplankton abundances in the littoral zone of Lake Texoma, Oklahoma–Texas, in relation to abiotic variables. *Environmental Biology of Fishes* 59: 271–283.

Mackie GL and Schloesser DW. 1996. Comparative biology of zebra mussels in Europe and North America: An overview. *American Zoologist* 36: 244–258.

Marsden JE. 1992. Standard protocols for monitoring and sampling zebra mussels. *Illinois Natural History Survey, Biological Notes* 138: 1–38.

Martel AF. 1992. Collector for veliger and drifting post-metamorphic zebra mussels. *Illinois Natural History Survey, Biological Notes* 138: 1–38.

Martel AF, Findlay CS, Nepszy SJ, and Leach JH. 1994. Daily settlement rates of the zebra mussel, *Dreissena polymorpha*, on an artificial substrate correlate with veliger abundance. *Canadian Journal of Fisheries and Aquatic Science* 54: 856–861.

McMahon RF. 1992. The zebra mussel—The biological basis of its macrofouling and potential for distribution in North America. Reprint # 342 from "Corrosion 92" (*Proceedings of the Annual Meetings of the National Association of Corrosion Engineers*). National Association of Corrosion Engineers, Houston, TX, 14pp.

McMahon RF and Bogan AE. 2001. Bivalves. In JH Thorp and AP Covich (eds.), *Ecology and Classification of North American Freshwater Invertebrates*, 2nd edn. Academic Press, New York, pp. 331–428.

McMahon RF and Lutey RW. 1996. Review of the effects of invertebrate macrofouling on microbiologically influenced corrosion in raw water systems. In NC Millhouse, P O'Boyle, and J Schubert (eds.), *Official Proceedings of the International Water Conference*, Paper IWC-96-70, Engineers Society of Western Pennsylvania, Pittsburgh. PA, pp. 650–658.

McMahon RF and Tsou JL. 1990. Impact of European zebra mussel infestation to the electric power industry. *Proceedings of the American Power Conference* 52: 988–997.

Miller AC, Payne BS, Neilson F, and McMahon RF. 1992. Control strategies for zebra mussel infestations at public facilities. Technical Report EL-92-25. Department of the Army, U.S. Army Corps of Engineers, Washington, DC, 24pp.

Morse JT. 2009. Thermal tolerance, physiological condition, and population genetics of dreissenid mussels (*Dreissena polymorpha* and *Dreissena rostriformis bugensis*) relative to their invasion of waters in the western United States. PhD dissertation. The University of Texas at Arlington, Arlington, TX, 279pp.

Mussalli YG, McMahon RF, Jenner H, Kema NV, Kasper JR, Secchia RF, Shuman SA, Adams TA, and Martin RL. 1992. Zebra mussel monitoring and control guide. RP-3052-03, Electric Power Research Institute, Palo Alto, CA, 280pp.

Nalepa TF. 2010. An overview of the spread, distribution and ecological impacts of the quagga mussel, *Dreissena rostriformis bugensis*, with possible implications to the Colorado River System. In TS Melis, JF Hamill, GE Bennett, LG Coggins Jr, PE Grams, TA Kennedy, DM Kubly, BE Ralston, (eds.), *Proceedings of the Colorado River Basin Science and Resource Management Symposium. Coming Together, Coordination of Science, Technology and Restoration Activities for the Colorado River Ecosystem*. U.S. Geological Survey Scientific Investigations Report 2010-5135, Reston, VA, pp. 113–121.

Nichols SJ and Black MG. 1994. Identification of larvae: The zebra mussel (*Dreissena polymorpha*), quagga mussel (*Dreissena rostriformis bugensis*), and Asian clam (*Corbicula fluminea*). *Canadian Journal of Zoology* 72: 406–417.

O'Neill CR Jr. 1996. The zebra mussel: Impacts and control. *Information Bulletin 238*, New York Sea Grant, Cornell University, Ithaca, NY, 62pp.

Petrie SA and Knapton RW. 1999. Rapid increase and subsequent decline of zebra and quagga mussels in Long Point Bay, Lake Erie: Possible influence of waterfowl predation. *Journal of Great Lakes Research* 25: 772–782.

Pimentel D, Zuniga R, and Morrison D. 2005. Update on the environmental and economic costs associated with alien-invasive species in the United States. *Ecological Economics* 52: 273–288.

Prince AM and Andrus L. 1992. PCR: How to kill unwanted DNA. *BioTechniques* 12: 358–360.

Qualls TM, Dolan DM, Reed T, Zorn ME, and Kennedy J. 2007. Analysis of the impacts of the zebra mussel, *Dreissena polymorpha*, on nutrients, water clarity, and the chlorophyll-phosphorus relationships in lower Green Bay. *Journal of Great Lakes Research* 33: 617–626.

Severson AM. 2010. Effects of zebra mussel (*Dreissena polymorpha*) invasion on the aquatic community of a Great Plains reservoir. MS Thesis. Kansas State University, Manhattan, KS, 73pp.

Spidle AP, Mills EL, and May B. 1995. Limits to tolerance of temperature and salinity in the quagga mussel (*Dreissena bugensis*) and the zebra mussel (*Dreissena polymorpha*). *Canadian Journal of Fisheries and Aquatic Science* 52: 2018–2119.

Stańczykowska A and Lewandowski K. 1993. Thirty years of studies of *Dreissena polymorpha* ecology in Marzurian Lakes of northeastern Poland. In TF Nalepa and WD Schloesser (eds.), *Zebra Mussels: Biology Impacts and Control*. Lewis Publishers, Boca Raton, FL, pp. 3–33.

Strayer DL. 2009. Twenty years of zebra mussels: Lessons from the mollusk that made headlines. *Frontiers in Ecology and the Environment* 7: 135–141.

Strayer DL and Malcom HM. 2006. Long-term demography of a zebra mussel (*Dreissena polymorpha*) population. *Freshwater Biology* 51: 117–130.

Strayer DL and Malcom HM. 2014. Long-term change in the Hudson River's bivalve populations: A history of multiple invasions (and recovery?). In TF Nalepa and DW Schlosser (eds.), *Quagga and Zebra Mussels: Biology Impacts, and Control*, 2nd edn. CRC Press, Taylor & Francis Group, Boca Raton, FL, pp. 71–81.

Surber T. 1912. Identification of the glochidia of freshwater mussels. Bureau of Fisheries document No. 771, United States Department of Labor and Commerce, Washington, DC, 16pp.

Texas Parks and Wildlife Department. 2009a. Lone zebra mussel found in Lake Texoma. Texas Parks and Wildlife Department, news release, April 21, 2009. Accessed February 6, 2014 at http://www.tpwd.state.tx.us/newsmedia/releases/print.phtml?req=20090421a.

Texas Parks and Wildlife Department. 2009b. Zebra mussels spreading in Texas: Invasive threat believed to be entering Trinity River via Lake Lavon. Texas Parks and Wildlife Department, news release, August 17, 2009. Accessed February 16, 2014 at http://www.tpwd.state.tx.us/newsmedia/releases/print.phtml?req=20090817a.

Texas Parks and Wildlife Department. 2012. Zebra mussels found in Lake Ray Roberts. Texas Parks and Wildlife Department, news release, July 18, 2012. Accessed February 6, 2014 at http://www.tpwd.state.tx.us/newsmedia/releases/print.phtml?req=20120718a.

Texas Parks and Wildlife Department. 2013a. Zebra mussels found in Lake Belton and suspected in Lakes Worth and Joe Pool. Texas Parks and Wildlife Department, news release, July 18, 2012. Accessed February 14, 2014 at http://www.tpwd.state.tx.us/newsmedia/releases/print.phtml?req=20130926a.

Texas Parks and Wildlife Department. 2013b. Zebra mussels confirmed in Lake Bridgeport. Texas Parks and Wildlife Department, news release, July 18, 2013. Accessed February 14, 2014 at http://www.tpwd.state.tx.us/newsmedia/releases/index.phtml?req=20130627a.

Texas Parks and Wildlife Department. 2013c. "Clean, drain and dry" regimen bad news for zebra mussels. Texas Parks and Wildlife Department, news release, May 24, 2013. Accessed February 16, 2014 at http://www.tpwd.state.tx.us/newsmedia/releases/print.phtml?req=20130524a.

Texas Parks and Wildlife Department. 2014a. Zebra mussels confirmed in Lake Lavon. Texas Parks and Wildlife Department, news release, January 21, 2014. Accessed February 6, 2014 at http://www.tpwd.state.tx.us/newsmedia/releases/print.phtml?req=20140121a.

Texas Parks and Wildlife Department. 2014b. Zebra mussel rules now expanded statewide. Texas Parks and Wildlife Department news release, May 22, 2014. Accessed June 24, 2014 at http://www.tpwd.state.tx.us/newsmedia/releases/?req=20140522f.

URS Group, Inc. 2009. Final summary report: Zebra mussel eradication project, Lake Offutt, Offutt Air Force Base, Nebraska. 55CES/CEV, Offutt Air force base, Belluve, NE, 51pp. Accessed April 8, 2010 at http://www.aquaticnuisance.org/wordpress/wp-content/uploads/2009/01/OAFB-ZM-Final-Summary-Report.pdf.

United States Army Corps of Engineers. 2013. Zebra mussel resource document: Trinity River basin, Texas. United States Army Corps of Engineers, Fort Worth District, Fort Worth, TX, 207pp. Accessed February 16, 2014 at http://www.swf.usace.army.mil/Portals/47/docs/Environmental/Water/Zebra_Mussel_Resource_Document_FINAL.pdf.

United States Geological Survey. 2010a. Progression of the zebra mussel (*Dreissena polymorpha*) distribution in North America. Accessed April 8, 2010 at http://fl.biology.usgs.gov/Nonindigenous_Species/ZM_Progression/zm_progression.html.

United States Geological Survey. 2010b. Quagga mussel locations in lakes and reservoirs outside the Great Lakes. Accessed April 8, 2010 at http://nas.er.usgs.gov/taxgroup/mollusks/zebramussel/QuaggaMusselLakeList.aspx.

United States Geological Survey. 2010c. Zebra mussel query by state. Accessed April 8, 2010 at http://nas.er.usgs.gov/queries/zmbyst.aspx.

United States Geological Survey. 2014a. NAS—Noindigenous aquatic species, specimen information, *Dreissena rostriformis bugensis* collection info. Accessed February 5, 2014 at http://nas.er.usgs.gov/queries/collectioninfo.aspx?SpeciesID=95.

United States Geological Survey. 2014b. NAS—Noindigenous aquatic species, specimen information, *Dreissena polymorpha* collection info. Accessed February 5, 2014 at http://nas.er.usgs.gov/queries/collectioninfo.aspx?SpeciesID=5.

United States National Park Service. 2007. Quagga/zebra mussel infestation prevention and response planning guide, Appendix H, Early detection monitoring. U.S. National Park Service, National Resources Program Center, Fort Collins, CO, 43pp.

van der Velde G, Paffen BGP, van den Brink FWB, bij de Vaate A, and Jenner HA. 1994. Decline of zebra mussel populations in the Rhine: Competition between two mass invaders (*Dreissena polymorpha* and *Corophium curvispinum*). *Naturwissenschaften* 81: 32–34.

Walz N. 1978. The energy balance of the freshwater mussel *Dreissena polymorpha* Pallas in laboratory experiments and in Lake Constance. III. Growth under standard conditions. *Archiv für Hydrobiologie-Supplement* 2: 121–141.

WFAA. 2011. Zebra mussels found at Lake Ray Hubbard. American Broadcasting Company (ABC) Television (Dallas-Fort Worth) news article. Accessed June 23, 2014 at http://www.wfaa.com/news/local/Zebra-mussels-found-at-Lake-Ray-Hubbard-122482064.html.

Whittier TR, Ringold PL, Herlihy AT, and Pierson S. 2008. A calcium-based invasion risk assessment for zebra and quagga mussels (*Dreissena* spp). *Frontiers in Ecology and the Environment* 6: 180–184. Accessed February 16, 2014 at http://www.esajournals.org/doi/pdf/10.1890/070073.

Wolnomiejski N and Woźniczka A. 2008. A drastic reduction in abundance of *Dreissena polymorpha* Pall. in the Skoszewska Cove (Szczecin Lagoon, River Odra estuary): Effects in the population and habitat. *Ecological Questions* 9: 103–111.

Wong WH and Gerstenberger S. 2011. Quagga mussels in the western United States: Monitoring and management. *Aquatic Invasions* 6: 125–129.

25

Monitoring Quagga Mussels, Dreissena bugensis, *in California: How, When, and Where*

Carolynn S. Culver, Andrew J. Brooks, and Daniel Daft

CONTENTS

ABSTRACT   Development of effective and efficient monitoring programs for quagga mussels, *Dreissena bugensis*, requires knowledge of settlement preferences, patterns of recruitment, and survivability. While these topics have been investigated extensively, additional studies are warranted in the recently invaded region of the southwestern United States where environmental conditions differ considerably from other invaded areas. In this study, we (1) evaluated the efficacy of four samplers by measuring mussel recruitment among samplers, (2) identified spatial and temporal patterns of mussel recruitment by quantifying monthly recruitment at eight water depths for 13 months, (3) explored potential relationships between mussel recruitment and environmental parameters (water transparency, water temperature, dissolved oxygen, and incoming veligers), and (4) assessed mussel survivorship within the stratification layers (epilimnion, metalimnion, and hypolimnion). Mussel densities were low where we investigated settlement preferences and recruitment.

Stacked polyvinyl chloride (PVC) plates and mesh scrub pads (Tuffy™) were significantly more effective at monitoring mussel recruitment than PVC pipes or acrylonitrile butadiene styrene (ABS) plates. One primary and one secondary recruitment event occurred during the summer and early fall, with little to no recruitment in the winter and early spring. The timing of recruitment events was highly correlated with veligers entering the system via aqueduct water, but was not significantly correlated with water temperature, dissolved oxygen, or water transparency. Mussel recruitment was highest at the shallowest depths (3–12 m), with extremely limited recruitment at the deeper depths (≥18 m). Few mussels recruited below the photic zone or at depths where dissolved oxygen was limited. Mussel survivorship was significantly lower within the hypolimnion versus the shallower layers at or above the thermocline.

The low mussel densities and presumably less competition at our study site likely enhanced the detection of mussel preferences and patterns that previously had not been reported. Additional studies evaluating spatial and temporal variations in mussel recruitment among water bodies with differing environmental conditions are needed to refine our recommendations. Nonetheless, our results are likely to improve the monitoring of quagga mussel populations, be it for long-term monitoring or early detection and evaluation of eradication and control efforts when mussel densities are low and difficult to detect.

Introduction

Quagga mussels, *Dreissena bugensis* (Andrusov 1897; Mackie and Claudi 2010), were first found in California in 2007, entering and then spreading through the southern part of the state via the Colorado Aqueduct. As of 2013, 25 Southern California water bodies have been infested by this invasive pest, almost all of which serve as drinking water sources. Annually, millions of dollars are being spent to control mussels to ensure continued water delivery for more than 20 million California residents in the infested region. These water sources also are highly valued for recreational activities, including sportfishing and boating. Mussels already have impacted these activities at some locations, with fishing lines and persons being cut by the sharp mussel shells (K. Kidd-Tackleberry, personal communication). Mussels are continually exposed due to water drawdowns, impacting the walkability of the area and also resulting in an unpleasant odor as the mussels desiccate and eventually die.

Upon the arrival of quagga mussels in California, and in some cases even before, many recognized the importance of early detection and long-term monitoring of this pest. This was in part driven by the impacts occurring elsewhere in the United States and Europe and the expected high success of the mussels in the mild climate of the southwestern United States. It was clear that early detection efforts were critically needed to determine whether other locations had become infested, particularly because of the large number of fishing tournaments occurring throughout the state that were attended by persons with their own, trailered watercraft—a major vector for spreading quagga and zebra mussels. Early detection also was needed to identify sites for potential eradication efforts. Long-term monitoring was equally important and needed for evaluating existing mussel populations in the state. This information was essential for not only identifying key parameters influencing the success of mussel populations in California, but also providing baseline data useful for evaluating the efficacy of management actions. Such actions were anticipated to minimize the likelihood of spread of the pest and reduce the impacts of the mussels on the water body.

With this in mind, early on state funds were directed toward monitoring. The California Department of Fish and Wildlife (CDFW) in collaboration with the California Sea Grant Extension Program (SGEP) developed a monitoring manual (Culver et al. 2009) and training course to help guide monitoring efforts. Trainings have since continued being led by the U.S. Fish and Wildlife Service, in collaboration with CDFW and SGEP. However, the development of these materials was hindered by a lack of information about quagga mussel population dynamics in California and the western United States more broadly. While California certainly benefited from the vast amount of existing literature and experiences of mussel monitoring elsewhere (mid-western and eastern United States, Europe; see reviews Neumann and Jenner 1992; Nalepa and Schloesser 1993; Claudi and Mackie 1994; D'Itri 1997; Mackie and Claudi 2010; van der Velde et al. 2010), much of this knowledge was collected from places with widely differing environmental conditions, and it often was in reference to zebra, not quagga, mussels. In addition, new artificial substrate samplers were being used by various groups in the southwestern United States, but comparative studies on the efficacy of these samplers were lacking. Taken together, studies were clearly needed to identify the most effective and efficient ways to monitor mussel infestations in California water bodies, including how, when, and where to monitor.

Various methods have been and continue to be used to monitor dreissenid mussels, including tactile and visual surveys, artificial substrates, and plankton sampling (Nalepa and Schloesser 1993; Boelman et al. 1997; Claudi and Mackie 1994, 2010; Culver et al. 2009). While it is best to use these methods in combination, limitations led to artificial substrates being the most practical choice for the monitoring of mussels in California. Tactile and visual surveys were limited by restrictions on body contact with the water at many locations; many lakes and reservoirs serve as drinking water sources, so swimming and diving is prohibited. Plankton sampling was limited by the need for technical training and special and costly equipment for sampling and processing. In contrast, sampling using artificial substrates was fairly inexpensive and could be conducted with minimal training and without any water body contact.

Designing a monitoring program using artificial substrates requires certain biological knowledge. First, artificial substrates are designed to facilitate settlement and attachment of postlarval and juvenile organisms by providing a *favorable* substrate, a surface that the organism would like to settle on and/or attach to. This method takes advantage of the settlement behavior of aquatic organisms where they search for, test, and eventually settle on or attach to a preferred substrate (see reviews Crisp 1984; Lindner 1984; Pawlik 1993). Many studies have examined substrate preferences of zebra and quagga mussels (Kilgour and Mackie 1992; Yankovich and Haffner 1993; Marsden and Lansky 2000; Czarnoleski et al. 2004; Kobak 2005). Based on these findings, different types of samplers have been developed. Yet, few studies have evaluated the efficacy of the different sampling designs, making it difficult to provide guidance on the best way to monitor quagga and zebra mussels. Second, knowledge of recruitment* dynamics of mussel populations can vastly improve the effectiveness and efficiency of monitoring, by identifying appropriate times and places (water depths) to monitor. Several studies have evaluated recruitment patterns of quagga and zebra mussels, but almost all of these have been conducted at locations with environmental conditions different from those prevalent in Southern California. Of exception is recent work done at Lake Mead, Nevada–Arizona (Mueting et al. 2010; Chen et al. 2011; Wong et al. 2012), but it too differs from systems in the coastal counties of California, including San Diego. As a result, it remains unclear whether similar patterns of mussel recruitment occur in California, making it difficult to determine when and where to monitor. Without this information, monitoring becomes much like finding a needle in a haystack.

The goal of our study was to improve the efficiency and effectiveness of mussel monitoring in California by evaluating not only sampler designs but also through the documentation of recruitment dynamics and survivability of mussels. We investigated five hypotheses:

1. Mussels do not exhibit a preference among different samplers.
2. Mussel recruitment occurs throughout the year, as environmental conditions are suitable year-round.

---

* We define recruitment as the settlement and attachment of postlarval and early juvenile mussel stages. We do not address the movement of larger juvenile and adult mussels.

3. Mussel recruitment varies among water depths, being higher in deeper, low light depths.
4. Mussel survivability differs among stratification zones (epilimnion, metalimnion, and hypolimnion).
5. Recruitment patterns are associated with certain environmental parameters (water transparency, water temperature, dissolved oxygen, and larval input).

Based on our results, we discuss how (type of sampler), when (time of year), and where (water depth) to monitor quagga mussels in California and the implications of our findings for monitoring mussels elsewhere.

Materials and Methods

Study Site

We conducted the main portion of this study at Miramar Reservoir, San Diego, California, 32°54.83′N, 117°5.99′W (Figure 25.1). The function of Miramar Reservoir is to serve as emergency and operational storage for the Miramar water treatment plant (WTP) and citywide emergency storage needs. Operational storage is between reservoir gauges 104 and 106; the spillway is at reservoir gauge 114. Miramar Reservoir is located on Big Sur Creek and impounds runoff from its one square mile watershed. It also impounds imported water from the San Diego County Water Authority (SDCWA) Aqueduct System carrying Colorado River and State Project water to the San Diego area. The reservoir has a storage capacity of 2178 million gallons (MGs), a surface area of 0.42 square miles at the spillway crest. Miramar Reservoir is a popular recreational area supporting fishing, boating, picnicking, and walking/running, but human contact with the water (e.g., swimming, water skiing) is prohibited. The reservoir is a warm monomictic lake that circulates freely in the winter and stratifies in the summer. Although low in nutrients, adult quagga mussels are present and initially were detected in this oligotrophic system on December 14, 2007.

We also conducted an additional portion of this study (survivability experiment) at Lower Otay Reservoir, San Diego, California (32°36′35.34 N, 116°55′50.43 W) (Figure 25.1). The function of Otay Reservoir is to serve as the primary emergency and operational storage required to meet Otay WTP and citywide emergency storage needs. Otay Reservoir is located on the Otay River and impounds runoff from the surrounding 100 square mile watershed (6427 acre ft/year). It also impounds water transferred from Morena and Barrett Reservoirs located in the Cottonwood Watershed via the Dulzura Conduit, and imported water from the SDCWA Aqueduct System carrying Colorado River and State Project water to

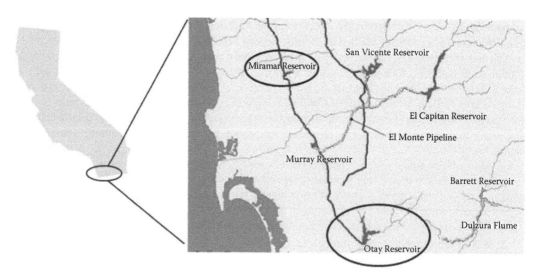

FIGURE 25.1   California State image with circle representing San Diego County. The expanded inlay shows study locations circled; Miramar and Otay Reservoirs.

the San Diego area. The reservoir has a storage capacity of 18,417 MGs at the top of the central over pour spillway flash gates, 16,243 MGs at the crest of the independent spillway, and 15,334 MGs at the crest of the central over pour spillway. The surface area of the reservoir is 1.92 square miles. Otay Reservoir is a popular recreational area supporting fishing, boating, picnicking, biking, and walking/running, but human contact with the water (e.g., swimming, water skiing) is prohibited. It is a warm monomictic lake that circulates freely in the winter and stratifies directly in the summer. However, it is higher in nutrients than Miramar and, as such, is a mesotrophic reservoir. Adult quagga mussels remain present, first detected in Otay Reservoir in September 2007.

Sampler Experiment

To evaluate the efficacy of different sampling designs for early detection monitoring of *Dreissena* mussels, we measured recruitment to four types of samplers at two different times of the year. In the initial trial, we evaluated the ability of two filamentous and two hard artificial samplers to serve as a substrate for newly recruiting/settling quagga mussels:

- Scrub brushes with wood handles (Figures 25.2(1) and 25.3a)
- Plastic mesh scrub pads (Tuffy) (Figures 25.2(2) and 25.3b)
- Four horizontally stacked gray PVC plates (15.0 cm × 15.0 cm × 0.6 cm), based on a design promoted by the CDFW (CDFW 2013) (Figures 25.2(3) and 25.3c)
- One white PVC pipe with holes based on samplers developed by researchers at Portland State University (PSU) (PSU 2013) (Figures 25.2(4) and 25.3d)

Both of the filamentous types of samplers are commonly used to monitor mussels in marine habitats along the west coast, while the hard substrate samplers typically are used to monitor quagga and zebra mussels in the western United States.

For the second trail, we evaluated one filamentous substrate and three hard substrate samplers:

- Plastic mesh scrub pads (Tuffy) (Figures 25.2(2) and 25.3b)
- Large ABS plastic plate (Figures 25.2(5) and 25.3e)
- Four horizontally stacked gray PVC plates (15.0 cm × 15.0 cm × 0.6 cm), based on a design promoted by the CDFW (CDFW 2013) (Figures 25.2(3) and 25.3c)
- One white and one black PVC pipe with holes based on samplers developed by researchers at PSU (PSU 2013) (Figures 25.2(4) and 25.3d)

We replaced the wood handled scrub brushes with a large ABS plastic plate because the brush was difficult to process and the ABS plastic plate was being used by other research groups to monitor quagga

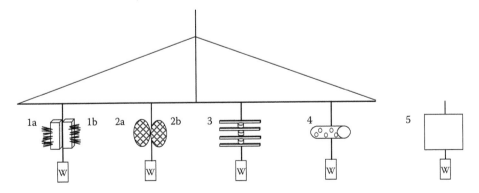

FIGURE 25.2 Experimental apparatus consisting of four test substrates attached to a PVC pipe. Test substrates— (1) scrub brush: a. 1 month; b. 2 months. (2) Plastic mesh scrub pad (Tuffy): a. 1 month; b. 2 months. (3) PVC horizontal plates with round separators in between each plate. (4) PVC pipe. (5) ABS vertical plate. W, weight.

FIGURE 25.3 Experimental artificial substrates. (a) Scrub brushes. (b) Mesh scrub pads (Tuffy). (c) PVC horizontal plates with round separators in between each plate. (d) PVC pipe. (e) ABS vertical plate.

mussels in California. We also used two PVC pipes of different colors (one white and one black) instead of a single white PVC pipe to conform with modifications made by PSU to their sampler design.

For both trials, the surfaces of all hard substrates were sanded prior to the samplers being assembled to increase rugosity in accordance with standard procedures for the sampler types. Each experimental line (Figure 25.2) consisted of a long PVC pole with a line at each end and in the center joined together at the top and used to attach to the dock. Attached to the PVC pole was the set of samplers. Duplicate filamentous samplers (brushes and scrub pads) were needed to allow removal and processing of a single sampler at the first sampling interval. Unlike the hard substrate samplers, filamentous substrate samplers must be removed from the line and processed in the laboratory; they are not typically visually examined while in the field. A small weight was added below each substrate to keep the substrates vertically oriented within the water column. Substrates were randomly assigned and connected to a horizontal position on the experimental line. Each line (Figure 25.2) was attached to the north side of a dock, such that the substrates were all submerged 10 m below the surface of the water, a depth known for high mussel recruitment (D. Daft, personal observation). Four replicate experimental lines were deployed at the boat docks during each trial. Activity at and around the docks was minimal. Access to the dock is restricted to city personnel and patrons renting boats, with only about 30 boat rentals per month. Further, fishing is restricted within 30 m of the docks. The depth of the water where the substrates were deployed (~40 m) minimized impacts due to turbidity.

Each trial ran for a period of 2 months, with substrates examined and mussels enumerated monthly. Samplers were examined using typical methods for the type of sampler under examination. Hard substrates were visually examined with the naked eye (no additional magnification) in the field. Filamentous substrates were removed and taken to the laboratory where they were frozen for a minimum of 24 h, then thawed and rinsed over a 150 µm mesh screen, a screen size appropriate for collecting pediveligers and larger mussels. The resulting collected material was washed off the screen into petri dishes and microscopically sorted and mussels enumerated. A single filamentous substrate of each type was removed from each line and processed after 1 month, with the other companion substrate removed and processed after 2 months. The experiments were conducted in the fall (September–November) of 2008 and the summer (July–September) of 2009.

Recruitment Study

Based on the results of our study to determine the optimal sampler design, we deployed a series of plastic mesh scrub pads (Tuffy) at the same reservoir to identify spatial and temporal patterns of mussel recruitment. Specifically, we deployed four replicate experimental lines at the boat docks as done before, but this time, each weighted line contained eight mesh scrub pads (Figure 25.4), with samplers spaced at 3 m intervals between water depths of 3 and 24 m. Samplers were collected and replaced monthly for 13 months. Collected samplers were processed as described previously.

Survivability Experiment

To assess the survivability of quagga mussels in anoxic conditions, we quantified mortality among groups of mussels deployed at different water depths in Otay and Miramar Reservoirs. Adult mussels (7–15 mm shell length) were collected and held in an aerated holding tank (20 L) in the laboratory for approximately 10 days prior to starting the experiment. Mussels were fed ground commercial fish food during this time. After about 1 week, groups of 10 mussels were placed into aquatic mesh bags (6" × 8") and returned to the holding tank for an additional 3 days. During this period, mussels in the mesh bags were examined and any that were dead or slow to close their shell when lightly touched were replaced.

Bagged mussels were kept moist and transported in a cooler to each reservoir and deployed at differing water depths within the reservoir. At Otay Reservoir, two groups of 10 mussels were deployed at three water depths, 4, 8, and 16 m (epilimnion, metalimnion, and hypolimnion, respectively), while at Miramar Reservoir, two groups of 10 mussels were deployed at two water depths, 11 and 20 m (epilimnion and hypolimnion, respectively). At Miramar Reservoir, two additional experimental lines containing 20 mussels

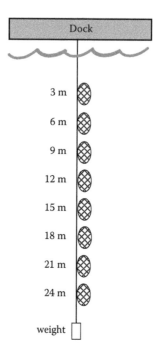

FIGURE 25.4   Experimental weighted line containing eight passive samplers (Tuffy plastic mesh scrub pads) attached at 3 m intervals.

each were deployed at each water depth and checked weekly to evaluate if daily exposure of the mussels to surface conditions affected survivorship. Mussels were deployed at Otay Reservoir on August 11, 2008, and at Miramar Reservoir on September 23, 2008.

Mussels were monitored approximately daily, weekly in the case of the additional lines deployed at Miramar, for mussel viability by retrieving the outplanted mussels and visually examining each individual for signs of mortality. Mussels with gaping valves were gently prodded at the posterior mantel edges. Mussels that failed to close their valves were considered dead. Dead mussels were removed, and the remaining mussels were returned to the appropriate experimental water depth. Exposure of mussels to surface conditions during examination did not exceed 2 min. The experiment concluded when 100% mortality occurred at one of the experimental depths.

Environmental Parameters

To explore potential relationships between mussel recruitment and environmental parameters, we measured several environmental variables during the recruitment study. We recorded water temperature and dissolved oxygen weekly using a YSI 6600 sonde at the identical eight water depths where recruitment samplers were deployed. We also measured water transparency (clarity) weekly using a Secchi disk to identify the photic and aphotic zones. To examine the correlation between recruitment and the source of veligers entering the reservoir via the aqueduct, we obtained data from the California Water Authority on the number of veligers per liter sampled from water taken from the aqueduct during the months of the study.

For the survivability study, we recorded water temperature, dissolved oxygen, and pH. These environmental parameters were measured almost daily at Otay and weekly at Miramar using a YSI 6600 sonde.

Data Analyses

Data were analyzed using the SAS 9.2 statistical package. Average monthly water temperature, dissolved oxygen, and water transparency were calculated for the 1-month recruitment sampling intervals.

Average values for the environmental parameters also were calculated for the survivability experiment as well as percent survivorship over time. For the sampler experiment, we used an analysis of variance (ANOVA) to evaluate mussel recruitment among substrate types for each month. Initial analyses found an interaction between substrates and month; thus, the analysis was run for each month separately. An ANOVA also was used for the recruitment study, evaluating mussel recruitment among water depths for the peak periods of recruitment. Each recruitment event was analyzed separately due to interactions between depth and recruitment periods. Data were log-transformed to help satisfy assumptions of homogeneity of variances between depths and recruitment events. To evaluate whether recruitment varied between the photic and aphotic zones, we compared mussel recruitment above and below the average monthly Secchi disk reading. Recruitment data were assigned to the photic or aphotic zone based on the Secchi depth measured each month and the depth of the deployed substrate. Recruitment was then summed for each zone for each month, and total recruitment was compared between zones using a repeated measures ANOVA for only those months when recruitment occurred (May–October). Data on the number of recruits were log-transformed to help satisfy assumptions of homogeneity of variances between zones and months. The correlation between recruitment and incoming veliger supply was analyzed using the Spearman's rank correlation. The Kaplan–Meier method as implemented in the SAS version 9.2 procedure PROC LIFETEST was used to compute lake-specific nonparametric estimates of the survivor function from the recorded survivorship data. This procedure also was used to determine if mussel survivorship differed between stratification layers within a lake. Where necessary, a Sidak adjustment of the model Chi-square value was performed to account for multiple comparisons of the different depths.

Results

Comparison of Samplers

Fall Trial

Recruitment of quagga mussels was virtually nonexistent during the fall trial, with only three mussels recruiting to the experimental samplers: all three of the mussels recruited to samplers containing filamentous substrates, not hard substrates, with recruitment occurring during each of the exposure times (1 month, 2 months). One mussel recruited to one mesh scrub pad sampler (Tuffy) in the first month, with a single mussel recruiting to both a mesh scrub pad and a scrub brush in the second month. All mussel recruitment occurred in the same area of the reservoir, with mussels recovered from samplers attached to the same experimental line.

Summer Trial

Recruitment of quagga mussels was significantly different among samplers for both the first ($p = 0.0188$, $F_{3,12} = 4.91$) and second ($p = 0.0039$, $F_{3,12} = 7.74$) months of exposure (Figure 25.5). The difference initially was less prominent, with mussel recruitment similar among the mesh scrub pad, PVC plates, and PVC pipes, as well as among the PVC plates, PVC pipe, and ABS plate. By the second month, significantly more mussels recruited to the mesh scrub pad samplers and the hard PVC plates than the other two hard substrates. Further, for the two samplers with higher mussel recruitment (mesh scrub pads and PVC plates), significantly more mussels recruited during the longer exposure time of 2 months (paired t-test between months: $t_3 = -0.47$, $p = 0.005$ for PVC plates; $t_3 = -8.40$, $p = 0.004$ for mesh scrub pads) (Figure 25.5). This was not the case for the other two samplers (PVC pipe and ABS plate, $t_3 = 0.46$, $p = 0.679$; $t_3 = 1.80$, $p = 0.171$, respectively); recruitment was similarly low for both exposure times. While the average number of mussels recruiting to the mesh scrub pads and PVC plates after 2 months was not found to be statistically different using a post hoc Ryan–Einot–Gabriel–Welsch multiple range test, the coefficient of variation (COV) between the number of recruits counted on the four replicate mesh scrub pads ($COV_{scrub\ pads} = 18.10$) after 2 months was less than that computed for the PVC plates ($COV_{ABS\ Plates} = 127.66$).

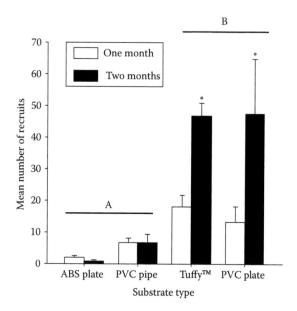

FIGURE 25.5 Average recruitment of quagga mussels to four types of samplers over two time periods. Bars with different letters (A, B) indicate a significant difference ($p < 0.05$) in the mean number of recruits on the samplers. An asterisks (*) denotes a significant difference ($p < 0.05$) in the mean number of recruits on each sampler after a one or two month deployment within each substrate type.

Recruitment Experiment

More than 2300 quagga mussels recruited to the passive samplers during the period the recruitment experiment was conducted. Mussel recruitment was highly seasonal, with two recruitment events of differing magnitudes: one large primary event (72% of all mussel recruitment) in the summer (June–July) and one smaller secondary event (24.5% of all mussel recruitment) in the fall (September–October) (Figures 25.6 and 25.7). No mussel recruitment was detected in the winter (December–February), with extremely limited recruitment (five mussels) in the early spring (March–April).

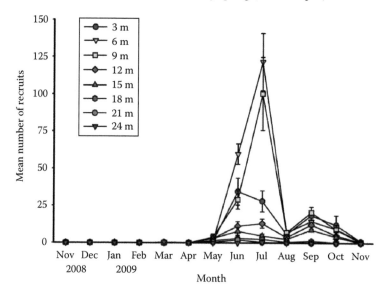

FIGURE 25.6 Average monthly recruitment of quagga mussels at various water depths at Miramar Reservoir, San Diego, California.

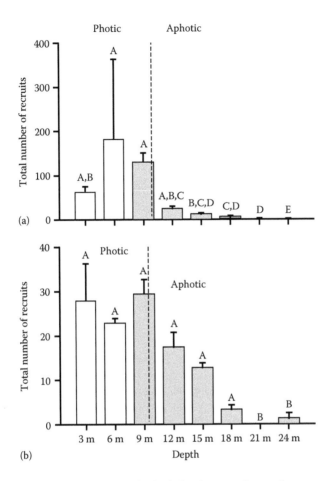

FIGURE 25.7 Mussel recruitment at various water depths during the two major recruitment events. (a) Primary recruitment period (June–July 2009). (b) Secondary recruitment period (September–October, 2009). Dashed line denotes Secchi disk transparency depth.

Mussel recruitment varied significantly by water depth for both the primary summer event ($F_{7,24} = 10.82$, $p < 0.0001$) and the secondary fall event ($F_{7,24} = 34.81$, $p < 0.0001$) (Figures 25.6 and 25.7). During the primary event, significantly more mussels recruited to 6 and 9 m than water depths ≥ 15 m. Recruitment at the other two shallow depths (3 and 12 m) was similar not only to recruitment at 6 and 9 m but also to recruitment between 15 and 21 m. A similar but more defined recruitment pattern occurred during the smaller secondary recruitment event. Mussel recruitment was significantly higher at the five shallowest depths (≤ 15 m), followed by 18 m, and then the two deepest depths (21 and 24 m). Overall, mussel recruitment was limited at the three deepest depths throughout the study, with only 39, 13, and 4 mussels recruiting to samplers at 18, 21, and 24 m water depths, respectively.

Environmental Parameters: Recruitment Experiment

Water Transparency

The Secchi depth in Miramar Reservoir averaged 9.5 m, ranging from 7.68 to 11.25 m during the recruitment experiment (Figure 25.8). In general, transparency varied seasonally, with the greater Secchi disk readings occurring in the winter and spring (January–June) and lesser readings in the summer (July–September).

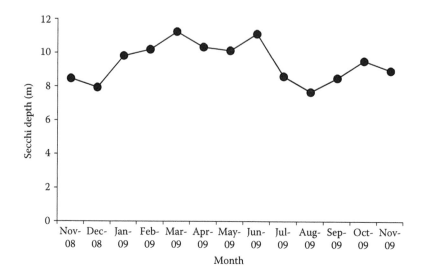

FIGURE 25.8   Monthly water transparency at Miramar Reservoir, San Diego, California.

Water Temperature

Water temperatures ranged from 13.0°C to 27.3°C across the eight experimental water depths and over the 13-month time period of the recruitment experiment (Figure 25.9). At the majority of water depths, water temperature increased in the late spring, peaking in the summer and then decreasing in the fall until reaching the lowest temperatures in the winter. An exception occurred at 24 m where the water temperature remained fairly constant and low (13°C–14°C) throughout the year. Water temperatures at all depths were ideal (10°C–28°C) for long-term survival of quagga mussels and within the range of thermal tolerance for quagga mussels (<2°C to >30°C; Mackie and Claudi 2010).

Dissolved Oxygen

Dissolved oxygen levels ranged from anaerobic (0.03%) to supersaturated (107.8%), varying among water depths (Figure 25.10). Levels remained quite high (>80% saturation) throughout the year at the four

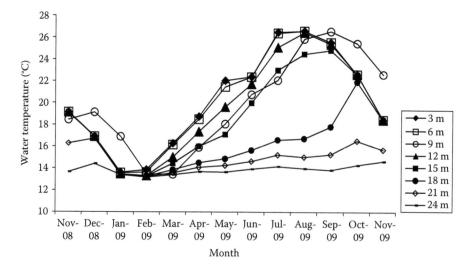

FIGURE 25.9   Average monthly water temperatures at Miramar Reservoir, San Diego, California. *Note:* scale starts at 10°C.

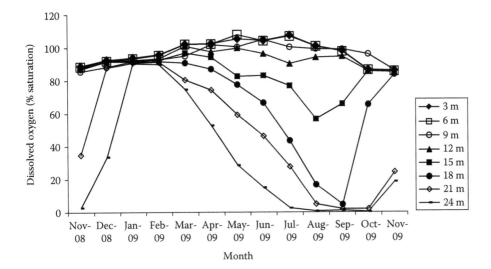

FIGURE 25.10   Average monthly dissolved oxygen levels at Miramar Reservoir, San Diego, California.

shallowest depths (≥12 m). Dissolved oxygen at the three deepest depths (≥18 m) decreased below the level required for mussel survival (<25% saturation; Mackie and Claudi 2010) in the summer and fall, remaining low for 2–5 months depending on the depth. While dissolved oxygen also decreased at the intermediate depth of 15 m in the summer, it remained well above (>55%) the 25% saturation level required for the survival of quagga mussels.

Incoming Veliger Source

Water was continuously added to Miramar Reservoir from the San Diego County Water Authority Aqueduct (carrying Colorado River and State Project water) during the recruitment experiment (Figure 25.11). The largest volumes of aqueduct water flowed in during the summer though early fall (July–October), averaging 2715 MGs per month. Almost half as much flowed into Miramar during mid-fall through most of the winter (November–February), averaging 1598 MGs per month, with intermediate volumes (2317 MGs per month) added throughout the spring (March–June). The incoming water

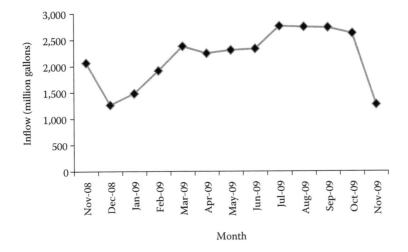

FIGURE 25.11   Monthly additions of water from the San Diego County Water Authority Aqueduct System to Miramar Reservoir, San Diego, California.

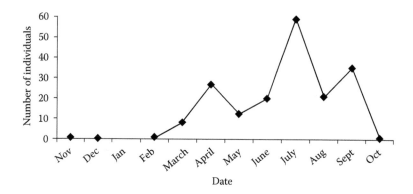

FIGURE 25.12   Monthly larval mussel source from aqueduct water entering Miramar Reservoir, San Diego, California.

contained veliger quagga mussels throughout the spring and summer (March–September) (Figure 25.12). Veligers were most abundant in July, about two to six times higher than in the other months.

Recruitment and Environmental Parameters

Recruitment may have been influenced by some, but not all of the considered environmental parameters (Figures 25.7, 25.13, and 25.14). First, significantly more mussels recruited to the passive samplers deployed in water shallower than the average depth of transparency ($F_{1,6}$ = 113.56, p < 0.0001) (Figure 25.7). This pattern was particularly evident when large numbers of mussels first began to settle. Second, mussel recruitment also was highly correlated with the density of incoming veligers (r = 0.64, p = 0.02) (Figure 25.13). In general, higher numbers of incoming veligers resulted in higher numbers of recruits. Third, and by contrast, recruitment was influenced little by water temperature or dissolved

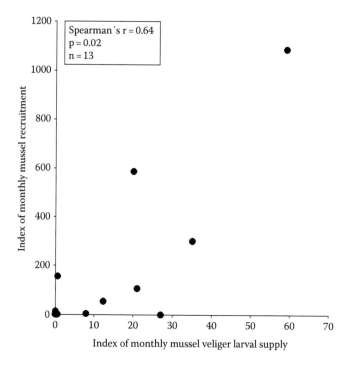

FIGURE 25.13   Correlation between incoming monthly larval mussel source from the aqueduct and mussel recruitment at Miramar Reservoir, San Diego, California.

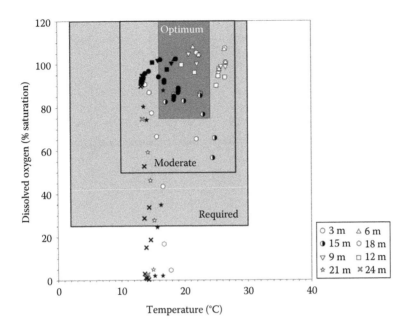

FIGURE 25.14 Relationship between recruitment of quagga mussels and two environmental parameters, water temperature and dissolved oxygen, at Miramar Reservoir, San Diego, California. Filled symbols, no mussel recruitment; open symbols, mussel recruitment. Dark gray box represents the optimum environmental conditions for the two environmental parameters for quagga mussel development and reproduction. Light gray boxes represent moderate and minimum requirements of mussels for the two environmental parameters.

oxygen (Figure 25.14). During some periods of the year, recruitment was lacking even though the water temperature and dissolved oxygen levels were favorable for mussels and veligers were present. Furthermore, some mussels settled in less than optimal and even unsuitable conditions (low dissolved oxygen levels). Notably, while the actual number of recruits may not have been determined by dissolved oxygen levels, oxygen saturation may have influenced wherein the water column mussels recruited. More mussels recruited to depths with high oxygen saturation, with very limited recruitment at depths with extremely low dissolved oxygen levels (Figure 25.15).

Environmental Parameters: Survivability Experiment

Otay Reservoir

Environmental parameters varied among the three experimental depths (Table 25.1). Water temperatures were fairly similar among the two shallow depths (4 and 8 m), ranging from 21.0°C to 27.23°C. By contrast, water temperatures were much colder at the deepest depth (16 m), averaging twice as cold as temperatures at the shallowest depth (4 m). Dissolved oxygen levels also were substantially lower at the deeper depth than the shallowest depth. Average oxygen saturation at the intermediate depth (8 m) was low and closer to that of the deeper depth (16 m). While it was at or just above the minimum required level (25% saturation) for the first 5 days, it fell below this critical level for the remainder of the experiment, falling as low as 7.0% saturation at Day 17 (Figure 25.16).

Miramar Reservoir

Environmental variables differed between the two experimental depths (Table 25.1). Average water temperature was almost 10°C higher, with greater variation at the shallowest depth than the deeper depth. A similar pattern occurred with dissolved oxygen, with levels being much higher with greater variability at the shallower depth.

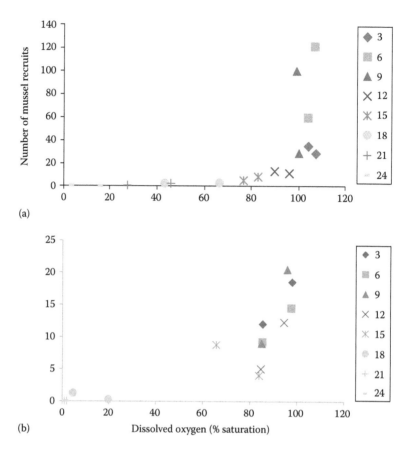

FIGURE 25.15 Relationship between mussel recruitment and dissolved oxygen levels. (a) Primary recruitment period (June–July, 2009). (b) Secondary recruitment period (September–October, 2009).

TABLE 25.1

Water Temperatures and Oxygen Saturation Levels for Otay and Miramar Reservoirs, San Diego, California, during Survivability Experiment

Depth (m)	Water Temperature (°C)		Dissolved Oxygen[a] (%)	
	Average	Range	Average	Range
Otay				
4	26.81 ± 0.21	26.47–27.23	112.7 ± 7.9	100.7–124.1
8	22.11 ± 0.69	21.0–23.36	19.0 ± 8.0	7.0–32.6
16	13.72 ± 0.12	13.64–14.07	2.3 ± 1.3	0.2–4.5
Miramar				
11	24.01 ± 1.09	22.41–24.77	89.2 ± 10.1	80.6–103.7
20	14.89 ± 0.40	14.30–15.11	1.4 ± 0.7	0.7–2.3

[a] The accuracy of the dissolved oxygen (DO) probe was ±2% of the reading.

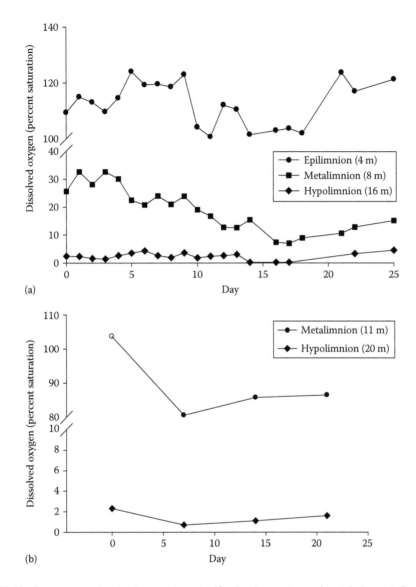

(a)

(b)

FIGURE 25.16 Oxygen saturation levels at various stratification layers at experimental sites. (a) Otay Reservoir. (b) Miramar Reservoir.

Survivability Experiment

Otay Reservoir

Mussel survivorship varied among stratification layers at Otay Reservoir (Figure 25.17). While mussel survivorship was high (85%) and similar at the epilimnion (4 m) and metalimnion (8 m) ($\chi^2_2 = 0.0713$, $p < 0.9907$), significantly more mussels died at the hypolimnion (16 m) as compared to both the shallower depths (4 m: $\chi^2_2 = 31.9703$, $p < 0.001$; 8 m: $\chi^2_2 = 33.8109$, $p < 0.001$). All but three mussels survived at both of the two shallower depths. The three mussel mortalities occurred within the first 5 days at 4 m, whereas two of the three mussels died after 21 days at 8 m. By contrast, no mussels survived at 16 m, with mean survivorship time 10.4 days (±1 SE 1.3 days).

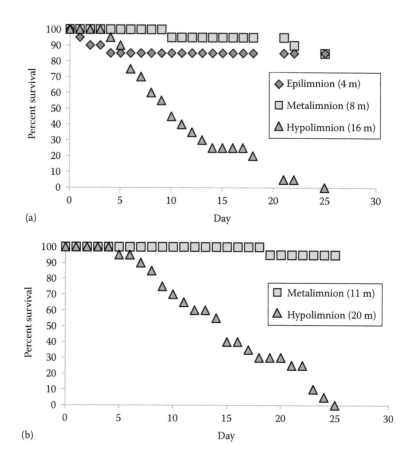

(a)

(b)

FIGURE 25.17   Survivorship of quagga mussels at the thermal stratification layers at two reservoirs in San Diego, California. (a) Otay Reservoir. (b) Miramar Reservoir.

Miramar Reservoir

Mussel survivorship was significantly different between the two experimental depths at Miramar Reservoir (χ_i^2 = 41.4867, p < 0.0001) (Figure 25.17). Survivorship was high throughout the experiment at 11 m, with only a single mussel dying at this depth. Comparatively, all mussels died within 25 days at 20 m. Mussel mortality gradually declined after the first 5 days, with mean survivorship time 14.2 days.

Discussion

Results from our study provide insight useful for developing effective and efficient early detection and long-term monitoring programs for quagga mussels in Southern California. Based on our examination of settlement preferences, patterns of recruitment, and survivability, we provide guidance on how, when, and where to monitor quagga mussels. We also discuss several implications resulting from our work.

How to Monitor

PVC plate and mesh scrub pad samplers were clearly more effective for monitoring quagga mussels than PVC pipe and ABS plate samplers. Significantly more mussels recruited to the PVC plate and mesh scrub pad than the pipe and ABS samplers. Further, the number of mussel recruits increased over time on the

PVC plate and mesh scrub pads, indicating continued high recruitment of mussels to these samplers. This was not the case for the PVC pipe and ABS plate samplers, where there was little to no increase in the number of recruits from the first to second month. Mussels either avoided the samplers altogether in the second month or the few initial recruits moved elsewhere or died after the first month and were replaced with a few new recruits during the second month. In either case, mussels did not readily or continually recruit to the PVC pipes and ABS plates, making them less desirable samplers for monitoring quagga mussels.

Several factors may have influenced the higher recruitment of mussels to the PVC plates and mesh scrub pads. Folino-Rorem et al. (2006) suggested that recruitment of zebra mussels was facilitated by the increased surface area of artificial filamentous substrates. Our results for the mesh scrub pads support this phenomenon for quagga mussels, as the mesh is a complex of interwoven flat threads that create a filamentous-type substrate. The PVC plate sampler although not filamentous also likely provided ample surface area for the small mussel recruits because it contained four separated stacked plates, providing several different surfaces for mussel attachment. In contrast, the ABS plate samplers contained just a single plate, one that was only slightly larger than a single PVC plate. Likewise, there were only two small diameter PVC pipes available on the PVC pipe samplers. However, while the latter two samplers may have had less total surface area available, there were still large areas of unoccupied space available for mussels to attach, suggesting characteristics other than open surface area promoted/inhibited recruitment. For the ABS plate sampler, orientation may have been a factor resulting in lower mussel recruitment. Marsden (1992) and Kilgour and Mackie (1993) reported higher recruitment to horizontal surfaces than vertical surfaces. The ABS plates were the only samplers with a vertical orientation. This orientation may have allowed greater light exposure on the ABS plates, as compared to the other samplers, which in turn also may have reduced the suitability of these samplers for mussels. Zebra and quagga mussels are known to avoid light, preferring shaded, dark areas (Marsden 1992; Marsden and Lansky 2000; Kobak 2001). Similarly, the quagga mussels may have found the cylindrical shape of the PVC pipe sampler less favorable to the flat and angular (e.g., corners, edges) surfaces of the other hard substrates, although this type of preference has not been studied for quagga and zebra mussels. Regardless of the mechanisms influencing mussel recruitment, a preference for the PVC plate and mesh scrub pad samplers was evident in this study.

Our results differ from studies that found no preference exhibited by mussels for various substrates, with mussels settling on all nontoxic surfaces (Marsden 1992; Mackie and Claudi 2010; Mueting et al. 2010; Chen et al. 2011). Unlike our study, these investigations were conducted at locations with high mussel infestations, not low infestations. It is likely that mussel preferences may be more readily detectable when mussel abundance is low due to less competition for space. This may explain the surprisingly low level of mussel recruitment to the PVC pipe collectors in our study. Mussels readily recruited to these and other similar samplers in other studies and monitoring efforts in the southwestern United States (Mueting et al. 2010; Wong et al. 2012), but mussels were highly abundant in these other locations. In our study, there were several different samplers for mussels to recruit to and space clearly was not limited as evidenced by the large amount of empty space remaining on several of the samplers. These findings illustrate how sampler type may not be as important for monitoring in highly infested systems, but that it could be critical for early detection and monitoring of new and low-level mussel infestations.

The limited recruitment of three mussels to the mesh scrub pads in our fall recruitment trial may illustrate an initial preference of mussels for filamentous substrates, as suggested by Folino-Rorem et al. (2012). This preference is further supported by our observations of mussels attaching to polypropylene and other plastic rope (but not nylon, coir, or cloth rope) used for deploying samplers, even when hard substrates are present. This has been noted during periods of low recruitment and at locations with low mussel infestations, when presumably preferences would be detected. If true, the use of the mesh scrub pad sampler and plastic rope may improve the likelihood of early detection and enhance monitoring efficacy in systems with low mussel infestations as long as the rope is inspected as part of the monitoring effort. The rope offers the added benefit of providing substrate throughout the water column instead of just at a single depth as occurs with the sampler substrates.

While the filamentous mesh scrub pad sampler may enhance the early detection of mussels, there are some disadvantages in using this method. In particular, instantaneous results are not usually obtained in the field because the samplers typically are processed in the laboratory. Such is not the case with the PVC plates that are examined in the field and results are known immediately. Further, processing of the filamentous mesh sampler is very labor intensive and it requires additional equipment. Because the mesh pads are rolled up, one has to unroll and rinse the mesh scrub pad, then filter and examine the extracted material under a dissecting microscope. Often the samples contain dirt, debris, and/or algae, making it even more difficult to process and examine under the microscope. One might be able to carefully open and visually inspect the sampler for attached mussels while in the field. However, this method would likely preclude examination of the material under a dissecting microscope, thereby eliminating the potential observation of the microscopic postlarval stages. As a result, the advantages of using this type of sampler for early detection may be traded off for the increased efficiency of other sampler types if it is necessary to obtain immediate results while still in the field.

When choosing a sampler type, it is important to consider the goal of the monitoring program—be it early detection, long-term monitoring, or efficacy of eradication and control tactics. For example, if early detection is the goal of the monitoring program, one might consider using a combination of effective sampler types in order to minimize efforts and costs while maximizing the chance for detecting mussels when few exist. In this case, mesh scrub pad samplers could be used only during certain times of the year when there is a high likelihood for mussel settlement (see next section) and in areas where infestations are most likely to occur (high-risk areas), with PVC plate samplers used at other times of the year and in lower-risk areas if more extensive monitoring is desired. This monitoring design would minimize the labor and time required for processing the mesh scrub pads year-round, but it would still allow more rapid detection of mussels at times and locations most susceptible to mussel infestations. It also likely would be a good way for monitoring the effectiveness of eradication efforts, maximizing the chance of mussel detection when few mussels may remain. Comparatively, if long-term monitoring is the goal of the monitoring effort, one might forego the labor and costs of using mesh scrub pads entirely and just use the PVC plate samplers to document mussel recruitment year-round. This same scenario also could be used to evaluate the efficacy of implemented control tactics, assuming mussels are at a level that continues to be readily detectable.

When to Monitor

The strong seasonal pattern of recruitment of quagga mussels at Miramar Reservoir suggests that monitoring could potentially be limited to a short period of time. For example, effort could be concentrated over 2 months in the summer (June and July) when there is a high likelihood of mussel recruitment, thereby maximizing the chance for early detection. Monitoring the rest of the summer and into the early fall also may be worthwhile, as conditions may vary year to year, possibly resulting in delayed and stronger recruitment a little later in the season. In terms of early detection, little would likely be gained if monitoring was conducted the rest of the year, as virtually no mussel recruitment is expected then. Year-round monitoring of established mussel populations may, however, be useful for detecting changes in recruitment over time, including *boom and bust* cycles that are known to occur with populations of invasive species.

Environmental conditions provided little insight as to when to expect mussel recruitment, and thus conduct monitoring. Although most of the recruitment events occurred during periods of acceptable and even favorable environmental conditions, these conditions were not enough to guarantee the presence of recruits. Such was evident during the winter and spring, when mussel recruitment was lacking despite environmental conditions being suitable, and even favorable, for mussel development and reproduction. In fact, dissolved oxygen levels were optimal (>75% saturation) at the shallow depths (≤12 m) throughout the study. Water temperatures also were optimal (16°C–24°C) for all but 2 months (January and February) when they were still *suitable*, but not optimal. Nonetheless, recruitment remained limited to just a few months of the year, indicating other factors influenced when the mussels recruited.

The lack of a predictive relationship between water temperature and dissolved oxygen on mussel recruitment, along with the strong correlation between mussel recruitment and the incoming source of veligers from the aqueduct, suggests that the established mussel population may not be reproducing and therefore not self-sustaining. Instead, the incoming aqueduct water is likely providing a source of veligers that is sustaining the mussel population at this reservoir. Knowing this can be useful not only for determining when to monitor, but also for control of the mussel population. In terms of monitoring, sampling the incoming water could be an early indicator of upcoming recruitment. One could work with the California Water Authority to identify times when incoming veligers are most abundant, and implement monitoring efforts at that time. This information also could be used to determine when to apply control tactics. Tactics that target veligers and/or juvenile mussels could be implemented when veligers begin entering the system and/or when mussels begin settling and attaching. Continuous monitoring would be required before, during, and after the application of such control measures to determine the efficacy of the actions.

Where to Monitor

Our study of spatial patterns of recruitment illustrates how monitoring at shallow (3–12 m) depths (particularly 6–9 m) may enhance the detection of mussel recruitment events of varying magnitudes. That is, while mussel recruitment was similar and highest among all but the deepest (21 and 24 m) depths during the secondary recruitment pulse, significantly more mussels recruited to the shallower depths (≤12 m) during the primary peak recruitment period. The recording of such a high density of mussel recruits would have been missed if monitoring was only conducted at depths greater than 12 m. Further, one or both recruitment events could have been missed all together if samplers were only at the deepest depths (21 and 24 m), as one or both recruitment events went undetected there. These findings support the deployment of samplers at depths ≤12 m to maximize the efficiency and effectiveness of detecting mussel recruitment.

The large mussel recruitment events at shallow depths were unexpected, as quagga mussels are typically found attached to substrates in shaded and dark places. Kobak (2001) points out that the anecdotal field observations of mussels in such locations could be due to factors other than light, including predation, water flow, and habitat complexity. However, Kobak's (2001) laboratory investigations documented zebra mussels selecting shadowed areas of experimental dishes over illuminated ones, suggesting phototactic behavior. In line with this, we expected recruitment to occur just below the Secchi disk depth, where light is more limited but other conditions are adequate for mussel survival. This was not the case: mussel recruitment was higher in the photic zone. Others also have documented mussel recruitment in shallow depths where light is readily available (Waker and von Elert 2003; Mueting et al. 2010; Wong et al. 2012). Thus, Secchi disk depth may be useful for determining where to place monitoring samplers, but samplers should be placed at and above the identified depth not below.

Water temperature likely had little influence on where mussels recruited. The water temperatures of the top five experimental depths (≤15 m) were all quite similar and favorable during the mussel recruitment events, but mussel recruitment differed among these shallower depths. Water temperatures were a bit cooler (14°C–15°C) at the bottom three depths (18, 21, and 24 m) during periods of mussel recruitment, potentially explaining why there was less recruitment there. However, the temperatures at those deeper depths were still quite suitable for quagga mussels, being well within the range required for reproduction and growth (≥10°C).

Dissolved oxygen levels may explain where within the water column mussels recruited, thus identifying the best depths to monitor. Karatayev et al. (1998) reviewed several studies that found dreissenid mussels to be sensitive to low dissolved oxygen levels, with recent field evidence of dissolved oxygen influencing settlement of quagga mussels in Lake Mead, Nevada–Arizona (Chen et al. 2011). While it is difficult to distinguish between the influence of dissolved oxygen, water temperature, and water depth using field experiments because they covary throughout the water column, the trend for limited mussel recruitment at deeper depths and high mussel recruitment at shallower depths in our study is more readily explained by the physiological tolerances of quagga mussels for dissolved oxygen, not water temperature and depth. We found high mussel recruitment at depths with high oxygen saturation levels (≥100%), with extremely limited recruitment at depths with dissolved oxygen levels below 50%, the lower limit supporting moderate mussel infestations. By contrast, water temperatures (as just discussed) remained at levels suitable for moderate-to-high mussel infestations. Further, quagga mussel recruitment occurred at water depths much shallower than the deepest depth (130 m) from which mussel populations have been observed (Mills et al. 1996). Taken together, our results further support the likely influence of dissolved oxygen on quagga mussel recruitment, and the usefulness of profiling dissolved oxygen levels of a system to identify water depths most appropriate for monitoring mussels.

Likewise, the stratification layers in a lake/reservoir system are important to consider when monitoring mussels. As found in this and many other studies, the hypolimnion is not suitable for mussel infestations. All adult mussels died within 25 days at both study sites. While water temperatures were still within the range required for mussel survival, development, and reproduction, dissolved oxygen levels were not. This finding further supports the use of dissolved oxygen profiles of a lake to determine where to monitor.

Although not studied here, water flow may have influenced where mussels recruited in our study, as quagga mussels are typically found in areas with low flow (<2 m/s). Chen et al. (2011) found higher mussel recruitment in low flow areas of Lake Mead, Nevada–Arizona, suggesting that wind effects and lake circulation may have influenced mussel settlement. Given the large influx of water (and veligers) entering Miramar from the aqueduct, it could be that the resulting water circulation leads to different water flows at different depths. Future traceability studies are needed to elucidate whether flow patterns and velocities may play a role in where mussel recruitment occurs within a system.

Implications

As this study largely was limited to a single reservoir and a single year, additional studies are needed to determine whether the results are broadly applicable over time and to other systems in California. Our observations indicate that the mussel infestation at Miramar may not be representative of infestations at other Southern California reservoirs as mussel populations are thriving at some of these other locations. For example, substrates examined at Miramar Reservoir had a high of 30 mussels, whereas those at Otay and El Capitan Reservoirs were covered with thousands of mussels (D. Daft, unpublished data). It remains unknown whether just infestation levels vary among these reservoirs with patterns of recruitment being similar to Miramar, or if recruitment patterns vary too. It may be that recruitment at other Southern California reservoirs is more similar to patterns documented at Lake Mead, Nevada–Arizona. There, quagga mussels recruit year-round, being highest in late fall (November–December) and lowest in the late summer/early fall (August–September) (Wong et al. 2012). This different pattern of mussel recruitment is likely due to two factors: (1) Lake Mead has a large self-sustaining population of quagga mussels, with recruitment magnitudes higher than in this study and (2) water temperatures are less suitable for quagga mussels during the summer, thereby presumably explaining the lower mussel recruitment at that time of year as compared to the fall. Such is not the case at Miramar Reservoir, but it could be at other California water bodies. If true, monitoring may need to be adjusted to cover longer periods of time or additional water depths. Comparative studies will help to determine whether recruitment patterns vary among freshwater systems in California.

Although not studied here, productivity of the system may be an important factor for invasion success in California. The highest densities of zebra mussels in Poland have been found in mesotrophic lakes (Stanczkowska and Lewandowski 1993) where nutrient levels are moderate. Miramar Reservoir is considered an oligotrophic system with low nutrient levels: yearly average values of total phosphorous <6.0 µg/L, total nitrogen 670 µg/L, chlorophyll a 1.0 µg/L, and Secchi disk depth >9.0 m (D. Daft, unpublished data). Obviously quagga mussels can survive there, as evident by the presence of large adult mussels, but they may not be reaching reproductive condition and thus proliferating there. Productivity at other nearby mesotrophic reservoirs (Otay and El Capitan) is two to four times higher (D. Daft, unpublished data), and mussels are significantly more abundant and robust (large, thick shells) (D. Daft, personal observation). Comparative studies would help to elucidate the role of productivity in the success of mussel infestations in Southern California.

Results from our survivability study may support a lower minimum dissolved oxygen tolerance level for quagga mussels than that reported by Mackie and Claudi (2010), and in closer agreement with the level reported by Shkorbatov et al. (1994, as cited by Karatayev et al. 2007). Mackie and Claudi (2010) estimated a tolerance of <25% saturation for quagga mussels based on findings for zebra mussels and indications that quagga mussels were more tolerant of low oxygen levels than zebra mussels. However, Shkorbatov et al. (1994, as cited by Karatayev et al. 2007) reported that quagga mussels require 1.5 mg/L of oxygen at 20°C or 16.3% saturation. Our findings suggest a tolerance level of around 19% saturation. However, two of the three mussel mortalities that occurred under the lower dissolved oxygen levels at the metalimnion occurred in the last 5 days of the experiment. It may be that the higher dissolved oxygen levels (25%–32% saturation) at the beginning of the experiment maintained higher survivorship early on, but as the levels dropped below 25% saturation, the mussels started to become stressed and began dying by the end of the experiment. Our observations indicate such events might have been occurring, as the mussels were not attached to the mesh bags or clumped together as they were at the epilimnion. Attachment and clumping are typical behaviors for nonstressed quagga mussels. Thus, additional studies are needed to determine whether defining a lower dissolved oxygen tolerance level for quagga mussels is warranted.

Management of mussel infestations in California would be improved through collaborations with Nevada and Arizona. Lake Mead, which is located in Arizona and Nevada, is providing Southern California water bodies with a continual source of mussel larvae via the Colorado aqueduct. Working together to reduce and control the source population will not only minimize impacts in one state but also the others. Without such efforts, California will have to put more effort into the management of each individual water body receiving mussel larvae from the aqueduct. Additional management actions would likely still be needed at some locations in California, as some populations are believed to be self-sustaining. However, targeting the source at a single location could greatly reduce infestations downstream in California, particularly those like Miramar that appear to be sustained by the larval source from the aqueduct.

Conclusions

When quagga mussels were first detected in Southern California, they were expected to proliferate in most areas because of the mild year-round environmental conditions that are favorable for mussel development and reproduction. This has not been the case; mussels have been quite successful at some, but not at all locations, despite suitable conditions. Multiple factors are undoubtedly influencing the success of mussels, including the extent of stratification and resulting dissolved oxygen levels, incoming sources of veligers, and productivity.

Comparative studies of mussel recruitment are critically needed to better manage mussel populations in California and the southwestern United States. As shown here, recruitment studies can help identify when (time of year) and where (water depth) to monitor mussels. This information can help to maximize the chance of detecting new mussel populations. It also can be extremely useful for evaluating mussel populations and the efficacy of control measures by identifying the best places and times to track changes in mussel recruitment. As our data and those from other studies suggest that recruitment dynamics may vary among systems, comparative studies also would be extremely beneficial for elucidating factors influencing the success of mussel infestations. Such information would improve the ability to identify locations at highest risk to mussel infestations in the southwestern United States.

Acknowledgments

We are grateful to Jason Tay, student intern and laboratory assistant at the University of California Santa Barbara, for his help with processing recruitment samples, and the staff at Miramar Reservoir who provided access and assistance with field activities. We thank Lisa Prus, San Diego County Water Authority, for providing veliger data for our analyses. In-kind funding was provided by the California Sea Grant Extension Program, City of San Diego Water Department, and the University of California Cooperative Extension, Santa Barbara and Ventura counties. The statements, findings, and conclusions reported here are those of the authors and are not necessarily those of the supporting agencies.

REFERENCES

Andrusov, N.I. 1897. Fossil and recent Dreissenidae of Eurasia. Trudy Sankt-Peterburgskago Obschestva Estestvoispitatelei. *Department of Geology and Mineralogy*. 25: 1–683.

Boelman, S.F., F.M. Neilson, E.A. Dardeau Jr., and T. Cross. 1997. Zebra mussel (*Dreissena polymorpha*) control handbook for facility operators. US Army Corps of Engineers, Washington, DC. Miscellaneous Paper EL-07-1. 78pp.

California Department of Fish and Wildlife. 2013. Zebra and quagga mussel artificial substrate monitoring protocol. http://www.dfg.ca.gov/invasives/quaggamussel/, accessed October 30, 2013.

Chen, D., S.L. Gerstenberger, S.A. Mueting, and W.H. Wong. 2011. Potential factors affecting settlement of quagga mussel (*Dreissena bugensis*) veligers in Lake Mead, Nevada-Arizona, USA. *Aquatic Invasions*. 6: 149–156.

Claudi, R. and G.L. Mackie. 1994. Zebra mussel monitoring and control. Boca Raton, FL: Lewis Publishers, Inc.

Crisp, D.J. 1984. Overview of research on marine invertebrate larvae, 1940–1980. In: J.D. Costlow and R.D. Tipper (eds.) *Marine Biodeterioration: An Interdisciplinary Study*. Annapolis, MD: U.S. Naval Institute Press, pp. 103–126.

Culver, C.S., S.L. Drill, M.R. Myers, and V.T. Borel. 2009. Early detection monitoring manual in quagga and zebra mussels. Technical Report T-069, California Sea Grant College Program, University of California San Diego, San Diego, CA.

Czarnoleski, M., L. Michalezyk, and A. Pajdak-Stós. 2004. Substrate preference in settling zebra mussels *Dreissena polymorpha*. *Arch Hydrobiology*. 159: 263–270.

D'Itri, F.M. (ed.) 1997. *Zebra Mussels and Aquatic Nuisance Species*. Chelsea, MI: Anne Arbor Press. 638 p.

Folino-Rorem, N., J. Stoeckel, E. Thorn, and L. Page. 2012. Effects of artificial filamentous substrate on zebra mussel (*Dreissena polymorpha*) settlement. *Biological Invasions*. 8(1): 89–96.

Karatayev, A., L. Burlakova, and D.K. Padilla. 1998. Physical factors that limit the distribution and abundance of *Dreissena polymorpha* (Pall.). *Journal of Shellfish Research*. 17: 1219–1235.

Karatayev, A., D. Padilla, D. Minchin, D. Boltovskoy, and L. Burlakova. 2007. Changes in global economies and trade: The potential spread of exotic freshwater bivalves. *Biological Invasions*. 9: 161–180.

Kilgour, B.W. and G.L. Mackie. 1993. Colonization of different construction materials by the zebra mussel (*Dreissena polymorpha*). In: T.F. Nalepa and D.W. Schloesser (eds.) *Zebra Mussels: Biology, Impacts and Control*. Boca Raton, FL: Lew Publishers, pp. 167–173.

Kobak, J. 2001. Light, gravity and conspecifics as cues to site selection and attachment behavior of juvenile and adult *Dreissena polymorpha* Pallas, 1771. *Journal of Molluscan Studies*. 67: 183–189.

Lindner, E. 1984. The attachment of macrofouling invertebrates. In: J.D. Costlow and R.D. Tipper (eds.) *Marine Biodeterioration: An Interdisciplinary Study*. Annapolis, MD: U.S. Naval Institute Press, pp. 183–201.

Mackie, G.L. and R. Claudi. 2010. *Monitoring and Control of Macrofouling Mollusks in Fresh Water Systems*. Boca Raton, FL: CRC Press. 508pp.

Marsden, J.E. 1992. *Standard Protocols for Monitoring and Sampling Zebra Mussels*. Biological Notes, vol. 138. Champaign, IL: Illinois Natural History Survey. 40pp.

Marsden, J.E. and D.M. Lansky. 2000. Substrate selection by settling zebra mussels, *Dreissena polymorpha*, relative to material, texture, orientation, and sunlight. *Canadian Journal of Zoology*. 78: 787–793.

Mills, E.L., G. Rosenberg, A.P. Spidle, M. Ludyanskiy, Y. Pligin, and B. May. 1996. A review of the biology and ecology of the quagga mussel (*Dreissena bugensis*), a second species of freshwater Dreissenid introduced to North America. *American Zoologist*. 36: 271–286.

Mueting, S.A., S.L. Gerstenberger, and W.H. Wong. 2010. An evaluation of artificial substrates for monitoring the quagga mussel (*Dreissena bugensis*) in Lake Mead, Nevada-Arizona. *Lake Reservoir Management*. 26: 283–292.

Nalepa, T.F. and D.W. Schloesser (eds.) 1993. *Zebra Mussels: Zebra Mussels Biology, Impact and Control*. Boca Raton, FL: Lewis Publishers.

Neumann, D. and H.A. Jenner (eds.) 1992. *The Zebra Mussel Dreissena polymorpha*. New York: Gustav Fischer.

Pawlik, J.R. 1992. Chemical ecology of the settlement of benthic marine invertebrates. *Oceanography and Marine Biology*. 30: 273–335.

Portland State University. 2013. Zebra and quagga mussel monitoring program. http://www.clr.pdx.edu/projects/volunteer/zebra.php, accessed October 30, 2013.

Shkorbatov, G.L., A.F. Karpevich, and P.I. Antonov. 1994. Ecological physiology. In: J.I. Starobogatov (ed.) *Freshwater Zebra Mussel Dreissena polymorpha (Pall.) (Bivalvia, Dreissenidae). Systematics, Ecology, Practical Meaning.* Moscow, Russia: Nauka Press, pp. 67–108 (in Russian).

Stanczykowska, A. and W. Lewandowski. 1993. Thirty years of studies of *Dreissena polymorpha* ecology in Mazurian lakes of northeastern Poland. In: T.F. Nalepa and D.W. Schloesser (eds.) *Zebra Mussels Biology, Impacts and Control.* Boca Raton, FL: Lewis Publishers, pp. 3–38.

Van der Velde, G., S. Rajagopal, and A. Bij de Vaate (eds.) 2010. *The Zebra Mussel in Europe.* Leiden, the Netherlands: Backhuys Publishers. 490pp.

Waker, A. and E. von Elert. 2003. Settlement pattern of the zebra mussel, *Dreissena polymorpha*, as a function of depth in Lake Constance. *Hydrobiologia.* 158(3): 289–301.

Wong, W.H., S. Gerstenberger, W. Baldwin, and B. Moore. 2012. Settlement and growth of quagga mussels (*Dreissena rostriformis bugensis* Andrusov, 1897) in Lake Mead, Nevada-Arizona, USA. *Aquatic Invasions.* 7(1): 7–19.

Yankovich, T.L. and G.D. Haffner. 1993. Habitat selectivity by the zebra mussel (*Dreissena polymorpha*) on artificial substrates in the Detroit River. In: T.F. Nalepa and D.W. Schloesser (eds.) *Zebra Mussels Biology, Impacts and Control.* Boca Raton, FL: Lewis Publishers, pp. 175–191.

26

Monitoring and Control of Quagga Mussels in Sweetwater Reservoir

Mark D. Hatcher and Scott W. McClelland

CONTENTS

ABSTRACT   The Sweetwater Authority (Authority) owns and operates the Robert A. Perdue Water Treatment Plant (WTP), and the adjacent source of supply, Sweetwater Reservoir (SWR) in southern San Diego County. In response to the quagga mussel (*Dreissena rostriformis bugensis*) infestation in SWR, first detected in April 2008, the Authority developed a Dreissenid Mussel Monitoring, Response, and Control Plan (Quagga Plan) in compliance with the California Department of Fish and Game Code 2301. This chapter will describe the development and efficacy of the main components of the Authority's Quagga Plan, which include a systematic monitoring program, inspection procedures, and methods of control. Chlorine dioxide, ferric salts, and cationic polymer have been effective in preventing impacts to WTP operations. In addition, seasonal stratification with hypolimnotic anoxia appears to be limiting the overall severity of infestation in SWR. Perhaps the most innovative control strategy evaluated by the Authority was the redear sunfish predation study conducted in SWR in 2009–2010.

Introduction and Background

Since the development of Sweetwater Reservoir (SWR) in 1888, now owned and operated by the Sweetwater Authority (Authority), many nonnative species have been introduced into the reservoir and in the surrounding watershed. These have included numerous fish species such as the common carp (*Cyprinus caprio*), channel catfish (*Ictalurus punctatus*), threadfin shad (*Dorosoma petenense*), and mollusks such as Asian clams (*Corbicula fluminea*). However, none of the previously introduced nonnative species has had the potential to fundamentally change the ecosystem of the reservoir in quite the same way as the quagga mussel (*Dreissena rostriformis bugensis*). SWR is a eutrophic water body, supporting a diverse population of plankton species. Cyanobacteria blooms are common during the spring and summer, corresponding with a seasonal thermocline as well as daily diurnal variations in pH and turbidity. Annual cycling of the reservoir level regularly occurs, averaging 13 ft over the past 5 years.

Quagga mussels originally infested the Great Lakes region in 1989, altering the ecology of the lakes, impacting fishing areas, and creating an esthetic nuisance for other recreational activities (Mills et al. 1993). On January 6, 2007, adult quagga mussels were discovered in Lake Mead (LaBounty and Roefer 2007). By the summer of 2007, the quagga mussel infestation had begun to spread (via the Colorado River Aqueduct) to San Diego County reservoirs that were connected to the aqueduct (USGS 2012). In April 2008, quagga mussel veligers were detected in SWR. On January 8, 2009, the first adult quagga mussel was detected at the intake tower in SWR.

Quagga mussels are prolific breeders and are currently impacting water conveyance, power generating, and water treatment facilities throughout the Colorado River Basin and in Southern California. Once established, the quagga mussels can create their own habitat by forming self-sustaining colonies that can clog water intake structures, pipelines, pumps, intake screens, and trash racks (DeLeon 2009). The quagga mussels' main mechanism of nutrient intake is through filtering the water. Each individual mussel can filter up to 1 L of water per day and produce up to 1 million eggs per season (Moy 2009). Dreissenid mussels may increase the clarity (transparency) of the water, which can increase the potential for severe algal blooms in the affected water body (Vanderploeg 2002).

The focus of this chapter is to describe how the Authority's Dresissenid Mussel Monitoring, Response, and Control Plan (Quagga Plan) was developed and implemented to address the quagga mussel infestation in Sweetwater Reservoir and to also document the effectiveness of the Authority's monitoring and control strategies that have been employed to limit the spread of the quagga mussels in the reservoir and infrastructure within the Perdue Water Treatment Plant (WTP).

The Quagga Plan fulfilled the Authority's regulatory obligation under Fish and Game Code Section 2301, which requires public or private entities that operate a water supply system to implement measures to avoid infestation by dreissenid mussels and to control or eradicate any infestation that may occur in a water supply system.

In order to provide an overall context of the development of the Authority's Quagga Plan, an operational snapshot of the Authority is provided later.

The Authority is located in San Diego County and is a publicly owned, joint-powers water agency with policies and procedures established by a seven-member Board of Directors. The Authority provides water service to approximately 187,000 people in National City, Bonita, and the western and central portions of Chula Vista. The 32 square mile service area (Figure 26.1) receives water from four sources: the SWR, deep freshwater wells in National City, semi-brackish groundwater from wells in Chula Vista and National City, and imported water that is drawn from the Colorado River or the State Water Project in Northern California.

The Authority's watershed encompasses approximately 230 square miles. The Sweetwater River is the primary water course through the watershed, flowing from the Cuyamaca Mountains in eastern San Diego County down to San Diego Bay. There are two impoundment facilities located along the Sweetwater River: Loveland Reservoir (25,400 acre-ft capacity), which is located in the eastern San Diego County community of Alpine; and SWR (28,100 acre-ft capacity), which borders Spring Valley. A map of the Authority's watershed is provided in Figure 26.2.

The 30 million gallon per day (MGD) Robert A. Perdue WTP, which is located adjacent to SWR, was originally constructed in 1959 as a direct filtration plant. Today, it is a conventional filtration plant consisting of chemical treatment, chemical mixing using pumped diffusion, flocculation, dissolved air flotation, and dual-media filters. As mentioned earlier, the Perdue Plant is capable of treating (pumping) raw water either from SWR or from the raw water aqueduct pipeline.

The Authority's Quagga Plan was developed, in part, to determine the impacts from veliger and adult mussel populations on infrastructure within SWR and at the Perdue WTP. Therefore, the first order of business in developing the Quagga Plan was to implement a systematic early warning quagga mussel monitoring program.

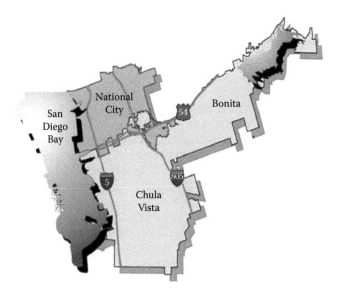

FIGURE 26.1 Sweetwater Authority service area.

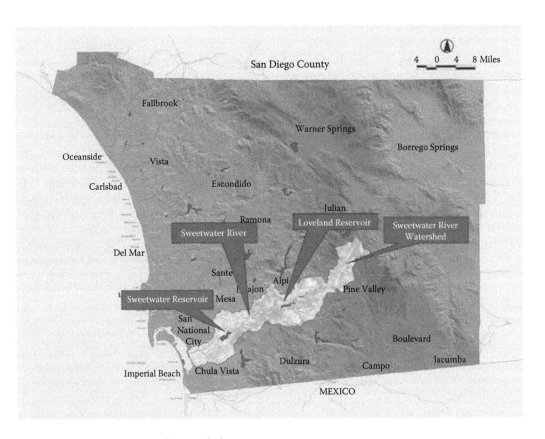

FIGURE 26.2 Sweetwater Authority watershed.

Development of Monitoring and Control Strategies

During the summer of 2007, the Authority started gathering information from the 100th Meridian website related to the unfolding quagga mussel infestation in the Southwest. The website also included a link to the U.S. Bureau of Reclamation (USBR), which was in the process of investigating the spread of veligers throughout the Colorado River Basin. The Authority contacted quagga experts at USBR, who were involved with tracking veliger and adult mussel densities along the Colorado River and in Lake Mead. USBR provided guidance and technical assistance to Authority staff with respect to quagga mussel tow net specifications and developing veliger and solid substrate monitoring protocols in SWR (Kelly et al. 2007). The Authority initiated veliger tows in SWR in October 2007. The samples were preserved and sent to USBR for analysis using polarizing microscopy and polymerase chain reaction (PCR), which detects the presence of quagga mussel veliger DNA.

In August 2007, the San Diego County Water Authority (SDCWA) formed a regional Quagga Mussel Control Plan Workgroup that consisted of member water agencies located in San Diego County and the Metropolitan Water District of Southern California (MWD). MWD, which wholesales water to the SDCWA, provided quagga mussel research and operational control recommendations to the SDCWA Workgroup (DeLeon 2007). The California Department of Fish and Game (DFG), now called California Department of Fish and Wildlife, staff attended the workgroup meetings, and, as the enforcement agency in California, provided valuable regulatory guidance relating to quagga mussel control and monitoring (Maxwell 2007–2012).

One of the main goals of the Quagga Workgroup was to develop a Draft San Diego Regional Dreissena Mussel Response and Control Plan (Prus et al. 2008). The Regional Plan was completed in June 2008. Each agency was then encouraged to use the general guidance provided in the Regional Plan to develop their own agency or site specific quagga mussel control plans.

Also attending the Workgroup meetings during that time were marine biologists from Scripps Institute of Oceanography. In 2008, the SDCWA teamed up with Scripps Institute to create a regional Agreement for performing PCR and polarizing microscopy analysis of San Diego County water agency veliger tow net samples (Burton et al. 2008). This arrangement enabled local agencies to drop off their tow samples without having to preserve them (preserving veliger samples makes it impossible to determine viability).

In the year following, the Authority experimented with several different types of veliger tows (i.e., vertical, horizontal, and pumped) and solid substrates (i.e., plexiglass and ABS plastic squares, perforated pipes, and cinder blocks) to determine which monitoring methods were optimal for SWR. In addition, Authority staff attended a hands-on workshop regarding early detection of quagga mussels (Culver et al. 2008). In April 2008, following guidance provided by DFG, strings of black ABS perforated pipes (1.5 in. diameter; 12 in. length) and square plates (6 in. × 6 in.) were deployed at all substrate monitoring locations in SWR. The substrate strings were deployed from fixed surface buoys and extended from approximately 5 ft below the surface to approximately 15–20 ft below the surface of the reservoir. The buoys were anchored with cement weights. A counterweight and pulley system allowed the monitoring buoys to automatically adjust to increasing or decreasing reservoir level. It was thought that the ABS substrates would provide a better settlement environment for the adult quagga mussels in comparison to the clear plexiglass substrates, which had been deployed since the summer of 2007.

Pictures of typical monitoring substrates employed by the Authority are provided in Figure 26.3. Solid substrate and monitoring buoy construction schematics are provided in Figures 26.4 and 26.5 and monitoring locations for adult quagga mussels are provided in Figure 26.6. Tables 26.1 and 26.2 summarize quagga mussel substrate and veliger monitoring locations in SWR. Veliger tow net specifications are provided in Table 26.3.

The quagga mussel monitoring program was developed and prioritized to target specific areas of SWR where the mussels were expected to colonize, such as the boulder field offshore from the SDCWA NCSB-3 (SB-3) raw aqueduct pipeline inflow to SWR and the treatment plant intake tower. Quagga mussel veligers were first detected in the raw aqueduct water flowing into SWR through the SB-3 pipeline

FIGURE 26.3 Monitoring substrate types.

in December 2007; however, veligers were not detected in SWR itself until April 2008. In January 2009, the first adult mussel was detected on monitoring substrates deployed at the intake tower and by September 2009, adult mussels had been detected at all solid substrate locations in SWR.

Because of the effectiveness of the early warning monitoring program, the Authority was able to detect the quagga mussel infestation in its earliest stages in SWR. This gave the Authority time to respond to the infestation by taking corrective actions to limit the spread of quagga mussels in the reservoir. The effectiveness of these strategies will be discussed in subsequent sections of this chapter. The Authority's Quagga Plan was approved by the DFG in June 2009, making it one of the first response and control plans to be approved in the state of California.

Quagga Mussel Control and Inspection

When developing quagga control strategies, it is advisable to have contingency control strategies in place to cover a worst-case infestation scenario. For example, in the event the population density of quagga mussels in SWR were to increase dramatically, it is very likely the intake screens could become clogged with quagga mussels, which could require that mechanical control methods be employed on a periodical basis. Because no single control strategy has been shown to completely eradicate quagga mussels in large water bodies, it is best to use an integrative approach to quagga mussel control (DeLeon 2007). Having a thorough, well-designed inspection program is critical in determining the extent of adult mussel colonization and can be used to determine which control strategies are most appropriate for current conditions.

Quagga mussel monitoring substrates

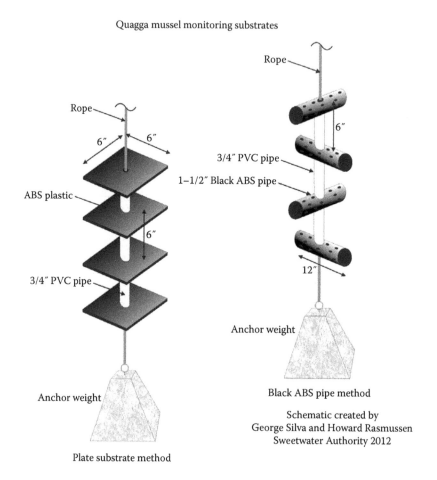

FIGURE 26.4 Substrate construction schematic.

Inspection Procedures

An initial vulnerability assessment was performed in order to develop an in-depth strategy for controlling the population of dreissenid mussels in SWR. This assessment included an evaluation of which assets were most likely to be operationally impacted by a quagga mussel infestation. Routine inspections of Perdue Plant infrastructure (such as raw water intake screens and spools [pictured later], pipelines, and raw water pump stations) were performed and were typically bundled with already scheduled plant maintenance activities.

It is advisable to always be prepared for ongoing (i.e., unexpected) opportunities for infrastructure inspection. For example, in September 2012, as a test, an intake spool (Figure 26.7) was retrieved from SWR and inspected after being submerged near the boat dock for several months. The spool had been infested with quagga mussels (estimated density 180 mussels/m²) with shell lengths ranging from 7 to 10 mm.

Underwater inspection methods employed in SWR have included using ROVs, scuba divers, cameras, and visual examinations of infested areas near the boulder field off shore from the SB-3 pipeline and at the intake tower. Visual inspections of the SWR dam face and intake tower are performed as water levels drop.

ROV inspections were performed in 2008 and 2009 but the effectiveness of this inspection technique is somewhat limited, depending on the visibility within the reservoir and the amount of algae growing on the underwater structures or substrates being inspected. In contrast, diving surveys (conducted 2009

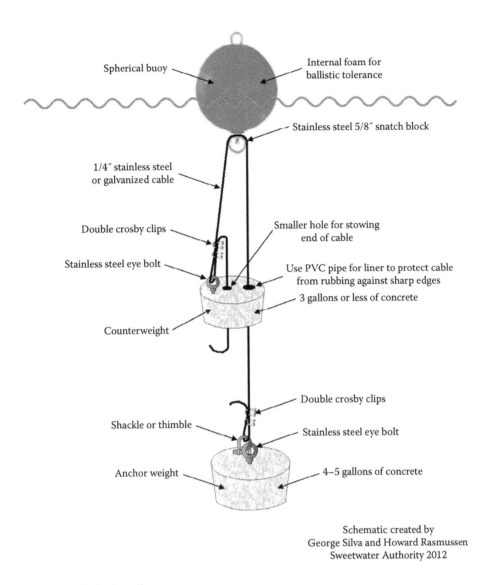

Spherical buoy

Internal foam for
ballistic tolerance

Stainless steel 5/8″ snatch block

1/4″ stainless steel
or galvanized cable

Double crosby clips

Smaller hole for stowing
end of cable

Stainless steel eye bolt

Use PVC pipe for liner to protect cable
from rubbing against sharp edges

3 gallons or less of concrete

Counterweight

Double crosby clips

Shackle or thimble

Stainless steel eye bolt

Anchor weight

4–5 gallons of concrete

Schematic created by
George Silva and Howard Rasmussen
Sweetwater Authority 2012

FIGURE 26.5 Monitoring buoy diagram.

and 2010) are especially effective during the early stages of quagga infestation because divers can move debris and algae and actually touch and harvest the mussels as well as providing estimates of in situ population densities. Areas of the reservoir with the highest colonization of quagga mussels can also be identified and targeted for control.

For example, an ROV examination of the boat dock, dam face, intake tower, and the boulder field offshore from the SB-3 raw aqueduct pipeline was performed on July 15, 2009. Due to the poor visibility of the lake water and the abundance of algae growing on the underwater substrate surfaces, the video images did not show the presence of quagga mussels in any of the areas examined.

Eight days later, on July 23, 2009, DFG divers performed a diving survey of quagga mussel colonization in SWR (Schrimsher and Lewis 2009). The DFG divers examined critical infrastructure on the intake tower (tower walls, intake cups, and screens), the dam face, the emergency release tunnel pipe, boat dock, and the boulder field offshore from the SB-3 raw aqueduct pipeline inflow to SWR. The findings indicated that low densities of adult quagga mussels had begun to colonize the intake tower

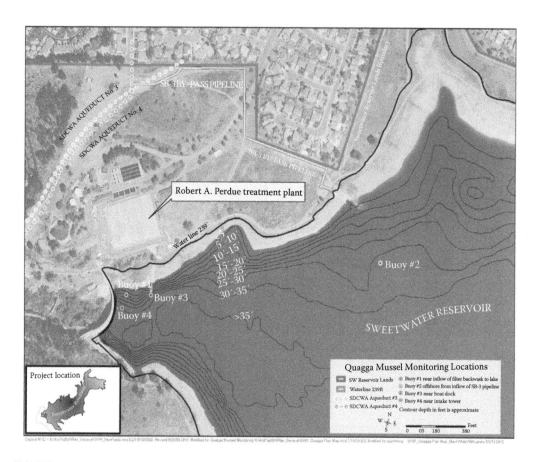

FIGURE 26.6 Quagga mussel monitoring locations.

TABLE 26.1

Substrate Monitoring Locations

Monitoring Location	Cinder Block	Plexiglass Plates	ABS Pipe and Plates	Inspection Frequency[a]
Intake tower (IT)		X	X	Monthly
SB-3 buoy	X	X	X	Monthly
Boat dock		X	X	Monthly
Filter backwash	X	X	X	Monthly

[a] Includes visual inspection and pictures.

TABLE 26.2

Plankton Tow Sites

Monitoring Location	Vertical Tow (Composite)	Horizontal Tow	Pumped	Sampling Frequency
Offshore from SB3	X			Monthly
Log boom to IT		X		As necessary
Log boom to SB3		X		As necessary
Intake tower	X			Monthly
Operator sink aqueduct			X	As necessary
Operator sink Sweetwater raw			X	As necessary

TABLE 26.3

Veliger Tow Net Specifications

Description	Diameter	Mesh Size (µm)	Length (in.)	Ring and Bridle	Dolphin Bucket
Wildco Part # 33-E28	20 in. (horizontal and vertical tows)	63	80	Yes	Yes
Wildco Part # 31-B28; 48-C60; 47-C28	8 in. (pumped samples)	63	24	Yes	Yes

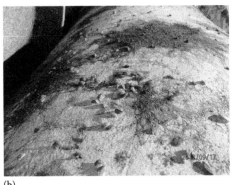

(a)   (b)

FIGURE 26.7 Intake spool September 2012: (a) side view and (b) top view.

(approximately 2 adults were seen on the tower wall near the intake cup and about 12 mussels were seen on the underside of the cup itself). The highest densities of quagga mussels in Sweetwater Lake occurred in the boulder field offshore from the SB-3 inlet to the lake (ranging up to an estimated density of 100–300 mussels/m^2 on the large boulders to only 5–10 quagga mussels per square meter on either side of the boulder field in the more sandy areas). No mussels were seen on the dam face, release tunnel, or on the boat dock. Examples of in situ colonization, shell lengths of the harvested mussels, and colonized monitoring substrates are provided in Figure 26.8. This illustrates the efficacy of performing a diving survey to thoroughly examine submerged critical infrastructure, especially in the early stages of infestation.

Boat Inspections and Decontamination

The Authority does not allow public boating, the use of live bait for fishing or human contact on either Loveland Reservoir or the Sweetwater Reservoir; however, the Authority does maintain several boats to perform maintenance and sampling functions, and these boats are occasionally transferred between the two reservoirs. The Authority does employ decontamination procedures, which are based upon general guidelines and recommendations provided by the 100th Meridian Project, whenever a boat is transferred from Sweetwater Reservoir to Loveland Reservoir. These procedures include application of dilute bleach solutions and/or hot water treatments. The boats are cleaned and inspected each time before being transferred between reservoirs.

Chemical Control

To mitigate the spread of quagga mussels in the Sweetwater Reservoir, the Authority has implemented chemical control strategies, which are summarized in Table 26.4. The goal in designing chemical control methods was to make maximum use of existing treatment processes and chemical addition points at the Perdue Plant to kill the quagga veligers and prevent adult mussel settlement within plant infrastructure.

    Chemical control included the application of chlorine dioxide, which is used as a primary oxidant (disinfectant) for normal operations at the water treatment facilities, to the SB-3 raw aqueduct water pipeline

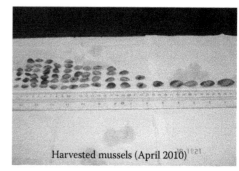

FIGURE 26.8   Substrate inspection and harvested mussels.

TABLE 26.4

Perdue Plant Chemical Control Points

Location	Source Water	Chemical Dose (mg/L)	Application Duration
SB-3 pipeline	Aqueduct raw	ClO_2 at 0.5	Continuous
SB-1 pipeline	Aqueduct raw	ClO_2 at 1 or Cl_2 at 3	Continuous
Intake tower	Sweetwater raw	ClO_2 at 1.6 or Cl_2 at 5	Continuous
Flash mixer	Aqueduct raw	Poly at 1–2 and $FeCl_3$ at 2–4	Continuous
Flash mixer	Sweetwater raw	Poly at 6–10 and $FeCl_2$ at 3–6	Continuous
Pump station	Sweetwater raw	$FeCl_3$ at 6–18	Continuous

(initiated in 2009 when storing water directly into SWR) and at the raw water inputs to the Perdue Plant including the intake tower, when treating water from SWR, and to the SDCWA NCSB-1 (SB-1) raw aqueduct water pipeline, prior to the flash mixer.

Ferric chloride ($FeCl_3$), ferrous chloride ($FeCl_2$), and cationic polymer are chemicals used in the treatment process to coagulate the suspended particles that are present in the raw water. The coagulated particle's size is increased using gentle mixing during the flocculation stage, and the particles are then floated to the surface and skimmed (removed) from the water using dissolved air flotation (DAF). Finally, dual-media filters (sand/anthracite) remove greater than 90% of the turbidity that was originally present in the raw water (before treatment).

The four gravity filters are typically cleaned (i.e., backwashed) every few days and the used backwash water is returned to the lake near the intake tower. The ferric chloride sludge generated from the DAF process is also recycled back to SWR at a different location than that of the filter backwash water.

In order to determine the effectiveness of the Perdue Plant chemical processes to inactivate quagga mussel veligers, pumped tow net samples were taken in January 2008 from process sample taps located in the Perdue Plant Operations Laboratory; the aqueduct raw water (no treatment), conditioned raw water (just after disinfection), and settled water (after sedimentation) were sampled. The microscopy test results showed a dramatic drop off in veliger density and viability between the raw aqueduct water (0.27 veligers/L) and the settled water sample locations (0.012 veligers/L). These preliminary quagga veliger sampling test results indicated that the Perdue Plant disinfection, coagulation, flocculation, and sedimentation processes had effectively removed and/or inactivated the quagga mussel veligers. Since the infestation began in 2008, no adult quagga mussels have been detected within the Perdue Plant infrastructure, with the exception of the intake tower screens as described later.

Mechanical Control and Desiccation

An initial evaluation of potential mechanical control measures was conducted, including the efficacy of placing screens, on the SB-1 and SB-3 raw water aqueduct pipelines to catch any adult quagga mussels that could break loose and travel through the pipelines. The downside of using this approach is that the screens would have to be inspected and maintained at regular intervals. This strategy has not been employed to control quagga mussels in SWR because of potential hydraulic and logistical concerns.

The durability of copper-based coatings (applied to the intake screen surfaces) may also be evaluated at some point in the future. Although copper-based coatings may leach copper into the water, the amount of copper ions leached may not impact current drinking water standards. Health departments may grant an exception to use such coatings in small areas even though most commercial products are not NSF 60 approved.

Initially, when designing mechanical control strategies, it was believed that if the adult quagga mussels were to become firmly established in SWR, the most likely mechanical control measure to be employed by the Authority would be mechanical removal of the mussels from critical infrastructure such as the intake tower screens. This approach would include scraping the mussels from the intake tower screen in situ or removing a clogged screen and replacing it with a clean screen and using a power washer to remove mussels as appropriate.

Mechanical removal by Authority staff was required for the first time in May 2011 as approximately eight adult mussels, ranging in size from 1.6 to 3.2 cm, were harvested from the SWR raw water sample pump screen. In September 2012, significant numbers of mussels were removed from a 3 ft diameter by 5 ft in length, intake tower spool with an approximate density equal to 180 mussels/m².

In April 2013, quagga mussels (approximate density 44 mussels/m²) were mechanically removed (scraped) from the intake tower screen during a change of the intake cup level in SWR. The mussels ranged in size from 1.2 to 2.4 cm in length. Quagga mussels that have been physically removed from the lake are digitally photographed (sized), desiccated, and disposed of properly in a sanitary landfill.

Mechanical removal of mussels is most effective on infrastructure that is readily accessible or removable. For situations where critical infrastructure is submerged, divers may be required to remove the mussels, making mechanical removal more of a logistical challenge.

As the reservoir level drops over the course of the year, mussels attached to critical infrastructure such as the intake tower may become exposed to desiccative conditions above the water line. Desiccated adult quagga mussels have been observed along the shoreline of the boulder field near the SB-3 pipeline and approximately one dozen mussels were observed above the water line on the intake tower in January 2013.

Historically, the highest water levels in Sweetwater Reservoir SWR occur in early spring, resulting from natural (storm) runoff (i.e., approximately 3000–4000 acre-ft annually) and/or periodic water transfers from Loveland Reservoir in winter (i.e., approximately 10,000 acre ft in January of 2008, 2010, and 2012). The Perdue Plant primarily treats raw aqueduct water from November through June and then switches sources to SWR over the summer/fall time frame, which accelerates the drop in water level in the reservoir during those months.

Over the 2008–2012 monitoring period, the reservoir (gage) level ranged from a low of 57 ft in November 2008 to a high of 82 ft in February 2012. The yearly change in lake (gage) level ranged from 5.7 ft in 2009 to 21 ft in 2012, with an average yearly level change of approximately 13 ft. Aside from the benefit of desiccation, the constantly changing water level in SWR may also assist in controlling the adult mussel population, altering their habitat by limiting the extent of oxygenated water within the reservoir as described later.

Limnological Control

Perhaps the most effective (and benign control) strategy for limiting the spread of quagga mussels in SWR has been using the lake's natural limnological processes to suppress quagga mussel spawning and settlement. Literature suggests that the colonization potential of dreissenid mussels should be lowered as the dissolved oxygen level drops below 4 mg/L (O'Neill 1996).

SWR typically stratifies in the late spring and summer, and an anoxic hypolimnetic layer forms in the lower reaches of the reservoir. SWR is a monomictic lake that typically destratifies in the fall each year. Figures 26.9 and 26.10 provide graphical profiles of dissolved oxygen and temperature under typical stratifying and nonstratifying conditions in SWR. Figure 26.11 provides a graphical summary of limnological data over the 2008–2012 monitoring period.

Figure 26.12 shows how some reservoir water quality parameters such as pH, manganese, and total phosphorus may change under stratified (summer) versus nonstratified (winter) conditions. The water quality monitoring data are representative of the raw water at the current intake cup level. Over the 2008–2012 monitoring period, the intake level ranged from 3 to 20 ft below the surface, with an

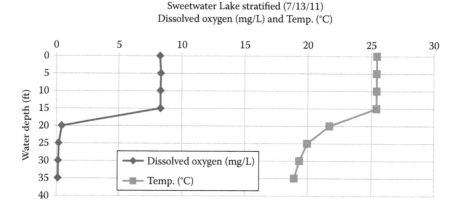

FIGURE 26.9 Graphical profiles of dissolved oxygen and temperature under typical stratifying conditions in Sweetwater Reservoir.

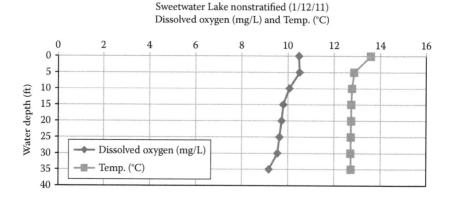

FIGURE 26.10 Graphical profiles of dissolved oxygen and temperature under typical nonstratifying conditions in Sweetwater Reservoir.

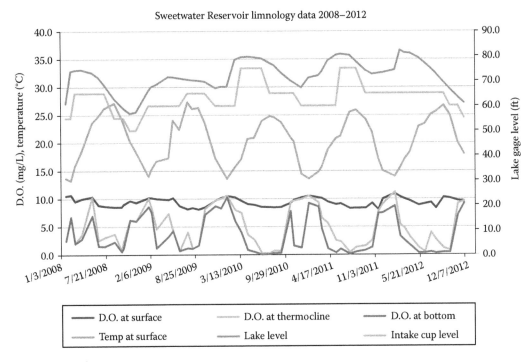

FIGURE 26.11 Graphical summary of limnological data over the 2008–2012 monitoring period.

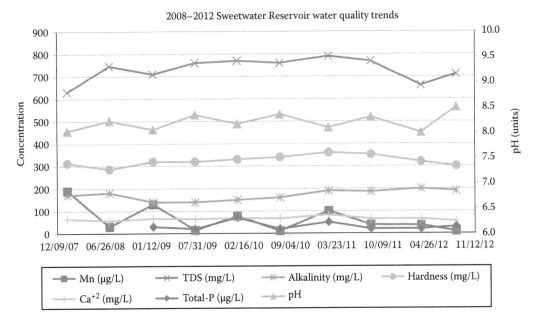

FIGURE 26.12 2008–2012 Sweetwater Reservoir water quality trends.

average depth of 9 ft (Figure 26.11). The intake cup is always positioned above the thermocline when treating water from SWR under stratifying conditions to ensure that oxygenated water is being supplied to the Perdue Plant.

With the exception of 2012, manganese concentrations peaked during the winter sampling events, ranging up to 190 μg/L in January 2008. Total phosphorous monitoring, which was initiated in 2009, exhibited a similar pattern, ranging up to 70 µg/L in January 2010. The water quality data indicate that the observed ranges for pH (8.0–8.5 units), calcium (54–82 mg/L), temperature (13°C–27°C), total hardness (286–360 mg/L as $CaCO_3$), and TDS (630–790 mg/L) should provide an optimal environmental for supporting quagga mussel growth (O'Neill 1996).

Most cyanobacteria blooms (i.e., *Pseudanabaena limnetica, Anabaena circinalis, Anabaena spiroides, Leptolyngbya* spp., and *Anabaena flos-aquae*) tend to occur from spring to early fall when the reservoir is stratified and will typically raise the pH above the thermocline. The concentrations of manganese and total phosphorus (above the thermocline) show the opposite trend as they tend to stay dissolved in the hypolimnion under anaerobic conditions and then mix throughout the water column when the reservoir destratifies or *turns over* in the fall and winter months. It is unknown at this time whether seasonal cyanobacterial blooms and increases in manganese and total phosphorus concentrations, in addition to dissolved oxygen, could potentially have an effect on quagga mussel reproduction in Sweetwater Reservoir.

As expected, thermal stratification in Sweetwater Reservoir was most pronounced during the summer months, corresponding with the time frame the lower reaches of the reservoir (i.e., hypolimnion) became almost completely anaerobic. The dissolved oxygen level at the surface was fairly consistent, ranging from approximately 8 to 10 mg/L. In contrast, under peak stratifying conditions, the dissolved oxygen levels at the thermocline ranged from less than 1 mg/L to approximately 4 mg/L and from less than 1 mg/L to approximately 2 mg/L in the lower reaches of the hypolimnion over the 5-year monitoring period.

With the exception of 2009, when veliger densities peaked in September at the intake tower (2.1 veligers/L), most of the time, veliger densities peaked in the April–May time frame. This trend could indicate that quagga mussels that had settled on natural substrates below the thermocline prior to the summer (peak) stratification were inhibited from spawning during the summer months because of low dissolved oxygen levels. However, mussels that may have settled on natural or man-made substrates above the thermocline, such as at the intake tower during 2009, would not be affected and would still be able to spawn. Also, in 2009, the reservoir level was relatively low (maximum gage height was 71.6 ft), which may not have allowed thermal (and dissolved oxygen) stratification to fully develop in comparison to other years with higher lake levels.

The three rather sharp increases in Sweetwater Reservoir level in January of 2008, 2010, and 2012 were due to transfers of water from Loveland Reservoir, which is located about 15 miles upstream of Sweetwater Reservoir. The deepest area of Sweetwater Reservoir is located near the intake tower. The maximum lake level during 2008–2012 was 82 ft. Over the years, sediment has accumulated near the intake tower, so a lake gage level of 82 ft would correspond to maximum water depth of approximately 60 ft near the intake tower.

As the lake level drops, a visual inspection of the intake tower and dam face are conducted. Perdue Plant operational considerations are always the main concern with respect to lake level management.

While limnological control of quagga mussels within Sweetwater Reservoir appears to be effective in limiting the extent of quagga mussel colonization, there are limitations with this approach. First, stratification is a seasonal condition that means quagga mussel spawning can still occur (below the thermocline) from October through April as the dissolved oxygen levels recover in the hypolimnion. Second, for mussels that have attached to natural or man-made substrates above the thermocline, reservoir stratification (i.e., anoxic conditions below the thermocline) would be expected to have little or no effect. Third, quagga mussels that have colonized in the more shallow areas of the reservoir, such as the portion of the boulder field, which extends from the shoreline to an approximate depth of 15–20 ft offshore from the SB-3 pipeline, may not be affected because the depth of the water column may not be large enough to create stratifying conditions. Under normal conditions, dissolved oxygen levels are stable down to 15 ft below the surface of the reservoir. Sweetwater Reservoir is relatively shallow, with a gently sloping lake bottom. In most years, the maximum depth in the center of the lake ranges from approximately 35 to 45 ft. Over the 5-year monitoring period, the (Secchi disk) clarity in Sweetwater Reservoir ranged from a depth of 1 ft in December 2010 to 25 ft in February 2012, with an average depth of 5.5 ft.

One key element of the implementation of a limnological control strategy is to avoid hypolimnotic lake aeration during the stratification time frame. This approach may be difficult for some water systems that employ lake aeration in order to minimize water quality issues such as dissolved manganese and algal blooms.

Biological Control

In September 2009, the Authority collaborated with the University of Nevada Las Vegas (UNLV) and DFG to undertake an innovative study to determine the effectiveness of using biological predation with redear sunfish (*Lepomis microlophus*) to control quagga mussels in Sweetwater Reservoir. The redear sunfish, also known as the *shellcracker*, is a mollusk-eating fish that should thrive on quagga mussels (Wong and Gerstenberger 2010). The total budget for this study was approximately $11,000, including fabrication of the boundary net and the mussel examinations conducted at UNLV. DFG provided diving services for the duration of the study.

As previously mentioned, a dive survey was conducted by DFG in July of 2009 to determine the extent of quagga mussel colonization in Sweetwater Reservoir and it was discovered that the highest densities of adult quagga mussels in the reservoir (approximately 100–300 mussels/m^2) were found in the boulder field offshore from the SB-3 pipeline inflow to the reservoir.

The study site consisted of a 25 ft by 45 ft area within the boulder field offshore from the SB-3 pipeline in Sweetwater Reservoir (Figure 26.13). The water depth ranged from approximately 12 to 18 ft. The study area was enclosed on all four sides with a weighted boundary net purchased from Christensen Networks (Everson, WA). The net was fabricated from black knotless nylon netting with floats every 0.45 m across the top and a 95/100 lead line. Anchor weights were attached to plastic rings at the bottom of the net. The DFG divers installed the boundary netting on November 5, 2009, making sure the weighted bottom (i.e., the lead line) was securely anchored to the bottom of the boulder field.

On November 10, 2009, the DFG divers measured the initial quagga mussel densities inside the study area and in the control area, which was located adjacent to the study area. Small sampling quadrats (5 in. × 5 in.) were employed to count the mussels (Figure 26.13). The original study design included fixed quadrats to be placed at five locations each in the study and control areas. However, due to poor visibility

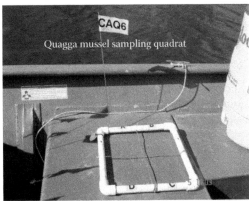

FIGURE 26.13 Redear predation study photographs.

in the reservoir, the divers used hand held quadrats that they placed in five random areas to count mussels in the control and study areas. In order to determine settlement rates during the 6-month study, five (4 in. diameter × 2 ft) perforated pipes were placed in the control and study areas.

On December 1, 2009, roughly 200 redear sunfish (obtained from International Bait and Supply; San Diego, CA) were stocked into the study area (Figure 26.13). The density of the fish within the study area was approximately 0.42 fish/m³, which was slightly higher than the highest naturally observed range. The redears averaged about 3 in. in length. Ten of these fish were sent to UNLV for analysis. Three subsequent mussel and fish sampling episodes were scheduled at 2-month intervals over the course of the 6-month study.

The results of the experiment were promising as on January 29, 2012, a noticeable decline in the population of mussels was observed inside the study area in comparison to the control area (Figure 26.14). The mussel population in the study area was only 8% of the initial density after 2 months (Wong and Gerstenberger 2010). The density of the mussels in the control area was significantly higher than that in the study area. Based upon the size frequency distribution of the mussel populations, the new recruitment of young juveniles was evident in the control area and was less significant in the enclosure where the redears were stocked (Figure 26.15) (Wong and Gerstenberger 2010). The eight redears harvested from the study area were observed to be thriving, increasing in length to an average of approximately 4 in. The contents of the redear's stomachs were examined and evidence of quagga shell fragments was discovered (Figure 26.16) (Wong and Gerstenberger 2010). Unfortunately, only the 2-month sampling episode was conducted because the water level of Sweetwater Reservoir had risen approximately 10 ft during the January–February 2009 time frame due to the transfer of approximately 10,000 acre-ft of water from Loveland Reservoir, causing the bottom of the boundary net to rise approximately 1 ft above the bottom of the reservoir, releasing some of the redears.

Based upon the experimental results, biological control of quagga mussels with redear sunfish in Sweetwater Reservoir could be an effective approach, provided it is applied strategically to critical infrastructure. For example, a boundary net could be placed around the intake tower or boat dock and stocked with redears above their naturally occurring population density to control mussels in those specific areas. Targeting specific infrastructure would be much more practical then attempting a whole lake approach because the costs of stocking and maintaining redears at population levels required for effective reservoir-wide quagga mussel control would be extremely high. Of course, it would be necessary to consider both logistical (i.e., boundary net construction and access to infrastructure) and regulatory (permitting) issues before employing any biological control strategy with redear sunfish.

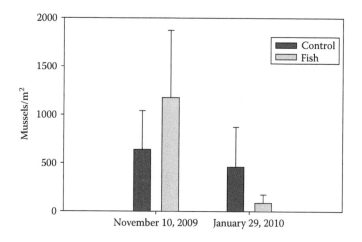

FIGURE 26.14 Quagga mussel densities in the boulder area of Sweetwater Lake (n = 5, mean ± standard deviation).

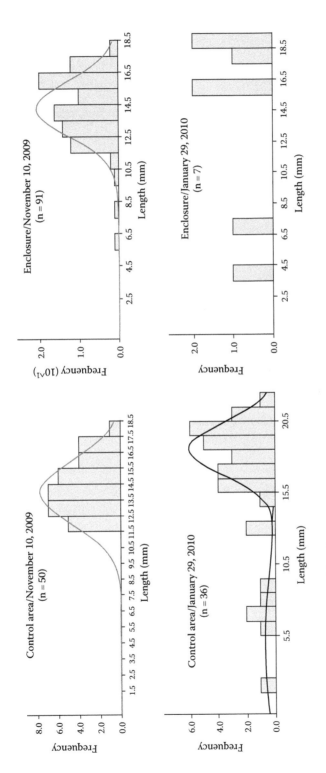

FIGURE 26.15 Frequency of quagga mussel shell length.

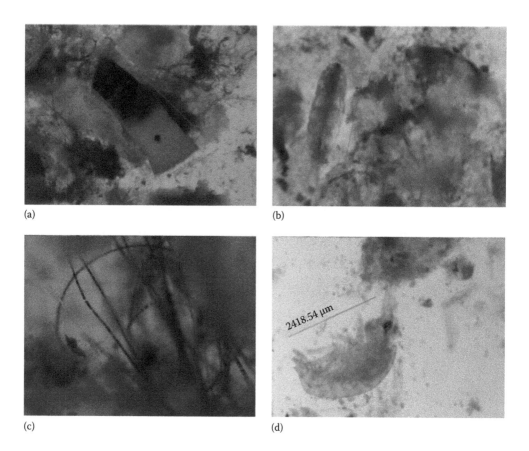

FIGURE 26.16   Broken quagga mussel shell (a), zooplankton (b, copepod is shown), microalgae (c), and benthic organism (d, an amphipod is shown) in the stomach of redear sunfish.

Quagga Mussel Monitoring and Population Trends

Monitoring Methods

The monitoring methods employed at Sweetwater Reservoir included using a large veliger tow net (20 in. diameter by 80 in. in length) to acquire monthly composited vertical tow samples from two locations in Sweetwater Reservoir as follows: (1) three composited vertical tow samples from near the intake tower and (2) three composited tows near the buoy offshore from the SB-3 pipeline. The composited tow volumes typically ranged from approximately 2500 to 6000 L. Each composited sample was transferred from the collection bucket to a 1 L poly bottle, labeled, and analyzed for the presence/absence and quantification of veligers using polarizing microscopy and for quagga DNA using PCR at Scripps Institute of Oceanography.

Monitoring substrates were inspected on a monthly basis; adult quagga mussels were counted on each substrate and were harvested periodically. Adult mussel population densities were determined in mussels per square meter and the data were stored in an Excel database. Digital photographs were taken of the substrates each month and served to document the extent of quagga colonization over time.

Quagga Mussel Population Trends

Population trends over the 2008–2012 monitoring period are provided for quagga mussels (maximum density values per substrate monitoring location) and veligers in Figures 26.17 and 26.18.

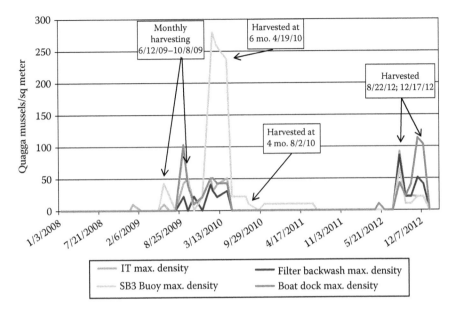

FIGURE 26.17 Sweetwater Reservoir quagga population trends 2008–2012.

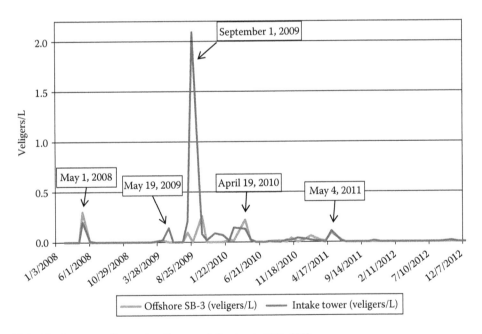

FIGURE 26.18 Sweetwater Reservoir veliger population trends 2008–2012.

Perhaps the largest influence in the creation of a sustainable population of quagga mussels in Sweetwater Reservoir was the inflow (i.e., storage) of untreated raw aqueduct water from the SB-3 pipeline into the lake. The first quagga mussel in the western United States was discovered in Lake Mead in January 2007, so it is likely that SWR was being infested with quagga mussel veligers during the February 2007 to July 2007 time frame (while storing an average of approximately 30 acre-ft/day of raw aqueduct water through the SB-3 pipeline).

A boulder field extends approximately 150 ft into Sweetwater Reservoir from the SB-3 pipeline outfall into the reservoir, thus providing an excellent habitat for quagga mussel colonization. This area of the reservoir had the highest adult mussel densities ever recorded on a solid substrate in Sweetwater Reservoir (i.e., 280 mussels/m^2 in February 2010). The population of quagga mussels in this area was stable and well established as DFG divers had estimated the average (in situ) adult mussel density to be approximately 100–300 mussels/m^2 7 months earlier in July 2009.

Polarizing microscopy was more sensitive than PCR in determining the presence of veligers in Sweetwater Reservoir over the 2008–2012 monitoring period. This trend was more pronounced during years when the frequency of veliger detections was low such as in 2008 (43% of microscopy positives confirmed by PCR) and 2012 (20% of microscopy positives confirmed by PCR). During years with a higher frequency of veliger detections such as in 2009 (69% of microscopy positives confirmed by PCR) and 2010 (79% of microscopy positives confirmed by PCR), there was a much higher correlation between the microscopy and PCR results. Since Sweetwater Reservoir is considered to be a eutrophic system (with abundant phytoplankton and zooplankton populations), it is possible there could have been sample matrix-related interferences with the PCR test.

Interestingly, no veligers were detected in Sweetwater Reservoir until April 2008. Veliger quantification was initiated in May 2008. With the exception of the September 2009 composite vertial tow near the intake tower, which had the highest veliger density ever recorded in Sweetwater Reservoir (2.1 veligers/L), veliger densities generally peaked during the April–May time frame from 2008 to 2011.

The fact that the veliger densities at the intake tower were roughly 10 times higher during the September 2009 sampling event, in comparison to any other composite tow sample, could indicate that the sampling happened to coincide with the peak of a quagga mussel spawning event in the area of Sweetwater Reservoir near the intake tower. The relatively high level of spawning near the intake tower in September 2009 correlates well with the observed increase in the adult quagga mussel population on substrates deployed at both the intake tower (43 mussels/m^2) and the boat dock (103 mussels/m^2), which is located nearby, during that same month.

In 2010, veliger tow net sampling in April again indicated a good correlation between the observed veliger densities offshore from SB-3 (0.2 veligers/L) and at the intake tower (0.1 veligers/L) in comparison to observed adult mussel populations offshore from the SB-3 pipeline (240 mussels/m^2), intake tower (51 mussels/m^2), at the boat dock (43 mussels/m^2), and at the filter backwash monitoring buoy (31 mussels/m^2).

From September 2010 through July 2012, only one adult mussel was detected on the solid substrates deployed in Sweetwater Reservoir and veliger spawning activity also appeared to be somewhat suppressed as the highest density of veligers detected both the intake tower and offshore from SB-3 was 0.1 veligers/L in May 2011.

In 2012, the maximum density of veligers detected near the intake tower was very low (0.012 veligers/L, October 2012) but it did correspond to an observed increase in the adult mussel population at the boat dock in that general time frame, ranging from 41 mussels/m^2 in October to 113 mussels/m^2 in November 2012.

The apparent drop off in both veliger densities and adult mussel colonization in 2010–2012 could be due to a combination of environmental factors. Although it is currently unknown whether cyanobacteria blooms in Sweetwater Reservoir over the past 5 years have had an effect on quagga mussel populations, it is important to note that during the almost 2-year time frame with minimal adult mussel settlement, two extended cyanobacteria blooms were observed (July 2010–December 2010 [*Pseudanabaena limnetica*] and May 2011–October 2011 [*P. limnetica/Leptolyngbya* spp.]). Whereas, during the period of maximum mussel settlement (i.e., June 2009–June 2010), cyanobacteria populations were at relatively low levels.

In 2011, the reservoir started mixing and *turning over* earlier than normal. During the summer and fall of 2011, diurnal changes in the turbidity of the raw water entering the Perdue Plant were observed, making the water difficult to treat. A spike in the raw water turbidity was observed each day around 5 pm. This increase in turbidity was thought to be due to dissolved manganese upwelling from the hypolimnion and entering the raw water intake. Manganese concentrations greater than 5 mg/L were measured entering the Perdue Plant. In contrast, the observed diurnal spikes in turbidity in 2012 were much less severe, coinciding with a late summer through fall recovery in adult mussel populations on the monitoring substrates. Although no direct correlation of quagga mussel mortality as a function of manganese concentration has been confirmed in Sweetwater reservoir, future study in this area may be warranted (Claudi and Prescott 2011).

From August 2007 through November 2008, no raw aqueduct water was stored in Sweetwater Reservoir from the SB-3 pipeline. The storage of raw aqueduct water in Sweetwater Reservoir through the SB-3 pipeline was reinitiated in December 2008 and continued (intermittently) through April 2009, with an average inflow of approximately 50–70 acre-ft/day. By this time, the SB-3 pipeline raw aqueduct water inflow to Sweetwater Reservoir was being treated with a dose of approximately 0.5 mg/L of chlorine dioxide to reduce the levels of viable veligers entering the reservoir. Subsequent to 2007, when untreated raw aqueduct water seeded Sweetwater Reservoir with veligers for approximately 6 months, the number of viable veligers entering the reservoir since then has been either eliminated (by not storing raw aqueduct water) or reduced (by treating the SB-3 flow with chlorine dioxide). The cumulative effect of reducing the loading of viable veligers flowing into the reservoir may have had a dampening effect on the increase in quagga mussel populations in Sweetwater Reservoir.

Quagga Mussel Growth and Resettlement Rates

In order to determine quagga mussel resettlement rates, the solid substrates were harvested periodically throughout the 2008–2012 monitoring period. From June 2009 through October 2009, despite mussels being harvested from the substrates each month, the maximum resettlement densities of the quagga mussels were still significant, especially at the boat dock (103 mussels/m^2; September), the intake tower (51 mussels/m^2; October), and at the buoy offshore from SB-3 (92 mussels/m^2; September).

To measure resettlement rates over a longer time frame, no mussels were harvested from November 2009 until after monitoring was completed in April 2010. During this time frame, a large increase in adult mussels was observed on the substrates located offshore from the SB-3 pipeline, ranging up to 280 mussels/m^2 in February 2010. This increase in population was most likely due to spawning activity from the quagga mussel population in the boulder field near the SB-3 monitoring buoy. However, 4 months later (August 2010), only one mussel was detected on the monitoring substrates deployed in Sweetwater Reservoir (offshore from SB-3). Over the next 2 years (through July 2012), a maximum density of only 10 mussels/m^2 was detected on the monitoring substrates.

By August 2012, adult quagga mussels were again detected at all solid substrate monitoring locations (Figure 26.17). The maximum mussel densities ranged from 43 mussels/m^2 at the boat dock to 92 mussels/m^2 at the intake tower. After harvesting in August 2012, only the substrates deployed at the boat dock showed a marked increase in colonization over the next 4 months as the maximum density more than doubled to 113 mussels/m^2 in November 2012. The adult quagga mussels were harvested from all substrates after the completion of monitoring in December 2012.

In September 2012, a spare intake spool was removed from Sweetwater Reservoir, where it had been submerged next to the boat dock for several months (Figure 26.7). The mussels were harvested from the intake spool and the density was determined to be approximately 180 mussels/m^2.

Figure 26.19 provides a comparison of the quagga mussel shell length (frequency) distribution for each of the three adult mussel harvesting events in 2012. As would be expected for a recent colonization event, the shell length distribution for the mussels harvested in August 2012 was clustered around 3–6 mm. Assuming both the solid substrates and the intake spool were initially colonized within the same time frame, the generational cohort shell length had increased by approximately 4 mm from August 2012 to September 2012.

At the end of the 4-month time period between the August 2012 and December 2012 harvesting events, the shell length distribution of the harvested mussels clustered around 17–20 mm, which would equate to an average shell length growth rate of approximately 4–5 mm/month. The lack of smaller generational cohorts in December 2012 indicates that the quagga mussel colonization on the solid substrates must have taken place in the August/September time frame, when conditions for spawning and settlement were optimal.

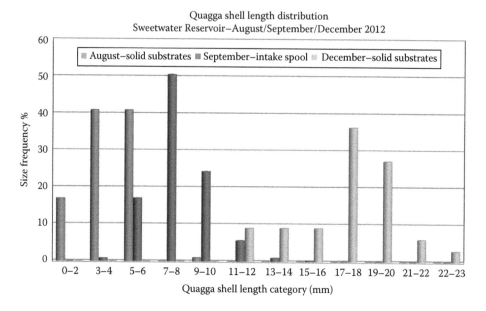

FIGURE 26.19   Quagga mussel shell length (frequency) distribution for each of the three adult mussel harvesting events in 2012.

Quagga Mussel Substrate Preferences

Figure 26.20 shows quagga mussel settlement preference by substrate type (i.e., black ABS perforated pipes vs. black ABS squares) and monitoring location in 2010. The most dramatic difference in quagga mussel substrate settling preference occurred at the SB-3 substrate monitoring location, where the average densities of mussels were three to eight times greater on the ABS squares versus the pipes over the 4-month period (January 2010–April 2010). The opposite trend occurred at the intake tower as the quagga mussels preferred the ABS pipes versus the squares by a factor of 4–5 in January and February and by even higher margins in March and April, where the average density of mussels on the ABS squares was zero.

The relatively strong water currents created as water from Sweetwater Reservoir is pumped from the intake tower to the Perdue Plant for treatment may make the inside of the pipes a more optimal

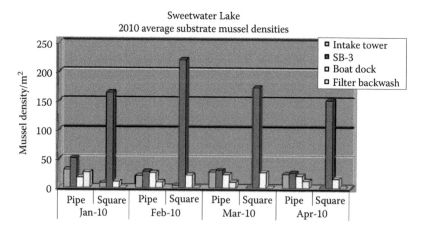

FIGURE 26.20   Quagga mussel settlement preference by substrate type (i.e., black ABS perforated pipes vs. black ABS squares) and monitoring location in 2010.

(i.e., protected) settling environment for the mussels. The perforations in the pipes would still allow the mussels to filter feed efficiently. The same effect may be occurring on the substrates located offshore from where the filter backwash flows enter the reservoir (a single filter backwash flow event can last for 15 min and use over 200,000 gal). In fact, 100% of the mussels settled on the perforated pipes—not a single mussel was detected on the ABS squares during the 4-month period. The quagga mussels that settled on the substrates hung from the boat dock exhibited only a slight preference for settling on the pipes versus the squares, which makes sense because the current at the boat dock, which is located several hundred feet from both the intake tower and the filter backwash monitoring buoy, would be expected to be much less pronounced.

Over the 2008–2012 monitoring period, no quagga mussels were detected on the cinder blocks tethered to the bottom of the reservoir at the Filter Backwash Buoy and to the buoy offshore from SB-3. This could be due to the soft layer of muddy sludge that exists at the bottom of the reservoir, coupled with the fact that anaerobic conditions exist in the lower reaches (i.e., hypolimnion) of the reservoir from spring to fall each year, possibly inhibiting the quagga mussels' ability to survive.

Conclusions

The quagga mussel monitoring and control protocols described in this chapter were implemented by the Authority in an attempt to prevent the population of quagga mussels from multiplying exponentially in Sweetwater Reservoir. While this chapter has focused on the efficacy of the strategies employed in Sweetwater Reservoir, the quagga mussel infestation in reservoirs in the western United States is a regional problem that requires a regional approach. The guidance, exchange of information, and education provided by the SDCWA Regional Quagga Workgroup, DFG, MWD, California Sea Grant, USGS, and USBR were invaluable in the development of the Authority's Quagga Plan. In addition, forming a cooperative partnership with DFG and UNLV created a unique opportunity to study the efficacy of redear predation of quagga mussels in a natural habitat within Sweetwater Reservoir.

It should be noted that the degree of infestation in Sweetwater Reservoir may not be typical of most infested lakes and reservoirs in the Southwest because the densities of both the adult and larval stages of the quagga mussel have remained relatively low over the past 5 years. However, this has also provided the Authority an opportunity to track long-term quagga mussel population trends and to proactively investigate control strategies that may help to limit the severity of the infestation in Sweetwater Reservoir in the years ahead.

One of the main limitations of tracking quagga mussel population trends in Sweetwater Reservoir has been that seasonal veliger populations have not always correlated with the degree of quagga mussel colonization occurring on the solid monitoring substrates. In addition, the only way to reliably determine whether the adult mussel settlement on the monitoring substrates is reflective of reservoir-wide quagga mussel population trends is to periodically perform a dive or ROV survey.

Ultimately, the best approach for controlling quagga mussels in infested water bodies may be to use a variety of control or eradication strategies that take advantage of local conditions (such as reservoir stratification) and existing or available (chemical) treatment processes to prevent the quagga mussels from obtaining a foot hold in critical infrastructure. Finally, it should be noted that the potential for controlled selective predation by redear sunfish shows promise as a control method and in some instances could be a viable option.

REFERENCES

Burton L, Wilson N, Moy G. 2008. Scripps institute of oceanography quagga mussel identification and quantification laboratory services, Scripps Institute of Oceanography (Monthly data reports), San Diego, CA.

Claudi R, Prescott T. 2011. Vulnerability assessment to quagga mussel infestation for Olivenhain, Hodges and San Dieguito Reservoirs and associated facilities, including a list of viable control strategies. San Diego County Water Authority, San Diego, CA.

Culver C, Drill S, Myers M. 2008. Early detection methods and protocols. *Invasive Eurasian Mussel (Quagga & Zebra) Early Detection Monitoring Training Workshop*. California Sea Grant, ANR University of California, UCCE, El Cajon, CA. 100th Meridian Initiative, 2007. General Decontamination Procedures. http://www.100thmeridian.org, accessed May 6, 2008.

DeLeon R. 2007. Elements of a rapid response plan for quagga mussels. *Member Agency Water Quality Managers Meeting Workshop Presentation*, Los Angeles, CA.

DeLeon R. 2009. Invasive quagga and zebra mussels in the west. *American Water Works Association Webcast Presentation*, San Diego, CA.

Kelly K, Hosler D, Nibling F. 2007. Collecting water samples for *Dreissena* spp. Veliger PCR analysis. U.S. Bureau of Reclamation Sampling Protocol, Technical Service Center, Denver, CO.

LaBounty JF, Roefer P. 2007. Quagga mussels invade Lake Mead. *Lakeline* 27:17–22.

Maxwell D, McAlexander B, Schrimsher D, Tavares E, Black D. 2007–2012. California Department of Fish and Game regulatory guidance. *SDCWA Regional Quagga Workgroup Meetings*, San Diego, CA.

Mills EL, Dermott RM, Roseman EF, Dustin D, Mellina E, Conn DB, Spidle AP. 1993. Colonization, ecology, and population-structure of the quagga mussel (Bivalvia, Dreissenidae) in the Lower Great Lakes. *Canadian Journal of Fisheries and Aquatic Sciences* 50(11):2305–2314.

Moy P. 2009. Dreissenid mussels—Their origin, spread, and effects. *American Water Works Association Webcast Presentation*, San Diego, CA.

O'Neill CR. 1996. The zebra mussel: Impacts and control. Cornell Cooperative Extension Information Bulletin 238. New York Sea Grant, Cornell University, State University of New York, Ithaca, NY.

Prus L, Eaton G, Wegand J et al. 2008. San Diego regional dreissena mussel response plan, San Diego County Water Authority, San Diego, CA.

Schrimsher D, Lewis R. 2009. CDFG Sweetwater Reservoir quagga mussel population survey (summary report), California Department of Fish and Game, Sacramento, CA.

U.S. Geological Survey (USGS). 2012. Quagga and zebra mussel sightings distribution in California, 2007–2012, U.S. Geological Survey, Virginia, U.S. http://www.dfg.ca.gov/invasives/quaggamussel/, accessed May 2013.

Vanderploeg H. 2002. The zebra mussel connection: Nuisance algal blooms, Lake Erie anoxia, and other water quality problems of the Great Lakes. National Oceanic and Atmospheric Administration, Great Lakes Environmental Research Laboratory, Ann Arbor, MI.

Wong D, Gerstenberger S. 2010. Density of invasive quagga mussels is reduced by redear sunfish (*Lepomis microlophus*) in Sweetwater Reservoir (final report), University of Nevada Las Vegas, Nevada.

27

Quagga Mussel Monitoring in the Central Arizona Project: 2009–2012

Scott Bryan

CONTENTS

ABSTRACT The Central Arizona Project (CAP) is a 336-mile aqueduct system that annually transports over 1.5 million acre-ft of Colorado River water to the central and southern portions of Arizona, including the Phoenix and Tucson metropolitan areas. In 2008, the CAP became infested with the invasive quagga mussel (*Dreissena bugensis*), which was first found in Lake Mead the previous year. A monitoring program was implemented in 2009 to determine the distribution of the invasive mussel within the CAP, identify environmental factors that may limit its success, and to help assess the risk of infestation at critical pumping plants along the system. Early results indicated that water temperature was the main environmental factor that limited the survival of veligers and settlement of adults; however, more recent data suggest that the limiting factors are much more complex. After 5 years in the system, quagga mussels have become established throughout the entire CAP and have the potential to negatively impact this vital water delivery system.

Introduction

The Central Arizona Project (CAP) is a multipurpose water resource development and management project that provides irrigation, municipal and industrial water, power, flood control, outdoor recreation, environmental enhancement, and sediment control. The primary means of water conveyance is a 336-mile concrete-lined aqueduct that transports water from the Colorado River (Lake Havasu), on Arizona's western border, across the state to Phoenix, and then southward to the aqueduct terminus near Tucson (Figure 27.1).

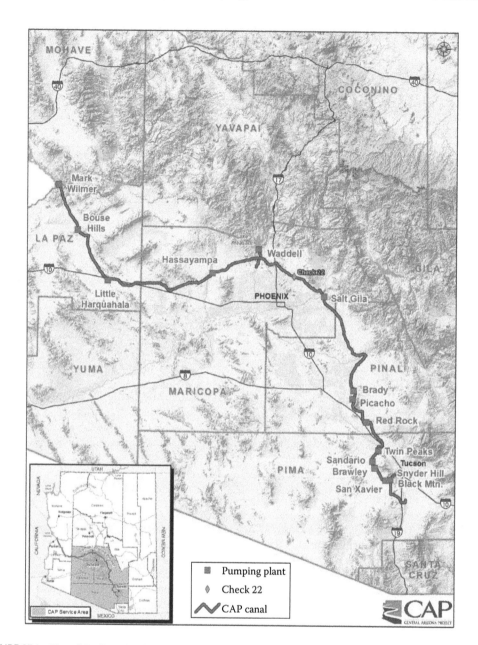

FIGURE 27.1   Map of the Central Arizona Project, including pumping plants.

Over 1.5 million acre-ft of water is delivered to customers each year. During low demand periods (typically mid-September through March), water is pumped directly from the Colorado River at a rate of approximately 3500 cubic feet per second (cfs). Customers receive their allotment of water and the remainder is used to fill Lake Pleasant, which is CAP's 10,000 surface acre storage reservoir located just north of Phoenix (Figure 27.1). During high demand periods (typically April through mid-September), water from Lake Pleasant is released into the aqueduct and mixed with Colorado River water before being delivered to downstream customers. Although highly variable, summer discharge from Lake Pleasant averages approximately 2000 cfs, with the Colorado River contributing approximately 1500 cfs. Lake Pleasant is rarely the sole water source and typically only occurs for short periods of time when upstream maintenance is required. There are 14 pumping plants throughout the system that collectively lift water approximately 2375 ft between Lake Havasu and the terminus near Tucson (Figure 27.1). The canal and its associated structures were completed in 1993.

Adult quagga mussels were first discovered in Colorado River reservoirs in 2007 (LaBounty and Roefer 2007; Whittmann et al. 2010; Turner et al. 2011). One year later (2008), live veligers were observed in plankton samples throughout the CAP aqueduct. Despite the large number of veligers found in the system, including Lake Pleasant, adult mussels only sporadically settled in the aqueduct and downstream pumping plants (Claudi and Prescott 2008). The most likely reason for the lack of settlement was hypothesized to be a combination of factors. First, it was believed that veligers suffered severe mortality and/or damage from single stage pumps that provide the initial lift of water (824 ft) from its origin at Lake Havasu to the beginning of the aqueduct (Claudi and Prescott 2008). Second, Claudi and Prescott (2009) hypothesized that settlement below Lake Pleasant was inhibited by the anoxic releases from the lake, high levels of inorganic suspended solids, high levels of algal production, and/or high summer water temperatures.

In 2009, CAP began to monitor the abundance of live veligers in the aqueduct and the settlement of adults at pumping plants. The primary goals of the monitoring effort were (1) to determine seasonal distribution of both veligers and settled adults throughout the canal, (2) to identify environmental factors that may limit the survival of the mussel or its ability to settle, and (3) to assess the level of risk for infestation in each pumping plant. Although the goals of the monitoring program have remained unchanged over the 4-year period, some of the sampling methods have been refined to provide the most beneficial information.

Methods

Water Temperature

From 2010 to 2012, a temperature logger (Onset Hobo TidbiT v2) was deployed in the forebay of Mark Wilmer Pumping Plant (Mark Wilmer). In 2011, a similar logger was also deployed approximately 21 miles below Lake Pleasant (check #22; Figure 27.1). Loggers were suspended at a depth of approximately 6 ft. Water temperatures were recorded every 6 h and loggers were downloaded annually.

Veliger Sampling

From 2009 to 2012, the forebays of Mark Wilmer and Waddell Pump/Generating Plant (Waddell) were sampled for veliger abundance. A volume of water (250 L) was pumped directly from raw water intakes through a 50 µm plankton net to obtain veliger samples. Samples were preserved in a buffered alcohol solution and transported to the lab for identification and enumeration. Five replicates were subsampled from the original sample and veligers were identified and counted using cross-polarized light microscopy.

During 2009–2010, samples were collected bi-weekly, in 2011 samples were collected daily, and in 2012 samples were collected weekly. Daily samples were collected in 2011 to ensure that "peaks and valleys" in veliger abundance were not being missed with bi-weekly samples. However, it was difficult to analyze the daily samples with limited manpower and the higher level of accuracy achieved with daily samples was determined to be unnecessary for CAP's monitoring goals.

Settling Plates

To monitor mussel settlement throughout the aqueduct system, settling plates were deployed on chains in the forebay at each pumping plant (13; excluding Black Mountain). Plates were 6″ × 6″ and made of a hard black plastic. Each chain had 2–5 plates spaced approximately 3 ft apart, depending on water depth. Upon retrieval, adult quagga mussels were scraped from the plate and counted. No determination of alive or dead was made; however, size of each individual was estimated. Chains from each pumping plant were retrieved annually each year from 2009 to 2012. A separate chain of plates was retrieved quarterly in 2012.

Bio-Boxes

Within each of the 14 pumping plants, mussel settlement and survival were monitored using a bio-box. The bio-boxes are aquaria that are installed in-line with the raw water system, so they receive the same water as pumping plant service water (e.g., cooling water and fire suppression water). Bio-boxes act as a "canary in the coal mine" and give an indication of quagga mussel settlement that may be occurring within the critical piping of the pumping plants.

From 2009 to 2011, only anecdotal data were collected from bio-boxes and very few conclusions could be drawn from the limited information. In 2012, bio-boxes were cleaned of all sediment and mussels in April and quagga mussels were allowed to settle for the remainder of the year. Temperature loggers (Onset Hobo Pendant) were installed in each bio-box, and the number of settled adults was counted each month to correlate quagga mussel density with water temperature.

Data Analysis

Analysis of variance (ANOVA) and Tukey's multiple comparison test (Tukey's HSD) were used to determine differences in mean monthly water temperature among years. A T-test was used to determine whether veliger density differed between Mark Wilmer and Waddell. Spearman rank correlation was used to determine the relationship between mussel settlement and ambient water temperature. All statistical tests were considered significant if $P < 0.05$.

Results

Water Temperature

Although monthly mean water temperatures varied significantly (ANOVA; $P < 0.05$) at the Mark Wilmer forebay (Table 27.1) from 2010 to 2012, overall summer mean temperature (April–October) varied by less than 0.25°C and was not significantly different among years (ANOVA; $P > 0.05$). The maximum temperature recorded at Mark Wilmer reached 31.4°C in both 2010 and 2011, while the

TABLE 27.1

Mean Monthly Water Temperatures (°C) in the Forebay of Mark Wilmer (Lake Havasu) and at Check #22 on the CAP

	Mark Wilmer			Check #22	
	2010	2011	2012	2011	2012
January		11.64[a]	10.95[b]		11.18
February		11.44[a]	12.19[b]		12.62
March	15.65[a]	15.04[b]	14.34[c]		14.98
April	18.01[a]	18.58[b]	18.00[a]	19.36	19.05
May	21.31[a]	21.15[a]	22.18[b]	20.66[a]	22.57[b]
June	24.87[a]	24.64[ab]	25.20[ac]	20.80[a]	21.70[b]
July	28.68[a]	28.45[a]	26.36[b]	17.35	17.03
August	28.63[a]	29.48[b]	27.77[c]	18.77[a]	22.29[b]
September	27.02[a]	27.58[b]	27.38[b]	24.53[a]	27.23[b]
October	23.78[a]	23.06[b]	24.12[a]	22.67[a]	18.15[b]
November	18.00[a]	17.79[a]	18.79[b]	16.25[a]	18.11[b]
December	13.04[a]	13.13[a]	15.27[b]	12.53[a]	14.39[b]
Mean (April–October)	24.63	24.71	24.44	20.73[a]	21.13[b]
Max. (April–October)	31.43	31.43	30.52	27.63	27.73
No. of days > 30°C	15	17	2	0	0

Superscript letters represent values that are statistically significant (Tukey's HSD; $P < 0.05$).

maximum temperature in 2012 was 30.5°C. In 2010, water temperatures were over 30°C on 15 separate days, 17 days in 2011, and just 2 days in 2012.

Below Lake Pleasant, water temperatures are influenced based on whether water comes solely from the Colorado River (typically mid-September through March) or if water is a mixture of Colorado River and Lake Pleasant water (typically April through mid-September). At check #22, approximately 21 miles downstream of Lake Pleasant, mean water temperatures varied significantly between years (ANOVA; $P < 0.05$), and mean monthly water temperatures in 2012 were typically higher than in 2011 (Table 27.1). In 2011, the maximum temperature reached was 27.6°C, while in 2012 the maximum temperature was 27.7°C.

Trends in Veliger Density

Mean annual veliger density in 2012 (4.06 veligers/L) was at its lowest level since 2009 (1.98 veligers/L) at Mark Wilmer. The highest mean density of veligers was observed in 2011 (15.78 veligers/L) when variable peaks in density were measured throughout the summer and extremely high peaks were measured in September and October (Figure 27.2). In 2012, there was a similar trend in late summer, but densities were much lower. The estimated number of veligers per liter never exceeded 25.6 in 2012 compared with

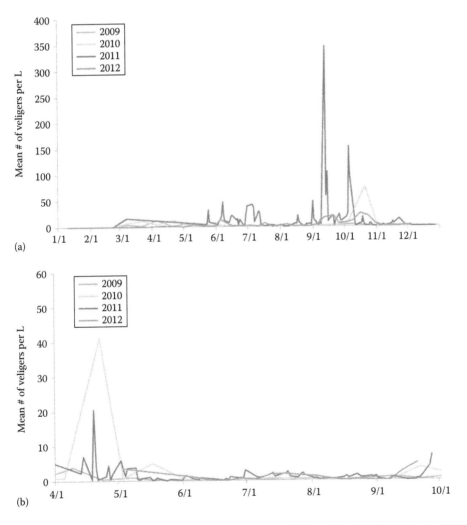

FIGURE 27.2 Trend in veliger abundance collected from the CAP at Mark Wilmer (a) and Waddell (b) during 2009–2012.

densities as high as 345.9 per liter in 2011. In 2009, veliger density reached only 13 per liter, while in 2010 there was a single peak in late October with a density of 76.2 veligers/L.

At Waddell, data are only presented for periods when water was being pumped out of the lake. Since water is not mixed at this point, veliger densities represent quagga mussel spawning in Lake Pleasant rather than the Colorado River. Compared to mean quagga mussel density at Mark Wilmer, mean densities at Waddell were significantly lower in 2009, 2011, and 2012 (T-test; $P < 0.05$). The only spikes in quagga mussel density at Waddell were observed when pumps were first started in early April (Figure 27.2). Throughout the remainder of the summer, mean densities were typically less than 2 veligers/L.

Trends in Mussel Settlement: Settling Plates

Annual settling plates show no discernible pattern in quagga mussel settlement from 2009 to 2012 (Figure 27.3). Settlement was always highest at Mark Wilmer, with numbers ranging from 30,700 per m² in 2010 to 49,050 per m² in 2011. Settlement rates were substantially lower at Waddell (140–3275 per m²). At Bouse Hills, settlement on annual plates was relatively low from 2009 to 2011, but in 2012, numbers increased to nearly 1500 per m². Quagga mussel settlement on plates at all other plants was virtually nonexistent over the 4-year period.

Throughout the 4-year monitoring period, the density of quagga mussels on annual plates at Mark Wilmer was higher on the deeper plates, while plates near the surface had fewer settled mussels (Table 27.2). Similar patterns were observed at Waddell and Salt Gila (Claudi and Prescott 2012; Bryan 2013).

Quagga mussels were found on quarterly settling plates at five pumping plants in 2012 (Figure 27.4). Although the plates showed variable settlement among the plants, numbers were typically higher in spring and summer than during winter. Estimated size of settled mussels follows an expected pattern of smaller individuals during spring, progressively growing into larger individuals during winter (Figure 27.5).

Trends in Mussel Settlement: Bio-boxes

In 2012, quagga mussels were consistently found in bio-boxes at Black Mountain, San Xavier, and Mark Wilmer (Figure 27.6), while mussels were more sporadically found in bio-boxes at Twin Peaks, Brady, and Waddell. Relative to observations at other pumping plants, there was an extremely high infestation

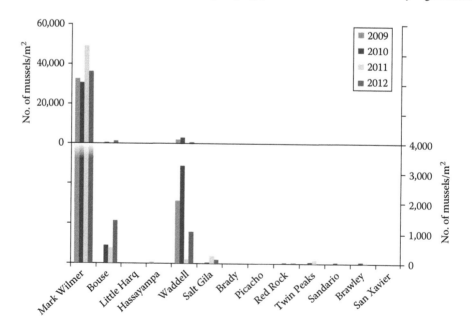

FIGURE 27.3   Trend in adult quagga mussel settlement on annual settling plates at CAP pumping plants from 2009 to 2012.

TABLE 27.2

Number of Quagga Mussels Settled on Annual Plates at Depths Ranging from Approximately 3 to 15 ft at Mark Wilmer from 2009 to 2012

Plate Depth (m)	2009	2010	2011	2012
1	1271	1289	49	28
2	1280	1120	294	848
3	1585	1520	236	476
4	1607	1200	684	920
5	1426	—	2428	1800

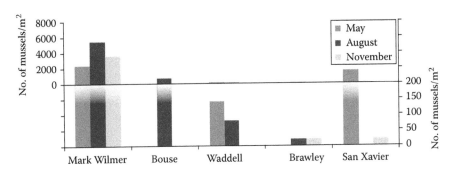

FIGURE 27.4 Settlement of quagga mussels on quarterly settling plates at CAP pumping plants in 2012.

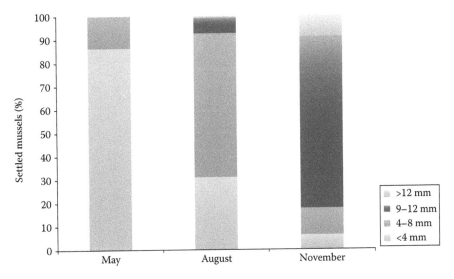

FIGURE 27.5 Percent of settled quagga mussels in each size category collected on quarterly settlement plates at CAP pumping plants in 2012.

of quagga mussels found in the Sandario bio-box during early summer in 2012. However, plant maintenance required water flow to be stopped into the bio-box during July and no further observations could be made. At Mark Wilmer, quagga mussel settlement was relatively low until October, when it increased significantly through December.

Temperatures in bio-boxes ranged from 13°C in the winter to over 31°C during summer. Although there were no significant statistical correlations between quagga settlement and water temperatures (Spearman's R; $P > 0.05$), the graph shows that quagga densities in the bio-boxes were lower with warmer water temperatures, and then increased when water temperatures fell below 27°C (Figure 27.7).

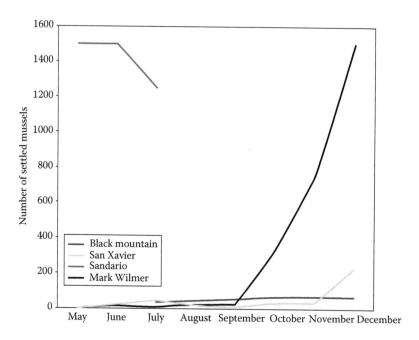

FIGURE 27.6   Number of settled quagga mussels found in monthly bio-box observations in 2012.

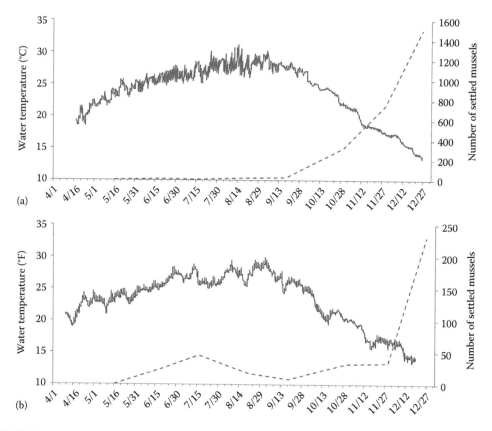

FIGURE 27.7   Water temperatures (solid line) and the number of settled quagga mussels (dashed line) in bio-boxes at Mark Wilmer (a) and San Xavier (b) pumping plants on the CAP.

Discussion

The three methods used to monitor the various life stages of quagga mussels in the CAP are commonly used throughout the United States and Canada (Mackie and Claudi 2010). However, it is evident that each method has its limitations. For example, settling plates were found to be largely ineffective at measuring the abundance of adult quagga mussels in specific areas, and in some cases, even the presence or absence of quagga mussels. During the 4 years of monitoring, quagga mussels have not been observed on settling plates at Brawley, but in 2012, pumps experienced high vibration due to quagga mussel infestation. Veliger sampling is extremely labor intensive and veliger densities have not correlated well with settlement densities, especially at Waddell. Bio-boxes do not represent the flows or piping materials used in pumping plants. However, despite their individual limitations, when used collectively the three methods can provide valuable insight into the degree of infestation that may be occurring.

Our findings of lower veliger densities in 2012 are consistent with findings reported by various agencies throughout the lower Colorado River (personal communications with the U.S. Bureau of Reclamation, Metropolitan Water District, Southern Nevada Water Authority, and University of Nevada, Las Vegas). Although there has been speculation that the food base throughout the lower Colorado River may be limiting, Baldwin et al. (2002) found that quagga mussels can survive, grow, and feed extremely well in waters with both high and low productivity.

Therefore, it is reasonable to assume that some other factor explains the annual fluctuations in veliger density. Spidle et al. (1995) found that biological processes of quagga mussels can be impacted when temperatures exceed 25°C, and the upper thermal limit for quagga mussels has been reported to be 30°C (Cohen 2007; Karatayev et al. 2007). However, Choi et al. (2013) found that quagga mussel veligers can survive temperatures greater than 35°C when acclimated in containers for short durations. Mastitsky and Claudi (2011) conducted a series of statistical analyses (generalized additive mixed-effects modeling) on 2009–2010 CAP data that attempted to correlate veliger abundance with various water quality parameters, including temperature, pH, salinity, chlorophyll, turbidity, and dissolved oxygen. Their report concluded that water temperature had the strongest statistical association with the abundance of quagga mussel veligers. The effect of temperature on the veliger density showed a steady increase in veliger numbers within the range from 10°C to 24°C, followed by a decline at higher temperatures.

Data collected from settling plates and bio-boxes appear to confirm that increased water temperature has a negative impact on quagga mussels. Annual settling plates that were positioned at depths of 10–15 ft below the water surface typically had higher quagga mussel infestation than plates nearer the surface, where water temperatures can be several degrees warmer. In bio-boxes, quagga mussel density increased in the late fall, after water temperatures cooled below 27°C.

However, mean water temperatures in the forebay of Mark Wilmer (Lake Havasu) did not differ from 2010 through 2012, and were even slightly higher in 2011. If water temperature negatively impacted veliger density, we would have expected to observe lower numbers in 2011. Instead, veliger density spiked at Mark Wilmer during early September when water temperatures were near the highest of the year (28°C–30°C). In addition, despite the high mean monthly water temperatures in 2011, mussel settlement rate (on annual settling plates) was higher than any other year.

Although it is reasonable to believe that increased water temperatures may have some negative impact on the quagga mussel population, our data suggest that the issue is much more complex and there must be other factors that combine to influence the survival and distribution of the invasive mussel.

Claudi and Prescott (2008) suggested that the pressures and forces exerted by the pumps at Mark Wilmer and the initial lift (824 vertical feet) through the Buckskin Tunnel cause a high level of damage and mortality to quagga mussels. The consistent infestation of quagga mussels in bio-boxes and settlement plates at the beginning of the system (Mark Wilmer) coupled with the lack of settlement at Bouse, Little Harquahala, and Hassayampa seems to support this theory. However, there have been recent observations of substantial quagga mussel settlement in the aqueduct just downstream of the Buckskin Tunnel (personal observation).

Other environmental factors, such as cold, anoxic releases from Lake Pleasant, high levels of inorganic suspended solids, and high levels of algal production, were thought to be limiting quagga mussel settlement in the southern reaches of the CAP (Claudi and Prescott 2009). However, bio-box data and visual inspections showed infestation in many of the southern pumping plants in 2012. It is evident that the limiting factor that previously minimized infestation below Lake Pleasant is either no longer present or quagga mussels have adapted to it, which has allowed them to expand their range in the aqueduct. Mills et al. (1996) found that quagga mussels acclimated to temperatures and salinity levels in North America that differed from those in the Ukraine. Quagga mussels have adapted to deeper water and sandy sediments in Lake Erie (Dermott and Munawar 1993). McMahon and Ussery (1995) and Choi et al. (2013) found that Dreissenids could adapt to elevated water temperatures when gradually acclimated.

Pimental et al. (2005) reported damage and associated control costs of zebra and quagga mussels at approximately $1 billion per year. His report was prior to the expansion of quagga mussels to the West, so costs have likely increased exponentially since 2005. As the quagga mussel population in the CAP continues to grow, the threat of damage to piping, pumps, and critical systems increases.

Although continued monitoring of the population will not protect the CAP from damage caused by quagga mussels, gaining a better understanding of the environmental conditions that limit the mussel will help ensure that effective control measures can be implemented. Future monitoring within the CAP will focus on developing new techniques, including visual observations using underwater cameras and remotely operated vehicles. Comprehensive water quality monitoring may also provide better insight into the conditions that limit quagga mussel infestation. In addition, water flow appears to affect colonization and growth and various times throughout the year in the CAP (personal observation). An in-depth study of the effect of water flow on quagga mussels may be warranted.

REFERENCES

Baldwin, B.S., M.S. Mayer, J. Dayton, N. Pau, J. Mendilla, M. Sullivan, A. Moore, A. Ma, and E.L. Mills. 2002. Comparative growth and feeding in zebra and quagga mussels (*Dreissena polymorpha* and *Dreissena bugensis*): Implications for North American lakes. *Canadian Journal of Fisheries and Aquatic Sciences* 59:680–694.

Bryan, S.D. 2013. 2012 Annual biology report. Report prepared for the Central Arizona Project, Phoenix, AZ. 49pp.

Choi, W.J., S. Gerstenberger, R.F. McMahon, and W.H. Wong. 2013. Estimating survival rates of quagga mussel (*Dreissena rostriformis bugensis*) veliger larvae under summer and autumn temperature regimes in residual water of trailered watercraft at Lake Mead, USA. *Management of Biological Invasions* 4:61–69.

Claudi, R. and T. Prescott. 2008. Assessment of the potential impact of quagga mussels on Central Arizona Project Facilities and Structures; Recommendations for Monitoring and Control. Report prepared for the Central Arizona Project, Contract #C0825. RNT Consulting, Inc., Ontario, Canada. 46pp.

Claudi, R. and T. Prescott. 2009. Assessment of the potential impact of quagga mussels on CAP Southern Canal Facilities and Structures; Recommendations for Monitoring and Control. Report prepared for the Central Arizona Project, Contract #C0825. RNT Consulting, Inc., Ontario, Canada. 58pp.

Claudi, R. and K. Prescott. 2012. Results of the 2011 quagga mussel monitoring program in the Central Arizona Project Aqueduct. Report prepared for the Central Arizona Project, Contract #0941. RNT Consulting, Inc., Ontario, Canada. 33pp.

Cohen, A.N. 2007. Potential distribution of zebra mussels (*Dreissena polymorpha*) and quagga mussels (*Dreissena bugensis*) in California. Phase 1 Report for the California Department of Fish and Game. San Francisco Estuary Institute, Oakland, CA.

Dermott, R. and M. Munawar. 1993. Invasion of Lake Erie offshore sediments by *Dreissena*, and its ecological implications. *Canadian Journal of Fisheries and Aquatic Sciences* 50:2298–2304.

Karatayev, A.Y., D.K. Padilla, D. Minchin, D. Boltovskoy, and L.E. Burlakova. 2007. Changes in global economies and trade: The potential spread of exotic freshwater bivalves. *Biological Invasions* 9:161–180.

LaBounty, J.F. and P. Roefer. 2007. Quagga mussels invade Lake Mead. *Lakeline* 27:17–22.

Mackie, G.L. and R. Claudi. 2010. *Monitoring and Control of Macrofouling Mollusks in Fresh Water Systems*, 2nd edn. CRC Press, Boca Raton, FL.

Mastitsky, S. and R. Claudi. 2011. Results of the 2010 quagga mussel monitoring program in the Central Arizona Project Aqueduct. Report prepared for the Central Arizona Project, Contract #0941. RNT Consulting, Inc., Ontario, Canada. 81pp.

McMahon, R.F. and T.A. Ussery. 1995. Thermal tolerance of zebra mussel (*Dreissena polymorpha*) relative to rate of temperature increase and acclimation temperature. U.S. Army Engineer Waterways Experiment Station, Vicksburg, MS. Technical Report No. EL-95-10. 28pp.

Mills, E.L., G. Rosenberg, A.P. Spidle, M. Ludyanskiy, Y. Pligin, and B. May. 1996. A review of the biology and ecology of the quagga mussel (*Dreissena bugensis*), a second species of freshwater Dreissenid introduce to North America. *Integrative and Comparative Biology* 36(3):271–286.

Pimental, D., R. Zuniga, and D. Morrison. 2005. Update on the environmental and economic costs associated with the alien-invasive species in the United States. *Ecological Economics* 52:273–288.

Spidle, A.P., E.L. Mills, and B. May. 1995. Limits to tolerance of temperature and salinity in the quagga mussel (*Dreissena bugensis*) and the zebra mussel (*Dreissena polymorpha*). *Canadian Journal of Fisheries and Aquatic Sciences* 52:2108–2119.

Turner, K., W.H. Wong, S.L. Gerstenberger, and J.M. Miller. 2011. Interagency monitoring action plan (I-MAP) for quagga mussels in Lake Mead, Nevada-Arizona, USA. *Aquatic Invasions* 6:195–204.

Whittmann, M.E., S. Chandra, A. Caires, M. Denton, M.R. Rosen, W.H. Wong, T. Tietjen, K. Turner, P. Roefer, and G.C. Holdren. 2010. Early invasion population structure of quagga mussel and associated benthic invertebrate community composition on soft sediment in a large reservoir. *Lake and Reservoir Management* 26:316–327.

Section VII

Control

28

Mussel Byssus and Adhesion Mechanism: Exploring Methods for Preventing Attachment

Bobbi Jo Merten, Allen D. Skaja, and David Tordonato

CONTENTS

Introduction

Zebra and quagga mussels, *Dreissena polymorpha* and *Dreissena bugensis*, respectively, received much attention since their introduction to United States waterways in the 1980s. The zebra mussel significantly populated the eastern half of the United States over a 10-year period. However, the quagga mussel was first discovered in the Colorado River system in 2007 and has since spread to other western waters.

Reclamation began a multifaceted research program in 2008 to mitigate the potential impacts of these mussels on hydraulic equipment. One of the tasks was to evaluate coatings that could prevent mussel attachment, such as antifouling and foul-release systems (*Coatings for Invasive Mussel Control— Colorado River Field Study*). Materials engineers and coatings researchers reviewed the biology and adhesion mechanism of mussel species in order to evaluate methods or materials to discourage attachment and colonization. This chapter presents literature on adhesion mechanisms for zebra and quagga mussels. Important factors include the mussel byssus physical structure and properties, identified mussel foot proteins and biochemical characterization, byssal plaque formation, and mechanics of adhesion.

The marine blue mussel, *Mytilus edulis*, is included where information is lacking on its freshwater counterparts because its research is the most comprehensive. Comparisons are provided where practical. Finally, possible methods for preventing mussel attachment and colonization are discussed with a focus on the development of nonfouling coating surfaces.

Mussel Byssus

The mussel animal occupies the interior of its shell with the exception of the exogenous byssus that allows for surface attachment. The root of the byssus organ anchors into muscle tissue within the shell. Its stem extends from this root, numerous threads emerge from the stem, and a disc-shaped plaque attaches to the surface as the distal end of each thread. The average diameter of byssal plaque is 200 μm, and its appearance is foam like, with vacuoles (voids) that contain the flocculent (adhesive) material (Bonner and Rockhill, 1994).

The mussel foot organ forms the byssal thread and plaque. The mussel places this foot against an underwater surface at which time it triggers the initiation of a series of protein secretions. The protein adhesive and cohesive cure processes take about 5 min. Afterward, the foot reveals a single thread connecting the mussel organism to the byssal plaque attached to the surface. The mussel repeats this process to give the animal multiple anchor points.

This mussel byssus structure and attachment mechanism is similar throughout all mussel species. However, slight differences occur in the number of proteins and their amino acid sequences. In addition, it is worth noting that freshwater and marine species have evolved to live in ecosystems with quite different water chemistries.

Permanent, Temporary, and the Belaying Byssal Threads

Several types of byssal threads have been observed in mussels; these include the permanent, temporary, and belaying. The elongated belaying byssal thread measures 20–30 times the length of the mussel and provides the initial attachment of the mussel to a surface (Rzepecki and Waite, 1993a). The belaying byssal thread has multiple plaques to increase the probability of attaching to a surface. The temporary and permanent byssal threads are much shorter in length (Frisina and Eckroat, 1993). One difference between them is that temporary threads are thinner, longer, and the plaques attach in a tripod pattern for greater stability. Permanent byssal plaques tend to align in straight rows and columns directly beneath the mussel (Frisina and Eckroat, 1993). However, recent observations contradict this alignment of the permanent plaques. Also, the mussel can release the temporary byssal plaque from a surface by secreting an enzyme to disrupt the adhesion mechanism (Gilbert, 2010).

Juvenile and smaller adult mussels utilize temporary byssal threads more often because they tend to move around more (Frisina and Eckroat, 1993). Large adult mussels can also produce temporary byssal threads when needed. Relocating adult or swimming juvenile mussels employ the belaying byssal thread as the initial point of attachment to an arriving surface.

Researchers have not yet identified the enzyme utilized by the mussel to detach temporary byssal thread. However, this may be a potentially significant discovery for developing materials for preventing mussel attachment.

Mechanical and Adhesive Strengths of Mussel Thread

The mechanical properties of the zebra mussel byssal threads were compared to other marine mussels (Bonner and Rockhill, 1994). Zebra and quagga mussels have shorter and thinner threads, but they have higher mechanical properties than all the marine mussels. The tensile strength and modulus (stiffness) are provided in Table 28.1 for zebra mussels as well as marine mussels (Brazee and Carrington, 2006). Zebra mussel threads also had the greatest elongation and extent of recovery. This was somewhat

TABLE 28.1

Mussel Mechanical and Adhesion Strength

Mussel Species	Tensile Strength (MPa)	Tensile Modulus (MPa)	Adhesion/Animal (N)
Zebra	48	137	1.6
Marine (range)	13–26	35–79	10–36

surprising since marine mussels in the tidal zone experience significant wave action, requiring added strength (Brazee and Carrington, 2006).

It is believed that the fibers in the threads of marine mussels resemble collagen, whereas the zebra and quagga mussels fibers are elastin (Bonner and Rockhill, 1994). The elastin fiber provides greater elongation prior to fracture compared to the collagen fibers, which are more rigid. In addition, zebra mussel byssal threads have lateral filament connections between the fibers, providing reinforcement to the fiber bundles (Bonner and Rockhill, 1994).

The adhesion strengths of zebra mussels and marine mussels have been measured on numerous substrates. Table 28.1 provides these data for smooth steel. It is a measure of relative adhesion strength because it does not include the number of byssal threads for each mussel. Zebra mussel's relative adhesion strength is weaker than marine mussels (Bonner and Rockhill, 1994). The reference does not mention the duration of attachment or number of byssal threads. The explanation for the disparity in adhesion strengths was that the zebra mussel plaque is not as porous as the marine species.

The mussel adhesive layer is approximately 10 nm thick, extending between the bottom of the plaque and the substrate surface (Farsad et al., 2009). The small dimensional scale is one reason that the adhesive layer is difficult to measure by nanoindentation and other mechanical techniques.

Biochemical Characterization of Adhesive and Cohesive Proteins

All mussel foot proteins contain amino acid protein 3,4-dihydroxyphenylalanine (DOPA), which is key to the animal's wet adhesion mechanism (Nicklisch and Waite, 2012; Yu et al., 2011b). DOPA forms adhesive and cohesive bonds through its reduced and oxidized (DOPA quinone) states, respectively. Figure 28.1 provides a schematic of protein-bound DOPA's diverse interactions.

Reduced DOPA, at top, leads to the strongest surface bonding (Lee et al., 2006). The reported routes for DOPA surface bonding at the byssal plaque interface include bidentate hydrogen bonding (Yu et al., 2011a), metal/metal oxide coordination (Lee et al., 2006), and oxidative cross-linking (Wilker, 2010). DOPA has been known to chelate with metal ions such as calcium, iron, and aluminum (Brazee and Carrington, 2006). Furthermore, the bidentate hydrogen bonding is twice as strong as a single hydrogen bond, making DOPA oxidation of the bidentate structure highly improbable (Hwang et al., 2013).

Oxidized DOPA, at bottom, can only lead to cohesive bonding. The only exception to this is if the quinone is reduced back to DOPA. This occurs through a thiol-containing reduction *partner protein* described in a later section (Yu et al., 2011a). Potential cohesive interactions include metal-mediated bonding, especially with iron, and intrinsic binding as well as oxidative covalent cross-links (Nicklisch and Waite, 2012).

Further details of the mussel redox chemistry appear in later sections of this chapter. Beforehand, it is important to describe biochemical advances in the characterization of mussel foot proteins to understand the locations and occurrences of DOPA and other important amino acids.

Researchers have characterized the unique amino acid compositions of mussel species' foot proteins. Six polyphenolic *foot proteins* contribute to the makeup of the byssal plaque and thread. They are numbered using acronyms that include the mussel genus and species. For example, the *zebra mussel foot protein number one* is Dpfp-1, where *Dp* denotes *D. polymorpha*. Three additional proteins, often collagen- or elastin-containing proteins, constitute the byssal thread (Bonner and Rockhill, 1994; Nicklisch and Waite, 2012). The focus of this chapter is on the mussel adhesion mechanism and, therefore, is largely limited to the byssal plaque features and formation beyond this point.

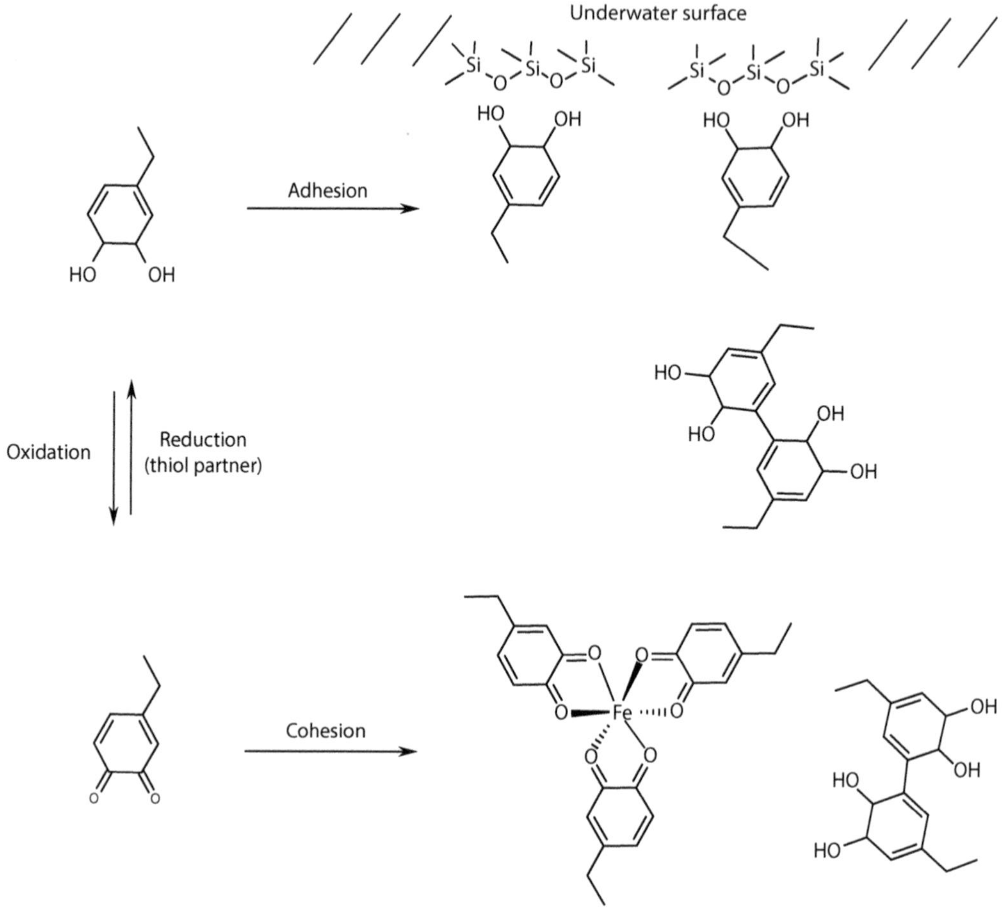

FIGURE 28.1   Adhesive and cohesive DOPA interactions. (Image adapted from Wilker, J.J., *Nat. Chem. Biol.*, 7, 579, 2011.)

Each foot protein is composed of unique amino acid sequences and molecular weights. These vary slightly from species to species and ocean to freshwater. The percentage of DOPA contributes significantly to each protein's structural role. The blue, zebra, and quagga mussel are described in the following using available information on each identified protein.

Blue Mussel, *Mytilus edulis*, Foot Proteins (Mefp-)

The blue mussel Mefp-1 was first characterized in 1981. It is a large protein containing 897 amino acids, of which approximately 60%–70% are hydroxylated amino acid (Silverman and Roberto, 2007). The main function of this protein is for the cuticle (sheath) around the byssal thread and plaque (Frank and Belfort, 2002). The protein is highly cross-linked with Fe^{+3} and Ca^{+2} metal-mediated bonds to provide a protective layer around the threads (Holten-Andersen et al., 2009). Table 28.2 characterizes each blue mussel protein by name, location, molecular weight, and DOPA content.

Mefp-2 is the solid foam structure found exclusively in the plaque that provides cohesiveness to the bulk matrix and is cross-linked by DOPA (Frank and Belfort, 2002; Silverman and Roberto, 2007). Mefp-3 is the smallest protein, and it primarily participates in substrate adhesion at the byssal plaque interface (Silverman and Roberto, 2007). Later, two distinct variants were named—Mefp-3f and Mefp-3s—which are outside the scope of this chapter (Zhao et al., 2006). Mefp-4 is also found within the plaque; its main function is to act as a coupling agent at the byssal thread/plaque junction (Silverman and Roberto, 2007). Mefp-5 accompanies Mefp-3 at the byssal plaque interface, and it contains phosphoserine, which binds

TABLE 28.2

Blue Mussel Foot Proteins, Location, Molecular Weight, and DOPA Content

Protein	Location (Role)	Molecular Weight (kDa)	DOPA (%)
Mefp-1	Thread and plaque outer layer (cohesion)	115	10–15
Mefp-2	Plaque matrix (25%–40% of total matrix, cohesion)	47	2–3
Mefp-3	Plaque interface (adhesion)	5–7	20–25
Mefp-4	Plaque/thread interface (cohesion)	79	4
Mefp-5	Plaque interface (adhesion)	9.5	27
Mefp-6	Plaque matrix (cohesion)	11.5	Small

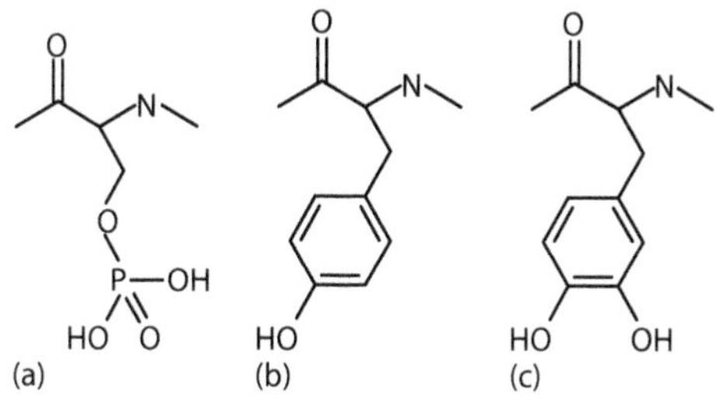

FIGURE 28.2 Chemical structure of protein-bound phosphoserine (a), tyrosine (b), and DOPA (c).

to calcareous materials in certain circumstances (Silverman and Roberto, 2007). From looking at the chemical structure of phosphoserine in Figure 28.2, it would be fair to assume that the protein-bound molecule would be capable of forming bidentate hydrogen bonds with substrate species. Mefp-6 is found in the plaque, containing a large amount of tyrosine and a small concentration of DOPA (Silverman and Roberto, 2007). Tyrosine, or 4-hydroxyphenylalanine, (also shown in Figure 28.2) is the precursor of DOPA, and it converts to DOPA in the presence of catechol oxidase (Silverman and Roberto, 2007).

Mefp-3 and Mefp-5 contain the highest concentrations of DOPA and are the main proteins involved in the adhesion to substrates. They are also the smallest proteins and, therefore, likely to have low viscosity to optimize the wetting of surfaces. Mefp-6 plays a vital role in the mussel adhesion mechanism and provides the link between these DOPA-rich proteins at the adhesive interface and the bulk of the plaque structure (Nicklisch and Waite, 2012; Silverman and Roberto, 2007; Yu et al., 2011a).

Zebra Mussel Foot Proteins (Dpfp-)

The foot proteins of the zebra mussel are not fully characterized. Two proteins are characterized and one is identified, but not characterized. The zebra mussel differs from the marine mussel in several ways. For instance, the proteins contain significant levels of carbohydrates, with galactosamine being most prevalent (Rzepecki and Waite, 1993a). This is not observed in marine mussels, and its significance is unknown at this time. The most common amino acid is asparagines, with much higher levels than marine mussels; alternatively, the glycine levels are lower (Rzepecki and Waite, 1993a). In addition, the zebra mussels lack hydroxyproline, indicating that there are no collagen fibers in the threads (Rzepecki and Waite, 1993a).

Table 28.3 provides molecular weight and DOPA content for the characterized zebra mussel proteins. Dpfp-1 has a high molecular weight and contains 8.5% tyrosine, which, as mentioned previously, can be oxidized to DOPA in the presence of catechol oxidase (Rzepecki and Waite, 1993a). Like the marine mussels, the main function of this protein is for the cuticle (sheath) around the thread and plaque and is highly cross-linked with Fe^{+3} and Ca^{+2} (Frank and Belfort, 2002; Holten-Andersen et al., 2009; Rzepecki and Waite, 1993a). Dpfp-2 has a lower molecular weight, a tyrosine concentration of 13%, and participates in the cohesion of the plaque's solid foam structure (Rzepecki and Waite, 1993b). The Dpfp-3 sequence has not been fully identified, but it has a very low molecular weight and participates in adhesion to substrates; the concentration of DOPA is not known (Rzepecki and Waite, 1993b). Further details on the actual amino acid concentration and sequencing can be found in the references by Rzepecki and Waite (1993a,b).

TABLE 28.3

Zebra Mussel Foot Proteins (Dpfp-4 through Dpfp-6 Not Available)

Proteins	Location (Role)	Molecular Weight (kDa)	DOPA (%)
Dpfp-1	Thread and plaque outer layer (cohesion)	76	6.6
Dpfp-2	Plaque matrix (adhesion)	26	7
Dpfp-3	Plaque interface (adhesion)	4.5–7	N/A

TABLE 28.4

Quagga Mussel Foot Proteins (Dbfp-3 through Dpfp-6 Not Available)

Protein	Location (Role)	Molecular Weight (kDa)	DOPA (%)
Dbfp-1	Thread and plaque outer layer (cohesion)	76	0.6
Dbfp-2	Plaque matrix (cohesion)	35	2

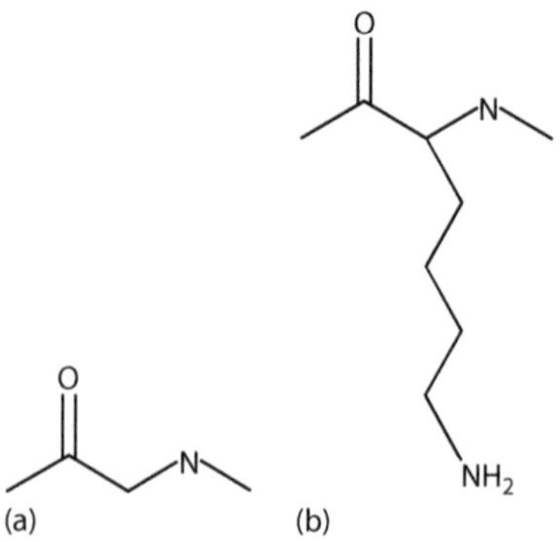

FIGURE 28.3   Chemical structure for protein-bound glycine (a) and lysine (b).

Quagga Mussel Foot Proteins (Dbfp-)

Table 28.4 provides the foot proteins characterized for the quagga mussel. Information on these proteins is given in Table 28.4. Dbfp-1 has a high molecular weight and a low DOPA concentration (Rzepecki and Waite, 1993b). Notably, however, it has a much higher concentration of tyrosine, glycine, and lysine compared to Dpfp-1 (Anderson and Waite, 2002; Rzepecki and Waite, 1993b). Figure 28.3 provides the chemical structure for glycine and lysine. As with the other mussel species, the main function of Dbfp-1 is for the sheath around the byssal thread and plaque, and it is highly cross-linked with Fe^{+3} and Ca^{+2} (Anderson and Waite, 2002; Holten-Andersen et al., 2009; Rzepecki and Waite, 1993b).

Dbfp-2 has a low molecular weight (Rzepecki and Waite, 1993b). Again, Dbfp-2 has a much higher concentration of tyrosine compared to Dpfp-2. Like the blue and zebra mussel, Dbfp-2 is a solid foam structure that participates in byssal plaque cohesion. Further details on the actual amino acid concentration and sequencing can be found in the references by Rzepecki and Waite (1993b) and Anderson and Waite (2002).

Mussel Adhesive Bonding Methods

Physical adhesion between two materials is achieved through a combination adsorption, mechanical interlocking, and molecular diffusion across an interface. It is vital to have intimate contact between the two materials. For byssal adhesion, microtopography, viscosity of adhesive, and wetting tendency are all important (Nalepa and Schloesser, 1993). The byssal adhesion mechanism of the zebra and quagga mussels is not completely understood, but mechanical interlocking is thought to play an important role. The mussel achieves this when the secreted adhesive flows into and wets the microscopic pores and crevices of surfaces. As stated previously, Mfp-3 and Mfp-5 are very low molecular weight, allowing deep wetting and diffusion to occur.

Mussel adhesive also provides chemical functionality to provide bidentate hydrogen bonding, chemical cross-linking, and covalent chelating bonding with metals (Nicklisch and Waite, 2012). The adhesive foot proteins, Mfp-3 and Mfp-5, are also primarily responsible for the chemical adhesion to substrates. Here, dynamic amino acids, including DOPA, tyrosine, and phosphoserine, seem to play significant roles by tightly bonding or chelating with substrate moieties.

It is most likely that a combination of mechanical and chemical adhesion mechanisms is utilized to form the characteristically strong byssal plaque/substrate bonds. Further details on the byssal plaque formation, including this adhesive interface, are provided in the following sections.

Experimental Observations on Byssal Thread and Plaque Formation

Direct observation of the zebra mussel plaque formation and attachment was conducted in 1990 (Nalepa and Schloesser, 1993). One interesting observation was that the mussel foot always swept across the substrate surface prior to attachment (Nalepa and Schloesser, 1993). It was unclear in this study if the mussel foot secretes an adhesive just prior to the thread and plaque formation, or if the foot was used to clean the surface. Recent discoveries indicate that the byssal plaque and thread form within the distal depression of the mussel foot. The number of byssal threads formed depends on several environmental factors, including water velocity, water temperature, salinity, and food availability, as well as substrate type (Marsden and Lansky, 2000; Rajagopal et al., 1996).

The process of plaque formation in the blue mussel was recently investigated by real-time spectroscopic study of the protein secretions (Yu et al., 2011b). These studies include the natural, unperturbed adhesive secretion as well as chemically induced secretions. Matrix-assisted laser desorption ionization-time of flight (MALDI-TOF) mass spectrometry was the protein detection method used in the perturbed and unperturbed studies. Utilizing the previous biochemical characterizations, this allowed for a step-by-step analysis of the byssal plaque and thread formation.

Researchers achieved induced plaque formation by injecting 0.55 M KCl solution into the mussel's pedal ganglion organ at the base of the foot (Yu et al., 2011b). The plaque formation initiates within 1 min of injection. Mefp-3 is the first detected protein, followed by Mefp-6 approximately 30 s later (Yu et al., 2011b). The instrument also detects Mefp-5 during this first minute; however, this particular protein requires higher laser power for improved resolution.

The researchers used a number of quality control methods to compare the results of induced plaques to natural depositions. The secreted foot proteins, themselves, are identical; however, it is unknown whether the process of deposition is the same. The induced plaque formation experiments have been valuable in that they confirm and improve our understanding of each foot protein's role. It is assumed, here, that the zebra and quagga mussel plaque and thread formation process is similar to the blue mussel.

Solution Chemistry for Plaque Formation

The water chemistry or solution conditions that occur under the foot during adhesive cure are vital to the formation of a strong, well-adhered plaque. Saltwater and freshwater are quite different, especially in regard to ionic strength. Ocean water has a pH ~8.2 with an ionic strength of 0.7 M. Previously, it seemed that marine and freshwater mussels could have different attachment mechanisms due to different water chemistries.

Recent studies of blue mussel discovered that the mussel foot isolates and facilitates a microenvironment during plaque formation. Yu et al. (2011b) utilized microelectrodes to reveal the pH and ionic strength conditions of ~5 and 0.1 M, respectively. The high viscosity and stickiness of the secreted proteins quickly foul the electrode; therefore, these estimates are somewhat conservative (Yu et al., 2011b).

Redox Chemistry for Plaque Formation

In addition to the mussel foot being responsible for the pH and ionic strength regulation, it performs vital redox cycling processes (Nicklisch and Waite, 2012). The discovery that the oxidation of DOPA to DOPA quinone results in reduced adhesion forces revealed the mussel's participation of redox cycling (Lee et al., 2006). The mussel carefully controls the redox chemistry of DOPA (and possibly other entities) to achieve proper plaque formation; too much oxidation leads to interfacial failure and too little oxidation causes cohesive failure. Yet more impressive is that the redox control is both positional and temporal during thread and plaque formation (Nicklisch and Waite, 2012). The discovery of this insight occurred during research on marine mussels; however, zebra and quagga mussels likely utilize solution chemistry regulation and redox chemistries to achieve strong, covalent cross-links.

Thiol-containing Mfp-6 is responsible for balancing the redox chemistry in mussel adhesive and cohesive processes. The oxidation of DOPA to DOPA quinone is favorable over a wide range of pH. Mfp-6 is cosecreted with Mfp-3 and Mfp-5 to maintain a reducing environment for DOPA. Studies utilize the diphenylpicryl hydrazyl (DPPH) free radical to measure the redox species and their locations during plaque formation. The results confirmed the role of thiol-containing proteins, such as Mfp-6, with DOPA quinone (Nicklisch and Waite, 2012; Wilker, 2010). This provides DOPA with added opportunities to form strong bidentate substrate bonds.

Figure 28.4 summarizes mussel adhesion from the macroscale to molecular scale, beginning with the mussel body, foot, thread, and plaque at the top left. The center image provides the theoretical organization of all mussel foot proteins within the plaque. And, finally, the top-right schematic demonstrates bidentate hydrogen bonding interactions of DOPA-containing foot proteins (Mfp-3 or Mfp-5) at the mussel plaque/underwater surface interface. Here, the central molecule represents the initial bidentate hydrogen bond adhesion of DOPA with the surface. The molecule at left shows the pendant DOPA

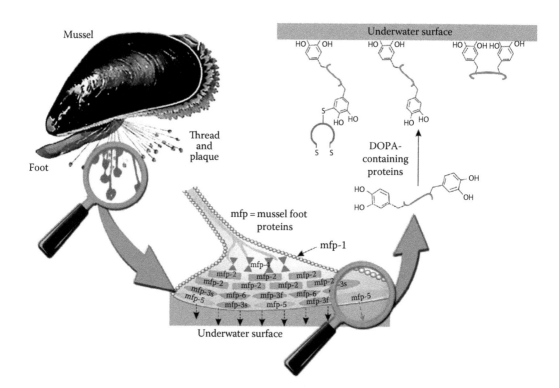

FIGURE 28.4 Mussel macroscale to molecular adhesion. (Images adapted from Hwang, D.S. et al., *J. Biol. Chem.*, 285, 25850, 2010; Yu, J. et al., *Nat. Chem. Biol.*, 7, 588, 2011a.)

cross-linking with thiol-containing Mfp-6 to form a cohesive bond with the bulk plaque. The final molecule at right shows the pendent DOPA forming an additional adhesive bond with the substrate.

Methods for Preventing Zebra and Quagga Mussel Attachment

Zebra and quagga mussels likely contain high concentrations of DOPA in Mfp-3 and Mfp-5 for adhesion although the proteins have not been fully characterized. DOPA's primary interactions are strong, bidentate hydrogen bonding and covalent bonding with metals. From a coatings perspective, mussel adhesive adheres strongly to epoxy and polyurethane coatings (Skaja, 2010). This is, in part, due to the oxygen and nitrogen within the coating chemistry, which are hydrogen bond acceptor sites. These functional groups are present at the surface in most organic coatings, allowing for facile mussel attachment. There are coating chemistries without hydrogen bond acceptor sites; however, materials such as polyethylene, polypropylene, and polytetrafluoroethylene have shown strong mussel attachment (Puzzuoll and Leitch, 1991; Race and Kelly, 1997). The ability for the mussel adhesive to wet and mechanically interlock with the surface microstructure and porosity of these systems becomes important. Thus far, the only nontoxic coatings to prevent mussel attachment are silicone foul-release coatings (FRCs) (Skaja, 2010). This unique combination of chemistry and physical properties prevents mussel attachment.

Foul-Release Coatings

The theory, as well as chemical and physical properties, required for formulating a biocide-free FRC system is discussed in this section. The adhesion mechanism at the point of contact between the mussel foot and the surface of the coating is critical. The plaque adhesive must be able to have intimate contact and wetting or spreading across the surface is essential. One physical property that can help prevent the wetting of the mussel adhesive on the surface is to minimize the critical surface energy (Brady, 1999). Low surface energy is a key property in an FRC system. Fluorine- and silicone-based polymers are the lowest surface energy polymers known at this time. However, fluorine is known to have porosity into which the bioadhesives adhere by mechanical interlocking (Brady, 2000).

The fracture mechanics between the marine fouling species and the coating surface is also an important property. There are three different types of fracture: tensile, shear, and peel, with peel requiring the lowest amount of energy (Brady, 2000). A low elastic modulus aids in the peel failure between the marine fouling species and the coating surface (Brady, 2000). This allows for flexibility and chain mobility, which is thought to prevent strong mechanical bond formation (Brady, 2000). A high elastic modulus provides a rigid surface for the mussel to form strong mechanical interlocking. The drawback of the low modulus is that the coatings are inherently low toughness and durability.

There is a difference between the results observed in the marine species and the freshwater species using the silicone FRCs. In the marine environment, barnacles, tube worms, and other fouling species attach to the silicone ship surface while docked, and the species are removed by drag force during ship movement, hence the name *FRCs*. However, quagga mussels have no attachment to silicone FRCs (Skaja, 2010).

Durable Foul-Release Coatings

In recent years, scientists have tried to develop more durable FRCs. There are few commercially available silicone–epoxy hybrid FRCs and fluorinated polyurethane FRCs. Initial test results show that these coatings experience mussel attachment (Skaja and Tordonato, 2011). However, these coatings were easier to clean than traditional epoxy or polyurethane coatings. Advances in polymer synthesis, coatings chemistry, and formulation are necessary to provide a durable FRC, which prevents mussel attachment.

REFERENCES

Anderson, K.E., Waite, J.H., Biochemical characterization of a byssal protein from *Dreissena bugensis*, *Biofouling*, 18(1), 37–45, 2002.

Bonner, T., Rockhill, R., Ultrastructure of the byssus of the zebra mussel, *Transactions of the American Microscopical Society*, 113(3), 302–315, 1994.

Brady, R.F., Properties which influence marine fouling resistance in polymers containing silicon and fluorine, *Progress in Organic Coatings*, 35, 31–35, 1999.

Brady, R.F., Clean hulls without poisons: Devising and testing nontoxic marine coatings, *Journal of Coatings Technology*, 72, 44–56, 2000.

Brazee, S.L., Carrington, E., Interspecific comparison of the mechanical properties of mussel byssus, *The Biological Bulletin*, 211(3), 263–274, 2006.

Farsad, N., Gilbert, T.W., Sone, E.D., Adhesive structure of the freshwater zebra mussel, *Dreissena polymorpha*, *Materials Research Society Symposia Proceedings*, 1187, 33–40, 2009.

Frank, B.P., Belfort, G., Adhesion of *Mytilus edulis* foot protein 1 on silica: Ionic effects on biofouling, *Biotechnology Progress*, 18, 580–586, 2002.

Frisina, A.C., Eckroat, L.R., Histological and morphological attributes of the byssus of the zebra mussel (*Dreissena polymorpha*) (Palla), *Journal of the Pennsylvania Academy of Science*, 66, 63–67, 1993.

Gilbert, T., Investigation of the protein components of the zebra mussel (*Dreissena polymorpha*) byssal adhesion apparatus. Masters thesis, University of Toronto, Toronto, Ontario, Canada, 2010.

Holten-Anderson, N., Mates, T.E., Toprak, M.S., Stucky, G.D., Zok, F.W., Waite, J.H., Metals and the integrity of a biological coating: The cuticle of mussel byssus, *Langmuir*, 25(6), 3323–3326, 2009.

Hwang, D., Wei, W., Rodriguez-Martinez, N., Danner, E., Waite, J., Chapter 8—A microcosm of wet adhesion: Dissecting protein interactions in mussel plaques, In Zeng, H. (ed.), *Polymer Adhesion, Friction, and Lubrication*, John Wiley & Sons Inc., Hoboken, NJ, pp. 319–350, 2013.

Hwang, D.S., Zeng, H., Masic, A., Harrington, M.J., Israelachvili, J.N., Waite, J.H., Protein- and metal-dependent interactions of a prominent protein in mussel adhesive plaques, *Journal of Biological Chemistry*, 285, 25850–25858, 2010.

Lee, H., Scherer, N.F., Messersmith, P.B., Single-molecule mechanics of mussel adhesion, *Proceedings of the National Academy of Sciences*, 103, 12999–13003, 2006.

Marsden, J.E., Lansky, D.M., Substrate selection by settling zebra mussels, *Dreissena polymorpha*, relative to material, texture, orientation, and sunlight, *Canadian Journal of Zoology*, 78, 787–793, 2000.

Nalepa, T., Schloesser, D., *Zebra Mussels: Biology, Impacts, and Control*, Lewis Publishers, Boca Raton, FL, pp. 239–282, 1993.

Nicklisch, S.C.T., Waite, J.H., Mini-review: The role of redox in Dopa-mediated marine adhesion, *Biofouling: The Journal of Bioadhesion and Biofilm Research*, 28, 865–877, 2012.

Puzzuoll, F.V., Leitch, E.G., Evaluation of coatings to control zebra mussel colonization—Year one—Interim report. Ontario Hydro Report 91–56-K, April 25, 1991.

Race, T.D., Kelly, M.A., Chapter 23—A summary of a three year evaluation effort of anti-zebra mussel coatings and materials, In *Zebra Mussels and Aquatic Nuisance Species*. Lewis Publishers, Boca Raton, FL, pp. 359–388, 1997.

Rajagopal, S., Van Der Velde, G., Jenner, H.A., Van Der Gaag, M., Kempers, A.J., Effect of temperature, salinity, and agitation on byssus thread formation of zebra mussel *Dreissena Polymorpha*, *Netherlands Journal of Aquatic Ecology*, 30(2–3), 187–195, 1996.

Rzepecki, L.M., Waite, J.H., The byssus of the zebra mussel, *Dreissena polymorpha*. I: Morphology and in situ protein processing during maturation, *Molecular Marine Biology and Biotechnology*, 2(5), 255–266, 1993a.

Rzepecki, L.M., Waite, J.H., The byssus of the zebra mussel, *Dreissena polymorpha*. II: Structure and polymorphism of byssal polyphenolic protein families, *Molecular Marine Biology and Biotechnology*, 2(5), 267–279, 1993b.

Silverman, H.G., Roberto, F.F., Understanding marine mussel adhesion, *Marine Biotechnology*, 9(6), 661–681, 2007.

Skaja, A.D., Testing coatings for zebra and quagga mussel control, *Journal of Protective Coatings and Linings*, 27(7), 57–65, 2010.

Skaja, A.D., Tordonato, D.S., Evaluating coatings to control zebra mussel fouling, *Journal of Protective Coatings and Linings*, 28(11), 46–53, 2011.

Wilker, J.J., The iron fortified adhesive system of marine mussels, *Angewandte Chemie*, 49, 8076–8078, 2010.

Wilker, J.J., Redox and adhesion on the rocks, *Nature Chemical Biology*, 7, 579–580, 2011.

Yu, J., Wei, W., Danner, E., Ashley, R.K., Israelachvili, J.N., Waite, J.H., Mussel protein adhesion depends on thiol-mediated redox modulation, *Nature Chemical Biology*, 7, 588–590, 2011a.

Yu, J., Wei, W., Danner, E., Israelachvili, J.N., Waite, J.H., Effects of interfacial redox in mussel adhesive protein films on mica, *Advanced Materials*, 23, 2362–2366, 2011b.

Zhao, H., Robertson, N.B., Jewhurst, S.A., Waite, J.H., Probing the adhesive footprints of *Mytilus californianus* byssus, *Journal of Biological Chemistry*, 280, 11090–11096, 2006.

29

Coatings for Invasive Mussel Control: Colorado River Field Study

Allen D. Skaja, David Tordonato, and Bobbi Jo Merten

CONTENTS

Introduction

Background

The first colonization of invasive quagga mussels in the western United States occurred in a marina on Lake Mead Reservoir in 2007 (McKinnon 2007). The US Department of the Interior, Bureau of Reclamation (Reclamation) is responsible for the operation and maintenance of many federally owned dams, powerplants, and water distribution systems in the western United States (including Hoover Dam) and thus has a vested interest in ensuring that invasive species fouling does not interfere with the continued reliability of its infrastructure.

In 2008, a test program was initiated by Reclamation's Materials Engineering and Research Laboratory (MERL) to review and evaluate the effectiveness of coating technologies that may prevent or deter attachment (colonization) by zebra/quagga mussels. This research was a subset of a multifaceted approach being taken by Reclamation to mitigate impacts from invasive mussels. MERL's research included both field and laboratory testing. Field testing directly evaluated antifouling efficacy, whereas laboratory testing measured other properties of interest such as durability, weathering resistance, and corrosion protection. This chapter will focus on the field test results and compare them to previous research by others.

Previous Coatings Research

Marine fouling provides the environmental test conditions for a majority of previous research. There are more than 6000 fouling species in the ocean, and it has been an ongoing challenge since marine vessels entered the seas (Jones 2009). All antifouling coatings currently manufactured were specifically designed for the marine market. Not surprisingly, some products are also useful for controlling zebra and quagga mussels. The major difference between the marine industry and the freshwater infrastructure is that the ships move through the water, while freshwater infrastructure is stationary with flowing water. With stationary equipment there are additional factors to be concerned about such as impact, abrasion, erosion, and gouging caused by debris, ice, and the occasional tree. In addition, freshwater infrastructure is usually inaccessible and may require removal and handling during the recoating process. The coating longevity is important to consideration for use due to the inaccessibility of infrastructure.

The US Army Corps of Engineers (USACE) and Ontario Hydro were pioneers in researching materials and coatings to prevent zebra mussel attachment in North America. Their research showed that some metal alloys, antifouling paints, zinc metallic coatings, and silicone foul release coatings (FRCs) were effective in preventing freshwater fouling (Puzzuoll and Leitch 1991; Race and Kelly 1997; Spencer 1998). The long-term studies created a foundation for identifying systems that prevent zebra and quagga mussel attachment (Race and Kelly 1997; Spencer 1998).

Another long-term study, by the Long Island Lighting Company, showed that silicone FRCs exhibited excellent performance in a marine environment for powerplant cooling water lines. The coatings did not allow *Mytilus edulis* (blue mussels) attachment, but a few barnacles attached to the surface (Gross 1997). This study and the Ontario Hydro study focused on silicone coatings, showing excellent foul release performance for mussels (Gross 1997; Spencer 1998). These products were new to the marketplace at the time. Delamination and blistering occurred due to the challenge of adhering low surface energy surface coats to corrosion-resistant primers (Gross 1997; Spencer 1998). Formulation advancements have corrected many of these delamination issues.

Materials Approach to Deterring Mussel Attachment

Copper and Copper Alloys

Wooden ships from as early as 200 BC through the 1900s have used copper cladding to prevent fouling (Hellio and Yebra 2009). The USACE study tested copper metal and copper alloys for freshwater fouling, finding the copper leach rates to be in the range of 0.2–3.8 µg/cm²/day (Race and Kelly 1997).

These results were within the range of most antifouling paints. Copper cations are toxic to small organisms, including invasive mussels, and therefore prevent settlement. Another construction material used in marine applications that prevents corrosion and biofouling is 90/10 copper/nickel (Powell 1976). The efficiency of this metal alloy is dependent on the corrosion rate of the metal in water.

Antifouling Paints

The shipping industry primarily uses antifouling paints to prevent biofouling since the 1860s (Hellio and Yebra 2009). Antifouling paints work by using active ingredients that are toxic to aquatic organisms, normally referred to as a *biocide*. Today, many different types of effective biocides are used; copper is the most widely used. Antifouling paints are formulated using one of three binders: nonablative, ablative, and self-polishing (Bressy et al. 2009; Kiil and Yebra 2009). Nonablative binders do not degrade, but instead, the biocide leaches into the water, leaving the coating with a honeycomb appearance. Once the surface biocides are consumed, effectiveness diminishes. Typical service life is 1–2 years. Ablative binders prevent fouling by controlled depletion. These coatings have two different erosion fronts: pigment liberation followed by binder dissolution. Typically, the ablative binders provide 2–3 years of service life. Self-polishing binders are designed to release biocide at a constant rate. The polymer and biocide leach rates are the same, allowing for a consistent antifouling performance. The self-polishing coatings have the longest service life of the antifouling paints for ships, typically 5–6 years.

The effectiveness of an antifouling paint is dependent on water chemistry, velocity, temperature, salinity, and pH (Yebra et al. 2004). A specific antifouling paint may prevent fouling in one water body and exhibit reduced performance in another. Antifouling paints must be evaluated for site-specific conditions in freshwater.

The U.S. Environmental Protection Agency (EPA) requires the registration of antifouling paints for each designated use. Most products are specifically approved for ship hulls. This may be an obstacle for many manufacturers because it requires them to either reclassify their products for freshwater infrastructure use or obtain special permits each time the coating is applied. Biocide-containing materials must be compliant with the Federal Insecticide, Fungicide and Rodent Act (FIFRA) for freshwater mussel control use. Furthermore, permitting and compliance with the National Environmental Policy Act (NEPA) is required for the installation of these coating systems to any water infrastructure.

Zinc Metals and Zinc-Rich Primers

Galvanized steel, zinc metallizing, and zinc-rich primers have been shown to reduce the attachment of mussels (Race and Kelly 1997). Zinc cations are also toxic to freshwater invasive mussels. The efficiency is dependent upon the corrosion rate (leach rate) of the zinc in water. Zinc has an advantage, because currently, there are no regulations for its use as an antifouling material.

Nontoxic Foul Release Coatings

Most FRCs rely on the material properties of the coating and do not contain leaching toxins or biocides. Therefore, these coating systems provide consistent performance regardless of water chemistry. Most FRCs are silicone-based materials that have low surface energy and low elastic modulus. The low surface energy prevents fouling species from wetting out the coating surface, thus preventing an adhesive bond. The low elastic modulus causes the attached species to be released by a peeling fracture mechanism, which requires less force than shearing (Brady 1999, 2001; Brady and Singer 2000; Berglin et al. 2003; Kohl and Singer 1999).

Reclamation's Coatings and Materials Research Program

The program was conducted in two phases; Phase I consisted of evaluating coatings and materials shown to work in the previous studies. Phase II focused on examining these and other technologies with the potential to provide the longest fouling-prevention service life.

Steel substrates were solvent cleaned according to SSPC SP1 and abrasive blasted to SSPC SP10/ NACE 2 near-white metal blast with a 3.0 mil surface profile. The coating application proceeded in accordance with manufacturer recommendations. For several products, steel substrates were shipped to the coating manufacturer for application.

Test Site, Conditions, and Sample Preparation

Reclamation's MERL identified Parker Dam on the lower Colorado River to be the ideal field test site. This facility provided quasistatic water (very low flow rate—0.03–0.15 ft/s) test locations, which is similar to the service conditions of previous research by Ontario Hydro and the USACE. Parker Dam also allowed for dynamic (0.30–3.00 ft/s) conditions, which better simulates the typical conditions for Reclamation water infrastructure. Most importantly, the facility already sustained heavily populations of invasive mussels. In addition, the water temperature remained above 52°F year-round, facilitating continuous mussel reproduction and fast test results.

For each coating system tested in quasistatic exposure, three 12-in. by 12-in. by 3/16-in. thick steel plates were tied off by a nylon rope and lowered to approximately 50 ft below the water surface, near the face of the dam. For the dynamic conditions, one 18-in. by 24-in. coated floor grate with 1-in. spacing was tied off with two nylon ropes (to prevent twisting) and lowered to a depth of 40 ft below the water surface. These samples were secured downstream from the forebay trash rack structure to receive continuous water flow. The flow rate varied depending upon the number of turbine units operating.

Phase I: Evaluation of Coatings and Materials from Previous Studies

Many of the coatings and materials evaluated in the first year were selected based on the previous research by the USACE, Ontario Hydro, and the Long Island Lighting Company. Since many different types of products were reported to prevent mussel fouling, a few of each type of technology were tested. Three zinc-rich primers, three zinc metallic coatings, two copper-based antifouling coatings, three chemical-biocide-based antifouling coatings, three FRCs, and four copper metal alloys were placed in the water at Parker Dam in May 2008.

From early in the study, it was evident that the water flow rate greatly influenced the mussel population and fouling behavior. Flowing water provided mussel attachment and colonization faster than the quasistatic environment. Figure 29.1 shows the contrast between quasistatic and dynamic exposure conditions on steel substrates. Some products performed better in quasistatic environment, but not well in the dynamic condition.

Reclamation developed a testing protocol and criteria to determine the products' efficacy in preventing mussel attachment. If a product accumulated 25% or greater reduction in flow (measured by the area reduction of grate spacings) for dynamic conditions, it is unacceptable. For quasistatic exposure, if the mussels attached to the surface and were not easily cleaned, that is, with the gentle brush of the hand, it was unacceptable. If any coating is blistered or delaminated, it was also unacceptable. The first-year results are shown in Table 29.1.

Zinc-Rich Primers

The USACE study reported that mussels attached to the inorganic and organic zinc-rich primers but had a low mussel density compared to epoxy-based systems (Race and Kelly 1997). Reclamation observations support this for the quasistatic environment. However, all (Devoe Cathacoat 304, 304L, and 313) had significant mussel colonization on the grates in flowing water with 75%–100% blockage after 7 months of exposure (Skaja 2010). The flowing water conditions create a more favorable environment for attached mussels to thrive. Another possibility is that the Colorado River water chemistry encourages zinc carbonate formation, which scavenges the free zinc cations key to deterring mussel attachment. The water hardness at Parker Dam is 260 mg/L compared to 92 mg/L at the USACE testing site (Race and Kelly 1997). These concepts were not examined further. In either case, the extent of fouling in flowing water was more aggressive than the quasistatic environment. All samples failed to meet the criteria and were removed after 7 months.

FIGURE 29.1 Uncoated ASTM A788 steel in dynamic conditions and quasistatic conditions after 7 months of exposure.

TABLE 29.1

First-Year Test Results

Manufacturer	Product Name	Material Type	Environmental Conditions Quasistatic 0.03–0.15 ft/s	Dynamic 0.3–3 ft/s	Date Tested
N/A	Steel	Metal	30% coverage	100% blocked	5-08 to 12-08
N/A	Copper	Metal	0% coverage	N/A	5-08 to 12-08
N/A	Brass	Metal	0% coverage	N/A	5-08 to 12-12
N/A	Bronze	Metal	0% coverage	N/A	5-08 to 12-12
N/A	90-10 Copper nickel	Metal	80% coverage	N/A	5-08 to 12-08
N/A	Galvanized steel	Metallic coating	10% coverage	25% blocked	5-08 to 12-08
N/A	100% zinc Metallizing	Metallic coating	10% coverage	50% blocked	5-08 to 12-08
N/A	85-15 zinc-A; metalizing	Metallic coating	30% coverage	75% blocked	5-08 to 12-08
Devoe Coatings	Cathoacoat 304	Inorganic zinc-rich primer	10% coverage	100% blocked	5-08 to 12-08
Devoe Coatings	Cathoacoat 304L	Inorganic zinc-rich primer	10% coverage	75% blocked	5-08 to 12-08
Devoe Coatings	Cathoacoat 313	Organic zinc-rich primer	30% coverage	100% blocked	5-08 to 12-08
Sealife	Sealife ZMP	Cuprous antifouling paint	1% coverage	25% blocked	5-08 to 12-08
LuminOre	LuminOre	Copper antifouling paint	0% coverage	5% blocked	5-08 to 12-12
E-Paint	Sunwave plus	Peroxide antifouling paint	1% coverage	25% blocked	5-08 to 5-09
E-Paint	SN-1	Chemical antifouling paint	1% coverage	25% blocked	5-08 to 12-08
E-Paint	ZO-HP	Chemical antifouling paint	1% coverage	25% blocked	5-08 to 12-08
Sherwin Williams	Sher-Release	Silicone foul release	0% coverage	0% blocked	5-08 to current
International Paint	Intersleek 970	Fluorinated silicone FRC	0% coverage	0% blocked	5-08 to current
Ecological Coatings	Wearlon	Silicone epoxy FRC	50% coverage	100% blocked	5-08 to 12-08

Zinc Metallic Coatings

The USACE found that the mussels attached to the zinc metallic coatings, but did not colonize in high densities (Race and Kelly 1997). Reclamation's study showed the same result for thermal spray 100% zinc, 85–15 zinc aluminum, and galvanized steel in the quasistatic environment (Skaja 2010). However, under dynamic conditions, high densities of quagga mussels attached to the metallic coated grates, causing 50%–100% blockage after 7 months (Skaja 2010). All samples were removed after 7 months of exposure.

Copper and Copper Alloys

Reclamation evaluated four different copper alloys in quasistatic condition only: copper, brass, bronze, and 90–10 copper-nickel. Evaluation in flowing water did not occur because floor grates were not available for these materials.

Copper performed very well with no mussels attached in 4 years of testing. However, unlike the Army Corps of Engineers test results, the corrosion rate, and thus the leach rate, of copper was significantly higher. The copper plate had lost over half of its thickness within the 4 years of testing. The original thickness was 0.125 in. and after 4 years of testing the thickness was 0.052 in., a 60% metal loss. The calculated leach rate was 473 µg/cm²/day. This is 2–3 orders of magnitude higher than the USACE study. Due to the high copper leach rates observed, the use of copper on the Colorado River should be approached with caution.

Brass and bronze also prevented mussel attachment. However, the brass and bronze leach rates were also very high, and they occasionally had an attached mussel. The 90–10 copper-nickel allowed mussel attachment, undoubtedly because the corrosion rate is very low. There is insufficient copper cation released to deter attachment. This material was removed after 4 months of testing.

Copper-Based Antifouling Paints

Reclamation tested two copper-based antifouling paints. One contained an ablative binder with cuprous oxide pigments (Sealife ZMP) and the other was a nonablative binder that contained metallic copper pigments (LuminOre) (Skaja 2010).

The ablative binder effectively deterred most mussels in quasistatic exposure. However, this product did not prevent mussel attachment in flowing water. The floor grate was 25% blocked with mussels after 7 months and was withdrawn from the test.

The nonablative system prevented mussel attachment in both quasistatic and dynamic environments. However, the coated floor grate in dynamic conditions had a service life of approximately 18 months before the mussels blocked the grate at 30%. In contrast, the quasistatic samples surpassed 4 years of service without mussel attachment. At 4 years, blistering was observed, which negatively impacts the coating's remaining service life.

Antifouling paints are designed for specific environmental conditions. Environmental conditions that affect the service life and performance are flow velocities, temperature, pH, salinity, and water chemistry (Yebra et al. 2004). Coating manufacturers recommend different products for end users frequently on the move, such as a cruise liner or cargo tanker, than pleasure craft or military ships that sit in harbor for longer periods.

Reclamation's test results indicate that zinc metallic coating systems are not suitable for the specific dynamic service environment tested. The coatings performed well in the quasistatic environment with lower flow rates; therefore, they may not have been designed for continuous water flow.

In marine applications, antifouling paints typically provide 60 months of service before ships are drydocked to receive a fresh coat of paint. This service life is not economical for Reclamation due to the inaccessibility of its infrastructure. Another problem for the ablative binder was that the paint became very soft in the 115°F temperatures in the Mohave Desert. The coating surfaces began to physically adhere to one another prior to installing them for testing. These temperatures are significantly higher than any shipyard.

Reclamation has decided to not use antifouling paints due to permitting and regulation compliance that would be required for use on infrastructure.

Chemical-Based Antifouling Paints

Reclamation tested three chemical-based antifouling paints. Two contained an ablative binder (E-Paint SN-1 and ZO-HP) and the other system was a cross-linked coating (E-Paint Sunwave Plus). The active biocide in SN-1 was Seanine 211®, which controls algae and soft fouling. Seanine 211 has a short half-life and biodegrades quickly. The SN-1 performed well in quasistatic conditions but allowed a few mussels to attach. The grate was 30% fouled in flowing water and was withdrawn after 7 months (Skaja 2010). These coatings also softened at the test site's ambient temperatures.

The ZO-HP active biocide is Zinc Omadine, the same active ingredient in antidandruff shampoos. The ZO-HP performed well in quasistatic conditions, allowing only a few mussels to attach. However, in flowing conditions, the grate was 25% blocked and was withdrawn after 7 months (Skaja 2010). The coatings also experienced melting.

Sunwave Plus is a unique coating that uses peroxide as the active biocide. It performed very well for the first 7 months with only 10% blockage but then allowed significant mussel attachment and was withdrawn at 1 year with 30% blockage (Skaja 2010). Long Island Lighting Company (Gross 1997) reported similar results.

Foul Release Coatings

Reclamation tested three FRCs, two silicone elastomers and one silicone epoxy hybrid. The first silicone coating (Fujifilm, DUPLEX, currently sold by Sherwin Williams as Sher-Release) is a 100% silicone that contained silicone oil as a release agent. This product prevented mussel attachment in both dynamic and quasistatic water conditions (Skaja 2010). Occasionally algae and bryozoans fouling attached and accumulated over time but were easily released by flowing water, that is, self-cleaning. This coating system is still performing well after 5 years of testing, shown in Figure 29.2.

The second silicone elastomer (International Paint Intersleek 970) is a fluorinated silicone that does not contain silicone oil and is inert. This system is amphiphilic: it contains both hydrophobic and hydrophilic domains. This product performed well in dynamic conditions but allowed some mussels to attach to the surface (Skaja 2010). The mussels were easily released with 0.05–0.20 lb of force, and they did not accumulate greater than 10% grate blockage before self-cleaning. In quasistatic conditions, the mussels did not attach.

FIGURE 29.2 Silicone FRC in dynamic conditions after 5 years of exposure.

The silicone epoxy hybrid (Ecological Coatings Wearlon) did not prevent mussel attachment and blocked the grate 100%. This product was removed from testing at 7 months (Skaja 2010). The Long Island Lighting Company reported similar results (Gross 1997).

Phase I Discussion

Reclamation test results are, in some instances, quite different than the previous results for antifouling paints, zinc-rich primers, and zinc metallic coatings in flowing water. Water flow velocity is one key difference between Reclamation's study and previous research sites. The USACE and Ontario Hydro studies were conducted in the facilities' forebay against a structure wall. Reclamation's results suggest that these products prevent mussel attachment more successfully in quasistatic water (face of dam) than flowing water conditions (anchored to a trash rack structure). Reclamation's test site provided the advantage to test in these higher velocity water environments. This better correlates to its existing infrastructure, such as trash racks, intakes, penstocks, and outlet works.

It is not surprising that flowing water provides a more severe fouling environment than quasistatic test sites. Mussels are filter feeders and prefer that food be brought to them. Products should be evaluated in the exact service conditions in order to accurately assess the anticipated performance. Reclamation has shown here that the quasistatic results would have been misleading for higher water velocity applications.

Copper and copper alloys performed well in Reclamation's and previous research studies. However, the leach rate of copper is significantly greater. The calculated rate was 473 $\mu g/cm^2/day$ at Parker Dam compared to the 0.2–3.8 $\mu g/cm^2/day$ reported by the USACE (Race and Kelly 1997). This is in part because copper corrodes at different rates in different water chemistries. Furthermore, the corrosion-resistant copper alloy, 90–10 copper-nickel, allowed mussels to attach to the surface. From an environmental perspective, caution should be used in applying copper or copper-based materials for mussel control.

The copper-based and chemical-based antifouling paints provided acceptable performance in quasistatic water for 1 year, but performance was unacceptable in flowing water. The leach rate is expected to be higher in the flowing water because the fresh ions are washed away from the surface. It is also possible that the binders erode at a different rate than the biocides and thus have insufficient biocide on the surface.

The silicone FRCs performed very well in both quasistatic and dynamic conditions. The 100% silicones prevented mussel attachment altogether. The fluorinated silicone allowed a few mussels to attach, but significant grate blockage did not occur. This was surprising because the manufacturer indicated that the mussels would attach in quasistatic conditions. This will be a point of discussion later in the chapter. It should be noted that these coatings are not very durable and lack abrasion or gouge resistance. However, laboratory testing by Reclamation found the silicone elastomers to be more erosion resistant than epoxy, probably due to its elastic nature (Tordonato and Skaja 2012).

The silicone epoxy hybrid (durable foul release) allowed mussels to attach and blocked the floor grate 100%. Previous research reported the same result (Gross 1997).

Phase II—Identifying Technologies for the Longest Service Life

The initial Phase I test results were quite alarming because many of the coating systems shown to prevent attachment in previous studies only prevented mussel attachment in Reclamation's quasistatic conditions. The only exception was the silicone elastomer FRCs, which performed well in both quasistatic and dynamic conditions. The results observed were attributed to the environmental conditions, but this is very important because Reclamation's infrastructure experiences both service conditions.

MERL decided to focus on silicone FRCs because they prevented mussel attachment in both water flow velocities and do not release biocides into the water. Antifouling product testing ceased unless the manufacturer could provide ecological impacts of their system in freshwater. This is important because

Reclamation water is used for municipality drinking water and agricultural irrigation. Silicone FRCs indicate the potential to provide the longest service life, absence of coating damage by debris or other mechanical impacts. MERL also began evaluating fluorinated powder coatings and durable FRCs, such as silicone epoxy and fluorinated polyurethane, to identify coating systems with increased durability while maintaining foul release performance.

Once the focus of Reclamations research shifted toward FRCs, specifically, slight modifications were made to the testing procedure. Stainless steel control substrates were attached to each sample to verify and quantify the extent of fouling during any test period. In addition, the force required to remove mussels from the surface was measured using a handheld force gauge (Shimpo Model FGV-5XY, maximum capacity of 5 lb). The procedure was modeled after ASTM D 5618–94, which is used to determine the attachment strength of barnacles. The major deviation from the test method is that no attempt was made to measure the attachment area due to the difficulty in performing this in the field. Image analysis was also used to quantify the percent coverage or blockage of the substrates. Finally, MERL measured flow velocities at dynamic testing locations as well as along the face of the dam (quasistatic), which varied depending on the number of turbine units in operation.

Reclamation has evaluated over 60 products using the aforementioned methods. Of these, only 15 technologies are marketed as FRCs. Reclamation evaluated an additional 7 products that were *"nonstick"* coatings that provided insight for the experiment. Table 29.2 provides a summary of the results for FRCs and other *nonstick* coatings.

Further discussion of these results appears in the following sections and separates these coatings by type.

TABLE 29.2
Results of FRCs and *Nonstick* Coatings

Manufacturer	Product Name	Product Description	Force to Remove Muscle	Percent Blockage of Grate	Date Tested
Sherwin Williams	Sher-Release	Silicone FRC	No mussels	Self-cleaning	5-08 to current
PPG	Sigmaglide	Silicone FRC	No mussels	Self-cleaning	10-09 to current
Chuguko Marine Paint	Bioclean	Silicone FRC	No mussels	Self-cleaning	10-09 to current
Nusil	9706	Silicone FRC	No mussels	Self-cleaning	5-12 to 12-12
Nusil	9707	Silicone FRC	Delaminated	Self-cleaning	5-12 to current
International Paint	Intersleek 970	Fluorinated silicone FRC	0.05–0.2 lb	Self-cleaning	5-08 to current
Ecological Coatings	Wearlon	Silicone epoxy FRC	Unknown	100%	5-08 to 12-08
Seacoat	Seaspeed V%	Silicone epoxy FRC	0.56 lb	72%	10-09 to 11-11
Duromar	HPL 2510FR	Silicone epoxy FRC	0.76 lb	41%	11-10 to 11-11
Greensfield Manufacturing	Hullspeed	Silicone epoxy FRC	0.69 lb	54%	11-11 to 12-12
SEI Chemical	SHC-500	Fluorinated polyurethane FRC	0.73 lb	97%	10-09 to 11-10
Telluride East Inc.	Aquafast	Silicate FRC	0.95 lb	50%	08-11 to 5-12
Telluride East Inc.	Aquafast V2	Silicate FRC	0.83 lb	50%	08-11 to 5-12
Durachemie	DuSlip	Silicone polyurea FRC	1.3 lb	71%	11-11 to 12-12
Rylar	Rylar	1K Polyurethane FRC	1.3 lb	37%	11-11 to 12-12
Arkema	PVDF Kynar	Fluorinated powder coating	0.41 lb	25%	05-09 to 11-09
Solvay Plastics	ECTFE Halar	Fluorinated powder coating	0.41 lb	44%	05-09 to 11-09
Daikin	ETFE (Neoflon)	Fluorinated powder coating	0.43 lb	51%	05-09 to 11-09
Daikin	FEP (Neoflon)	Fluorinated powder coating	0.47 lb	45%	05-09 to 11-09
Daikin	PFA (Neoflon)	Fluorinated powder coating	0.2 lb	48%	05-09 to 11-11
Sherwin Williams	Sher-Release Tiecoat	Tiecoat	0.02–0.2 lb	53%	06-10 to 11-11
Fujifilm Hunt Smart Surfaces	Fujifilm oil-free	Silicone FRC oil-free	No mussels	Self-cleaning	07-12 to current

Silicone Foul Release Coatings

Reclamation has evaluated five silicone FRCs that perform very well in quasistatic and dynamic conditions. Silicone FRCs contain silicone oil that slowly exudes to act as a release agent. There are additional silicone FRCs available in the market, and they are assumed to have similar performance.

    With all observations to date, no mussel has directly attached to the silicone FRC surface. However, over time slime, sponges, algae, and bryozoans weakly attach to the coating surface, and the mussels may attach to these organisms. The slow buildup of fouling eventually creates enough drag force that it releases from or shears off the coating surface, that is, self-cleaning. Figure 29.3 illustrates the degree to which the percent of grate blockage changed with each inspection. It was observed that darker colored coatings have more algae and bryozoans attached than white- or light-colored coatings (Skaja and Tordonato 2011). This phenomenon was also observed in the ocean (Finlay et al. 2008; Swain et al. 2006). The silicone FRC results for several products can be summarized as follows:

- Sherwin Williams Sher-Release has been in the study for 5 years with no mussel attachment.
- PPG Sigmaglide 890 has been in the study for 3.5 years with no mussel attachment.
- Chugoku Bioclean SPG-H has been in the study for 3.5 years with no mussel attachment.
- Nusil 9706 was removed from the study after 7 months because the topcoat delaminated from the epoxy primer—no mussels were observed on the intact topcoat.
- Nusil 9707 has been in the study for 1 year with no mussel attachment.

Previous research seems to suggest that marine fouling conditions are significantly more aggressive than freshwater (Clare and Aldred 2009; Gross 1997; Puzzoull and Leitch 1991; Race and Kelly 1997; Skaja and Tordonato 2011; Spencer 1998). The ocean has more than 6000 different fouling species (Jones 2009). Barnacles, tubeworms, and other marine fouling species attach to the silicone FRCs (Clare and Aldred 2009). For freshwater, the primary goal is to prevent two species of mussels from attaching to a surface.

Fluorinated Silicone Foul Release Coatings

Reclamation evaluated a fluorinated silicone FRC product named International Intersleek 970. This is a unique product because it does not rely on silicone oil to assist in its release mechanism; it is an amphiphilic coating with hydrophilic and hydrophobic domains. In the marine environment, this product has

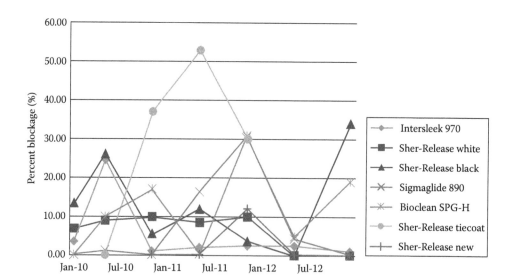

FIGURE 29.3   Percent blockage vs. time of silicone FRCs, showing self-cleaning.

shown better release characteristics for all fouling species. Reclamation has observed that a few mussels attach to the coating surface but are released with very little force—between 0.05 and 0.20 lb. This coating is self-cleaning and mussel colonies do not achieve large colonization. This coating also prevents slime and algae fouling more effectively than the silicone FRCs tested.

Silicone Epoxy Foul Release Coatings

Reclamation evaluated four silicone epoxy FRCs: Ecological Coatings Wearlon, Seacoat Seaspeed V5, Duromar HPL 2510FR, and Greenfield Manufacturing Hullspeed. These products are all very different but classified similarly here for ease of discussion. The products are hard like an epoxy but have a very slick surface. Unfortunately, mussels attach to the coating surface and continue to build up until the floor grate is 100% blocked. These coatings are not self-cleaning under Reclamations testing conditions. It required varying degrees of force to remove mussels between 0.50 and 0.76 lb of force. This is less force than a traditional epoxy or coal tar epoxy, which is 1.2–1.3 lb. Of these types of coatings, Seacoat Seaspeed V5 required the least amount of force to remove the mussels. However, Reclamation conducted a water jet cleanability study and the mussels adhered strong enough that the byssal threads severed leaving the mussel plaque attached to the coating surface (Mortensen 2013). There was no appreciable difference between the traditional epoxy and the Seacoat product following water jet cleaning.

Fluorinated Polyurethane Foul Release Coatings

Reclamation evaluated a fluorinated polyurethane FRC, SEI Chemical SHC-500. The product is durable and hard like traditional polyurethanes. However, the product allowed mussel attachment and required moderately high force to remove the mussels: 0.73 lb. The grate in flowing water was 100% blocked within 1 year.

Silicate Foul Release Coatings

Reclamation evaluated two silicate FRCs, Telluride East Inc. Aquafast and Aquafast V2. Both products are hard coatings that are abrasion resistant. However, both products allowed mussel attachment and required significant force to remove mussels, 0.83–0.95 lb of force. Within 9 months, the grates were 50% blocked.

Silicone Polyurea Foul Release Coatings

Reclamation evaluated a silicone polyurea FRC, Durachemie DuSlip. This product is hard and abrasion resistant. However, the product allowed mussel attachment and required 1.30 lb of force to remove a mussel, similar to a traditional epoxy coating. The floor grate was 71% blocked after 1 year.

Polyurethane Foul Release Coating

Reclamation evaluated a polyurethane FRC, Rylar. This product was hard and durable. However, the product allowed mussel attachment and required 1.3 lb to remove a mussel. The floor grate was 37% blocked after 1 year of service.

Fluorinated Powder Coatings

Reclamation evaluated five fluorinated powder coatings. These coatings are not designed as FRCs but rather as *nonstick* coatings. The coatings evaluated were polyvinylidene fluoride, Kynar (PVDF), polyethylenechlorotrifluoro ethylene, Halar (ECTFE), polyethylenetetrafluoroethylene, Neoflon (ETFE), polyfluoroethylene propylene, Dyneon (FEP), and polyperfluoroalkoxy, Neoflon (PFA). Mussels attached to all of the coating systems. PVDF, ECTFE, ETFE, and FEP all required about 0.40 lb of force to remove a mussel. PFA was the best performing of the fluorinated powder coatings and only required 0.2 lb to remove a mussel. Reclamation terminated the study prior to seeing if the coating was self-cleaning. After 1 year, the grate was 48% blocked and the test was terminated.

Sher-Release Tiecoat

Reclamation evaluated the tiecoat from Sherwin Williams as a stand-alone product. The tiecoat is slightly more durable than its foul release topcoat. Unfortunately, the mussels attach to the tiecoat but require minimal force to remove a mussel, 0.02–0.20 lb. Even with the low force, the mussels seemed to be able to slowly build up to create 50% blockage through the grate. At the subsequent inspection, the percent blockage was reduced to 30%. This tiecoat has some self-cleaning characteristics. Mussels released from the surface sometime during the summer months. The dam typically operates at full capacity during this season to deliver water for irrigation and meet increased power generation demands. As a result, the water velocities are higher; the increased hydrodynamic drag force most likely resulted in the release of mussels.

The Sher-Release tiecoat and Intersleek 970 were the only coatings evaluated that allow mussel attachment but released the mussels once a critical drag force was meet. Intersleek 970 appears to have a lower critical drag force than the Sher-Release tiecoat. The entire byssal thread and plaque released from the coating surfaces in both cases.

Fujifilm Silicone Oil-Free Foul Release Coating

Reclamation desired to better understand the silicone FRC's method for preventing mussel attachment. Another goal was to determine the meaning or usefulness of the Sher-Release tiecoat and Intersleek 970 self-cleaning properties. One of the components in the silicone FRCs is silicone oil, used as a release agent (Anderson et al. 2003; Milne 1977; Stein et al. 2003; Truby et al. 2000). Reclamation proceeded with an investigation of the necessity of this ingredient for preventing mussel attachment. The coating surface is free from mussels following 16 months in flowing water, as shown in Figure 29.4. However, the stainless steel controls and zip ties were significantly fouled with mussels. This observation contradicts previously accepted marine fouling and FRCs' theory, suggesting that freshwater mussel fouling prevention may be achieved without the use of oils.

FIGURE 29.4   Fujifilm silicone oil-free version after 16 months of exposure.

Discussion on Foul Release Coatings

Effects of Hydrogen Bonding on Freshwater Mussel Attachment

Mussel adhesives contain numerous polyphenolic amino acid proteins. These proteins have high concentrations of 3,4-dihydroxyphenylalanine (DOPA), which is considered to be a critical component in wet adhesion (Farsad and Sone 2012; Nicklisch and Waite 2012). The adhesion occurs by DOPA-mediated bidentate hydrogen bonding, metal/metal oxide coordination, or oxidative cross-linking (Nicklisch and Waite 2012), allowing the mussel to attach to almost any surface. Chapter 28 is dedicated to further explaining the mechanisms for zebra and quagga mussel attachment.

Reclamation has shown that silicone FRCs prevent quagga mussel attachment (Skaja 2010). These coatings are based on polydimethylsiloxane (PDMS); the chemical structure is shown in Figure 29.5. This structure prevents mussel attachment because the methyl groups shield the mussel adhesive intermolecular attraction, that is, hydrogen bonding with oxygen's unpaired electrons. In addition, PDMS has a flexible backbone containing high energy and partially ionic siloxane bonds (Owen 1990). PDMS also has low intermolecular forces occurring between these methyl groups and its surface is very smooth (Brady 2000), minimizing the penetration or mechanical interlocking of the fouling specie's adhesive. The foundation of FRC theory requires low surface energy and low elastic modulus. PDMS satisfies these requirements and, in addition, does not contain hydrogen bond–accepting sites.

Other silicon-oxygen-based materials (silicates) experience strong mussel attachment because these oxygen atoms have no shielding methyl groups or otherwise. Silicates can be fabricated materials such as glass, inorganic coatings (silicate FRC and inorganic zinc), etc., or can be minerals such as sand, rocks, clays, talc, mica, etc. The silicate chemical structure is shown in Figure 29.6. The high elastic modulus and microporosity of these materials also contributes to the adhesion of these materials.

Coatings containing hydrogen bond–accepting sites, such as epoxy, urethane, or urea linkages, allow mussel adhesives to form a bidentate hydrogen bond, which is twice as strong as traditional hydrogen bonds (Hwang et al. 2013).

The durable FRCs evaluated thus far have a high elastic modulus and may contain hydrogen bond–accepting sites on the coating surface due to epoxy, urethane, or urea linkages. The PDMS concentration within the formulation is not sufficient to prevent the mussel adhesive from forming a bidentate hydrogen bond at these linkages. Another option is that the high elastic modulus does not allow easy release. However, it is likely that some percentage of silicone is on the coating surface because the mussel removal force is lower than a traditional epoxy. Therefore, at this time, it is unknown if a

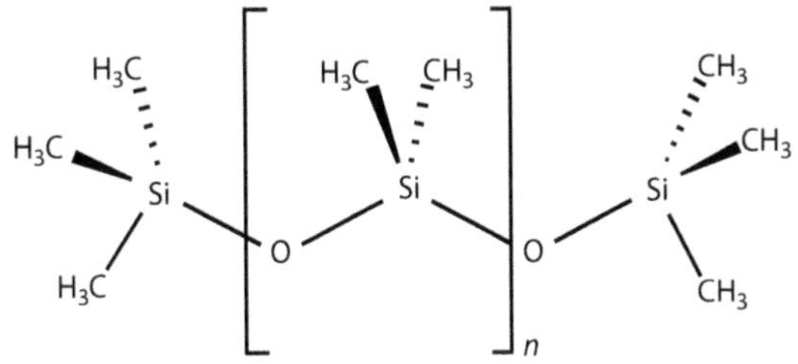

FIGURE 29.5   Chemical structure of PDMS.

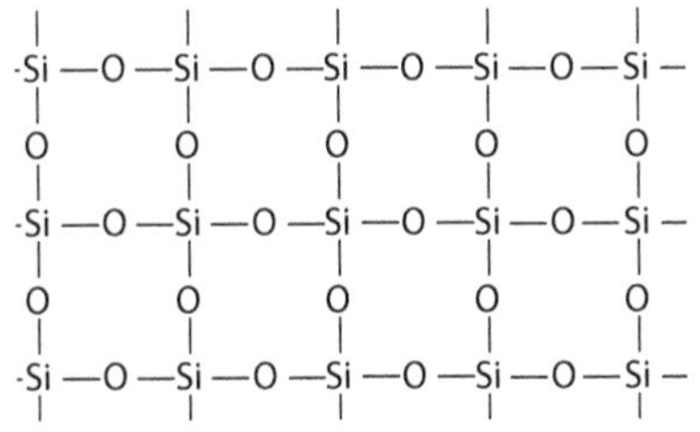

FIGURE 29.6   Chemical structures of silicates.

durable (high elastic modulus) coating could be formulated that would prevent mussel attachment or if hydrogen bond–accepting sites are the limiting stipulation.

Mussels attach to and release from the Sher-Release tiecoat and Intersleek 970. Both coatings are PDMS silicone-based materials with some organic functionality in the coating formulation. The Sher-Release tiecoat contains small concentrations of nitrogen and oxygen in order to bond to the epoxy primers, both of which are hydrogen bond–accepting sites. Therefore, mussels attach weakly to the surface. Intersleek 970 is an amphiphilic coating, and it is believed that the hydrophilic domains contain hydrogen bond–accepting sites. The mussels attach to these hydrogen bond–accepting sites, but the adjacent PDMS segments ensure that the attachment surface area, and therefore force, is minimal. The mussels are released from the surface once the critical drag force is met. The main difference between Sher-Release tiecoat and Intersleek 970 is likely the concentration of hydrogen bond–accepting sites or the strength of the bidentate hydrogen bonds, and therefore, Sher-Release tiecoat fouling was greater than Intersleek 970 as shown in Figure 29.3.

Discussion of Fluorinated Powder Coatings

Reclamation evaluated five fluorinated powder coatings with the intent of finding a durable coating that prevents mussels from adhering to the surface. Fluorine, when bonded with carbon, is electron withdrawing, which is the reason for the low surface energy. Most fluoropolymers have an elastic modulus around 500 MPa, compared to the elastic modulus of silicone FRCs 0.5–5 MPa. The fluoropolymers PVDF, ECTFE, ETFE, and FEP all required approximately the same force to remove a mussel, 0.40 lb. It appeared that the slight changes in chemistry did not affect the mussel attachment mechanism. However, the mussels attached with a lower force, 0.20 lb, to PFA. The difference in chemistry is a pendant perfluoro alkoxy chain. The ether group provides some flexibility, which may assist in the release of the mussel. Again, the oxygen in the ether group is shielded by the trifluoro methyl group, hindering hydrogen bonding for the mussel. Reclamation tested this product for 1 year, and it was discarded. This was done prior to understanding the mussel attachment mechanism and prior to collecting force measurements on the durable FRCs. PFA fluoropolymers could be examined further as fouling release coatings.

Conclusion

Previous research has shown that metallic zinc coatings, copper alloys, antifouling paints, and FRCs prevent mussel attachment. Reclamation has found that the service environment greatly influences the coating's performance properties. Reclamation has extensive data on various coating systems for preventing invasive mussel attachment in both dynamic and quasistatic conditions. FRCs have been shown to prevent mussel attachment in both dynamic and static water conditions. Examinations of freshwater mussel attachment as well as the chemical and physical properties of coating surfaces have been revisited to build a foundation for freshwater fouling-prevention technologies. Reclamation continues to evaluate new products for mussel control with the hopes of identifying a durable FRC that prevents mussel attachment similarly to the silicone FRCs.

Disclaimer

Information in this chapter may not be used for advertising or promotional purposes. The enclosed data and findings should not be construed as an endorsement of any product or firm by the Bureau of Reclamation (Reclamation), US Department of the Interior, or the Federal Government. The products in this report were evaluated in environmental conditions and for purposes specific to Reclamation's mission. Most of these products were originally developed for the marine environment and not necessarily for use in freshwater. The data should be viewed as site specific and not necessarily applicable to all freshwater exposure conditions. Reclamation gives no warranties or guarantees, expressed or implied, for the products evaluated in this report, including merchantability or fitness for a particular purpose.

Acknowledgments

We would like to thank the Bureau of Reclamation Research and Development Office—Science and Technology (S&T) Program—for funding this research over the past 5 years. We would also like to thank the Bureau of Reclamation Lower Colorado (LC) Region, the LC Dams Office, and the Parker Dam staff for all of their in-kind contributions, access, and support.

REFERENCES

Anderson C, Atlar M, Callow M, Candries M, Milne A, Townsin RL. 2003. The development of foul-release coatings for seagoing vessels. *Proceedings of the Institute of Marine Engineering, Science, and Technology. Part B, Journal of Marine Design and Operations* B4:11–23.

ASTM D 5618–94. 2011. Standard test method for measurement of barnacle adhesion strength in shear, ASTM International, West Conshohocken, PA, http://www.astm.org.

Berglin M, Lonn N, Gatenholm P. 2003. Coating modulus and barnacle bioadhesion. *Biofouling* 19(S):63–69.

Brady RF Jr. 1999. Properties which influence marine fouling resistance in polymers containing silicon and fluorine. *Progress in Organic Coatings* 35:31–35.

Brady RF Jr. 2000. No more tin. What now for fouling control? *Journal of Protective Coatings and Linings* 17:42–46.

Brady RF Jr. 2001. A fracture mechanical analysis of fouling release from nontoxic antifouling coatings. *Progress in Organic Coatings* 43:188–192.

Brady RF Jr., Singer IL. 2000. Mechanical factors favoring release from fouling coatings. *Biofouling* 15:73–81.

Bressy C, Margaillan A, Fay F, Linossier I, Rehel K. 2009. Chapter 18, Tin-free self-polishing marine antifouling coatings. In Hellio, C. and Yebra, D. (eds.), *Advances in Marine Antifouling Coatings and Technologies*, Oxford, U.K.: Woodhead Publishing Limited, pp. 445–491.

Clare A, Aldred N. 2009. Chapter 3, Surface colonization by marine organisms and its impact on antifouling research. In Hellio, C. and Yebra, D. (Eds.), *Advances in Marine Antifouling Coatings and Technologies*, Oxford, U.K.: Woodhead Publishing Limited, pp. 46–79.

Farsad N, Sone ED. 2012. Zebra mussel adhesion: Structure of the byssal adhesive apparatus in the freshwater mussel, *Dreissena polymorpha. Journal of Structural Biology* 177:613–620.

Finlay J, Fletcher B, Callow M, Callow J. 2008. Effect of background colour on growth and adhesion strength of ulva sporelings. *Biofouling* 22:219–225.

Gross AC. 1997. Chapter 21, Long term experience with non-fouling coatings and other means to control macrofouling. In Frank M. D'Itri (ed.), *Zebra Mussels and Aquatic Nuisance Species*. Boca Raton, FL: Lewis Publishers, pp. 329–342.

Hellio C, Yebra D. 2009. Chapter 1, Introduction. In Hellio, C. and Yebra, D. (Eds.), *Advances in Marine Antifouling Coatings and Technologies*. Oxford, U.K.: Woodhead Publishing Limited, pp. 1–17.

Hwang D, Wei W, Rodriguez-Martinez N, Danner E, Waite J. 2013. Chapter 8, A microcosm of wet adhesion: Dissecting protein interactions in mussel plaques. In Zeng, H. (ed.), *Polymer Adhesion, Friction, and Lubrication*. Hoboken, NJ: John Wiley & Sons Inc., pp. 319–350.

Jones G. 2009. Chapter 2, The battle against marine biofouling: A historical review. In Hellio, C. and Yebra, D. (Eds.), *Advances in Marine Antifouling coatings and Technologies*. Oxford, U.K.: Woodhead Publishing Limited, pp. 19–45.

Kiil S, Yebra DM. 2009. Chapter 14, Modelling the design and optimization of chemically active marine antifouling coatings. In Hellio, C. and Yebra, D. (Eds.), *Advances in Marine Antifouling Coatings and Technologies*. Oxford, U.K.: Woodhead Publishing Limited, pp. 334–364.

Kohl JG, Singer IL. 1999. Pull-off behavior of epoxy bonded to silicone duplex coatings. *Progress in Organic Coatings* 36:15–20.

McKinnon S. 2007 Jan 23. Mussels invading Arizona waterways. *The Arizona Republic.*

Milne A, inventor; the International Paint Co., Ltd., assignee. May 24 1977. Anti-fouling marine compositions. United States patent US 4,025,693.

Mortensen J. 2013. Resistance of protective coatings to high pressure water jets for invasive mussel removal, Reclamation Hydraulic Laboratory Technical Memorandum PAP-1074.

Nicklisch SCT, Waite JH. 2012. Mini-review: The role of redox in dopa-mediated marine adhesion. *Biofouling* 28:865–877.

Owen MJ. 1990. Chapter 40, Siloxane surface activity. In John, M.Z. and Fearon, F.W.G. (Eds.), *Silicon-Based Polymer Science: A Comprehensive Resource*. Washington, DC: American Chemical Society, pp. 705–739.

Powell M. 1976. Resistance of copper-nickel expanded metal to fouling and corrosion in mariculture operations. *The Progressive Fish Culturist* 38(1):58–59.

Puzzuoll FV, Leitch EG. 1991. Evaluation of coatings to control zebra mussel colonization—Year one—Interim report. Ontario Hydro Report 91–56-K, April 25.

Race TD, Kelly MA. 1997. Chapter 23, A summary of a three year evaluation effort of anti-zebra mussel coatings and materials. In Frank M. D'Itri (ed.), *Zebra Mussels and Aquatic Nuisance Species*. Boca Raton, FL: Lewis Publishers, pp. 359–388.

Skaja AD. 2010. Testing coatings for zebra and quagga mussel control. *Journal of Physical Chemistry Letters* 27(7):57–65.

Skaja A, Tordonato D. 2011. Evaluating coatings to control zebra mussel fouling. *Journal of Physical Chemistry Letters* 28(11):47–53.

Spencer FS. 1998. Evaluation of coatings to control zebra mussel colonization: Year 8 update. Ontario Hydro Report 344-000-1998-RA-0001-R00, April 1998.

SSPC SP1. Solvent Cleaning, Society of Protective Coatings, Pittsburgh PA, http://www.sspc.org.

SSPC SP10/NACE 2. Near white blast cleaning, Society of Protective Coatings, Pittsburgh PA, National Association of Corrosion Engineers, Houston TX, http://www.sspc.org, http://www.nacce.org.

Stein J, Truby K, Wood CD, Stein J, Gardner M, Swain G, Kavanagh C et al. 2003. Silicone foul release coatings: Effect of the interaction of oil and coating functionalities on the magnitude of macrofouling attachment strengths. *Biofouling* 19:71–82.

Swain G, Herpe S, Ralston E, Tribou M. 2006. Short term testing of antifouling surfaces: The importance of colour. *Biofouling* 24:425–429.

Tordonato D, Skaja A. 2012. Durability assessment of foul release coatings. *Journal of Protective Coatings and Linings* 29(8):34–41.

Truby K, Wood C, Stein J, Cella J, Carpenter J, Kavanagh C, Swain G et al. 2000. Evaluation of the performance enhancement of silicone biofouling-release coatings by oil incorporation. *Biofouling* 15:141–150.

Yebra D, Kiil S, Dam-Johansen K. 2004. Antifouling technology—Past, present and future steps towards efficient and environmentally friendly antifouling coatings. *Progress in Organic Coatings* 50:75–104.

30

Efficacy of Two Approaches for Disinfecting Surfaces and Water Infested with Quagga Mussel Veligers

Christine M. Moffitt, Amber Barenberg, Kelly A. Stockton, and Barnaby J. Watten

CONTENTS

ABSTRACT   Disinfection tools and protocols are needed to reduce the probability of transferring invasive mollusk species as hitchhikers. Applications are needed to provide rapid disinfection of recreational equipment, boats, and tankers and other materials used in fire suppression. Moreover, tools are needed that are safe for hatchery and aquaculture operations. In two separate laboratory trials, we tested the lethality of elevated pH or a commonly used aquaculture disinfectant on quagga mussel (*Dreissena rostriformis bugensis*) veligers. We found that all reagents were effective in killing veligers. Aqueous solutions of pH 12 were created with NaOH or Ca(OH)2 and tested at 16°C and 20°C, and three aqueous concentrations of Virkon® Aquatic were tested at 20°C. We observed 100% mortality within a 10 min exposure in solutions of pH 12 prepared with Ca(OH)2 and within a 30 min exposure in solutions prepared with NaOH. We found that solutions of 5 g/L of Virkon Aquatic killed all veligers within a 10 min exposure. We concluded that all chemicals show promise as disinfectants, and use of $Ca(OH)_2$ or NaOH to elevate the pH of disinfecting solutions may provide a more economical and environmentally acceptable way to disinfect large surfaces or tanks.

Introduction

Global transportation of commodities and human movements pose challenges to natural resource managers, as the risks of transporting nonnative, invasive organisms are compounded with each vehicle and vector. In addition, many altered ecosystems are more vulnerable to invasion as natural processes can be in flux (Norbury et al. 2013; Thorp 2014). Mollusks are among the most highly successful invasive aquatic organisms, as they are armored with shells that protect them from desiccation, and have a great capacity as engineers of systems in which they establish because of their capacity for filtration (Cuhel and Aguilar 2013). The vectors and vehicles that transport and distribute invasive mollusks to new locations include fish, birds and mammals, aquarium traders, ship ballast water, recreationalists, agency personnel, and other water users (Alanso and Castro-Diez 2008; Bowler 1991; Bruce and Moffitt 2010).

Quagga mussels, native to eastern Europe, were accidentally introduced into the Laurentian Great Lakes in North America in the 1980s in ballast water (Strayer 2009; Vanderploeg et al. 2002). The major pathway of quagga mussel introduction from the Great Lakes to other nonhydrologically connected locations in North America has been assumed to be recreational boaters (Johnson et al. 2001, 2006; Leung et al. 2006; Mueting and Gerstenberger 2011). It is hypothesized that quagga mussels were introduced into Lake Mead sometime before 2007 in bilge water or with bait or live wells in boats from the Great Lakes region (Wong and Gerstenberger 2011).

In addition to transport on recreational vessels, activities such as fire suppression and fish stocking are of concern in western North America as they move large quantities of raw water resulting in high risks of transfer of invasive mollusks (Britton and Dingman 2011; Bruce et al. 2009; Culver et al. 2000; Edwards et al. 2000). Fish hatcheries in Colorado, Arizona, and Nevada are now infested with quagga mussels, and such facilities are constrained to reduce the risk of transporting veligers with movement of fish, or equipment between locations. Western states transfer water from nearby open water sources via helibucket, pumps, and other operations for fire suppression. The U.S. Forest Service, Bureau of Land Management, and interagency fire management teams are concerned that raw water from infested areas could introduce invasive mollusks into new areas (Britton and Dingman 2011).

Once established, quagga mussels can affect ecosystems and damage infrastructure of cooling pipes, water intake pipes, head gates, and hatchery facilities (Peyer et al. 2009; Pimentel et al. 2000, 2005; Wong and Gerstenberger 2011). There is a growing demand for tools that can be used to address both small- and large-scale disinfection needs using methods that are effective and safe to humans and the environment. Despite the need for innovative disinfection tools, economic constraints for purchase of chemicals and infrastructure used in the disinfection process can be limiting factors (Karatayev et al. 2007). Often highly effective disinfectants are expensive or they have substantial environmental consequences if used in large quantities. Chemicals that can be used in large-scale applications must be less expensive and should have attributes that would allow for safe neutralization and disposal.

Developmental stages of quagga mussel veligers have different survival times, but veligers are reported to remain alive for weeks in moist environments depending on temperature (Choi et al. 2013). Britton and Dingman (2011) found that a 10 min exposure to a 3% concentration of Sparquat 256® was sufficient to kill 100% of tested veligers held 60 min post treatment. These and similar quaternary ammonia reagents have been reported effective in killing quagga mussel veligers and other invasive species in and around fish hatchery environments (Mitchell and Cole 2008; Oplinger and Wagner 2009; Waller et al. 1996). However, widespread use of quaternary ammonia compounds has caused environmental concerns as the treatments can affect aquatic and soil systems (Garcia et al. 2001; Li and Brownawell 2010; Sarkara et al. 2010), and their persistence has increased concerns regarding genotoxic effects (Ferk et al. 2007). Other compounds have been considered to reduce the risk of transporting dreissenid mussels and mitigate the effects of their biofouling abilities. Edwards et al. (2000) used KCl and formalin to reduce risks of invasive zebra mussels (*Dreissena polymorpha*). Antifouling compounds are also of great interest to protect infrastructure in aquaculture and industry but environmental consequences must be considered (Guardiola et al. 2012). Chlorine and other oxidizing reagents are effective tools for disinfection (Brady et al. 1996), but the residual compounds have environmental consequences for water supplies and aquatic organisms (Watson et al. 2012). Recent studies reported the efficacy of a proprietary copper chemical EarthTec® to kill and prevent settlement of veligers (Watters et al. 2013), but dosages tested were above those considered safe for most aquatic organisms.

Ship ballast is well recognized as a source of introduction of invasive organisms throughout the globe (Briski et al. 2012). Watten et al. (2007) have explored the use of elevated pH as a means of killing invasive species in ship ballast water and residual ballast solids. Their early studies demonstrated that pH in the range of 11–12.4 was effective in killing zooplankton with relatively short exposure requirements and that pH could be returned readily to neutral levels through the use of dilution or carbonation. The end product of the reaction with CO_2 (bicarbonate alkalinity) can be environmentally benign. The amount of reagent required, either lye (NaOH) or hydrated lime [$Ca(OH)_2$], for specific pH targets increased with increasing ballast salinity. This application has been particularly attractive given that elevated pH avoids ship corrosion concerns associated with certain alternative acid/oxidant ballast treatments. Reagent cost estimates for freshwater applications were attractive, which led to further testing in cooperation with the Great Ships Initiative group, Duluth Minnesota, in a series of bench-scale, pilot-scale, and shipboard-scale studies (Cangelosi 2009, 2011a,b; Cangelosi et al. 2013) conducted with support from the U.S. National Park Service, the U.S. Environmental Protection Agency, the U.S. Department of Transportation, and the American Steamship Company. Other studies support the use of hydroxide alkalinity as a tool to reduce microbial populations including fish pathogens (Starliper and Watten 2013). NaOH has long been successfully used as a dairy disinfection tool for milking equipment as an alternative to chlorine (Gleeson et al. 2013). Hydrated lime is a reagent successfully used in treatment of wastewater (Grabow et al. 1978; Scanar et al. 2001).

Virkon Aquatic (reformulated from Virkon S in 2007) is one of very few U.S. Environmental Protection Agency–registered disinfectants labeled for use in aquaculture facilities for use on bacterial, fungal, and viral pathogens (DuPont 2011; Mainous et al. 2010). Active ingredients of Virkon Aquatic are potassium monopersulfate and sodium chloride, 21.9% (Mitchell and Cole 2008). Its mode of action is oxidizing proteins and other components of cell protoplasm, resulting in the inhibition of enzyme systems and loss of cell-wall integrity (Curry et al. 2005). The recommended concentration for disinfecting most surfaces for a majority of the organisms is a 10 g/L concentration with an exposure time of at least 5 min (Western Chemical 2008).

The objectives of our study were to compare the mortality response of quagga mussel veligers in disinfectant strength solutions of aqueous pH 12 made with $Ca(OH)_2$ or NaOH and in solutions of the approved hatchery disinfectant, Virkon Aquatic. In addition, we explore the relative costs, environmental risks and benefits, and suitability for each disinfectant for use on gear, boats, tanks, and other applications.

Methods

Study Location and Source of Organisms

All tests were conducted at Willow Beach National Fish Hatchery, Willow Beach, Arizona. Tests were conducted from October 13 to 22, 2009 (Virkon Aquatic trials) and from November 17 to 25, 2012 (elevated pH). Veligers for tests were collected with a 30 μm mesh plankton tow net hung from the grating system of a raceway head box for 30–60 min. The contents of the cod end were then filtered through a 500 μm nylon mesh screen into individual 500 mL Nalgene sample bottles. The contents of each sample bottle were allowed to settle at test temperatures for 1–3 h acclimation. Condition of veligers was assessed, and plankton tow samples with greater than 90% living active veligers were retained for testing. The veligers were concentrated by collecting settled filtrate on ~100 μm mesh nylon fabric and gently placing this into a 1.8 L Nalgene container and rinsing with squirt bottles of aerated well water at test temperatures.

Test Chemicals

We weighed ~0.6 g/L NaOH (Fisher Chemical, Pittsburgh, PA, Lot 060432) or $Ca(OH)_2$ (J.T. Baker Chemical, Phillipsburg, NJ, Lot G43364), dissolved into aerated well water to reach a final solution pH of 12. The test solutions were filtered to remove particulates, and placed into sealed Nalgene carboys to acclimate to test temperature overnight. The pH of each solution was verified with a Hach sensION platinum portable pH meter, or YSI 556 multiprobe. Virkon Aquatic (lot #2258523) (Western Chemical, Ferndale, WA) was measured (0.01 g) and mixed in volumetric flasks with well water at 18°C. Solutions were placed at test temperatures for at least 1 h to acclimate and activate. The test concentrations of 2.5, 5, and 20 g/L were verified with Virkon Aquatic test strips (Western Chemical, Ferndale, WA).

Experimental Design

Elevated pH: A 2 mL sample from the bottom of the Nalgene container with concentrated veligers acclimated to test temperature was removed with a disposable plastic pipette. Each veliger sample was placed into a separate 120 mL beaker (4 test replicates per time interval). We then added 100 mL of test solution rapidly and recorded the starting time of exposure. Static exposures were conducted at a cool temperature (~16°C) and at room temperature (~20°C). To provide the cool temperatures, test and control beakers were maintained in a water bath (raceway trough), and temperatures were recorded continuously with data loggers (mean ± SD = 15.9°C ± 0.14°C). Room temperature tests were conducted on a laboratory bench (20.1°C ± 0.84°C). Groups of beakers were terminated at intervals of 2, 5, 10, 20, or 30 min. At the end of each time interval, the contents of each beaker were poured through a stainless steel mesh spoon lined with ~100 µm nylon mesh to collect veligers. We used well water acclimated to the test temperature to rinse residual test solution from the veligers. After rinsing, the nylon mesh with veligers was placed into a labeled glass Petri dish and rinsed to remove all veligers. Veligers were maintained in the glass Petri dish and allowed to recover for 24 h at test temperature before final assessment of mortality. As a control, we held and handled replicated beakers of veligers not exposed to test compounds at each interval. To determine surviving veligers, the glass Petri dishes were placed under a dissecting microscope and the number of live and dead veligers was counted. Veligers were considered alive if cilia was moving and moving or spinning organs and food contents could be seen inside the shell. Veligers were considered dead if shells were open or empty, body parts had crystallized, or veligers showed no reaction when disturbed.

Virkon Aquatic trials: We used a serological pipette to transfer 1 mL of concentrated veligers acclimated to room temperature from the Nalgene bottle into a 150 mL glass beaker. We then added 9 mL of stock Virkon Aquatic concentration to achieve a final concentration of 2.5, 5, or 20 g/L. Tests were conducted with two replicate beakers for each concentration and time interval. Replicate groups of veligers were assessed after 5, 10, 15, or 20 min of exposure. To evaluate survival in each test replicate, we removed 3 mL of test solution and veligers from the bottom of a test beaker and placed the contents into a 30 mm diameter Petri dish. A dissecting microscope was used to locate and enumerate veligers. After a rapid assessment of mortality, all veligers were placed in aerated well water for recovery, and final assessment of survival was recorded after 24 h with the aid of a Sedgewick rafter microscope slide and compound microscope. No movement of cilia, visible organs, and darkening or crystallization of the veliger, and lack of reaction when disturbed constituted mortality. All tests with Virkon Aquatic were conducted at room temperatures that ranged between 19°C and 22°C.

Data Analysis

We analyzed the proportion of live veligers in tests of elevated pH with a global logistic regression model that included test chemical, temperature, and exposure interval. Estimates of model fit and maximum likelihood estimates were evaluated with type III Wald chi-square effects. To discern more details of the response within each test intervals, we compared the proportions of live and dead veligers by the duration of exposure to observe chemical and temperature effects using log-linear categorical models (Stokes et al. 2000). For test intervals where only one chemical was tested, we report simple exact frequency probabilities to compare the outcome of live and dead by test temperature. In tests of Virkon Aquatic, no temperature comparisons were made, and we compared the frequency of live and dead veligers in concentrations of 2.5 and 5 g/L at exposure times of 5 and 10 min. We reported the results of exposure to 20 g/L in tabular format only. We used a probit model to estimate the LT_{50} in exposures to 2.5 g/L as there were adequate data for model convergence. All statistical analyses were conducted in SAS version 9.2 (SAS Institute, Cary, NC).

Results

We observed rapid mortality of all veligers exposed to elevated pH (Figure 30.1). We found significant effects attributed to test compound, temperature, and exposure time (all Wald χ^2 $P < 0.001$; Table 30.1). The mortality response to $Ca(OH)_2$ was more rapid and evident best in comparisons of

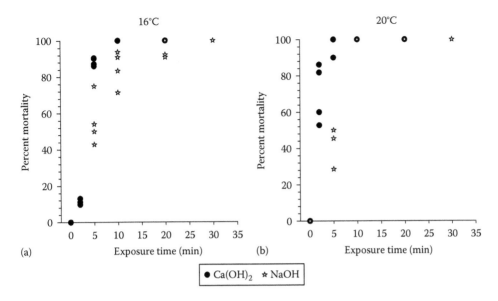

FIGURE 30.1 Summary of percent mortality of veligers versus time of exposure in replicated trials of Ca(OH)₂ and NaOH at pH 12 at two temperatures: 16°C Panel (a) and 22°C Panel (b).

TABLE 30.1

Summary of Logistic Regression Analysis of Proportion Live in Replicate Tests of Elevated pH 12 with Two Chemicals [Ca(OH)₂ or NaOH]

Parameter	DF	Estimate	Wald Chi-Square	P
Intercept	1	7.35	172.88	0.001
Chemical	1	−1.148	91.91	0.001
Temperature	1	−0.231	88.90	0.001
Interval	1	−0.814	225.71	0.001

Note: Maximum likelihood estimates of Wald chi-square and P values are provided.

responses in 10 and 12 min exposures. We found no effect of test temperature after 2 min exposure in solutions of Ca(OH)₂ ($\chi^2 = 1.7$, DF = 1; $P > 0.180$; Table 30.2), but the response was observed at the 5 min exposure ($P < 0.001$; Tables 30.2 and 30.3). After 10 min, the differences between test temperature were not as pronounced ($P = 0.069$), but we detected a significant test chemical by temperature interaction. The likelihood ratios of survival after a 20 min exposure to either chemical were highly significant between temperature and test chemical. In veligers exposed to pH 12 created with Ca(OH)₂, we recorded complete mortality at both water temperatures, but 3% of veligers held at 15°C in NaOH survived (Tables 30.2 and 30.3). Throughout our trials, we observed little to no mortality in control tests of veligers that were handled similarly, but not exposed to elevated pH (Table 30.2).

Veligers exposed to 2.5 g/L Virkon Aquatic achieved 100% mortality between 10 and 15 min (Table 30.4). Higher concentrations (5 g/L) achieved 100% mortality within 10 min. We found significant differences between the frequency of live and dead veligers between concentrations of 2.5 and 5 g/L after 5 min of exposure ($\chi^2 = 30.49$; $P < 0.001$), but no significant difference in frequency of live and dead veligers after a 10 min exposure ($\chi^2 = 2.52$; $P > 0.285$). Using the results of exposure to 2.5 g/L Virkon Aquatic, we fit a significant probit model for mortality versus exposure time (Wald $\chi^2 = 15.68$; $P = 0.001$) and estimated an LC₅₀ of 6.1 min (95% CI = 4.1 – 7.3). Our test system had no effect on veliger survival, as all controls were alive during the testing and recovery time (Table 30.4).

TABLE 30.2

Summary of the Number of Veligers Dead or Alive and Percent Survival
(in Parentheses) in Tests of Ca(OH)$_2$ and NaOH and Controls by Time
Interval, Temperature, and Test Compound, Replicates Combined

	16°C		20°C	
Chemical	**Dead**	**Live**	**Dead**	**Live**
2 min				
Ca(OH)$_2$	21	185	116	56
	(11)	(89)	(67)	(32)
Control	0	117	4	82
	(0)	(100)	(5)	(95)
5 min				
Ca(OH)$_2$		182	25	150
	(88)	(12)	(97)	(3)
Control	3	60	3	56
	(5)	(95)	(5)	(95)
NaOH	45	39	24	31
	(54)	(46)	(43)	(56)
Control	3	64	5	35
	4	96	88	12
10 min				
Ca(OH)$_2$	137	0	84	0
		(100)	(0)	(100)
Control	0	45	2	29
		(100)	(6)	(94)
NaOH	70	12	97	0
		(85)	(15)	(100)
Control	0	41	6	97
		(100)	(6)	(94)
20 min				
Ca(OH)$_2$	60	0	110	0
		(100)		(100)
Control	0	34	1	72
		(100)	(1)	(99)
NaOH	68	2	72	0
		(97)	(3)	(100)
Control	2	51	2	122
	(4)	(96)	(2)	(98)
30 min				
NaOH	53	0	78	0
		(100)		(100)
Control	1	40	4	92
	(2)	(98)	(4)	(96)

Note: Veligers exposed to NaOH were not scored at 2 min, and veligers exposed to
Ca(OH)$_2$ were not scored at 30 min. All final assessments were made after 24 h
recovery period.

TABLE 30.3

Summary of Maximum Likelihood Analysis of Variance Log-Linear Comparisons of Independence for Frequency of Live and Dead Veligers, Tested Individually for Three Intervals of Exposure Modeled by Differences between Chemical Tested and Test Temperatures at pH 12

Source	DF	Chi-square	*P*
5 min			
Temperature	1	12.28	0.0005
Chemical	1	90.79	<0.0001
Chemical*Temp	1	0.44	0.5094
Likelihood ratio	4	306.44	<0.0001
10 min			
Temperature	1	3.32	0.0685
Chemical	1	90.79	<0.0001
Chemical*Temp	1	0.44	<0.0001
Likelihood ratio	1	45.4	<0.0001
20 min			
Temperature	1	32.67	<0.0001
Chemical	1	17.19	<0.0001
Chemical*Temp	1	0.25	0.6199
Likelihood ratio	1	78.88	<0.001

TABLE 30.4

Summary of Live and Dead Quagga Mussel Veligers in Trials of Three Concentrations of Virkon Aquatic for Durations of 5–20 min Exposure

Concentration (g/L)	Exposure (min)	Number Dead	Number Alive	Percent Mortality
2.5	5	3	7	30
	10	31	1	96.9
	15	20	0	100
	20	20	0	100
5.0	5	110	7	94
	10	80	0	100
20	5	30	0	100
0 (control)	5	0	60	0
	10	0	20	0
	15	0	20	0
	20	0	30	0

Note: Trials were conducted at room temperature, ~20°C.

Discussion

Solutions of pH 12 made with NaOH or $Ca(OH)_2$ were effective in killing veligers, although more time was needed at the cooler water temperature to achieve 100% mortality. Similar success with laboratory trials of disinfection testing a suite of microorganisms was achieved by Starliper and Watten (2013). They compared colony-forming units as an endpoint of efficacy in solutions of NaOH at pH 10–12. They achieved 100% mortality of all Gram-negative and Gram-positive cultures. In recently reported shipboard trials conducted by the Great Ships Initiative, tests with NaOH pH 12 showed large reductions in the density of live organisms ≥50 µm in the treatment discharge samples over samples analyzed in control discharge (Cangelosi et al. 2013).

Claudi et al. (2012) evaluated the effect of a range of pH on mussel settlement and mortality and determined that dreissenid mussels have a relatively narrow range of pH tolerance, with the optimum range for settlement of pH 7.5–9.3. Other factors such as calcium, flow velocity, water temperature, dissolved oxygen, conductivity, total organic carbon, and the surface roughness scan affect the settlement of mussels (Chen et al. 2011). The rapid mortality in our tests with veligers is likely also related to the size of the organisms tested when compared with time to mortality in adult-sized mussels. The veligers in our trials ranged from approximately 180 to 220 μm in length and were characteristic of the sizes reported for Lake Mead in October and November (Gerstenber et al. 2011). Mitchell and Cole (2008) found that solutions of pH 12–13 made with hydrated lime or sodium hydroxide were not effective in killing adult-sized faucet snails after a 1 h exposure.

Our tests with Virkon Aquatic were also very effective in killing veligers. A short exposure of 5–10 min at 5 g/L killed all veligers. Our quagga mussel veliger test results support the report by O'Connor et al. (2008) that found 5 mg/L of Virkon Aquatic caused 100% abnormal embryo development in Sydney rock oysters (*Saccostrea glomerata*). Stockton and Moffitt (2013) found that a 15–20 min bath application of 20 g/L Virkon Aquatic was a reliable tool to disinfect boot surfaces infested with New Zealand mudsnails (*Potamopyrgus antipodarum*) and other aquatic invertebrates. Mitchell et al. (2007) reported 100% mortality in red-rim melania (*Melanoides tuberculata*) at concentrations greater than 1 g/L for 24 h (Mitchell et al. 2007). However, faucet snails (*Bithynia tentaculata*) exposed to concentrations of 2 g/L Virkon for 1–24 h showed little mortality (Mitchell and Cole 2008).

Our tests with Virkon Aquatic were conducted at one temperature, and additional tests are needed to establish the relationship between Virkon Aquatic efficacy and water temperature. The active ingredient in Virkon Aquatic is potassium permonosulfate and sodium chloride (Mitchell and Cole 2008; Western Chemical 2008). Potassium and sodium gradients control the neuron response in humans (Starr and Taggert 1998), and have been associated with neural responses in mollusks. A neuron response was first identified in the California sea hare (*Aplysia californica*) (Russell and Brown 1972). Fisher et al. (1991) reported that zebra mussels had a low tolerance for elevated potassium concentrations, and potassium can destroy the integrity of the mussel gill epithelium, leading to asphyxiation.

Trials conducted in our laboratory indicated that Virkon Aquatic was safe for use in and around vertebrates in aquaculture environments (Stockton 2011). Further testing of the efficacy of Virkon Aquatic on other aquatic invasive species is recommended to enable broad-spectrum use. The use of Virkon Aquatic as a disinfectant was comparable to the results obtained using Sparquat 256, but Virkon Aquatic is easily deactivated by organics into simple salts that pose little environmental risk (Stockton and Moffitt 2013).

Determining the appropriate tool for disinfection requires understanding the costs, risks, and environmental consequences of each of these treatments. At present, high-pressure hot water spray has been recommended as an effective means for boat decontamination of attached juvenile and adult mussels (Comeau et al. 2011; Zook and Phillips 2012). The addition of an elevated pH treatment could improve the effectiveness of treatment, as elevated pH residues could be effective in areas that were not adequately heated with the hot water sprays. The use of elevated pH could also be effective in treating water for fire suppression activities. The solutions were easily prepared and used approximately 0.6 g/L dissolved in well water to achieve the pH target. Natural resource interest groups and regulatory agencies have made it clear that safe and nonchemical alternatives for controlling mussel fouling are preferred (Britton and Dingman 2011).

The costs of treatment must be considered as part of any evaluation of chemical disinfection choices. The cost estimates for lye assume a market price of $250/ton for a 75% by weight solution. The cost for hydrated lime assumes a market price of $81/ton. A simple cubic meter solution of aqueous pH 12 would require approximately $0.17 of lye and $0.04 of lime. A kg of Virkon Aquatic sells from aquaculture suppliers for approximately $30, and it would take 5 kg or $150 to make a 5 g/L solution.

Acknowledgments

We thank Mark Olson, Thomas Frew, Giovanni Cappellii, and Andrew Flaten at Willow Beach National Fish Hatchery, Arizona, for their support and assistance with sampling and accommodations for these trials. This research was supported with funding from the Fish and Wildlife Service, Paul Heimowitz,

Project officer, and the U.S. Geological Survey, Work Order 153 in collaboration with the National Park Service. We are grateful to Dr. Andrew Ray, Bob Kibler, and two anonymous reviewers for their reviews of earlier drafts of this manuscript. The use of trade names or products does not constitute endorsement by the U.S. Government.

REFERENCES

Alanso, A. and P. Castro-Diez. 2008. What explains the invading success of the aquatic mud snail *Potamopyrgus antipodarum* (Hydrobiidae, Mollusca)? *Hydrobiologia* 614:107–116.

Bowler, P.A. 1991. The rapid spread of the freshwater Hydrobiid snail *Potamopyrgus antipodarum* in the middle Snake River, southern Idaho. *Proceedings of the Desert Fishes Council* 21:173–179.

Brady, T.J., J.E. Van Benschoten, and J.N. Jensen. 1996. Chlorination effectiveness for zebra and quagga. *Journal of the American Water Works Association* 88:107–110.

Briski, E., S. Ghabooli, S.A. Bailey, and H.J. MacIsaac. 2012. Invasion risk posed by macroinvertebrates transported in ships' ballast tanks. *Biological Invasions* 14:1843–1850.

Britton, D. and S. Dingman. 2011. Use of quaternary ammonium to control the spread of aquatic invasive species by wildland fire equipment. *Aquatic Invasions* 6:169–173.

Bruce, R.L. and C.M. Moffitt. 2010. Quantifying risks of volitional consumption of New Zealand mudsnails by steelhead and rainbow trout. *Aquaculture Research* 41:552–558.

Bruce, R.L., C.M. Moffitt, and B. Dennis. 2009. Survival and passage of ingested New Zealand mudsnails through the intestinal tract of rainbow trout. *North American Journal of Aquaculture* 71:287–301.

Cangelosi, A. 2009. Great Ships Initiative bench-scale test findings, Technical Report, Sodium Hydroxide (NaOH). Great Ships Initiative, Northeast-Midwest Institute, Washington, DC, 20001.

Cangelosi, A. 2011a. Great Ships Initiative bench-scale test findings, Technical Report, Hydrated Lime, Ca(OH)$_2$. Great Ships Initiative, Northeast-Midwest Institute, Washington, DC, 20001.

Cangelosi, A. 2011b. Final report of the land-based, freshwater testing of the lye (NaOH) ballast water treatment system. Great Ships Initiative, Northeast-Midwest Institute, Washington, DC, 20001.

Cangelosi, A., M. Balcer, L. Fanberg, D. Fobbe, S. Hagedorn, T. Mangan, A. Marksteiner et al. 2013. Final report of the shipboard testing of the sodium hydroxide (NaOH) ballast water treatment system onboard the MV Indiana Harbor. Great Ships Initiative, Northeast-Midwest Institute, Washington, DC, 20001.

Chen, D., S.L. Gerstenberger, S.A. Mueting, and W.H. Wong. 2011. Potential factors affecting settlement of quagga mussel (*Dreissena bugensis*) veligers in Lake Mead, Nevada-Arizona, USA. *Aquatic Invasions* 6:149–156.

Choi, W.J., S. Gerstenberger, R.F. McMahon, and W.H. Wong. 2013. Estimating survival rates of quagga mussel (*Dreissena rostriformis bugensis*) veliger larvae under summer and autumn temperature regimes in residual water of trailered watercraft at Lake Mead, USA. *Management of Biological Invasions* 4:61–69.

Claudi, R., A. Graves, A.C. Taraborelli, R.J. Prescott, and S.E. Mastitsky. 2012. Impact of pH on survival and settlement of dreissenid mussels. *Aquatic Invasions* 7(1):21–28.

Comeau, S., S. Rainville, B. Baldwin, E. Austin, S.L. Gerstenberger, C. Cross, and W.H. Wong. 2011. Susceptibility of quagga mussels (*Dreissena rostriformis bugensis* Andrusov) to hot-water sprays as a means of watercraft decontamination. *Biofouling* 27:267–274.

Cuhel, R.L. and C. Aguilar. 2013. Ecosystem transformations of the Laurentian Great Lake Michigan by nonindigenous biological invader. *Annual Review Marine Science* 5:289–320.

Culver, D.A., W.J. Edwards, and L. Babcock-Jackson. 2000. Preventing the introduction of zebra mussels during aquaculture, and fish stocking activities. *Internationale Vereinigung fuer Theoretische und Angewandte Limnologie* 27:1809–1811.

Curry, C.H., J.S. McCarthy, H.M. Darragh, R.A. Wake, S.E. Chruchill, A.M. Robins, and R.J. Lowen. 2005. Identification of an agent suitable for disinfecting boots of visitors to the Antarctic. *Polar Record* 41:39–45.

DuPont. 2011. Virkon® Aquatic efficacy against specific fish pathogens. Available: http://www2.dupont.com/DAHS_EMEA/en_GB/ahb/fish/efficacy_data.html (accessed January 4, 2015).

Edwards, W.J., L. Babcock-Jackson, and D.A. Culver. 2000. Prevention of the spread of zebra mussels during fish hatchery and aquaculture activities. *North American Journal of Aquaculture* 62:229–236.

Ferk, F., M. Mišík, C. Hoelz, M. Uhl, M. Fuerhacker, B. Grillitsch, W. Parzefall et al. 2007. Benzalkonium chloride (BAC) and dimethyldioctadecyl-ammonium bromide (DDAB), two common quaternary ammonium compounds, cause genotoxic effects in mammalian and plant cells at environmentally relevant concentrations. *Mutagenesis* 22:363–370.

Fisher, S.W., P. Stromberg, K.A. Bruner, and L.D. Boulet. 1991. Molluscicidal activity of potassium to the zebra mussel, *Dreissena polymorpha*: Toxicity and mode of action. *Aquatic Toxicology* 20:219–234.

Garcia, M.T., I. Ribosa, T. Guindulain, J. Sanchez-Leal, and J. Vives-Rego. 2001. Fate and effect of monoalkyl quaternary ammonium surfactants in the aquatic environment. *Environmental Pollution* 111:169–175.

Gerstenberger, S., S.A. Mueting, and W.H. Wong. 2011. Abundance and size of quagga mussels (*Dreissena bugensis*) veligers in Lake Mead, Nevada-Arizona. *Journal of Shellfish Research* 30:933–938.

Gleeson, D., B. O'Brien, and K. Jordan. 2013. The effect of using nonchlorine products for cleaning and sanitising milking equipment on bacterial numbers and residues in milk. *International Journal of Dairy Technology* 66:182–188.

Grabow, W.O.K., I.G. Middendorff, and N.C. Basson. 1978. Role of lime treatment in the removal of bacteria, enteric viruses, and coliphages in a wastewater reclamation plant. *Applied and Environmental Microbiology* 35:663–669.

Guardiola, F.A., A. Cuesta, J. Meseguer, and M.A. Esteban. 2012. Risks of using antifouling biocides in aquaculture. *International Journal of Molecular Sciences* 13:1541–1560.

Johnson, L.E., J.M. Bossenbroek, and C.E. Kraft. 2006. Patterns and pathways in the post-establishment spread of nonindigenous aquatic species: The slowing invasion of North American inland lakes by the zebra mussel. *Biological Invasions* 8:475–489.

Johnson, L.E., A. Ricciardi, and J.T. Carlton. 2001. Overland dispersal of aquatic invasive species: A risk assessment of transient recreational boating. *Ecological Applications* 11:1789–1799.

Karatayev, A.Y., D.K. Padilla, D. Minchin, D. Boltovskoy, and L.E. Burlakova. 2007. Changes in global economies and trade: The potential spread of exotic freshwater bivalves. *Biological Invasions* 9:161–180.

Li, X. and B.J. Brownawell. 2010. Quaternary ammonium compounds in urban estuarine sediment environments—A class of contaminants in need of increased attention? *Environmental Science and Technology* 44:7561–7568.

Leung, B., J.M. Bossenbroek, and D.M. Lodge. 2006. Boats, pathways, and aquatic biological invasions: Estimating dispersal potential with gravity models. *Biological Invasions* 8:241–254.

Mainous, M.E., S.A. Smith, and D.D. Kuhn. 2010. Effect of common aquaculture chemicals against *Edwardsiella ictaluri* and *E. tarda*. *Journal of Aquatic Animal Health* 22:224–228.

Mitchell, A.J. and R.A. Cole. 2008. Survival of the faucet snail after chemical disinfection, pH extremes, and heated water bath treatments. *North American Journal of Fisheries Management* 28:1597–1600.

Mitchell, A.J., M.S. Hobbs, and T.M. Brandt. 2007. The effect of chemical treatments on red-rim melania *Melanoides tuberculata*, an exotic aquatic snail that serves as a vector of trematodes to fish and other species in the USA. *North American Journal of Fisheries Management* 27:1287–1293.

Mueting, S.A. and S.L. Gerstenberger. 2011. The 100th Meridian Initiative at the Lake Mead National Recreation Area, NV, USA: Differences between boater behaviors before and after a quagga mussel, *Driessena rostiformis bugensis*, invasion. *Aquatic Invasions* 6:223–229.

Norbury, G., A. Bryon, R. Pech, J. Smith, D. Clarke, D. Anderson, and G. Forrester. 2013. Invasive mammals and habitat modification interact to generate unforeseen outcomes for indigenous fauna. *Ecological Applications* 23:1707–1721.

O'Connor, W.A., M. Dove, and B. Finn. 2008. Sydney rock oysters: Overcoming constraints to commercial scale hatchery and nursery production. NSW Department of Primary Industries Fisheries Final Report Series No. 104.

Oplinger, R.W. and E.J. Wagner. 2009. Toxicity of common aquaculture disinfectants to New Zealand mud snails and mud snail toxicants to rainbow trout eggs. *North American Journal of Aquaculture* 71:229–237.

Peyer, S.M., A.J. McCarthy, and C.E. Lee. 2009. Zebra mussels anchor byssal threads faster and tighter than quagga mussels in flow. *The Journal of Experimental Biology* 212:2027–2036.

Pimentel, D., L. Lach, R. Zuniga, and D. Morrison. 2000. Environmental and economic costs of nonindigenous species in the United States. *BioScience* 50:53–65.

Pimentel, D., R. Zuniga, and D. Morrison. 2005. Update on the environmental and economic costs associated with alien-invasive species in the United States. *Ecological Economics* 52:273–288.

Russell, J.M. and A.M. Brown. 1972. Active transport of potassium by the giant neuron of the *Aplysia* abdominal ganglion. *The Journal of General Physiology* 60:519–532.

Sarkara, B., M. Megharaja, Y. Xia, G.S.R. Krishnamurtia, and R. Naidua. 2010. Sorption of quaternary ammonium compounds in soils: Implications to the soil microbial activities. *Journal of Hazardous Materials* 184:448–456.

Scanar, J., R. Milacic, M. Strazar, O. Burica, and P. Bukovec. 2001. Environmentally safe sewage sludge disposal: The impact of liming on the behaviour of Cd, Cr, Cu, Fe, Mn, Ni, Pb, and Zn. *Journal of Environmental Monitoring* 3:226–231.

Starliper, C.E. and B.J. Watten. 2013. Bactericidal efficacy of elevated pH on fish pathogenic and environmental bacteria. *Journal of Advanced Research* 4:345–353.

Starr, C. and R. Taggart. 1998. *Biology: The Unity and Diversity of Life*, 8th edn. Wadsworth Publishing Company, Belmont, CA.

Stockton, K.A. 2011. Methods to assess, control, and manage risks for two invasive mollusks in fish hatcheries. University of Idaho, Master's thesis. Moscow, ID.

Stockton, K.A. and C.M. Moffitt. 2013. Disinfection of three wading boot surfaces infested with New Zealand mudsnails. *North American Journal of Fisheries Management* 33:529–538.

Stokes, M.E., C.S. Davis, and G.G. Koch. 2000. *Categorical Data Analysis Using the SAS System*, 2nd edn. SAS Institute Inc., Cary, NC.

Strayer, D.L. 2009. Twenty years of zebra mussels: Lessons from the mollusk that made headlines. *Frontiers in Ecology and the Environment* 7:135–141.

Thorp, J.H. 2014. Metamorphosis in river ecology: From reaches to macrosystems. *Freshwater Biology* 59:200–210.

Vanderploeg, H.A., T.F. Nalepa, D.J. Jude, E.L. Mills, K.T. Holeck, J.R. Liebig, I.A. Grigorovich, and H. Ojaveer. 2002. Dispersal and emerging ecological impacts of Ponto-Caspian species in the Laurentian Great Lakes. *Canadian Journal of Fisheries and Aquatic Sciences* 59:1209–1228.

Waller, D.L., S.W. Fisher, and H. Dabrowska. 1996. Prevention of zebra mussel infestation and dispersal during aquaculture operations. *The Progressive Fish-Culturist* 58:77–84.

Watson, K., M.J. Farré, and N. Knight. 2012. Strategies for the removal of halides from drinking water sources, and their applicability in disinfection by-product minimisation: A critical review. *Journal of Environmental Management* 110:276–298.

Watten, B.J., P.L. Sibrell, C.E. Starliper, and K.L. Ritenour. 2007. New method for treating ship ballast solids to prevent future introductions of aquatic invasive species: Effect of lime stabilization/recarbonation on survivorship, water chemistry and process economics. Published Abstract, *15th International Conference on Aquatic Invasive Species*, Nijmegen, the Netherlands.

Watters, A., S.L. Gerstenberger, and W.H. Wong. 2013. Effectiveness of EarthTec for killing invasive quagga mussels (*Dreissena rostriformis bugensis*) and preventing their colonization in the western United States. *Biofouling* 29:21–29.

Western Chemical. 2008. Virkon® Aquatic brochure. Available: http://www.wchemical.com/Assets/File/virkonAquatic_brochure.pdf (accessed December 2010).

Wong, W.H. and S.L. Gerstenberger. 2011. Quagga mussels in the western United States: Monitoring and management. *Aquatic Invasions* 6:125–129.

Zook, B. and S. Phillips. 2012. Recommended uniform minimum protocols and standards for watercraft interception programs for Dreissenid mussels in the western United States. (UMPS II) Updated version of the original 2009 document. Pacific States Marine Fisheries Commission. Portland, OR. http://www.aquaticnuisance.org/.

31

Effectiveness of the SafeGUARD Ultraviolet Radiation System as a System to Control Quagga Mussel Veligers (Dreissena rostriformis bugensis)

Patricia Delrose, Shawn L. Gerstenberger, and Wai Hing Wong

CONTENTS

ABSTRACT   The quagga mussel (*Dreissena rostriformis bugensis*) has quickly spread to Lake Mohave and further down the lower Colorado River drainage since its January 2007 discovery in Lake Mead. The microscopic sizes (70 μm or larger) of the veliger life stages make them impossible to see with the unaided eye and difficult to remove from water delivery systems and fish stocking trucks. The purpose of this study is to determine if exposure to different doses of ultraviolet radiation can irreparably damage or kill quagga mussel veligers. The ultraviolet radiation (UVR) exposure doses were 1, 3, 6, and 12 cycles through the SafeGUARD UVR system. After exposure, 50 veligers were sampled and observed at 0, 24, 48, 72, and 96 h. Results indicate that veligers exposed to 12 cycles through the UVR system experienced 100% mortality after 96 h. Results also show a significant difference in mortality of veligers between one cycle and multiple cycles of UVR exposure ($p < 0.05$), while there is no statistical difference between cycles 3, 6, and 12 ($p > 0.05$).

Introduction

In January 2007, the invasive quagga mussel (*Dreissena rostriformis bugensis*) was found in Lake Mead. This became the first known dreissenid species inhabiting the Southwest, and the only time a large aquatic ecosystem was first infested by the quagga mussel rather than the zebra mussel (*Dreissena polymorpha*) (Gerstenberger et al., 2011; LaBounty and Roefer, 2007). The rapidly expanding quagga mussel population, due to their fouling nature, impacted water circulation and associated delivery infrastructure at lower Colorado River dams and their associated water distribution systems, along with two fish hatcheries operated by Nevada Department of Wildlife and the US Fish and Wildlife Service. Facility operators have discussed the use of ultraviolet radiation (UVR) as an improved control method for dreissenid mussels, but few studies have been conducted on the ability of UVR to negatively impact dreissenid adult or veliger life states (Chalker-Scott et al., 1994a,b; Seaver et al., 2009). Thus, the purpose of this study is to test the Emperor Aquatics, Inc. (Pottstown, PA) SafeGUARD Ultraviolent Radiation System (SafeGUARD UVR System), assessing the number of passage cycles through it that would be needed to irreparably damage or kill quagga mussel veligers.

The Lake Mead National Recreational Area (LMNRA) is situated along the Colorado River adjacent, but downstream to Grand Canyon National Park. The LMNRA, administered by the National Park Service, includes Lake Mead and Lake Mohave. Hoover Dam and Davis Dam, respectively, impound both lakes. Hoover Dam is located 56.3 km outside of Las Vegas, Nevada, while Davis Dam is located 107.8 km downstream near Laughlin, Nevada (NPS, 2010). Lake Mead is the largest reservoir by volume (3.5×10^{10} m^3) in the United States (LaBounty and Burns, 2005). Lake Mohave, although smaller, impounds 1,818,330 acre-ft of water. The LMNRA covers about 1.5 million acres and is important for the development of the Southwest. The LMNRA supplies drinking water to the Las Vegas area, electricity to the Southwest, recreational activities for millions of visitors, and irrigation water to farmlands (Holdren and Turner, 2010).

LaBounty and Roefer (2007) report that the zebra and quagga mussels have become the most serious nonnative, biofouling pests introduced into North American freshwater systems. In a short amount of time, these species have caused severe economic, ecological, and human health impacts to the Southwest. *Dreissenid* mussels are very efficient filter feeders that are capable of filtering large volumes of water in a very short amount of time (Karatayev et al., 1997). Through filtering the water, they have the ability to reduce the biomass and change the structure of phytoplankton and zooplankton communities (Wong et al., 2011). This increases the water clarity and reduces the amount of suspended solids and oxygen in the water column, allowing aquatic plants to grow more rapidly (Wong et al., 2011). Dreissenid mussels have a rapid filtration rate, a planktonic veliger stage, high fecundity, and the ability to attach easily to surfaces, allowing them to spread easily throughout North America (Gerstenberger et al., 2011; Herbert et al., 1989; Wong et al., 2011). These mussels have the ability to attach to surfaces using their strong byssal threads, allowing them to clog water pipes, damage boat motors, and destroy recreational equipment. The Metropolitan Water District of Southern California spends $10–15 million a year to deal with quagga mussel damage caused to the 390 km Colorado River aqueduct and reservoir system (Fonseca, 2009; Gerstenberger et al., 2011). It is estimated that 1 billion dollars are spent annually in the Great Lakes region and throughout other areas of North America to monitor and control dreissenid populations (Pimentel et al., 2005; Wong et al., 2011).

The sun is a natural and major source of UVR, and manufactured lamps can also emit it. According to the National Science Foundation, UVR is high in energy; therefore, it has the ability to change the chemical structure of a DNA molecule and cause mutations in the genetic code. This change in the chemical structure can cause cell damage and deformities in living organisms. UVR is divided into three categories that are based on the wavelength band, the amount of energy it contains, and the effects it has on biological material as follows:

> The shortest UVR wavelength band, UV-C (200–280 nm), is the most energetic of the three, potentially the most lethal, but the least harmful, because the radiation is absorbed by the ozone layer and does not reach the Earth. Man-made lamps can emit UV-C radiation, but water absorbs most of its rays, so only the aquatic organisms in the immediate area of a lamp are affected (Chalker-Scott et al., 1994a). Researchers have also found that UV-C radiation has the ability to change veliger behavior and increased mortality (Chalk-Scott et al., 1994b). Exposure to UV-C rays has also been linked to major human health hazards in occupational settings, such as welders (Chalker-Scott et al., 1994a; http://uv.biosphereical.com, 2012).

> The second type of UVR, midrange UV-B (280–320 nm), is able to pass through the ozone layer and reach the Earth's surface. Studies have shown this type is the most damaging to biological systems under natural conditions. *Dreissena polymorpha* veligers have shown sensitivity to UV-B, experiencing up to 100% mortality, although mortality decreases with increasing larval age (Chalker-Scott et al., 1994a).

Radiation from the longest UVR wavelength band, UV-A (320–400 nm), dependent upon cloud cover, can deliver up to 95% of its rays to the Earth's surface (Biospherical Instruments, 2012). Black lights and fluorescent lights are manufactured ways of producing UV-A rays. UV-A does not damage DNA directly, but it produces chemicals such as hydroxyl and oxygen radicals that can cause damage to an organism's DNA.

Materials and Methods

Design of the System

A SafeGUARD UVR System is currently in place at the Willow Beach National Fish Hatchery (WBNFH), Willow Beach, AZ. The system was used to determine the number of cycles for water containing live veligers to be exposed to UVR radiation to cause irreparable damage or death (Figure 31.1). The SafeGUARD UVR System contains three 80-watt UVR lights that are encased in a metal vessel and are arranged to maximize their output potential. The quartz sleeve, made from transparent hard quartz glass, thermally protects each lamp, which allows the highest UVR transmittance to ensure maximum UVR energy output (Emperor Aquatics Inc., 2008). The spectral power distribution (SPD) for the unit is 180,000 μWs/cm^2. The rays emitted are UV-C, which have been found to cause irreparable damage to veliger DNA, behavior changes, and increased mortality (Chalker-Scott et al., 1994a,b). The system's owner manual states that the low pressure, mercury arc germicidal lamp produces about 90% of its radiation energy at 253.7 nm, which approximates the most lethal wavelength (265 nm) to microorganisms (265 nm) (Emperor Aquatics Inc., 2008).

Collection and Enumeration of Veligers

On each sampling day, a batch of live quagga mussel veligers was collected from the B8 raceway at WBNFH. This was done by placing a 64 μm plankton net under the water flowing out of the head box. After a 20 min collection time, the batch of veligers and water in the net's collection cup was placed in a 300 mL glass beaker. Using a pipette and a 64 μm sieve, water was decanted from the batch until

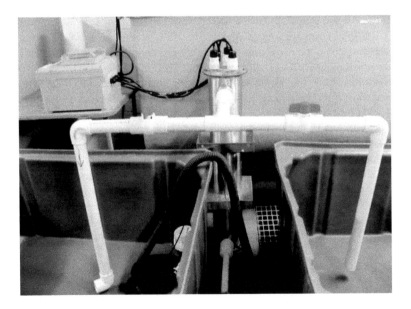

FIGURE 31.1 Design for the SafeGUARD UVR System at Willow Beach National Fish Hatchery used to test its damage caused to quagga veligers.

a volume of 50 mL was reached. From the thoroughly mixed 50 mL sample, 5 mL as a subsample was removed and placed in a glass petri dish. The 5 mL subsample was observed under an Olympus SZX7 dissecting scope (Olympus, Valley Center, PA), and the number of veligers was counted to ensure that a sufficient number of veligers were present. The 5 mL subsample was returned to the sample. The 50 mL sample was then added to 227 L of well water and pumped through the SafeGUARD UVR System using a standard bilge pump (Atwood V1250, Model 4212). Emperor Aquatics Inc. (2008) suggests a flow rate of 6–8 gallon per minute (GPM) past the UVR lamps, so our test was run at the slower 6 GPM to optimize veliger exposure to UVR. By moving the bilge pump back and forth between the two holding tanks, the veliger-laden water was cycled through the system for the desired number of times (1, 3, 6, and 12). An identical number of control sample cycles were run through the system with its UVR lights turned off.

In order to ensure that veligers were not getting trapped or lost within the bilge pump, holding tank, or the SafeGUARD UVR System, all samples, including the controls, were re-enumerated following their first run through the system. An identical procedure to the aforementioned enumeration, comparing preveliger/postveliger numbers, confirmed no loss.

Following the desired number of cycles (1, 3, 6, and 12) for UVR treatments and controls, 64 μm plankton net was placed under the outflow pipe to re-collect the veligers from each group. Two plankton nets were used: one for control groups and one for UVR exposed groups. A 5 mL subsample from each cycle group was examined at the selected observation times of 0, 24, 48, 72, and 96 h, assessing 50 veligers per subsample for 3 min to detect any movement (presumed alive), no movement (presumed dead), or structural damage. The data were recorded, and the 5 mL subsample per observation period was added back into the sample for its respective group along with fresh well water to a volume of 300 mL. In addition, a sample of veligers was collected, enumerated, and placed in 300 mL of well water for assessment at 0, 24, 48, 72, and 96 h. All samples were placed in a 16°C well water bath until the next observation time.

Between sampling days, the plankton net, including its collection cup, and the utensils, petri dish, pipette, and sieve were rinsed with well water and disinfected by placing them in 5% acetic acid for a 24 h period. The SafeGUARD UVR System and holding tanks were rinsed with well water, then drained and dried between sampling days too. Within a sampling day, all equipment was rinsed with well water between taking samples.

Statistical Analysis and Data Interpretation

To determine if there was a significant difference between the numbers of UVR exposures, an analysis of covariance (ANCOVA) was performed. Before the UVR testing began, well water was collected and observed under stereoscope, verifying that no veligers were present in the water. Observation of 50 veligers from a sample placed in well water held at 16°C and examined under an Olympus SZX7 dissecting microscope after 0, 24, 48, 72, and 96 h showed that all life stages were actively swimming and feeding. From this, it can be concluded that the well water at WBNFH does not kill veligers within 96 h. Before the experiment and after each UVR treatment was conducted, 100 mL water samples were collected for Emperor Aquatics, Inc. to determine the UVR transmittance percentage (%UVT). The %UVT is the total amount of UVR energy available to treat the water. The higher the percent value, the greater the UVR dose will be. Before treatments, the source water was analyzed and determined to have a %UVT reading of 93%. UVT readings for the various cycles examined ranged from 94% to 96%. At a flow rate of 6 GPM and a 95% UVT reading with a 10% safety factor included, Emperor Aquatics determined the fluence (UVR dose) to be 700.11 mJ/cm². The fluence calculation is proprietary information; therefore, the dose at 94% and 96% UVT can only be estimated to be 700.11 mJ/cm².

Results

For the controls, we determined that passing veligers through the SafeGUARD UVR System multiple times without the UVR lights on did not damage or kill them at the beginning (Tables 31.1 and 31.2). As the UVR exposure cycles increased from 1, 3, 6, and 12 so did the number of veligers that appeared not

TABLE 31.1

Quagga Mussel Veliger Survival (M = Movement Observed; Presumed Alive) and Mortality (N = No Movement Observed; Presumed Dead) Data for the Effects of Increasing Cycles of UVR Treatment (1, 3, 6, and 12) on 50-Veliger Observations at 0, 24, 48, 72, and 96 h in the SafeGUARD UVR System Experiment

		0		24		48		72		96	
	Time (h)	M	N	M	N	M	N	M	N	M	N
Control	# of cycles										
	1	50	0	48	2	50	0	49	1	47	1
	3	50	0	47	3	49	1	49	1	45	5
	6	50	0	47	3	48	2	44	6	43	7
	12	50	0	49	1	39	11	18	32	17	33
UVR	# of cycles										
	1	17	33	40	10	39	11	6	44	2	48
	3	7	43	15	35	7	43	5	45	5	45
	6	6	44	2	48	7	43	7	43	1	49
	12	0	50	12	38	7	43	3	47	0	50

TABLE 31.2

Percent of Quagga Mussel Veligers Showing No Movement (Presumed Dead) Following Increasing Cycles of Exposure for UVR Treatments at Time 0, 24, 48, 72, and 96 h

	Control					UVR Exposure				
Cycles of Exposure	0 (h)	24 (h)	48 (h)	72 (h)	96 (h)	0 (h)	24 (h)	48 (h)	72 (h)	96 (h)
1	0%	4%	0%	2%	2%	66%	20%	22%	88%	96%
3	0%	6%	2%	2%	10%	86%	70%	86%	90%	90%
6	0%	6%	4%	12%	14%	88%	96%	86%	86%	98%
12	0%	2%	22%	64%	66%	100%	76%	86%	94%	100%

to be moving; thus, these were presumed to be dead (Table 31.1). Veligers observed after 24 h of UVR exposure showed signs of recovery, but as the observation times increased, so did the number of veligers not moving. After being exposed to UVR, veligers initially showed higher percentages of no movement (Table 31.2). After 96 h of observation, all UVR treatments had an increase in the percentage of veligers not moving (Table 31.2). With a treatment of 12 times through the SafeGUARD UVR System at a period of 96 h, 100% of the veligers observed were not moving. Under the same conditions without the UVR lights on, there was a 66% chance of veligers not moving. Therefore, UVR increased the likelihood of killing veligers, and the more times they are exposed to UVR, the greater is the chance that they will die.

There is a statistically significant difference between veligers being exposed once to UVR compared to the other treatment cycles (ANCOVA, $p < 0.05$). The more the veligers are exposed to UVR, the more significant the difference between the cycles becomes: 1:3, $p = 0.0153$; 1:6, $p = 0.0032$; and 1:12 $p = 0.0029$. When comparing 3:6 ($p = 0.5322$), 3:12 ($p = 0.5071$), and 6:12 ($p = 0.9688$), there is no significant difference between the cycles. To get the highest percent mortality of veligers, the maximum number of exposure cycles should be used (Table 31.2). The longer the exposure to UVR, the more damaging it is to veligers.

Discussion

Since quagga mussels were found in Lake Mead in January 2007, they quickly spread throughout the lower Colorado River drainage. *Dreissenid* mussels are considered to be the most serious nonnative biofouling pest introduced into a large North American freshwater ecosystem (LaBounty and Roefer, 2007).

Quagga mussels have caused severe economic, ecological, and human health impacts to the Southwest. There have been many efforts such as the introduction of an enemy species or the application of toxic chemicals directed toward the eradication and control of this invasive species. However, these efforts often result in more ecological harm such as the excessive poisoning of nontarget organisms, the transfer of poisons up the food chain, or a population explosion of introduced enemy species (Simberloff et al., 2005). Thus, research should focus on ways to eradicate this invasive species without causing harm to the aquatic environment.

Immediately after the initial exposure (1 cycle) to the SafeGUARD UVR System, most (66%–100%) veligers appeared to be dead. But, after 24 h, many began to recover. Assessing 1 cycle of UVR exposure, the recovery from what initially appeared to be 66% mortality to only 20% and 22% after 24 and 48 h, respectively, took 72 h to see increasing rates of veliger mortality: 88% by 72 h, increasing to 96% by 96 h. So, it is clear that 1 cycle of exposure to UVR is not enough to immediately kill quagga mussel veligers; multiple exposures are needed.

After 3 cycles of UVR, the percent mortality increased when comparing it to one exposure cycle. The longer the veligers are exposed to UVR, the higher the mortality rate becomes. Under the laboratory conditions at WBNFH, 100% mortality was reached at 12 exposure cycles with an observation time of 96 h. To ensure increased mortality, veligers should be exposed to UVR for a minimum of 3 cycles and held for a minimum of 5 days.

This study confirms the findings by Chalker-Scott et al. (1994a,b) that veligers are sensitive to multiple exposures of UV-C radiation and they have potential effectiveness as a control strategy. It has also been suggested that adult mussels are able to survive higher doses of UV-C radiation due to the thickness of their outer shell (Chalker-Scott et al., 1994a), which would explain why 100% mortality was not seen until the highest exposure cycle. Chalker-Scott et al. (1994b) report that the water absorbs UV-C rays, so only the aquatic organisms in the immediate area of the UVR source are affected. This may explain why percent mortality varies among the treatment cycles. To reduce the length of time it takes to obtain 100% mortality, veligers should be passed under UVR multiple times and at a flow rate of 6 GPM or slower.

Limitations

For the UVR treatment, using UV-C lamps that emit a range of wavelengths at 240–280 nm instead of exactly 264 nm, the wavelength that kills most biological organisms, could have caused a longer time period for veligers to die. Because these lamps emit a range of wavelengths, there is the chance that the lower end of the wavelength was being emitted and the veligers were not receiving the wavelength that is most damaging to their systems. There is also the chance that not all the veligers observed were exposed to UVR during the multiple passes through the system due to their location in the water column when passing the UVR lamps.

Recommendations

The flow rate of 6 GPM through the SafeGUARD UVR System showed 30% mortality in the control samples, although they received no UVR since the lights were turned off. Thus, a reduced flow rate should be used to ensure that only UVR is killing the veligers and not the pressure of the water or abrasion of system elements against veligers going through the system.

Since it took 30 min to pass veligers through UV radiation system and it did not kill them immediately, doing more treatment cycles could give a better measurement of how long exposure to UVR is needed to kill veligers immediately. Studies could be performed using more UV-C lamps and longer exposure times to determine if the length of hours between exposure and death could be reduced.

Samples should be rechecked at intervals longer than 96 h, which will determine if percent mortality increased, since veligers could have been irreparably damaged, but simply had not died during the observation period. Also, veligers should be observed for longer than 3 min to ensure they are dead, because veligers have the ability to appear dead when they really are not. In addition, tripling the test cycles for the UVR study and increasing the number of veligers per sample observed would help to clarify observed differences in percent mortality between the treatment cycles and observation times. Testing more numbers of cycles through the SafeGUARD UVR System would give a more accurate determination for the number of cycles to which veligers need to be exposed in order to kill them sooner.

REFERENCES

Biospherical Instruments Inc. Ultraviolet radiation. Available at http://uv.biospherical.com. Retrieved on June 12, 2012.

Chalker-Scott, L., Scott, J.D., Carnevale, R., and Smith, K. (1994a). Comparison of acute and chronic mid-range ultraviolet radiation (280–320 nm) effects on adult zebra mussels. *Zebra Mussel Information Clearinghouse* 5(1), 1–4.

Chalker-Scott, L., Scott, J.D., Titus, J., and Scalia, J. (1994b). Influence of wide-range ultraviolet radiation upon behavior and mortality of *Dreissena polymorpha. Proceedings of the Fourth International Zebra Mussel Conference*, Madison, WI, March 1994, pp. 161–177.

Emperor Aquatics Inc. (2008). SafeGUARD UV system Hoss operator's manual.

Fonsca, F. 2009. Invasive mussels imperil western water system. *New York Daily News.* Available at http://articles.nydailynews.com/2009–07–21/news/17927572_1_quagga-mussel-zebra-mussel-lake-mead. Accessed on April 20, 2011.

Gerstenberger, S.L., Mueting, S.A., and Wong, W.H. (2011). Veligers of invasive quagga mussels (*Dreissena rostriformis bugensis,* Andrusov 1897) in Lake Mead, Nevada-Arizona. *Journal of Shellfish Research* 30(3), 933–938.

Herbert, P.D.N., Muncaster, B.W., and Mackie, G.L. (1989). Ecological and genetic studies on *Dreissena polymorpha* (Pallas): A new mollusk in the Great Lakes. *Canadian Journal of Fisheries and Aquatic Sciences* 46, 1587–1591.

Holdren, G.C. and Turner, K. (2010). Characteristics of Lake Mead, Arizona–Nevada. *Lake and Reservoir Management* 26, 230–239.

Karatayev, A.Y., Burlakova, L.E., and Padilla, D.K. (1997). The effects of *Dreissena polymorpha* (Pallas) invasion on aquatic communities in eastern Europe. *Journal of Shellfish Research* 16, 187–203.

LaBounty, J.F. and Burns, N.M. (2005). Characterization of Boulder Basin, Lake Mead, Nevada-Arizona, USA-Based on analysis of 34 limnological parameters. *Lake and Reservoir Management* 21, 277–307.

LaBounty, J.F and Roefer P. (2007). Quagga mussels invade Lake Mead. *LakeLine* 27, 17–22.

NPS (National Park Service). Overview of Lake Mohave. (December 13, 2010). Available at http://www.nps.gov/lake/naturescience/overview-of-lake-mohave.htm. Retrieved June 16, 2012, from National Park Service

Pimentel, D., Zuniga, R., and Morison, D. (2005). Update on the environmental and economic costs associated with alien-invasive species in the United States. *Ecological Economics* 52, 73–288.

Seaver, R.W., Ferguson, G.W., Gehrmann, W.H., and Misamore, M.J. (2009). Effects of ultraviolet radiation on gametic function during fertilization in zebra mussel (*Dreissenapolymorpha). Journal of Shellfish Research* 28(3), 625–633.

Simberloff, D., Parker, I.M., and Windle, P.N. (2005). Introduced species policy, management, and future research needs. *Frontiers in Ecology and the Environment* 3, 12–20.

Wong, W.H., Gerstenberger, S.L., Miller, J.M., Palmer, C.J., and Moore, B. (2011). A standardized design for quagga mussel monitoring in Lake Mead, Nevada-Arizona. *Aquatic Invasions* 6(2), 205–215.

32

Challenges of Developing a Molluscicide for Use on Dreissena rostriformis bugensis Veligers in Fish Transport Tanks

Catherine L. Sykes and Wade D. Wilson

CONTENTS

ABSTRACT   The transfer of fish from mussel-positive waters is a potential vector for moving quagga mussel *Dreissena rostriformis bugensis* and zebra mussel *Dreissena polymorpha* veligers to new water bodies. A multitude of chemicals have been published as effective molluscicides, but most are used over an extended period of time and are highly toxic to fish. The goal of our study was to develop a time-limited treatment that will ensure 100% mortality of *D. bugensis* veligers while not negatively impacting fish exposed simultaneously to the treatment. The efficacy of 18 chemicals as molluscicides was tested at multiple concentrations and treatment times on *D. bugensis* veligers and native fish species under water conditions of the southwestern United States. The acute toxicity data were highly variable, but one consistent result was that *D. bugensis* veligers exhibited greater tolerance to the chemicals than native fish species. Six chemicals produced 100% veliger mortality, but five (Peraclean® 15, magnesium chloride, menthol, propylene phenoxytol, and a clove oil/menthol mix) were at concentrations that cannot be tolerated by native fish. The sixth chemical, potassium pyrophosphate, did not appear to negatively impact the fish, but further testing is required because of the formation of a precipitate in the dilution water. Future research on the survivability and viability of veligers exposed to sublethal concentrations of these chemicals would provide greater insight into their potential use as molluscicides.

Introduction

Water bodies of the southwestern United States have some of the most highly used and visited aquatic recreational areas found in the nation. Areas along the Colorado River such as Lake Powell and Lake Mohave provide scenic and warm weather destinations for people to fish, camp, and boat throughout the year. This high use has come at a cost however, with the introduction of aquatic invasive species such as quagga mussel *Dreissena rostriformis bugensis*. *Dreissena bugensis* larvae are divided into four basic developmental stages identified by morphology and behavior: trochophore (40–100 µm in diameter), straight-hinged or D-stage veliger (97–112 µm), umbonal (112–347 µm), and pediveliger (231–462 µm). Several life stages of the species have been found within or on boats that have visited infested water bodies and, when not properly disinfected, passively transport *D. bugensis* to new water bodies.

In the cases of Lake Mohave, Mead, and Havasu, *D. bugensis* was detected in 2007 (McMahon 2011). These water bodies are rich in aquatic recreational activities, but they are also areas of extensive ichthyofaunal conservation efforts. The Colorado River basin and its major tributaries, including the Green River, provide habitat for several endemic fish species including the bonytail *Gila elegans*, humpback chub *Gila cypha*, roundtail chub *Gila robusta*, Colorado pikeminnow *Ptychocheilus lucius*, razorback sucker *Xyrauchen texanus*, flannelmouth sucker *Catostomus latipinnis*, and bluehead sucker *Catostomus discobolus*. All of these species are protected at the state or federal level with *G. elegans*, *G. cypha*, *P. lucius*, and *X. texanus* federally listed as endangered. Recovery plans and species conservation efforts for these four species include conservation tasks associated with captive breeding or refuge population programs that require the transfer of individuals from wild population to state, federal, or tribal facilities. For example, wild larval *X. texanus* were transported to Willow Beach National Fish Hatchery (WBNFH; Willow Beach, AZ) for captive rearing and were later transferred as fingerlings to Bubbling Ponds Fish Hatchery (BPFH; Cornville, AZ) and the Southwestern Native Aquatic Resources and Recovery Center (Southwestern Native ARRC; Dexter, NM, known as Dexter National Fish Hatchery and Technology Center prior to 2012). Currently, the only captive broodstock in the lower Colorado River basin (below Hoover Dam) is located at Southwestern Native ARRC and is used for the production of individuals to augment the Colorado River population below Lake Mohave. This broodstock was developed, in part, with the transport of Lake Mohave larval fish from a variety of localities to the Southwestern Native ARRC between 1999 and 2004. The intent of this process was capturing the genetic diversity of the Lake Mohave and lower Colorado River basin (Dowling et al. 1996).

After the discovery of *D. bugensis*, transfer of larval fish from Lake Mohave to BPFH and Southwestern Native ARRC was no longer an option. Mitigation efforts included restricting the movement of fish to waters within the lower Colorado River basin where the presence of *D. bugensis* was already confirmed. This quarantine of the infested portion of Colorado River has impacted the recovery efforts of *X. texanus* in Lake Mohave and above Hoover Dam. The only facility that currently receives larval fish for recovery activities in Lake Mohave is WBNFH. As *D. bugensis* spreads within the Colorado River system, it is anticipated that other native fish species will be impacted in similar ways.

In response to these recovery challenges, the Bureau of Reclamation's Lower Colorado River Multi-Species Conservation Program (LCR MSCP) funded mitigation research to investigate the efficacy of chemicals as potential molluscicides for *D. bugensis* veligers in fish transport tanks with minimal impact on the native fish held in the same container. This is a novel and challenging aspect in regard to *D. bugensis* mitigation research. Most prior research has been industrial based, focusing on the control and management of zebra mussel (*Dreissena polymorpha*) in power plants and water treatment facilities through equipment disinfection protocols. Research on the effects of chemical treatments to kill or control *D. bugensis* has rarely included consideration of acute or chronic effects of simultaneous chemical exposure to nontarget species, including native fish.

This chapter examines the challenges of developing a molluscicide treatment for *D. bugensis* veligers while not impacting native fish species. A decision tree (Figure 32.1) was used to select 18 compounds for acute toxicity evaluation on *D. bugensis* veligers at WBNFH and three native fish species at Southwestern Native ARRC. The chemicals were selected based on published data showing a high toxicity to bivalves, low toxicity to fish, low risk to humans, and availability for use in the United States. Twelve chemicals

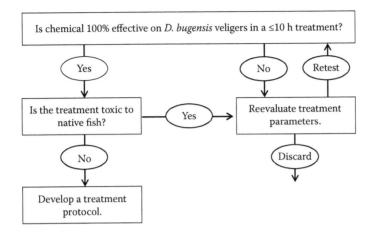

FIGURE 32.1   Decision tree used for progressive stages in testing each of the chemicals as a potential molluscicide for *D. bugensis* veligers within the guidelines established for fish transport tanks.

were evaluated in multiple combinations of concentrations and exposure times, including potassium chloride (KCl) plus formalin, a common treatment protocol for *D. polymorpha* (Edwards et al. 2000). The anesthetic properties of six additional chemicals were evaluated as a pretreatment to prevent veligers from closing their shells, a defense response to toxicant exposure. To explore as many options as possible, four biocides commonly used in aquaculture were tested in addition to other potassium-based compounds.

Methods

Veliger Sampling Protocol

Dreissena bugensis veligers (straight-hinged to pediveliger larvae) were collected from raceway head box inflows at WBNFH using a 35 μm plankton net fitted with a 1 L capacity cod end jar. Collection times were a minimum of a 15 min sampling period. Collected material was first passed through a 250 μm stainless steel sieve to remove large algal debris, then passed through a 35 μm mesh net to collect veligers and finally rinsed into a 250 mL sample jar with a final volume of 50–100. To eliminate temperature changes as a potential confounding factor, jars were stored in a water bath consistent with raceway water temperatures until veligers could be counted into test plates on the day of collection. Sykes (2012) examined the effect of collection methods on veliger survival during a 24 h posthandling period and showed that no veliger mortality occurred during the collection or transfer method.

Acute Toxicity Tests

Potassium Chloride and Formalin

Dreissena bugensis veliger exposure tests to a range of KCl (presumed anesthetic) and formalin (molluscicide) concentrations were conducted in either 100 mL beakers with 50 mL test solution or 1000 mL beakers with 500 mL test solution (Table 32.1). Potassium chloride solutions were prepared using WBNFH well water (pH 7.9; conductivity 1404 μS/cm; hardness, 380 mg/L CaCO$_3$; alkalinity, 182 mg/L CaCO$_3$) with concentrations based on percent active ingredient (99.8%, Cargill Salt, MN) and verified by the measurement of conductivity (Symphony SP90M5 meter and four cell conductivity electrode; VWR International, Inc., Brisbane, CA). Traceable conductivity standards (66.1, 664, and 6637 mg/L KCl) were used to create a standardized curve. Formalin concentrations were based on the complete product formula (Formacide-B, B.L. Mitchell, Inc., MS). Condition of veligers was assessed at the end of the KCl and formalin treatment with, and without, a 2 h recovery period (see Table 32.1). It was determined that

TABLE 32.1

Condition of *D. bugensis* Veligers Following a 1 h Exposure to KCl and 2 h Exposure to Formalin

KCl (mg/L)	Formalin (mg/L)	Test Temperature (°C)	Replicates per Test	Number of Veligers per Test	Percent Unresponsive after Treatment	Percent Mortality after Recovery
0	0	19	—	25	1	n/a
1500	25	19	—	67	0	n/a
1500	50	19	—	32	0	n/a
1500	100	17	—	18	72	n/a
2000	25	19	—	34	3	n/a
2000	50	19	—	43	2	n/a
2250	25	17	2	15	27	n/a
2250	50	17	2	12	25	n/a
2250	100	17	2	17	100	n/a
3500	25	17	2	10	20	n/a
3500	50	17	2	16	13	n/a
3500	100	17	2	15	60	n/a
0	0	17	2	≥20	0	0
4250	25	17	2	18	39	0
4250	50	17	2	15	93	0
4250	100	17	2	22	50	0

Note: Percent mortality was recorded after a 2 h recovery period except where noted (n/a). Initial tests did not include a recovery period.

adding a recovery period was necessary for accurate assessment of veliger condition (dead vs. dormant). All assessments were conducted under a stereomicroscope and used two criteria to determine mortality (Edwards et al. 2000): (1) lack of motion in velum cilia and internal organs and (2) lack of response to touch. To evaluate the effect of transferring veligers from Lake Mohave water to well water, a negative control was included.

Molluscicides: Cutrine®-Ultra, Peraclean 15, and Spectrus™ CT1300

Three additional molluscicides were selected for testing based on application rates and toxicity levels (see Sprecher and Getsinger 2000):

- Cutrine-Ultra (Applied Biochemists, Germantown, WI), an algaecide based on a 9% copper-ethanolamine complex (Kennedy et al. 2006)
- Peraclean 15 (Evonik Industries, Parsippany, NJ), a biocide based on 15% peracetic acid (de Lafontaine et al. 2008; Fuchs and de Wilde 2004; Verween et al. 2009)
- Spectrus CT1300 (formerly packaged as Clamtrol CT-2; GE Betz, Trevose, PA), a molluscicide based on 50% n-alkyl dimethylbenzyl ammonium chloride (Fisher et al. 1994; Waller et al. 1993)

All test solutions were prepared using Lake Mohave water supplied to raceways (pH 7.8–8.1; conductivity 949–1000 µS/cm; hardness, 266–310 mg/L $CaCO_3$; alkalinity, 120–138 mg/L $CaCO_3$). All water was filtered prior to use. Tests were conducted in six-well plastic tissue culture plates (well volume = 10 mL; Corning Inc., Corning, NY). A different plate was used for each of the three chemicals tested. Veligers were first consolidated into petri dishes using a stereomicroscope and volumetric pipette (total volume = 0.02 mL); then a minimum of 10 veligers were transferred to each well containing 1.98 mL of river water for a final well volume of 2 mL. Chemical stock solutions were prepared daily and added to wells containing veligers at the appropriate volume for the desired test concentration. Toxicity tests were conducted in two stages: (1) preliminary tests were conducted to determine potential ranges of

TABLE 32.2

Dreissena bugensis Veliger Mortality Observed Following a Time-Series Exposure to
Candidate Molluscicides (mg/L Based on Percent Active Ingredient), No Replication per Well

		Treatment Time (Hours)					
		1	2	3	4	5	6
Chemical	(mg/L)	Percent Mortality per Treatment Time (Veligers per Well)					
Cutrine-Ultra	15	0 (10)	20 (10)	70 (10)	36 (11)	80 (10)	80 (10)
	20	30 (10)	60 (10)	60 (10)	100 (12)	100 (11)	83 (12)
Peraclean 15	35	73 (11)	100 (10)	90[a] (10)	100 (10)	100 (10)	100 (10)
	50	60 (10)	100 (11)	100 (10)	100 (10)	100 (9)	100 (10)
Spectrus CT1300	30	0 (10)	50 (10)	90 (10)	60 (10)	55 (9)	90 (10)
	37.5	27 (11)	27 (11)	60 (10)	50 (10)	90 (10)	91 (11)

Note: Percent mortality was recorded after a recovery period of ≥ 1 h.

[a] One veliger exhibited movement during the first hour in fresh water but was dead by the second hour.

efficacy for each chemical; and (2) time series (1–6 h) tests were conducted to determine potential effective treatment times. After treatment exposure, veligers were transferred from individual wells to fresh water on an hourly basis (e.g., well 1 after 1 h exposure, well 2 after 2 h exposure, etc.) up to a 6 h time point (Table 32.2). Recovery periods ranged from 1 to 6 h. Test plates were placed in a raceway water bath to maintain consistent water temperatures and oxygen levels.

The acute toxicity of Peraclean 15 and Spectrus CT1300 to juvenile *G. elegans* (approximate total length 30 mm) was evaluated using concentrations of 15 and 25 mg/L, respectively. Tests were conducted in 38 L aquaria without replication ($n = 10$ fish per aquarium) using well water (temperature 20°C, hardness 3325 mg/L $CaCO_3$, alkalinity 162 mg/L $CaCO_3$, pH 6.9) with aeration and were terminated when 100% mortality was observed.

Anesthetic Pretreatments

Veligers close their shells in a defense response to being exposed to toxicants. Because this response may interfere with the delivery of the tested toxicants, the anesthetic properties of additional chemicals that may prevent veliger shell closure prior to toxicant exposure (pretreatment) were evaluated. Chemical selection was restricted to those having been tested on both (1) bivalves—documenting an anesthetic response (see Table 32.A.1) and (2) fish (see Table 32.A.2). In total, six chemicals were chosen for testing (all chemicals purchased from Sigma-Aldrich, St. Louis, MO): benzocaine, clove oil, magnesium chloride ($MgCl_2$), menthol, propylene phenoxytol, and a clove oil/menthol mix (6 mL clove oil and 1 g menthol in 3 mL ethanol; Saydmohammed and Pal 2009). Multiple treatment and exposure times were performed with each chemical either alone or as a pretreatment followed by a treatment of 50 mg/L formalin (Table 32.3). Additional tests were conducted with 5 g/L $MgCl_2$ alone or combined with 50 or 75 mg/L formalin (Table 32.4). All tests were performed in six-well tissue culture plates according to established protocols.

Acute toxicity of menthol (1g/L), clove oil/menthol mix (400 µL/L), and propylene phenoxytol (3 mL/L) was evaluated on juvenile *G. cypha* (range of total length 50–70 mm) in 150 L aquaria without replication (due to a limited number of test animals, $n = 5$ fish per aquarium) using well water (previously described) with aeration. Tests were terminated when 100% mortality was observed. Two treatments of $MgCl_2$ (5 and 10 mg/L) were also evaluated on *G. cypha* following the aforementioned protocols.

In addition, the acute toxicity of $MgCl_2$ was evaluated on *G. elegans* (range of total length 51–58 mm) through exposure to two treatments: (1) 5 g/L $MgCl_2$ plus 50 mg/L formalin for 7 h and (2) 5 g/L $MgCl_2$ only for 10 h. *Xyrauchen texanus* range of total length 55–63 mm were exposed to treatment 2 only. The tests were conducted in 38 L aquaria with three replicates per treatment ($n = 10$ fish per aquarium) using 18°C well water (other water quality parameters previously described) and aeration.

TABLE 32.3

Dreissena bugensis Veliger Mortality Observed after Exposure to Pretreatment with a Candidate Anesthetic Followed by a Treatment of 50 mg/L Formalin, Two Replicates per Treatment

Chemical	Concentration	Pretreatment Time (h)	50 mg/L Formalin Treatment Time (h)	Number of Veligers per Test	Percent Mortality
Magnesium chloride	3 g/L	4	2	20	5
	3 g/L	6	2	23	65
	5 g/L	2	2	21	33
	5 g/L	4	2	20	50
	5 g/L	6	n/a	19	53
Menthol	0.75 g/L	6	n/a	23	74
	1 g/L	2	2.5	18	100
	1 g/L	3	n/a	20	100
	1 g/L	4	n/a	20	100
Menthol/clove oil mix	800 uL/L	2	n/a	30[a]	100
Propylene phenoxytol	3 mL/L	2	2	25	60
	3 mL/L	3	2	22	59
	3 mL/L	5	2	33[a]	79
	3 mL/L	6	n/a	34[a]	82
	4 mL/L	2	n/a	19	95
	4 mL/L	3	n/a	19	100
	4 mL/L	4	n/a	14	100
	5 mL/L	2	n/a	21	100
	5 mL/L	3	n/a	22	100
	5 mL/L	4	n/a	22	100

Note: Percent mortality was recorded after a minimum 24 h recovery period.
[a] Denotes three replicates.

TABLE 32.4

Dreissena bugensis Veliger Mortality Observed Following Toxicity Range-Finding Tests with 5 g/L $MgCl_2$ Combined with Formalin, Two Replicates per Treatment

$MgCl_2$ Concentration (g/L)	Formalin Concentration (mg/L)	Treatment Time (h)	Number of Veligers per Test	Percent Mortality
5	0	10	22	100[a]
5	50	6	22	91
5	75	6	26	92
5	50	7	22	100
5	75	7	24	96
5	50	8	25	100
5	75	8	23	96

Note: Percent mortality was recorded after a minimum 48 h recovery period.
[a] Denotes recovery period ended at 18 h due to time constraints.

Biocides: Chloramine-T, Dimilin® 2L, and Praziquantel

Biocides commonly used in aquaculture were also used in veliger toxicity tests following the methodology used for molluscicides. These four biocides included chloramine-T (20 mg/L), Dimilin 2L (1.25–5 mg/L based on 22% active ingredient), and three treatments of praziquantel (20 mg/L): praziquantel only or in combination with formalin at either 50 or 100 mg/L.

TABLE 32.5

Dreissena bugensis Veliger Mortality Observed after Exposure to a 2000 mg/L Potassium-Based Compound for 4 h Followed by 338 mg/L Formalin (125 mg/L Active Ingredient) for 1 h, No Treatment Replication

Chemical	Number of Veligers per Test	Percent Unresponsive after Treatment	Percent Mortality after		
			24 h Recovery	72 h Recovery	96 h Recovery
Control	23	0	0	0	0
KH_2PO_4	19	100	0	68	74
$K_4P_2O_7$	17	100	n/a	n/a	n/a
Langbeinite	22	81	14	53	76
Potash	16	81	81	81	81
$KCl + MgCl_2$	35	77	43	43	79

Notes: The KCl + $MgCl_2$ mix was 2000 mg/L of each chemical. Percent mortality was recorded immediately following the treatment and after 24, 72, and 96 h recovery periods.
n/a, veligers too deteriorated to transfer to a recovery plate.

Potassium-Based Compounds

Potassium chloride (KCl), as stated, is a presumed anesthetic; however, other potassium compounds are reported to have similar effects. Using formalin as the molluscicide, five potassium compounds were used in toxicity tests on *D. bugensis* veligers:

- Potassium phosphate monobasic (KH_2PO_4; Sigma-Aldrich, St. Louis, MO)
- Potassium pyrophosphate ($K_4P_2O_7$; Sigma-Aldrich, St. Louis, MO)
- Langbeinite (a potassium magnesium sulfate compound; Intrepid Potash-New Mexico, LLC, Carlsbad, NM)
- Potash (KCl as 62% K_2O; Intrepid Potash-New Mexico, LLC, Carlsbad, NM), and
- KCl + $MgCl_2$ mix (KCl, Fisher Scientific, Pittsburg, PA; $MgCl_2$, Sigma-Aldrich, St. Louis, MO)

Using the established protocol, veligers were exposed to one of the five potassium compounds for 4 h at a concentration of 2000 mg/L (for the KCl+MgCl2 mix, it was 2000 mg/L of each component) followed by 338 mg/L formalin (125 mg/L active ingredient) for 1 h. At the end of the 5 h treatment, veligers were transferred to fresh water and mortality was assessed at 24 h posttreatment (Table 32.5). Based on the preliminary results, additional tests were conducted with the $K_4P_2O_7$ and formalin (50 mg/L) to refine the range of concentrations and exposure times (Table 32.6).

A test was conducted to evaluate the acute toxicity of $K_4P_2O_7$ to *G. elegans* (range of total length 42–70 mm) in duplicate ($n = 10$ fish per aquarium) using 18.5°C well water as previously described. The fish were exposed to a concentration of 1050 mg/L $K_4P_2O_7$ for 4 h followed by 25 mg/L formalin for an additional 2 h and were then moved to fresh water and held for a 5 d observation period.

Results and Discussion

Bivalves have sensitive chemoreceptors that allow them to detect and respond to chemicals in the water by closing their shells, thereby avoiding exposure to potentially toxic substances (Kennedy et al. 2006; Sprecher and Getsinger 2000). Potassium chloride has been reported as an intoxicant (anesthetic) to mussels and has been used as a pretreatment to prevent shell closure when a molluscicide is subsequently added (Edwards et al. 2000; Fisher et al. 1994; O'Donnell et al. 1996; Waller et al. 1996; Wildridge et al. 1998). Edwards et al. (2000, 2002) developed a treatment protocol with KCl and formalin for use in hatchery settings to kill *D. polymorpha* veligers. The treatment (750 mg/L KCl for 1 h followed by 25 mg/L formalin for 2 h), used by many fish culture facilities in the United States, had not been tested on *D. bugensis* veligers. The purpose of the first toxicity test was to evaluate the efficacy of the

TABLE 32.6

Dreissena bugensis Veliger Mortality Observed after Toxicity Range-Finding Tests with $K_4P_2O_7$ Followed by a 50 mg/L Formalin Treatment (Unless Noted by n/a), Two Replicates per Treatment

K4P2O7 Treatment		50 mg/L Formalin Treatment	Number of	Percent Mortality after		
mg/L	Time (h)	Time (h)	Veligers per Test	24 h Recovery	48 h Recovery	72 h Recovery
750	2.5	3	24	75	83	83
750	3	2	20	80	70	75
750	4	2	33[a]	82[b]	—	—
750	6	2	36[a]	97[b]	—	—
1000	2.5	3	23	100	100	100
1000	3	2	22	100	100	100
1000	4	1	20	95[b]	—	—
1000	4	2	≥30[a]	100[b]	—	—
1500	2.5	n/a	21	52	81	81
1500	3	n/a	21	100	100	100

Note: Percent mortality was recorded after 24, 48, and 72 h recovery periods.
[a] Denotes three treatment replicates.
[b] Denotes mortality recorded at 36 h recovery.

KCl and formalin treatment on *D. bugensis* veligers under the fish culture conditions at WBNFH. The results from that test revealed that the treatment is ineffective at WBNFH; no physiological effect was observed in the *D. bugensis* veligers. To determine if the lack of treatment effect was due to a component of WBNFH water or the quality of the test chemicals, additional tests were conducted with the same protocol using well water from Southwestern Native ARRC and new lots of KCl and formalin, but the same results were observed.

The additional tests conducted with higher concentrations of KCl ranging from 1500 to 3500 mg/L and formalin concentrations of 25, 50, or 100 mg/L resulted in varying levels of estimated mortality (recovery period not included) at the end of the treatments (Table 32.1). Within that range of chemical concentrations, the physical condition of treated veligers was variable and two notable observations were made: (1) there was a contrast in physical condition of veligers recorded as dead based on published protocols (lack of internal movement or response to stimuli), and (2) there appeared to be less physical distress in the higher KCl concentration treatments. In terms of physical condition, shells of some veligers were gapping with their velum exposed (Figure 32.2a), while other veligers were unresponsive, but their shells remained closed (Figure 32.2b). To address the first observation and further assess their condition, treated veligers were transferred to fresh water and all were observed to exhibit some level of recovery within 2 h. From those results, it became apparent that a recovery period is critical for toxicity studies using *D. bugensis* veligers and, most likely, *D. polymorpha* veligers. Wildridge et al. (1998) reported that a recovery period is critical for adult *D. polymorpha* after observations of a fivefold lower estimate of lethality following a 96 h recovery period, suggesting that mortality could not be accurately assessed by lack of movement or response to tactile stimulation only. The second observation suggests that there is an inverse relationship between KCl concentration and veliger distress, that is, as the concentration increased, physical distress decreased. Pucherelli et al. (2010) observed the same effect, suggesting that higher KCl concentrations may stimulate veliger chemoreceptors, triggering a response to close their shells rather than acting as an anesthetic.

Three subsequent tests of 4250 mg/L KCl for 1 h followed by 25, 50, or 100 mg/L formalin for 2 h were conducted with the inclusion of a recovery period after the treatments. Estimated mortalities in the 25, 50, and 100 mg/L formalin treatments were 39%, 93%, and 50%, respectively, immediately following the exposure period, but all veligers showed signs of movement when transferred to fresh water.

Edwards et al. (2000, 2002) reported 100% morality of *D. polymorpha* veligers exposed to 750 mg/L KCl for 1 h followed by 25 mg/L formalin for 2 h. The difference between their results and the current study's findings may suggest a difference in the level of sensitivity between *D. polymorpha* and *D. bugensis*. This hypothesis needs to be tested though as they did not include recovery periods citing the

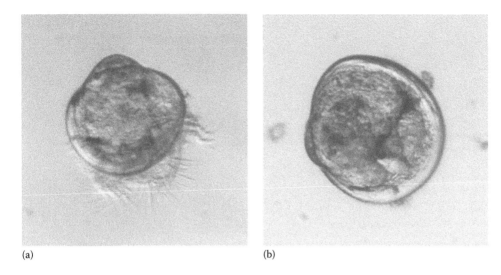

(a)   (b)

FIGURE 32.2 (a) *D. bugensis* veliger recorded as dead with gaping velum. (b) Veliger recorded as dead with no observable motion but shell tightly closed.

inability to transfer *D. polymorpha* veligers to fresh water without damaging them. Fisher et al. (1994) reported that 3-day-old *D. polymorpha* preveligers were tolerant of handling but D-stage and post-D veligers experienced high mortality during handling. This may be another characteristic that differs between *D. polymorpha* and *D. bugensis* as all veliger stages used in this study were resilient to handling.

Given the inefficacy of the KCl and formalin treatments on *D. bugensis* veligers, a search was undertaken to explore alternative molluscicide treatments. There are a multitude of chemicals listed as effective molluscicides; however, most are (1) used over extended periods of time, (2) highly toxic to fish (Sprecher and Getsinger 2000), (3) not regulated for use in the United States, (4) tested predominantly with *D. polymorpha* mussels, or (5) have not been tested under the water conditions of the southwestern United States. From the Zebra Mussel Control Guide (Sprecher and Getsinger 2000), three commercially available compounds of copper (Cutrine-Ultra), peracetic acid (Peraclean 15), and quaternary ammonium chloride (Spectrum CT1300) were chosen for testing.

Cutrine-Ultra was selected because of the potential increase in toxicity to veligers from an added emulsifier, surfactant, and solvent (ethanolamine). It also contains limonene, a reported molluscicide (Kumar and Singh 2006). Copper toxicity is dependent on water hardness, but the monoethanolamine and triethanolamine in Cutrine-Ultra aid in preventing the precipitation of copper in hard water, allowing the copper to remain active as a biological control. Kennedy et al. (2006) reported a 24 h LC_{50} of 0.012 mg Cu/L for preveliger stage *D. polymorpha* larvae. Preliminary trials were conducted, exposing *D. bugensis* for 1–6 h to Cutrine-Ultra concentrations, ranging from 0.5 to 6.25 mg Cu/L. The highest mortality rate observed was 50% (data not shown). Cutrine-Ultra concentrations were increased to 15 and 20 mg Cu/L and at the end of the 6 h treatment, 80% mortality was recorded in the 15 mg/L and 84% in the 20 mg/L after a recovery period (Table 32.2). Hamilton and Buhl (1997) reported a 24 h LC_{50} of 0.305 mg Cu/L for larval *P. lucius* and 0.408 mg/L for larval *X. texanus*. Based on the sensitivity of those fish species to copper compared to the high concentration required to produce veliger mortality, copper does not appear to be a time-limited treatment option under the hard water conditions of southwestern United States.

The second chemical tested, Peraclean 15 (15% peracetic acid, 21% hydrogen peroxide), is a disinfectant for controlling microorganisms in sewage treatment plants. It easily hydrolyzes into acetic acid and hydrogen peroxide, which are biodegradable by-products (Christiani 2005). Peracetic acid also has undergone extensive testing and has been found effective as a treatment for controlling aquatic species introduced through ship ballast waters in both salt water and fresh water (de Lafontaine et al. 2008; Fuchs and de Wilde 2004). Verween et al. (2009) tested the toxicity of peracetic acid on preveliger *D. polymorpha* larvae and reported 95% mortality with a 15 min exposure to 3 mg/L. Initial tests on *D. bugensis* veligers using concentrations of 1.25–10 mg/L peracetic acid for 7 h produced a maximum

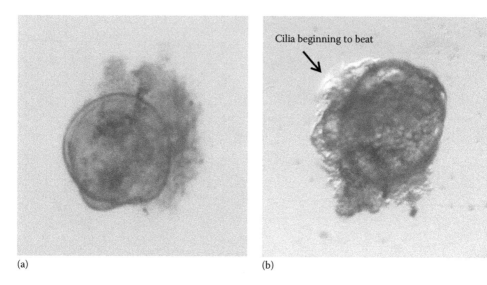

(a)   (b)

FIGURE 32.3   (a) Disintegrated *D. bugensis* veliger after exposure to peracetic acid. (b) Veliger exhibiting signs of recovery in fresh water after exposure to peracetic acid.

mortality rate of 70% (data not shown). Many of the veligers were disintegrated from the peracetic acid exposure by the end of the treatment (Figure 32.3a), but during the recovery period, they began showing signs of cilia movement (Figure 32.3b) and erratic swimming. This further supports the importance of not relying on physical characteristics to determine mortality levels and including recovery periods with all toxicity work involving *D. bugensis* veligers. It also reveals the resilient nature of *D. bugensis*. Time series tests were then conducted with 35 and 50 mg/L peracetic acid to determine the time to achieve 100% mortality. Following a recovery period, 100% mortality was observed after 2 h of exposure at both 35 mg/L and 2 h of exposure to 50 mg/L, although one veliger from the 3 h 35 mg/L treatment did exhibit brief signs of movement during the recovery period.

The third chemical tested was Spectrus CT1300, a quaternary ammonium chloride compound (QUAT) approved by the Environmental Protection Agency (EPA) as a mollusk control agent. Clamtrol CT-1, which contains 8% alkyl dimethylbenzyl ammonium chloride compared to 50% in Spectrus CT1300, has been used in toxicity tests with *D. polymorpha* (Fisher et al. 1994; Waller et al. 1993). Fisher et al. (1994) reported a 24 h LC_{50} value of 0.179 mg/L QUAT for a composite of umbonal and pediveliger stage veligers and 8.8 mg/L for plantigrades (early settled juvenile stage), and greater than 13 mg/L for adults (>5 mm). Waller et al. (1993) found that 5–8 mm *D. polymorpha* had a 48 h LC_{50} of 0.290 mg/L and 0.738 for 20–25 mm mussels. An initial dose of 10 mg/L QUAT was chosen in the current study for a preliminary test, but little adverse response was observed in *D. bugensis* veligers. Subsequent time series tests were conducted using 30 and 37.5 mg/L QUAT with observed mortality rates of 90% and 91%, respectively, at 6 h (Table 32.2).

Given the relatively high mortalities, the toxicity of Peraclean 15 and Spectrus CT1300 to juvenile *G. elegans* was tested to determine if further investigation was warranted. Fish exposed to 15 mg/L peracetic acid died within 30 min and those exposed to 25 mg/L QUAT died within 5 min. These results clearly demonstrate that *D. bugensis* veligers are more resistant to the toxic effects of these chemicals than native fish and should not be considered effective molluscicides under the specific test conditions presented in this study.

The focus of this study was then shifted to evaluating potential pretreatments to prevent veligers from closing their shells when exposed to a molluscicide (Table 32.A.1). Based on the observed ability of *D. bugensis* veligers to withstand relatively high concentrations of selected chemicals for lengthy periods of time, most likely due to shell closure, it was determined that a pretreatment to relax *D. bugensis* veligers similar to the reported effect of KCl on *D. polymorpha* would be beneficial. Reducing the veliger's ability to close its shell in response to a chemical may allow application of lower doses of molluscicides, reducing the toxicity to fish. A literature search was conducted focusing on responses of shell gaping, sedation, and intoxication in bivalves and crustacean exposed to reported anesthetics.

Twenty-six chemicals were identified that had been used on aquatic invertebrates (Table 32.A.1) and tested on fish (Table 32.A.2), but most were known to be more toxic to fish than bivalves or were not regulated for use in the United States. Six compounds were chosen from the list for testing (Table 32.3).

Multiple toxicity tests with the six anesthetics were run simultaneously using a range of exposure times and concentrations with 50 mg/L formalin as the molluscicide. Table 32.3 lists only the concentrations and times that resulted in at least a partial response in veligers. In pretreatments where complete mortality appeared to have been reached within the specified time frame, veligers were transferred to fresh water without a formalin treatment. Four of the six chemicals (menthol, clove oil/menthol mix, $MgCl_2$, and propylene phenoxytol) produced 100% mortality, but only at relatively high concentrations (Table 32.3). Benzocaine had no effect, and the effect of clove oil was minimal. Within the range of $MgCl_2$ treatments, a high degree of physical deterioration was observed among the surviving veligers.

The toxicity tests conducted to evaluate the effect of menthol, the clove oil/menthol mix, and propylene phenoxytol on juvenile *G. cypha* demonstrated that the chemicals were highly toxic, killing the fish in less than 30 min in the concentrations determined to kill *D. bugensis* veligers. Magnesium chloride was observed to be less toxic. *Gila cypha* were able to tolerate 5 g/L $MgCl_2$ for 72 h without visible signs of stress. In the 10 g/L $MgCl_2$ tests, fish were visibly stressed within 4 h, lost equilibrium by 7 h, and first mortality occurred at 8 h.

From those results, additional tests were designed to determine the treatment time required to achieve 100% veliger mortality using 5 g/L $MgCl_2$ with (as combined treatment) and without the addition of formalin. The lowest effective treatments were a 7 h treatment with 5g/L $MgCl_2$ and 50 mg/L formalin and a 10 h treatment with 5g/L $MgCl_2$ and no formalin (Table 32.4).

Toxicity tests were conducted to evaluate the effect of the 7 h 5g/L $MgCl_2$ and 50 mg/L formalin treatment on juvenile *G. elegans* and *X. texanus*. *Gila elegans* exhibited the lowest tolerance, with 10% mortality ($n = 30$) observed at the end of the treatment and a total of 63% at the end of a 10-day posttreatment period. In *X. texanus*, no mortality was observed during the treatment, but 50% ($n = 30$) mortality was recorded by 8-day posttreatment. The effect of the 10 h 5g/L $MgCl_2$ only treatment was also evaluated on *G. elegans*, resulting in 3% mortality during treatment and 43% by 7-day posttreatment. Based on the results from the *G. elegans* and *X. texanus* toxicity tests, $MgCl_2$ was not considered for further testing.

In the toxicity tests on *D. bugensis* veligers conducted with the biocides chloramine-T, Dimilin 2L, and praziquantel, mortality was observed only in the chloramine-T tests (80% mortality within 2.5 h). The surviving veligers exhibited a high degree of deterioration, which increased through the 24 h recovery period. Currently, chloramine-T is regulated by the U.S. Fish and Wildlife Service (USFWS) Aquatic Animal Drug Approval Partnership (AADAP) Program under the chloramine-T Investigational New Animal Drug (INAD) 9321 exemption. Treatment dosage is limited to a maximum of 20 mg/L for 60 min on three consecutive days. Increased dosages and exposure periods have been tested on multiple fish species with varied, but promising survival rates (Altinok 2004; Gaikowski et al. 2008, 2009). However, further testing on mussel veligers and fish is necessary to determine the efficacy of chloramine-T as a molluscicide as well as the acute and chronic effects on fish.

In the continuing search for an effective molluscicide to be used on *D. bugensis* veligers in fish transport tanks, attention was then given to the other potassium-based compounds (in addition to KCl) that have been reported to work as molluscicides (Fisher et al. 1991; Waller et al. 1993; O'Donnell et al. 1996; VDGIF 2005; Wildridge et al. 1998). Based on published results, new toxicity experiments were developed to evaluate the effect of five potassium-based molluscicides (monopotassium phosphate, potassium pyrophosphate, langbeinite, potash, and a KCl/$MgCl_2$ mix) on *D. bugensis* veligers. The initial toxicity test conducted with the five potassium compounds provided preliminary data for the assessment of each chemical's potential effect on veligers (Table 32.5). The only treatment with 100% veliger mortality was the 4 h potassium pyrophosphate ($K_4P_2O_7$) and formalin treatment. A recovery period for the $K_4P_2O_7$ treatment was not included because the deteriorated condition of the veligers made it difficult to pipette them to recovery wells. Based on those results, only $K_4P_2O_7$ was chosen for further testing. To determine the lowest concentration and treatment time that would produce 100% veliger mortality, two series of experiments using various combinations of $K_4P_2O_7$, formalin concentrations, and exposure times were conducted (series one, Table 32.6; series two, Table 32.7). The lowest effective treatment combination was a 3 h exposure to 1000 mg/L $K_4P_2O_7$ followed by a 2 h exposure to 25 mg/L formalin.

TABLE 32.7

Dreissena bugensis Veliger Mortality Observed after Toxicity Range-Finding Tests with 1000 mg/L $K_4P_2O_7$ Followed by a Formalin Treatment, Two Replicates per Treatment

$K_4P_2O_7$ (1000 mg/L) Treatment Time (h)	Formalin		Total Treatment Time (h)	Number of Veligers per Test	Percent Mortality
	mg/L	Treatment Time (h)			
3	0	0	3	24	63
3	25	2	5	19	100
3	50	1	4	19	74
4	0	0	4	28	75
4	25	2	6	21	100
4	50	1	5	22	91
5	0	0	5	20	100
5	25	2	7	21	100
5	50	1	6	21	100

Note: Percent mortality was recorded after a 24 h recovery period.

An interesting note from the 3 and 4 h $K_4P_2O_7$ treatments was that 100% mortality was observed in the 2 h 25 mg/L formalin treatment but not in the 1 h 50 mg/L formalin treatment (Table 32.7). These results suggest that exposure time is a more important factor with formalin than concentration.

To evaluate the effect of $K_4P_2O_7$ on native fish, one test was conducted on juvenile *G. elegans* with a treatment of 1000 mg/L $K_4P_2O_7$ for 4 h followed by 25 mg/L formalin for 2 h. When the chemical was mixed with well water (hardness 3325 mg/L $CaCO_3$) and dechlorinated domestic water (hardness 618 mg/L $CaCO_3$), a strong precipitate was formed. In contrast, the amount of precipitate that formed in the river water at WBNFH (hardness 266 mg/L $CaCO_3$) during the veliger tests was minimal. There was no visible stress apparent in the fish during the treatment. All fish were moved to fresh water after the treatment and monitored for 5 days, during which no mortality was observed. Because of the presence of the precipitate, it is unknown whether $K_4P_2O_7$ is nontoxic to fish at that dose in all levels of water hardness or if the toxicity was reduced because components of the compound were bound in the precipitate and no longer available in solution. In addition, the lack of an acute toxic effect on *G. elegans* does not provide insight into possible chronic impacts the treatment may have on the fish. Further evaluation of the toxicity of $K_4P_2O_7$ may be warranted to determine its effectiveness under a range of water conditions.

Summary

The presence of invasive mussel species impacts all aspects of aquaculture from sport fish stockings to endangered species recovery programs. Finding a chemical that cannot be detected by *D. bugensis* veligers, is lethal to veligers in a relatively short time frame, and can be tolerated by fish species has proven to be a monumental challenge. Results from this research show that *D. bugensis* veligers are for more tolerant of many published molluscicides than native fish species. One area of future research that needs to be considered is monitoring the survivability and viability of veligers after exposure to a chemical treatment. Although many natural resource managers have expressed reluctance to accept a treatment protocol that did not achieve 100% veliger mortality at the conclusion of the treatment, agencies working with sensitive fish species such as *G. elegans* may need to consider the long-term impact of continued restrictions on the movement of endangered species because of *D. bugensis*. Although 100% mortality was not observed immediately following exposure to lower concentrations of $MgCl_2$, $K_4P_2O_7$, and chloramine-T, the deteriorated state of the veligers exposed to each of those chemicals may warrant further investigation into long-term viability of veligers and the impact the chemicals may have on their ability to develop into settling juveniles. In conjunction with continued veliger testing, evaluating the chronic effects after exposure to these chemicals on multiple life stages of fish species is imperative. One last aspect that should be considered when evaluating the toxicity of a chemical to *D. polymorpha* and *D. bugensis* veligers is the efficacy of the treatment under various water quality parameters.

32.A Appendix

TABLE 32.A.1

Results from Literature Search for Chemicals Tested as Anesthetics on Molluscs, Crustaceans, and Other Invertebrates

Chemical	Species	References
2-phenoxyethanol	Abalone	Bilbao et al. (2010); White et al. (1996)
	Oysters	Mamangkey et al. (2009); Norton et al. (1996)
	Prawns	Coyle et al. (2005)
	Queen conch	Acosta-Salmon and Davis (2007)
	Sea urchins	Hagen (2003)
Aqui-S	Prawns	Coyle et al. (2005)
Aspirin	Scallops	Heasman et al. (1995)
Benzocaine	Abalone	Aquilina and Roberts (2000); Hooper et al. (2011)
	Oysters	Acosta-Salmon et al. (2005); Mamangkey et al. (2009); Norton et al. (1996); Suquet et al. (2009)
	Queen conch	Acosta-Salmon and Davis (2007)
	Scallops	Heasman et al. (1995)
	Sea snail	Noble et al. (2009)
	Sea urchins	Hagen (2003)
Chloral hydrate	Oysters	Culloty and Mulcahy (1992); Norton et al. (1996)
	Scallops	Heasman et al. (1995)
Clove oil/eugenol	Abalone	Bilbao et al. (2010)
	Oysters	Mamangkey et al. (2009); Norton et al. (1996); Suquet et al. (2009)
	Prawns	Coyle et al. (2005); Saydmohammed and Pal (2009)
EDTA	Abalone	White et al. (1996)
Ethanol	Scallops	Heasman et al. (1995)
	Sea snail	Noble et al. (2009)
Isobutanol	Scallops	Heasman et al. (1995)
Ketamine	Clam	Jamieson and Lander (1984)
	Snails	Martins-Sousa et al. (2001)
Magnesium chloride	Cephalopods	Messenger et al. (1985)
	Oysters	Culloty and Mulcahy (1992); Norton et al. (1996); Suquet et al. (2009)
	Queen conch	Acosta-Salmon and Davis (2007)
	Rock oyster	Butt et al. (2008)
	Scallops	Heasman et al. (1995)
	Sea snail	Noble et al. (2009)
	Sea urchin	Arafa et al. (2007); Hagen (2003)
Magnesium sulfate	Abalone	White et al. (1996)
	Oysters	Culloty and Mulcahy (1992)
	Scallops	Heasman et al. (1995)
Menthol	Oysters	Mamangkey et al. (2009); Norton et al. (1996)
	Prawns	Saydmohammed and Pal (2009)
	Queen conch	Acosta-Salmon and Davis (2007)
Metomidate	Scallops	Heasman et al. (1995)
MS222	Abalone	Aquilina and Roberts (2000)
	Oysters	Norton et al. (1996)
	Queen conch	Acosta-Salmon and Davis (2007)
	Scallops	Heasman et al. (1995)
Phenoxy ethanol	Scallops	Heasman et al. (1995)

(Continued)

TABLE 32.A.1 (*Continued*)

Results from Literature Search for Chemicals Tested as Anesthetics on Molluscs, Crustaceans, and Other Invertebrates

Chemical	Species	References
Procaine hydrochloride	Abalone	White et al. (1996)
Propylene phenoxetol	Abalone	Aquilina and Roberts (2000)
	Benthic invertebrates	McKay and Hartzband (1970)
	Oysters	Acosta-Salmon et al. (2005); Mamangkey et al. (2009)
Quinaldine	Prawns	Coyle et al. (2005)
	Scallops	Heasman et al. (1995)
Serotonin (5-HT)	Zebra mussel adults	Fong (1998); Kennedy et al. (2006); Ram et al. (1991)
Sodium bicarbonate	Oysters	Norton et al. (1996)
Sodium pentobarbital	Abalone	Aquilina and Roberts (2000); Sharma et al. (2003)
	Oysters	Culloty and Mulcahy (1992); Norton et al. (1996)
	Scallops	Heasman et al. (1995)
	Sea snail	Noble et al. (2009)
	Snails	Martins-Sousa et al. (2001)
Tertiary amyl alcohol	Scallops	Heasman et al. (1995)
Valium	Scallops	Heasman et al. (1995)

TABLE 32.A.2

Results from Literature Search for Chemicals Tested as Anesthetics on Fish Species

Chemical	Species	References
2-phenoxyethanol	Carp	Dziaman et al. (2010)
	Cod	Zahl et al. (2009)
	Dusky kob	Bernatzeder et al. (2008)
	Perch	Velisek et al. (2009)
	Rainbow trout	Gilderhus and Marking (1987)
	Sea bass	Basaran et al. (2007); King et al. (2005); Mylonas et al. (2005)
	Sea bream	Mylonas et al. (2005); Tsantilas et al. (2006)
	Sole	Weber et al. (2009)
	Trout	Ucar and Atamanalp (2010); Valisek et al. (2011)
Aqui-S	Atlantic salmon	Iversen et al. (2003)
	Channel catfish	Bosworth (2007); Small (2004); Small and Chatakondi (2005)
	Striped bass	Davis and Griffin (2004); Woods III et al. (2008)
	Western rainbowfish	Young (2009)
Benzocaine	Atlantic salmon	Kiessling et al. (2009)
	Carp	Hasan and Bart (2007); Heo and Shin (2010)
	Catfish	Hayton et al. (1996)
	Cod	Zahl et al. (2009)
	Codling	Bolasina (2006)
	Halibut	Zahl et al. (2010)
	Rainbow trout	Cotter and Rodnick (2006); Gilderhus and Marking (1987); Stehly et al. (1998)
	Salmon	Iversen et al. (2003)
Chloral hydrate	Mullet fry	Durve (1975)
	Rohu	Sharma and Sharma (1996)
	Tilapia	Lanzing (1971)
Clove oil/isoeugenol	Atlantic salmon	Iversen et al. (2003); Kiessling et al. (2009)
	Caspian salmon	Ghazilou et al. (2010)
	Halibut	Zahl et al. (2010)
	Largemouth bass	Cooke et al. (2004)
	Perch	Velisek et al. (2009)
	Rainbow trout	Cotter and Rodnick (2006); Sattari et al. (2009)
	Reef fishes	Cunha and Rosa (2006)
	Rock bream	Park et al. (2009)
	Rockpool fishes	Griffiths (2000)
	Sea bass	King et al. (2005); Mylonas et al. (2005)
	Sea bream	Mylonas et al. (2005)
	Sole	Weber et al. (2009)
	Steelhead fry	Woolsey et al. (2004)
	Striped bass	Davis and Griffin (2004); Sink and Neal (2009)
	Trout	Ucar and Atamanalp (2010); Velisek et al. (2011)
Dimilin	Freshwater organisms	Fischer and Hall (1992)
	Mosquitofish	Draredja-Beldi and Soltani (2003)
Etomidate	Carp	Dziaman et al. (2010)
	Rainbow trout	Gilderhus and Marking (1987)
Ketamine	Carp	Al-Hamdani et al. (2010)
Magnesium chloride	Daphnia	Dowden and Bennett (1965)
	Gambusia	Wallen et al. (1957)

(Continued)

TABLE 32.A.2 (*Continued*)

Results from Literature Search for Chemicals Tested as Anesthetics on Fish Species

Chemical	Species	References
Menthol	Guppy	Pickering et al. (1983)
	Tambaqui	Facanha and Gomes (2005)
	Tilapia	Simoes and Gomes (2009)
Metomidate	Atlantic salmon	Iversen et al. (2003)
	Cod	Zahl et al. (2009)
	Halibut	Zahl et al. (2010)
	Koi	Crosby et al. (2010)
	Ornamental fish	Kilgore et al. (2009)
	Rainbow trout	Gilderhus and Marking (1987)
	Sea bass	King et al. (2005)
	Sole	Weber et al. (2009)
	Striped bass	Davis and Griffin (2004)
Pentobarbital	Goldfish	Greizerstein (1979)
Praziquantel	Carp/shiners	Mitchell and Hobbs (2007)
Propylene phenoxetol	Fish	Bagenal (1963)
	Sockeye salmon	Sehdev et al. (1963)
	Tilapia	Lanzing (1971)
Quinaldine	Carp, rohu	Hasan and Bart (2007)
	Striped bass	Davis and Griffin (2004)
	Tilapia	Lanzing (1971)
Quinaldine sulfate	Rainbow trout	Gilderhus and Marking (1987)
	Striped bass	Davis and Griffin (2004)

REFERENCES

Acosta-Salmon, H. and M. Davis. 2007. Inducing relaxation in the queen conch *Strombus gigas* (L.) for cultured pearl production. *Aquaculture* 262:73–77.

Acosta-Salmon, H., E. Martinez-Fernandez, and P. C. Southgate. 2005. Use of relaxants to obtain saibo tissue from the blacklip pearl oyster (*Pinctada margaritifera*) and the Akoya pearl oyster (*Pinctada fucata*). *Aquaculture* 246:167–172.

Al-Hamdani, A. H., S. K. Ebrahim, and F. K. Mohammad. 2010. Experimental xylazine-ketamine anesthesia in the common Carp (*Cyprinus carpio*). *Journal of Wildlife Diseases* 46:596–598.

Altinok, Ilhan. 2004. Toxicity and therapeutic effects of chloramine-T for treating *Flavobacterium columnare* infection of goldfish. *Aquaculture* 239:47–56.

Aquilina, B. and R. Roberts. 2000. A method for inducing muscle relaxation in the abalone, *Haliotis iris*. *Aquaculture* 190:403–408.

Arafa, S., S. Sadok, and A. El Abed. 2007. Assessment of magnesium chloride as an anaesthetic for adult sea urchins (*Paracentrotus lividus*): Incidence on mortality and spawning. *Aquaculture Research* 38:1673–1678.

Bagenal, T. B. 1963. Propylene phenoxetol as a fish anaesthetic. *Nature* 197:1222–1223.

Basaran, F., H. Sen, and S. Karabulut. 2007. Effects of 2-phenoxyethanol on survival of normal juveniles and malformed juveniles having lordosis or nonfunctional swimbladders of European sea bass (*Dicentrarchus labrax* L., 1758). *Aquaculture Research* 38:933–939.

Bernatzeder, A. K., P. D. Cowley, and T. Hecht. 2008. Effect of short term exposure to the anaesthetic 2-phenoxyethanol on plasma osmolality of juvenile dusky kob, *Argyrosomus japonicus* (Sciaenidae). *Journal of Applied Ichthyology* 24:303–305.

Bilbao, A., G. Courtois de Vicose, M. del Pino Viera, B. Sosa, H. Fernandez-Palacios, and M. del Carmen Hernandez. 2010. Efficiency of clove oil as anesthetic for abalone (*Haliotis tuberculata coccinea*, Revee). *Journal of Shellfish Research* 29:679–682.

Bolasina, S. N. 2006. Cortisol and hematological response in Brazilian codling, *Urophycis brasiliensis* (Pisces, Phycidae) subjected to anesthetic treatment. *Aquaculture International* 14:569–575.

Bosworth, B. G., B. C. Small, D. Gregory, J. Kim, S. Black, and A. Jerrett. 2007. Effects of rested-harvesting using the anesthetic AQUI-S™ on channel catfish, *Ictalurus punctatus*, physiology and fillet quality. *Aquaculture* 262:302–318.

Butt, D., S. J. O'Connor, R. Kuchel, W. A. O'Connor, and D. A. Raftos. 2008. Effects of the muscle relaxant, magnesium chloride, on the Sydney rock oyster (*Saccostrea glomerata*). *Aquaculture* 275:342–346.

Christiani, P. 2005. Solutions to fouling in power station condensers. *Applied Thermal Engineering* 25:2630–2640.

Cooke, S. J., C. D. Suski, K. G. Ostrand, B. L. Tufts, and D. H. Wahl. 2004. Behavioral and physiological assessment of low concentrations of clove oil anaesthetic for handling and transporting largemouth bass (*Micropterus salmoides*). *Aquaculture* 239:509–529.

Cotter, P. A. and K. J. Rodnick. 2006. Differential effects of anesthetics on electrical properties of the rainbow trout (*Oncorhynchus mykiss*) heart. *Comparative Biochemistry and Physiology A—Molecular & Integrative Physiology* 145:158–165.

Coyle, S. D., S. Dasgupta, J. H. Tidwell, T. Beavers, L. A. Bright, and D. K. Yasharian. 2005. Comparative efficacy of anesthetics for the freshwater prawn *Macrobrachium rosenbergii*. *Journal of the World Aquaculture Society* 36:282–290.

Crosby, T. C., B. D. Petty, H. J. Hamlin, L. J. Guillette Jr., J. E. Hill, K. H. Hartman, and R. P. E. Yanong. 2010. Plasma cortisol, blood glucose, and marketability of koi transported with metomidate hydrochloride. *North American Journal of Aquaculture* 72:141–149.

Culloty, S. C. and M. F. Mulcahy. 1992. An evaluation of anaesthetics for *Ostrea edulis* (L.). *Aquaculture* 107:249–252.

Cunha, F. E. A. and I. L. Rosa. 2006. Anaesthetic effects of clove oil on seven species of tropical reef teleosts. *Journal of Fish Biology* 69:1504–1512.

Davis, K. B. and B. R. Griffin. 2004. Physiological responses of hybrid striped bass under sedation by several anesthetics. *Aquaculture* 233:531–548.

de Lafontaine, Y., S. Despatie, and C. Wiley. 2008. Effectiveness and potential toxicological impact of the PERACLEAN® Ocean ballast water treatment technology. *Ecotoxicology and Environmental Safety* 71:355–369.

Dowden, B. F. and H. J. Bennett. 1965. Toxicity of selected chemicals to certain animals. *Journal (Water Pollution Control Federation)* 37:1308–1316.

Dowling, T. E., W. L. Minckley, and P. C. Marsh. 1996. Mitochondrial DNA diversity within and among populations of razorback sucker (*Xyrauchen texanus*) as determined by restriction endonuclease analysis. *Copeia* 1996:542–550.

Draredja-Beldi, H. and N. Soltani. 2003. Laboratory evaluation of dimilin on growth and glutathione activity in mosquitofish, a non-target species. *Communications in Agricultural and Applied Biological Sciences* 68:299–305.

Durve, V. S. 1975. Anaesthetics in the transport of mullet seed. *Aquaculture* 5:53–63.

Dziaman, R., G. Hajek, and B. Klyszejko. 2010. Effect of 2-phenoxyethanol and etomidate on cardiac and respiratory functions and behaviour of common carp, *Cyprinus carpio* L. (Actinopterygii, Cypriniformes, Cyprinidae), during general anaesthesia. *Acta Ichthyologica Et Piscatoria* 40:37–43.

Edwards, W. J., L. Babcock-Jackson, and D. A. Culver. 2000. Prevention of the spread of zebra mussels during hatchery and aquaculture activities. *North American Journal of Aquaculture* 62:229–236.

Edwards, W. J., L. Babcock-Jackson, and D. A. Culver. 2002. Field testing of protocols to prevent the spread of zebra mussels *Dreissena polymorpha* during fish hatchery and aquaculture activities. *North American Journal of Aquaculture* 64:220–223.

Facanha, M. F. and L. C. Gomes. 2005. Efficacy of menthol as an anesthetic for tambaqui (*Colossoma macropomum*, Characiformes: Characidae). *Acta Amazonica* 35:71–75.

Fischer, S. A. and L. W. Hall. 1992. Environmental concentrations and aquatic toxicity data on diflubenzuron (Dimilin). *Critical Reviews in Toxicology* 22:45–79.

Fisher, S. W., H. Dabrowska, D. L. Waller, L. Babcock-Jackson, and X. Zhang. 1994. Sensitivity of zebra mussel (*Dreissena polymorpha*) life stages to candidate molluscicides. *Journal of Shellfish Research* 13:373–377.

Fisher, S. W., P. Stromberg, K. A. Bruner, and L. D. Boulet. 1991. Molluscicidal activity of potassium to the zebra mussel, *Dreissena polymorphia*: Toxicity and mode of action. *Aquatic Toxicology* 20:219–234.

Fong, P. P. 1998. Zebra mussel spawning is induced in low concentrations of putative serotonin reuptake inhibitors. *The Biological Bulletin* 194:143–149.

Fuchs, R. and I. de Wilde. 2004. PERACLEAN Ocean: A potentially environmentally friendly and effective treatment option for ballast water. In *Second International Symposium on Ballast Water Treatment*, ed. J. T. Matheickal and S. Raaymakers, pp. 175–180. London, U.K.: International Maritime Organization.

Gaikowski, M. P., C. L. Densmore, and V. S. Blazer. 2009. Histopathology of repeated, intermittent exposure of chloramine-T to walleye (*Sander vitreum*) and (*Ictalurus punctalus*) channel catfish. *Aquaculture* 287:28–34.

Gaikowski, M. P., W. J. Larson, and W. H. Gingerich. 2008. Survival of cool and warm freshwater fish following chloramine-T exposure. *Aquaculture* 275:20–25.

Ghazilou, A., H. S. Hasankandi, F. Chenary, A. Nateghi, N. Haghi, and M. R. Sahraeean. 2010. The anesthetic efficiency of clove oil in Caspian salmon, *Salmo trutta caspius* K., smolts in dosage-salinity-pH linked approach. *Journal of the World Aquaculture Society* 41:655–660.

Gilderhus, P. A. and L. L. Marking. 1987. Comparative efficacy of 16 anesthetic chemicals on rainbow trout. *North American Journal of Fisheries Management* 7:288–292.

Greizerstein, H. B. 1979. Development of functional tolerance to pentobarbital in goldfish. *Journal of Pharmacological and Experimental Therapeutics* 208:123–127.

Griffiths, S. P. 2000. The use of clove oil as an anaesthetic and method for sampling intertidal rockpool fishes. *Journal of Fish Biology* 57:1453–1464.

Hagen, N. T. 2003. KCl induced paralysis facilitates detachment of hatchery reared juvenile green sea urchins, *Strongylocentrotus droebachiensis*. *Aquaculture* 216:155–164.

Hamilton, S. J. and K. J. Buhl. 1997. Hazard assessment of inorganics, individually and in mixtures, to two endangered fish in the San Juan River, New Mexico. *Environmental Toxicology and Water Quality* 12:195–209.

Hasan, M. and A. N. Bart. 2007. Improved survival of rohu, *Labeo rohita* (Hamilton-Buchanan) and silver carp, *Hypophthalmichthys molitrix* (Valenciennes) fingerlings using low-dose quinaldine and benzocaine during transport. *Aquaculture Research* 38:50–58.

Hayton, W. L., A. Szoke, B. H. Kemmenoe, and A. M. Vick. 1996. Disposition of benzocaine in channel catfish. *Aquatic Toxicology* 36:99–113.

Heasman, M. P., W. A. O'Connor, and A. W. J. Frazer. 1995. Induction of anaesthesia in the commercial scallop, *Pecten fumatus* Reeve. *Aquaculture* 131:231–238.

Heo, G. J. and G. Shin. 2010. Efficacy of benzocaine as an anaesthetic for Crucian carp (*Carassius carassius*). *Veterinary Anaesthesia and Analgesia* 37:132–135.

Hooper, C., R. Day, R. Slocombe, K. Benkendorff, and J. Handlinger. 2011. Effect of movement stress on immune function in farmed Australian abalone (hybrid *Haliotis laevigata* and *Haliotis rubra*). *Aquaculture* 315:348–354.

Iversen, M., B. Finstad, R. S. McKinley, and R. A. Eliassen. 2003. The efficacy of metomidate, clove oil, Aqui-S™ and Benzoak® as anaesthetics in Atlantic salmon (*Salmo salar* L.) smolts, and their potential stress-reducing capacity. *Aquaculture* 221:549–566.

Jamieson, D. D. and J. Lander. 1984. Investigation of the mode of action of the anaesthetic agent, ketamine, in the heart of the bivalve mollusc *Tapes watlingi**. *Comparative Biochemistry and Physiology C-Molecular & Integrative Physiology* 77:109–114.

Kennedy, A. J., R. N. Millward, J. A. Steevens, J. W. Lynn, and K. D. Perry. 2006. Relative sensitivity of zebra mussel (*Dreissena polymorpha*) life-stages to two copper sources. *Journal of Great Lakes Research* 32:596–606.

Kiessling, A., D. Johansson, I. H. Zahl, and O. B. Samuelsen. 2009. Pharmacokinetics, plasma cortisol and effectiveness of benzocaine, MS-222 and isoeugenol measured in individual dorsal aorta-cannulated Atlantic salmon (*Salmo salar*) following bath administration. *Aquaculture* 286:301–308.

Kilgore, K. H. and J. E. Hill. 2009. Investigational use of metomidate hydrochloride as a shipping additive for two ornamental fishes. *Journal of Aquatic Animal Health* 21:133–139.

King, W. V., B. Hooper, S. Hillsgrove, C. Benton, and D. L. Berlinsky. 2005. The use of clove oil, metomidate, tricaine methanesulphonate and 2-phenoxyethanol for inducing anaesthesia and their effect on the cortisol stress response in black sea bass (*Centropristis striata* L.). *Aquaculture Research* 36:1442–1449.

Kumar, P. and D. K. Singh. 2006. Molluscicidal activity of *Ferula asafoetid*a, *Syzygium aromaticum* and *Carum carvi* and their active components against the snail *Lymnaea acuminate*. *Chemosphere* 63:1568–1574.

Lanzing, W. J. R. 1971. Effects of some anaesthetics on laboratory-reared *Tilapia mossambica* (Cichlidae). *Copeia* 1971:182–185.

Mamangkey, N. G. F., H. Acosta-Salmon, and P. C. Southgate. 2009. Use of anaesthetics with the silver-lip pearl oyster, *Pinctada maxima* (Jameson). *Aquaculture* 288:280–284.

Martins-Sousa, R. L., D. Negrao-Correa, F. S. M. Bezerra, and P. M. Z. Coelho. 2001. Anesthesia of *Biomphalaria* spp. (Mollusca, Gastropoda): Sodium pentobarbital is the drug of choice. *Memorias do Instituto Oswaldo Cruz* 96:391–392.

McKay, C. R. and D. J. Hartzband. 1970. Propylene phenoxetol: Narcotic agent for unsorted benthic inverte-brates. *Transactions of the American Microscopical Society* 89:53–54.

McMahon, R. F. 2011. Quagga mussel (*Dreissena rostriformis bugensis*) population structure during the early invasion of Lakes Mead and Mohave January-March 2007. *Aquatic Invasions* 6:131–140.

Messenger, J. B., M. Nixon, and K. P. Ryan. 1985. Magnesium chloride as an anesthetic for cephalopods. *Comparative Biochemistry and Physiology C-Molecular & Integrative Physiology* 82:203–205.

Mills, D., A. Tlili and J. Norton. 1997. Large-scale anesthesia of the silver-lip pearl oyster, *Pinctada maxima* Jameson. *Journal of Shellfish Research* 16:573–574.

Mitchell, A. J. and M. S. Hobbs. 2007. The acute toxicity of praziquantel to grass carp and golden shiners. *North American Journal of Aquaculture* 69:203–206.

Mylonas, C. C., G. Cardinaletti, I. Sigelaki, and A. Polzonetti-Magni. 2005. Comparative efficacy of clove oil and 2-phenoxyethanol as anesthetics in the aquaculture of European sea bass (*Dicentrarchus labrax*) and gilthead sea bream (*Sparus aurata*) at different temperatures. *Aquaculture* 246:467–481.

Noble, W. J., R. R. Cocks, J. O. Harris, and K. Benkendorff. 2009. Application of anaesthetics for sex identi-fication and bioactive compound recovery from wild *Dicathais orbita*. *Journal of Experimental Marine Biology and Ecology* 380:53–60.

Norton, J. H., M. Dashorst, T. M. Lansky, and R. J. Mayer. 1996. An evaluation of some relaxants for use with pearl oysters. *Aquaculture* 144:39–52.

O'Donnell, J. M., M. E. Durand, P. L. Robitaille, S. W. Fisher, and P. C. Stromberg. 1996. 31P-NMR analysis of lethal and sublethal lesions produced by KCl-intoxication in the zebra mussel, *Dreissena polymorpha*. *The Journal of Experimental Zoology* 276:53–62.

Park, M. O., S. Y. Im, D. W. Seol, and I. S. Park. 2009. Efficacy and physiological responses of rock bream, *Oplegnathus fasciatus* to anesthetization with clove oil. *Aquaculture* 287:427–430.

Pickering, Q. H., E. P. Hunt, G. L. Phipps, T. H. Roush, W. E. Smith, D. L. Spehar, C. E. Stephan, and D. K. Tanner. 1983. Effects of pollution on freshwater fish and amphibians. *Journal (Water Pollution Control Federation)* 55:840–863.

Pucherelli, S., D. Hosler, and D. E. Portz. 2010. Assessment of quagga mussel veliger treatments for Pueblo State Fish Hatchery transport. Report of Bureau of Reclamation to Colorado Division of Wildlife, Denver, CO.

Ram, J. L., D. Moore, S. Putchakayala, A. A. Paredes, D. Ma, and R. P. Croll. 1999. Serotonergic responses of the siphons and adjacent mantle tissue of the zebra mussel, *Dreissena polymorpha*. *Comparative Biochemistry and Physiology C-Molecular & Integrative Physiology* 124:211–220.

Sattari, A., S. S. Mirzargar, A. Abrishamifar et al. 2009. Comparison of electroanesthesia with chemical anes-thesia (MS222 and clove oil) in rainbow trout (*Oncorhynchus mykiss*) using plasma cortisol and glucose responses as physiological stress indicators. *Asian Journal of Animal and Veterinary Advances* 4:306–313.

Saydmohammed, M. and A. K. Pal. 2009. Anesthetic effect of eugenol and menthol on handling stress in *Macrobrachium rosenbergii*. *Aquaculture* 298:162–167.

Sehdev, H. S., J. R. McBride, and U. H. M. Fagerlund. 1963. 2-Phenoxyethanol as a general anaesthetic for sockeye salmon. *Journal of the Fisheries Research Board of Canada* 20:1435–1440.

Sharma, P. D., H. H. Nollens, J. A. Keogh, and P. K. Probert. 2003. Sodium pentobarbitone-induced relaxation in the abalone *Haliotis iris* (Gastropoda): Effects of animal size and exposure time. *Aquaculture* 218:589–599.

Sharma, S. K. and L. L. Sharma. 1996. Effect of chloral hydrate on metabolic rate of *Labeo rohita* (Ham) and *Poecilia reticulata* (Peters). *Journal of the Indian Fisheries Association* 26:121–125.

Simoes, L. N. and L. C. Gomes. 2009. Eficácia do mentol como anestésico para juvenis de tilápia-do-nilo (*Oreochromis niloticus*). *Arquivo Brasileiro De Medicina Veterinaria E Zootecnia* 61:613–620.

Sink, T. D. and J. W. Neal. 2009. Stress response and posttransport survival of hybrid striped bass transported with or without clove oil. *North American Journal of Aquaculture* 71:267–275.

Small, B. C. 2004. Effect of isoeugenol sedation on plasma cortisol, glucose, and lactate dynamics in channel catfish *Ictalurus punctatus* exposed to three stressors. *Aquaculture* 238:469–481.

Small, B. C. and N. Chatakondi. 2005. Routine measures of stress are reduced in mature channel catfish during and after AQUI-S anesthesia and recovery. *North American Journal of Aquaculture* 67:72–78.

Sprecher, S. L. and K. D. Getsinger. 2000. Zebra mussel chemical control guide. ERDC/EL TR-00–1, U.S. Army Engineer Research and Development Center, Vicksburg, MS.

Stehly, G. R., J. R. Meinertz, and W. H. Gingerich. 1998. Effect of temperature on the pharmacokinetics of benzocaine in rainbow trout (*Oncorhynchus mykiss*) after bath exposures. *Journal of Veterinary Pharmacology and Therapeutics* 21:121–127.

Suquet, M., G. de Kermoysan, R. G. Araya et al. 2009. Anesthesia in Pacific oyster, *Crassostrea gigas. Aquatic Living Resources* 22:29–34.

Sykes, C. L. 2012. Development of an efficient method for removal of quagga mussel veligers from transport tanks at Willow Beach National Fish Hatchery: 2009–2012 comprehensive report. Report of U.S. Fish and Wildlife Service to Bureau of Reclamation, Boulder City, NV.

Tsantilas, H., A. D. Galatos, F. Athanassopoulou, N. N. Prassinos, and K. Kousoulaki. 2006. Efficacy of 2-phenoxyethanol as an anaesthetic for two size classes of white sea bream, *Diplodus sargus* L., and sharp snout sea bream, *Diplodus puntazzo* C. *Aquaculture* 253:64–70.

Ucar, A. and M. Atamanalp. 2010. The effects of natural (clove oil) and synthetical (2-phenoxyethanol) anesthesia substances on hematology parameters of rainbow trout (*Oncorhynchus mykiss*) and brown trout (*Salmo trutta fario*). *Journal of Animal and Veterinary Advances* 9:1925–1933.

VDGIF (Virginia Department of Game and Inland Fisheries). 2005. Millbrook Quarry zebra mussel and quagga mussel eradication. Final Environmental Assessment Report of Virginia Department of Game and Inland Fisheries to U.S. Fish and Wildlife Service, Hadley, MA.

Velisek, J., A. Stara, Z.-H. Li, S. Silovska, and J. Turek. 2011. Comparison of the effects of four anaesthetics on blood biochemical profiles and oxidative stress biomarkers in rainbow trout. *Aquaculture* 310:369–375.

Velisek, J., V. Stejskal, J. Kouril, and Z. Svobodova. 2009. Comparison of the effects of four anaesthetics on biochemical blood profiles of perch. *Aquaculture Research* 40:354–361.

Verween, A., M. Vincx, and S. Degraer. 2009. Comparative toxicity of chlorine and peracetic acid in the biofouling control of *Mytilopsis leucophaeata* and *Dreissena polymorpha* embryos (Mollusca, Bivalvia). *International Biodeterioration and Biodegradation* 63:523–528.

Wallen, I. E., W. C. Greer, and R. Lasater. 1957. Toxicity to *Gambusia affinis* of certain pure chemicals in turbid waters. *Sewage and Industrial Wastes* 29:695–711.

Waller, D. L., S. W. Fisher, and H. Dabrowska. 1996. Prevention of zebra mussel infestation and dispersal during aquaculture operations. *The Progressive Fish-Culturist* 58:77–84.

Waller, D. L., J. J. Rach, W. G. Cope, L. L. Marking, S. W. Fisher, and H. Dabrowska. 1993. Toxicity of candidate molluscicides to zebra mussels (*Dreissena polymorpha*) and selected nontarget organisms. *Journal of Great Lakes Research* 19:695–702.

Weber, R. A., J. B. Peleteiro, L. O. Garcia Martin, and M. Aldegunde. 2009. The efficacy of 2-phenoxyethanol, metomidate, clove oil and MS-222 as anaesthetic agents in the Senegalese sole (*Solea senegalensis* Kaup 1858). *Aquaculture* 288:147–150.

White, H. I., T. Hecht, and B. Potgieter. 1996. The effect of four anaesthetics on *Haliotis midae* and their suitability for application in commercial abalone culture. *Aquaculture* 140:145–151.

Wildridge, P. J., R. G. Werner, F. G. Doherty, and E. F. Neuhauser. 1998. Acute effects of potassium on filtration rates of adult zebra mussels, *Dreissena polymorpha. Journal of Great Lakes Research* 24:629–636.

Woods, L. C. III, D. D. Theisen, and S. He. 2008. Efficacy of Aqui-S as an anesthetic for market-sized striped bass. *North American Journal of Aquaculture* 70:219–222.

Woolsey, J., M. Holcomb, and R. L. Ingermann. 2004. Effect of temperature on clove oil anesthesia in steelhead fry. *North American Journal of Aquaculture* 66:35–41.

Young, M. J. 2009. The efficacy of the aquatic anaesthetic AQUI-S® for anaesthesia of a small freshwater fish, *Melanotaenia australis. Journal of Fish Biology* 75:1888–1894.

Zahl, I. H., A. Kiessling, O. B. Samuelsen, and M. K. Hansen. 2009. Anaesthesia of Atlantic cod (*Gadus morhua*)—Effect of pre-anaesthetic sedation, and importance of body weight, temperature and stress. *Aquaculture* 295:52–59.

Zahl, I. H., A. Kiessling, O. B. Samuelsen, and R. E. Olsen. 2010. Anesthesia induces stress in Atlantic salmon (*Salmo salar*), Atlantic cod (*Gadus morhua*) and Atlantic halibut (*Hippoglossus hippoglossus*). *Fish Physiology and Biochemistry* 36:719–730.

33

Use of a Molluscicide on Preventing Quagga Mussel Colonization

Ashlie Watters, Shawn L. Gerstenberger, and Wai Hing Wong

CONTENTS

ABSTRACT   Quagga mussels (*Dreissena rostriformis bugensis*) continue to pose a threat to water delivery systems because of their ecological and biofouling impacts. Chemical control protocols are a commonly used method for preventing the colonization of quagga mussels because they are versatile, easy to implement, and cost effective. The objective of the study is to evaluate the effectiveness of EarthTec QZ®, a new molluscicide with copper sulfate pentahydrate ($CuSO_4 \cdot 5H_2O$) as an active ingredient, on preventing quagga mussel veliger colonization during fall, winter, spring, and summer seasons at Lake Mead. The following four Cu^{2+} concentrations were used in the study: 0 (control), 0.06, 0.12, and 0.18 ppm. The results of the study show that 0.18 ppm of $CuSO_4 \cdot 5H_2O$ is effective in preventing veliger colonization on fiberglass panels during all four seasons, and 0.12 ppm of $CuSO_4 \cdot 5H_2O$ reduces colonization rates by 59% for the fall season and 93% for the winter, spring, and summer seasons. This yearlong study shows that $CuSO_4 \cdot 5H_2O$ can be useful as an added option for the management of quagga mussel biofouling in water treatment facilities and hydraulic power plants.

Introduction

The zebra (*Dreissena polymorpha*) and the quagga mussel (*Dreissena rostriformis bugensis*) have arguably become the most impactful, nonindigenous biofouling pests introduced into North American freshwater systems (LaBounty and Roefer 2007). The ecological, recreational, and economic impacts of these mussels are profound, in that they are difficult to control and eradicate. Dreissenid mussels were accidently introduced into the Laurentian Great Lakes in North America in the 1980s via ballast water of large cargo ships (Ludyanskiy et al. 1993; Carlton 2008; Van der Velde et al. 2010). Quagga mussels were first discovered in the Boulder Basin of Lake Mead (Nevada, USA) on January 6, 2007 (LaBounty and Roefer 2007). This is the first confirmed introduction of a dreissenid species in the western United States, and it is also the first time that a large ecosystem was infested by quagga mussels without a previous infestation by zebra mussels. It has been postulated that quagga mussels were introduced into Lake Mead through bilge water (i.e., bait or live wells) carried by a recreational boat from the Great Lakes region (LaBounty and Roefer 2007; Hickey 2010; McMahon 2011; Wong and Gerstenberger 2011).

Shortly after the discovery of quagga mussels in Lake Mead, they were found colonized in the lower Colorado River, as well as other lakes in Arizona, California, and Nevada. Quagga mussels can easily spread at an unprecedented rate throughout the United States because of their high fecundity, efficient larval dispersal, lack of natural controls, and the strength of their byssal threads, which allows for attachment to a variety of indiscriminate substrates (Wong and Gerstenberger 2011).

Since the discovery of dreissenids in North America, billions of dollars have been expended for monitoring and control efforts (Connelly et al. 2007). For the first couple decades, the financial focus was centered on the Great Lakes region. Now, however, the focus is shared with the Southwest region with infestations occurring down the Colorado River, which provides water to 27 million people in Arizona, California, and Nevada (Wong et al. 2013). The quagga mussel infestation in the Southwest has negatively impacted water delivery systems, water treatment facilities, hydroelectric power plants, fisheries, and recreational boating activities. Once a quagga mussel infestation occurs, it is difficult to completely eradicate the problem.

There are several, albeit limited methods for controlling quagga mussels that are not only effective and cost efficient, but ecologically sound. Measures to control quagga mussels include mechanical, thermal, desiccation, biological alternatives, and chemical protocols. These methods are commonly used in closed systems and have been used alone or as combined methods. The most common and widely used method in both the United States and Europe is chemical control with the use of chlorine as chlorine gas or as liquid sodium hypochlorite (Claudi and Mackie 1994; Rajagopal et al. 1996; Sprecher and Getsinger 2000). Chlorine is effective at low concentrations and efficient against all life stages of the quagga mussel (Jenner and Janssen-Mommen 1993). However, when chlorine reacts with organic or inorganic material already present in the water being treated, trihalomethanes (THMs), which are potential carcinogens, are produced (Cotruvo and Regelski 1989). THMs are regulated by the U.S. Environmental Protection Agency and are monitored closely. Because large amounts of continuous chlorine dosing can cause health effects, other chemical options are being evaluated.

Recently, another form of chemical control being evaluated for preventing quagga mussel colonization is copper. Historically, copper has not been used as a means of preventing nor killing quagga mussels in municipal settings. Rather, copper has been added to antifouling coatings on ships to prevent barnacle growth (Claudi and Mackie 1994) and applied as an antifouling agent on underwater pipes (Dormon et al. 1996). The general toxicity of copper has been found to be successful in mussel control; however, copper ions leach from the coatings and result in unacceptable copper concentrations in water systems. Copper ions have been shown to be effective in killing adult and juvenile quagga mussels, and preventing veliger colonization with limited exposure time in a closed system, during the winter season at Lake Mead (Watters et al. 2013). The tested copper product is called EarthTec® and copper sulfate pentahydrate ($CuSO_4 \cdot 5H_2O$) is the active ingredient (Watters et al. 2013). It is generally used as an algaecide, but recently, it has been approved as a molluscicide because it can be used for preventing quagga mussel veligers from colonizing. It also has a new registered label, EarthTec QZ (www.earthtecqz.com). Copper treatments may be more effective in the summer season at Lake Mead, when water temperatures are higher (28°C–30°C). With increasing water temperatures, the respiration rate of the mussel increases; hence, the copper has a greater effect on the mussels in the summer (Rao and Khan 2000).

Quagga mussel veligers are present year-round in Lake Mead with the percentage of settlement of competent veligers peaking at greater than 60% during the fall and declining to less than 5% in February when surface water temperatures are at their lowest (10°C–12°C) (Gerstenberger et al. 2013). The abundance of veligers in Lake Mead is associated with the metalimnion water temperature. Because veliger dynamics in Lake Mead vary by season, it is essential to have a more accurate assessment of the effectiveness of $CuSO_4 \cdot 5H_2O$ over a wide range of seasonal temperatures at which veliger settlement occurs. The objective of this study is to evaluate the effectiveness of EarthTec QZ, on preventing quagga mussel colonization on fiberglass panels during fall, winter, spring, and summer seasons in a laboratory setting at Lake Mead. Because $CuSO_4 \cdot 5H_2O$ is the active ingredient of this molluscicide, it will be specifically used to represent this commercial product in this manuscript. The results of the study will show that $CuSO_4 \cdot 5H_2O$ can be useful as an added option for the management of quagga mussel biofouling in water treatment facilities and hydraulic power plants.

Methods

Experimental Design

The experiment was performed in four phases coinciding with each of the four seasons. In September 2011, December 2011, March 2012, and June 2012 (Table 33.1), veliger specimens of *D. rostriformis bugensis* were collected from Lake Mead (36°1′27.52′N; 114°46′18.21′W) at a depth of 30 m using a 64 μm pore size plankton net (Watters et al. 2013). A National Park Service permit was obtained, granting permission to collect quagga mussel veligers. Immediately following collection, samples were brought back to the Nevada Department of Wildlife's (NDOW) fish hatchery in Boulder City, NV. They were divided into 12, 10 gallon aquarium tanks which were aerated and filled with 25 L of raw Lake Mead water. Colonization was measured using 24 (two per tank) fiberglass panels (79 × 68 × 1.66 mm), suspended from a shelf above the tank. Ten days prior to experimentation, the panels were immersed in raw Lake Mead water to develop a biofilm. The following four concentrations of $CuSO_4 \cdot 5H_2O$ solution were tested for effectiveness on preventing veliger colonization on the fiberglass panels: 0 (control), 0.06, 0.12, and 0.18 ppm. Veligers were fed twice daily with 0.375 mL of Instant Algae® *Isochrysis* 1800 (Reed Mariculture, Campbell, California) (1.54×10^8 cells). Each week, half of the water in each tank was exchanged and replaced with fresh Lake Mead water. To prevent the loss of the veligers, a 64 μm filter was used during the water exchange and the veligers were placed back into the corresponding tank. Each tank received a minimum of 25 veligers per liter of water ($N = 625$). After veligers were added to each aquarium, the appropriate amount of $CuSO_4 \cdot 5H_2O$ was added to the medium to attain the appropriate test concentration. The pH, water temperature, and dissolved oxygen (DO) were recorded (Table 33.2).

Panel Analysis

After 57 days for fall, spring, and summer seasons, and 63 days for the winter season, the fiberglass panels were removed from all tanks and were brought back to the University of Nevada, Las Vegas' Environmental Health Laboratory. Colonization status of attached quagga mussel settlers was assessed using cross-polarized light microscopy. Each mussel was then recorded and photographed with the Zeiss Discovery V8 stereo microscope (Carl Zeiss, Inc., Peabody, MA). To measure the amount of colonization, all six surfaces of each panel were observed. To determine the number of mussels per square meter, the total number of settlers was divided by 0.01 (Watters et al. 2013).

TABLE 33.1

Duration of Colonization Experiment per Season

Season	Dates	Days
Fall	September–November 2011	57
Winter	December–February 2012	63
Spring	March–May 2012	57
Summer	June–August 2012	57

TABLE 33.2

Average pH, Water Temperature, and DO per Season in All Treatment Tanks

Season	pH (Range)	Water Temperature (C°) (Range)	DO (mg/L) (Range)
Fall	8.3 (8.2–8.3)	23.2 (22.3–24.4)	8.9 (7.8–9.3)
Winter	8.4 (8.3–8.5)	12.2 (11.2–12.9)	9.9 (9.9–10)
Spring	8.3 (8.3–8.4)	17.2 (14.9–21.6)	8.5 (7.9–8.9)
Summer	8.3 (8.2–8.3)	24.2 (22.8–24.9)	7.4 (6.9–7.9)

Statistical Analysis

A two-way analysis of variance (ANOVA) was conducted to examine the effect of the four seasons (fall, winter, spring, and summer) and the four concentrations (0, 0.06, 0.12, and 0.18 ppm) of $CuSO_4 \cdot 5H_2O$ on the rate of veliger colonization. A Tukey *post hoc* multiple comparisons test was conducted to determine if there was a significant difference among the four seasons and the four concentrations of $CuSO_4 \cdot 5H_2O$. A linear regression was created to estimate the theoretically lowest concentration of $CuSO4 \cdot 5H2O$ at which a zero colonization rate could be reached for the four seasons. The significance criterion was set at $\alpha = 0.05$. All statistical analyses were performed using SPSS Statistics software (version 20.0, IBM SPSS Statistics, Inc. Armonk, New York).

Results

The results show that the colonization rate of veligers is significantly affected by the $CuSO_4 \cdot 5H_2O$ concentration, and the season during which it was administered ($p < 0.001$). Higher concentrations of $CuSO_4 \cdot 5H_2O$ resulted in less veliger colonization. Treatments with 0.18 ppm $CuSO_4 \cdot 5H_2O$ had a zero colonization rate in all four seasons (Figure 33.1). The groups treated with 0.12 and 0.06 ppm were less colonized compared to the control groups (ANOVA, $p < 0.01$) (Figure 33.1). There is a significant difference between each treatment of $CuSO_4 \cdot 5H_2O$ ($p < 0.01$).

The season during which the $CuSO_4 \cdot 5H_2O$ treatments were administered had a significant effect on veliger colonization (ANOVA, $p < 0.01$). There is a significant difference in the fall season (September–November, 2011) and the other three seasons, winter (December 2011–February 2012), spring (March–May 2012), and summer (June–August 2012), as well as a significant difference between winter and spring and summer (ANOVA, $p < 0.05$); however, there is no significant difference between spring and summer seasons (ANOVA, $p > 0.05$).

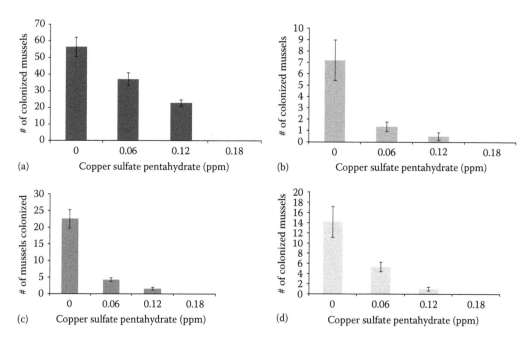

FIGURE 33.1   Number of colonized quagga mussel veligers treated with 0, 0.06, 0.12, and 0.18 ppm of copper sulfate pentahydrate for fall (a), winter (b), spring (c), and summer (d) seasons. Mean ± 1 SD; values shown on top of the bars are means. (Note the Y-axis scales are different for different seasons.)

TABLE 33.3

Quagga Mussel Veliger Colonization Rates for Fall, Winter, Spring, and Summer

Season	CuSO$_4$·5H$_2$O Dose	Colonization Density	Colonization Rate (%)	Reduced Colonization Rate (%)
Fall	0	5469	100	0
	0.06	3592	66	44
	0.12	2216	41	59
	0.18	0	0	100
Winter	0	696	100	0
	0.06	129	19	81
	0.12	49	7	93
	0.18	0	0	100
Spring	0	2184	100	0
	0.06	405	19	81
	0.12	146	7	93
	0.18	0	0	100
Summer	0	1375	100	0
	0.06	518	38	62
	0.12	97	7	93
	0.18	0	0	100

TABLE 33.4

Theoretical Minimal CuSO$_4$·5H$_2$O Concentration Leading to Zero Colonization for Fall, Winter, Spring, and Summer Seasons Using the Regression between Mussel Colonization (Y) and Concentration of CuSO$_4$·5H$_2$O (X)

Season	Regression Equation	CuSO$_4$·5H$_2$O Concentration (ppm)
Fall	Y = −305.3 X + 56.5 (R^2 = 0.99, P < 0.01)	0.185
Winter	Y = −37.2 X + 5.6, (R^2 = 0.63, P > 0.05[a])	0.18
Spring	Y = −116.9 X + 17.6 (R^2 = 0.62, P > 0.05[a])	0.18
Summer	Y = −78.1 X + 12.2 (R^2 = 0.82, P > 0.05[a])	0.18

[a] Linear model is not significant; 0.18 ppm of CuSO$_4$·5H$_2$O is recommended, which resulted in zero colonization.

Assuming the control (0 ppm) treatment had a 100% colonization rate in the fall season, an average 66% colonization rate was found for 0.06 ppm, 41% colonization rate was found for 0.12 ppm, and 0% colonization rate was found for 0.18 ppm of CuSO$_4$·5H$_2$O (Table 33.3). The same assumption was applied to the winter, spring, and summer seasons. In winter, the colonization rates for 0.06, 0.12, and 0.18 ppm were 19%, 7%, and 0%, respectively; in the spring season, the colonization rates for 0.06, 0.12, and 0.18 ppm were 19%, 7%, and 0%, respectively; and in the summer season, the colonization rates for 0.06, 0.12, and 0.18 ppm were 38%, 7%, and 0% (Table 33.3). Therefore, in all four seasons, 0.06, 0.12, and 0.18 ppm of CuSO$_4$·5H$_2$O were effective in reducing the colonization rates (Table 33.3). Because the control treatments in all four seasons had a 100% colonization rate, least square fit regression was used to determine the relationship between colonization rate (%) and dose (ppm). Therefore, to result in zero colonization, it is estimated that 0.185 ppm of CuSO$_4$·5H$_2$O is necessary for fall months (September–November), and 0.18 ppm of CuSO$_4$·5H$_2$O is necessary for winter months (December–February), spring months (March–May), and summer months (July–August) (Table 33.4).

Discussion

This study shows that the colonization rate of quagga mussel veligers is affected by not only the CuSO$_4$·5H$_2$O treatment but also the season in which it is administered. In all four seasons, 0.06 and 0.12 ppm of CuSO$_4$·5H$_2$O greatly reduced quagga mussel veliger colonization on fiberglass panels when compared to the control groups ($p < 0.01$) and 0.18 ppm prevented all quagga mussel colonization. The data are in agreement with a previous study that tested the effectiveness of EarthTec in preventing veliger colonization on fiberglass panels during the winter season (December–February). The results of the study found that 3 ppm of EarthTec (0.18 ppm of $CuSO_4 \cdot 5H_2O$) was effective in deterring quagga mussel veliger settlement (Watters et al. 2013).

Colonization rate in each season varied within the control groups. This can be caused by food quantity and quality, water temperature, and/or hydrodynamics of the water column. These environmental factors can affect the abundance of live and competent quagga mussel veligers (Gerstenberger et al. 2011). In this study, quagga mussel veligers had the highest colonization rates in the control group in the fall and spring seasons. The mean annual temperature ranges in the epilimnion, metalimnion, and hypolimnion of Boulder Basin of Lake Mead, where the samples were collected, are 12°C–27°C, 12°C–18°C, and 12°C–12.5°C, respectively (LaBounty and Burns 2005). Quagga mussel veligers can be found alive and competent year-round in this area of Lake Mead because the water temperature is above the threshold for spawning and recruitment (Gerstenberger et al. 2011).

Chemicals are important components in controlling for quagga mussels. For reducing established infestations or controlling new ones, chemicals are versatile, easy to implement, and cost effective (Spreecher and Getsinger 2000). There are numerous studies that examine the effectiveness of chlorine on killing quagga mussels and preventing veliger colonization (Claudi and Mackie 1994; Rajagopal et al. 1996; Sprecher and Getsinger 2000; Watters et al. 2013). While chlorine is effective, THM production has been shown to have adverse health effects if humans are exposed to high or chronic doses (Cotruvo and Regelski 1989). Copper is an effective metal for killing quagga mussel veligers and deterring settlement. However, it may not be applicable to all scenarios. Studies have shown that copper alloys may be useful on boat hulls and immersed infrastructure. A study by Dormon et al. (1996) examined copper alloys as antifoulants for zebra and quagga mussels and found that using plates of copper and copper alloy (copper-nickel 90:10) was effective up to 37 months. However, when a biofilm was able to form, as the copper leached into the water, settlers were found, suggesting that this is not a permanent solution.

Currently, there is not just one solution that will eradicate quagga mussels. If chemical control is the chosen solution for water treatment facilities and hydropower managers, it is important to focus on a preventive plan by deterring quagga mussel veligers from attaching to the infrastructure as opposed to killing an established adult population, as it is more difficult to remove and kill an established adult population than preventing veligers from settling. A viable tool for managing quagga mussel infestations and for preventing veligers from colonizing infrastructure is 0.18 ppm of $CuSO_4 \cdot 5H_2O$. The current study was performed in a laboratory setting. To use $CuSO_4 \cdot 5H_2O$ in an open water system, further research is needed to test the risk to nontarget species and the surrounding ecosystem.

REFERENCES

Carlton J.T. 2008. The zebra mussel *Dreissena polymorpha* found in North America in 1986 and 1987. *Journal of Great Lakes Research* 34:770–773.

Claudi R., Mackie G.L. 1994. *Practical Manual for Zebra Mussel Monitoring and Control*, 1st edn. Boca Raton, FL: Lewis Publishing.

Connelly N.A., O'Neill C.R., Knuth B.A., Brown T.L. 2007. Economic impacts of zebra mussels on drinking water treatment and electric power generation facilities. *Environmental Management* 40:105–112.

Cotruvo J.A., Regelski M. 1989. Issues in developing national primary drinking water regulations for disinfection and disinfection by-products. In: Calabrese E.J., Gilbert C.E., Pastids H., eds. *Safe Drinking Water Act: Amendments, Regulations, and Standards*. Chelsea, MI: Lewis Publishers, pp. 57–69.

Dormon J.M., Cottrell C.M., Allen D.G., Ackerman J.D., Spelt J.K. 1996. Copper and copper-nickel alloys as zebra mussel antifoulants. *Journal of Environmental Engineering* 122:276–283.

Gerstenberger S.L., Meuting S.A., Wong W.H. 2011. Veligers of invasive quagga mussels (*Dreissena rostriformis bugensis*, Andrusov 1897) in Lake Mead, Nevada-Arizona. *Journal of Shellfish Research* 30:933–938.

Hickey V. 2010. Quagga mussel crisis at Lake Mead National Recreation Area, Nevada (U.S.A.). *Conservation Biology* 24:931–937.

Jenner H.A., Janssen-Mommen J.P.M. 1993. Monitoring and control of *Dreissena polymorpha* and other macrofouling bivalves in the Netherlands. In: Nalepa TF, Schloesser DW, eds. *Zebra Mussels: Biology, Impacts, and Control.* Boca Raton, FL: Lewis Publishers, pp. 537–554.

LaBounty J.F., Roefer P. 2007. Quagga mussels invade Lake Mead. *LakeLine* 27:17–22.

Ludyanskiy M.L., McDonald D., Macneill D. 1993. Impact of the zebra mussel, a bivalve invader. *Dreissena polymorpha* is rapidly colonizing hard surfaces throughout waterways of the United States and Canada. *Bioscience* 43:533–544.

McMahon R.F. 2011. Quagga mussel (*Dreissena rostriformis bugensis*) population structure during the early invasion of Lakes Mead and Mohave January-March 2007. *Aquatic Invasions* 6:131–140.

Rajagopal S., Nair K.V.K., Azariah J., van der Velde G., Jenner H.A. 1996. Chlorination and mussel control in the cooling conduits of a tropical coastal power station. *Marine Environmental Research* 41:201–221.

Rao D.G.V., Khan M.A.Q. 2000. Zebra mussels: Enhancement of copper toxicity by high temperatures and its relationship with respiration and metabolism. *Water Environment Research* 72:175–178.

Sprecher S.L., Getsinger K.D. 2000. Zebra mussel chemical control guide. US Army Corps of Engineers, Washington, DC. ERDC/EL TR-00-1.

Van der velde G., Rajagopal S., Bij de Vaate A. 2010. *The Zebra Mussel in Europe.* Leiden, the Netherlands: Backhuys Publishers, 490pp.

Watters A., Gerstenberger S.L., Wong W.H. 2013. Effectiveness of EarthTec® for killing invasive quagga mussels (*Dreissena rostriformis bugensis*) and preventing their colonization in the western United States. *Biofouling* 29:21–28.

Wong W.H., Gerstenberger S.L. 2011. Quagga mussels in the western United States: Monitoring and management. *Aquatic Invasions* 6:125–129.

Wong W.H., Gerstenberger S.L., Hatcher M.D., Thompson D.R., Schrimsher D. 2013. Invasive quagga mussels can be attenuated by redear sunfish (*Lepomismicrolophus*) in the southwestern United States. *Biological Control* 64:276–282.

34

Zequanox®:
Bio-Based Control of Invasive Dreissena *Mussels*

Sarahann Rackl and Carolyn Link

CONTENTS

ABSTRACT   Despite the continued spread of zebra and quagga mussel populations throughout Europe and the United States, regulatory pressure is mounting to limit the use of pesticidal chemicals for mussel control. Previously, nonselective biocidal chemicals, including both oxidizing (e.g., chlorine) and nonoxidizing (e.g., polyquaternary ammonium compounds) chemicals, were the primary choice for mussel control. Power and industrial facilities have historically elected to use these biocides, because there were no other available control options that were highly effective, easy to apply, required limited capital expenditures, and could be injected at the inlet while still offering control throughout the entire system. However, nonselective biocides used for invasive mussel control are toxic to a broad spectrum of aquatic organisms and present safety issues for workers and water quality. For open water treatments, such as within lakes and rivers, the concentration of currently approved molluscicides required to achieve zebra and quagga mussel control is generally sufficient to cause mortality in other organisms, including fish and native bivalve species. Consequently, there is a need for an environmentally safe option to control zebra and quagga mussels for environmental and recreational habitat restoration, as well as a more environmentally compatible product for power and industrial facilities.

Research for almost two decades has led to the development of this alternative, Zequanox®, the first biopesticide for the control of invasive zebra and quagga mussels (*Dreissena polymorpha* and *Dreissena rostriformis bugensis*). Efficacy trials suggest that this biologically derived product can effectively control invasive mussels while mitigating risks and addressing regulatory agency concerns with using chemicals. In this chapter, we discuss how Zequanox was developed to maintain the advantages of chemical control while addressing many of the disadvantages to the current chemical control practices. Included in the chapter is a product overview—discovery and science behind this innovation, treatment programs and use, registration, and safety. We conclude the chapter with five unique case studies that demonstrate the versatility of Zequanox and the ability to tailor treatments according to customer needs in a variety of settings and offer an equivalent level of control to Zequanox's chemical counterparts. Collectively, the case studies for confined systems demonstrate that Zequanox can provide greater than 90% control in both flowing and static infrastructure in high volumes of water. In addition, settlement management treatments provide additional flexibility for customers in need of enhanced control of mussel growth and shell debris management by achieving greater than 85% settlement control.

Introduction and Product Overview

In the early 1990s, power and industrial facility operators in the United States and Canada were employing a variety of methods to try and stave off impending damage in their water systems caused by the ever-expanding colonies of quagga and zebra mussels. The control methods that commercial and public entities implemented included various aqueous or chemical controls, antifouling coatings, physical removal, and mechanical controls. Each of these methods has significant drawbacks. While some methods were labor intensive and/or turned out to be completely ineffective, others posed health and safety risks to the facilities' employees and the surrounding ecosystems. Without a better alternative, operators had to tolerate these shortcomings. However, the discovery of the activity against zebra and quagga mussels from *Pseudomonas fluorescens* CL145A (*Pf* Cl145A) led to the development of Zequanox (Molloy et al. 2013a).

In this section, we discuss how Zequanox was developed to maintain the advantages of chemical control while addressing many of the disadvantages of the current chemical control practices—including the discovery and science behind this innovation, discussed in a different section, and the treatment programs and uses.

Overview

The most common approach to mussel control is the use of oxidizing (e.g., chlorine) and nonoxidizing chemicals (e.g., polyquaternary ammonium compounds). The use of chemicals, such as hypochlorite, chlorine gas, chlorine dioxide, and quaternary ammonia compounds, necessarily involves careful practices to ensure that the chemicals are safely stored and utilized. Further, most chemical products used for invasive mussel control are general, nonselective biocides and are therefore toxic to other aquatic organisms. Because of this nontarget toxicity, facilities using chlorine and other chemical-based molluscicides may be required to deactivate or detoxify the treated water before discharge to meet environmental requirements (U.S. Department of Energy, National Energy Technology Laboratory 2006). Bisulfate or similar salts are used to help prevent the release of chlorine into the environment and reduce the impact on other aquatic organisms, contributing to salt loading in water bodies. Many polyquaternary molluscicides require the addition of clay to a treated water system to quench or deactivate the chemicals' toxicity before discharge into the environment. The ultimate fate and transport of the clay-bound molluscicides once discharged is unknown; many of these substances are nonbiodegradable and stay in the ecosystem long after discharge.

Some chemical molluscicides, chlorine in particular, exacerbate the increased corrosion and pitting that invasive mussels initiate. An additional disadvantage of using chlorine is that the mussels perceive the chlorinated water as a threat, causing them to shut their siphons, an avoidance behavior that necessitates extensive application times in order to achieve target mortality. The formation of harmful byproducts is yet another area of concern; when chlorine combines with organic compounds in water, potentially carcinogenic substances, such as trihalomethanes, haloacetic acids, and dioxins, are formed (Thornton 2000; U.S. Environmental Protection Agency [EPA] 1999).

Power and industrial facilities have historically elected to use these nonselective biocides, with chlorine being the most popular choice, because there were no other available control options that were highly effective, relatively easy to apply, required limited capital expenditures, and could be injected at the inlet of a system while still offering control throughout the entire system. It was the desire to maintain these important advantages that the chemical control options provide, while eliminating the environmental and safety concerns and other disadvantages, which drove the extensive research and ultimate discovery of *Pf* CL145A, the active ingredient in Zequanox. Faced with the threat of zebra mussels fouling electric power facilities within New York State, a research consortium of New York State's electric power generation companies contracted with the New York State Museum's Field Research Laboratory in 1991 for the screening of bacteria as potential biological control agents. The use of microbial, natural product compounds already had a clear record of commercial success and environmental safety in the control of invertebrate pests in North America as well as globally (Rodgers 1993). Extensive laboratory screening trials identified a North American bacterial isolate, strain CL145A of *P. fluorescens*, to be lethal to zebra and quagga mussels (Molloy et al. 2013a). A patent for this purpose was issued in both the United States (Molloy 2001, patent number 6,194,194) and Canada (Molloy 2004, patent number 2,225,436). *P. fluorescens* is worldwide in distribution and is present in all North American water bodies. In nature, it is a bacterial species that is found protecting the roots of plants from diseases. In 2007, Marrone Bio Innovations (MBI) entered into a commercial partnership with the New York State Museum to bring this naturally occurring soil microorganism to market. The result was Zequanox, manufactured by MBI, the industry's first aqueous, environmentally compatible molluscicide. Zebra and quagga mussels filter Zequanox from the water, and toxins associated with the cellular membrane of the inactivated *Pf* CL145A cells disrupt the epithelial cells lining the mussel's digestive system, leading to mortality (Molloy et al. 2013b). A killed, or nonviable, form of the *Pf* CL145A cells is formulated to make Zequanox.

Zequanox was developed to maintain the benefits of current chemical control options while eliminating the worker and environmental health and safety concerns for the control of zebra and quagga mussels in cooling water (and other similar infrastructure) systems within power and industrial facilities. Zequanox is applied as a liquid, just like its chemical counterparts. Therefore, Zequanox can directly replace current chemical treatment practices with no to minimal additional capital expenditures. The ability to apply Zequanox as a liquid, like other chemicals, provides the added benefit of being able to reach and treat even the smallest of crevices in the water system, whereas mechanical control tools offer control only at a fixed accessible location. Zequanox poses very limited to no risk to workers, nontarget species, and the environment. As a reduced-risk pesticide, Zequanox is safe to store, handle, and apply; only minimal personal protective equipment is needed. Additionally, complicated detoxification is not needed. In contrast, chlorine and other chemical pesticides are toxic to aquatic life and the environment (i.e., they often fall into the level 1 pesticide, or other high-risk category as opposed to Zequanox, which is a level 4 pesticide). The alternative products require special handling, safety warning placards, sophisticated permitting, tracking, and monitoring. If not properly managed, chlorine and other hazardous chemicals can cause serious (even fatal) harm to humans, and can cause irreparable harm to the environment. Zequanox is noncorrosive, and therefore will not negatively impact equipment or infrastructure like the oxidizing chemicals, as mentioned earlier.

Applications of Zequanox are less labor intensive and less operationally disruptive than chemical methods. Zequanox treatments can be done during normal facility operations and typically occur within a 2–8 h period. This time frame is in contrast to chlorine treatments, which can require several weeks of around-the-clock treatment, and often require special procedures to ensure worker safety during the treatments. Zequanox treatment regimens can also be adjusted to achieve the balance of mussel control, application frequency, and shell debris management desired by facilities.

Treatment Programs and Application

Zequanox can be used to control adult populations, prevent new settlement, and target the free-floating veliger stage to aid in the prevention of new infestations. At the time of publication, MBI was working with customers to implement two standard treatment programs in compliance with product regulatory approval; these include attached mussel and settlement management treatment programs.

Attached mussel treatments focus on the treatment of adult mussels and typically occur once a year, approximately at the end of the mussel's peak reproductive season after a majority of seasonal settlement has occurred. These treatments are designed for facilities with a tolerance for moderate-to-large shell sizes—larger than 4 mm in length. In warmer climatic regions of the United States where the water temperatures are higher year-round, such as the Colorado River, treatments may occur more than once a year due to longer reproduction periods and more rapid growth rates. After a single day of treatment, mortality occurs over time and typically is observed between 4 and 21 days after the treatment. Mortality can be monitored by assessing the mussel health in bioboxes as described further in Case Study 1 and Case Study 3.

Settlement management treatments are designed to minimize the number of settled mussels that will grow and exceed 4 mm in length. Because these treatments target a smaller and less mature mussel, they are conducted at significantly lower concentrations and for shorter durations compared to adult treatments. This treatment is ideal for sensitive systems and equipment where debris from shells greater than 4 mm in length can cause significant damage or blockage. Settlement management treatments can be performed as frequently as every other week, up to approximately monthly, throughout the settlement season. Treatment efficacy is measured via reduction in settled mussels either by counting or by measuring the biomass of mussels accumulating in specifically defined areas. Bioboxes can also be used to quantify the degree of settlement control as described further in Case Study 4.

For both treatment programs, concentrations, treatment times, and frequency of treatments are based on the directions for use, the specific location of the treatment, and the degree of mussel infestation.

Veliger treatments target the free-floating planktonic stage of both zebra and quagga mussels. At the time of publication, Zequanox has been demonstrated to be highly effective for veliger control. Providing defined treatment programs and uses for veliger control is challenging, because currently there are limited peer-reviewed methodologies for assessing veliger mortality. Alongside the development of veliger treatment and application programs for veligers, MBI is developing methodologies for assessing absolute veliger mortality.

At time of publication, the two developed Zequanox treatment programs are further divided into two application types: (1) in-pipe and infrastructure applications with defined flow or volumes and (2) applications in open waters for environmental and recreational rehabilitation. The following sections provide a general description of how Zequanox can be used to provide invasive mussel control in these application types; however, there are many other situations in which Zequanox can be used to control invasive mussels.

Confined System (In-Pipe and Other Water-Related Infrastructure) Treatments

In general, the confined system treatment category applies to all types of in-pipe applications, as well as applications to infrastructure where the flow and/or use of fresh surface water supplies is impeded by mussel infestation. Historically, the majority of invasive mussel control options have targeted the negative impacts and problems mussels cause to these systems. Confined system applications include the treatment of any type of conduit that conveys water, including enclosed or semienclosed pipes such as cooling, service, and fire suppression systems in power and industrial facilities and water distribution systems for irrigation and municipalities.

Aqueous products are ideal for this type of control, because they can be applied at the beginning of a pipe to control mussels throughout the length of the system to the outlet or be mixed uniformly into a large volume of water. In the case of water distribution in pipes, mussel treatment is applied to remove mussel growth occluding the pipes and to maintain water flow. With Zequanox, either adult mussel treatments or settlement management treatment programs can be employed and designed specifically depending on the system to be treated and the system's sensitivity to mussel shells.

Facility managers also treat infrastructure that holds defined volumes of water, such as wet wells, intake bays, and raw water storage facilities, under static conditions. Aqueous products are ideal for these treatment situations, because the product can be applied into the entire volume of water, providing control on all surfaces in a timely manner.

Results for infrastructure type applications are described in Case Studies 1 through 3. The results of these studies represent product efficacy in various waters, with different zebra and quagga mussel populations, and in different geographies.

Environmental and Recreational Rehabilitation Treatments

Where invasive mussels are present, the composition of native organisms can be altered. Growth of nuisance weeds and algae can increase, negatively impacting fisheries, recreational life, facility operations, and property values. Invasive mussels are voracious filter feeders that grow rapidly, eliminating food sources and altering native habitat critical to other aquatic organisms (e.g., see Caraco et al. 1997; Strayer 2008). In addition, zebra and quagga mussels often attach directly to many benthic organisms and are frequently found in dense drusses on endangered and threatened unionid species. Currently, the approved chemical pesticide and control options available for zebra and quagga mussels are not target specific, and the treatment concentrations required to kill zebra and quagga mussels are toxic to most other organisms, including fish and native bivalve species. There is a need for an environmentally safe option to control zebra and quagga mussels in these sensitive environments. However, there are very few methodologies and standard practices for application of treatment options in open waters, because there have been limited options. As a result, it was necessary for MBI to develop methodologies, with our partners and peers, for open water applications with Zequanox.

Due to its low environmental toxicity concerns, Zequanox is a highly flexible mussel control tool that can achieve a variety of open water treatment objectives, including

- *Rapid response*: Treatments with Zequanox can occur quickly when point sources of infestation are identified early to prevent establishment of new populations.
- *Shoreline restoration*: Shorelines support many recreational and educational uses, and many waterways heavily infested with mussels have lost their usable shorelines due to invasive mussel accumulation and debris. Reducing mussels to controllable levels can, in time, lead to the restoration of these water resources.
- *Habitat restoration support*: Invasive mussels can have widespread impacts on habitats for sensitive native species, including species with limited mobility, such as native unionid mussels. Control of invasive mussel populations can support programs to reestablish native populations. Reduction or removal of an invasive mussel population may prevent endangered benthic species from being outcompeted by the invasive mussels for habitat, and can support overall ecosystem restoration, with benefits to native mussels, algae, and fish.
- *Protection of infrastructure*: Applying Zequanox to open water sources can help to protect the intake, transmission, and transport equipment and infrastructure that draw water from them.
- *Control of source populations*: A large invasive mussel population in one area of a flowing water body may not cause problems at its own location, but can act as a source of reproducing mussels, affecting downstream locations heavily if the veliger source is uncontrolled. Applying Zequanox and controlling the mussels upstream at the source of the veliger population can reduce mussel populations and their impacts downstream.

Case Study 4 provides information and results from one of the completed open water treatments in Deep Quarry Lake, Illinois.

Additional work on native unionid habitat restoration and recreation restoration is being completed by Dr. Denise Mayer with the New York State Museum and James Luoma and Mark Gaikowski with the United States Geological Survey.

Zequanox Registration

Because Zequanox is used to control a pest or nuisance species, pesticide registration is required. In general, after identification and characterization of the product, the primary objective of the registration process is to allow regulatory bodies to conduct a human and ecological risk assessment. Thus, the fundamental components of a registration submission include the following: (1) an effects dose–response profile for standard species (typically species of interest that represent environments likely to be impacted) and (2) an exposure profile for representative species relevant to the endpoints being considered (Touart and Maciorowski 1997). To meet these guidelines, MBI completed a wide range of mammalian and nontarget species testing. MBI also conducted additional nontarget ecotoxicological studies and other studies beyond those required for registration as a biopesticide to address product-specific environmental concerns and to meet state and local requirements for discharge.

Zequanox was first registered in the United States, in July 2011, and in Canada, in November 2012, through the United States Environmental Protection Agency (USEPA) and Health Canada Pesticide Management Regulatory Agency (PMRA), respectively, as a microbial-based biopesticide for use in pipes and infrastructure. The portion of the product, Zequanox, that is effective towards invasive mussels, *Pf* CL145A, is called the active ingredient (a.i.) in regulatory terms. The maximum allowable application concentration was 200 mg of active ingredient per liter (mg a.i./L) for treatments of adult mussels (50 mg a.i./L for younger life stages) with a maximum treatment duration of up to 12 h. In June 2014, the USEPA approved an expanded master label for Zequanox that included uses in open waters to address invasive mussel control for environmental and recreational restoration. Between the 2011 and 2014 USEPA registrations, advances in product application allowed for a lowering of the maximum treatment conditions without impacting efficacy, further protecting the environment. The maximum application concentration was reduced to 100 mg a.i./L for adult mussels (50 mg a.i./L was kept the same for younger life stages), and the maximum treatment duration was reduced to 8 h. At the time of this publication, PMRA was reviewing additional ecotoxicology studies and product use descriptions (or application types) to expand the product use label to include environmental and recreational restoration applications as well, and the required information was being reviewed for registration of Zequanox in the European Union.

Zequanox Human and Environmental Safety

As described in the earlier section, as part of the state and federal registration process, a wide variety of testing has been conducted to assess the potential for Zequanox to adversely impact nontarget organisms. Testing was generally conducted by certified contract laboratories according to Good Laboratory Practices (GLP; 40 CFR part 160), and some selected testing with nonstandard organisms was conducted by academic experts. Resulting studies were submitted for regulatory review as described, and each regulatory agency conducted their own (publicly available) environmental risk assessment on the product, such as the "Biopesticides Registration Action Document (BRAD): *P. fluorescens* strain CL 145A" (USEPA, Office of Pesticide Programs Biopesticides and Pollution Prevention Division 2011) and "Revised risk assessment for Zequanox (EPA Reg No. 84059-15) *(P. fluorescens* strain CL 145A killed cells)" (USEPA Office of Chemical Safety and Pollution Prevention 2014).

Because nontarget reports and data can be used by other companies to register another product with the same active ingredient, the reports and data are deemed confidential; therefore, the publicly available risk assessments are the best resources to review submitted nontarget data. Two peer-reviewed manuscripts are available that include the methods and data for a number of studies on a variety of nontarget species: "Ecotoxicological impact of Zequanox®, a novel biocide, on selected non-target Irish aquatic species" by Meehan et al. (2014), and "Non-target trials with *P. fluorescens* strain CL145A, a lethal control agent of dreissenid mussels (Bivalvia: Dreissenidae)" by Molloy et al. (2013c).

Molloy et al. (2013c) evaluated the impact of a single 72 h exposure of *Pf* CL145A at concentrations between 50 and 200 mg a.i./L to six species; *Colpidium colpoda, Daphnia magna, Hylella azteca,*

Pimephales promelas, Lepomis macrochirus, and *Salmo trutta.* The duration Molloy et al. evaluated the nontarget organisms post exposure varied among the different species tested. No mortality was seen in five of the six species evaluated. *H. azteca* was the only organism for which Molloy et al. observed low mortality, between 3% and 27%, but additional trials suggested that most, if not all, of the morality could be attributed to some other unidentified factor (e.g., possibly particle load or a water quality issue) rather than the zebra and quagga mussel chemistry of *Pf* CL145A. In addition to these six species, Molloy et al. also evaluated the impact of *Pf* CL145A to other nontarget bivalve species. Overall, Molloy et al. concluded that exposing these nontarget organisms under aerated conditions to live *Pf* CL145A cells clearly points to the potential for high host specificity.

Meehan et al. (2014) exposed *Austropotatamobius pallipes, Chironomus plumosus,* and *Andonta* species to Zequanox in 72 h static renewal studies, with renewals every 12 or 24 h. Lethal concentrations were calculated for 10%, 50%, and 100% mortality of the organisms when statistically possible. These organisms were selected, because they are representative nontarget organisms native to the Irish ecosystem for the manner in which Zequanox is used. The resulting LC50 values were *Anodonta*: ≥500 mg a.i./L; *C. plumosus*: 1075 mg a.i./L; *A. pallipes*: ≥750 mg a.i./L. The authors concluded that these results demonstrate that Zequanox does not negatively affect these organisms at the concentration required for greater than 80% zebra mussel mortality and the maximum allowable registered treatment concentration in the United States* and the studies further demonstrate the high species specificity of Zequanox, and support its use in commercial facilities and open waters.

Nontarget Bivalve Studies

In order to investigate the toxicity of Zequanox to nontarget mussels, testing was conducted using several freshwater mussels native to the United States, as well as one marine bivalve native to northern coastal areas of the Atlantic Ocean. In this section, we describe three bodies of work testing different nontarget mussel species. Because control of zebra and quagga mussels is frequently occurring in the habitats of these native species, it is important that control options do not further impact the native species' survival. In addition, because these organisms exhibit similar filter feeding behaviors to zebra and quagga mussels, it was important to determine the target specificity in relation to other bivalve filter feeders. We elected to specifically present the body of data on the other nontarget mussel species in this chapter, because the methodologies and results are available in the public domain and the similar exposure pathway for the target and nontarget mussels. *Mytilus edulis* is frequently used in nontarget testing where there is concern for marine aquatic life; however, there are no standard or one representative freshwater bivalve species for nontarget testing.

Molloy et al. (2013c) evaluated the ecotoxicological impact of unformulated live *Pf* CL145A cells to other bivalve species in static single exposure nontarget tests at concentrations of 100 and 200 mg a.i./L, a total of six native freshwater unionid species, *Pyganodon grandis, Pyganodon cataracta, Lasmigona compressa, Strophitus undulates, Elliptio complanata, Lampsilis radiata* and, one saltwater species, *M. edulis* mussel. The nontarget mussel species were exposed for 72 h and then rinsed and placed in freshwater or salt water for the *M. edulis* species, and monitored up to 28 days post exposure. In these studies, no mortality was observed in all the nontarget mussels tested at concentrations that are effective at controlling zebra and quagga mussels.

Meehan et al. (2014) evaluated the impact of Zequanox to an *Anodonta* species collected in Ireland at concentrations from 100 to 500 mg a.i./L in 12 h static renewal tests with a 72 h exposure. The *Anodonta* species was selected by Meehan et al., because in Ireland, *Anodonta* are widespread in freshwater systems, but have been largely extirpated in waters where zebra mussels have invaded, due to the attachment of zebra mussels to *Anodonta* shells and competition for food (Lucy et al. in press; Minchin et al. 1998). No *Anodonta* mortality was observed, and the authors concluded that the data indicated that Zequanox could be a valuable tool in *Anodonta* restoration efforts.

In a third independent set of nontarget bivalve exposure studies completed by Pletta (2013), freshwater species within the *Lampsilis, Megalonaias,* and *Margaritifera* genera were selected for testing,

---

* 200 mg a.i./L for adult mussels from 2011 to 2014 in the United States. Lowered to 100 mg a.i./L in June 2014 due to developmental improvements (Marrone 2012).

TABLE 34.1

Zequanox 72 h Exposure LC50 Endpoint Values for Subadult and Juvenile Freshwater Bivalve Species

Species	Exposure	72 h LC$_{50}$[b] (mg a.i./L)
Lampsilis siliquoidea (subadult)	12 h static renewal with circulation	235[a]
Lampsilis abrupta (subadult)	12 h static renewal with circulation	207[a]
Megalonaias nervosa (subadult)	12 h static renewal with circulation	>500[a]
Lampsilis abrupta (juvenile)	12 h static Renewal[c]	157[a]
Margaritifera falcate (juvenile)	12 h static Renewal[c]	126[a]

Source: Pletta, M.E., Particle capture by freshwater bivalves: Implications for feeding ecology and biopesticide delivery (published Masters thesis), Missouri State University, Springfield, MO, 2013.

[a] Endpoint calculated using nominal active ingredient concentrations.

[b] Mortality endpoints assessed 10 days after completion of the 72 h exposure period.

[c] Study conducted in petri dishes without active aeration.

because these species are a representative range within the unionid family of native freshwater mussels, representing different tribes, and are species that have been impacted by invasive mussels both indirectly (through habitat destruction) and directly (by attachment to, and growth upon, shells). For each of the species selected in this study, 30 organisms (3 replicates of 10 organisms) were exposed to Zequanox at concentrations between 1 and 500 mg a.i./L for a total of 72 h according to ASTM E2455-06 (ASTM 2006). Following completion of the exposure period, test mussels were transferred to Zequanox-free cultivation tanks and monitored for mortality over 10 days. The LC50 values reported in this study account for delayed mortality effects following a shorter-term acute exposure. As shown in Table 34.1, resulting LC50 values were higher for all tested adult mussel species than the maximum allowed Zequanox application concentration. Zequanox exposure to subadult *Lampsilis siliquoidea, Lampsilis abrubta,* and *Megalonaias nervosa* mussels at the maximum allowable Zequanox application concentration of 100 mg a.i./L for a total of 72 h resulted in very low levels of mussel mortality (between 0% and 7% for each species). Although 72 h exposures to juvenile *L. abrubta* and *Margaritifera falcate* at 100 mg a.i./L Zequanox resulted in higher total mortality (60% ± 20% and 50% ± 10%, respectively), it is important to note that the maximum allowable Zequanox treatment time is 8 h, significantly lower than the static renewal 72 h exposures presented here. The higher mortality in the test conducted with juveniles may also be experimental artifacts, as the testing configuration employed for juvenile studies allowed for dissolved oxygen (DO) concentrations to deplete below 1 mg/L in the 100 mg a.i./L treatments.* Likewise, the toxicity of Zequanox to *L. abrubta* and *M. falcate* juveniles was considerably lower when concentrations were reduced to 10 mg a.i./L (mean 72 h mortality of 13% and 23%, respectively) or 1 mg a.i./L (mean 72 h mortality of 0% and 3%, respectively). Reduction of Zequanox concentrations to these levels is likely to occur rapidly in most receiving waters due to dilution and biodegradation. Importantly, treatment management practices can be employed to further reduce risks posed to juvenile mussel life stages. These results suggest that Zequanox is not especially toxic to a variety of nontarget mussel species, and demonstrates the potential applicability of Zequanox for habitat and native unionid species restoration.

The results from the three independent studies presented in this section suggest that Zequanox does not present an acute toxicity risk to a wide variety of different nontarget mussel species in the adult and subadult life stages. Although study results suggest that Zequanox may be slightly more impactful to mussels in the juvenile life stage, this effect is unlikely to be detrimental at the population level.

---

* It is common to include aeration in most aquatic nontarget toxicology studies because the small static chambers in which the studies are conducted can exaggerate impacts to water quality. These impacts are not always observed in the untreated controls due to differences in the microbial communities after the product is introduced. However, due to necessary study design, aeration was not included.

According to Dr. Chris Barnhart, a professor at Missouri State University and expert on the ecology of freshwater mussels native to the United States

> When looking at the non-target toxicity of a product like Zequanox and its potential to be used for environmental restoration, the more critical life stages to consider for native freshwater mussels in ecotoxicity assessments are subadults and adults. While mortality in juveniles shouldn't be ignored, the natural background juvenile recruitment and survival is so low that the additional negative impact of a product like Zequanox is less of a concern than maintaining the health of subadults and reproducing adult populations (personal communication).

Treatment and Efficacy Assessment Methodologies for Presented Case Studies

In order to demonstrate the efficacy of the Zequanox product under field conditions, we have selected five representative case studies to describe in this chapter. The general methodology for the case studies was to apply the product to an infested system and monitor mussel mortality after the treatment to determine the efficacy. This section describes the efficacy of monitoring and application techniques employed for each of the case studies.

Product Concentration Monitoring

To monitor Zequanox product concentration in treated water, a linear correlation between the turbidity and the active ingredient of the product, *Pf* CL145A cells, is developed for each specific treatment. Using water from the treated systems, field scientists determine a site-specific target turbidity measured as milligrams of active ingredient per liter (mg a.i./L) and turbidity in nephelometric turbidity units (NTU).

Biobox Efficacy Monitoring

Bioboxes are aquaria-like structures that receive a continuous flow to support live invasive mussels that can be monitored to demonstrate the presence of mussel infestations and efficacy of control practices (Mackie and Claudi 2010). They are also commonly referred to as side stream samplers. Bioboxes are used to monitor efficacy for most Zequanox applications in pipes and other infrastructures with defined flows and volumes. An example biobox is presented in Figure 34.1. Water enters the biobox via the supply pipe on the left side of the unit. The water then travels through the biobox with serpentine flow over or below the baffles (dark gray plates within the biobox) before exiting through a standpipe on the right. Flow-through enclosures containing live mussels are placed into the biobox sections for adult treatment

FIGURE 34.1 Example of a baffled biobox for efficacy monitoring (see Mackie and Claudi 2010 for additional description).

efficacy monitoring. For monitoring settlement management treatments, settlement plate assemblies are placed into the biobox (see section Settlement Management Treatment Assessment).

Bioboxes have been used by the industry for more than 20 years to monitor invasive mussel control treatments by either applying products directly into the bioboxes or to monitor efficacy from facility treatments by receiving a steady flow of the treated water during treatment programs. Water flows continuously through the bioboxes before, during, and after treatments, continuing until final mortality is measured. Because the treated mussels in a biobox are exposed to water with the same properties as the water in the treated system, observed mussel mortality in the bioboxes is considered to be an approximation of the mussel mortality inside the treated system.

For Case Study 1, bioboxes received a constant flow of water from the treated system and a control biobox was set up to receive untreated water. For Case Study 3, bioboxes received continuous flow of water before and after treatment, but the mussels were treated under static conditions within the biobox. For the settlement management case study presented in Case Study 4, concentrated product solution was applied directly into the bioboxes, which received 1–2 gpm (gallons per minute) of system water flow. For all treatments, injection rates were metered to achieve the target treatment concentration as measured by a turbidimeter.

Adult Mussel Collection and Efficacy Monitoring within Bioboxes

For each case study conducted at a facility, mussels were collected from the facility's system water source—within close proximity of the testing location—or from a nearby lake. The mussels were collected at least 3 days prior to treatments. Collection was done manually by gently removing the mussels from surfaces with a paint scraper or long-handled scraper. Scraping, as opposed to pulling, ensures that the mussels are removed by the release of the byssal threads from the structure and not by tearing the threads from the mussels. Mussels were then sorted by size, and intact, siphoning adult mussels (15–25 mm) were selected for use in the treatments. Fifty mussels were placed into each enclosure. The enclosures consisted of acrylic pipe sections with mesh caps that allowed water to flow through them while keeping the mussels contained.

A minimum of three enclosures containing mussels (150 mussels total) was placed in each biobox. After each treatment, scientists measured mussel mortality at least weekly, by counting the number of mussels that were alive (closed valves) and those that were dead (gaping and unresponsive to prodding). The field scientists removed the dead mussels, leaving only live mussels within the enclosures. Mortality was scored from 2 weeks post treatment up to 1 month post treatment, depending on the water temperature. The number of living mussels remaining within each enclosure at the end of the posttreatment period was compared with the original number of live mussels (50), and the percent mortality (the percent of killed mussels) was calculated. The percent mortality from each enclosure of mussels was then averaged with the other enclosures within the same treatment to calculate a mean mortality and standard deviation (n = 3). The mortality of the treated mussels was compared with that of the untreated mussels (in a untreated control biobox) to confirm that mortality in the treated system was the result of the treatment, and not the result of handling, old age, or other reasons unrelated to the efficacy of the product.

Settlement Management Treatment Assessment

The general method for settlement management treatments is to apply low concentrations of Zequanox throughout a system for a few hours on a frequent basis (up to once every 2 weeks), and determine the treatment strategy's effectiveness by comparing mussel density on both treated and untreated sample plates. For application and monitoring in Case Study 4, the methods and setup were similar to those used for adult treatments, but the equipment used was smaller and required less space, because the product needed per treatment was decreased.

The assessment of the level of control (or reduction in mussel accumulation) was evaluated by comparing the density of mussels on sample plates within treated systems to the density on untreated control sample plates (Link 2013). Paired sets of ABS plates were placed in bioboxes facing one another, parallel to the flow of water moving through the biobox to have equivalent settlement occurring on each set of plates (Figure 34.2).

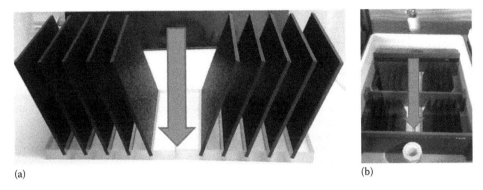

(a)　　　　　　　　　　　　　　　　　　　　(b)

FIGURE 34.2 Paired treated and untreated ABS settlement plates as positioned in bioboxes in between Zequanox settlement management treatments. The arrows indicate direction of water flow. (a) Close up of plates outside of biobox and (b) plates within biobox.

One set of each pair was treated with 10 mg a.i./L of Zequanox for 2 h once every 14 days throughout the 3.5-month study, while the untreated plates are stored in an untreated biobox. Both sets were handled in equal amounts and stored in the same biobox between treatments. Mussel densities from the treated plates are compared with the densities on the untreated partner set of plates. The evaluation techniques included both digital microscope imaging and visual plate counts to verify the imaging method. Mussel density, and thereby percent control or reduction of the density, was monitored biweekly, prior to treatments.

Environmental and Recreational Rehabilitation Application Assessments

For open water treatments, MBI followed similar collection, handling, and selection procedures employed for our in-pipe treatments. For the open water case study, Case Study 5, 50 siphoning mussels were placed into each of 18 mussel containment chambers and groupings of 3 chambers were secured to a weight to keep them stable, then groupings of 3 chambers were randomly placed among the benthos in each treated and control site (Whitledge et al. in press). After the treatments, all the chambers were relocated to one location where they could be easily accessed for posttreatment mortality assessments (Figure 34.3).

FIGURE 34.3 Mussel containment chambers used in monitoring Zequanox efficacy treatments conducted in open waters such as lakes.

In addition to the MBI monitoring, collaborators on the project, Southern Illinois University (SIU), completed an independent evaluation of the efficacy of the treatments. The day following treatment and again, 1 week post treatment, SIU conducted mortality monitoring by randomly selecting three 1 m² sample plots within each treated and control site (Whitledge et al. in press). Samples of the benthic substrate within these plots were collected and analyzed for dead and live mussels. Mortality monitoring of naturally settled mussels provided representative data for mussels that were not handled or otherwise disturbed by the collection process and was complementary to the analysis completed by MBI.

Case Studies

This section presents an overview of demonstration and commercial treatments completed in 2011 and 2012 throughout North America and in Europe. The five case studies that we present demonstrate the product efficacy in different water bodies, different geographies, and with different invasive mussel populations. We have included case studies that cover both the adult annual and the settlement management treatment programs for applications within pipes and other infrastructure and one example of an open water treatment.

Case Study 1—Adult Mussel Control within a Hydropower Facility in Canada

- Canada Hydropower Facility—Nameplate Capacity 100–500 Mega Watts (MW)
- Treated cooling water for two turbines with a 1940 gpm cooling water system
- Achieved 94% mortality with a single 6 h treatment

In previous years, the customer treated the cooling water system by continuously applying chlorine for approximately 10–14 days. The annual treatments required extensive health and safety training review each year for all personnel involved in the treatments and a minimum of two personnel to monitor the treatments around the clock. Over the years of treating with chlorine, the customer had optimized the chlorine treatment regime to reduce the total amount of chlorine used while achieving greater than 90% mussel control. However, the customer remained concerned about the worker and environmental safety with these chlorine treatments and was interested in identifying a more environmentally compatible solution to their mussel control needs.

A single approximately 6 h treatment was performed in August 2012 to control a mixed population of adult zebra and quagga mussels within the facility's cooling water system. Zequanox was mixed on site and transferred into the facility's existing day tank used previously for chlorine applications. From the day tank, Zequanox was applied to the cooling water system using the pumps and chlorine injection system equipment already in place for chlorine treatments. Mussel mortality within bioboxes indicated that the treatment resulted in mussel mortality greater than 90% (Figure 34.4). Mortality reached 91% by day 7; final mortality was calculated as 94% at day 28.

Case Study 2—Adult Mussel Control within Intake Bays at a Coal-Powered Plant in Oklahoma

- Oklahoma, United States, Coal Power Plant—Nameplate Capacity 10,000 MW
- Achieved 95% mussel mortality in treatment of two 214,000 gal intake bays to cooling water system, static treatment

In Case Study 2, cooling water intake bays were treated rather than the flowing cooling water system. The customer determined that the primary problem they wanted to address was to reduce the population and size of the mussels in the intake bays so that the mussels would not slough off and become sucked into the cooling water system. Due to the high velocities, the customer concluded that the cooling water

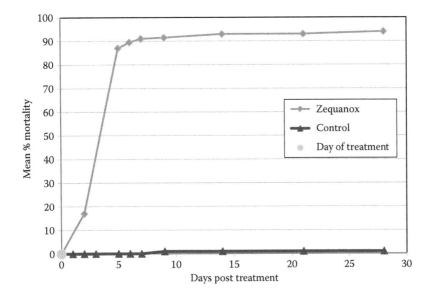

FIGURE 34.4 Average mussel morality observed in bioboxes after an approximately 6 h treatment of the two cooling water systems of a Canadian hydropower facility, August 2012.

system itself was not heavily fouled by mussels at the time of treatment. In previous years, the customer sent divers into each intake bay to scrap the walls. The task was difficult, and the customer was concerned about employee safety.

On the day of treatment, the inlet and the outlet of the intake bays were blocked and flow to the cooling water system was stopped. Rather than using bioboxes for testing adult mussel mortality, slightly modified methods were used. Adult mussels were collected according to the methods described earlier. Fifty mussels were enclosed in hard plastic mesh enclosures on rope that was then directly suspended into the intake bays. Three enclosures were placed in the treated bays, and three enclosures were placed in an untreated bay. After treatment, mortality was monitored over 25 days and compared to the untreated control. Zequanox was mixed on site, and the concentrated solution was pumped directly into the intake bays for approximately 1 h. MBI scientists used air compressors attached to diffusers to completely mix the product in the intake bays. The facility staff held the treated water static in each intake bay for approximately 12 h.

The customer quantified the final mussel mortality in both intake bays, via the placed mussels, to be greater than 95% (Figure 34.5). The customer also visibly observed a significant reduction in the treated bays versus the untreated bays and determined that it was not necessary to send divers into the intake bays to scrap the mussels from the walls.

Case Study 3—Zequanox Compared to Chlorine at a Drinking Water Treatment Facility in Sligo, Ireland

The main objective of Case Study 3 was to demonstrate the efficacy of Zequanox in controlling zebra mussels in comparison to the current practice of chlorination at a drinking water treatment facility in Sligo, Ireland. This case study is described in detail in Meehan et al. (2013).

The Sligo drinking water treatment plant began using chlorine on an annual basis to control zebra mussels in three square concrete raw water chambers in 2009. Live mussels and mussel attachment was not a concern in the facility after the raw water chambers, because the drinking water treatment processes downstream of the raw water chambers sufficiently control and remove the mussels. However, mussel shells and debris were clogging pipes and impacting the downstream drinking water treatment processes. Therefore, it was necessary to address the mussel population on an annual basis to limit shell debris and mussel growth.

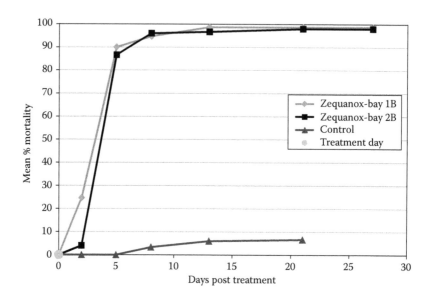

FIGURE 34.5 Quagga mussel mortality of seeded mussels in intake bays at an Oklahoma coal plant after a Zequanox application.

From 2009 to 2011, chlorine treatments were conducted at the end of the reproductive and primary growth seasons to treat all the mussels that colonized that year. These treatments achieved between 80% and 100% mussel mortality. During a typical chlorine treatment, the plant was forced to shut down drinking water treatment for approximately 7 days to isolate the water in the raw water chambers. The length of the shutdown was disruptive to the water supply, and a shorter treatment time was desirable.

Three 200 L bioboxes were placed on a flow through system in the Sligo drinking water treatment plant on July 13, 2011 (Figure 34.6). These tanks received water from the water treatment plant's main chambers via gravity flow, with a total flow of 287,000 L over 13 weeks until October 11, 2011. Of these three tanks, one was established to serve as the experimental control and the other two received Zequanox treatments. The tanks were covered with heavy plastic with weights on each side to protect from any

FIGURE 34.6   Bioboxes outside of Sligo drinking water treatment plant.

harsh weather exposure or interference. PVC plates were placed in each of the three tanks to allow for natural mussel settlement. For additional information on the methods and results from the mussels settled on the plates, see Meehan et al. (2013).

Prior to treatment on October 10, 2011, the bioboxes were moved from the water treatment plant to the research facility at IT Sligo. The bioboxes were then no longer on a flow-through system. Twenty-four hours prior to Zequanox treatment, the seeded mussels were checked for mortality and any dead mussels were replaced with healthy, live mussels. One week after treatment of the bioboxes with Zequanox, on October 17, 2011, the chlorine treatment of the raw water chambers at the drinking water treatment plant took place. The same methods for assessing adult mortality use for the Zequanox treatments were also used for the chlorine treatments.

Zequanox was applied directly to bioboxes through a peristaltic pump and flexible tubing. The product was gently mixed in the biobox with an overhead mixer. The Zequanox treatment was compared to the chlorine treatment completed in 2011 according to the facilities standard practices (Table 34.2). The facility operators maintained chlorine residual concentration of 2 mg/L for 7 days. Meehan et al. (2013) observed approximately 80% total mortality for both treatments; however, the high chlorine concentrations were maintained for 168 h versus 8 h of treatment with Zequanox (Figure 34.7). The results of this case study demonstrated that Zequanox was as effective at controlling zebra mussels as the previously implemented chlorine treatment program. Meehan et al. (2013) concluded that Zequanox provided a suitable alternative to chlorine and would reduce the downtime of the drinking water treatment plant from greater than 7 days to only 1 day.

TABLE 34.2

Chlorine and Zequanox Treatment in Sligo, Ireland: Treatment Summary and Results

Product	Concentration	Treatment Duration (h)	Treatment Date	Final Mean Percent Mortality
Zequanox	200 mg a.i./L	8	October 10	80
Chlorine	2 mg/L	168	October 17	80

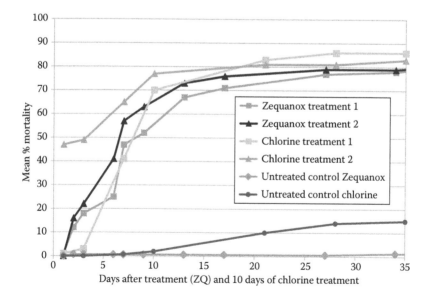

FIGURE 34.7 Efficacy of 8 h 200 mg a.i./L Zequanox treatment compared to 7 day 2 mg/L residual chlorine treatment at a drinking water treatment facility in Sligo, Ireland. (From Meehan, S. et al., *Manag. Biol. Invasions*, 4(2), 113, 2013. With permission.)

Case Study 4—Mussel Settlement Management within a Hydropower Facility on the Colorado River

In the summer and fall of 2012, frequent, low concentration Zequanox treatments were evaluated for the control of newly settling young invasive mussels. Settlement management treatments provide a mussel control strategy for facilities in which mussels greater than 4 mm in size must be minimized, or in facilities where the amount of debris that may occur within their system must be minimized to prevent the occlusion of strainer baskets or other infrastructure. These trials were conducted at a hydropower facility on the Colorado River, a location highly infested with quagga mussels.

To evaluate the treatment strategy, paired sets of ABS plates were placed in storage bioboxes receiving raw water from the intake bay of the facility. One set of each pair of plates was treated in a treatment biobox with 10 mg a.i./L of Zequanox for 2 h once every 14 days throughout the 3.5-month study, while the partner set of plates was placed in an untreated control biobox. Mussel density from the treated plates was compared with the density on the untreated partner set of plates (both sets were handled equal amounts, and stored in the same biobox between treatments) once every 14 days.

The results of the study indicated that the biweekly 2 h applications with a treatment concentration of 10 mg a.i./L of Zequanox achieved controlled of 87% of mussel settlement. During the biweekly treatments, the mussel density in the untreated control increased to approximately 160 mussels per plate, depicted by the darker line and triangles in Figure 34.8, while the Zequanox treated plates remained less than 10 mussels per plate, shown by a lighter line and diamonds in Figure 34.8.

In conclusion, the biweekly Zequanox treatment strategy successfully controlled the settling mussel population. Also significant, the biweekly treatments reduced the total amount of product needed annually by approximately one half when compared to higher concentration, single annual adult treatments, which can drastically reduce the overall annual cost to treat and protect a flowing system from mussel problems. In addition, the biweekly Zequanox treatment strategy requires little space and limited effort to complete, and often requires no, to minimal, changes to a system, providing a valuable treatment option for facilities that are sensitive to mussel debris.

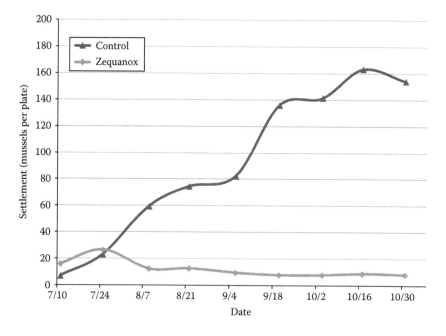

FIGURE 34.8   Mussel densities during the course of a 3-month treatment series on untreated plates (triangles) and plates treated with Zequanox at biweekly intervals for 2 h periods at 10 mg a.i./L (diamonds).

Case Study 5—Open Water Treatments within Deep Quarry Lake, Illinois

Deep Quarry Lake is within the West Branch Forest Preserve near Bartlett, Illinois. Deep Quarry Lake is a popular urban fishing location containing bass, sunfish, channel and flathead catfish, carp, bluegill, walleye, northern pike, and crappie. Zebra mussels were first discovered in Deep Quarry Lake in 2009 and have since developed into a well-established population within the lake.

Treatments in Deep Quarry Lake were conducted during July 2012 in partnership with Professor Gregory Whitledge and his graduate student assistants from Southern Illinois University and the staff of the Forest Preserve District of DuPage County. Three sets of paired, 24-square-meter (m²) treatment and control sites were established within Deep Quarry Lake (Whitledge 2015). Zequanox treatments in 2012 sought to evaluate the efficacy of the product in open water setting. Treatment sites were enclosed by PVC barriers. The study team selected sites based on the presence of settled zebra mussels throughout the site, as well as accessibility, and similar depth within the paired sites.

Zequanox was applied through a rigid pipe to the bottom one meter (Figure 34.9). Treatment concentration and water quality were monitored both at the surface and at the bottom with a long-cabled probe. Treatment concentrations ranged between 93 and 124 mg a.i./L. Barriers were removed 24 h after application.

DO, pH, conductivity, and temperature were monitored by MBI staff during treatment, and no lasting effect was observed between parameters within treated and untreated control sites. The pH values ranged from 8.4 to 8.6 during the treatment and after the treatment. The DO and turbidity values for treated and control site 2 are presented in Figure 34.10. During the treatment, the turbidity values increased as expected. Temporary but substantial reductions in DO were observed in treatment locations during the morning following the Zequanox treatment, likely due to the presence of the barriers that prevented well-oxygenated water from circulating into treatment zones from adjacent areas in the lake. DO concentrations quickly rebounded to levels consistent with control sites upon removal of barriers (Figure 34.10). Results from grab samples sent to an independent lab and analyzed for ammonia, total nitrogen, total phosphorus, biochemical oxygen demand, and chlorophyll *a* showed no impact for all parameters (Table 34.3). Mussel mortality was visually observed by the study team (as gaping shells) in treated sites the day following treatment. Additionally, 24 h after treatment application, prior to removal of the treatment barriers, fish were observed swimming in the treated barriers.

Mortality monitoring conducted by MBI staff showed that the Zequanox treatment was highly effective against invasive zebra mussels in all three treated sites by 14 days after treatment with an average of 97% mortality across all three sites and a control mortality of 11% (Figure 34.11). As a result of the partnerships developed in this project, private and academic institutions assessed Zequanox treatment efficacy utilizing different methods. Figure 34.12 shows the results from SIU assessments, which support the results found by MBI with average treated site mortality of greater than 90% compared to a control site mortality of less than 10% one week after treatment. Results and data collection are described further in Whitledge et al. (2015).

(a)

(b)

FIGURE 34.9 Example of treated site: during application with rigid piping (a) and after application (b). Treatment concentration ranged between 93 and 124 mg a.i./L. (From Whitledge, G.W. et al., *Manag. Biol. Invasions*, 2015. in press. With permission.)

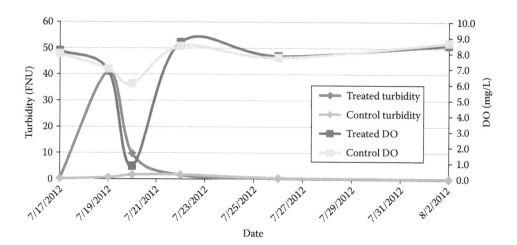

FIGURE 34.10 Dissolved oxygen (DO) concentrations and turbidity values in Deep Quarry at site 2 during and 14 days post Zequanox treatment in Deep Quarry. Site 1 and 3 values are nearly identical. (From Whitledge, G.W. et al., *Manag. Biol. Invasions*, 2015. in press. With permission.)

TABLE 34.3

Independent Water Quality Results from Deep Quarry Lake Zequanox Treatment

Parameter	Day before Treatment	Day after Treatment
BOD (mg/L)	<2	<2
Chlorophyll *a* (mg/L)	Nondetected	Nondetected
Total phosphorus (mg/L)	0.032	0.032
Total nitrogen (mg/L)	Nondetected	Nondetected
Ammonia (mg/L)	Nondetected	Nondetected

Source: Whitledge, G.W. et al., *Manag. Biol. Invasions*, 2015. in press. With permission.

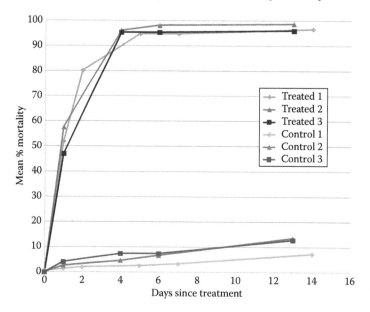

FIGURE 34.11 Average mussel mortality in treated and control plots of collected mussels placed in enclosures and placed within plots over the 14 days after a Zequanox treatment at Deep Quarry Lake. (MBI assessment and described in Whitledge, G.W. et al., *Manag. Biol. Invasions*, 2015. in press. With permission.)

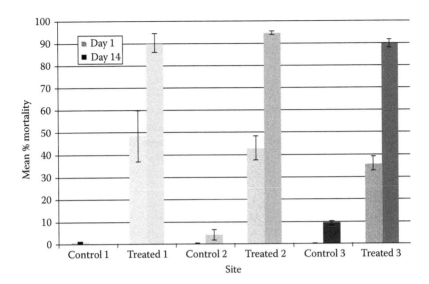

FIGURE 34.12 Average mussel mortality in treated and control plots at 1 day and 14 days after a Zequanox treatment in Deep Quarry Lake during July 2012. (Assessment conducted by Dr. Whitledge and graduate student team (SIU) and described in Whitledge, G.W. et al., *Manag. Biol. Invasions*, 2015. in press. With permission.)

In 2013, one large Zequanox treatment (324 m²), inclusive of shoreline, was conducted with the same partners at Deep Quarry Lake with an objective of evaluating larger (area) treatments and improved application techniques (among other objectives). This study also resulted in similarly high mussel control (see Whitledge et al. 2015 for additional details).

Overall, the results of the Zequanox studies conducted at Deep Quarry Lake strongly support the use of Zequanox in open water settings. The high level of mussel control, in combination with the lack of lasting water quality impact, indicate Zequanox can meet many of the challenging demands of open water treatments, offering a registered treatment option where there has historically been no alternative options.

Summary

Collectively, the product overview and the presented case studies demonstrate that Zequanox provides a highly effective, safer mussel control option. As is summarized by the nontarget studies completed to date and the reviews of these studies available from the EPA, at treatment concentrations and durations used for zebra and quagga mussel control, Zequanox treatments result in minimal impact to nontarget species. Zequanox thereby greatly reduces the need for nonselective chemical products and their associated neutralization, extended application times, and the addition of potentially carcinogenic by products from chlorine and its derivatives to natural waters while increasing environmental protection and worker safety.

The presented case studies demonstrate the versatility of Zequanox and the ability to tailor treatments according to customer needs in a variety of settings and offer an equivalent level of control to Zequanox's chemical counterparts. Collectively, the case studies for confined systems demonstrate that Zequanox can provide greater than 90% control in both flowing and static infrastructure in high volumes of water. In addition, settlement management treatments provide flexibility for customers in need of enhanced control of mussel growth and shell debris management by achieving greater than 85% settlement control. Further, optimized treatment strategies for various life stages, as well as for differing application settings, provide flexibility for custom treatment plan development based on facility conditions (flowing vs. static, growth rate of mussels) and needs (tolerance for mussel size and debris).

In natural habitats and recreational waters that have been degraded by mussel infestation, Zequanox provides a tool for rehabilitation and management with limited impact to nontarget organisms where there have previously been few options. MBI has been able to effectively demonstrate that invasive mussel control in open waters is an option without harming the beneficial uses of the water body. MBI plans to continue to expand open water applications to address specific needs in open water recreation and environmental restoration.

Because of the human and environmental safety of Zequanox, MBI continues to explore other applications for invasive mussel control where previously there have been limited options. These categories include native mussel restoration, fish spawning bed restoration, and the aquaculture industry including both infrastructure and transportation of fish from infested water bodies to uninfested water bodies. Because Zequanox is effective at controlling the veliger life stage, MBI plans to pursue treatment techniques and applications for treating and control of the veliger life stage that meet customer needs.

REFERENCES

ASTM. 2006. Standard guide for conducting laboratory toxicity tests with freshwater mussels (E2455-06).

Caraco, N.F., Cole, J.J., Raymond, P.A., Strayer, D.L., Pace, M.L., Findlay, S.E.G., and Fischer, D.T. 1997. Zebra mussel invasion in a large, turbid river: Phytoplankton response to increased grazing. *Ecology* 78: 588–602.

Lucy, F.E., Burlakova, L.E., Karatayev, A.Y., Mastitsky, S.E., and Zanatta, D.T. 2014. Zebra mussel impacts on unionids. In: Nalepa T and Schloesser D (eds.), *Quagga and Zebra Mussels,* 2nd edn. Taylor & Francis, Oxford, U.K.

Mackie, G.L. and Claudi, R. 2010. *Monitoring and Control of Macrofouling Mollusks in Freshwater Systems,* 2nd edn. CRC Press, Boca Raton, FL, 508pp.

Marrone Bio Innovations. 2012. Product label: MBI-401 SDP. EPA Reg. No. 84059-15. http://www.epa.gov/pesticides/chem_search/ppls/084059-00015-20120309.pdf. Accessed on March 9, 2012.

Meehan, S., Lucy, F.E., Gruber, B., and Rackl, S. 2013. Comparing a microbial biocide and chlorine as zebra mussel control strategies in an Irish drinking water treatment plant. *Management of Biological Invasions* 4(2): 113–122.

Meehan, S., Lucy, F.E., Gruber, B., Rackl, S., and Shannon, A. 2014. Ecotoxicological impact of Zequanox®, a novel biocide, on selected non-target Irish aquatic species. *Ecotoxicology and Environmental Safety* 107: 148–153.

Molloy, D.P. 2001. A method for controlling *Dreissena* species. US Patent and Trademark Office, US Department of Commerce. Patent No. 6,194,194, filed December 17, 1997 and issued February 27, 2001.

Molloy, D.P. 2004. A method for controlling *Dreissena* species. Canadian Intellectual Property Office, Industry Canada. Patent No. 2,225,436, filed December 27, 1997 and issued December 21, 2004.

Molloy, D.P., Mayer, D.A., Gaylo, M.J., Burlakova, L.E., Karatayev, A.Y., Presti, K.T., Sawyko, P.M., Morse, J.T., Paul, E.A. 2013c. Non-target trials with *Pseudomonas fluorescens* strain CL145A, a lethal control agent of dreissenid mussels (Bivalvia: Dreissenidae). *Management of Biological Invasions* 4(1): 71–79.

Molloy, D.P., Mayer, D.A., Gaylo, M.J., Morse, J.T., Presti, K.T., Sawyko, P.M., Karatayev, A.Y. et al. 2013a. *Pseudomonas fluorescens* strain CL145A—A biopesticide for the control of zebra and quagga mussels (Bivalvia: Dreissenidae). *Journal for Invertebrate Pathology* 113: 104–114.

Molloy, D.P., Mayer, D.A., Giamberini, L., Gaylo, M.J. 2013b. Mode of action of *Pseudomonas fluorescens* strain CL145A, a lethal control agent of dreissenid mussels (Bivalvia: Dreissenidae). *Journal for Invertebrate Pathology* 113: 115–121.

Minchin, D. and Moriarty, C. 1998. Zebra mussels in Ireland. Fisheries leaflet 177. Marine Institute, Dublin, Ireland, 11pp.

Pletta, M.E. 2013. Particle capture by freshwater bivalves: Implications for feeding ecology and biopesticide delivery (published Masters thesis). Missouri State University, Springfield, MO.

Rodgers, P.B. 1993. Potential of biopesticides in agriculture. *Pesticide Science* 39: 117–129.

Strayer, D.L. 2008. Twenty years of zebra mussels: Lessons from the mollusk that made headlines. *Frontiers in Ecology and the Environment* 7: 135–141.

Thornton, J. 2000. *Pandora's Poison: Chlorine, Health, and a New Environmental Strategy.* MIT Press, Cambridge, MA.

Tourt, L.W. and Maciorowski, A.F. 1997. Information needs for pesticide registration in the United States. *Ecological Applications* 7: 1086–1093.

U.S. Department of Energy, National Energy Technology Laboratory (NETL). 2006. Effectiveness of a micro-bial control agent method of controlling zebra mussel fouling compared to chlorine injection. Draft report. Prepared by WorleyParsons Group, Inc., Report No. EJ-2004-06. February 17, 2006.

U.S. Environmental Protection Agency, 2011. Biopesticides registration action document. *Pseudomonas fluo-rescens* strain CL145A/pesticide chemical (PC) Code: 006533. Version #2. Office of Pesticide Programs Biopesticides and Pollution Prevention Division. July 29.

U.S. Environmental Protection Agency, (DRAFT). 2004. The globally harmonized system of classification and labeling of chemicals (GHS): Implementation planning issues for the Office of Pesticide Programs. http://www.epa.gov/oppfead1/international/global/globa-whitepaper.pdf. Accessed on August 7, 2014.

U.S. Environmental Protection Agency (EPA). 1999. Wastewater technology fact sheet: Chlorine disinfection. U.S. Environmental Protection Agency, Washington, DC. EPA/832-F99-062.

Whitledge, G.W., Weber, M.M., DeMartini, J., Oldenburg, J., Roberts, D., Link, C., Rackl, S.M. et al. 2015. An evaluation of Zequanox® efficacy and application strategies for targeted control of zebra mussels in shallow-water habitats in lakes. *Management of Biological Invasions*.

Index

A

Acrylonitrile butadiene styrene (ABS) plates, 379, 383, 392–393
Acute toxicity
    anesthetic pretreatments, 491–492, 497
    biocides, 492–493, 497
    Cutrine-Ultra, 490–491, 495
    formalin, 489–490, 498
    Peraclean 15, 490–491, 495–496
    potassium chloride, 489–490, 493–494
    Spectrus CT1300, 490–491, 496
ACWG, *see* Asian Clam Working Group (ACWG)
AERF, *see* Aquatic Ecosystem Restoration Foundation (AERF)
AIS, *see* Aquatic invasive species (AIS)
Analysis of covariance (ANCOVA), 107
Analysis of variance (ANOVA), 107, 383
ANSTF, *see* Aquatic Nuisance Species Task Force (ANSTF)
Aquatic Ecosystem Restoration Foundation (AERF), 344
Aquatic invasive species (AIS)
    boat decontamination, 161–162
    equipment inspection and cleaning
        aging water management, 227–228
        clusters, 231, 233
        declining budgets, 227–228
        definition, 228–229
        Deming's observation, 234–236
        drought and climate change, 227–228
        High Park Fire, 230
        human factors, 235, 237–238
        hydroelectric power, 231
        vs. IPM, 231, 233–234
        law enforcement, 264
        mountain pine beetle, 229–231
        nonhuman factors, 235, 237
        policymakers, 231
        protocols and standards, 177, 184–185
        self-certification process, 263
        traffic routes, 231–232
    firefighting operation
        cleaning internal tanks, 158
        equipment contracts, 158–159
        NWCG, 155–156, 158
        pathways, 155
        quaternary ammonium compounds, 157–158
        Redbook operational guidelines, 156–157
        risks, 155
        vectors, 154–155
    WFDSS, 159
Aquatic Nuisance Species Task Force (ANSTF), 302
Army Corps of Engineers method, 21
Asian Clam Working Group (ACWG), 326–327

B

Boat decontamination
    AIS, 161–162
    easy access surface areas
        characteristics, 162–163
        field validation, 170–172
    field tests, 164–165
    hard access surface areas
        characteristics, 162–163
        evaluation, 165
        field validation, 170–172
    heat-sensitive areas
        characteristics, 162–163
        evaluation, 167–168
    hot water spray, 163–164
    mortality, 165–167
    validation results
        summer validation, 168–169
        winter validation, 169–170
Boater management, *see* Utah
Bonneville Power Administration (BPA), 336
Box models
    detritus (D), 78
    ecosystem-level models, 77
    mussels (M), 78
    nutrients (N), 77–78
    phytoplankton (P), 77–78
    Simile software, 78
    simplified box models, 78–79
    simulation results, 80
    variables and measurements, 80
    zooplankton, 77
BPA, *see* Bonneville Power Administration (BPA)

C

CAISMP, *see* California AIS Management Plan (CAISMP)
CALFED, 275–276
California
    CDFW, 377
    data analyses, 382–383
    dissolved oxygen levels, 386–387
    environmental parameters, 382, 388–389
    experimental artificial substrates, 379–381
    fall trial, 383
    hypotheses investigation, 377–378
    long-term monitoring, 376
        ABS, 392–393
        dissolved oxygen levels, 395–396
        environmental conditions, 394
        goal, 394

9 780367 575755